Grundlehren der mathematischen Wissenschaften 243

A Series of Comprehensive Studies in Mathematics

Editors

M. Artin S. S. Chern J. L. Doob A. Grothendieck
E. Heinz F. Hirzebruch L. Hörmander
S. Mac Lane W. Magnus C. C. Moore J. K. Moser
M. Nagata W. Schmidt D. S. Scott J. Tits
B. L. van der Waerden

Managing Editors

B. Eckmann S. R. S. Varadhan

B. Huppert   N. Blackburn

Finite Groups III

Springer-Verlag
Berlin Heidelberg New York 1982

Bertram Huppert
Mathematisches Institut der Universität
Saarstraße 21
D-6500 Mainz

Norman Blackburn
Department of Mathematics
The University
GB-Manchester M13 9 PL

ISBN 3-540-10633-2 Springer-Verlag Berlin Heidelberg New York
ISBN 0-387-10633-2 Springer-Verlag New York Heidelberg Berlin

Library of Congress Cataloging in Publication Data. Huppert, Bertram, 1927–. Finite groups III. (Grundlehren der mathematischen Wissenschaften; 243). Bibliography: p. Includes index. 1. Finite groups. I. Blackburn, N. (Norman). II. Title. III. Series. QA 171.B578. 512'.22. 81-2288. ISBN 0-387-10633-2 (U.S.). AACR 2

Printed in Germany

Typesetting: Asco Trade Typesetting Limited, Chai Wan, Hong Kong.
Printing and bookbinding: Konrad Triltsch, Würzburg
2141/3140-543210

Preface

Und dann erst kommt der „Ab - ge - sang," daß der nicht kurz und nicht zu lang,

From "Die Meistersinger von Nürnberg", Richard Wagner

This final volume is concerned with some of the developments of the subject in the 1960's. In attempting to determine the simple groups, the first step was to settle the conjecture of Burnside that groups of odd order are soluble. The proof that this conjecture was correct is much too long and complicated for presentation in this text, but a number of ideas in the early stages of it led to a local theory of finite groups, some aspects of which are discussed in Chapter X. Much of this discussion is a continuation of the theory of the transfer (see Chapter IV), but we also introduce the generalized Fitting subgroup, which played a basic role in characterization theorems, that is, in descriptions of specific groups in terms of group-theoretical properties alone. One of the earliest and most important such characterizations was given for Zassenhaus groups; this is presented in Chapter XI. Characterizations in terms of the centralizer of an involution are of particular importance in view of the theorem of Brauer and Fowler. In Chapter XII, one such theorem is given, in which the Mathieu group $\mathfrak{M}_{11}$ and $PSL(3, 3)$ are characterized. This last chapter is mainly concerned with some aspects of multiply transitive permutation groups loosely connected with the Mathieu groups or with sharp n-fold transitivity, and several results from Chapter XI are used in it. The two last chapters are, however, independent of Chapter X.

Again we wish to acknowledge our indebtedness to the many colleagues who have assisted us with this work. In addition to those named in the preface to Volume II, thanks are due to George Glauberman, who read an earlier version of Chapter X. The contributions of all have done a great deal to improve this volume, and it is with the greatest pleasure that we express our gratitude to them.

January, 1982

Bertram Huppert, Mainz
Norman Blackburn, Manchester

Contents

Index of Symbols

Chapter X

Local Finite Group Theory

The word *local* is used in finite group-theory in relation to a fixed prime p; thus properties of p-subgroups or their normalisers, for example, are regarded as local. In the case of a soluble group, then, everything is local, but an insoluble group also has global aspects. Now the local behaviour influences the global, that is, there are theorems in which the hypothesis involves only p-subgroups and their normalisers, but the conclusion involves the whole group. This chapter is an introduction to theorems of this sort.

Some such theorems are already known from Chapter IV; for example, Burnside's transfer theorem, which asserts that if the centre of the normaliser of a Sylow p-subgroup $\mathfrak{S}$ contains $\mathfrak{S}$, then the whole group is p-nilpotent. This is proved by showing that the transfer into $\mathfrak{S}$ is an epimorphism. An essential lemma (IV, 2.5) states that two $\mathfrak{S}$-invariant subsets of $\mathfrak{S}$ are conjugate in $\mathfrak{G}$ if and only if they are conjugate in $\mathbf{N}_{\mathfrak{G}}(\mathfrak{S})$. This has many other applications, being a link between the global and local properties. More generally, the situation in which two subsets A, B of $\mathfrak{S}$ are conjugate in $\mathfrak{G}$ frequently arises; such sets A, B are often described as *fused*, particularly when they are not conjugate in $\mathfrak{S}$. In general, fusion can be reduced not to one but to a sequence of local transformations. This is the subject matter of § 4, where the precise way in which A can be transformed into B is investigated. It is shown that if $A^g = B$, then $g = g_1 g_2 \cdots g_n$ where g_i normalises some subgroup $\mathfrak{P}_i$ of $\mathfrak{S}$ and $A^{g_1 \cdots g_{i-1}} \subseteq \mathfrak{P}_i$. Moreover there are certain sets $\mathscr{F}$ of subgroups of $\mathfrak{S}$ for which the additional condition that $\mathfrak{P}_i \in \mathscr{F}$ may be imposed. These sets $\mathscr{F}$ are called *conjugation families*.

Another theorem with a local hypothesis but a global conclusion is the theorem of Thompson (IV, 6.2) that, for p odd, if $\mathbf{C}_{\mathfrak{G}}(\mathbf{Z}(\mathfrak{S}))$ and $\mathbf{N}_{\mathfrak{G}}(\mathbf{J}_0(\mathfrak{S}))$ are p-nilpotent, so is $\mathfrak{G}$. Here $\mathbf{J}_0(\mathfrak{S})$ denotes a characteristic subgroup of $\mathfrak{S}$. Certain similarly defined characteristic subgroups are very useful for establishing non-simplicity criteria; this is shown in § 2, where a character-free proof of the solubility of groups of order $p^a q^b$ is given. In § 3, it is shown that there is such a characteristic group $\mathbf{ZJ}(\mathfrak{S})$ which is always normal in $\mathfrak{G}$ whenever $\mathbf{O}_p(\mathfrak{G}) \geq \mathbf{C}_{\mathfrak{G}}(\mathbf{O}_p(\mathfrak{G}))$ and $\mathfrak{G}$ is p-stable. This can be used to give another proof of the above theorem of Thompson; it was also used by BENDER [3] to simplify greatly a section of the proof of the solubility of groups of odd order. For all such applications, criteria for p-stability are of course required, such as those given in Chapter IX.

Now $\mathbf{J}(\mathfrak{P})$ is defined for any p-group $\mathfrak{P}$ by certain rules. To analyse these, we consider first, in § 5, completely general rules, supposing only that there is defined in each p-group $\mathfrak{P}$ a subgroup $\mathbf{W}(\mathfrak{P})$ and that whenever α is an isomorphism of $\mathfrak{P}$ onto $\tilde{\mathfrak{P}}$, $\mathbf{W}(\mathfrak{P})\alpha = \mathbf{W}(\tilde{\mathfrak{P}})$. Such a $\mathbf{W}$ is called a *characteristic p-functor*. In order to study fusion, a conjugation family is defined in § 5 corresponding to any characteristic p-functor $\mathbf{W}$. This enables us to prove, for example, that $\mathfrak{G}$ and $\mathbf{N}_{\mathfrak{G}}(\mathbf{W}(\mathfrak{S}))$ have isomorphic maximal p-factor groups if and only if the same is true in the normaliser of any non-identity p-subgroup (Theorem 7.3). By combining this with some results about the transfer developed in § 6, a commutator condition is obtained which implies that $\mathfrak{G}$ and $\mathbf{N}_{\mathfrak{G}}(\mathbf{W}(\mathfrak{S}))$ have isomorphic maximal p-factor groups. In § 8, two characteristic p-functors $\underline{\mathbf{K}}$, $\overline{\mathbf{K}}$ are defined, and, completely within the context of p-groups, a complementary commutator condition is established (Theorem 8.10). Putting the two together, a theorem of Glauberman, which states that for $p \geq 5$, $\mathfrak{G}$, $\mathbf{N}_{\mathfrak{G}}(\overline{\mathbf{K}}(\mathfrak{S}))$ and $\mathbf{N}_{\mathfrak{G}}(\underline{\mathbf{K}}(\mathfrak{S}))$ all have isomorphic maximal p-factor groups, follows. Grün's second theorem (IV, 3.7) makes a similar assertion, but requires that $\mathfrak{G}$ be p-normal. Glauberman's theorem, however, has no such hypothesis. Among its consequences are the fact that if $\mathfrak{G}$ is not a p-group, there exists a Sylow subgroup $\mathfrak{S}$ of $\mathfrak{G}$ for which $\mathbf{N}_{\mathfrak{G}}(\mathfrak{S}) > \mathfrak{S}$. In § 9, it is shown that $\underline{\mathbf{K}}$, $\overline{\mathbf{K}}$ could be used in place of $\mathbf{J}_0$ in the theorem of Thompson, and before this, it is shown that when every section of $\mathfrak{G}$ is p-stable, $\overline{\mathbf{K}}$, $\underline{\mathbf{K}}$ and, for p odd, $\mathbf{ZJ}$ have a property which is described as *strongly controlling fusion*: whenever A, A^g are contained in $\mathfrak{S}$, there exists $h \in \mathbf{N}_{\mathfrak{G}}(\mathbf{W}(\mathfrak{S}))$ such that $a^g = a^h$ for all $a \in A$.

In § 10, we consider another property of $\mathbf{J}$. If $\mathfrak{G}$ is p-soluble, $\mathbf{O}_{p'}(\mathfrak{G}) = 1$ and $\mathfrak{S} \in S_p(\mathfrak{G})$, then the equation

$$\mathfrak{G} = \mathbf{N}_{\mathfrak{G}}(\mathbf{J}(\mathfrak{S}))\mathbf{C}_{\mathfrak{G}}(\mathbf{Z}(\mathfrak{S}))$$

holds under many circumstances; certainly if $p > 3$. This kind of factorization is of considerable importance and made its first appearance (implicitly) in Thompson's theorem. Conditions for its validity when p is 2 or 3 are found in § 10, and in § 11 these are applied to prove a theorem on fixed point free automorphism groups. It is conjectured that if $\mathfrak{A}$ is a fixed point free group of automorphisms of $\mathfrak{G}$ and $(|\mathfrak{A}|, |\mathfrak{G}|) = 1$, then $\mathfrak{G}$ is soluble; in § 11 this is proved when $\mathfrak{A}$ is elementary Abelian.

Since $\mathfrak{G}/\mathfrak{G}'$ is in duality with $\mathrm{Hom}(\mathfrak{G}, \mathbb{C}^\times) = \mathbf{H}^1(\mathfrak{G}, \mathbb{C}^\times)$, where $\mathbb{C}^\times$ is regarded as a trivial $\mathbb{Z}\mathfrak{G}$-module, the transfer of $\mathfrak{G}$ into a subgroup $\mathfrak{H}$ gives rise to a homomorphism of $\mathbf{H}^1(\mathfrak{H}, \mathbb{C}^\times)$ into $\mathbf{H}^1(\mathfrak{G}, \mathbb{C}^\times)$. This is a special case of the corestriction homomorphism of $\mathbf{H}^n(\mathfrak{H}, \mathsf{M})$ into $\mathbf{H}^n(\mathfrak{G}, \mathsf{M})$ described in I, 16.17. It is shown in 12.8 that $\mathbf{H}^n(\mathfrak{G}, \mathsf{M})$ and $\mathbf{H}^n(\mathbf{N}_{\mathfrak{G}}(\mathbf{W}(\mathfrak{S})), \mathsf{M})$ have isomorphic Sylow p-subgroups if M is a trivial $\mathfrak{G}$-module and $\mathbf{W}$ is a characteristic p-functor which strongly controls fusion in $\mathfrak{G}$—Grün's second theorem is a special case of this. This is applied to the Schur multiplier of $\mathfrak{G}$ in 12.17; if $\mathfrak{S} \in S_p(\mathfrak{G})$ and the class of $\mathfrak{S}$ is at most $\frac{1}{2}p$, the Sylow p-subgroups of the Schur multipliers of $\mathfrak{G}$ and of $\mathbf{N}_{\mathfrak{G}}(\mathfrak{S})$ are isomorphic.

In addition to the transfer, a number of results, which have become very familiar in finite group theory, are frequently used in proving these theorems; these include the properties of the centralizers of the Fitting subgroup and $\mathbf{O}_{p',p}(\mathfrak{G})$ and a number of other facts which are collected in § 1. In § 13 and § 14, some of these results are generalized in such a way that solubility hypotheses are removed. In doing this, the role of the nilpotent group is taken by the quasinilpotent group (13.2) and that of the p'-group by the p^*-group (14.2). It is shown, for example, that every group $\mathfrak{G}$ has a unique maximal normal quasinilpotent subgroup $\mathbf{F}^*(\mathfrak{G})$ and that $\mathbf{C}_{\mathfrak{G}}(\mathbf{F}^*(\mathfrak{G})) \le \mathbf{F}(\mathfrak{G})$; again, every group $\mathfrak{G}$ has a generalized p'-core $\mathbf{O}_{p^*}(\mathfrak{G})$, and if $\mathfrak{P}$ is a p-subgroup of $\mathfrak{G}$, $\mathbf{O}_{p^*}(\mathbf{C}_{\mathfrak{G}}(\mathfrak{P})) \le \mathbf{O}_{p^*}(\mathfrak{G})$. Finally, in § 16, another aspect of local properties is briefly considered; this involves the relationship between the various soluble p'-subgroups of a group $\mathfrak{G}$ which are normalised by a fixed Abelian p-subgroup of $\mathfrak{G}$.

§ 1. Elementary Lemmas

In this chapter a number of elementary results will be used frequently. Some of these have already appeared in the previous chapter; the remainder are collected together in this section.

First, we establish a lemma for characteristic subgroups of p-groups analogous to the theorem (III, 7.3) that a maximal normal Abelian subgroup of a p-group is its own centralizer.

1.1 Lemma. *Let $\mathfrak{G}$ be a p-group and let $\mathfrak{A}$ be a characteristic Abelian subgroup of $\mathfrak{G}$. Then there exists a characteristic subgroup $\mathfrak{B}$ of $\mathfrak{G}$ such that*

(i) $\mathfrak{B} \geq \mathbf{C}_{\mathfrak{G}}(\mathfrak{B}) = \mathbf{Z}(\mathfrak{B}) \geq \mathfrak{A}$, *and*

(ii) $\mathfrak{B}/\mathbf{Z}(\mathfrak{B})$ *is an elementary Abelian subgroup of* $\mathbf{Z}(\mathfrak{G}/\mathbf{Z}(\mathfrak{B}))$.

In particular, the class of $\mathfrak{B}$ is at most 2.

Proof. Let $\mathscr{X}$ be the set of characteristic subgroups $\mathfrak{X}$ of $\mathfrak{G}$ such that $\mathbf{Z}(\mathfrak{X}) \geq \mathfrak{A}$ and $\mathfrak{X}/\mathbf{Z}(\mathfrak{X})$ is an elementary Abelian subgroup of $\mathbf{Z}(\mathfrak{G}/\mathbf{Z}(\mathfrak{X}))$. Thus $\mathfrak{A} \in \mathscr{X}$. Let $\mathfrak{B}$ be a maximal element of $\mathscr{X}$. If $\mathfrak{B} \geq \mathbf{C}_{\mathfrak{G}}(\mathfrak{B})$, then $\mathfrak{B}$ has all the required properties. Suppose then that $\mathfrak{B} \ngeq \mathbf{C}_{\mathfrak{G}}(\mathfrak{B})$, that is, $\mathbf{Z}(\mathfrak{B}) < \mathbf{C}_{\mathfrak{G}}(\mathfrak{B})$. Let $\mathfrak{D}/\mathbf{Z}(\mathfrak{B})$ be the set of elements of order at most p in $(\mathbf{C}_{\mathfrak{G}}(\mathfrak{B})/\mathbf{Z}(\mathfrak{B})) \cap \mathbf{Z}(\mathfrak{G}/\mathbf{Z}(\mathfrak{B}))$. Thus $\mathfrak{D} \leq \mathbf{C}_{\mathfrak{G}}(\mathfrak{B})$, $\mathfrak{D}^p \leq \mathbf{Z}(\mathfrak{B})$ and $[\mathfrak{D}, \mathfrak{G}] \leq \mathbf{Z}(\mathfrak{B})$. By III, 7.2, $\mathfrak{D} > \mathbf{Z}(\mathfrak{B})$, so $\mathfrak{D} \nleq \mathfrak{B}$. But $\mathfrak{D}\mathfrak{B} \in \mathscr{X}$, since $[\mathfrak{D}\mathfrak{B}, \mathfrak{G}] = [\mathfrak{D}, \mathfrak{G}][\mathfrak{B}, \mathfrak{G}] \leq \mathbf{Z}(\mathfrak{B})$ and $(\mathfrak{D}\mathfrak{B})^p = \mathfrak{D}^p\mathfrak{B}^p \leq \mathbf{Z}(\mathfrak{B})$. This contradicts the maximality of $\mathfrak{B}$. **q.e.d.**

1.2 Lemma. *Suppose that $\mathfrak{G}$ is a p-group and that α is an automorphism of $\mathfrak{G}$ of order prime to p. If there exists a subgroup $\mathfrak{H}$ of $\mathfrak{G}$ for which $\mathfrak{H}\mathbf{C}_{\mathfrak{G}}(\mathfrak{H}) \leq \mathbf{C}_{\mathfrak{G}}(\alpha)$, then α is the identity automorphism.*

Proof. This is proved by induction on $|\mathfrak{G}|$. There is nothing to prove if $\mathfrak{H}\mathbf{C}_{\mathfrak{G}}(\mathfrak{H}) = \mathfrak{G}$. Otherwise, there exists a maximal α-invariant proper subgroup $\mathfrak{M}$ of $\mathfrak{G}$ such that $\mathfrak{M} \geq \mathfrak{H}\mathbf{C}_{\mathfrak{G}}(\mathfrak{H})$. Since $\mathbf{N}_{\mathfrak{G}}(\mathfrak{M})$ is α-invariant, $\mathbf{N}_{\mathfrak{G}}(\mathfrak{M}) = \mathfrak{G}$, by III, 2.3. Thus $\mathfrak{M} \trianglelefteq \mathfrak{G}$. By the inductive hypothesis, $\mathfrak{M} \leq \mathbf{C}_{\mathfrak{G}}(\alpha)$. By IX, 6.3, α induces the identity mapping on $\mathfrak{G}/\mathbf{C}_{\mathfrak{G}}(\mathfrak{M})$. But $\mathbf{C}_{\mathfrak{G}}(\mathfrak{M}) \leq \mathbf{C}_{\mathfrak{G}}(\mathfrak{H}) \leq \mathfrak{M}$, so α induces the identity mapping on $\mathfrak{G}/\mathfrak{M}$. By I, 4.4, α is the identity mapping. **q.e.d.**

1.3 Lemma. *Suppose that $\mathfrak{G}$ is a p-group and $\mathbf{C}_{\mathfrak{G}}(\mathfrak{N}) \leq \mathfrak{N} \trianglelefteq \mathfrak{G}$. Suppose that*

$$\mathfrak{N} = \mathfrak{N}_0 \geq \mathfrak{N}_1 \geq \cdots \geq \mathfrak{N}_k = 1,$$

where $\mathfrak{N}_i \trianglelefteq \mathfrak{G}$ $(i = 0, 1, \ldots, k)$. Let $\mathfrak{A}$ be a group of automorphisms α of $\mathfrak{G}$ for which $\mathfrak{N}_i\alpha = \mathfrak{N}_i$ $(i = 0, 1, \ldots, k)$. Let

$$\mathfrak{B} = \{\beta | \beta \in \mathfrak{A}, (x\mathfrak{N}_i)\beta = x\mathfrak{N}_i \quad \textit{for all} \quad x \in \mathfrak{N}_{i-1}\ (i = 1, \ldots, k)\}.$$

Then $\mathfrak{B}$ is a normal p-subgroup of $\mathfrak{A}$.

In particular, any non-identity p'-element of $\mathfrak{A}$ induces a non-identity automorphism on $\mathfrak{N}_{i-1}/\mathfrak{N}_i$ for at least one i.

Proof. If $\alpha \in \mathfrak{A}$, let $\rho(\alpha)$ denote the automorphism of $\mathfrak{N}$ induced by α. Then ρ is a homomorphism of $\mathfrak{A}$ into the group of automorphisms of $\mathfrak{N}$. Let $\mathfrak{K} = \ker \rho$, $\mathfrak{A}_0 = \operatorname{im} \rho$, $\mathfrak{B}_0 = \rho(\mathfrak{B})$. Then there is an isomorphism between $\mathfrak{A}/\mathfrak{K}$ and $\mathfrak{A}_0$ in which $\mathfrak{B}/\mathfrak{K}$ and $\mathfrak{B}_0$ correspond. Now

$$\mathfrak{B}_0 = \{\beta | \beta \in \mathfrak{A}_0, (x\mathfrak{N}_i)\beta = x\mathfrak{N}_i \quad \text{for all} \quad x \in \mathfrak{N}_{i-1}\ (i = 1, \ldots, k)\}.$$

By IX, 7.3, $\mathfrak{B}_0$ is a normal p-subgroup of $\mathfrak{A}_0$. Hence $\mathfrak{B} \trianglelefteq \mathfrak{A}$ and it only remains to show that $\mathfrak{K}$ is a p-group. If $\gamma \in \mathfrak{K}$, choose r such that the order of $\delta = \gamma^{p^r}$ is prime to p. Then $\mathfrak{N}\mathbf{C}_{\mathfrak{G}}(\mathfrak{N}) = \mathfrak{N} \leq \mathbf{C}_{\mathfrak{G}}(\delta)$, so $\delta = 1$ by 1.2. Thus $\mathfrak{K}$ is a p-group. **q.e.d.**

1.4 Lemma. *Suppose that α is an automorphism of the Abelian p-group $\mathfrak{A}$ and that α leaves fixed every element of $\mathfrak{A}$ of order p. Then the order of α is a power of p.*

Proof. Let $\mathfrak{A}_n = \Omega_n(\mathfrak{A})$ $(n = 0, 1, \ldots)$. Thus $\mathfrak{A}_n\alpha = \mathfrak{A}_n$ and

$$1 = \mathfrak{A}_0 \leq \mathfrak{A}_1 \leq \cdots \leq \mathfrak{A}_m = \mathfrak{A}$$

for some m. If $x \in \mathfrak{A}_n$ and $x\alpha = xy$,

$$y^{p^{n-1}} = x^{-p^{n-1}}(x\alpha)^{p^{n-1}} = x^{-p^{n-1}}(x^{p^{n-1}})\alpha = 1,$$

since the order of $x^{p^{n-1}}$ is at most p. Thus $y \in \mathfrak{A}_{n-1}$, and $(x\mathfrak{A}_{n-1})\alpha = x\mathfrak{A}_{n-1}$. By 1.3, the order of α is a power of p. **q.e.d.**

1.5 Lemma (THOMPSON). *Suppose that $\mathfrak{P}$, $\mathfrak{K}$ are subgroups of the group $\mathfrak{G}$, $\mathfrak{P}$ is a p-group, $\mathfrak{K} = \mathbf{O}^p(\mathfrak{K})$ and $[\mathfrak{K}, \mathfrak{P}] = 1$. Suppose that $\mathfrak{B}$ is a p-subgroup of $\mathfrak{G}$ and $\mathfrak{P}\mathfrak{K} \leq \mathbf{N}_{\mathfrak{G}}(\mathfrak{B})$. If $[\mathfrak{K}, \mathbf{C}_{\mathfrak{B}}(\mathfrak{P})] = 1$, then $[\mathfrak{K}, \mathfrak{B}] = 1$.*

Proof. Let $\mathfrak{H} = \{(u, v) | u \in \mathfrak{P}, v \in \mathfrak{B}\}$. Then $\mathfrak{H}$ is a group if we put

$$(u, v)(u', v') = (uu', v^{u'}v')$$

for all u, u' in $\mathfrak{P}$ and v, v' in $\mathfrak{B}$. If x is an element of $\mathfrak{K}$ of order prime to p, then $x \in \mathbf{C}_{\mathfrak{G}}(\mathfrak{P})$, so

$$(u, v^x)(u', v'^x) = (uu', v^{xu'}v'^x) = (uu', (v^{u'}v')^x).$$

Hence if we put $(u, v)\alpha = (u, v^x)$, α is an automorphism of $\mathfrak{H}$. Note that $\mathfrak{H}$ is a p-group and the order of α is prime to p. If $\mathfrak{Q} = \{(u, 1) | u \in \mathfrak{P}\}$,

$\mathfrak{Q} \leq \mathfrak{H}$ and $\mathbf{C}_\mathfrak{H}(\mathfrak{Q}) = \{(u, v) | u \in \mathbf{Z}(\mathfrak{P}), v \in \mathbf{C}_\mathfrak{B}(\mathfrak{P})\}$. Since $[\mathfrak{K}, \mathbf{C}_\mathfrak{B}(\mathfrak{P})] = 1$, $\mathfrak{Q}\mathbf{C}_\mathfrak{H}(\mathfrak{Q}) \leq \mathbf{C}_\mathfrak{H}(\alpha)$. By 1.2, α is the identity automorphism. Thus $x \in \mathbf{C}_\mathfrak{G}(\mathfrak{B})$ for every element x of $\mathfrak{K}$ of order prime to p. Since $\mathfrak{K} = \mathbf{O}^p(\mathfrak{K})$, it follows that $\mathfrak{B}$ centralizes $\mathfrak{K}$. **q.e.d.**

1.6 Lemma. *If $\mathfrak{P}$ is a p-subgroup of the p-constrained group $\mathfrak{G}$,*

$$\mathbf{O}_{p'}(\mathbf{N}_\mathfrak{G}(\mathfrak{P})) \leq \mathbf{O}_{p'}(\mathfrak{G}).$$

Proof. Let $\overline{\mathfrak{G}} = \mathfrak{G}/\mathbf{O}_{p'}(\mathfrak{G})$, $\overline{\mathfrak{P}} = \mathfrak{P}\mathbf{O}_{p'}(\mathfrak{G})/\mathbf{O}_{p'}(\mathfrak{G})$, $\overline{\mathfrak{N}} = \mathbf{N}_{\overline{\mathfrak{G}}}(\overline{\mathfrak{P}})$, $\overline{\mathfrak{K}} = \mathbf{O}_{p'}(\overline{\mathfrak{N}})$. Then $\overline{\mathfrak{K}} = \mathbf{O}^p(\overline{\mathfrak{K}})$ since $\overline{\mathfrak{K}}$ is a p'-group, and $[\overline{\mathfrak{P}}, \overline{\mathfrak{K}}] = 1$ since $\overline{\mathfrak{P}}$, $\overline{\mathfrak{K}}$ are normal subgroups of $\overline{\mathfrak{N}}$ of coprime orders. Also $\overline{\mathfrak{P}}\overline{\mathfrak{K}} \leq \mathbf{N}_{\overline{\mathfrak{G}}}(\mathbf{O}_p(\overline{\mathfrak{G}}))$ and

$$[\overline{\mathfrak{K}}, \mathbf{C}_{\mathbf{O}_p(\overline{\mathfrak{G}})}(\overline{\mathfrak{P}})] \leq \mathbf{O}_p(\overline{\mathfrak{G}}) \cap [\overline{\mathfrak{K}}, \overline{\mathfrak{N}}] \leq \mathbf{O}_p(\overline{\mathfrak{G}}) \cap \overline{\mathfrak{K}} = 1.$$

Hence by 1.5, $[\overline{\mathfrak{K}}, \mathbf{O}_p(\overline{\mathfrak{G}})] = 1$. But $\mathbf{O}_{p'}(\overline{\mathfrak{G}}) = 1$ and $\overline{\mathfrak{G}}$ is p-constrained (VII, 13.3), so $\overline{\mathfrak{K}} \leq \mathbf{O}_p(\overline{\mathfrak{G}})$. Since $\overline{\mathfrak{K}}$ is a p'-group, $\overline{\mathfrak{K}} = 1$. Now by IX, 6.11, $\overline{\mathfrak{N}} = \mathbf{N}_\mathfrak{G}(\mathfrak{P})\mathbf{O}_{p'}(\mathfrak{G})/\mathbf{O}_{p'}(\mathfrak{G})$, so $\overline{\mathfrak{N}} \cong \mathbf{N}_\mathfrak{G}(\mathfrak{P})/\mathfrak{M}$, where $\mathfrak{M} = \mathbf{N}_\mathfrak{G}(\mathfrak{P}) \cap \mathbf{O}_{p'}(\mathfrak{G})$ is a normal p'-subgroup of $\mathbf{N}_\mathfrak{G}(\mathfrak{P})$. Since $\mathbf{O}_{p'}(\overline{\mathfrak{N}}) = \overline{\mathfrak{K}} = 1$, it follows that $\mathbf{O}_{p'}(\mathbf{N}_\mathfrak{G}(\mathfrak{P})) = \mathfrak{M} \leq \mathbf{O}_{p'}(\mathfrak{G})$. **q.e.d.**

1.7 Corollary. *Suppose that $\mathfrak{P}$ is a p-subgroup of the p-constrained group $\mathfrak{G}$. If $\mathfrak{P}$ contains every p-element of $\mathbf{C}_\mathfrak{G}(\mathfrak{P})$, then*

$$\mathbf{C}_\mathfrak{G}(\mathfrak{P}) = \mathbf{Z}(\mathfrak{P}) \times (\mathbf{C}_\mathfrak{G}(\mathfrak{P}) \cap \mathbf{O}_{p'}(\mathfrak{G})).$$

Proof. By hypothesis, $\mathbf{Z}(\mathfrak{P})$ is the only Sylow p-subgroup of $\mathbf{C}_\mathfrak{G}(\mathfrak{P})$. Since $\mathbf{Z}(\mathfrak{P}) \leq \mathbf{Z}(\mathbf{C}_\mathfrak{G}(\mathfrak{P}))$, $\mathbf{C}_\mathfrak{G}(\mathfrak{P}) = \mathbf{Z}(\mathfrak{P}) \times \mathfrak{N}$ for some $\mathfrak{N}$, by IV, 2.6. Thus $\mathfrak{N} = \mathbf{O}_{p'}(\mathbf{C}_\mathfrak{G}(\mathfrak{P})) \leq \mathbf{O}_{p'}(\mathbf{N}_\mathfrak{G}(\mathfrak{P})) \leq \mathbf{O}_{p'}(\mathfrak{G})$ by 1.6, and $\mathfrak{N} = \mathbf{C}_\mathfrak{G}(\mathfrak{P}) \cap \mathbf{O}_{p'}(\mathfrak{G})$. **q.e.d.**

We now turn to a generalization of part of IX, 6.11.

1.8 Lemma. *Suppose that $\mathfrak{A}$ is a group of operators on a group $\mathfrak{G}$ and that either $\mathfrak{A}$ or $\mathfrak{G}$ is soluble.*

a) *If $\mathfrak{N}$ is a normal $\mathfrak{A}$-invariant subgroup of $\mathfrak{G}$ and* $(|\mathfrak{N}|, |\mathfrak{A}|) = 1$, *then*

$$\mathbf{C}_{\mathfrak{G}/\mathfrak{N}}(\mathfrak{A}) = \mathbf{C}_\mathfrak{G}(\mathfrak{A})\mathfrak{N}/\mathfrak{N}.$$

b) *If* $(|[\mathfrak{G}, \mathfrak{A}]|, |\mathfrak{A}|) = 1$, *then* $\mathfrak{G} = \mathbf{C}_\mathfrak{G}(\mathfrak{A})[\mathfrak{G}, \mathfrak{A}]$.

Proof. a) Obviously $\mathbf{C}_{\mathfrak{G}}(\mathfrak{A})\mathfrak{N}/\mathfrak{N} \leq \mathbf{C}_{\mathfrak{G}/\mathfrak{N}}(\mathfrak{A})$. If $x\mathfrak{N} \in \mathbf{C}_{\mathfrak{G}/\mathfrak{N}}(\mathfrak{A})$, then by I, 18.6, there exists $y \in \mathbf{C}_{\mathfrak{G}}(\mathfrak{A})$ such that $y\mathfrak{N} = x\mathfrak{N}$. Hence $x \in \mathbf{C}_{\mathfrak{G}}(\mathfrak{A})\mathfrak{N}$.

b) By III, 1.6b), $\mathfrak{M} = [\mathfrak{G}, \mathfrak{A}]$ is a normal $\mathfrak{A}$-invariant subgroup of $\mathfrak{G}$. Clearly $\mathbf{C}_{\mathfrak{G}/\mathfrak{M}}(\mathfrak{A}) = \mathfrak{G}/\mathfrak{M}$. But by a),

$$\mathbf{C}_{\mathfrak{G}/\mathfrak{M}}(\mathfrak{A}) = \mathbf{C}_{\mathfrak{G}}(\mathfrak{A})\mathfrak{M}/\mathfrak{M}.$$

Thus $\mathfrak{G} = \mathbf{C}_{\mathfrak{G}}(\mathfrak{A})\mathfrak{M} = \mathbf{C}_{\mathfrak{G}}(\mathfrak{A})[\mathfrak{G}, \mathfrak{A}]$. **q.e.d.**

Note that on account of the solubility of groups of odd order, the hypothesis that either $\mathfrak{A}$ or $\mathfrak{G}$ be soluble is unnecessary.

1.9 Lemma. *Suppose that $\mathfrak{G}$ is a p'-group and $\mathfrak{A}$ is an Abelian p-group of operators on $\mathfrak{G}$. Then*

$$\mathfrak{G} = \langle \mathbf{C}_{\mathfrak{G}}(\mathfrak{B}) \mid \mathfrak{B} \leq \mathfrak{A}, \mathfrak{A}/\mathfrak{B} \textit{ is cyclic} \rangle.$$

If also $\mathfrak{A}$ is not cyclic,

$$\mathfrak{G} = \langle \mathbf{C}_{\mathfrak{G}}(x) \mid x \in \mathfrak{A}, x \neq 1 \rangle.$$

Proof. The first assertion is proved by induction on $|\mathfrak{G}|$. If q is any prime other than p, $\mathfrak{G}$ possesses an $\mathfrak{A}$-invariant Sylow q-subgroup $\mathfrak{Q}$, by IX, 1.11, and $\mathfrak{G}$ is generated by all such Sylow subgroups. Thus it suffices to prove that $\mathfrak{Q} = \mathfrak{C}$, where

$$\mathfrak{C} = \langle \mathbf{C}_{\mathfrak{Q}}(\mathfrak{B}) \mid \mathfrak{B} \leq \mathfrak{A}, \mathfrak{A}/\mathfrak{B} \text{ is cyclic} \rangle.$$

Suppose that $\mathfrak{C} < \mathfrak{Q}$. Then $\Phi(\mathfrak{Q})\mathfrak{C} < \mathfrak{Q}$. Let $\mathfrak{M}$ be a maximal $\mathfrak{A}$-invariant subgroup of $\mathfrak{Q}$ for which $\Phi(\mathfrak{Q})\mathfrak{C} \leq \mathfrak{M} < \mathfrak{Q}$. Then $\mathfrak{Q}/\mathfrak{M}$ is irreducible under $\mathfrak{A}$. If $\mathfrak{B}_0$ is the kernel of the representation of $\mathfrak{A}$ on $\mathfrak{Q}/\mathfrak{M}$, then $\mathfrak{A}/\mathfrak{B}_0$ is cyclic, by II, 3.10. Also

$$\mathfrak{Q}/\mathfrak{M} = \mathbf{C}_{\mathfrak{Q}/\mathfrak{M}}(\mathfrak{B}_0) = \mathbf{C}_{\mathfrak{Q}}(\mathfrak{B}_0)\mathfrak{M}/\mathfrak{M},$$

by 1.8a), so

$$\mathfrak{Q} = \mathbf{C}_{\mathfrak{Q}}(\mathfrak{B}_0)\mathfrak{M} = \mathfrak{M},$$

a contradiction.

Suppose that $\mathfrak{A}$ is not cyclic. If $\mathfrak{A}/\mathfrak{B}$ is cyclic, then $\mathfrak{B} \neq 1$ and $\mathbf{C}_{\mathfrak{G}}(\mathfrak{B}) \leq \mathbf{C}_{\mathfrak{G}}(x)$ for some $x \in \mathfrak{A} - \{1\}$. Hence
$\mathfrak{G} = \langle \mathbf{C}_{\mathfrak{G}}(x) \mid x \in \mathfrak{A}, x \neq 1 \rangle$. **q.e.d.**

1.10 Lemma. a) *If $\mathfrak{S} \in S_p(\mathfrak{G})$ and $\mathfrak{A}$ is a maximal normal Abelian subgroup of $\mathfrak{S}$, $\mathbf{C}_{\mathfrak{G}}(\mathfrak{A}) = \mathfrak{A} \times \mathfrak{D}$ for some p'-subgroup $\mathfrak{D}$.*

b) *If p is odd, $\mathfrak{S} \in S_p(\mathfrak{G})$ and $\mathfrak{A}$ is a maximal normal elementary Abelian subgroup of $\mathfrak{S}$, every element of order p in $\mathbf{C}_{\mathfrak{G}}(\mathfrak{A})$ lies in $\mathfrak{A}$.*

c) *Suppose that $\mathfrak{A}$ is a p-subgroup of $\mathfrak{G}$ and every element of order p in $\mathbf{C}_{\mathfrak{G}}(\mathfrak{A})$ lies in $\mathfrak{A}$. If $\mathfrak{K} \leq \mathfrak{G}$, $\mathfrak{A} \leq \mathbf{N}_{\mathfrak{G}}(\mathfrak{K})$ and $\mathfrak{K} \cap \mathfrak{A} = 1$, $\mathfrak{K}$ is a p'-group.*

Proof. a) See IX, 5.9.

b) Since $\mathfrak{S} \in S_p(\mathbf{N}_{\mathfrak{G}}(\mathfrak{A}))$ and $\mathbf{C}_{\mathfrak{G}}(\mathfrak{A}) \trianglelefteq \mathbf{N}_{\mathfrak{G}}(\mathfrak{A})$, $\mathbf{C}_{\mathfrak{S}}(\mathfrak{A}) \in S_p(\mathbf{C}_{\mathfrak{G}}(\mathfrak{A}))$. Hence if x is an element of $\mathbf{C}_{\mathfrak{G}}(\mathfrak{A})$ of order p, $x^c \in \mathbf{C}_{\mathfrak{S}}(\mathfrak{A})$ for some $c \in \mathbf{C}_{\mathfrak{G}}(\mathfrak{A})$. But x^c is of order p, so by III, 12.1, $x^c \in \mathfrak{A}$. Since $c \in \mathbf{C}_{\mathfrak{G}}(\mathfrak{A})$, it follows that $x \in \mathfrak{A}$.

c) Suppose $\mathfrak{P} \in S_p(\mathfrak{A}\mathfrak{K})$ and $\mathfrak{P} \geq \mathfrak{A}$. Then $\mathfrak{P} = \mathfrak{A}\mathfrak{P}_0$, where $\mathfrak{P}_0 = \mathfrak{P} \cap \mathfrak{K} \in S_p(\mathfrak{K})$. Also $\mathfrak{P}_0 \trianglelefteq \mathfrak{P}$, so if $\mathfrak{P}_0 \neq 1$, $\mathfrak{P}_0$ contains an element z of $\mathbf{Z}(\mathfrak{P})$ of order p. But then $z \in \mathbf{C}_{\mathfrak{G}}(\mathfrak{A})$, so by hypothesis, $z \in \mathfrak{A}$. Since $\mathfrak{A} \cap \mathfrak{K} = 1$, it follows that $z = 1$, a contradiction. Thus $\mathfrak{P}_0 = 1$ and $\mathfrak{K}$ is a p'-group. **q.e.d.**

1.11 Theorem. *Suppose that $\mathfrak{G}$ is p-constrained, $\mathfrak{S} \in S_p(\mathfrak{G})$ and $\mathfrak{A}$ is a maximal normal Abelian subgroup of $\mathfrak{S}$. If $\mathfrak{K}$ is a subgroup of $\mathfrak{G}$ for which $\mathfrak{A} \leq \mathbf{N}_{\mathfrak{G}}(\mathfrak{K})$ and $\mathfrak{A} \cap \mathfrak{K} = 1$, then $\mathfrak{K} \leq \mathbf{O}_{p'}(\mathfrak{G})$.*

Proof. Suppose that this is false and that $\mathfrak{G}$ is a counterexample of minimal order. By 1.10a) and c), $\mathfrak{K}$ is a p'-group.

a) $\mathbf{O}_{p'}(\mathfrak{G}) = 1$.

Let $\overline{\mathfrak{G}} = \mathfrak{G}/\mathbf{O}_{p'}(\mathfrak{G})$, $\overline{\mathfrak{S}} = \mathfrak{S}\mathbf{O}_{p'}(\mathfrak{G})/\mathbf{O}_{p'}(\mathfrak{G})$; thus $\overline{\mathfrak{S}} \in S_p(\overline{\mathfrak{G}})$ and $\overline{\mathfrak{A}} = \mathfrak{A}\mathbf{O}_{p'}(\mathfrak{G})/\mathbf{O}_{p'}(\mathfrak{G})$ is a maximal normal Abelian subgroup of $\overline{\mathfrak{S}}$. Also $\overline{\mathfrak{A}} \leq \mathbf{N}_{\overline{\mathfrak{G}}}(\overline{\mathfrak{K}})$, where $\overline{\mathfrak{K}} = \mathfrak{K}\mathbf{O}_{p'}(\mathfrak{G})/\mathbf{O}_{p'}(\mathfrak{G})$. Since $\mathfrak{K}$ is a p'-group, $\overline{\mathfrak{A}} \cap \overline{\mathfrak{K}} = 1$. But $\overline{\mathfrak{K}} \neq 1 = \mathbf{O}_{p'}(\overline{\mathfrak{G}})$, so $\overline{\mathfrak{G}}$ is a counterexample to the theorem, and $\mathbf{O}_{p'}(\mathfrak{G}) = 1$ on account of the minimality of $|\mathfrak{G}|$.

Let $\mathfrak{T}$ be a minimal non-identity $\mathfrak{A}$-invariant subgroup of $\mathfrak{K}$. Thus $\mathfrak{T}$ is a p'-group.

b) $\mathfrak{G} = \mathbf{O}_p(\mathfrak{G})\mathfrak{T}\mathfrak{A}$.

Let $\mathfrak{H} = \mathbf{O}_p(\mathfrak{G})\mathfrak{T}\mathfrak{A}$. Thus $\mathfrak{H} \leq \mathfrak{G}$ and $\mathbf{O}_p(\mathfrak{G})\mathfrak{A} \in S_p(\mathfrak{H})$. Since $\mathbf{O}_p(\mathfrak{G})\mathfrak{A} \leq \mathfrak{S}$ and $\mathbf{C}_{\mathfrak{S}}(\mathfrak{A}) = \mathfrak{A}$, $\mathfrak{A}$ is a maximal normal Abelian subgroup of $\mathbf{O}_p(\mathfrak{G})\mathfrak{A}$. Since $\mathbf{O}_{p'}(\mathfrak{G}) = 1$ and $\mathfrak{G}$ is p-constrained, $\mathbf{O}_p(\mathfrak{G}) \geq \mathbf{C}_{\mathfrak{G}}(\mathbf{O}_p(\mathfrak{G}))$, so $[\mathbf{O}_p(\mathfrak{G}), \mathfrak{T}] \neq 1$ and $\mathfrak{T} \not\leq \mathbf{O}_{p'}(\mathfrak{H})$. Thus $\mathfrak{H}$ is a counterexample to the theorem, and $\mathfrak{H} = \mathfrak{G}$ on account of the minimality of $|\mathfrak{G}|$.

c) $[\mathfrak{T}, \mathfrak{A}] \neq 1$.

This is clear if $\mathfrak{T} \not\leq \mathbf{N}_{\mathfrak{G}}(\mathfrak{A})$, so we suppose that $\mathfrak{T} \leq \mathbf{N}_{\mathfrak{G}}(\mathfrak{A})$. Then $\mathfrak{A} \trianglelefteq \mathfrak{G}$ since $\mathfrak{G} = \mathfrak{S}\mathfrak{T}$ by b). By 1.10a), $\mathbf{C}_{\mathfrak{G}}(\mathfrak{A}) = \mathfrak{A} \times \mathfrak{D}$ for some

p'-subgroup $\mathfrak{D}$. Thus $\mathfrak{D}$ is a characteristic subgroup of $\mathbf{C}_{\mathfrak{G}}(\mathfrak{A})$ and $\mathbf{C}_{\mathfrak{G}}(\mathfrak{A}) \trianglelefteq \mathfrak{G}$. Hence $\mathfrak{D} \trianglelefteq \mathfrak{G}$ and $\mathfrak{D} \leq \mathbf{O}_{p'}(\mathfrak{G}) = 1$. Thus $\mathbf{C}_{\mathfrak{G}}(\mathfrak{A}) = \mathfrak{A}$ and $[\mathfrak{T}, \mathfrak{A}] \neq 1$.

d) Let $\mathsf{V} = \mathbf{O}_p(\mathfrak{G})/\Phi(\mathbf{O}_p(\mathfrak{G}))$. Then $\mathfrak{G}$ is a group of operators on V, and by III, 13.4b),

$$\mathsf{V} = [\mathsf{V}, \mathfrak{T}] \times \mathbf{C}_{\mathsf{V}}(\mathfrak{T}).$$

Now $[\mathsf{V}, \mathfrak{T}]$ is $\mathfrak{A}$-invariant, so $[[\mathsf{V}, \mathfrak{T}], \mathfrak{A}] \leq [\mathsf{V}, \mathfrak{T}]$. But also $[[\mathsf{V}, \mathfrak{T}], \mathfrak{A}] \leq [\mathsf{V}, \mathfrak{A}] \leq \mathbf{C}_{\mathsf{V}}(\mathfrak{T})$, since

$$[\mathbf{O}_p(\mathfrak{G}), \mathfrak{A}, \mathfrak{T}] \leq [\mathfrak{A}, \mathfrak{T}] \cap \mathbf{O}_p(\mathfrak{G}) \leq \mathfrak{T} \cap \mathbf{O}_p(\mathfrak{G}) = 1.$$

Hence $[[\mathsf{V}, \mathfrak{T}], \mathfrak{A}] = 1$ and $[\mathsf{V}, \mathfrak{T}] \leq \mathbf{C}_{\mathsf{V}}(\mathfrak{A})$. Since $[\mathsf{V}, \mathfrak{T}]$ is $\mathfrak{T}$-invariant, it follows that $[\mathsf{V}, \mathfrak{T}] \leq \mathbf{C}_{\mathsf{V}}(\mathfrak{A}^t)$ for all $t \in \mathfrak{T}$; hence

$$[\mathsf{V}, \mathfrak{T}] \leq \mathbf{C}_{\mathsf{V}}([\mathfrak{T}, \mathfrak{A}]).$$

Now $[\mathfrak{T}, \mathfrak{A}]$ is an $\mathfrak{A}$-invariant subgroup of $\mathfrak{T}$, and by c), $[\mathfrak{T}, \mathfrak{A}] \neq 1$. By minimality of $\mathfrak{T}$, $[\mathfrak{T}, \mathfrak{A}] = \mathfrak{T}$. Thus $[\mathsf{V}, \mathfrak{T}] \leq \mathbf{C}_{\mathsf{V}}(\mathfrak{T})$. Hence $[\mathsf{V}, \mathfrak{T}] = 1$ and

$$\mathfrak{T} \leq \mathbf{C}_{\mathfrak{G}}(\mathbf{O}_p(\mathfrak{G})/\Phi(\mathbf{O}_p(\mathfrak{G}))).$$

By IX, 1.6, $\mathfrak{T} \leq \mathbf{O}_p(\mathfrak{G})$, contrary to $\mathfrak{T} \neq 1$. **q.e.d.**

We show that for p odd, the conclusion of 1.11 holds under weaker assumptions.

1.12 Theorem (THOMPSON-BENDER). *For p odd, suppose that $\mathfrak{A}$ is a p-subgroup of the p-constrained group $\mathfrak{G}$ and that every element of order p in $\mathbf{C}_{\mathfrak{G}}(\mathfrak{A})$ lies in $\mathfrak{A}$. If $\mathfrak{K} \leq \mathfrak{G}$, $\mathfrak{A} \leq \mathbf{N}_{\mathfrak{G}}(\mathfrak{K})$ and $\mathfrak{K} \cap \mathfrak{A} = 1$, then $\mathfrak{K} \leq \mathbf{O}_{p'}(\mathfrak{G})$.*

Proof. Suppose that this is false and let $\mathfrak{G}$ be a counterexample of minimal order. By 1.10, $\mathfrak{K}$ is a p'-group.

a) $\mathbf{O}_{p'}(\mathfrak{G}) = 1$.

Suppose that $\mathfrak{N} = \mathbf{O}_{p'}(\mathfrak{G}) \neq 1$. By IX, 6.11, $\mathbf{C}_{\mathfrak{G}/\mathfrak{N}}(\mathfrak{A}\mathfrak{N}/\mathfrak{N}) = \mathbf{C}_{\mathfrak{G}}(\mathfrak{A})\mathfrak{N}/\mathfrak{N}$. Thus any element of order p in $\mathbf{C}_{\mathfrak{G}/\mathfrak{N}}(\mathfrak{A}\mathfrak{N}/\mathfrak{N})$ is of the form $x\mathfrak{N}$ with $x \in \mathbf{C}_{\mathfrak{G}}(\mathfrak{A})$ and $x^p \in \mathfrak{N}$. If $x = x_1x_2 = x_2x_1$, where x_1 is a p-element and x_2 is a p'-element, then $x_2 \in \mathfrak{N}$ and $x_1^p = 1$. Since $x_1 \in \mathbf{C}_{\mathfrak{G}}(\mathfrak{A})$, it follows from the hypothesis that $x_1 \in \mathfrak{A}$. Hence $x\mathfrak{N} \in \mathfrak{A}\mathfrak{N}/\mathfrak{N}$. Thus $\mathfrak{G}/\mathfrak{N}$

satisfies the conditions of the theorem. Since $\mathfrak{G}$ is a minimal counterexample, it follows that $\mathfrak{K}\mathfrak{N}/\mathfrak{N} \leq \mathbf{O}_{p'}(\mathfrak{G}/\mathfrak{N}) = 1$ and $\mathfrak{K} \leq \mathfrak{N} = \mathbf{O}_{p'}(\mathfrak{G})$.

Let $\mathscr{S}$ be the set of subgroups $\mathfrak{X}$ of $\mathbf{O}_p(\mathfrak{G})$ such that $\mathfrak{A}\mathfrak{K} \leq \mathbf{N}_{\mathfrak{G}}(\mathfrak{X})$ but $\mathfrak{K} \nleq \mathbf{C}_{\mathfrak{G}}(\mathfrak{X})$. Since $\mathbf{O}_{p'}(\mathfrak{G}) = 1$, $\mathfrak{G}$ is p-constrained and $\mathfrak{K}$ is a non-identity p'-group, $\mathbf{O}_p(\mathfrak{G}) \in \mathscr{S}$. Thus $\mathscr{S}$ is non-empty. Let $\mathfrak{H}$ be a minimal element of $\mathscr{S}$. Thus

b) $\mathfrak{H}$ is a subgroup of $\mathbf{O}_p(\mathfrak{G})$, $\mathfrak{A}\mathfrak{K} \leq \mathbf{N}_{\mathfrak{G}}(\mathfrak{H})$, $\mathfrak{K} \nleq \mathbf{C}_{\mathfrak{G}}(\mathfrak{H})$.

We prove next that

c) the class of $\mathfrak{H}$ is at most 2 (cf. III, 13.5).

Indeed, by minimality of $\mathfrak{H}$, $\mathfrak{K} \leq \mathbf{C}_{\mathfrak{G}}(\mathfrak{H}')$. Thus $[\mathfrak{H}', \mathfrak{H}, \mathfrak{K}] = [\mathfrak{K}, \mathfrak{H}', \mathfrak{H}] = 1$. By III, 1.10, $[\mathfrak{H}, \mathfrak{K}, \mathfrak{H}'] = 1$. Now since $\mathfrak{K} \nleq \mathbf{C}_{\mathfrak{G}}(\mathfrak{H})$, there exists $y \in \mathfrak{K}$ such that y induces a non-trivial automorphism η on $\mathfrak{H}$. In fact η is not of order a power of p, since $\mathfrak{K}$ is a p'-group. But $\mathfrak{K} \leq \mathbf{N}_{\mathfrak{G}}([\mathfrak{H}, \mathfrak{K}])$, so η leaves $[\mathfrak{H}, \mathfrak{K}]$ fixed and induces the identity automorphism on $\mathfrak{H}/[\mathfrak{H}, \mathfrak{K}]$. It follows from I, 4.4 that η induces a non-identity automorphism on $[\mathfrak{H}, \mathfrak{K}]$. Thus $\mathfrak{K} \nleq \mathbf{C}_{\mathfrak{G}}([\mathfrak{H}, \mathfrak{K}])$. But $\mathfrak{A}\mathfrak{K} \leq \mathbf{N}_{\mathfrak{G}}([\mathfrak{H}, \mathfrak{K}])$, so $[\mathfrak{H}, \mathfrak{K}] \in \mathscr{S}$. By minimality of $\mathfrak{H}$, $[\mathfrak{H}, \mathfrak{K}] = \mathfrak{H}$. Thus

$$[\mathfrak{H}, \mathfrak{H}'] = [\mathfrak{H}, \mathfrak{K}, \mathfrak{H}'] = 1$$

and the class of $\mathfrak{H}$ is at most 2.

Since $|\mathfrak{H}|$ is odd, each element of $\mathfrak{H}$ has a unique square root. Thus by c) and VIII, 9.16, there exists an addition on $\mathfrak{H}$ with respect to which $\mathfrak{H}$ is an Abelian group $\tilde{\mathfrak{H}}$ and $\mathfrak{A}\mathfrak{K}$ is a group of operators on $\tilde{\mathfrak{H}}$. By III, 13.4b),

$$\tilde{\mathfrak{H}} = [\tilde{\mathfrak{H}}, \mathfrak{K}] \oplus \mathbf{C}_{\tilde{\mathfrak{H}}}(\mathfrak{K});$$

here $[\tilde{\mathfrak{H}}, \mathfrak{K}]$ is understood in the sense of the additive structure of $\tilde{\mathfrak{H}}$ and $\mathbf{C}_{\tilde{\mathfrak{H}}}(\mathfrak{K}) = \mathbf{C}_{\mathfrak{H}}(\mathfrak{K})$. Thus $[\tilde{\mathfrak{H}}, \mathfrak{K}]$ is an $\mathfrak{A}$-invariant subgroup of $\tilde{\mathfrak{H}}$, and by b), $[\tilde{\mathfrak{H}}, \mathfrak{K}] \neq 0$. Since $[\tilde{\mathfrak{H}}, \mathfrak{K}]$ and $\mathfrak{A}$ are p-groups, there exists an element u of order p in $[\tilde{\mathfrak{H}}, \mathfrak{K}]$ such that $u \in \mathbf{C}_{\tilde{\mathfrak{H}}}(\mathfrak{A}) = \mathbf{C}_{\mathfrak{H}}(\mathfrak{A})$. But then, by hypothesis, $u \in \mathfrak{A}$. Thus if $g \in \mathfrak{K}$ and the commutator $[u, g]$ is now understood in the ordinary sense,

$$[u, g] \in \mathfrak{H} \cap [\mathfrak{A}, \mathfrak{K}] \leq \mathfrak{H} \cap \mathfrak{K} = 1.$$

Thus $u \in \mathbf{C}_{\mathfrak{H}}(\mathfrak{K}) = \mathbf{C}_{\tilde{\mathfrak{H}}}(\mathfrak{K})$ and

$$u \in \mathbf{C}_{\tilde{\mathfrak{H}}}(\mathfrak{K}) \cap [\tilde{\mathfrak{H}}, \mathfrak{K}] = 0,$$

a contradiction. **q.e.d.**

§ 2. Groups of Order Divisible by at Most Two Primes

The main aim of this section is a proof of Burnside's theorem on the solubility of groups $\mathfrak{G}$ for which $|\mathfrak{G}|$ is divisible by at most two primes. This was proved in V, 7.3, following Burnside's original method (1904), which uses the theory of group characters. In 1961, Thompson gave a proof in the case when $|\mathfrak{G}|$ is odd, which made no use of character-theory. About 10 years later a number of improvements were made by various authors, and the proof given here is a culmination of these.

We begin with a rather technical lemma.

2.1 Lemma. *Suppose that $\mathfrak{G}$ is a p-soluble group of odd order, $\mathbf{O}_{p'}(\mathfrak{G}) = 1$ and the Sylow p-subgroups of $\mathfrak{G}/\mathbf{O}_{p,p'}(\mathfrak{G})$ are cyclic. Suppose also that if $\mathfrak{Z} = \Omega_1(\mathbf{Z}(\mathbf{O}_p(\mathfrak{G})))$, $\mathbf{C}_{\mathfrak{G}}(\mathfrak{Z}) = \mathbf{O}_p(\mathfrak{G})$. Let $\mathfrak{S}$ be a Sylow p-subgroup of $\mathfrak{G}$ and let $\mathfrak{A}$ be an elementary Abelian subgroup of $\mathfrak{S}$ of order as large as possible. Then $\mathfrak{A} \leq \mathbf{O}_p(\mathfrak{G})$.*

Proof. Let $\mathfrak{P} = \mathbf{O}_p(\mathfrak{G})$, $\mathfrak{N} = \mathbf{O}_{p,p'}(\mathfrak{G})$. Since $\mathfrak{S}\mathfrak{N}/\mathfrak{N}$ is cyclic and $\mathfrak{A}\mathfrak{N}/\mathfrak{N}$ is elementary Abelian, $|\mathfrak{A}\mathfrak{N}/\mathfrak{N}| \leq p$. Since $\mathfrak{A} \cap \mathfrak{N} = \mathfrak{A} \cap \mathfrak{P}$, it follows that $|\mathfrak{A} : \mathfrak{A} \cap \mathfrak{P}| \leq p$. But $(\mathfrak{A} \cap \mathfrak{P})\mathfrak{Z}$ is an elementary Abelian subgroup of $\mathfrak{S}$, since $\mathfrak{Z} \leq \mathbf{Z}(\mathfrak{P})$. Hence $|(\mathfrak{A} \cap \mathfrak{P})\mathfrak{Z}| \leq |\mathfrak{A}|$ and

$$|\mathfrak{Z} : \mathfrak{A} \cap \mathfrak{Z}| \leq |\mathfrak{A} : \mathfrak{A} \cap \mathfrak{P}| \leq p.$$

But $\mathfrak{A} \cap \mathfrak{Z} \leq \mathbf{Z}(\mathfrak{A}\mathfrak{Z})$, so $\mathfrak{Z}/(\mathfrak{A} \cap \mathfrak{Z})$ is a normal subgroup of $\mathfrak{A}\mathfrak{Z}/(\mathfrak{A} \cap \mathfrak{Z})$ of order at most p. Thus $[\mathfrak{Z}, \mathfrak{A}] \leq \mathfrak{A} \cap \mathfrak{Z}$ and $[\mathfrak{Z}, \mathfrak{A}, \mathfrak{A}] = 1$. But by IX, 7.4, $\mathfrak{G}$ is p-stable, since $|\mathfrak{G}|$ is odd; (since $|SA(2, p)|$ is even, this also follows from IX, 7.10). Since $\mathbf{C}_{\mathfrak{G}}(\mathfrak{Z}) = \mathfrak{P}$, $\mathbf{O}_p(\mathbf{N}_{\mathfrak{G}}(\mathfrak{Z})/\mathbf{C}_{\mathfrak{G}}(\mathfrak{Z})) = \mathbf{O}_p(\mathfrak{G}/\mathbf{O}_p(\mathfrak{G})) = 1$, so it follows that $\mathfrak{A} \leq \mathfrak{P}$. **q.e.d.**

To use this, we make the following definition.

2.2 Definition. For any p-group $\mathfrak{X}$, let $\mathbf{J}_e(\mathfrak{X})$ be the subgroup of $\mathfrak{X}$ generated by all elementary Abelian subgroups $\mathfrak{A}$ which are of order as large as possible. Thus $\mathbf{J}_e(\mathfrak{X})$ is a characteristic subgroup of $\mathfrak{X}$.

2.3 Lemma. *Suppose that $\mathfrak{G}$ is a p-soluble group of odd order, $\mathbf{O}_{p'}(\mathfrak{G}) = 1$ and the Sylow p-subgroups of $\mathfrak{G}/\mathbf{O}_{p,p'}(\mathfrak{G})$ are cyclic. Suppose also that if $\mathfrak{Z} = \Omega_1(\mathbf{Z}(\mathbf{O}_p(\mathfrak{G})))$, $\mathbf{C}_{\mathfrak{G}}(\mathfrak{Z}) = \mathbf{O}_p(\mathfrak{G})$. Then if $\mathfrak{S} \in S_p(\mathfrak{G})$, $\mathbf{J}_e(\mathfrak{S}) \trianglelefteq \mathfrak{G}$.*

Proof. Let $\mathfrak{P} = \mathbf{O}_p(\mathfrak{G})$. It follows from 2.1 that $\mathbf{J}_e(\mathfrak{S}) \leq \mathfrak{P}$. Hence the elementary Abelian subgroups of $\mathfrak{P}$ of greatest possible order are the same as those of $\mathfrak{S}$. Thus $\mathbf{J}_e(\mathfrak{S}) = \mathbf{J}_e(\mathfrak{P})$. Hence $\mathbf{J}_e(\mathfrak{S})$ is a characteristic subgroup of $\mathfrak{P}$ and hence of $\mathfrak{G}$. **q.e.d.**

The significance of the conclusion of 2.3 lies in the fact that a normal subgroup of $\mathfrak{G}$ is constructed from $\mathfrak{S}$ alone. In applications, it leads to a situation in which the following version of I, 8.8 can be applied.

2.4 Lemma. *Suppose that $\mathfrak{P}$ is a p-subgroup of a group $\mathfrak{G}$, $\mathfrak{U}$ is a characteristic subgroup of $\mathfrak{P}$ and $\mathfrak{P} \in S_p(\mathbf{N}_{\mathfrak{G}}(\mathfrak{U}))$. Then $\mathfrak{P} \in S_p(\mathfrak{G})$.*

Proof. If $\mathfrak{P} < \mathfrak{S} \in S_p(\mathfrak{G})$, then $\mathfrak{P}_0 = \mathbf{N}_{\mathfrak{S}}(\mathfrak{P}) > \mathfrak{P}$, by I, 8.8. But $\mathfrak{P} \lhd \mathfrak{P}_0$ and $\mathfrak{U}$ is a characteristic subgroup of $\mathfrak{P}$, so $\mathfrak{U} \lhd \mathfrak{P}_0$. Thus $\mathfrak{P}_0 \leq \mathbf{N}_{\mathfrak{G}}(\mathfrak{U})$ and $\mathfrak{P}$ cannot be a Sylow p-subgroup of $\mathbf{N}_{\mathfrak{G}}(\mathfrak{U})$. **q.e.d.**

2.5 Lemma. *Suppose that $\mathfrak{G}$ is a simple group, $\mathfrak{B} \leq \mathfrak{A} < \mathfrak{G}$ and $\mathfrak{G} = \mathfrak{A}\mathbf{N}_{\mathfrak{G}}(\mathfrak{B})$. Then $\mathfrak{B} = 1$.*

Proof. If $g \in \mathfrak{G}$, then $g = yx$ with $y \in \mathbf{N}_{\mathfrak{G}}(\mathfrak{B})$, $x \in \mathfrak{A}$ and $\mathfrak{B}^g = \mathfrak{B}^{yx} = \mathfrak{B}^x \leq \mathfrak{A}$. Thus every conjugate of $\mathfrak{B}$ is contained in $\mathfrak{A}$, and if $\overline{\mathfrak{B}}$ is the normal closure of $\mathfrak{B}$ in $\mathfrak{G}$, $\overline{\mathfrak{B}} \leq \mathfrak{A}$. Since $\overline{\mathfrak{B}} \lhd \mathfrak{G}$ and $\mathfrak{G}$ is simple, $\overline{\mathfrak{B}} = 1$ and $\mathfrak{B} = 1$. **q.e.d.**

2.6 Lemma (MATSUYAMA [1]). *Suppose that $\mathfrak{G} = \mathfrak{A}\mathfrak{B}$, $\mathfrak{A}_0 \leqq \mathfrak{A}$ and $\mathfrak{B}_0 \leqq \mathfrak{B}$. Suppose that $\mathfrak{K} = \langle \mathfrak{B}_0^c \mid c \in \mathfrak{A}_0 \rangle$ is a p-group. Then there exists $\mathfrak{S} \in S_p(\mathfrak{G})$ such that $\mathfrak{A}_0$ normalises the weak closure*

$$\langle \mathfrak{B}_0^g \mid g \in \mathfrak{G}, \mathfrak{B}_0^g \leq \mathfrak{S} \rangle.$$

Proof. Let $\mathscr{X}$ be the set of p-subgroups $\mathfrak{X}$ of $\mathfrak{G}$ such that $\mathfrak{X}$ is generated by a set of subgroups of the form $\mathfrak{K}^a$ with $a \in \mathfrak{A}$. Since $\mathfrak{K} \in \mathscr{X}$, $\mathscr{X}$ is non-empty. Let $\mathfrak{J}$ be a maximal element of $\mathscr{X}$ and choose $\mathfrak{S} \in S_p(\mathfrak{G})$ such that $\mathfrak{J} \leq \mathfrak{S}$. Let

$$\mathfrak{J}_1 = \langle \mathfrak{B}_0^g \mid g \in \mathfrak{G}, \mathfrak{B}_0^g \leq \mathfrak{S} \rangle.$$

Since $\mathfrak{J}$ is generated by conjugates of $\mathfrak{K}$ and $\mathfrak{K}$ by conjugates of $\mathfrak{B}_0$, $\mathfrak{J} \leq \mathfrak{J}_1$. Also $\mathfrak{A}_0 \leq \mathbf{N}_{\mathfrak{G}}(\mathfrak{K})$, so for all $a \in \mathfrak{A}$, $\mathfrak{A}_0 = \mathfrak{A}_0^a \leq \mathbf{N}_{\mathfrak{G}}(\mathfrak{K}^a)$. Hence $\mathfrak{A}_0 \leq \mathbf{N}_{\mathfrak{G}}(\mathfrak{J})$. We show that $\mathfrak{J} = \mathfrak{J}_1$. First we make the following observations.

a) Every conjugate of $\mathfrak{B}_0$ in $\mathfrak{G}$ is of the form $\mathfrak{B}_0^a$ for some $a \in \mathfrak{A}$.

For $\mathfrak{G} = \mathfrak{B}\mathfrak{A}$ and $\mathfrak{B}_0 \trianglelefteq \mathfrak{B}$.

b) If $a \in \mathfrak{A}$ and $\mathfrak{B}_0^a \leq \mathbf{N}_{\mathfrak{G}}(\mathfrak{J})$, then $\mathfrak{B}_0^a \leq \mathfrak{J}$.

For

$$\mathfrak{K}^a = \langle \mathfrak{B}_0^{ca} \mid c \in \mathfrak{A}_0 \rangle = \langle \mathfrak{B}_0^{ac'} \mid c' \in \mathfrak{A}_0 \rangle \leq \langle \mathfrak{B}_0^a, \mathfrak{A}_0 \rangle.$$

Thus if $\mathfrak{B}_0^a \leq \mathbf{N}_{\mathfrak{G}}(\mathfrak{J})$, $\mathfrak{K}^a \leq \mathbf{N}_{\mathfrak{G}}(\mathfrak{J})$. But then $\mathfrak{K}^a\mathfrak{J} \in \mathscr{X}$, so $\mathfrak{K}^a \leq \mathfrak{J}$ by maximality of $\mathfrak{J}$. Hence $\mathfrak{B}_0^a \leq \mathfrak{J}$.

Now suppose that $\mathfrak{J} < \mathfrak{J}_1$. Then by a) and the definition of $\mathfrak{J}_1$, $\mathfrak{B}_0^a \leq \mathfrak{S}$ and $\mathfrak{B}_0^a \not\leq \mathfrak{J}$ for some $a \in \mathfrak{A}$. Thus by b), $\mathfrak{B}_0^a \not\leq \mathbf{N}_{\mathfrak{G}}(\mathfrak{J})$. Hence $\mathfrak{S} \not\leq \mathbf{N}_{\mathfrak{G}}(\mathfrak{J})$ and $\mathbf{N}_{\mathfrak{S}}(\mathfrak{J}) < \mathfrak{S}$. By I, 8.8, there exists $s \in \mathfrak{S}$ such that $s \notin \mathbf{N}_{\mathfrak{S}}(\mathfrak{J})$ but $\mathbf{N}_{\mathfrak{S}}(\mathfrak{J})^s = \mathbf{N}_{\mathfrak{S}}(\mathfrak{J})$. Thus $\mathfrak{J}^s \neq \mathfrak{J}$ and for some $g \in \mathfrak{G}$, $\mathfrak{B}_0^g \leq \mathfrak{J}$ but $\mathfrak{B}_0^{gs} \not\leq \mathfrak{J}$. By a), $\mathfrak{B}_0^{gs} = \mathfrak{B}_0^{a'}$ for some $a' \in \mathfrak{A}$. Thus $\mathfrak{B}_0^{a'} \not\leq \mathfrak{J}$, so by b), $\mathfrak{B}_0^{a'} \not\leq \mathbf{N}_{\mathfrak{G}}(\mathfrak{J})$. However,

$$\mathfrak{B}_0^{a'} = \mathfrak{B}_0^{gs} \leq \mathfrak{J}^s \leq \mathbf{N}_{\mathfrak{S}}(\mathfrak{J})^s = \mathbf{N}_{\mathfrak{S}}(\mathfrak{J}),$$

a contradiction. Thus $\mathfrak{J} = \mathfrak{J}_1$ and $\mathfrak{A}_0 \leq \mathbf{N}_{\mathfrak{G}}(\mathfrak{J}_1)$. **q.e.d.**

2.7 Corollary. *Let $\mathfrak{M}$ be a maximal subgroup of the simple group $\mathfrak{G}$ and suppose that $\mathfrak{M}$ is soluble. If $\mathfrak{G}$ possesses a subgroup $\mathfrak{Q}$ such that $|\mathfrak{G} : \mathfrak{Q}|$ is a power of the prime p and $\mathbf{O}_{p'}(\mathfrak{M}) = 1$, then $\mathfrak{M} \cap \mathbf{Z}(\mathfrak{Q}) = 1$.*

(Note that this follows trivially from V, 7.2. It appears, however, that no character-free proof of V, 7.2 is known.)

Proof. We suppose that $\mathfrak{M} \neq 1$. Since $\mathbf{O}_{p'}(\mathfrak{M}) = 1$ and $\mathfrak{M}$ is soluble, $\mathbf{O}_p(\mathfrak{M}) \neq 1$. By IX, 1.4, $\mathbf{O}_p(\mathfrak{M}) \geq \mathbf{C}_{\mathfrak{M}}(\mathbf{O}_p(\mathfrak{M}))$. If $\mathbf{O}_p(\mathfrak{M}) \leq \mathfrak{P} \in S_p(\mathfrak{G})$, $\mathbf{Z}(\mathfrak{P}) \leq \mathbf{C}_{\mathfrak{G}}(\mathbf{O}_p(\mathfrak{M})) \leq \mathbf{N}_{\mathfrak{G}}(\mathbf{O}_p(\mathfrak{M}))$. But since $\mathfrak{G}$ is simple and $\mathfrak{M}$ is maximal, $\mathbf{N}_{\mathfrak{G}}(\mathbf{O}_p(\mathfrak{M})) = \mathfrak{M}$. Thus $\mathbf{Z}(\mathfrak{P}) \leq \mathfrak{M}$ and

$$\mathbf{Z}(\mathfrak{P}) \leq \mathbf{C}_{\mathfrak{M}}(\mathbf{O}_p(\mathfrak{M})) \leq \mathbf{O}_p(\mathfrak{M}).$$

Hence $\mathbf{Z}(\mathfrak{P})^y \leq \mathbf{O}_p(\mathfrak{M})$ for all $y \in \mathfrak{M}$. Let

$$\mathfrak{K} = \langle \mathbf{Z}(\mathfrak{P})^y \mid y \in \mathfrak{M} \cap \mathbf{Z}(\mathfrak{Q}) \rangle.$$

Thus $\mathfrak{K} \leq \mathbf{O}_p(\mathfrak{M})$ and $\mathfrak{K}$ is a p-group. The conditions of 2.6 are therefore satisfied, with

$$(\mathfrak{A}, \mathfrak{A}_0, \mathfrak{B}, \mathfrak{B}_0) = (\mathfrak{Q}, \mathfrak{M} \cap \mathbf{Z}(\mathfrak{Q}), \mathfrak{P}, \mathbf{Z}(\mathfrak{P})).$$

Thus there exists $\mathfrak{S} \in S_p(\mathfrak{G})$ such that $\mathfrak{M} \cap \mathbf{Z}(\mathfrak{Q}) \leq \mathbf{N}_{\mathfrak{G}}(\mathfrak{J})$, where

$$\mathfrak{J} = \langle \mathbf{Z}(\mathfrak{P})^g \,|\, g \in \mathfrak{G}, \mathbf{Z}(\mathfrak{P})^g \leq \mathfrak{S} \rangle.$$

But clearly, $\mathfrak{S} \leq \mathbf{N}_{\mathfrak{G}}(\mathfrak{J}) < \mathfrak{G}$, so $\mathfrak{G} = \mathfrak{S}\mathfrak{Q} = \mathbf{N}_{\mathfrak{G}}(\mathfrak{J})\mathbf{N}_{\mathfrak{G}}(\mathfrak{M} \cap \mathbf{Z}(\mathfrak{Q}))$. By 2.5, $\mathfrak{M} \cap \mathbf{Z}(\mathfrak{Q}) = 1$. **q.e.d.**

2.8 Theorem (BURNSIDE). *A group of order divisible by at most two primes is soluble.*

Proof. Suppose that this is false and let $\mathfrak{G}$ be a counterexample of minimal order. We shall obtain a contradiction in several steps.

2.9. a) *$\mathfrak{G}$ is simple and is not of prime-power order.*

b) *Every proper subgroup of $\mathfrak{G}$ is soluble.*

c) *If p is a prime divisor of $|\mathfrak{G}|$ and $\mathfrak{S} \in S_p(\mathfrak{G})$, $\mathfrak{S}$ does not normalise any non-identity p'-subgroup of $\mathfrak{G}$.*

Proof. a) This is trivial.

b) This follows at once from the minimality of $|\mathfrak{G}|$.

c) Suppose that $\mathfrak{K}$ is a p'-subgroup and $\mathfrak{S} \leq \mathbf{N}_{\mathfrak{G}}(\mathfrak{K})$. Then $\mathfrak{K}$ is a q-group, where q is the prime divisor of $|\mathfrak{G}|$ other than p. Thus $\mathfrak{K}$ is contained in a Sylow q-subgroup $\mathfrak{Q}$ of $\mathfrak{G}$ and $\mathfrak{G} = \mathfrak{Q}\mathfrak{S} = \mathfrak{Q}\mathbf{N}_{\mathfrak{G}}(\mathfrak{K})$. By 2.5, $\mathfrak{K} = 1$. **q.e.d.**

2.10. *If $\mathfrak{M}$ is a maximal subgroup of $\mathfrak{G}$, $\mathbf{F}(\mathfrak{M})$ is of prime-power order.*

Proof. This is proved in three steps.

a) Let $\mathfrak{M}$ be a maximal subgroup of $\mathfrak{G}$, and put $\mathfrak{F} = \mathbf{F}(\mathfrak{M})$. If $\mathfrak{F}$ is not of prime-power order, there exists a prime p such that $\mathfrak{F}$ has more than one subgroup of order p.

Suppose that this is false. Then by III, 8.2, for any prime p, either $\mathbf{O}_p(\mathfrak{M})$ is cyclic or $p = 2$ and $\mathbf{O}_2(\mathfrak{M})$ is non-Abelian.

Suppose first that $\mathbf{O}_p(\mathfrak{M})$ is cyclic for all p. Let p, q be the prime divisors of $|\mathfrak{G}|$ with $p > q$. Then p does not divide $|\mathbf{Aut}\,\mathbf{O}_q(\mathfrak{M})|$, by I, 4.6. Thus if $\mathfrak{P} \in S_p(\mathfrak{M})$, $\mathfrak{P}$ centralizes $\mathbf{O}_q(\mathfrak{M})$. Since $\mathbf{O}_p(\mathfrak{M}) \leq \mathfrak{P}$ and $\mathfrak{F} = \mathbf{O}_p(\mathfrak{M}) \times \mathbf{O}_q(\mathfrak{M})$, it follows that $\mathbf{Z}(\mathfrak{P}) \leq \mathbf{C}_{\mathfrak{M}}(\mathbf{O}_p(\mathfrak{M})\mathbf{O}_q(\mathfrak{M})) = \mathbf{C}_{\mathfrak{M}}(\mathfrak{F})$. By III, 4.2, $\mathbf{C}_{\mathfrak{M}}(\mathfrak{F}) \leq \mathfrak{F}$. Thus $\mathbf{Z}(\mathfrak{P}) \leq \mathfrak{F}$ and $\mathbf{Z}(\mathfrak{P}) \leq \mathbf{O}_p(\mathfrak{M})$. Since $\mathbf{O}_p(\mathfrak{M})$ is cyclic, $\mathbf{Z}(\mathfrak{P})$ is a characteristic subgroup of $\mathbf{O}_p(\mathfrak{M})$ and hence of $\mathfrak{M}$. By 2.9a), it follows that $\mathfrak{M} = \mathbf{N}_{\mathfrak{G}}(\mathbf{Z}(\mathfrak{P}))$, since $\mathfrak{M}$ is a maximal subgroup of $\mathfrak{G}$. By 2.4, $\mathfrak{P} \in S_p(\mathfrak{G})$. By 2.9c), $\mathbf{O}_q(\mathfrak{M}) = 1$. Thus $\mathfrak{F}$ is a p-group, contrary to hypothesis.

Now suppose that $\mathbf{O}_2(\mathfrak{M})$ is non-Abelian, and let p be the odd prime divisor of $|\mathfrak{G}|$. Suppose $\mathfrak{Q} \in S_2(\mathfrak{M})$. Then $\mathfrak{Q}/\mathbf{C}_{\mathfrak{Q}}(\mathbf{O}_p(\mathfrak{M}))$ is isomorphic to a group of automorphisms of the cyclic group $\mathbf{O}_p(\mathfrak{M})$ and is therefore Abelian. Hence $\mathbf{C}_{\mathfrak{Q}}(\mathbf{O}_p(\mathfrak{M})) \geq \mathfrak{Q}'$. Thus $\mathbf{Z}(\mathfrak{Q}) \cap \mathfrak{Q}' \leq \mathbf{C}_{\mathfrak{M}}(\mathbf{O}_p(\mathfrak{M})\mathbf{O}_2(\mathfrak{M})) = \mathbf{C}_{\mathfrak{M}}(\mathfrak{F}) \leq \mathfrak{F}$, as in the previous case, and $\mathbf{Z}(\mathfrak{Q}) \cap \mathfrak{Q}' \leq \mathbf{O}_2(\mathfrak{M})$. But since $\mathfrak{Q} \geq \mathbf{O}_2(\mathfrak{M})$, $\mathfrak{Q}' \neq 1$. Thus $\mathbf{Z}(\mathfrak{Q}) \cap \mathfrak{Q}' \neq 1$. Let $\mathfrak{U}$ be a subgroup of $\mathbf{Z}(\mathfrak{Q}) \cap \mathfrak{Q}'$ of order 2. Since $\mathbf{O}_2(\mathfrak{M})$ has only one subgroup of order 2, $\mathfrak{U}$ is a characteristic subgroup both of $\mathbf{Z}(\mathfrak{Q}) \cap \mathfrak{Q}'$ and of $\mathbf{O}_2(\mathfrak{M})$. Hence $\mathfrak{U}$ is a characteristic subgroup of $\mathfrak{Q}$ and of $\mathfrak{M}$. Thus $\mathfrak{M} = \mathbf{N}_{\mathfrak{G}}(\mathfrak{U})$ and by 2.4, $\mathfrak{Q} \in S_2(\mathfrak{G})$. By 2.9c), $\mathbf{O}_p(\mathfrak{M}) = 1$ and $\mathfrak{F}$ is a 2-group, contrary to hypothesis.

b) Suppose that $\mathfrak{M}$ is a maximal subgroup of $\mathfrak{G}$ and $\mathfrak{Z} = \mathbf{Z}(\mathbf{F}(\mathfrak{M}))$. Suppose that $\mathbf{F}(\mathfrak{M})$ is not of prime-power order. If $\mathfrak{L}$ is a maximal subgroup of $\mathfrak{G}$ and $\mathfrak{L}$ contains $\mathfrak{Z}$, then $\mathbf{F}(\mathfrak{L})$ is not of prime-power order and $\mathbf{F}(\mathfrak{L}) \leq \mathbf{F}(\mathfrak{M})$.

Denote the prime divisors of $|\mathfrak{G}|$ by p, q. Then $\mathfrak{Z} = \mathfrak{Z}_p \times \mathfrak{Z}_q$, where $\mathfrak{Z}_p \in S_p(\mathfrak{Z})$, $\mathfrak{Z}_q \in S_q(\mathfrak{Z})$. By hypothesis, $\mathfrak{Z}_p \neq 1$, $\mathfrak{Z}_q \neq 1$. Hence, since $\mathfrak{M}$ is a maximal subgroup of $\mathfrak{G}$, $\mathfrak{M} = \mathbf{N}_{\mathfrak{G}}(\mathfrak{Z}_p) = \mathbf{N}_{\mathfrak{G}}(\mathfrak{Z}_q)$. Thus $\mathbf{N}_{\mathfrak{L}}(\mathfrak{Z}_q) \leq \mathfrak{M}$. But $\mathfrak{Z}_p \trianglelefteq \mathfrak{M}$ and $\mathfrak{Z}_p \leq \mathbf{N}_{\mathfrak{L}}(\mathfrak{Z}_q)$; thus $\mathfrak{Z}_p \trianglelefteq \mathbf{N}_{\mathfrak{L}}(\mathfrak{Z}_q)$ and $\mathfrak{Z}_p \leq \mathbf{O}_{q'}(\mathbf{N}_{\mathfrak{L}}(\mathfrak{Z}_q))$. By 2.9b) and 1.6, $\mathbf{O}_{q'}(\mathbf{N}_{\mathfrak{L}}(\mathfrak{Z}_q)) \leq \mathbf{O}_{q'}(\mathfrak{L}) = \mathbf{O}_p(\mathfrak{L})$, so $\mathfrak{Z}_p \leq \mathbf{O}_p(\mathfrak{L})$. Hence $[\mathfrak{Z}_p, \mathbf{O}_q(\mathfrak{L})] \leq [\mathbf{O}_p(\mathfrak{L}), \mathbf{O}_q(\mathfrak{L})] = 1$ and $\mathbf{O}_q(\mathfrak{L}) \leq \mathbf{N}_{\mathfrak{G}}(\mathfrak{Z}_p) = \mathfrak{M}$. Similarly $\mathbf{O}_p(\mathfrak{L}) \leq \mathfrak{M}$. Also $\mathbf{O}_q(\mathfrak{L}) \leq \mathbf{N}_{\mathfrak{M}}(\mathbf{O}_p(\mathfrak{L}))$. But since $\mathfrak{Z}_p \leq \mathbf{O}_p(\mathfrak{L})$, $\mathbf{O}_p(\mathfrak{L}) \neq 1$, so $\mathbf{N}_{\mathfrak{G}}(\mathbf{O}_p(\mathfrak{L})) = \mathfrak{L}$. It follows that $\mathbf{O}_q(\mathfrak{L}) \trianglelefteq \mathbf{N}_{\mathfrak{M}}(\mathbf{O}_p(\mathfrak{L}))$ and $\mathbf{O}_q(\mathfrak{L}) \leq \mathbf{O}_{p'}(\mathbf{N}_{\mathfrak{M}}(\mathbf{O}_p(\mathfrak{L})))$. Using 1.6, $\mathbf{O}_q(\mathfrak{L}) \leq \mathbf{O}_{p'}(\mathfrak{M}) \leq \mathbf{F}(\mathfrak{M})$. Similarly, $\mathbf{O}_q(\mathfrak{L}) \neq 1$ and $\mathbf{O}_p(\mathfrak{L}) \leq \mathbf{F}(\mathfrak{M})$. Thus $\mathbf{F}(\mathfrak{L})$ is not of prime-power order and $\mathbf{F}(\mathfrak{L}) \leq \mathbf{F}(\mathfrak{M})$.

c) We now prove 2.10.

Suppose that it is false. By a), there exists a prime p such that $\mathbf{F}(\mathfrak{M})$ has more than one subgroup of order p. Hence $\mathbf{O}_p(\mathfrak{M})$ possesses an elementary Abelian subgroup $\mathfrak{B}$ of order p^2. Let $\mathfrak{Z} = \mathbf{Z}(\mathbf{F}(\mathfrak{M}))$.

If x is an element of $\mathfrak{B} - \{1\}$, $\mathbf{C}_{\mathfrak{G}}(x) \neq \mathfrak{G}$. Let $\mathfrak{L}$ be a maximal subgroup of $\mathfrak{G}$ containing $\mathbf{C}_{\mathfrak{G}}(x)$. Since $x \in \mathfrak{B} \leq \mathbf{F}(\mathfrak{M})$, $\mathfrak{Z} \leq \mathbf{C}_{\mathfrak{G}}(x) \leq \mathfrak{L}$. By b), $\mathbf{F}(\mathfrak{L})$ is not of prime-power order and $\mathbf{F}(\mathfrak{L}) \leq \mathbf{F}(\mathfrak{M})$. Thus if $\mathfrak{Z}_1 = \mathbf{Z}(\mathbf{F}(\mathfrak{L}))$, $\mathfrak{Z}_1 \leq \mathfrak{M}$. Hence by b) again (with $\mathfrak{L}$, $\mathfrak{M}$ the other way round), $\mathbf{F}(\mathfrak{M}) \leq \mathbf{F}(\mathfrak{L})$. Thus $\mathbf{F}(\mathfrak{M}) = \mathbf{F}(\mathfrak{L})$. Since $\mathfrak{M} = \mathbf{N}_{\mathfrak{G}}(\mathbf{F}(\mathfrak{M}))$ and $\mathfrak{L} = \mathbf{N}_{\mathfrak{G}}(\mathbf{F}(\mathfrak{L}))$, it follows that $\mathfrak{L} = \mathfrak{M}$. Thus $\mathfrak{M} \geq \mathbf{C}_{\mathfrak{G}}(x)$ for any element x of $\mathfrak{B} - \{1\}$. It follows from 1.9 that $\mathfrak{M}$ contains any p'-subgroup of $\mathfrak{G}$ normalised by $\mathfrak{B}$.

Suppose that $\mathfrak{P} \in S_p(\mathfrak{M})$ and $g \in \mathbf{N}_{\mathfrak{G}}(\mathfrak{P})$. Then $\mathfrak{P} = \mathfrak{P}^g$ normalises $\mathbf{O}_{p'}(\mathfrak{M})^g$, and $\mathbf{O}_{p'}(\mathfrak{M})^g$ is a p'-subgroup of $\mathfrak{G}$ normalised by $\mathfrak{B}$, since $\mathfrak{P} \geq \mathbf{O}_p(\mathfrak{M}) \geq \mathfrak{B}$. Hence $\mathbf{O}_{p'}(\mathfrak{M})^g \leq \mathfrak{M}$. By IX, 1.5b), $\mathbf{O}_{p'}(\mathfrak{M})^g \leq \mathbf{O}_{p'}(\mathfrak{M})$. Hence $\mathbf{N}_{\mathfrak{G}}(\mathfrak{P}) \leq \mathbf{N}_{\mathfrak{G}}(\mathbf{O}_{p'}(\mathfrak{M}))$. But if $\mathbf{O}_{p'}(\mathfrak{M}) \neq 1$, $\mathbf{N}_{\mathfrak{G}}(\mathbf{O}_{p'}(\mathfrak{M})) = \mathfrak{M}$. Thus

$\mathbf{N}_{\mathfrak{G}}(\mathfrak{P}) \leq \mathfrak{M}$ and, by 2.4, $\mathfrak{P} \in S_p(\mathfrak{G})$. But then by 2.9c), $\mathbf{O}_{p'}(\mathfrak{M}) = 1$. Hence $\mathbf{F}(\mathfrak{M})$ is a p-group. **q.e.d.**

2.11. *If* $\mathfrak{S} \in S_p(\mathfrak{G})$ *and* $x \in \mathbf{Z}(\mathfrak{S}) - \{1\}$, *x normalises no non-identity* p'*-subgroup of* $\mathfrak{G}$.

Proof. Let q be the other prime divisor of $|\mathfrak{G}|$ and suppose that $x \in \mathbf{N}_{\mathfrak{G}}(\mathfrak{Q}_0)$, where $\mathfrak{Q}_0$ is a non-identity q-subgroup of $\mathfrak{G}$. Suppose $\mathfrak{Q}_0 \leq \mathfrak{Q} \in S_q(\mathfrak{G})$ and let $\mathfrak{M}$ be a maximal subgroup of $\mathfrak{G}$ containing $\mathbf{N}_{\mathfrak{G}}(\mathfrak{Q}_0)$. Then $x \in \mathfrak{M}$, so $\mathfrak{M} \cap \mathbf{Z}(\mathfrak{S}) \neq 1$. Also $\mathbf{Z}(\mathfrak{Q}) \leq \mathbf{N}_{\mathfrak{G}}(\mathfrak{Q}_0) \leq \mathfrak{M}$, so $\mathfrak{M} \cap \mathbf{Z}(\mathfrak{Q}) \neq 1$. By 2.7, $\mathbf{O}_p(\mathfrak{M}) \neq 1$ and $\mathbf{O}_q(\mathfrak{M}) \neq 1$, contrary to 2.10. **q.e.d.**

2.12. a) $|\mathfrak{G}|$ *is odd.*

b) *If* $\mathfrak{P}$ *is a p-subgroup of* $\mathfrak{G}$, *then* $\mathbf{O}_{p'}(\mathbf{N}_{\mathfrak{G}}(\mathfrak{P})) = 1$ *and* $\mathbf{O}_{p'}(\mathbf{C}_{\mathfrak{G}}(\mathfrak{P})) = 1$.

c) *If* $\mathfrak{P}$ *is a non-identity p-subgroup of* $\mathfrak{G}$, $\mathbf{C}_{\mathfrak{G}}(\Omega_1(\mathbf{Z}(\mathbf{O}_p(\mathbf{C}_{\mathfrak{G}}(\mathfrak{P})))))$ *is a p-group.*

Proof. a) If $|\mathfrak{G}|$ is even, there is an involution t in the centre of some Sylow 2-subgroup of $\mathfrak{G}$. Then $t \notin \mathbf{O}_2(\mathfrak{G})$, so by IX, 7.8, there exists $a \in \mathfrak{G}$ such that $\langle t, t^a \rangle$ is not a 2-group. But $\langle t, t^a \rangle$ is a dihedral group and therefore possesses a non-identity cyclic normal subgroup of odd order. By 2.11, this is impossible, so $|\mathfrak{G}|$ is odd.

b) Suppose $\mathfrak{P} \leq \mathfrak{S} \in S_p(\mathfrak{G})$. Then $\mathbf{Z}(\mathfrak{S}) \leq \mathbf{N}_{\mathfrak{G}}(\mathfrak{P})$, so $\mathbf{O}_{p'}(\mathbf{N}_{\mathfrak{G}}(\mathfrak{P}))$ is normalised by $\mathbf{Z}(\mathfrak{S})$. By 2.11, $\mathbf{O}_{p'}(\mathbf{N}_{\mathfrak{G}}(\mathfrak{P})) = 1$. Since $\mathbf{C}_{\mathfrak{G}}(\mathfrak{P}) \trianglelefteq \mathbf{N}_{\mathfrak{G}}(\mathfrak{P})$, $\mathbf{O}_{p'}(\mathbf{C}_{\mathfrak{G}}(\mathfrak{P})) \leq \mathbf{O}_{p'}(\mathbf{N}_{\mathfrak{G}}(\mathfrak{P})) = 1$.

c) By b), $\mathbf{F}(\mathbf{C}_{\mathfrak{G}}(\mathfrak{P}))$ is a p-group; suppose $\mathfrak{P}\mathbf{F}(\mathbf{C}_{\mathfrak{G}}(\mathfrak{P})) \leq \mathfrak{S} \in S_p(\mathfrak{G})$. Then $\Omega_1(\mathbf{Z}(\mathfrak{S}))$ centralizes $\mathfrak{P}$ and $\mathbf{F}(\mathbf{C}_{\mathfrak{G}}(\mathfrak{P}))$. Hence if $\mathfrak{C} = \mathbf{C}_{\mathfrak{G}}(\mathfrak{P})$, $\mathfrak{C}$ is soluble and $\Omega_1(\mathbf{Z}(\mathfrak{S})) \leq \mathbf{C}_{\mathfrak{C}}(\mathbf{F}(\mathfrak{C}))$. By III, 4.2b), $\mathbf{C}_{\mathfrak{C}}(\mathbf{F}(\mathfrak{C})) \leq \mathbf{F}(\mathfrak{C})$, so $\Omega_1(\mathbf{Z}(\mathfrak{S}))$ is contained in and centralizes $\mathbf{F}(\mathfrak{C})$. Hence $\Omega_1(\mathbf{Z}(\mathfrak{S})) \leq \Omega_1(\mathbf{Z}(\mathbf{F}(\mathbf{C}_{\mathfrak{G}}(\mathfrak{P}))))$. Thus by 2.11, $\Omega_1(\mathbf{Z}(\mathbf{F}(\mathbf{C}_{\mathfrak{G}}(\mathfrak{P}))))$ normalises no non-identity p'-subgroup of $\mathfrak{G}$. Hence $\mathbf{C}_{\mathfrak{G}}(\Omega_1(\mathbf{Z}(\mathbf{F}(\mathbf{C}_{\mathfrak{G}}(\mathfrak{P})))))$ has no p'-element other than 1. **q.e.d.**

2.13. *Suppose that p, q are the prime divisors of* $|\mathfrak{G}|$ *and* $q < p$. *If* $\mathfrak{Q}$ *is a non-identity q-subgroup of* $\mathfrak{G}$, *the Sylow p-subgroups of* $\mathbf{N}_{\mathfrak{G}}(\mathfrak{Q})$ *are cyclic.*

Proof. Suppose that this is false. Since p is odd, it then follows from III, 7.6 that there exists an elementary Abelian subgroup $\mathfrak{B}$ of order p^2 such that $\mathfrak{B} \leq \mathbf{N}_{\mathfrak{G}}(\mathfrak{Q})$ for some non-identity q-subgroup $\mathfrak{Q}$. Then $\mathfrak{B}$ normalises $\Omega_1(\mathbf{Z}(\mathfrak{Q}))$. Hence if $\mathscr{A}$ is the set of all ordered pairs $(\mathfrak{A}, \mathfrak{B})$, where $\mathfrak{A}$ is an elementary Abelian group of order p^2, $\mathfrak{B} \neq 1$ and $\mathfrak{B}$ is a maximal element of the set of elementary Abelian q-subgroups of $\mathfrak{G}$

normalised by $\mathfrak{A}$, $\mathscr{A}$ is non-empty. Choose an element $(\mathfrak{A}, \mathfrak{B})$ of $\mathscr{A}$ for which $|\mathbf{C}_{\mathfrak{B}}(\mathfrak{A})|$ is maximal.

Since $\mathfrak{B} \leq \Omega_1(\mathbf{Z}(\mathbf{O}_q(\mathbf{C}_{\mathfrak{G}}(\mathfrak{B}))))$, it follows from the maximality of $\mathfrak{B}$ that $\mathfrak{B} = \Omega_1(\mathbf{Z}(\mathbf{O}_q(\mathbf{C}_{\mathfrak{G}}(\mathfrak{B}))))$. By 2.12c), $\mathbf{C}_{\mathfrak{G}}(\mathfrak{B})$ is a q-group. Thus $\mathbf{C}_{\mathfrak{B}}(\mathfrak{A}) < \mathfrak{B}$. But by 1.9,

$$\mathfrak{B} = \langle \mathbf{C}_{\mathfrak{B}}(x) \mid 1 \neq x \in \mathfrak{A} \rangle.$$

Thus there exists $x \in \mathfrak{A}$ such that $x \neq 1$ and $\mathbf{C}_{\mathfrak{B}}(x) > \mathbf{C}_{\mathfrak{B}}(\mathfrak{A})$. Put $\mathfrak{U} = \mathbf{C}_{\mathfrak{B}}(x)$ and $\mathfrak{A} = \langle x, y \rangle$. Then y induces a non-identity automorphism η on $\mathfrak{U}$; indeed η is not a q-automorphism. But η operates trivially on $\mathbf{C}_{\mathfrak{B}}(\mathfrak{A})$, so by I, 4.4, η induces a non-trivial automorphism on $\mathfrak{U}/\mathbf{C}_{\mathfrak{B}}(\mathfrak{A})$. Since $q < p$, p does not divide the order of the automorphism group of a cyclic group of order q. But $\eta^p = 1$, so $|\mathfrak{U} : \mathbf{C}_{\mathfrak{B}}(\mathfrak{A})| \geq q^2$.

Let $\mathfrak{Z} = \Omega_1(\mathbf{Z}(\mathbf{O}_p(\mathbf{C}_{\mathfrak{G}}(x))))$. Note that $x \in \mathfrak{Z}$. Since $\mathfrak{U} \leq \mathbf{C}_{\mathfrak{G}}(x)$, $\mathfrak{U}$ normalises $\mathfrak{Z}$. But by 2.12c), $\mathbf{C}_{\mathfrak{G}}(\mathfrak{Z})$ is a p-group, so $\mathfrak{U}$ operates non-trivially on $\mathfrak{Z}$. By 1.9,

$$\mathfrak{Z} = \langle \mathbf{C}_{\mathfrak{Z}}(\mathfrak{U}_0) \mid |\mathfrak{U} : \mathfrak{U}_0| = q \rangle,$$

so there exists $\mathfrak{U}_0 < \mathfrak{U}$ such that $|\mathfrak{U} : \mathfrak{U}_0| = q$ and $\mathbf{C}_{\mathfrak{Z}}(\mathfrak{U}) < \mathbf{C}_{\mathfrak{Z}}(\mathfrak{U}_0)$. Since $x \in \mathbf{C}_{\mathfrak{Z}}(\mathfrak{U})$, it follows that $|\mathbf{C}_{\mathfrak{Z}}(\mathfrak{U}_0)| \geq p^2$. Let $\mathfrak{C}$ be a subgroup of $\mathbf{C}_{\mathfrak{Z}}(\mathfrak{U}_0)$ of order p^2. Since $|\mathfrak{U}| \geq q^2$, $\mathfrak{U}_0 \neq 1$ and so $(\mathfrak{C}, \mathfrak{U}_1) \in \mathscr{A}$ for some subgroup $\mathfrak{U}_1$ containing $\mathfrak{U}_0$. By the choice of $(\mathfrak{A}, \mathfrak{B})$,

$$|\mathbf{C}_{\mathfrak{B}}(\mathfrak{A})| \geq |\mathbf{C}_{\mathfrak{U}_1}(\mathfrak{C})| \geq |\mathfrak{U}_0| = q^{-1}|\mathfrak{U}| \geq q|\mathbf{C}_{\mathfrak{B}}(\mathfrak{A})|,$$

a contradiction. **q.e.d.**

We are now in a position to apply 2.3.

2.14. *Suppose that* $\mathfrak{M}$ *is a maximal subgroup of* $\mathfrak{G}$, $\mathbf{O}_{p'}(\mathfrak{M}) = 1$ *and* $\mathfrak{S} \in S_p(\mathfrak{M})$. *Then* $\mathfrak{M} = \mathbf{N}_{\mathfrak{G}}(\mathbf{J}_e(\mathfrak{S}))$ *and* $\mathfrak{S} \in S_p(\mathfrak{G})$.

Proof. We show that the Sylow p-subgroups of $\mathfrak{M}/\mathbf{O}_{p,p'}(\mathfrak{M})$ are cyclic. This is clear if $\mathfrak{M}$ is a p-group, so we suppose that this is not the case. Thus if q is the other prime divisor of $|\mathfrak{G}|$ and $\mathfrak{Q} \in S_q(\mathbf{O}_{p,p'}(\mathfrak{M}))$, $\mathfrak{Q} \neq 1$. Hence by 2.13, either the Sylow p-subgroups of $\mathbf{N}_{\mathfrak{G}}(\mathfrak{Q})$ are cyclic, or $q > p$. In the former case, the Frattini argument gives

$$\mathfrak{M} = \mathbf{O}_{p,p'}(\mathfrak{M})\mathbf{N}_{\mathfrak{M}}(\mathfrak{Q})$$

and $\mathfrak{M}/\mathbf{O}_{p,p'}(\mathfrak{M}) \cong \mathbf{N}_{\mathfrak{M}}(\mathfrak{Q})/\mathfrak{K}$ for some $\mathfrak{K}$, so the assertion is clear. Suppose then that $q > p$. In this case, 2.13 shows that the Sylow q-subgroups of $\mathbf{N}_{\mathfrak{G}}(\mathbf{O}_p(\mathfrak{M})) = \mathfrak{M}$ are cyclic. Hence $\mathfrak{R} = \mathbf{O}_{p,p'}(\mathfrak{M})/\mathbf{O}_p(\mathfrak{M})$ is a cyclic q-group. Since q is odd, it follows that **Aut** $\mathfrak{R}$ is cyclic. But by IX, 1.4, $\mathfrak{M}/\mathbf{O}_{p,p'}(\mathfrak{M})$ is isomorphic to a subgroup of **Aut** $\mathfrak{R}$, so the assertion is clear in this case also.

Put $\mathfrak{P} = \mathbf{O}_p(\mathfrak{M})$, $\mathfrak{Z} = \Omega_1(\mathbf{Z}(\mathfrak{P}))$. Since $\mathfrak{P} \trianglelefteq \mathfrak{M}$, $\mathbf{C}_{\mathfrak{G}}(\mathfrak{P}) \leq \mathbf{N}_{\mathfrak{G}}(\mathfrak{P}) = \mathfrak{M}$, so $\mathbf{C}_{\mathfrak{G}}(\mathfrak{P}) \trianglelefteq \mathfrak{M}$ and $\mathbf{O}_p(\mathbf{C}_{\mathfrak{G}}(\mathfrak{P})) \leq \mathbf{O}_p(\mathfrak{M}) = \mathfrak{P}$. Thus $\mathbf{O}_p(\mathbf{C}_{\mathfrak{G}}(\mathfrak{P})) \leq \mathbf{Z}(\mathfrak{P})$ and $\Omega_1(\mathbf{Z}(\mathbf{O}_p(\mathbf{C}_{\mathfrak{G}}(\mathfrak{P})))) \leq \Omega_1(\mathbf{Z}(\mathfrak{P})) = \mathfrak{Z}$. Hence $\mathbf{C}_{\mathfrak{G}}(\mathfrak{Z}) \leq \mathbf{C}_{\mathfrak{G}}(\Omega_1(\mathbf{Z}(\mathbf{O}_p(\mathbf{C}_{\mathfrak{G}}(\mathfrak{P}))))),$ so by 2.12c), $\mathbf{C}_{\mathfrak{G}}(\mathfrak{Z})$ is a p-group. But $\mathfrak{Z} \trianglelefteq \mathfrak{M}$, so $\mathbf{C}_{\mathfrak{M}}(\mathfrak{Z}) \trianglelefteq \mathfrak{M}$ and $\mathbf{C}_{\mathfrak{M}}(\mathfrak{Z}) \leq \mathbf{O}_p(\mathfrak{M}) = \mathfrak{P}$. Hence $\mathbf{C}_{\mathfrak{M}}(\mathfrak{Z}) = \mathfrak{P}$.

It follows from 2.3 that $\mathbf{J}_e(\mathfrak{S}) \trianglelefteq \mathfrak{M}$. Since $\mathbf{J}_e(\mathfrak{S}) \neq 1$, $\mathfrak{M} = \mathbf{N}_{\mathfrak{G}}(\mathbf{J}_e(\mathfrak{S}))$. By 2.4, $\mathfrak{S} \in S_p(\mathfrak{G})$. **q.e.d.**

2.15. *Let $\mathfrak{M}$ be a maximal subgroup of $\mathfrak{G}$. If p is a prime and $\mathbf{O}_{p'}(\mathfrak{M}) = 1$, then $\mathfrak{M} \cap \mathfrak{M}^g$ is a p'-group for any $g \in \mathfrak{G} - \mathfrak{M}$.*

Proof. Choose $g \in \mathfrak{G} - \mathfrak{M}$ such that $|\mathfrak{M} \cap \mathfrak{M}^g|$ is divisible by as large a power of p as possible. Suppose $1 \neq \mathfrak{P} \in S_p(\mathfrak{M} \cap \mathfrak{M}^g)$. Then $\mathbf{N}_{\mathfrak{G}}(\mathfrak{P}) < \mathfrak{G}$ and $\mathbf{N}_{\mathfrak{G}}(\mathfrak{P}) \leq \mathfrak{L}$ for some maximal subgroup $\mathfrak{L}$ of $\mathfrak{G}$. If $\mathfrak{P} \leq \mathfrak{S} \in S_p(\mathfrak{G})$, $\mathbf{Z}(\mathfrak{S}) \leq \mathbf{N}_{\mathfrak{G}}(\mathfrak{P}) \leq \mathfrak{L}$, so by 2.11, $\mathbf{O}_{p'}(\mathfrak{L}) = 1$.

By 2.14, $\mathfrak{L} = \mathbf{N}_{\mathfrak{G}}(\mathbf{J}_e(\mathfrak{P}_1))$ and $\mathfrak{M} = \mathbf{N}_{\mathfrak{G}}(\mathbf{J}_e(\mathfrak{P}_2))$ for Sylow p-subgroups $\mathfrak{P}_1$, $\mathfrak{P}_2$ of $\mathfrak{G}$. Since $\mathfrak{P}_1$, $\mathfrak{P}_2$ are conjugate in $\mathfrak{G}$, it follows that $\mathfrak{L} = \mathfrak{M}^h$ for some $h \in \mathfrak{G}$. If $\mathfrak{L} = \mathfrak{M}$, then $\mathbf{N}_{\mathfrak{G}}(\mathfrak{P}) \leq \mathfrak{M}$, $\mathbf{N}_{\mathfrak{M}^g}(\mathfrak{P}) \leq \mathfrak{M} \cap \mathfrak{M}^g$, $\mathfrak{P} \in S_p(\mathbf{N}_{\mathfrak{M}^g}(\mathfrak{P}))$ and $\mathfrak{P} \in S_p(\mathfrak{M}^g)$ by 2.4; thus $\mathfrak{P} \in S_p(\mathfrak{M})$. This is also the case if $\mathfrak{L} \neq \mathfrak{M}$. For in this case it follows from the choice of g that $|\mathfrak{L} \cap \mathfrak{M}|$ is not divisible by a higher power of p than $|\mathfrak{P}|$.
Thus $\mathfrak{P} \in S_p(\mathfrak{L} \cap \mathfrak{M})$, $\mathfrak{P} \in S_p(\mathbf{N}_{\mathfrak{M}}(\mathfrak{P}))$ and $\mathfrak{P} \in S_p(\mathfrak{M})$ by 2.4. Hence by 2.14, $\mathfrak{M} = \mathbf{N}_{\mathfrak{G}}(\mathbf{J}_e(\mathfrak{P})) = \mathfrak{M}^g$. But then $g \in \mathfrak{M}$, a contradiction. **q.e.d.**

2.16. *If $\mathfrak{S} \in S_p(\mathfrak{G})$, $|\mathfrak{S}|^2 \leq |\mathfrak{G}|$.*

Proof. Let $\mathfrak{M}$ be a maximal subgroup of $\mathfrak{G}$ such that $\mathfrak{M} \geq \mathfrak{S}$. By 2.9c), $\mathbf{O}_{p'}(\mathfrak{M}) = 1$, so by 2.15, $\mathfrak{M} \cap \mathfrak{M}^g$ is a p'-group for any $g \in \mathfrak{G} - \mathfrak{M}$. Thus $|\mathfrak{M} : \mathfrak{M} \cap \mathfrak{M}^g| \geq |\mathfrak{S}|$ and

$$|\mathfrak{S}|^2 \leq \frac{|\mathfrak{M}|\,|\mathfrak{M}^g|}{|\mathfrak{M} \cap \mathfrak{M}^g|} = |\mathfrak{M}\mathfrak{M}^g| \leq |\mathfrak{G}|. \qquad \textbf{q.e.d.}$$

But 2.16 cannot be true for both prime divisors of $|\mathfrak{G}|$, so we have a contradiction and 2.8 is proved.

2.17 Remark. Burnside also proved that if $|\mathfrak{G}| = p^a q^b$, where p, q are primes and $p^a > q^b$, then $\mathbf{O}_p(\mathfrak{G}) \neq 1$ except possibly when a) $p = 2$ and q is a Fermat prime, or b) $q = 2$ and p is a Mersenne prime.

GLAUBERMAN [8] proved the following analogous result. Suppose that $|\mathfrak{G}| = p^a q^b$, where p, q are primes. If $\mathfrak{G}$ has a p-subgroup of class at most 2 of order greater than that of any q-subgroup of class at most 2, $\mathbf{O}_p(\mathfrak{G}) \neq 1$.

§ 3. The J-Subgroup

A crucial point in the proof in § 2 was the use of the subgroup $\mathbf{J}_e(\mathfrak{X})$. A similarly defined subgroup was used in IV, § 6; it was denoted there by $\mathbf{J}(\mathfrak{X})$. In this chapter it will be denoted by $\mathbf{J}_0(\mathfrak{X})$, and we shall use $\mathbf{J}(\mathfrak{X})$ for yet another subgroup defined in a similar way. The properties of it depend on a replacement theorem of Thompson (3.3). We begin with the following lemma.

3.1 Lemma. *Suppose that $\mathfrak{P}$ is a p-group and $\mathfrak{A}$ is an Abelian subgroup of $\mathfrak{P}$. Let x be an element of $\mathfrak{P}$ for which $\mathfrak{M} = [x, \mathfrak{A}]$ is Abelian, and put*

$$\mathfrak{B} = \mathfrak{M}\mathbf{C}_{\mathfrak{A}}(\mathfrak{M}).$$

a) *$\mathfrak{B}$ is an Abelian subgroup of $\mathfrak{P}$, and $|\mathfrak{B} \cap \mathfrak{N}| \geq |\mathfrak{A} \cap \mathfrak{N}|$ for any $\mathfrak{N} \trianglelefteq \mathfrak{P}$. In particular, $|\mathfrak{B}| \geq |\mathfrak{A}|$.*

b) *If also $\mathfrak{P} = \mathfrak{P}_1 \geq \mathfrak{P}_2 \geq \cdots \geq \mathfrak{P}_k = 1$ is a central series of $\mathfrak{P}$ such that $|\mathfrak{B} \cap \mathfrak{P}_i| = |\mathfrak{A} \cap \mathfrak{P}_i|$ $(i = 1, \ldots, k)$, then $\mathfrak{B} = \mathfrak{A}$.*

Proof. a) Since $\mathfrak{M}$, $\mathfrak{A}$ are Abelian and $[\mathfrak{M}, \mathbf{C}_{\mathfrak{A}}(\mathfrak{M})] = 1$, $\mathfrak{B}$ is an Abelian subgroup of $\mathfrak{P}$.

Suppose that a, b are elements of $\mathfrak{A}$ and $[x, a] \equiv [x, b] \bmod \mathfrak{A} \cap \mathfrak{M}$. Then $x^a \equiv x^b$ and $x^{ab^{-1}} \equiv x \bmod \mathfrak{A} \cap \mathfrak{M}$, whence $[x, ab^{-1}] \in \mathfrak{A} \cap \mathfrak{M}$. It follows that for any element c of $\mathfrak{A}$, $[x, ab^{-1}]$ commutes with c and $[x, c]$; hence

$$\begin{aligned}[x, c]^{ab^{-1}} &= [x^{ab^{-1}}, c^{ab^{-1}}] = [x[x, ab^{-1}], c] \\ &= [x, c]^{[x, ab^{-1}]}[x, ab^{-1}, c] = [x, c].\end{aligned}$$

Since $\mathfrak{M} = [x, \mathfrak{A}] = \langle [x, c] | c \in \mathfrak{A} \rangle$, it follows that $ab^{-1} \in \mathbf{C}_{\mathfrak{A}}(\mathfrak{M})$ and $a \equiv b \bmod \mathbf{C}_{\mathfrak{A}}(\mathfrak{M})$.

If $|(\mathfrak{N} \cap \mathfrak{A})\mathbf{C}_{\mathfrak{A}}(\mathfrak{M}) : \mathbf{C}_{\mathfrak{A}}(\mathfrak{M})| = m$, there exist elements $a_1, \ldots, a_m$ in $\mathfrak{N} \cap \mathfrak{A}$ such that $a_i \not\equiv a_j \bmod \mathbf{C}_{\mathfrak{A}}(\mathfrak{M})$ $(i \neq j)$. It follows that if $b_i = [x, a_i]$ $(i = 1, \ldots, m)$, $b_i \in \mathfrak{N} \cap \mathfrak{M}$ and $b_i \not\equiv b_j \bmod \mathfrak{A} \cap \mathfrak{M}$ $(i \neq j)$. Since $b_i \in \mathfrak{M}$, $b_i \not\equiv b_j \bmod \mathbf{C}_{\mathfrak{A}}(\mathfrak{M})$ $(i \neq j)$, and $b_i \not\equiv b_j \bmod \mathfrak{N} \cap \mathbf{C}_{\mathfrak{A}}(\mathfrak{M})$. Since $b_i \in \mathfrak{N} \cap \mathfrak{B}$, it follows that

$$\begin{aligned} |\mathfrak{N} \cap \mathfrak{B} : \mathfrak{N} \cap \mathbf{C}_{\mathfrak{A}}(\mathfrak{M})| \geq m &= |(\mathfrak{N} \cap \mathfrak{A})\mathbf{C}_{\mathfrak{A}}(\mathfrak{M}) : \mathbf{C}_{\mathfrak{A}}(\mathfrak{M})| \\ &= |\mathfrak{N} \cap \mathfrak{A} : \mathfrak{N} \cap \mathbf{C}_{\mathfrak{A}}(\mathfrak{M})|. \end{aligned}$$

Hence $|\mathfrak{N} \cap \mathfrak{B}| \geq |\mathfrak{N} \cap \mathfrak{A}|$.

To prove b), suppose that $\mathfrak{B} \neq \mathfrak{A}$. Since $|\mathfrak{B}| = |\mathfrak{B} \cap \mathfrak{P}_1| = |\mathfrak{A} \cap \mathfrak{P}_1| = |\mathfrak{A}|$, it follows that $\mathfrak{A} \not\leq \mathfrak{B}$. Thus there exists a greatest integer $i \leq k$ such that $\mathfrak{P}_i \cap \mathfrak{A} \not\leq \mathfrak{P}_i \cap \mathfrak{B}$. Then $i < k$, $[\mathfrak{P}_i, \mathfrak{P}] \leq \mathfrak{P}_{i+1}$, and $\mathfrak{P}_{i+1} \cap \mathfrak{A} \leq \mathfrak{P}_{i+1} \cap \mathfrak{B}$. By hypothesis, $|\mathfrak{P}_{i+1} \cap \mathfrak{A}| = |\mathfrak{P}_{i+1} \cap \mathfrak{B}|$, so $\mathfrak{P}_{i+1} \cap \mathfrak{A} = \mathfrak{P}_{i+1} \cap \mathfrak{B}$. Thus

$$[x, \mathfrak{P}_i \cap \mathfrak{A}] \leq \mathfrak{P}_{i+1} \cap \mathfrak{M} \leq \mathfrak{P}_{i+1} \cap \mathfrak{B} = \mathfrak{P}_{i+1} \cap \mathfrak{A} \leq \mathfrak{P}_i \cap \mathfrak{A},$$

or $x \in \mathbf{N}_{\mathfrak{P}}(\mathfrak{P}_i \cap \mathfrak{A})$. Thus $\mathfrak{A}$ and $\mathfrak{A}^x$ both centralize $\mathfrak{P}_i \cap \mathfrak{A}$. But $\mathfrak{M} = [x, \mathfrak{A}] \leq \langle \mathfrak{A}, \mathfrak{A}^x \rangle$, so $\mathfrak{M}$ centralizes $\mathfrak{P}_i \cap \mathfrak{A}$. Thus $\mathfrak{P}_i \cap \mathfrak{A} \leq \mathbf{C}_{\mathfrak{A}}(\mathfrak{M}) \leq \mathfrak{B}$. Hence $\mathfrak{P}_i \cap \mathfrak{A} \leq \mathfrak{P}_i \cap \mathfrak{B}$, contrary to the choice of i. Thus $\mathfrak{B} = \mathfrak{A}$.

q.e.d.

3.2 Corollary. *If $\mathfrak{P}$ is a p-group, there exists an Abelian subgroup $\mathfrak{A}$ of $\mathfrak{P}$ with the following properties.*

a) *No Abelian subgroup of $\mathfrak{P}$ is of order greater than $|\mathfrak{A}|$.*

b) $\mathbf{N}_{\mathfrak{P}}(\mathfrak{A}) = \{x \mid x \in \mathfrak{P}, [x, \mathfrak{A}]$ *is Abelian*$\}$.

In particular, if $\mathfrak{B}$ is an Abelian subgroup of $\mathfrak{P}$ and $\mathfrak{A} \leq \mathbf{N}_{\mathfrak{P}}(\mathfrak{B})$, then $\mathfrak{B} \leq \mathbf{N}_{\mathfrak{P}}(\mathfrak{A})$.

Proof. Let $\mathscr{A}(\mathfrak{P})$ be the set of Abelian subgroups $\mathfrak{A}$ of $\mathfrak{P}$ which satisfy a). Each element of $\mathscr{A}(\mathfrak{P})$ is thus a maximal Abelian subgroup of $\mathfrak{P}$. Let $\mathfrak{P} = \mathfrak{P}_1 \geq \mathfrak{P}_2 \geq \cdots \geq \mathfrak{P}_k = 1$ be a fixed central series of $\mathfrak{P}$ and choose $\mathfrak{A} \in \mathscr{A}(\mathfrak{P})$ such that $\prod_{i=1}^{k} |\mathfrak{A} \cap \mathfrak{P}_i|$ is maximal. We show that if $x \in \mathfrak{P}$ and $[x, \mathfrak{A}]$ is Abelian, then $x \in \mathbf{N}_{\mathfrak{P}}(\mathfrak{A})$. For let $\mathfrak{M} = [x, \mathfrak{A}]$, $\mathfrak{B} = \mathfrak{M}\mathbf{C}_{\mathfrak{A}}(\mathfrak{M})$. By 3.1a), $\mathfrak{B}$ is an Abelian subgroup of $\mathfrak{P}$ and $|\mathfrak{B} \cap \mathfrak{P}_i| \geq |\mathfrak{A} \cap \mathfrak{P}_i|$ $(i = 1, \ldots, k)$. In particular $|\mathfrak{B}| \geq |\mathfrak{A}|$, so since $\mathfrak{A} \in \mathscr{A}(\mathfrak{P})$, $|\mathfrak{B}| = |\mathfrak{A}|$ and $\mathfrak{B} \in \mathscr{A}(\mathfrak{P})$. By choice of $\mathfrak{A}$, $\prod_{i=1}^{k} |\mathfrak{B} \cap \mathfrak{P}_i| \leq \prod_{i=1}^{k} |\mathfrak{A} \cap \mathfrak{P}_i|$. It follows that $|\mathfrak{B} \cap \mathfrak{P}_i| = |\mathfrak{A} \cap \mathfrak{P}_i|$ for all i. By 3.1b), $\mathfrak{B} = \mathfrak{A}$. Thus $[x, \mathfrak{A}] = \mathfrak{M} \leq \mathfrak{B} = \mathfrak{A}$ and $x \in \mathbf{N}_{\mathfrak{P}}(\mathfrak{A})$. Hence $\mathfrak{A}$ satisfies b).

q.e.d.

It will be noted that there does *not* necessarily exist a normal Abelian subgroup $\mathfrak{A}$ of $\mathfrak{P}$ such that no Abelian subgroup of $\mathfrak{P}$ is of order greater than $|\mathfrak{A}|$ (III, Aufg. 31).

3.3 Theorem *(Replacement theorem). Suppose that $\mathfrak{P}$ is a p-group, $\mathfrak{Q} \trianglelefteq \mathfrak{P}$, $\mathfrak{A}$ is an Abelian subgroup of $\mathfrak{P}$, $\mathfrak{A} \cap \mathfrak{Q} \geq \mathfrak{Q}'$ and $\mathfrak{Q} \not\leq \mathbf{N}_{\mathfrak{P}}(\mathfrak{A})$. If $p = 2$, suppose also that $\mathfrak{Q}$ is Abelian. Then there exists an Abelian subgroup $\mathfrak{B}$ of $\mathfrak{P}$ with the following properties.*

a) $\mathfrak{B} \cap \mathfrak{Q} > \mathfrak{A} \cap \mathfrak{Q}$.

b) $|\mathfrak{B}| \geq |\mathfrak{A}|$.

c) $\mathfrak{B} \leq \mathbf{N}_{\mathfrak{P}}(\mathfrak{A})$.

d) $\mathfrak{B} \leq \mathfrak{Q}(\mathfrak{A} \cap \mathfrak{B})$.

e) *If p is odd and $\mathfrak{A}$ is elementary Abelian, $\mathfrak{B}$ is elementary Abelian.*

Proof. Let $\mathfrak{N} = \mathbf{N}_{\mathfrak{A}\mathfrak{Q}}(\mathfrak{A})$. By hypothesis, $\mathfrak{Q} \not\leq \mathfrak{N}$, so $\mathfrak{N} < \mathfrak{A}\mathfrak{Q}$. Thus there exists a subgroup $\mathfrak{R}$ of $\mathfrak{A}\mathfrak{Q}$ such that $\mathfrak{N} \trianglelefteq \mathfrak{R}$ and $|\mathfrak{R} : \mathfrak{N}| = p$. Since $\mathfrak{R} \geq \mathfrak{A}$, $\mathfrak{R} = \mathfrak{A}(\mathfrak{Q} \cap \mathfrak{R})$, so $\mathfrak{Q} \cap \mathfrak{R} > \mathfrak{Q} \cap \mathfrak{N}$. Let x be an element of $\mathfrak{Q} \cap \mathfrak{R}$ which does not lie in $\mathfrak{N}$, and let $\mathfrak{M} = [\mathfrak{A}, x]$. Thus $\mathfrak{M} \leq \mathfrak{Q}$. Since $x \in \mathfrak{R}$, $\mathfrak{N}^x = \mathfrak{N}$ and $\mathfrak{A}$, $\mathfrak{A}^x$ are normal subgroups of $\mathfrak{N}$. Let $\mathfrak{U} = \mathfrak{A}\mathfrak{A}^x$. Then $\mathfrak{M} \leq \mathfrak{U}$ and

$$\mathfrak{U}' = [\mathfrak{A}, \mathfrak{A}^x] \leq \mathfrak{A} \cap \mathfrak{A}^x \leq \mathbf{Z}(\mathfrak{U}).$$

We prove that $\mathfrak{M}$ is Abelian. For $p = 2$, this is clear since $\mathfrak{Q}$ is Abelian. Suppose then that p is odd. We must show that $[x, a]$ and $[x, b]$ commute for any elements a, b of $\mathfrak{A}$. First observe that

$$[x, a, b]^x = [[x, a]^x, b^x] = [u[x, a], b^x],$$

where $u = [x, [a, x]] \in \mathfrak{Q}' = \mathfrak{Q}'^x \leq (\mathfrak{A} \cap \mathfrak{Q})^x \leq \mathfrak{A}^x$. Thus

$$[x, a, b]^x = [x, a, b^x].$$

Hence

$$\begin{aligned}[x, a, b]^x &= [a^{-x}a, b^x] = [a, b^x] \\ &= [b^{-x}, a]^{b^x} = [b^{-x}, a],\end{aligned}$$

since $\mathfrak{U}' \leq \mathbf{Z}(\mathfrak{U})$. Thus

$$[x, a, b]^x = [[x, b]b^{-1}, a] = [x, b, a]^{b^{-1}} = [x, b, a],$$

since $[x, b] \in \mathfrak{U}$ and $[x, b, a] \in \mathbf{Z}(\mathfrak{U})$. Hence x^2 commutes with $[x, a, b]$. Since p is odd, it follows that x commutes with $[x, a, b]$. Therefore

$$\begin{aligned}[x, a, b] &= [x, a, b]^x \\ &= [x, a, b^x] \\ &= [x, a, b[b, x]] \\ &= [[x, a], [b, x]][x, a, b]^{[b,x]} \\ &= [[x, a], [b, x]][x, a, b].\end{aligned}$$

It follows that $[x, a]$ and $[x, b]$ commute.

Let $\mathfrak{B} = \mathfrak{M}\mathbf{C}_{\mathfrak{A}}(\mathfrak{M})$. Then $\mathfrak{A} \cap \mathfrak{A}^x \leq \mathfrak{A} \cap \mathbf{Z}(\mathfrak{U}) \leq \mathbf{C}_{\mathfrak{A}}(\mathfrak{M}) \leq \mathfrak{B}$. But since $\mathfrak{A} \cap \mathfrak{Q} \geq \mathfrak{Q}'$, $x \in \mathbf{N}_{\mathfrak{P}}(\mathfrak{A} \cap \mathfrak{Q})$ and

$$\mathfrak{A} \cap \mathfrak{Q} = (\mathfrak{A} \cap \mathfrak{Q})^x = \mathfrak{A}^x \cap \mathfrak{Q} \leq \mathfrak{A} \cap \mathfrak{A}^x.$$

Hence $\mathfrak{A} \cap \mathfrak{Q} \leq \mathfrak{B} \cap \mathfrak{Q}$. Since $x \notin \mathfrak{N}$, $x \notin \mathbf{N}_{\mathfrak{P}}(\mathfrak{A})$, so $[\mathfrak{A}, x] \not\leq \mathfrak{A}$. Thus $\mathfrak{M} \not\leq \mathfrak{A} \cap \mathfrak{Q}$, although $\mathfrak{M} \leq \mathfrak{B} \cap \mathfrak{Q}$. Hence $\mathfrak{A} \cap \mathfrak{Q} < \mathfrak{B} \cap \mathfrak{Q}$, and a) is proved. b) follows from 3.1. Since $\mathfrak{M} \leq \mathfrak{U}$, $\mathfrak{B} \leq \mathfrak{U} \leq \mathbf{N}_{\mathfrak{P}}(\mathfrak{A})$. Since $\mathfrak{M} \leq \mathfrak{Q}$ and $\mathbf{C}_{\mathfrak{A}}(\mathfrak{M}) \leq \mathfrak{A} \cap \mathfrak{B}$, $\mathfrak{B} \leq \mathfrak{Q}\ (\mathfrak{A} \cap \mathfrak{B})$. If p is odd and $\mathfrak{A}$ is elementary Abelian, then $\mathfrak{U}$ is of exponent p, since $\mathfrak{U}$ is of class at most 2 and is therefore regular (III, 10.2). Since $\mathfrak{B} \leq \mathfrak{U}$, it follows that $\mathfrak{B}$ is elementary Abelian. **q.e.d.**

We prove the following related theorem.

3.4 Theorem (THOMPSON [7]). *Suppose that $p^r > 2$, $\mathfrak{P}$ is a p-group and $|\mathfrak{P} : \mathfrak{P}'\mathfrak{P}^{p^r}| > |\mathfrak{Q} : \mathfrak{Q}'\mathfrak{Q}^{p^r}|$ for every proper subgroup $\mathfrak{Q} < \mathfrak{P}$. Then the class of $\mathfrak{P}$ is at most 2.* (We remark that $|\mathfrak{P}: \mathfrak{P}'\mathfrak{P}^p| = |\mathfrak{P}: \Phi(\mathfrak{P})| = p^{d(\mathfrak{P})}$, where $d(\mathfrak{P})$ is the minimal number of generators of $\mathfrak{P}$.)

Proof. Suppose that this is false and let $\mathfrak{P}$ be a counterexample of minimal order. We observe that $\mathfrak{P}^{p^r} \leq \mathfrak{P}'$. For otherwise the exponent of $\mathfrak{P}/\mathfrak{P}'$ is greater than p^r. Hence if $\mathfrak{Q}/\mathfrak{P}' = \Omega_r(\mathfrak{P}/\mathfrak{P}')$, $\mathfrak{Q} < \mathfrak{P}$. By I, Aufg. 51, $|\mathfrak{Q}/\mathfrak{P}'| = |\mathfrak{P}/\mathfrak{P}' : \mathfrak{P}^{p^r}\mathfrak{P}'/\mathfrak{P}'| = |\mathfrak{P} : \mathfrak{P}^{p^r}\mathfrak{P}'| > |\mathfrak{Q} : \mathfrak{Q}'\mathfrak{Q}^{p^r}|$. Since $\mathfrak{Q}^{p^r}\mathfrak{Q}' \leq \mathfrak{P}'$, this gives a contradiction.

Let $\mathfrak{M} = [\mathfrak{P}', \mathfrak{P}]$. Since $\mathfrak{P}' \not\leq \mathbf{Z}(\mathfrak{P})$, $\mathfrak{M} \neq 1$. Hence there exists a normal subgroup $\mathfrak{M}_0$ of $\mathfrak{P}$ such that $|\mathfrak{M}_0| = p$ and $\mathfrak{M}_0 \leq \mathfrak{M}$. Since $\mathfrak{P}/\mathfrak{M}_0$ satisfies the hypotheses of the theorem but is not a counterexample, $[\mathfrak{P}'/\mathfrak{M}_0, \mathfrak{P}/\mathfrak{M}_0] = 1$; thus $\mathfrak{M} = [\mathfrak{P}', \mathfrak{P}] \leq \mathfrak{M}_0$, $\mathfrak{M} = \mathfrak{M}_0$ and $|\mathfrak{M}| = p$. Hence $\mathfrak{M} \leq \mathbf{Z}(\mathfrak{P})$, and by III, 2.12, $\mathfrak{P}'' \leq [\mathfrak{P}', \mathfrak{P}, \mathfrak{P}] = 1$. Thus $\mathfrak{P}'$ is Abelian. Let $\mathfrak{C} = \mathbf{C}_{\mathfrak{P}}(\mathfrak{P}')$; then $\mathfrak{P}' \leq \mathfrak{C}$ and $\mathfrak{C} \trianglelefteq \mathfrak{P}$. Note also that

$\mathfrak{P}/\mathfrak{C}$ is elementary Abelian. For if $x \in \mathfrak{P}$ and $y \in \mathfrak{P}'$, we have $[x, y] \in \mathfrak{M} \leq \mathbf{Z}(\mathfrak{P})$, so by III, 1.3, $[x^p, y] = [x, y]^p = 1$ and $x^p \in \mathbf{C}_{\mathfrak{P}}(\mathfrak{P}') = \mathfrak{C}$. Similarly, if $x \in \mathfrak{P}$ and $y \in \mathfrak{P}'$, $[x, y^p] = [x, y]^p = 1$, so $\mathfrak{P}'^p \leq \mathbf{Z}(\mathfrak{P})$. Hence if $\mathfrak{Z} = \mathbf{Z}(\mathfrak{P}) \cap \mathfrak{P}'$, $\mathfrak{P}'/\mathfrak{Z}$ is elementary Abelian.

By VIII, 6.1, there exists a mapping γ of $(\mathfrak{P}/\mathfrak{C}) \times (\mathfrak{P}'/\mathfrak{Z})$ into $\mathfrak{M}$, such that

$$(x\mathfrak{C}, y\mathfrak{Z})\gamma = [x, y] \quad (x \in \mathfrak{P}, y \in \mathfrak{P}');$$

also γ is bilinear over $\mathbb{Z}$. Since $\mathfrak{M}$, $\mathfrak{P}/\mathfrak{C}$ and $\mathfrak{P}'/\mathfrak{Z}$ are elementary Abelian, $\mathfrak{M}$, $\mathfrak{P}/\mathfrak{C}$ and $\mathfrak{P}'/\mathfrak{Z}$ can be regarded as vector spaces over $GF(p)$, and γ is bilinear over $GF(p)$. γ is non-singular, on account of the definitions of $\mathfrak{C}$ and $\mathfrak{Z}$. It follows that $|\mathfrak{P}/\mathfrak{C}| = |\mathfrak{P}'/\mathfrak{Z}|$. Hence $|\mathfrak{P} : \mathfrak{P}'\mathfrak{P}^{p^r}| = |\mathfrak{P} : \mathfrak{P}'| = |\mathfrak{C} : \mathfrak{Z}|$.

Now by III, 1.10b),

$$[\mathfrak{C}', \mathfrak{P}] = [\mathfrak{C}, \mathfrak{C}, \mathfrak{P}] \leq [\mathfrak{C}, \mathfrak{P}, \mathfrak{C}] \leq [\mathfrak{P}', \mathbf{C}_{\mathfrak{P}}(\mathfrak{P}')] = 1.$$

Hence $\mathfrak{C}' \leq \mathbf{Z}(\mathfrak{P})$. We show also that $\mathfrak{C}^{p^r} \leq \mathbf{Z}(\mathfrak{P})$. For suppose that $x \in \mathfrak{C}$ and $y \in \mathfrak{P}$. Since $[x, y]$ and x commute,

$$(x[x, y])^{p^r} = x^{p^r}[x, y]^{p^r}.$$

Thus

$$[x, y]^{p^r} = x^{-p^r}(x^y)^{p^r} - x^{-p^r}(x^{p^r})^y = [x^{p^r}, y].$$

But also $y^{p^r} \in \mathfrak{P}^{p^r} \leq \mathfrak{P}'$, so $[x, y^{p^r}] = 1$. Hence

$$1 = (y^{-p^r})^x y^{p^r} = (y^{-x})^{p^r} y^{p^r} = ([x, y]y^{-1})^{p^r} y^{p^r}.$$

Since $[x, y, y^{-1}] \in \mathfrak{M} \leq \mathbf{Z}(\mathfrak{P})$, it follows from III, 1.3b) that

$$([x, y]y^{-1})^{p^r} = [x, y]^{p^r} y^{-p^r} [y^{-1}, [x, y]]^{\binom{p^r}{2}}.$$

Since $p^r > 2$ and $\mathfrak{M}$ is of exponent p, it follows that

$$([x, y]y^{-1})^{p^r} = [x, y]^{p^r} y^{-p^r}$$

and $[x, y]^{p^r} = 1$. Hence $[x^{p^r}, y] = 1$ and $\mathfrak{C}^{p^r} \leq \mathbf{Z}(\mathfrak{P})$.

Thus $\mathfrak{C}'\mathfrak{C}^{p^r} \leq \mathbf{Z}(\mathfrak{P}) \cap \mathfrak{P}' = \mathfrak{Z}$. Thus

$$|\mathfrak{C} : \mathfrak{C}'\mathfrak{C}^{p'}| \geq |\mathfrak{C} : \mathfrak{Z}| = |\mathfrak{P} : \mathfrak{P}'\mathfrak{P}^{p'}|,$$

contrary to hypothesis. **q.e.d.**

3.5 Corollary. *Suppose that p is odd and that* $\mathfrak{A}$ *is a maximal element of the set of normal elementary Abelian subgroups of the p-group* $\mathfrak{P}$. *If* $|\mathfrak{A}| = p^n$, $\mathfrak{P}$ *can be generated by* $\frac{1}{2}n(n+1)$ *elements.*

Proof. Let $\mathfrak{C} = \mathbf{C}_{\mathfrak{P}}(\mathfrak{A})$. We prove by induction on $|\mathfrak{H}|$ that if $\mathfrak{H} \leq \mathfrak{C}$, then $|\mathfrak{H} : \Phi(\mathfrak{H})| \leq p^n$. First of all, if the class of $\mathfrak{H}$ is at most 2, then $\mathfrak{H}$ is regular, since p is odd (III, 10.2). Hence the elements of $\mathfrak{H}$ of order at most p form a subgroup $\mathfrak{H}_1$ of order $|\mathfrak{H} : \mathfrak{H}^p|$ (III, 10.5 and 10.7). Since $\mathfrak{H} \leq \mathfrak{C} = \mathbf{C}_{\mathfrak{P}}(\mathfrak{A})$, $\mathfrak{H}_1 \leq \mathfrak{A}$ by III, 12.1 since p is odd. Thus

$$|\mathfrak{H} : \Phi(\mathfrak{H})| \leq |\mathfrak{H} : \mathfrak{H}^p| = |\mathfrak{H}_1| \leq |\mathfrak{A}| = p^n,$$

as required. Secondly, if the class of $\mathfrak{H}$ is greater than 2, then by 3.4, there exists a subgroup $\mathfrak{H}_2 < \mathfrak{H}$ such that

$$|\mathfrak{H} : \Phi(\mathfrak{H})| = |\mathfrak{H} : \mathfrak{H}'\mathfrak{H}^p| \leq |\mathfrak{H}_2 : \mathfrak{H}_2\mathfrak{H}_2^p| = |\mathfrak{H}_2 : \Phi(\mathfrak{H}_2)|.$$

Since $|\mathfrak{H}_2 : \Phi(\mathfrak{H}_2)| \leq p^n$ by the inductive hypothesis, we again have $|\mathfrak{H} : \Phi(\mathfrak{H})| \leq p^n$.

Taking $\mathfrak{H} = \mathfrak{C}$, $|\mathfrak{C} : \Phi(\mathfrak{C})| \leq p^n$, so $\mathfrak{C}$ can be generated by n elements. By I, 4.3, $\mathfrak{P}/\mathfrak{C}$ is isomorphic to a group of automorphisms of $\mathfrak{A}$. Since the order of the group of all automorphisms of $\mathfrak{A}$ is $p^{\frac{1}{2}n(n-1)}q$ for some integer q with $(q, p) = 1$, $\mathfrak{P}/\mathfrak{C}$ can be generated by $\frac{1}{2}n(n-1)$ elements. Thus $\mathfrak{P}$ can be generated by $\frac{1}{2}n(n-1) + n = \frac{1}{2}n(n+1)$ elements. **q.e.d.**

3.6 Notation. a) For any p-group $\mathfrak{P}$, we define $\mathscr{A}(\mathfrak{P})$ to be the set of Abelian subgroups $\mathfrak{A}$ of $\mathfrak{P}$ of maximal order.

It is clear that if $\mathfrak{A} \in \mathscr{A}(\mathfrak{P})$, $\mathfrak{A}$ is a maximal Abelian subgroup of $\mathfrak{P}$; that is, $\mathbf{C}_{\mathfrak{P}}(\mathfrak{A}) = \mathfrak{A}$. In particular $\mathfrak{A} \geq \mathbf{Z}(\mathfrak{P})$.

The automorphisms of $\mathfrak{P}$ permute the elements of $\mathscr{A}(\mathfrak{P})$.

b) For any p-group $\mathfrak{P}$, write

$$\mathbf{J}(\mathfrak{P}) = \langle \mathfrak{A} \mid \mathfrak{A} \in \mathscr{A}(\mathfrak{P}) \rangle, \quad \mathbf{ZJ}(\mathfrak{P}) = \bigcap_{\mathfrak{A} \in \mathscr{A}(\mathfrak{P})} \mathfrak{A}.$$

Since $\mathbf{C}_{\mathfrak{P}}(\mathbf{J}(\mathfrak{P})) = \bigcap_{\mathfrak{A} \in \mathscr{A}(\mathfrak{P})} \mathbf{C}_{\mathfrak{P}}(\mathfrak{A}) = \bigcap_{\mathfrak{A} \in \mathscr{A}(\mathfrak{P})} \mathfrak{A} = \mathbf{ZJ}(\mathfrak{P})$, $\mathbf{ZJ}(\mathfrak{P})$ is the centre of $\mathbf{J}(\mathfrak{P})$.

$\mathbf{J}(\mathfrak{P})$ and $\mathbf{ZJ}(\mathfrak{P})$ are characteristic subgroups of $\mathfrak{P}$.

So far the theorems in this section have all been about p-groups. We now use them to prove an important theorem about p-stable groups.

3.7 Lemma. *Suppose that p is odd and that* $\mathfrak{G}$ *is a p-stable group. If* $\mathfrak{S} \in S_p(\mathfrak{G})$, $\mathbf{ZJ}(\mathfrak{S}) \cap \mathbf{O}_p(\mathfrak{G}) \trianglelefteq \mathfrak{G}$.

Proof. Suppose that this is false; write $\mathfrak{Z} = \mathbf{ZJ}(\mathfrak{S})$. The set $\mathscr{B}$ of normal p-subgroups $\mathfrak{B}$ of $\mathfrak{G}$ for which $\mathfrak{B} \cap \mathfrak{Z}$ is not normal in $\mathfrak{G}$ is non-empty, since $\mathbf{O}_p(\mathfrak{G}) \in \mathscr{B}$. Let $\mathfrak{Q}$ be an element of $\mathscr{B}$ of minimal order. Thus $\mathfrak{Q} \cap \mathfrak{Z} \ntrianglelefteq \mathfrak{G}$.

(1) $\mathfrak{Q}$ is the normal closure of $\mathfrak{Q} \cap \mathfrak{Z}$ in $\mathfrak{G}$.

Let $\mathfrak{Q}_1$ be the normal closure of $\mathfrak{Q} \cap \mathfrak{Z}$ in $\mathfrak{G}$. Since $\mathfrak{Q} \trianglelefteq \mathfrak{G}$, $\mathfrak{Q} \cap \mathfrak{Z} < \mathfrak{Q}_1 \le \mathfrak{Q}$. Thus $\mathfrak{Q}_1 \cap \mathfrak{Z} = \mathfrak{Q} \cap \mathfrak{Z}$ and $\mathfrak{Q}_1 \in \mathscr{B}$. By minimality of $|\mathfrak{Q}|$, $\mathfrak{Q}_1 = \mathfrak{Q}$.

(2) $\mathfrak{Q}' \le \mathfrak{Z}$.

Since $\mathfrak{Q}' < \mathfrak{Q}$, $\mathfrak{Q}' \notin \mathscr{B}$. Thus $\mathfrak{Q}' \cap \mathfrak{Z} \trianglelefteq \mathfrak{G}$. Since $\mathfrak{Z} = \mathbf{ZJ}(\mathfrak{S}) \trianglelefteq \mathfrak{S}$ and $\mathfrak{Q} \le \mathfrak{S}$, $[\mathfrak{Q} \cap \mathfrak{Z}, \mathfrak{Q}] \le \mathfrak{Q}' \cap [\mathfrak{Z}, \mathfrak{S}] \le \mathfrak{Q}' \cap \mathfrak{Z}$. Thus if $\mathfrak{C}/(\mathfrak{Q}' \cap \mathfrak{Z}) = \mathbf{C}_{\mathfrak{G}/(\mathfrak{Q}' \cap \mathfrak{Z})}(\mathfrak{Q}/\mathfrak{Q}' \cap \mathfrak{Z})$, $\mathfrak{Q} \cap \mathfrak{Z} \le \mathfrak{C}$. But $\mathfrak{C} \trianglelefteq \mathfrak{G}$. Hence $\mathfrak{Q} \le \mathfrak{C}$, by (1). Thus $\mathfrak{Q}' \le [\mathfrak{Q}, \mathfrak{C}] \le \mathfrak{Q}' \cap \mathfrak{Z}$.

Let $\mathfrak{N}$ be the intersection of all the conjugates of $\mathbf{N}_{\mathfrak{G}}(\mathfrak{Q} \cap \mathfrak{Z})$ in $\mathfrak{G}$. Thus $\mathfrak{N}$ is a proper normal subgroup of $\mathfrak{G}$, and $\mathfrak{S} \cap \mathfrak{N} \in S_p(\mathfrak{N})$. Write

$$\mathfrak{J} = \mathbf{J}(\mathfrak{S} \cap \mathfrak{N}), \quad \mathfrak{X} = \mathbf{Z}(\mathfrak{J}).$$

(3) $\mathfrak{G} = \mathfrak{N}\mathbf{N}_{\mathfrak{G}}(\mathfrak{S} \cap \mathfrak{N}) = \mathfrak{N}\mathbf{N}_{\mathfrak{G}}(\mathfrak{X}) = \mathfrak{N}\mathbf{N}_{\mathfrak{G}}(\mathfrak{Q} \cap \mathfrak{X})$.

The first equality is a consequence of the Frattini argument applied to the Sylow p-subgroup $\mathfrak{S} \cap \mathfrak{N}$ of $\mathfrak{N}$. Since $\mathfrak{X}$ is a characteristic subgroup of $\mathfrak{S} \cap \mathfrak{N}$, $\mathbf{N}_{\mathfrak{G}}(\mathfrak{S} \cap \mathfrak{N}) \le \mathbf{N}_{\mathfrak{G}}(\mathfrak{X})$. And $\mathbf{N}_{\mathfrak{G}}(\mathfrak{X}) \le \mathbf{N}_{\mathfrak{G}}(\mathfrak{Q} \cap \mathfrak{X})$ since $\mathfrak{Q} \trianglelefteq \mathfrak{G}$.

(4) There exists $\mathfrak{A} \in \mathscr{A}(\mathfrak{S})$ such that $\mathfrak{A} \not\le \mathfrak{N}$.

Since $\mathfrak{N} \le \mathbf{N}_{\mathfrak{G}}(\mathfrak{Q} \cap \mathfrak{Z}) \ne \mathfrak{G}$, $\mathfrak{G} \ne \mathfrak{N}\mathbf{N}_{\mathfrak{G}}(\mathfrak{Q} \cap \mathfrak{Z})$. By (3), $\mathfrak{Q} \cap \mathfrak{Z} \ne \mathfrak{Q} \cap \mathfrak{X}$, so $\mathfrak{Z} \ne \mathfrak{X}$ and $\mathbf{J}(\mathfrak{S}) \ne \mathfrak{J} = \mathbf{J}(\mathfrak{S} \cap \mathfrak{N})$. Thus there exists $\mathfrak{A} \in \mathscr{A}(\mathfrak{S})$ such that $\mathfrak{A} \not\le \mathfrak{S} \cap \mathfrak{N}$.

By (4), the set $\mathscr{A}^* = \{\mathfrak{A} \mid \mathfrak{A} \in \mathscr{A}(\mathfrak{S}), \mathfrak{A} \not\le \mathfrak{N}\}$ is non-empty. Let $\mathfrak{A}$ be an element of $\mathscr{A}^*$ for which $\mathfrak{A} \cap \mathfrak{Q}$ is maximal.

(5) $[\mathfrak{Q}, \mathfrak{A}, \mathfrak{A}] \ne 1$.

Suppose $[\mathfrak{Q}, \mathfrak{A}, \mathfrak{A}] = 1$. Since $\mathfrak{G}$ is p-stable, it follows that $\mathfrak{A} \le \mathfrak{M}$, where $\mathfrak{M}/\mathbf{C}_{\mathfrak{G}}(\mathfrak{Q}) = \mathbf{O}_p(\mathfrak{G}/\mathbf{C}_{\mathfrak{G}}(\mathfrak{Q}))$. Since $\mathfrak{M} \trianglelefteq \mathfrak{G}$, $\mathfrak{M} \cap \mathfrak{S} \in S_p(\mathfrak{M})$. Since $\mathfrak{M}/\mathbf{C}_{\mathfrak{G}}(\mathfrak{Q})$ is a p-group, it follows that $\mathfrak{M} = (\mathfrak{M} \cap \mathfrak{S})\mathbf{C}_{\mathfrak{G}}(\mathfrak{Q})$. But since $\mathfrak{S} \le \mathbf{N}_{\mathfrak{G}}(\mathfrak{Z})$ and $\mathfrak{S} \le \mathbf{N}_{\mathfrak{G}}(\mathfrak{Q})$, $\mathfrak{S} \le \mathbf{N}_{\mathfrak{G}}(\mathfrak{Q} \cap \mathfrak{Z})$. Also $\mathbf{C}_{\mathfrak{G}}(\mathfrak{Q}) \le \mathbf{N}_{\mathfrak{G}}(\mathfrak{Q} \cap \mathfrak{Z})$. Thus $\mathfrak{M} \le \mathbf{N}_{\mathfrak{G}}(\mathfrak{Q} \cap \mathfrak{Z})$. Since $\mathfrak{M} \trianglelefteq \mathfrak{G}$, it follows from the definition of $\mathfrak{N}$ that $\mathfrak{M} \le \mathfrak{N}$. Thus $\mathfrak{A} \le \mathfrak{N}$. contrary to $\mathfrak{A} \in \mathscr{A}^*$.

(6) By (2), $\mathfrak{Q}' \leq \mathfrak{Z} \leq \mathfrak{A}$, and by (5), $\mathfrak{Q} \not\leq \mathbf{N}_{\mathfrak{S}}(\mathfrak{A})$. Hence by 3.3, there exists an Abelian subgroup $\mathfrak{B}$ of $\mathfrak{S}$ such that $|\mathfrak{B}| \geq |\mathfrak{A}|$, $[\mathfrak{B}, \mathfrak{A}, \mathfrak{A}] = 1$ and $\mathfrak{B} \cap \mathfrak{Q} > \mathfrak{A} \cap \mathfrak{Q}$. Since $\mathfrak{B}$ is Abelian and $|\mathfrak{B}| \geq |\mathfrak{A}|$, $\mathfrak{B} \in \mathscr{A}(\mathfrak{S})$. Since $\mathfrak{B} \cap \mathfrak{Q} > \mathfrak{A} \cap \mathfrak{Q}$, $\mathfrak{B} \notin \mathscr{A}^*$. Thus $\mathfrak{B} \leq \mathfrak{N}$. Hence $\mathscr{A}(\mathfrak{S} \cap \mathfrak{N}) \subseteq \mathscr{A}(\mathfrak{S})$. Also $\mathfrak{Q} \cap \mathfrak{Z}$ centralizes every element of $\mathscr{A}(\mathfrak{S})$ and hence every element of $\mathscr{A}(\mathfrak{S} \cap \mathfrak{N})$. Since $\mathfrak{Q} \cap \mathfrak{Z} \leq \mathfrak{B} \leq \mathfrak{S} \cap \mathfrak{N}$, it follows that $\mathfrak{Q} \cap \mathfrak{Z} \leq \mathbf{Z}(\mathfrak{J}) = \mathfrak{X}$. Now if $g \in \mathfrak{G}$, there exist $a \in \mathfrak{N}$ and $b \in \mathbf{N}_{\mathfrak{G}}(\mathfrak{X})$ such that $g = ab$, by (3); thus $(\mathfrak{Q} \cap \mathfrak{Z})^g = (\mathfrak{Q} \cap \mathfrak{Z})^{ab} = (\mathfrak{Q} \cap \mathfrak{Z})^b$, since $\mathfrak{N} \leq \mathbf{N}_{\mathfrak{G}}(\mathfrak{Q} \cap \mathfrak{Z})$. Hence $(\mathfrak{Q} \cap \mathfrak{Z})^g \leq \mathfrak{X}^b = \mathfrak{X}$ for all $g \in \mathfrak{G}$. By (1), $\mathfrak{Q} \leq \mathfrak{X}$. Since $\mathfrak{B} \in \mathscr{A}(\mathfrak{S} \cap \mathfrak{N})$, $\mathfrak{X} \leq \mathfrak{B}$. Hence $\mathfrak{Q} \leq \mathfrak{B}$. Since $[\mathfrak{B}, \mathfrak{A}, \mathfrak{A}] = 1$, $[\mathfrak{Q}, \mathfrak{A}, \mathfrak{A}] = 1$. This contradicts (5). **q.e.d.**

3.8 Theorem (GLAUBERMAN [3]). *Suppose that p is an odd prime and that $\mathfrak{G}$ is a group which is p-stable and p-constrained. If $\mathfrak{S} \in S_p(\mathfrak{G})$,*

$$\mathbf{ZJ}(\mathfrak{S})\mathbf{O}_{p'}(\mathfrak{G}) \trianglelefteq \mathfrak{G},$$

and $\mathfrak{G} = \mathbf{N}_{\mathfrak{G}}(\mathbf{ZJ}(\mathfrak{S}))\mathbf{O}_{p'}(\mathfrak{G})$.

Proof. Let $\mathfrak{P} = \mathfrak{S} \cap \mathbf{O}_{p',p}(\mathfrak{G})$.

If $\mathfrak{A}$ is any normal Abelian subgroup of $\mathfrak{S}$, then $[\mathfrak{P}, \mathfrak{A}] \leq \mathfrak{P}$ and

$$[\mathfrak{P}, \mathfrak{A}, \mathfrak{A}] \leq [\mathfrak{S}, \mathfrak{A}, \mathfrak{A}] \leq [\mathfrak{A}, \mathfrak{A}] = 1.$$

Since $\mathfrak{G}$ is p-stable, it follows that $\mathfrak{A} \leq \mathfrak{D}$, where $\mathfrak{D}/\mathfrak{C} = \mathbf{O}_p(\mathfrak{N}/\mathfrak{C})$, $\mathfrak{C} = \mathbf{C}_{\mathfrak{G}}(\mathfrak{P})$ and $\mathfrak{N} = \mathbf{N}_{\mathfrak{G}}(\mathfrak{P})$. Since $\mathfrak{G}$ is p-constrained and $\mathbf{O}_{p',p}(\mathfrak{G}) = \mathfrak{P}\mathbf{O}_{p'}(\mathfrak{G})$, $\mathfrak{C} \leq \mathbf{O}_{p',p}(\mathfrak{G})$. Thus $\mathfrak{D}\mathbf{O}_{p',p}(\mathfrak{G})/\mathbf{O}_{p',p}(\mathfrak{G})$ is a p-group. But $\mathfrak{G} = \mathfrak{N}\mathbf{O}_{p',p}(\mathfrak{G})$ by the Frattini argument, and $\mathfrak{D} \trianglelefteq \mathfrak{N}$. Hence $\mathfrak{D}\mathbf{O}_{p',p}(\mathfrak{G}) \trianglelefteq \mathfrak{G}$. Since $\mathfrak{G}/\mathbf{O}_{p',p}(\mathfrak{G})$ has no non-identity normal p-subgroup, it follows that $\mathfrak{D} \leq \mathbf{O}_{p',p}(\mathfrak{G})$. Thus $\mathfrak{A} \leq \mathfrak{P}$.

Applying this with $\mathfrak{A} = \mathbf{ZJ}(\mathfrak{S})$, we have $\mathbf{ZJ}(\mathfrak{S}) \leq \mathfrak{P}$. Thus $\mathbf{ZJ}(\overline{\mathfrak{S}}) \leq \mathbf{O}_p(\overline{\mathfrak{G}})$, where $\overline{\mathfrak{S}} = \mathfrak{S}\mathbf{O}_{p'}(\mathfrak{G})/\mathbf{O}_{p'}(\mathfrak{G})$ and $\overline{\mathfrak{G}} = \mathfrak{G}/\mathbf{O}_{p'}(\mathfrak{G})$, for $\mathbf{ZJ}(\overline{\mathfrak{S}}) = \mathbf{ZJ}(\mathfrak{S})\mathbf{O}_{p'}(\mathfrak{G})/\mathbf{O}_{p'}(\mathfrak{G})$. By 3.7, $\mathbf{ZJ}(\overline{\mathfrak{S}}) \trianglelefteq \overline{\mathfrak{G}}$, whence

$$\mathbf{ZJ}(\mathfrak{S})\mathbf{O}_{p'}(\mathfrak{G}) \trianglelefteq \mathfrak{G}.$$

By the Frattini argument, $\mathfrak{G} = \mathbf{N}_{\mathfrak{G}}(\mathbf{ZJ}(\mathfrak{S}))\mathbf{O}_{p'}(\mathfrak{G})$. **q.e.d.**

3.9 Corollary. *Suppose that $\mathfrak{G}$ is p-soluble, where p is odd. Suppose also that if $p = 3$, the Sylow 2-subgroups of $\mathfrak{G}$ are Abelian. If $\mathfrak{S} \in S_p(\mathfrak{G})$,*

$$\mathbf{ZJ}(\mathfrak{S})\mathbf{O}_{p'}(\mathfrak{G}) \trianglelefteq \mathfrak{G}$$

and $\mathfrak{G} = \mathbf{N}_{\mathfrak{G}}(\mathbf{ZJ}(\mathfrak{S}))\mathbf{O}_{p'}(\mathfrak{G})$.

Proof. By IX, 1.4, $\mathfrak{G}$ is p-constrained, and by IX, 7.4 or 7.10, $\mathfrak{G}$ is p-stable. The assertions thus follow from 3.8. **q.e.d.**

3.10 Remark. As an example of the use of 3.8, we show how the proof of 2.8 can be simplified by using it. Suppose that we have reached 2.12a) in this proof; we may then proceed as follows.

(1) Every maximal subgroup $\mathfrak{M}$ of $\mathfrak{G}$ contains a Sylow subgroup of $\mathfrak{G}$.

By 2.10, $\mathbf{O}_{p'}(\mathfrak{M}) = 1$ for some p. If $\mathfrak{P} \in S_p(\mathfrak{M})$, $\mathbf{ZJ}(\mathfrak{P}) \trianglelefteq \mathfrak{M}$ by 3.9, since $|\mathfrak{G}|$ is odd. Thus $\mathfrak{M} = \mathbf{N}_{\mathfrak{G}}(\mathbf{ZJ}(\mathfrak{P}))$. By 2.4, $\mathfrak{P} \in S_p(\mathfrak{G})$.

(2) Any Sylow subgroup of $\mathfrak{G}$ is contained in only one maximal subgroup of $\mathfrak{G}$.

If $\mathfrak{S} \in S_p(\mathfrak{G})$, $\mathfrak{S} \leq \mathfrak{M}$ and $\mathfrak{M}$ is maximal, then $\mathbf{O}_{p'}(\mathfrak{M}) = 1$ by 2.9c). Hence $\mathfrak{M} = \mathbf{N}_{\mathfrak{G}}(\mathbf{ZJ}(\mathfrak{S}))$ and $\mathfrak{M}$ is determined by $\mathfrak{S}$.

Now write $|\mathfrak{G}| = p^a q^b$, where p, q are primes and $p^a < q^b$. Suppose that $\mathfrak{M}$ is a maximal subgroup of $\mathfrak{G}$ containing a Sylow q-subgroup of $\mathfrak{G}$; thus q^b divides $|\mathfrak{M}|$. Since $\mathfrak{G}$ is simple, $\mathfrak{M}$ does not contain every Sylow q-subgroup of $\mathfrak{G}$. Suppose that $\mathfrak{Q}^* \in S_q(\mathfrak{G})$ and $\mathfrak{Q}^* \not\leq \mathfrak{M}$. Let $\mathfrak{D} = \mathfrak{Q}^* \cap \mathfrak{M}$ and choose $\mathfrak{Q} \in S_q(\mathfrak{M})$ such that $\mathfrak{Q} \geq \mathfrak{D}$. Then $\mathfrak{Q} \cap \mathfrak{Q}^* = \mathfrak{D}$. Thus $|\mathfrak{Q}\mathfrak{Q}^*| = q^{2b}/|\mathfrak{D}| \leq |\mathfrak{G}|$, so $\mathfrak{D} \neq 1$. Hence $\mathbf{N}_{\mathfrak{G}}(\mathfrak{D}) < \mathfrak{G}$. Also $\mathfrak{D} < \mathfrak{Q}^*$, so $\mathbf{N}_{\mathfrak{Q}^*}(\mathfrak{D}) > \mathfrak{D} = \mathfrak{Q}^* \cap \mathfrak{M}$ and $\mathbf{N}_{\mathfrak{Q}^*}(\mathfrak{D}) \not\leq \mathfrak{M}$. Let $\mathfrak{M}^*$ be a maximal subgroup of $\mathfrak{G}$ containing $\mathbf{N}_{\mathfrak{G}}(\mathfrak{D})$; then $\mathbf{Z}(\mathfrak{Q}) \leq \mathfrak{M}^*$. Thus the set $\mathscr{M}$ of maximal subgroups $\mathfrak{K}$ of $\mathfrak{G}$ for which $\mathfrak{K} \neq \mathfrak{M}$ and $\mathbf{Z}(\mathfrak{Q}) \leq \mathfrak{K}$ is not empty.

Let $\mathfrak{L}$ be an element of $\mathscr{M}$ for which the order q^c of a Sylow q-subgroup of $\mathfrak{L} \cap \mathfrak{M}$ is maximal. Choose $\mathfrak{R} \in S_q(\mathfrak{L} \cap \mathfrak{M})$ such that $\mathfrak{R} \geq \mathbf{Z}(\mathfrak{Q})$. By (2), $\mathfrak{R} \notin S_q(\mathfrak{G})$; thus $\mathfrak{R} \notin S_q(\mathfrak{M})$. Suppose that $\mathfrak{R} < \mathfrak{Q}_0 \in S_q(\mathfrak{M})$; then $\mathfrak{R} < \mathbf{N}_{0}(\mathfrak{R}) \leq \mathbf{N}_{\mathfrak{G}}(\mathfrak{R}) \leq \mathfrak{K}$ for some maximal subgroup $\mathfrak{K}$. By maximality of q^c, $\mathfrak{K} \notin \mathscr{M}$. Since $\mathbf{Z}(\mathfrak{Q}) \leq \mathfrak{K}$, $\mathfrak{K} = \mathfrak{M}$. Thus $\mathbf{N}_{\mathfrak{G}}(\mathfrak{R}) \leq \mathfrak{M}$ and $\mathbf{N}_{\mathfrak{L}}(\mathfrak{R}) \leq \mathfrak{L} \cap \mathfrak{M}$. Hence $\mathfrak{R} \in S_q(\mathfrak{L})$ and q^b does not divide $|\mathfrak{L}|$. By (1), $\mathfrak{L}$ contains a Sylow p-subgroup $\mathfrak{P}$ of $\mathfrak{G}$. Thus $\mathfrak{G} = \mathfrak{P}\mathfrak{Q} = \mathfrak{L}\mathbf{N}_{\mathfrak{G}}(\mathbf{Z}(\mathfrak{Q}))$. Since $\mathbf{Z}(\mathfrak{Q}) \leq \mathfrak{L}$, 2.5 gives $\mathbf{Z}(\mathfrak{Q}) = 1$, a contradiction.

3.11 Remark. ARAD and GLAUBERMAN [1] have shown that if $\mathfrak{G}$ is soluble of odd order, $\mathfrak{H}$ is a Hall π-subgroup of $\mathfrak{G}$ and $\mathbf{O}_{\pi'}(\mathfrak{G}) = 1$, then $\mathbf{ZJ}(\mathfrak{H}) \trianglelefteq \mathfrak{G}$. Here $\mathbf{ZJ}(\mathfrak{H})$ is the intersection of the Abelian subgroups of $\mathfrak{H}$ of maximal order. More precisely, the following hold.

a) For every prime $p \in \pi - \{3\}$ and for every Abelian subgroup $\mathfrak{A}$ of $\mathfrak{H}$ of maximal order, $\mathbf{O}_p(\mathfrak{A}) \leq \mathbf{O}_p(\mathfrak{G})$.

b) The sets of prime divisors of $|\mathfrak{A}|$, $|\mathbf{ZJ}(\mathfrak{H})|$ and $|\mathbf{F}(\mathfrak{G})|$ are all the same.

c) The maximal orders of Abelian subgroups of $\mathfrak{H}$ and of $\mathfrak{G}$ are the same.

d) $\mathbf{ZJ}(\mathfrak{G}) = \mathbf{ZJ}(\mathfrak{H})$.

§ 4. Conjugate p-Subgroups

If $A \subseteq \mathfrak{P}_1 \leq \mathfrak{S} \in S_p(\mathfrak{G})$ and $g_1 \in \mathbf{N}_{\mathfrak{G}}(\mathfrak{P}_1)$, then $A^{g_1} \subseteq \mathfrak{P}_1^{g_1} = \mathfrak{P}_1 \leq \mathfrak{S}$. Repeating this, if $A^{g_1} \subseteq \mathfrak{P}_2 \leq \mathfrak{S}$ and $g_2 \in \mathbf{N}_{\mathfrak{G}}(\mathfrak{P}_2)$, then $A^{g_1 g_2} \subseteq \mathfrak{S}$. This may be repeated further, and we shall see that all A^g which are contained in $\mathfrak{S}$ may be obtained in this way. Moreover, some restrictions may be placed on $\mathfrak{P}_1, \mathfrak{P}_2, \ldots$. We therefore make the following definition.

4.1 Definition. Suppose that $\mathfrak{S} \in S_p(\mathfrak{G})$. A *conjugation family* for $\mathfrak{S}$ is a set $\mathscr{F}$ of subgroups of $\mathfrak{S}$ such that whenever A is a non-empty subset of $\mathfrak{S}$, g is an element of $\mathfrak{G}$ and $A^g \subseteq \mathfrak{S}$, there exist subgroups $\mathfrak{P}_1, \ldots, \mathfrak{P}_n$ in $\mathscr{F}$ and elements $g_i \in \mathbf{N}_{\mathfrak{G}}(\mathfrak{P}_i)$ $(i = 1, \ldots, n)$ such that $g = g_1 \cdots g_n$ and

$$A^{g_1 \cdots g_{i-1}} \subseteq \mathfrak{P}_i \quad (i = 1, \ldots, n).$$

Since $g_i \in \mathbf{N}_{\mathfrak{G}}(\mathfrak{P}_i)$, it follows at once that

$$A^{g_1 \cdots g_i} \subseteq \mathfrak{P}_i \quad (i = 1, \ldots, n);$$

in particular, $A \subseteq \mathfrak{P}_1$ and $A^g \subseteq \mathfrak{P}_n$.

Note that for such a family, $\mathfrak{G} = \langle \mathbf{N}_{\mathfrak{G}}(\mathfrak{P}) | \mathfrak{P} \in \mathscr{F} \rangle$. For taking $A = \{1\}$, we find that any element g of $\mathfrak{G}$ is of the form $g_1 \cdots g_n$ with $g_i \in \mathbf{N}_{\mathfrak{G}}(\mathfrak{P}_i)$ and $\mathfrak{P}_i \in \mathscr{F}$. And $\mathfrak{S}$ lies in any conjugation family, as is seen by taking $A = \mathfrak{S}$.

The fact mentioned above amounts to the assertion that there always exists a conjugation family for $\mathfrak{S}$. This will be proved by giving a characterization of conjugation families in terms of the following equivalence relation.

4.2 Definition. Suppose that $\mathfrak{S} \in S_p(\mathfrak{G})$ and $\mathfrak{P} \leq \mathfrak{S}$. Let

$$\mathsf{S}(\mathfrak{P}) = \{g | g \in \mathfrak{G}, \mathfrak{P}^g \leq \mathfrak{S}\}.$$

Note that $\mathsf{S}(\mathfrak{P})$ is a union of double cosets $\mathbf{N}_{\mathfrak{G}}(\mathfrak{P}) x \mathbf{N}_{\mathfrak{G}}(\mathfrak{S})$, and if $\mathfrak{P} = \mathfrak{S}$, $\mathsf{S}(\mathfrak{P}) = \mathbf{N}_{\mathfrak{G}}(\mathfrak{S})$.

For elements a, b in $\mathsf{S}(\mathfrak{P})$, we write $a \underset{\mathfrak{P}}{\approx} b$ if either (i) $\mathfrak{P} = \mathfrak{S}$ and $a = b$, or (ii) $\mathfrak{P} < \mathfrak{S}$ and there exists a sequence $a_1, \ldots, a_n$ $(n \geq 1)$ of elements of $\mathsf{S}(\mathfrak{P})$ such that $a = a_1$, $a_n = b$ and

$$\mathfrak{S}^{a_{i-1}^{-1}} \cap \mathfrak{S}^{a_i^{-1}} > \mathfrak{P} \quad (i = 2, \ldots, n).$$

Thus $\underset{\mathfrak{P}}{\approx}$ is an equivalence relation on $\mathsf{S}(\mathfrak{P})$, and, if $\mathfrak{P} < \mathfrak{S}$, each equivalence class is a union of left cosets of $\mathbf{N}_{\mathfrak{G}}(\mathfrak{S})$.

4.3 Lemma. *Suppose that* $\mathfrak{S} \in S_p(\mathfrak{G})$ *and* $\mathfrak{P} \leq \mathfrak{S}$.

a) *If* $\mathfrak{P}^g \leq \mathfrak{S}$, $\mathsf{S}(\mathfrak{P}^g) = g^{-1}\mathsf{S}(\mathfrak{P})$, *and for* a, b *in* $\mathsf{S}(\mathfrak{P}^g)$, $a \underset{\mathfrak{P}^g}{\approx} b$ *if and only if* $ga \underset{\mathfrak{P}}{\approx} gb$.

b) *Given* $a \in \mathsf{S}(\mathfrak{P})$, *there exists* $b \in \mathsf{S}(\mathfrak{P})$ *and* $c \in \mathbf{N}_{\mathfrak{G}}(\mathfrak{P})$ *such that* $\mathbf{N}_{\mathfrak{S}}(\mathfrak{P}^b) \in S_p(\mathbf{N}_{\mathfrak{G}}(\mathfrak{P}^b))$ *and* $a \underset{\mathfrak{P}}{\approx} b \underset{\mathfrak{P}}{\approx} c$. *Thus each equivalence class of* $\underset{\mathfrak{P}}{\approx}$ *contains an element of* $\mathbf{N}_{\mathfrak{G}}(\mathfrak{P})$.

c) *If* S *is the equivalence class of* $\underset{\mathfrak{P}}{\approx}$ *containing* 1, *then for any* $a \in \mathbf{N}_{\mathfrak{G}}(\mathfrak{P})$, *the equivalence class containing* a *is* $a\mathsf{S}$.

d) *If* $a \in \mathsf{S}(\mathfrak{P})$, *there exist* $c \in \mathbf{N}_{\mathfrak{G}}(\mathfrak{P})$ *and* $d \in \mathsf{S}(\mathfrak{P})$ *such that* $d \underset{\mathfrak{P}}{\approx} 1$ *and* $a = cd$.

Proof. a) This is obvious.

b) If $\mathfrak{P} = \mathfrak{S}$, take $b = c = a$. Suppose that $\mathfrak{P} < \mathfrak{S}$. Then since $\mathfrak{P} \leq \mathfrak{S}^{a^{-1}}$, the p-subgroup $\mathbf{N}_{\mathfrak{S}^{a^{-1}}}(\mathfrak{P})$ is contained in a Sylow p-subgroup $\mathfrak{S}_1$ of $\mathbf{N}_{\mathfrak{G}}(\mathfrak{P})$, and $\mathfrak{S}_1 \leq \mathfrak{S}^{b^{-1}}$ for some $b \in \mathfrak{G}$. Then $\mathbf{N}_{\mathfrak{S}^{b^{-1}}}(\mathfrak{P})$ is a p-subgroup of $\mathbf{N}_{\mathfrak{G}}(\mathfrak{P})$ containing the Sylow p-subgroup $\mathfrak{S}_1$ and

$$\mathbf{N}_{\mathfrak{S}^{b^{-1}}}(\mathfrak{P}) = \mathfrak{S}_1 \in S_p(\mathbf{N}_{\mathfrak{G}}(\mathfrak{P})).$$

Since $\mathfrak{P} \leq \mathfrak{S}_1 \leq \mathfrak{S}^{b^{-1}}$, $b \in \mathsf{S}(\mathfrak{P})$. Since $\mathfrak{P} < \mathfrak{S}$, we have $\mathfrak{P} < \mathfrak{S}^{a^{-1}}$ and by I, 8.8, $\mathfrak{P} < \mathbf{N}_{\mathfrak{S}^{a^{-1}}}(\mathfrak{P})$. Then

$$\mathfrak{S}^{a^{-1}} \cap \mathfrak{S}^{b^{-1}} \geq \mathfrak{S}^{a^{-1}} \cap \mathfrak{S}_1 \geq \mathbf{N}_{\mathfrak{S}^{a^{-1}}}(\mathfrak{P}) > \mathfrak{P},$$

so $a \underset{\mathfrak{P}}{\approx} b$. By Sylow's theorem, $\mathbf{N}_{\mathfrak{S}}(\mathfrak{P}) \leq \mathfrak{S}_1^c$ for some $c \in \mathbf{N}_{\mathfrak{G}}(\mathfrak{P})$. Thus $\mathfrak{S}^{b^{-1}} \cap \mathfrak{S}^{c^{-1}} \geq \mathbf{N}_{\mathfrak{S}}(\mathfrak{P})^{c^{-1}} > \mathfrak{P}^{c^{-1}} = \mathfrak{P}$, so $b \underset{\mathfrak{P}}{\approx} c$.

c) By a), $x \underset{\mathfrak{P}}{\approx} 1$ if and only if $ax \underset{\mathfrak{P}}{\approx} a$. Thus $y \underset{\mathfrak{P}}{\approx} a$ if and only if $a^{-1}y \in \mathsf{S}$.

d) By b), there exists $c \in \mathbf{N}_{\mathfrak{G}}(\mathfrak{P})$ such that $a \underset{\mathfrak{P}}{\approx} c$. By c), $a = cd$ for some $d \underset{\mathfrak{P}}{\approx} 1$. **q.e.d.**

4.4 Theorem (DOLAN). *Suppose that* $\mathfrak{G} \neq 1$ *and* $\mathfrak{S} \in S_p(\mathfrak{G})$, *and let* $\mathscr{F}$ *be a set of subgroups of* $\mathfrak{S}$. *Then* $\mathscr{F}$ *is a conjugation family if and only if given any subgroup* $\mathfrak{P}$ *of* $\mathfrak{S}$ *for which* $\underset{\mathfrak{P}}{\approx}$ *has more than one equivalence class, there exists* $a \in \mathsf{S}(\mathfrak{P})$ *such that* $\mathfrak{P}^a \in \mathscr{F}$.

Proof. First suppose that $\mathscr{F}$ is not a conjugation family. Then we may choose a subset A of $\mathfrak{S}$ with $|A|$ maximal for which there exists $g \in \mathfrak{G}$ such that $A^g \subseteq \mathfrak{S}$ and

(1) whenever $g = g_1 \cdots g_n$ with $g_i \in \mathbf{N}_{\mathfrak{G}}(\mathfrak{P}_i)$ and $\mathfrak{P}_i \in \mathscr{F}$, then $A^{g_1 \cdots g_{i-1}} \not\subseteq \mathfrak{P}_i$ for some i $(1 \leq i \leq n)$.

Let $\mathfrak{P} = \mathfrak{S} \cap \mathfrak{S}^{g^{-1}}$. Then $A \subseteq \mathfrak{P}$, $\mathfrak{P}^g \leq \mathfrak{S}$ and (1) holds with A replaced by $\mathfrak{P}$. It follows from the maximality of $|A|$ that $A = \mathfrak{P}$.

First suppose that $\mathfrak{P} = \mathfrak{S}$. By (1), $\mathfrak{S} \notin \mathscr{F}$. But since $\mathfrak{G} \neq 1$, $\underset{\mathfrak{S}}{\approx}$ has more than one equivalence class, so the condition of the theorem does not hold for $\mathfrak{S}$.

Suppose then that $\mathfrak{P} < \mathfrak{S}$. We prove the following.

(2) If $x \underset{\mathfrak{P}}{\approx} y$, there exist $\mathfrak{P}_r \in \mathscr{F}$ $(r = 1, \ldots, m)$ and $g_r \in \mathbf{N}_{\mathfrak{G}}(\mathfrak{P}_r)$ such that $x^{-1}y = g_1 \cdots g_m$ and $\mathfrak{P}^{xg_1 \cdots g_{j-1}} \subseteq \mathfrak{P}_j$ $(j = 1, \ldots, m)$.

To see this, observe that since $x \underset{\mathfrak{P}}{\approx} y$, there exist elements $a_1, \ldots, a_n$ of $\mathsf{S}(\mathfrak{P})$ such that $x = a_1$, $y = a_n$ and

$$\mathfrak{S}^{a_{i-1}^{-1}} \cap \mathfrak{S}^{a_i^{-1}} > \mathfrak{P} \quad (i = 2, \ldots, n).$$

Let $\mathfrak{Q}_i = \mathfrak{S}^{a_i^{-1}a_{i-1}} \cap \mathfrak{S}$. Then $\mathfrak{Q}_i > \mathfrak{P}^{a_{i-1}}$, so $|\mathfrak{Q}_i| > |A|$. Since $\mathfrak{Q}_i^{a_{i-1}^{-1}a_i} \leq \mathfrak{S}$, it follows from the maximality of $|A|$ that there exist $\mathfrak{P}_{ij} \in \mathscr{F}$ $(j = 1, \ldots, m_i)$ and $b_{ij} \in \mathbf{N}_{\mathfrak{G}}(\mathfrak{P}_{ij})$ such that $a_{i-1}^{-1}a_i = b_{i1} \cdots b_{im_i}$ and

$$\mathfrak{Q}_i^{b_{i1} \cdots b_{i,j-1}} \leq \mathfrak{P}_{ij} \quad (j = 1, \ldots, m_i).$$

Thus

$$\begin{aligned} x^{-1}y &= a_1^{-1}a_2 \cdot a_2^{-1}a_3 \cdots a_{n-1}^{-1}a_n \\ &= b_{21} \cdots b_{2m_2} \cdots b_{n1} \cdots b_{nm_n}, \end{aligned}$$

and

$$\mathfrak{P}^{xb_{21} \cdots b_{2m_2} \cdots b_{i1} \cdots b_{i,j-1}} = \mathfrak{P}^{a_{i-1}b_{i1} \cdots b_{i,j-1}} \leq \mathfrak{Q}_i^{b_{i1} \cdots b_{i,j-1}} \leq \mathfrak{P}_{ij}.$$

Thus (2) is proved.

It follows from (1) and (2) that g, 1 are not equivalent under $\underset{\mathfrak{P}}{\approx}$. Thus $\underset{\mathfrak{P}}{\approx}$ has more than one equivalence class. Suppose that $a \in \mathsf{S}(\mathfrak{P})$ and $\mathfrak{P}^a \in \mathscr{F}$. By 4.3d), $a = cd$, where $c \in \mathbf{N}_{\mathfrak{G}}(\mathfrak{P})$ and $d \underset{\mathfrak{P}}{\approx} 1$. Thus $\mathfrak{P}^a = \mathfrak{P}^d \in \mathscr{F}$. By 4.3b), $g \underset{\mathfrak{P}}{\approx} f$ for some $f \in \mathbf{N}_{\mathfrak{G}}(\mathfrak{P})$, and by 4.3c), $f \underset{\mathfrak{P}}{\approx} fd$. Thus $g \underset{\mathfrak{P}}{\approx} fd$. Since $1 \underset{\mathfrak{P}}{\approx} d$, it follows from (2) that there exist $\mathfrak{P}_i \in \mathscr{F}$ $(i = 1, \ldots, m)$ and $g_i \in \mathbf{N}_{\mathfrak{G}}(\mathfrak{P}_i)$ such that $d = g_1 \cdots g_m$ and $\mathfrak{P}^{g_1 \cdots g_{i-1}} \subseteq \mathfrak{P}_i$ $(i = 1, \ldots, m)$. Put $\mathfrak{P}_{m+1} = \mathfrak{P}^d$ and $g_{m+1} = f^d$; thus $\mathfrak{P}_{m+1} \in \mathscr{F}$ and $g_{m+1} \in \mathbf{N}_{\mathfrak{G}}(\mathfrak{P}_{m+1})$. Since $fd \underset{\mathfrak{P}}{\approx} g$, there exist $\mathfrak{P}_j \in \mathscr{F}$ $(j = m + 2, \ldots, n)$ and $g_j \in \mathbf{N}_{\mathfrak{G}}(\mathfrak{P}_j)$ such that $(fd)^{-1}g = g_{m+2} \cdots g_n$ and $\mathfrak{P}^{fdg_{m+2} \cdots g_{j-1}} \subseteq \mathfrak{P}_j$ $(j = m + 2, \ldots, n)$. Then $fd = df^d = g_1 \cdots g_m g_{m+1}$ and $g = g_1 \cdots g_n$.

Also

$$\mathfrak{P}^{g_1 \cdots g_{i-1}} \leq \mathfrak{P}_i \quad (i = 1, \ldots, n).$$

This contradicts (1). Hence $\mathfrak{P}^a \notin \mathscr{F}$ for every $a \in \mathsf{S}(\mathfrak{P})$.

Conversely, suppose that $\mathscr{F}$ is a conjugation family, $\mathfrak{P} \leq \mathfrak{S}$ and $\underset{\mathfrak{P}}{\approx}$ has more than one equivalence class. Thus some element g of $\mathsf{S}(\mathfrak{P})$ is not equivalent to 1 in $\underset{\mathfrak{P}}{\approx}$. Since $\mathfrak{P}^g \leq \mathfrak{S}$ and $\mathscr{F}$ is a conjugation family, there exist subgroups $\mathfrak{P}_1, \ldots, \mathfrak{P}_n$ in $\mathscr{F}$ and elements $g_i \in \mathbf{N}_{\mathfrak{G}}(\mathfrak{P}_i)$ $(i = 1, \ldots, n)$ such that $g = g_1 \cdots g_n$ and

$$\mathfrak{P}^{g_1 \cdots g_{i-1}} \leq \mathfrak{P}_i \quad (i = 1, \ldots, n).$$

If $|\mathfrak{P}_i| > |\mathfrak{P}|$, then $\mathfrak{S} \cap \mathfrak{S}^{g_i^{-1}} \geq \mathfrak{P}_i > \mathfrak{P}^{g_1 \cdots g_{i-1}}$, so $g_1 \cdots g_{i-1} \underset{\mathfrak{P}}{\approx} g_1 \cdots g_i$. Thus if $|\mathfrak{P}_i| > |\mathfrak{P}|$ for all i,

$$1 \underset{\mathfrak{P}}{\approx} g_1 \underset{\mathfrak{P}}{\approx} g_1 g_2 \underset{\mathfrak{P}}{\approx} \cdots \underset{\mathfrak{P}}{\approx} g_1 \cdots g_n = g,$$

contrary to the definition of g. Hence $|\mathfrak{P}_i| = |\mathfrak{P}|$ for some i, and $\mathfrak{P}^{g_1 \cdots g_{i-1}} = \mathfrak{P}_i \in \mathscr{F}$. **q.e.d.**

4.5 Definition. Suppose that $\mathfrak{S} \in S_p(\mathfrak{G})$. The subgroup $\mathfrak{P}$ of $\mathfrak{S}$ is said to be *extremal* in $\mathfrak{S}$ if $\mathbf{N}_{\mathfrak{S}}(\mathfrak{P}) \in S_p(\mathbf{N}_{\mathfrak{G}}(\mathfrak{P}))$.

For example, any normal subgroup of $\mathfrak{S}$ is extremal in $\mathfrak{S}$.

4.6 Corollary. *Suppose that $\mathfrak{S} \in S_p(\mathfrak{G})$. The set $\mathscr{F}$ of extremal subgroups of $\mathfrak{S}$ is a conjugation family for $\mathfrak{S}$.*

Proof. Suppose that $\mathfrak{P} \leq \mathfrak{S}$. By 4.3b), there exists $b \in \mathsf{S}(\mathfrak{P})$ such that $\mathbf{N}_{\mathfrak{S}}(\mathfrak{P}^b) \in S_p(\mathbf{N}_{\mathfrak{G}}(\mathfrak{P}^b))$, or $\mathfrak{P}^b \in \mathscr{F}$. Hence by 4.4, $\mathscr{F}$ is a conjugation family for $\mathfrak{S}$. **q.e.d.**

The existence of conjugation families is thus established. The following is useful in applications.

4.7 Theorem (ALPERIN). *Suppose that $\mathfrak{S} \in S_p(\mathfrak{G})$ and that $\mathscr{F}$ is a conjugation family for $\mathfrak{S}$ such that every element of $\mathscr{F}$ is extremal in $\mathfrak{S}$. If A is a non-empty subset of $\mathfrak{S}$, $g \in \mathfrak{G}$ and $A^g \subseteq \mathfrak{S}$, there exist $\mathfrak{P}_i \in \mathscr{F}$ $(i = 1, \ldots, n)$ and $g_i \in \mathbf{N}_{\mathfrak{G}}(\mathfrak{P}_i)$ such that $g = g_1 \cdots g_n$, $A^{g_1 \cdots g_{i-1}} \subseteq \mathfrak{P}_i$ and, for each $i = 1, \ldots, n$, either $\mathfrak{P}_i \geq \mathbf{C}_{\mathfrak{S}}(\mathfrak{P}_i)$ or $g_i \in \mathbf{C}_{\mathfrak{G}}(\mathfrak{P}_i)$.*

Proof. This is proved by induction on $|\mathfrak{S} : \langle A \rangle|$. Since $\mathscr{F}$ is a conjugation family, there exist $\mathfrak{Q}_1, \ldots, \mathfrak{Q}_m$ in $\mathscr{F}$ and $h_i \in \mathbf{N}_{\mathfrak{G}}(\mathfrak{Q}_i)$ such that $g = h_1 \cdots h_m$ and $A^{h_1 \cdots h_{i-1}} \subseteq \mathfrak{Q}_i$ $(i = 1, \ldots, m)$.

For a fixed i, let $B = A^{h_1 \cdots h_{i-1}}, h = h_i, \mathfrak{Q} = \mathfrak{Q}_i$. It is clearly sufficient to prove the theorem for (B, h) in the place of (A, g). Since $B \subseteq \mathfrak{Q}$ and $h \in \mathbf{N}_{\mathfrak{G}}(\mathfrak{Q})$, there is nothing more to prove if $\mathfrak{Q} \geq \mathbf{C}_{\mathfrak{S}}(\mathfrak{Q})$. We therefore suppose that this is not the case; in particular, $\langle B \rangle \neq \mathfrak{S}$. Let $\mathfrak{P} = \mathfrak{Q}\mathbf{C}_{\mathfrak{S}}(\mathfrak{Q})$; thus $\mathfrak{P} > \mathfrak{Q}$ and

$$\mathfrak{P} = \mathbf{N}_{\mathfrak{S}}(\mathfrak{Q}) \cap \mathfrak{Q}\mathbf{C}_{\mathfrak{G}}(\mathfrak{Q}).$$

Since $\mathfrak{Q} \in \mathscr{F}$, $\mathfrak{Q}$ is extremal in $\mathfrak{S}$, so $\mathbf{N}_{\mathfrak{S}}(\mathfrak{Q}) \in S_p(\mathbf{N}_{\mathfrak{G}}(\mathfrak{Q}))$. Since $\mathfrak{Q}\mathbf{C}_{\mathfrak{G}}(\mathfrak{Q}) \trianglelefteq \mathbf{N}_{\mathfrak{G}}(\mathfrak{Q})$, it follows that $\mathfrak{P} \in S_p(\mathfrak{Q}\mathbf{C}_{\mathfrak{G}}(\mathfrak{Q}))$. Also $\mathfrak{P}^h \leq \mathfrak{Q}\mathbf{C}_{\mathfrak{G}}(\mathfrak{Q})$ since $h \in \mathbf{N}_{\mathfrak{G}}(\mathfrak{Q})$, so $\mathfrak{P}^h = \mathfrak{P}^x$ for some $x \in \mathfrak{Q}\mathbf{C}_{\mathfrak{G}}(\mathfrak{Q})$. If $x = yz$ with $y \in \mathfrak{Q}$, $z \in \mathbf{C}_{\mathfrak{G}}(\mathfrak{Q})$, $\mathfrak{P}^h = \mathfrak{P}^z$ since $y \in \mathfrak{P}$. Then $\mathfrak{P}^{hz^{-1}} = \mathfrak{P}$, and since $|\mathfrak{P}| > |\mathfrak{Q}| \geq |\langle A \rangle|$, it follows from the inductive hypothesis that there exist $\mathfrak{P}_1, \ldots, \mathfrak{P}_{n-1}$ in $\mathscr{F}$ and $g_j \in \mathbf{N}_{\mathfrak{G}}(\mathfrak{P}_j)$ $(j = 1, \ldots, n-1)$ such that $hz^{-1} = g_1 \cdots g_{n-1}$, $\mathfrak{P}^{g_1 \cdots g_{i-1}} \leq \mathfrak{P}_i$ $(i = 1, \ldots, n-1)$ and, for each $j = 1, \ldots, n-1$, either $\mathfrak{P}_j \geq \mathbf{C}_{\mathfrak{S}}(\mathfrak{P}_j)$ or $g_j \in \mathbf{C}_{\mathfrak{G}}(\mathfrak{P}_j)$. Putting $g_n = z$ and $\mathfrak{P}_n = \mathfrak{Q} \in \mathscr{F}$, we have $g_n \in \mathbf{C}_{\mathfrak{G}}(\mathfrak{P}_n)$, $h = g_1 \cdots g_n$ and

$$B^{g_1 \cdots g_{i-1}} \subseteq \mathfrak{P}_i \quad (i = 1, \ldots, n),$$

since

$$B^{g_1 \cdots g_{n-1}} = B^{hz^{-1}} \subseteq \mathfrak{Q}^{hz^{-1}} = \mathfrak{Q}. \qquad \textbf{q.e.d.}$$

4.8 Corollary (ALPERIN). *Suppose that $\mathfrak{S} \in S_p(\mathfrak{G})$ and $\mathscr{F}$ is a conjugation family for $\mathfrak{S}$ such that every element of $\mathscr{F}$ is extremal in $\mathfrak{S}$. Let $\mathscr{F}^* = \{\mathfrak{P} | \mathfrak{P} \in \mathscr{F}, \mathfrak{P} \geq \mathbf{C}_{\mathfrak{S}}(\mathfrak{P})\}$. If A, B are non-empty subsets of $\mathfrak{S}$ conjugate in $\mathfrak{G}$, there exist $\mathfrak{P}_1, \ldots, \mathfrak{P}_n$ in $\mathscr{F}^*$ and $g_i \in \mathbf{N}_{\mathfrak{G}}(\mathfrak{P}_i)$ $(i = 1, \ldots, n)$ such that $A^{g_1 \cdots g_{i-1}} \subseteq \mathfrak{P}_i$ and $A^{g_1 \cdots g_n} = B$.*

Proof. Suppose that $A^g = B$. By 4.7, there exist $\mathfrak{Q}_1, \ldots, \mathfrak{Q}_m$ in $\mathscr{F}$ and $h_i \in \mathbf{N}_{\mathfrak{G}}(\mathfrak{Q}_i)$ such that $g = h_1 \cdots h_m$, $A^{h_1 \cdots h_{i-1}} \subseteq \mathfrak{Q}_i$ and, for each $i = 1, \ldots, m$, either $\mathfrak{Q}_i \in \mathscr{F}^*$ or $h_i \in \mathbf{C}_{\mathfrak{G}}(\mathfrak{Q}_i)$. In the latter case, $A^{h_1 \cdots h_{i-1}} = A^{h_1 \cdots h_i}$. Let $\{i_1, \ldots, i_n\}$ be the set of the suffixes for which $\mathfrak{Q}_{i_j} \in \mathscr{F}^*$. Then

$$B = A^g = A^{h_1 \cdots h_m} = A^{g_1 \cdots g_n},$$

where $g_j = h_{i_j}$, and

$$A^{g_1 \cdots g_{j-1}} = A^{h_1 \cdots h_{i_j - 1}} \subseteq \mathfrak{Q}_{i_j}. \qquad \textbf{q.e.d.}$$

To find smaller conjugation families, we investigate the condition in 4.4 involving $\underset{\mathfrak{P}}{\approx}$.

4.9 Lemma. *Suppose that $\mathfrak{S} \in S_p(\mathfrak{G})$, $\mathfrak{P} \leq \mathfrak{S}$ and $\underset{\mathfrak{P}}{\approx}$ has more than one equivalence class. Then there exist elements a, b of $\mathsf{S}(\mathfrak{P})$ such that $b \underset{\mathfrak{P}}{\approx} 1$, $\mathfrak{S}^{a^{-1}} \cap \mathfrak{S}^{b^{-1}} = \mathfrak{P}$ and $\mathfrak{P}$ is extremal in both $\mathfrak{S}^{a^{-1}}$ and $\mathfrak{S}^{b^{-1}}$.*

Proof. Let g be an element of $\mathsf{S}(\mathfrak{P})$ such that g, 1 are not equivalent in $\underset{\mathfrak{P}}{\approx}$. By 4.3b), there exist elements a, b in $\mathsf{S}(\mathfrak{P})$ such that $g \underset{\mathfrak{P}}{\approx} a$, $1 \underset{\mathfrak{P}}{\approx} b$ and $\mathfrak{P}^a$, $\mathfrak{P}^b$ are both extremal in $\mathfrak{S}$. Thus a, b are not equivalent in $\underset{\mathfrak{P}}{\approx}$, so $\mathfrak{S}^{a^{-1}} \cap \mathfrak{S}^{b^{-1}} = \mathfrak{P}$. Clearly $\mathfrak{P}$ is extremal in $\mathfrak{S}^{a^{-1}}$ and in $\mathfrak{S}^{b^{-1}}$. **q.e.d.**

4.10 Definition. Suppose that $\mathfrak{S}$, $\tilde{\mathfrak{S}}$ are Sylow p-subgroups of $\mathfrak{G}$. We put $\mathfrak{S} \sim \tilde{\mathfrak{S}}$ if there exist Sylow p-subgroups $\mathfrak{S}_1, \ldots, \mathfrak{S}_n$ of $\mathfrak{G}$ such that $\mathfrak{S} = \mathfrak{S}_1$, $\tilde{\mathfrak{S}} = \mathfrak{S}_n$ and $\mathfrak{S}_{i-1} \cap \mathfrak{S}_i > 1$ $(i = 2, \ldots, n)$. Then $\sim$ is an equivalence relation on $S_p(\mathfrak{G})$. We say that $\mathfrak{G}$ is *p-isolated* if $\sim$ has more than one equivalence class.

4.11 Theorem. a) *$\mathfrak{G}$ operates transitively on the set of equivalence classes defined by the relation given in 4.10, and if $\mathfrak{H}$ is the stabiliser of the class $\mathscr{C}$, $\mathscr{C} = S_p(\mathfrak{H})$.*

b) *Suppose that p divides $|\mathfrak{G}|$. Then $\mathfrak{G}$ is p-isolated if and only if $\mathfrak{G}$ has a proper subgroup $\mathfrak{H}$ such that p divides $|\mathfrak{H}|$ and $\mathfrak{H} \cap \mathfrak{H}^g$ is a p'-group whenever $g \in \mathfrak{G} - \mathfrak{H}$.*

c) *Suppose that $\mathfrak{S} \in S_p(\mathfrak{G})$ and $\mathfrak{P} < \mathfrak{S}$. If $\underset{\mathfrak{P}}{\approx}$ has more than one equivalence class, $\mathbf{N}_{\mathfrak{G}}(\mathfrak{P})/\mathfrak{P}$ is p-isolated.*

Proof. a) If $\mathfrak{S}_1 \sim \mathfrak{S}_2$, then $\mathfrak{S}_1^g \sim \mathfrak{S}_2^g$ for any $g \in \mathfrak{G}$; thus $\mathfrak{G}$ operates on the set of equivalence classes defined by $\sim$. By Sylow's theorem, $\mathfrak{G}$ operates transitively.

Let $\mathfrak{H}$ be the stabiliser of the class $\mathscr{C}$. If $\mathfrak{S} \in \mathscr{C}$ and $x \in \mathfrak{S}$, then $\mathscr{C}^x$ is the class containing $\mathfrak{S}^x = \mathfrak{S}$. Thus $\mathscr{C}^x = \mathscr{C}$ and $x \in \mathfrak{H}$. Hence $\mathfrak{S} \in S_p(\mathfrak{H})$, and $\mathscr{C} \subseteq S_p(\mathfrak{H}) \subseteq S_p(\mathfrak{G})$. Suppose, conversely, that $\mathfrak{P} \in S_p(\mathfrak{H})$. Choose any $\mathfrak{P}_0 \in \mathscr{C}$; then $\mathfrak{P} = \mathfrak{P}_0^h$ for some $h \in \mathfrak{H}$ and $\mathfrak{P} \in \mathscr{C}^h = \mathscr{C}$. Thus $\mathscr{C} = S_p(\mathfrak{H})$.

b) Suppose that $\mathfrak{G}$ is p-isolated. Let $\mathscr{C}$ be an equivalence class and let $\mathfrak{H}$ be the stabiliser of $\mathscr{C}$. Then $\mathfrak{H} < \mathfrak{G}$ and p divides $|\mathfrak{H}|$. Suppose that

$\mathfrak{H} \cap \mathfrak{H}^g$ is not a p'-group and that $\mathfrak{P} \in S_p(\mathfrak{H} \cap \mathfrak{H}^g)$. Then $\mathfrak{P} \neq 1$, and $\mathfrak{P} \leq \mathfrak{S}_1 \in S_p(\mathfrak{H})$, $\mathfrak{P} \leq \mathfrak{S}_2 \in S_p(\mathfrak{H}^g)$. By a), $\mathfrak{S}_1 \in \mathscr{C}$ and $\mathfrak{S}_2 \in \mathscr{C}^g$. But $\mathfrak{S}_1 \cap \mathfrak{S}_2 \geq \mathfrak{P} > 1$, so $\mathfrak{S}_1 \sim \mathfrak{S}_2$. Hence $\mathscr{C} = \mathscr{C}^g$ and $g \in \mathfrak{H}$. Hence $\mathfrak{H} \cap \mathfrak{H}^g$ is a p'-group whenever $g \in \mathfrak{G} - \mathfrak{H}$.

Conversely, suppose that $\mathfrak{G}$ has a proper subgroup $\mathfrak{H}$ such that p divides $|\mathfrak{H}|$ and $\mathfrak{H} \cap \mathfrak{H}^g$ is a p'-group whenever $g \in \mathfrak{G} - \mathfrak{H}$. Suppose that $\mathfrak{P} \in S_p(\mathfrak{H})$; thus $\mathfrak{P} \neq 1$. If $g \in \mathbf{N}_{\mathfrak{G}}(\mathfrak{P})$, $\mathfrak{P} \leq \mathfrak{H} \cap \mathfrak{H}^g$, so $g \in \mathfrak{H}$. Thus $\mathbf{N}_{\mathfrak{G}}(\mathfrak{P}) \leq \mathfrak{H}$, and by 2.4, $\mathfrak{P} \in S_p(\mathfrak{G})$. Thus $S_p(\mathfrak{H})$ is a non-empty subset of $S_p(\mathfrak{G})$. Since $\mathfrak{H}$ cannot contain a normal subgroup of $\mathfrak{G}$ of order divisible by p, $S_p(\mathfrak{H}) \neq S_p(\mathfrak{G})$. We observe that if $\mathfrak{S}$, $\mathfrak{S}_1$ are Sylow p-subgroups of $\mathfrak{G}$, $\mathfrak{S} \cap \mathfrak{S}_1 \neq 1$ and $\mathfrak{S} \leq \mathfrak{H}$, then $\mathfrak{S}_1 \leq \mathfrak{H}$. For if $\mathfrak{S}_1 = \mathfrak{S}^g$, $\mathfrak{H} \cap \mathfrak{H}^g \geq \mathfrak{S} \cap \mathfrak{S}_1 \neq 1$, so $g \in \mathfrak{H}$ and $\mathfrak{S}_1 \leq \mathfrak{H}$. It follows that if $\mathfrak{S}$, $\mathfrak{S}_1$ are Sylow p-subgroups of $\mathfrak{G}$, $\mathfrak{S} \sim \mathfrak{S}_1$ and $\mathfrak{S} \leq \mathfrak{H}$, then $\mathfrak{S}_1 \leq \mathfrak{H}$. Thus $S_p(\mathfrak{H})$ contains an equivalence class and $\mathfrak{G}$ is p-isolated.

c) By 4.3b), there exists $a \in \mathbf{N}_{\mathfrak{G}}(\mathfrak{P})$ such that a, 1 are not equivalent in $\underset{\mathfrak{P}}{\approx}$. Suppose $\mathbf{N}_{\mathfrak{S}}(\mathfrak{P}) \leq \mathfrak{P}_0 \in S_p(\mathbf{N}_{\mathfrak{G}}(\mathfrak{P}))$. Thus $\mathfrak{P}_0^{a^{-1}} \in S_p(\mathbf{N}_{\mathfrak{G}}(\mathfrak{P}))$. Suppose that $\mathfrak{P}_0/\mathfrak{P} \sim \mathfrak{P}_0^{a^{-1}}/\mathfrak{P}$. Then there exist Sylow p-subgroups $\mathfrak{P}_1, \ldots, \mathfrak{P}_n$ of $\mathbf{N}_{\mathfrak{G}}(\mathfrak{P})$ such that $\mathfrak{P}_{i-1} \cap \mathfrak{P}_i > \mathfrak{P}$ $(i = 1, \ldots, n)$ and $\mathfrak{P}_n = \mathfrak{P}_0^{a^{-1}}$. If $\mathfrak{P}_i^{x_i} \leq \mathfrak{S}$ $(x_i \in \mathfrak{G})$, $x_{i-1} \underset{\mathfrak{P}}{\approx} x_i$. Thus $x_0 \underset{\mathfrak{P}}{\approx} x_n$. But $\mathfrak{S}^{x_0^{-1}} \cap \mathfrak{S} \geq \mathbf{N}_{\mathfrak{S}}(\mathfrak{P}) > \mathfrak{P}$, so $1 \underset{\mathfrak{P}}{\approx} x_0$, and $\mathfrak{S}^{x_n^{-1}} \cap \mathfrak{S}^{a^{-1}} \geq (\mathfrak{P}_0 \cap \mathfrak{S})^{a^{-1}} \geq \mathbf{N}_{\mathfrak{S}}(\mathfrak{P})^{a^{-1}} > \mathfrak{P}$, so $x_n \underset{\mathfrak{P}}{\approx} a$. Thus $1 \underset{\mathfrak{P}}{\approx} a$, a contradiction. Hence $\mathbf{N}_{\mathfrak{G}}(\mathfrak{P})/\mathfrak{P}$ is p-isolated.

q.e.d.

We now obtain the following generalization of 4.6.

4.12 Theorem (ALPERIN, GOLDSCHMIDT [1]). *Suppose that $\mathfrak{S} \in S_p(\mathfrak{G})$ and that $\mathscr{F}$ is the set of all subgroups $\mathfrak{P}$ of $\mathfrak{S}$ such that*

(i) *there exists $\mathfrak{S}_0 \in S_p(\mathfrak{G})$ such that $\mathfrak{P} = \mathfrak{S} \cap \mathfrak{S}_0$ and $\mathfrak{P}$ is extremal in both $\mathfrak{S}$ and $\mathfrak{S}_0$, and*

(ii) *either $\mathfrak{P} = \mathfrak{S}$ or $\mathbf{N}_{\mathfrak{G}}(\mathfrak{P})/\mathfrak{P}$ is p-isolated.*

Then $\mathscr{F}$ is a conjugation family for $\mathfrak{S}$.

Proof. Suppose that $\mathfrak{Q} \leq \mathfrak{S}$ and that $\underset{\mathfrak{Q}}{\approx}$ has more than one equivalence class. By 4.9, there exist elements a, b of $\mathsf{S}(\mathfrak{Q})$ such that $b \underset{\mathfrak{Q}}{\approx} 1$, $\mathfrak{S}^{a^{-1}} \cap \mathfrak{S}^{b^{-1}} = \mathfrak{Q}$ and $\mathfrak{Q}$ is extremal in both $\mathfrak{S}^{a^{-1}}$ and $\mathfrak{S}^{b^{-1}}$. Let $\mathfrak{P} = \mathfrak{Q}^b = \mathfrak{S}^{a^{-1}b} \cap \mathfrak{S}$. Then $\mathfrak{P}$ is extremal in both $\mathfrak{S}^{a^{-1}b}$ and $\mathfrak{S}$. By 4.11c), either $\mathfrak{Q} = \mathfrak{S}$ or $\mathbf{N}_{\mathfrak{G}}(\mathfrak{Q})/\mathfrak{Q}$ is p-isolated. Hence either $\mathfrak{P} = \mathfrak{S}$ or $\mathbf{N}_{\mathfrak{G}}(\mathfrak{P})/\mathfrak{P}$ is p-isolated. Thus $\mathfrak{P} \in \mathscr{F}$. By 4.4, $\mathscr{F}$ is a conjugation family. **q.e.d.**

4.13 Corollary (ALPERIN). *Suppose that $\mathfrak{S}$, $\mathfrak{S}^*$ are Sylow p-subgroups of $\mathfrak{G}$. Then there exist Sylow p-subgroups $\mathfrak{S}_1, \ldots, \mathfrak{S}_n$ of $\mathfrak{G}$ and $g_i \in \mathbf{N}_{\mathfrak{G}}(\mathfrak{S} \cap \mathfrak{S}_i)$ $(i = 1, \ldots, n)$ such that*

a) $\mathfrak{S}^{*g_1 \cdots g_n} = \mathfrak{S}$,

b) $\mathfrak{S} \cap \mathfrak{S}_i$ *is extremal in both* $\mathfrak{S}$ *and* $\mathfrak{S}_i$,

c) $(\mathfrak{S} \cap \mathfrak{S}^*)^{g_1 \cdots g_{i-1}} \leq \mathfrak{S} \cap \mathfrak{S}_i$ $(i = 1, \ldots, n)$.

Proof. Let $\mathscr{F}$ be the conjugation family of 4.12. By Sylow's theorem, $\mathfrak{S}^{*g} = \mathfrak{S}$ for some $g \in \mathfrak{G}$. Hence $(\mathfrak{S}^* \cap \mathfrak{S})^g \leq \mathfrak{S}$. Hence there exist $\mathfrak{P}_1, \ldots, \mathfrak{P}_n$ in $\mathscr{F}$ and $g_i \in \mathbf{N}_{\mathfrak{G}}(\mathfrak{P}_i)$ $(i = 1, \ldots, n)$ such that $g = g_1 \cdots g_n$ and $(\mathfrak{S}^* \cap \mathfrak{S})^{g_1 \cdots g_{i-1}} \leq \mathfrak{P}_i$. But $\mathfrak{P}_i = \mathfrak{S} \cap \mathfrak{S}_i$ for some $\mathfrak{S}_i \in S_p(\mathfrak{G})$ and $\mathfrak{P}_i$ is extremal in both $\mathfrak{S}$ and $\mathfrak{S}_i$. **q.e.d.**

4.14 Remarks. BENDER [5] has proved that if $\mathfrak{G}$ is a 2-isolated group, either (i) the Sylow 2-subgroups of $\mathfrak{G}$ are either cyclic or generalized quaternion, or (ii) $\mathfrak{G}$ has a series of normal subgroups $1 \leq \mathfrak{M} < \mathfrak{L} \leq \mathfrak{G}$ such that $\mathfrak{M}$ and $\mathfrak{G}/\mathfrak{L}$ are of odd order and $\mathfrak{L}/\mathfrak{M}$ is isomorphic to one of the simple groups $PSL(2, q)$, $Sz(q)$ or $PSU(3, q)$, where q is a power of 2 and $q \geq 4$.

According to 4.11b), such groups are characterized by the existence of a subgroup $\mathfrak{H}$ of even order such that $|\mathfrak{H} \cap \mathfrak{H}^g|$ is odd for all $g \in \mathfrak{G} - \mathfrak{H}$. Such a subgroup $\mathfrak{H}$ is called *strongly embedded*.

§ 5. Characteristic p-Functors

5.1 Definition. Let p be a prime. A *characteristic p-functor* is a function $\mathbf{W}$, defined on the class of all finite p-groups, such that

a) for any finite p-group $\mathfrak{P}$, $\mathbf{W}(\mathfrak{P})$ is a subgroup of $\mathfrak{P}$, and

b) if $\mathfrak{P}_1$, $\mathfrak{P}_2$ are finite p-groups and α is an isomorphism of $\mathfrak{P}_1$ onto $\mathfrak{P}_2$, then $\mathbf{W}(\mathfrak{P}_1)\alpha = \mathbf{W}(\mathfrak{P}_2)$.

Typical of the characteristic p-functors which will be considered are **J**, **ZJ** (3.6).

5.2 Lemma. *Let* $\mathbf{W}$ *be a characteristic p-functor.*

a) *If* $\mathfrak{P}$ *is a p-group,* $\mathbf{W}(\mathfrak{P})$ *is a characteristic subgroup of* $\mathfrak{P}$.

b) *If* $\mathfrak{G}$ *is a finite group,* $\mathfrak{P}$ *is a p-subgroup of* $\mathfrak{G}$ *and* $g \in \mathfrak{G}$, *then* $\mathbf{W}(\mathfrak{P})^g = \mathbf{W}(\mathfrak{P}^g)$.

c) *If* $\mathfrak{N}$ *is a normal p'-subgroup of the finite group* $\mathfrak{G}$ *and* $\mathfrak{P}$ *is a p-subgroup of* $\mathfrak{G}$, *then* $\mathbf{W}(\mathfrak{P}\mathfrak{N}/\mathfrak{N}) = \mathbf{W}(\mathfrak{P})\mathfrak{N}/\mathfrak{N}$, *and* $\mathbf{N}_{\mathfrak{G}/\mathfrak{N}}(\mathbf{W}(\mathfrak{P}\mathfrak{N}/\mathfrak{N})) = \mathbf{N}_{\mathfrak{G}}(\mathbf{W}(\mathfrak{P}))\mathfrak{N}/\mathfrak{N}$.

Proof. a) By 5.1b), $\mathbf{W}(\mathfrak{P})\alpha = \mathbf{W}(\mathfrak{P})$ for every automorphism α of $\mathfrak{P}$, so $\mathbf{W}(\mathfrak{P})$ is a characteristic subgroup of $\mathfrak{P}$.

b) Since the mapping $x \to x^g$ $(x \in \mathfrak{P})$ is an isomorphism of $\mathfrak{P}$ onto $\mathfrak{P}^g$, $\mathbf{W}(\mathfrak{P})^g = \mathbf{W}(\mathfrak{P}^g)$ by 5.1b).

c) The mapping $x \to x\mathfrak{N}$ $(x \in \mathfrak{P})$ is an isomorphism α of $\mathfrak{P}$ onto $\mathfrak{P}\mathfrak{N}/\mathfrak{N}$. Thus $\mathbf{W}(\mathfrak{P}\mathfrak{N}/\mathfrak{N}) = \mathbf{W}(\mathfrak{P}\alpha) = \mathbf{W}(\mathfrak{P})\alpha = \mathbf{W}(\mathfrak{P})\mathfrak{N}/\mathfrak{N}$. By IX, 6.11, $\mathbf{N}_{\mathfrak{G}/\mathfrak{N}}(\mathbf{W}(\mathfrak{P}\mathfrak{N}/\mathfrak{N})) = \mathbf{N}_{\mathfrak{G}/\mathfrak{N}}(\mathbf{W}(\mathfrak{P})\mathfrak{N}/\mathfrak{N}) = \mathbf{N}_{\mathfrak{G}}(\mathbf{W}(\mathfrak{P}))\mathfrak{N}/\mathfrak{N}$. **q.e.d.**

5.3 Definitions. a) Suppose that $\mathbf{W}$ is a characteristic p-functor and that $\mathfrak{X}$ is a finite p-group. For each subgroup $\mathfrak{P}$ of $\mathfrak{X}$ and each integer $n \geq 0$, we define a subgroup $\omega_n(\mathfrak{P}) = \omega_{n,\mathfrak{X}}(\mathfrak{P})$ by induction on n. For $n = 0$, define $\omega_0(\mathfrak{P}) = \mathbf{N}_{\mathfrak{X}}(\mathfrak{P})$, and for $n > 0$,

$$\omega_n(\mathfrak{P}) = \mathbf{N}_{\mathfrak{X}}(\mathbf{W}(\omega_{n-1}(\mathfrak{P}))).$$

For $n > 0$, $\mathbf{W}(\omega_{n-1}(\mathfrak{P}))$ is a characteristic subgroup of $\omega_{n-1}(\mathfrak{P})$, by 5.2a). Hence $\mathbf{W}(\omega_{n-1}(\mathfrak{P})) \trianglelefteq \mathbf{N}_{\mathfrak{X}}(\omega_{n-1}(\mathfrak{P}))$ and $\mathbf{N}_{\mathfrak{X}}(\omega_{n-1}(\mathfrak{P})) \leq \mathbf{N}_{\mathfrak{X}}(\mathbf{W}(\omega_{n-1}(\mathfrak{P}))) = \omega_n(\mathfrak{P})$. Hence

$$\mathfrak{P} \leq \omega_0(\mathfrak{P}) < \omega_1(\mathfrak{P}) < \cdots < \omega_k(\mathfrak{P}) = \mathfrak{X} = \omega_{k+1}(\mathfrak{P}) = \cdots$$

for some integer $k \geq 0$.

b) Suppose that $\mathbf{W}$ is a characteristic p-functor, $\mathfrak{G}$ is a finite group, $\mathfrak{S} \in S_p(\mathfrak{G})$ and $\mathfrak{P} \leq \mathfrak{S}$. We say that $\mathfrak{P}$ is $\mathbf{W}$-*extremal* in $\mathfrak{S}$ if $\mathfrak{P}$ and all $\mathbf{W}(\omega_{n,\mathfrak{S}}(\mathfrak{P}))$ $(n \geq 0)$ are extremal in $\mathfrak{S}$.

Thus, *extremal* means the same as $\mathbf{T}$-*extremal*, where $\mathbf{T}$ is the trivial functor given by $\mathbf{T}(\mathfrak{P}) = 1$. Any normal subgroup of $\mathfrak{S}$ is $\mathbf{W}$-extremal for all $\mathbf{W}$.

5.4 Lemma. *Suppose that* $\mathbf{W}$ *is a characteristic p-functor,* $\mathfrak{G}$ *is a finite group and* $\mathfrak{S} \in S_p(\mathfrak{G})$. *If* $\mathfrak{Q} \leq \mathfrak{S}$, *there exists* $g \in \mathfrak{G}$ *such that* $\mathfrak{Q}^g \leq \mathfrak{S}$ *and* $\mathfrak{Q}^g$ *is* $\mathbf{W}$-*extremal in* $\mathfrak{S}$.

Proof. We define a sequence $(\mathfrak{Q}_n)$ $(n \geq 0)$ of p-subgroups of $\mathfrak{G}$ by induction on n. Define $\mathfrak{Q}_0$ to be a Sylow p-subgroup of $\mathbf{N}_{\mathfrak{G}}(\mathfrak{Q})$, and for $n > 0$ define $\mathfrak{Q}_n$ to be a Sylow p-subgroup of $\mathbf{N}_{\mathfrak{G}}(\mathbf{W}(\mathfrak{Q}_{n-1}))$ containing $\mathfrak{Q}_{n-1}$. Thus

$$\mathfrak{Q} \leq \mathfrak{Q}_0 \leq \mathfrak{Q}_1 \leq \cdots.$$

Let $\mathfrak{P} = \bigcup_{n \geq 0} \mathfrak{Q}_n$. Thus $\mathfrak{P}$ is a p-subgroup of $\mathfrak{G}$. By Sylow's theorem, there exists $g \in \mathfrak{G}$ such that $\mathfrak{P}^g \leq \mathfrak{S}$. Let $\mathfrak{P}_n = \mathfrak{Q}_n^g$ $(n \geq 0)$. Then $\mathfrak{P}_n \leq \mathfrak{S}$ and

$$\mathfrak{Q}^g \leq \mathfrak{P}_0 \leq \mathfrak{P}_1 \leq \cdots.$$

Further, $\mathfrak{P}_0 \in S_p(\mathbf{N}_{\mathfrak{G}}(\mathfrak{Q}^g))$ and $\mathfrak{P}_n \in S_p(\mathbf{N}_{\mathfrak{G}}(\mathbf{W}(\mathfrak{P}_{n-1})))$. Hence $\mathfrak{P}_0 = \mathbf{N}_{\mathfrak{S}}(\mathfrak{Q}^g)$ and, for $n > 0$, $\mathfrak{P}_n = \mathbf{N}_{\mathfrak{S}}(\mathbf{W}(\mathfrak{P}_{n-1}))$, since, for example, $\mathbf{N}_{\mathfrak{S}}(\mathfrak{Q}^g)$ is a p-subgroup of $\mathbf{N}_{\mathfrak{G}}(\mathfrak{Q}^g)$ containing the Sylow p-subgroup $\mathfrak{P}_0$. Thus $\mathfrak{Q}^g$ and each $\mathbf{W}(\mathfrak{P}_n)$ is extremal in $\mathfrak{S}$. We prove by induction on n that $\mathfrak{P}_n = \omega_n(\mathfrak{Q}^g)$; this is clear for $n = 0$ and, for $n > 0$,

$$\mathfrak{P}_n = \mathbf{N}_{\mathfrak{S}}(\mathbf{W}(\mathfrak{P}_{n-1})) = \mathbf{N}_{\mathfrak{S}}(\mathbf{W}(\omega_{n-1}(\mathfrak{Q}^g))) = \omega_n(\mathfrak{Q}^g).$$

Thus $\mathfrak{Q}^g$ and each $\mathbf{W}(\omega_n(\mathfrak{Q}^g))$ is extremal in $\mathfrak{S}$ and $\mathfrak{Q}^g$ is $\mathbf{W}$-extremal.

q.e.d.

5.5 Theorem. *Let* $\mathbf{W}$ *be a characteristic p-functor. If* $\mathfrak{S} \in S_p(\mathfrak{G})$, *the set of* $\mathbf{W}$*-extremal subgroups of* $\mathfrak{S}$ *is a conjugation family for* $\mathfrak{S}$.

Proof. This follows at once from 4.4 and 5.4. **q.e.d.**

We shall study the fusion of sequences of elements of a subgroup.

5.6 Definition. Suppose that $\mathfrak{X}$ is a subgroup of a finite group $\mathfrak{G}$.

a) Denote by $\mathscr{S}(\mathfrak{X})$ the set of all finite sequences $(x_1, \ldots, x_m)$ of elements x_i of $\mathfrak{X}$.

b) If $\mathbf{x} = (x_1, \ldots, x_m) \in \mathscr{S}(\mathfrak{X})$ and $g \in \mathfrak{G}$, write $\mathbf{x}^g = (x_1^g, \ldots, x_m^g)$.

c) Two elements $\mathbf{x}, \mathbf{y}$ of $\mathscr{S}(\mathfrak{X})$ are said to be *conjugate* in $\mathfrak{G}$ if there exists $g \in \mathfrak{G}$ such that $\mathbf{x}^g = \mathbf{y}$.

Conjugacy in $\mathfrak{G}$ is an equivalence relation on $\mathscr{S}(\mathfrak{X})$.

5.7 Definition. Suppose that $\mathfrak{G}$ is a finite group, $\mathfrak{H} \leq \mathfrak{G}$ and $\mathfrak{S} \in S_p(\mathfrak{G})$. Let $\sim$ be an equivalence relation on $\mathscr{S}(\mathfrak{S})$. We say that $\sim$ *contains fusion in* $\mathfrak{H}$ if $\mathbf{x} \sim \mathbf{y}$ whenever $\mathbf{x}, \mathbf{y}$ are elements of $\mathscr{S}(\mathfrak{S} \cap \mathfrak{H})$ conjugate in $\mathfrak{H}$.

It is clear that if $\sim$ contains fusion in $\mathfrak{H}$, then $\sim$ contains fusion in any subgroup of $\mathfrak{H}$.

5.8 Definition. The characteristic p-functor $\mathbf{W}$ is called *positive* if $\mathbf{W}(\mathfrak{P}) \neq 1$ for every p-group $\mathfrak{P} \neq 1$.

For example, $\mathbf{Z}$, $\mathbf{J}$, $\mathbf{ZJ}$ are all positive.

Roughly speaking, the following theorem shows that if $\sim$ contains fusion in $\mathbf{N}_{\mathfrak{G}}(\mathbf{W}(\mathfrak{S}))$ but not in $\mathfrak{G}$, then the same is true in the normaliser of some p-subgroup.

5.9 Theorem. *Suppose that* $\mathbf{W}$ *is a positive characteristic p-functor,* $\mathfrak{G}$ *is a finite group and* $\mathfrak{S} \in S_p(\mathfrak{G})$. *Suppose that* $\sim$ *is an equivalence relation* *on $\mathscr{S}(\mathfrak{S})$, that $\sim$ contains fusion in $\mathbf{N}_{\mathfrak{G}}(\mathbf{W}(\mathfrak{S}))$ and that $\sim$ does not contain fusion in $\mathfrak{G}$. Then there exists a subgroup $\mathfrak{P}$ of $\mathfrak{S}$ having the following properties.*

a) *$\mathfrak{P} \neq 1$ and $\mathfrak{P}$ is $\mathbf{W}$-extremal in $\mathfrak{S}$.*

b) *$\sim$ does not contain fusion in $\mathbf{N}_{\mathfrak{G}}(\mathfrak{P})$.*

c) *$\sim$ contains fusion in $\mathbf{N}_{\mathfrak{G}}(\mathbf{W}(\mathbf{N}_{\mathfrak{S}}(\mathfrak{P})))$ and hence in $\mathbf{N}_{\mathfrak{G}}(\mathbf{W}(\mathbf{N}_{\mathfrak{S}}(\mathfrak{P}))) \cap \mathbf{N}_{\mathfrak{G}}(\mathfrak{P})$.*

d) *Either*

(i) *there exists a subgroup $\mathfrak{Q}$ of $\mathfrak{S}$ such that $\mathfrak{P} = \mathbf{W}(\mathfrak{Q})$, $\mathfrak{Q} > \mathbf{O}_p(\mathfrak{G})$ and $\mathfrak{Q} \geq \mathbf{C}_{\mathfrak{S}}(\mathfrak{Q})$, or*

(ii) *$\mathfrak{P} \geq \mathbf{O}_p(\mathfrak{G})$ and there exist elements $\mathbf{x}$, $\mathbf{y}$ of $\mathscr{S}(\mathfrak{P})$ such that $\mathbf{x} \not\sim \mathbf{y}$ but $\mathbf{x}, \mathbf{y}$ are conjugate in $\mathbf{N}_{\mathfrak{G}}(\mathfrak{P})$.*

(Note that the condition (ii) of d) is only a slight strengthening of b)).

Proof. Since $\sim$ does not contain fusion in $\mathfrak{G}$, there exist $\mathbf{x} \in \mathscr{S}(\mathfrak{S})$ and $g \in \mathfrak{G}$ such that $\mathbf{x}^g \in \mathscr{S}(\mathfrak{S})$ but $\mathbf{x}^g \not\sim \mathbf{x}$. If $\mathbf{x} = (x_1, \ldots, x_m)$, put $\mathfrak{R} = \langle x_1, \ldots, x_m, \mathbf{O}_p(\mathfrak{G})\rangle$. Thus $\mathfrak{R} \neq 1$ and $\mathfrak{R}^g \leq \mathfrak{S}$. Hence by 5.5, there exist $\mathbf{W}$-extremal subgroups $\mathfrak{P}_1, \ldots, \mathfrak{P}_n$ and elements $g_i \in \mathbf{N}_{\mathfrak{G}}(\mathfrak{P}_i)$ such that $g = g_1 \cdots g_n$ and $\mathfrak{R}^{g_1 \cdots g_{i-1}} \leq \mathfrak{P}_i$ $(i = 1, \ldots, n)$. In particular,

$$\mathbf{O}_p(\mathfrak{G}) = \mathbf{O}_p(\mathfrak{G})^{g_1 \cdots g_{i-1}} \leq \mathfrak{R}^{g_1 \cdots g_{i-1}} \leq \mathfrak{P}_i \quad (i = 1, \ldots, n).$$

Now $\sim$ is transitive and $\mathbf{x}^{g_1 \cdots g_n} \not\sim \mathbf{x}$. Hence $\mathbf{x}^{g_1 \cdots g_{j-1}} \not\sim \mathbf{x}^{g_1 \cdots g_j}$ for some j, where $1 \leq j \leq n$. Since $\mathbf{x}^{g_1 \cdots g_{j-1}} \in \mathscr{S}(\mathfrak{P}_j)$ and $g_j \in \mathbf{N}_{\mathfrak{G}}(\mathfrak{P}_j)$, we also have $\mathbf{x}^{g_1 \cdots g_j} \in \mathscr{S}(\mathfrak{P}_j)$; hence $\mathfrak{P}_j$ satisfies the conditions d) (ii) and b). Of course $\mathfrak{P}_j$ also satisfies a), since $\mathfrak{P}_j \geq \mathfrak{R}^{g_1 \cdots g_{j-1}} \neq 1$.

Let $\mathscr{T}$ be the set of subgroups $\mathfrak{P}$ of $\mathfrak{S}$ which satisfy a), b) and d). Since $\mathfrak{P}_j \in \mathscr{T}$, $\mathscr{T}$ is non-empty. Let $\mathfrak{P}$ be an element of $\mathscr{T}$ for which $|\mathbf{N}_{\mathfrak{S}}(\mathfrak{P})|$ is maximal. Let $\mathfrak{P}_0 = \mathbf{N}_{\mathfrak{S}}(\mathfrak{P})$. If $\mathfrak{P}_0 = \mathfrak{S}$, $\mathfrak{P}$ satisfies c) by hypothesis, and there is nothing further to prove. Suppose then that $\mathfrak{P}_0 < \mathfrak{S}$. It then follows, since $\mathfrak{P}$ satisfies d), that $\mathfrak{P}_0 > \mathbf{O}_p(\mathfrak{G})$. For in case (i), $\mathfrak{P} \trianglelefteq \mathfrak{Q}$, so $\mathfrak{P}_0 = \mathbf{N}_{\mathfrak{S}}(\mathfrak{P}) \geq \mathfrak{Q} > \mathbf{O}_p(\mathfrak{G})$. And in case (ii), $\mathfrak{P} \leq \mathfrak{P}_0 < \mathfrak{S}$, so $\mathbf{O}_p(\mathfrak{G}) \leq \mathfrak{P} < \mathbf{N}_{\mathfrak{S}}(\mathfrak{P}) = \mathfrak{P}_0$. Note also that

$$\mathfrak{P}_0 = \mathbf{N}_{\mathfrak{S}}(\mathfrak{P}) \geq \mathbf{C}_{\mathfrak{S}}(\mathfrak{P}) \geq \mathbf{C}_{\mathfrak{S}}(\mathfrak{P}_0).$$

It follows that $\mathbf{W}(\mathfrak{P}_0)$ satisfies d) (i) (with $\mathfrak{P}_0$ in the place of $\mathfrak{Q}$).

We observe that $\mathbf{W}(\mathfrak{P}_0)$ satisfies a). Since $\mathfrak{P}_0 \neq 1$ and $\mathbf{W}$ is positive, $\mathbf{W}(\mathfrak{P}_0) \neq 1$. Since $\omega_0(\mathfrak{P}) = \mathfrak{P}_0$ and $\mathfrak{P}$ is $\mathbf{W}$-extremal,

$$\omega_0(\mathbf{W}(\mathfrak{P}_0)) = \mathbf{N}_{\mathfrak{S}}(\mathbf{W}(\omega_0(\mathfrak{P}))) = \omega_1(\mathfrak{P}) \in S_p(\mathbf{N}_{\mathfrak{G}}(\mathbf{W}(\omega_0(\mathfrak{P})))).$$

It is easy to deduce by induction on n that if $n \geq 0$,

$$\omega_n(\mathbf{W}(\mathfrak{P}_0)) = \omega_{n+1}(\mathfrak{P}),$$

for if $n > 0$,

$$\omega_n(\mathbf{W}(\mathfrak{P}_0)) = \mathbf{N}_{\mathfrak{S}}(\mathbf{W}(\omega_{n-1}(\mathbf{W}(\mathfrak{P}_0)))) = \mathbf{N}_{\mathfrak{S}}(\mathbf{W}(\omega_n(\mathfrak{P}))) = \omega_{n+1}(\mathfrak{P}).$$

Hence since $\mathfrak{P}$ is $\mathbf{W}$-extremal,

$$\begin{aligned}\mathbf{N}_{\mathfrak{S}}(\mathbf{W}(\omega_n(\mathbf{W}(\mathfrak{P}_0)))) &= \omega_{n+1}(\mathbf{W}(\mathfrak{P}_0)) \\ &= \mathbf{N}_{\mathfrak{G}}(\mathbf{W}(\omega_{n+1}(\mathfrak{P}))) \in S_p(\mathbf{N}_{\mathfrak{G}}(\mathbf{W}(\omega_{n+1}(\mathfrak{P})))) \\ &= S_p(\mathbf{N}_{\mathfrak{G}}(\mathbf{W}(\omega_n(\mathbf{W}(\mathfrak{P}_0))))).\end{aligned}$$

Thus $\mathbf{W}(\mathfrak{P}_0)$ is $\mathbf{W}$-extremal.

Since $\mathbf{W}(\mathfrak{P}_0)$ is a characteristic subgroup of $\mathfrak{P}_0$, $\mathbf{N}_{\mathfrak{S}}(\mathbf{W}(\mathfrak{P}_0)) \geq \mathbf{N}_{\mathfrak{S}}(\mathfrak{P}_0)$. Since $\mathfrak{P}_0 < \mathfrak{S}$, $\mathbf{N}_{\mathfrak{S}}(\mathfrak{P}_0) > \mathfrak{P}_0$; thus

$$\mathbf{N}_{\mathfrak{S}}(\mathbf{W}(\mathfrak{P}_0)) > \mathfrak{P}_0 = \mathbf{N}_{\mathfrak{S}}(\mathfrak{P}).$$

It follows from the definition of $\mathfrak{P}$ that $\mathbf{W}(\mathfrak{P}_0) \notin \mathscr{T}$. Since $\mathbf{W}(\mathfrak{P}_0)$ satisfies a) and d), it follows that $\mathbf{W}(\mathfrak{P}_0)$ does not satisfy b). Thus $\sim$ contains fusion in $\mathbf{N}_{\mathfrak{G}}(\mathbf{W}(\mathfrak{P}_0))$; that is, $\mathfrak{P}$ satisfies c). **q.e.d.**

§ 6. Transfer Theorems

To apply the results of § 5 to particular characteristic p-functors, we shall need some results proved by using the transfer (Chapter IV). The following is equivalent to IV, 2.2.

6.1 Lemma. *Suppose that $\mathfrak{P}$ is a normal p-subgroup of $\mathfrak{G}$ and $\mathfrak{S} \in S_p(\mathfrak{G})$. Then $\mathfrak{P} \cap \mathfrak{G}' = [\mathfrak{P}, \mathfrak{G}](\mathfrak{P} \cap \mathfrak{S}')$.*

Proof. Let $\mathfrak{Z}/[\mathfrak{P}, \mathfrak{G}] = \mathbf{Z}(\mathfrak{G}/[\mathfrak{P}, \mathfrak{G}])$. Obviously $\mathfrak{P} \leq \mathfrak{Z}$. By IV, 2.2 applied to $\mathfrak{G}/[\mathfrak{P}, \mathfrak{G}]$,

$$(\mathfrak{S}/[\mathfrak{P}, \mathfrak{G}]) \cap (\mathfrak{G}'/[\mathfrak{P}, \mathfrak{G}]) \cap (\mathfrak{Z}/[\mathfrak{P}, \mathfrak{G}]) \leq (\mathfrak{S}/[\mathfrak{P}, \mathfrak{G}])'.$$

Thus

$$\mathfrak{S} \cap \mathfrak{G}' \cap \mathfrak{Z} \leq [\mathfrak{P}, \mathfrak{G}]\mathfrak{S}'.$$

Since $\mathfrak{P} \leq \mathfrak{Z}$, it follows that

$$\mathfrak{P} \cap \mathfrak{G}' \leq \mathfrak{P} \cap [\mathfrak{P}, \mathfrak{G}]\mathfrak{S}' = [\mathfrak{P}, \mathfrak{G}](\mathfrak{P} \cap \mathfrak{S}'),$$

whence the assertion. **q.e.d.**

Next we prove the important "focal subgroup" theorem.

6.2 Theorem (D. G. HIGMAN [1]). *If* $\mathfrak{S} \in S_p(\mathfrak{G})$,

$$\mathfrak{S} \cap \mathfrak{G}' = \langle x^{-1}x^g \,|\, x \in \mathfrak{S}, g \in \mathfrak{G}, x^g \in \mathfrak{S} \rangle.$$

Proof. Let $X = \{x^{-1}x^g \,|\, x \in \mathfrak{S}, g \in \mathfrak{G}, x^g \in \mathfrak{S}\}$. Since $x^{-1}x^g = [x, g]$, $X \subseteq \mathfrak{G}'$; hence $\langle X \rangle \leq \mathfrak{S} \cap \mathfrak{G}'$. Let τ be the transfer of $\mathfrak{G}$ into $\mathfrak{S}/\mathfrak{S}'$. By IV, 1.7, if $g \in \mathfrak{G}$, $\tau(g)$ is a product of elements of the form $sg^ms^{-1}\mathfrak{S}'$, where $s \in \mathfrak{G}$, $sg^ms^{-1} \in \mathfrak{S}$ and the sum of the exponents m occurring in these factors is $|\mathfrak{G} : \mathfrak{S}| = n$, say. Now if $g \in \mathfrak{S}$, then $sg^ms^{-1}\mathfrak{S}'$ is the product of the elements $g^m\mathfrak{S}'$ and $g^{-m}(g^m)^{s^{-1}}\mathfrak{S}'$ of the Abelian group $\mathfrak{S}/\mathfrak{S}'$. Thus $\tau(g)$ is the product of $g^n\mathfrak{S}'$ and a number of elements of the form $x\mathfrak{S}'$ with $x \in X$.

Suppose, then, that $a \in \mathfrak{S} \cap \mathfrak{G}'$. Since $(n, p) = 1$ and $\mathfrak{S} \cap \mathfrak{G}'$ is a p-group, there exists $b \in \mathfrak{S} \cap \mathfrak{G}'$ such that $b^n = a$. Since $b \in \mathfrak{S}$, $\tau(b)$ is the product of $a\mathfrak{S}'$ and various elements $x\mathfrak{S}'$ with $x \in X$. But since $b \in \mathfrak{G}'$, $\tau(b)$ is the identity element of $\mathfrak{S}/\mathfrak{S}'$. Hence $a\mathfrak{S}' \in \langle x\mathfrak{S}' \,|\, x \in X \rangle$ and $a \in \langle X \rangle \mathfrak{S}'$. Thus $\mathfrak{S} \cap \mathfrak{G}' \leq \langle X \rangle \mathfrak{S}'$. But clearly $\mathfrak{S}' \leq \langle X \rangle$, since if $g \in \mathfrak{S}$, $h \in \mathfrak{S}$, then $[g, h] = g^{-1}g^h \in X$. Hence $\mathfrak{S} \cap \mathfrak{G}' \leq \langle X \rangle$. Since $\langle X \rangle \leq \mathfrak{S} \cap \mathfrak{G}'$, $\mathfrak{S} \cap \mathfrak{G}' = \langle X \rangle$. **q.e.d.**

The following concept is frequently used in transfer theory.

6.3 Definition. Suppose that $\mathfrak{S} \in S_p(\mathfrak{G})$. A non-empty subset A of $\mathfrak{S}$ is said to be *weakly closed* in $\mathfrak{S}$ (with respect to $\mathfrak{G}$) if, whenever $g \in \mathfrak{G}$ and $A^g \subseteq \mathfrak{S}$, then $A^g = A$.

If A is weakly closed in $\mathfrak{S}$ and $g \in \mathfrak{S}$, then $A^g \subseteq \mathfrak{S}$, so $A^g = A$. Hence A is necessarily a union of conjugacy classes of $\mathfrak{S}$. Thus a weakly closed subgroup of $\mathfrak{S}$ is normal in $\mathfrak{S}$ and a weakly closed element of $\mathfrak{S}$ lies in the centre of $\mathfrak{S}$.

Many applications of weak closure make use of the following.

6.4 Lemma. *Suppose that* $\mathfrak{S} \in S_p(\mathfrak{G})$, $\mathfrak{B} \trianglelefteq \mathfrak{S}$, $\mathbf{N}_{\mathfrak{G}}(\mathfrak{B}) \leq \mathfrak{M} \leq \mathfrak{G}$ and $t \in \mathfrak{G}$.

a) *There exists* $g \in \mathfrak{M}t$ *such that* $\mathfrak{B} \cap \mathfrak{S}^g = \mathfrak{B} \cap \mathfrak{M}^g$.

b) *If* $\mathfrak{B}$ *is weakly closed in* $\mathfrak{S}$ *with respect to* $\mathfrak{G}$ *and* $t \notin \mathfrak{M}$, *then* $\mathfrak{B} \cap \mathfrak{M}^g < \mathfrak{B}$.

Proof. a) We have $\mathfrak{S} \leq \mathbf{N}_{\mathfrak{G}}(\mathfrak{B}) \leq \mathfrak{M}$. Now $t\mathfrak{B}t^{-1} \cap \mathfrak{M}$ is a p-subgroup of $\mathfrak{M}$, so by Sylow's theorem, there exists $a \in \mathfrak{M}$ such that $t\mathfrak{B}t^{-1} \cap \mathfrak{M} \leq \mathfrak{S}^a$. Thus if $g = at$, we have $g \in \mathfrak{M}t$ and $\mathfrak{S}^g = \mathfrak{S}^{at} \geq \mathfrak{B} \cap \mathfrak{M}^t = \mathfrak{B} \cap \mathfrak{M}^g$, so $\mathfrak{B} \cap \mathfrak{S}^g = \mathfrak{B} \cap \mathfrak{M}^g$.

b) Since $t \notin \mathfrak{M}$, $g \notin \mathfrak{M}$ and $\mathfrak{B}^{g^{-1}} \neq \mathfrak{B}$. Since $\mathfrak{B}$ is weakly closed in $\mathfrak{S}$, it follows that $\mathfrak{B}^{g^{-1}} \not\leq \mathfrak{S}$, so $\mathfrak{B} \cap \mathfrak{M}^g = \mathfrak{B} \cap \mathfrak{S}^g < \mathfrak{B}$. **q.e.d.**

Next we obtain the very simple transfer result that will be needed.

6.5 Lemma. *Suppose that* $\mathfrak{G}$ *is a group and* V *is a* $\mathbb{Z}\mathfrak{G}$*-module. Suppose that* $\mathfrak{M}$ *is a subgroup of finite index in* $\mathfrak{G}$ *and* $v \in \mathbf{C}_{\mathsf{V}}(\mathfrak{M})$. *Put*

$$\bar{v} = \sum_{t \in \mathsf{T}} vt,$$

where T *is a transversal of* $\mathfrak{M}$ *in* $\mathfrak{G}$. *Then* $\bar{v}$ *is independent of the choice of* T *and* $\bar{v} \in \mathbf{C}_{\mathsf{V}}(\mathfrak{G})$.

Proof. Any transversal of $\mathfrak{M}$ in $\mathfrak{G}$ is of the form $\{x_t t \mid t \in \mathsf{T}\}$ for certain elements x_t of $\mathfrak{M}$. Since $v \in \mathbf{C}_{\mathsf{V}}(\mathfrak{M})$,

$$\sum_{t \in \mathsf{T}} vx_t t = \sum_{t \in \mathsf{T}} vt = \bar{v}.$$

Thus $\bar{v}$ is independent of the choice of T. If $g \in \mathfrak{G}$, $\mathsf{T}g$ is also a transversal of $\mathfrak{M}$ in $\mathfrak{G}$; hence

$$\bar{v} = \sum_{t \in \mathsf{T}} vtg = \bar{v}g.$$

Thus $\bar{v} \in \mathbf{C}_{\mathsf{V}}(\mathfrak{G})$. **q.e.d.**

To use this, we shall make a rather careful choice of the transversal, using the following lemmas.

6.6 Lemma. *Suppose that* $\mathfrak{A}$, $\mathfrak{B}$ *are subgroups of a group* $\mathfrak{G}$ *and that* $\mathfrak{B}$ *is subnormal in* $\mathfrak{G}$. *Then there exists a subset* T *of* $\mathfrak{G}$ *such that every element of* $\mathfrak{G}$ *is uniquely expressible in the form* xt *with* $x \in \mathfrak{A}\mathfrak{B}$, $t \in \mathsf{T}$.

Proof (Blessenohl). Since $\mathfrak{B}$ is subnormal in $\mathfrak{G}$ (see VII, 16.1), there exists a series

$$\mathfrak{B} = \mathfrak{B}_0 \leq \mathfrak{B}_1 \leq \cdots \leq \mathfrak{B}_k = \mathfrak{G}$$

such that $\mathfrak{B}_{i-1} \trianglelefteq \mathfrak{B}_i$ ($i = 1, \ldots, k$), and the lemma will be proved by induction on k. For $k = 0$, it is trivial. If $k > 0$, then by the inductive hypothesis, there exists a subset T_1 of $\mathfrak{G}$ such that every element of $\mathfrak{G}$ is uniquely expressible in the form yt_1 with $y \in \mathfrak{A}\mathfrak{B}_1$, $t_1 \in \mathsf{T}_1$. Now $\mathfrak{A}\mathfrak{B}_1$ is a union of double cosets $\mathfrak{A}u\mathfrak{B}$, and there exists a subset U of $\mathfrak{B}_1$ such that every element of $\mathfrak{A}\mathfrak{B}_1$ lies in $\mathfrak{A}u\mathfrak{B}$ for a unique $u \in \mathsf{U}$. We show that $\mathsf{T} = \mathsf{U}\mathsf{T}_1$ has the required property.

Since $\mathfrak{B} \trianglelefteq \mathfrak{B}_1$, $\mathsf{U}\mathfrak{B} = \mathfrak{B}\mathsf{U}$, so $\mathfrak{G} = \mathfrak{A}\mathfrak{B}_1\mathsf{T}_1 = \mathfrak{A}\mathsf{U}\mathfrak{B}\mathsf{T}_1 = \mathfrak{A}\mathfrak{B}\mathsf{U}\mathsf{T}_1 = \mathfrak{A}\mathfrak{B}\mathsf{T}$. Now suppose that $abut_1 = a'b'u't'_1$, with a, a' in $\mathfrak{A}$, b, b' in $\mathfrak{B}$, u, u' in U and t_1, t'_1 in T_1. Then abu, $a'b'u'$ are in $\mathfrak{A}\mathfrak{B}_1$, so $t_1 = t'_1$. Hence $abu = a'b'u'$ and $aub^u = a'u'b'^{u'}$, where $b^u \in \mathfrak{B}$, $b'^{u'} \in \mathfrak{B}$. Thus $u = u'$ and $ab = a'b'$. Hence every element of $\mathfrak{G}$ is uniquely expressible in the form xt with $x \in \mathfrak{A}\mathfrak{B}$, $t \in \mathsf{T}$. **q.e.d.**

Lemma 6.6 does not remain valid if the condition involving subnormality is omitted, for $|\mathfrak{A}\mathfrak{B}|$ need not divide $|\mathfrak{G}|$. But since every subgroup of a p-group is subnormal, we have the following.

6.7 Corollary. *Suppose that $\mathfrak{A}$, $\mathfrak{B}$ are subgroups of the p-group $\mathfrak{G}$. Then there exists a subset T of $\mathfrak{G}$ such that every element of $\mathfrak{G}$ is uniquely expressible in the form xt with $x \in \mathfrak{A}\mathfrak{B}$, $t \in \mathsf{T}$.*

6.8 Lemma. *Suppose that $\mathfrak{A} \leq \mathfrak{G}$, $\mathfrak{B} \leq \mathfrak{G}$, $t \in \mathfrak{G}$.*

a) *If U is a transversal of $\mathfrak{A}^t \cap \mathfrak{B}$ in $\mathfrak{B}$, each coset of $\mathfrak{A}$ in $\mathfrak{A}t\mathfrak{B}$ is of the form $\mathfrak{A}tu$ for a unique $u \in \mathsf{U}$.*

b) *If V is a transversal of $\mathfrak{A} \cap \mathfrak{B}$ in $\mathfrak{B}$, each coset of $\mathfrak{A}$ in $\mathfrak{A}\mathfrak{B}t$ is of the form $\mathfrak{A}vt$ for a unique $v \in \mathsf{V}$.*

Proof. a) Since $t(\mathfrak{A}^t \cap \mathfrak{B}) \subseteq \mathfrak{A}t$,

$$\mathfrak{A}t\mathfrak{B} = \mathfrak{A}t(\mathfrak{A}^t \cap \mathfrak{B})\mathsf{U} = \mathfrak{A}t\mathsf{U}.$$

And if $\mathfrak{A}tu = \mathfrak{A}tu'$ with u, u' in U, $u'u^{-1} \in \mathfrak{A}^t \cap \mathfrak{B}$, so $u' = u$.

b) V^t is a transversal of $\mathfrak{A}^t \cap \mathfrak{B}^t$ in $\mathfrak{B}^t$, so by a), each coset of $\mathfrak{A}$ in $\mathfrak{A}t\mathfrak{B}^t = \mathfrak{A}\mathfrak{B}t$ is of the form $\mathfrak{A}tv^t = \mathfrak{A}vt$ for a unique $v \in \mathsf{V}$. **q.e.d.**

This brings us to our first main lemma.

6.9 Lemma. *Suppose that $\mathfrak{G}$ is a group of operators on the finite Abelian p-group $\mathfrak{V}$ and that $\mathfrak{S} \in S_p(\mathfrak{G})$. Let $\mathfrak{B}$ be a weakly closed subgroup in $\mathfrak{S}$ with respect to $\mathfrak{G}$ and suppose that $\mathbf{N}_{\mathfrak{G}}(\mathfrak{B}) \leq \mathfrak{M}$. Let $\mathscr{A}$ be a set of subgroups which generates $\mathfrak{B}$. If $\mathbf{C}_{\mathfrak{V}}(\mathfrak{M}) > \mathbf{C}_{\mathfrak{V}}(\mathfrak{G})$, there exist $\mathfrak{A} \in \mathscr{A}$, $g \in \mathfrak{G} - \mathfrak{M}$ and $v \in \mathbf{C}_{\mathfrak{V}}(\mathfrak{A} \cap \mathfrak{S}^g)$ such that $\mathfrak{A} \cap \mathfrak{S}^g = \mathfrak{A} \cap \mathfrak{M}^g < \mathfrak{A}$ and*

$$\prod_{r \in \mathsf{R}} (vr) \notin \mathbf{C}_{\mathfrak{V}}(\mathfrak{G})$$

for any transversal R of $\mathfrak{A} \cap \mathfrak{S}^g$ in $\mathfrak{A}$.

Proof. Choose $u \in \mathbf{C}_{\mathfrak{V}}(\mathfrak{M}) - \mathbf{C}_{\mathfrak{V}}(\mathfrak{G})$. By 6.5,

$$\prod_{g \in \mathsf{Q}} ug \in \mathbf{C}_{\mathfrak{V}}(\mathfrak{G}),$$

where Q is any transversal of $\mathfrak{M}$ in $\mathfrak{G}$. But the set of right cosets of $\mathfrak{M}$ in $\mathfrak{G}$ is the disjoint union of the sets of those lying in the double cosets $\mathfrak{M}t\mathfrak{B}$ and $\mathfrak{M} = \mathfrak{M}1\mathfrak{B}$ is one of these double cosets. Since $u \notin \mathbf{C}_{\mathfrak{V}}(\mathfrak{G})$, it follows that there exists $t \notin \mathfrak{M}$ such that

$$(1) \qquad u_t = \prod_{h \in \mathsf{Q}_t} uh \notin \mathbf{C}_{\mathfrak{V}}(\mathfrak{G}),$$

where Q_t is a complete set of representatives of the right cosets of $\mathfrak{M}$ in $\mathfrak{M}t\mathfrak{B}$. Note that since $u \in \mathbf{C}_{\mathfrak{V}}(\mathfrak{M})$, u_t is independent of the choice of Q_t.

By 6.4, there exists $g \in \mathfrak{M}t$ such that $\mathfrak{B}_0 \lessdot \mathfrak{B}$, where

$$\mathfrak{B}_0 = \mathfrak{B} \cap \mathfrak{S}^g = \mathfrak{B} \cap \mathfrak{M}^g.$$

Since $\mathfrak{B}$ is generated by $\mathscr{A}$, there exists $\mathfrak{A} \in \mathscr{A}$ such that $\mathfrak{A} \not\leq \mathfrak{B}_0$. Thus

$$\mathfrak{A} \cap \mathfrak{B}_0 = \mathfrak{A} \cap \mathfrak{S}^g = \mathfrak{A} \cap \mathfrak{M}^g < \mathfrak{A}.$$

Let R be any transversal of $\mathfrak{A} \cap \mathfrak{B}_0$ in $\mathfrak{A}$. By 6.7, there exists a subset S of $\mathfrak{B}$ such that every element of $\mathfrak{B}$ is uniquely expressible in the form xs with $x \in \mathfrak{B}_0\mathfrak{A}$, $s \in \mathsf{S}$. Thus each coset of $\mathfrak{B}_0$ in $\mathfrak{B}$ is in $\mathfrak{B}_0\mathfrak{A}s$ for a unique $s \in \mathsf{S}$ and is therefore, by 6.8b), of the form $\mathfrak{B}_0rs$ for a unique $r \in \mathsf{R}$. By 6.8a), each coset of $\mathfrak{M}$ in $\mathfrak{M}g\mathfrak{B}$ is of the form $\mathfrak{M}grs$ for unique elements $r \in \mathsf{R}$, $s \in \mathsf{S}$. Since $\mathfrak{M}g\mathfrak{B} = \mathfrak{M}t\mathfrak{B}$, it follows from (1) that

$$\prod_{s \in \mathsf{S}} \prod_{r \in \mathsf{R}} ugrs \notin \mathbf{C}_{\mathfrak{V}}(\mathfrak{G}).$$

Hence

$$\prod_{r \in \mathsf{R}} ugr \notin \mathbf{C}_{\mathfrak{V}}(\mathfrak{G}).$$

But

$$ug \in \mathbf{C}_{\mathfrak{V}}(\mathfrak{M})g = \mathbf{C}_{\mathfrak{V}}(\mathfrak{M}^g) \leq \mathbf{C}_{\mathfrak{V}}(\mathfrak{B}_0) \leq \mathbf{C}_{\mathfrak{V}}(\mathfrak{A} \cap \mathfrak{S}^g),$$

so the stated condition holds with $v = ug$. **q.e.d.**

Our next aim is to remove the condition in 6.9 that $\mathfrak{V}$ is Abelian. For this we need a lemma similar to 1.8.

6.10 Lemma. *Suppose that $\mathfrak{G}$ is a group of operators on a group $\mathfrak{V}$. Let $\mathfrak{U}$ be an Abelian $\mathfrak{G}$-invariant subgroup of $\mathfrak{V}$ and suppose that $\mathfrak{G}$ has a subgroup $\mathfrak{M}$ such that $(|\mathfrak{G}:\mathfrak{M}|, |\mathfrak{U}|) = 1$. If $v \in \mathbf{C}_{\mathfrak{V}}(\mathfrak{M})$ and $v\mathfrak{U}$ is $\mathfrak{G}$-invariant, then $v\mathfrak{U}$ contains a $\mathfrak{G}$-invariant element.*

Proof. For each $g \in \mathfrak{G}$, there exists $f(g) \in \mathfrak{U}$ such that $vg = vf(g)$, since $v\mathfrak{U}$ is $\mathfrak{G}$-invariant. If also $h \in \mathfrak{G}$,

$$vf(gh) = (vg)h = (vf(g))h = (vh)(f(g)h) = (vf(h))(f(g)h),$$

so

(2) $$f(gh) = f(h)(f(g)h) \quad (g \in \mathfrak{G}, h \in \mathfrak{G}).$$

Since $v \in \mathbf{C}_{\mathfrak{V}}(\mathfrak{M})$, $f(x) = 1$ for all $x \in \mathfrak{M}$. Thus by (2),

(3) $$f(xg) = f(g) \quad (x \in \mathfrak{M}, g \in \mathfrak{G}).$$

Let $m = |\mathfrak{G}:\mathfrak{M}|$ and let T be a transversal of $\mathfrak{M}$ in $\mathfrak{G}$. By hypothesis, $ml \equiv 1$ $(|\mathfrak{U}|)$ for some positive integer l. Put

$$u = \left(\prod_{t \in \mathsf{T}} f(t)\right)^l.$$

(The order of the factors in this product need not be specified, since $\mathfrak{U}$ is Abelian). Then $u \in \mathfrak{U}$, and by (3), u is independent of the choice of the transversal T. But if $g \in \mathfrak{G}$, $\mathsf{T}g$ is a transversal of $\mathfrak{M}$ in $\mathfrak{G}$, so

$$u = \left(\prod_{t \in \mathsf{T}} f(tg)\right)^l.$$

By (2), it follows that

$$u = \left(\prod_{t\in T} f(g)(f(t)g)\right)^l = f(g)^{ml} \prod_{t\in T} (f(t)^l g) = f(g)(ug).$$

Thus

$$(vu)g = (vg)(ug) = vf(g)(ug) = vu$$

for all $g \in \mathfrak{G}$. **q.e.d.**

6.11 Lemma. *Suppose that $\mathfrak{B}$ is a finite p-group, $\mathfrak{G}$ is a group of operators on $\mathfrak{B}$ and $\mathfrak{M}$ is a subgroup of $\mathfrak{G}$ of index prime to p. Suppose that* $\mathbf{C}_{\mathfrak{U}}(\mathfrak{M}) = \mathbf{C}_{\mathfrak{U}}(\mathfrak{G})$ *for any $\mathfrak{G}$-composition factor $\mathfrak{U}$ (see I, 11.5) of $\mathfrak{B}$. Then* $\mathbf{C}_{\mathfrak{B}}(\mathfrak{M}) = \mathbf{C}_{\mathfrak{B}}(\mathfrak{G})$.

Proof. This is proved by induction on $|\mathfrak{B}|$. If $\mathfrak{B} = 1$, there is nothing to prove. Suppose that $\mathfrak{B} \neq 1$ and that $\mathfrak{Z}$ is a minimal $\mathfrak{G}$-invariant subgroup of the centre of $\mathfrak{B}$. Thus $\mathfrak{Z} \trianglelefteq \mathfrak{B}$ and $\mathfrak{Z}$ is a $\mathfrak{G}$-composition factor of $\mathfrak{B}$. Hence by hypothesis, $\mathbf{C}_{\mathfrak{Z}}(\mathfrak{M}) = \mathbf{C}_{\mathfrak{Z}}(\mathfrak{G})$. And by the inductive hypothesis, $\mathbf{C}_{\mathfrak{B}/\mathfrak{Z}}(\mathfrak{M}) = \mathbf{C}_{\mathfrak{B}/\mathfrak{Z}}(\mathfrak{G})$.

It is to be proved that any element v of $\mathbf{C}_{\mathfrak{B}}(\mathfrak{M})$ lies in $\mathbf{C}_{\mathfrak{B}}(\mathfrak{G})$. Since $v\mathfrak{Z} \in \mathbf{C}_{\mathfrak{B}/\mathfrak{Z}}(\mathfrak{M}) = \mathbf{C}_{\mathfrak{B}/\mathfrak{Z}}(\mathfrak{G})$, $v\mathfrak{Z}$ is $\mathfrak{G}$-invariant. By 6.10, there exists $z \in \mathfrak{Z}$ such that $vz \in \mathbf{C}_{\mathfrak{B}}(\mathfrak{G})$. Thus $vz \in \mathbf{C}_{\mathfrak{B}}(\mathfrak{M})$, and so $z \in \mathbf{C}_{\mathfrak{Z}}(\mathfrak{M}) = \mathbf{C}_{\mathfrak{Z}}(\mathfrak{G})$. Hence $v = (vz)z^{-1} \in \mathbf{C}_{\mathfrak{B}}(\mathfrak{G})$. **q.e.d.**

We also need a weak form of the dual of this.

6.12 Lemma. *Suppose that $\mathfrak{B}$ is a finite p-group, $\mathfrak{G}$ is a group of operators on $\mathfrak{B}$ and $\mathfrak{M}$ is a subgroup of $\mathfrak{G}$ of index prime to p. Suppose that for any elementary Abelian $\mathfrak{G}$-invariant section $\mathfrak{U}$ of $\mathfrak{B}$, $[\mathfrak{U}, \mathfrak{G}] = [\mathfrak{U}, \mathfrak{M}]$. Then* $[\mathfrak{B}, \mathfrak{G}] = [\mathfrak{B}, \mathfrak{M}]$.

Proof. We use induction on $|\mathfrak{B}|$ and suppose that $\mathfrak{B} \neq 1$. Let $\mathfrak{Z} = \Omega_1(\mathbf{Z}(\mathfrak{B})) \neq 1$. By hypothesis, $[\mathfrak{Z}, \mathfrak{G}] = [\mathfrak{Z}, \mathfrak{M}]$. If $[\mathfrak{Z}, \mathfrak{G}] \neq 1$, the inductive hypothesis may be applied to $\mathfrak{B}/[\mathfrak{Z}, \mathfrak{G}]$. This shows that $[\mathfrak{B}, \mathfrak{G}] = [\mathfrak{B}, \mathfrak{M}][\mathfrak{Z}, \mathfrak{G}]$. Hence

$$[\mathfrak{B}, \mathfrak{G}] = [\mathfrak{B}, \mathfrak{M}][\mathfrak{Z}, \mathfrak{M}] = [\mathfrak{B}, \mathfrak{M}],$$

as required. Suppose then that $[\mathfrak{Z}, \mathfrak{G}] = 1$.

If $[\mathfrak{V}, \mathfrak{M}] \neq 1$, then since $[\mathfrak{V}, \mathfrak{M}] \trianglelefteq \mathfrak{V}$, $\mathfrak{Y} = [\mathfrak{V}, \mathfrak{M}] \cap \mathfrak{Z} \neq 1$ (III, 7.2). Since $\mathfrak{Y} \leq \mathfrak{Z}$, $\mathfrak{Y}$ is normal and $\mathfrak{G}$-invariant. It follows from the inductive hypothesis that $[\mathfrak{V}, \mathfrak{G}]\mathfrak{Y} = [\mathfrak{V}, \mathfrak{M}]\mathfrak{Y}$, whence $[\mathfrak{V}, \mathfrak{G}] = [\mathfrak{V}, \mathfrak{M}]$.

Finally, suppose that $[\mathfrak{V}, \mathfrak{M}] = 1$. Then $[\mathfrak{V}/\mathfrak{Z}, \mathfrak{M}] = 1$, so by the inductive hypothesis, $[\mathfrak{V}/\mathfrak{Z}, \mathfrak{G}] = 1$ and $[\mathfrak{V}, \mathfrak{G}] \leq \mathfrak{Z}$. Also $[\mathfrak{Z}, \mathfrak{G}] = 1$. Let $\mathfrak{K} = \mathbf{C}_{\mathfrak{G}}(\mathfrak{V})$. Then $\mathfrak{K} \trianglelefteq \mathfrak{G}$ and $\mathfrak{G}/\mathfrak{K}$ is isomorphic to a group of automorphisms of $\mathfrak{V}$ which induce the identity automorphism on $\mathfrak{V}/\mathfrak{Z}$ and $\mathfrak{Z}$. By I, 4.4, $\mathfrak{G}/\mathfrak{K}$ is a p-group. But since $[\mathfrak{V}, \mathfrak{M}] = 1$, $\mathfrak{M} \leq \mathfrak{K}$. Since $|\mathfrak{G} : \mathfrak{M}|$ is prime to p, it follows that $\mathfrak{K} = \mathfrak{G}$ and $[\mathfrak{V}, \mathfrak{G}] = 1$. **q.e.d.**

6.13 Lemma. *Suppose that $\mathfrak{G}$ is a group of operators on the finite p-group $\mathfrak{V}$ and that $\mathfrak{S} \in S_p(\mathfrak{G})$. Let $\mathfrak{B}$ be a weakly closed subgroup in $\mathfrak{S}$ with respect to $\mathfrak{G}$ and suppose that $\mathbf{N}_{\mathfrak{G}}(\mathfrak{B}) \leq \mathfrak{M}$. Suppose that $\mathscr{A}$ is a set of subgroups which generates $\mathfrak{B}$.*

a) *Suppose that whenever $\mathfrak{A} \in \mathscr{A}$, $\mathfrak{A}_0 < \mathfrak{A}$, $\mathfrak{U}$ is a $\mathfrak{G}$-composition factor of $\mathfrak{V}$ and $u \in \mathbf{C}_{\mathfrak{U}}(\mathfrak{A}_0)$, then $\prod_{r \in \mathsf{R}} (ur) = 1$ for some transversal R of $\mathfrak{A}_0$ in $\mathfrak{A}$. Then $\mathbf{C}_{\mathfrak{V}}(\mathfrak{M}) = \mathbf{C}_{\mathfrak{V}}(\mathfrak{G})$.*

b) *Suppose that whenever $\mathfrak{A} \in \mathscr{A}$, $\mathfrak{A}_0 < \mathfrak{A}$, $\mathfrak{U}$ is a $\mathfrak{G}$-composition factor of $\mathfrak{V}$ and $u \in \mathfrak{U}$, then $\prod_{r \in \mathsf{R}} (ur^{-1}) \in [\mathfrak{U}, \mathfrak{A}_0]$ for some transversal R of $\mathfrak{A}_0$ in $\mathfrak{A}$. Then $[\mathfrak{V}, \mathfrak{M}] = [\mathfrak{V}, \mathfrak{G}]$.*

Proof. a) If $\mathbf{C}_{\mathfrak{V}}(\mathfrak{M}) > \mathbf{C}_{\mathfrak{V}}(\mathfrak{G})$, there is a $\mathfrak{G}$-composition factor $\mathfrak{U}$ of $\mathfrak{V}$ such that $\mathbf{C}_{\mathfrak{U}}(\mathfrak{M}) > \mathbf{C}_{\mathfrak{U}}(\mathfrak{G})$, by 6.11. By 6.9, there exist $\mathfrak{A} \in \mathscr{A}$ and $g \in \mathfrak{G}$ such that $\mathfrak{A}_0 = \mathfrak{A} \cap \mathfrak{S}^g$ does not satisfy the hypothesis of a).

b) By 6.12, it is sufficient to prove that $[\mathfrak{W}, \mathfrak{M}] = [\mathfrak{W}, \mathfrak{G}]$ for every elementary Abelian $\mathfrak{G}$-invariant section $\mathfrak{W}$ of $\mathfrak{V}$. Now $\mathfrak{W}$ can be regarded as a $\mathsf{K}\mathfrak{G}$-module, where $\mathsf{K} = GF(p)$, and by VII, 8.3, $\mathfrak{W}$ is the dual module W^* of some $\mathsf{K}\mathfrak{G}$-module W. We show that W satisfies the conditions of a). Suppose then that $\mathfrak{A} \in \mathscr{A}$, $\mathfrak{A}_0 < \mathfrak{A}$, U is a $\mathfrak{G}$-composition factor of W and $u \in \mathbf{C}_{\mathsf{U}}(\mathfrak{A}_0)$. Write $\mathsf{U} = \mathsf{W}_1/\mathsf{W}_2$ and $u = w + \mathsf{W}_2$, where $w \in \mathsf{W}_1$. By VII, 8.3, $\mathsf{W}_2^{\perp}/\mathsf{W}_1^{\perp}$ is a $\mathfrak{G}$-composition factor of $\mathsf{W}^* = \mathfrak{W}$, so by hypothesis*

$$\sum_{r \in \mathsf{R}} fr^{-1} \in [\mathsf{W}_2^{\perp}, \mathfrak{A}_0] + \mathsf{W}_1^{\perp}$$

for all $f \in \mathsf{W}_2^{\perp}$ and some transversal R of $\mathfrak{A}_0$ in $\mathfrak{A}$. By VII, 8.3g), $[\mathsf{W}_2^{\perp}, \mathfrak{A}_0] = \mathsf{W}_3^{\perp}$, where $\mathsf{W}_3/\mathsf{W}_2 = \mathbf{C}_{\mathsf{W}/\mathsf{W}_2}(\mathfrak{A}_0)$. Since $u \in \mathbf{C}_{\mathsf{U}}(\mathfrak{A}_0)$, $w \in \mathsf{W}_3$. Thus $w \in \mathsf{W}_1 \cap \mathsf{W}_3$. Since

* The additive notation is used in W^*.

$$\sum_{r\in\mathsf{R}} fr^{-1} \in \mathsf{W}_3^\perp + \mathsf{W}_1^\perp = (\mathsf{W}_1 \cap \mathsf{W}_3)^\perp,$$

$$0 = w\sum_{r\in\mathsf{R}} fr^{-1} = \left(\sum_{r\in\mathsf{R}} wr\right)f$$

for all $f \in \mathsf{W}_2^\perp$. Thus $\sum_{r\in\mathsf{R}} wr \in \mathsf{W}_2$ and $\sum_{r\in\mathsf{R}} ur = 0$. Hence W satisfies the conditions of a).

By a), $\mathbf{C}_\mathsf{W}(\mathfrak{M}) = \mathbf{C}_\mathsf{W}(\mathfrak{G})$. Using VII, 8.3g),

$$[\mathfrak{W}, \mathfrak{M}] = \mathbf{C}_\mathsf{W}(\mathfrak{M})^\perp = \mathbf{C}_\mathsf{W}(\mathfrak{G})^\perp = [\mathfrak{W}, \mathfrak{G}]. \qquad \textbf{q.e.d.}$$

6.14 Theorem. *Suppose that $\mathfrak{S} \in S_p(\mathfrak{G})$ and $\mathfrak{P}$ is a normal p-subgroup of $\mathfrak{G}$. Let $\mathfrak{B}$ be a weakly closed subgroup in $\mathfrak{S}$ with respect to $\mathfrak{G}$ and suppose that $\mathbf{N}_\mathfrak{G}(\mathfrak{B}) \leq \mathfrak{M}$. Suppose also that $\mathfrak{B}$ is generated by a set A having the property that whenever $a \in A$, $\mathfrak{X}/\mathfrak{Y}$ is a chief factor of $\mathfrak{G}$ and $\mathfrak{X} \leq \mathfrak{P}$, then $[\mathfrak{X}, \underbrace{a, \ldots, a}_{p-1}] \leq \mathfrak{Y}$. Then $\mathfrak{M} \geq \mathfrak{S}$ and*

$$\mathbf{Z}(\mathfrak{M}) \cap \mathfrak{P} = \mathbf{Z}(\mathfrak{G}) \cap \mathfrak{P}, \quad [\mathfrak{P}, \mathfrak{M}] = [\mathfrak{P}, \mathfrak{G}], \quad \mathfrak{P} \cap \mathfrak{M}' = \mathfrak{P} \cap \mathfrak{G}'.$$

Proof. Since $\mathfrak{B}$ is weakly closed, $\mathfrak{B} \trianglelefteq \mathfrak{S}$ and $\mathfrak{S} \leq \mathfrak{M}$.

We shall apply 6.13 to the p-group $\mathfrak{P}$ and to $\mathfrak{G}$, regarded as a group of operators on $\mathfrak{P}$. Let $\mathscr{A} = \{\langle a\rangle | a \in A\}$; thus $\mathfrak{B} = \langle \mathfrak{A} | \mathfrak{A} \in \mathscr{A}\rangle$.

Suppose that $\mathfrak{A} \in \mathscr{A}$, $\mathfrak{A}_0 < \mathfrak{A}$ and $\mathfrak{X}/\mathfrak{Y}$ is a $\mathfrak{G}$-composition factor of $\mathfrak{P}$. Thus $\mathfrak{X}$, $\mathfrak{Y}$ are $\mathfrak{G}$-invariant and are normal subgroups of $\mathfrak{G}$; also $\mathfrak{X}/\mathfrak{Y}$ is a chief factor of $\mathfrak{G}$. Let R be a transversal of $\mathfrak{A}_0$ in $\mathfrak{A}$. The conditions a) and b) of 6.13 will be satisfied if it is shown that for any $x \in \mathfrak{X}$,

$$\prod_{r\in\mathsf{R}} x^r \in \mathfrak{Y}, \quad \prod_{r\in\mathsf{R}} x^{r^{-1}} \in \mathfrak{Y}.$$

Since $\mathfrak{A}$ is Abelian, R^{-1} is also a transversal of $\mathfrak{A}_0$ in $\mathfrak{A}$; thus it is sufficient to prove that

$$\prod_{r\in\mathsf{R}} x^r \in \mathfrak{Y}$$

for all $x \in \mathfrak{X}$. Suppose that $\mathfrak{A} = \langle a\rangle$ and $a \in A$. If $|\mathfrak{A} : \mathfrak{A}_0| = p^k$, R consists of a set of elements of the form $b_i a^i$, where $b_i \in \mathfrak{A}_0$ $(i = 0, 1, \ldots, p^k - 1)$.

Since $\mathfrak{A}_0 < \mathfrak{A}$ and $\mathfrak{A}$ is cyclic, $\mathfrak{A}_0 \leq \mathfrak{A}^p$. Hence it suffices to prove that $x^{a^p} \equiv x \bmod \mathfrak{Y}$ and

$$\prod_{i=0}^{p^k-1} x^{a^i} \in \mathfrak{Y},$$

for all $x \in \mathfrak{X}$.

We regard $\mathfrak{X}/\mathfrak{Y}$ as a vector space over $GF(p)$ and denote by α the linear transformation $t\mathfrak{Y} \to t^a\mathfrak{Y}$ of $\mathfrak{X}/\mathfrak{Y}$. By IX, 1.8,

$$(t\mathfrak{Y})(\alpha - 1)^j = [t, a, \underset{j}{\ldots}, a]\mathfrak{Y}.$$

for any $t \in \mathfrak{X}$. Since $\mathfrak{X}/\mathfrak{Y}$ is a chief factor of $\mathfrak{G}$ and $a \in A$, $[\mathfrak{X}, a, \underset{p-1}{\ldots}, a] \leq \mathfrak{Y}$ by hypothesis. Thus $(\alpha - 1)^{p-1} = 0$. Since the characteristic of the ground-field is p, $(\alpha - 1)^p = \alpha^p - 1$. Hence $\alpha^p = 1$ and $x\mathfrak{Y} = (x\mathfrak{Y})\alpha^p = x^{a^p}\mathfrak{Y}$. Hence $x^{a^p} \equiv x \bmod \mathfrak{Y}$. Also

$$1 + \alpha + \alpha^2 + \cdots + \alpha^{p-1} = (\alpha - 1)^{p-1} = 0,$$

so

$$(x\mathfrak{Y})(x^a\mathfrak{Y}) \cdots (x^{a^{p-1}}\mathfrak{Y}) = 1,$$

or $\bar{x} = xx^a \cdots x^{a^{p-1}} \in \mathfrak{Y}$. It follows that

$$\prod_{i=0}^{p^k-1} x^{a^i} = \bar{x}\bar{x}^{a^p} \cdots \bar{x}^{a^{p^k-p}} \in \mathfrak{Y}.$$

Thus the conditions of 6.13 are satisfied.

By 6.13, $\mathbf{C}_{\mathfrak{P}}(\mathfrak{M}) = \mathbf{C}_{\mathfrak{P}}(\mathfrak{G})$, or $\mathfrak{P} \cap \mathbf{Z}(\mathfrak{M}) = \mathfrak{P} \cap \mathbf{Z}(\mathfrak{G})$. Also $[\mathfrak{P}, \mathfrak{M}] = [\mathfrak{P}, \mathfrak{G}]$. It follows from 6.1 that $\mathfrak{P} \cap \mathfrak{M}' = \mathfrak{P} \cap \mathfrak{G}'$. **q.e.d.**

We shall also require a generalization of IV, 4.7. First, we mention a few elementary facts about $\mathbf{O}^p(\mathfrak{G})$.

6.15 Lemma. *Suppose that* $\mathfrak{S} \in S_p(\mathfrak{G})$ *and* $\mathfrak{S} \leq \mathfrak{H} \leq \mathfrak{G}$.

a) $\mathfrak{G}/\mathbf{O}^p(\mathfrak{G}) \cong \mathfrak{H}/(\mathfrak{H} \cap \mathbf{O}^p(\mathfrak{G}))$ *and* $\mathfrak{H} \cap \mathbf{O}^p(\mathfrak{G}) \geq \mathbf{O}^p(\mathfrak{H})$.

b) *The following assertions are equivalent.*

(i) $\mathfrak{G}/\mathbf{O}^p(\mathfrak{G}) \cong \mathfrak{H}/\mathbf{O}^p(\mathfrak{H})$.

(ii) $|\mathfrak{G}/\mathbf{O}^p(\mathfrak{G})| = |\mathfrak{H}/\mathbf{O}^p(\mathfrak{H})|$.

(iii) $\mathfrak{H} \cap \mathbf{O}^p(\mathfrak{G}) = \mathbf{O}^p(\mathfrak{H})$.

c) $\mathfrak{G}/\mathbf{O}^p(\mathfrak{G})\mathfrak{G}' \cong \mathfrak{H}/(\mathfrak{H} \cap \mathbf{O}^p(\mathfrak{G})\mathfrak{G}')$ *and* $\mathfrak{H} \cap \mathbf{O}^p(\mathfrak{G})\mathfrak{G}' \geq \mathbf{O}^p(\mathfrak{H})\mathfrak{H}'$. *Also* $\mathfrak{G}/\mathfrak{G}' = (\mathfrak{S}\mathfrak{G}'/\mathfrak{G}') \times (\mathbf{O}^p(\mathfrak{G})\mathfrak{G}'/\mathfrak{G}')$.

d) *The following assertions are equivalent.*

(i) $\mathfrak{G}/\mathbf{O}^p(\mathfrak{G})\mathfrak{G}' \cong \mathfrak{H}/\mathbf{O}^p(\mathfrak{H})\mathfrak{H}'$.

(ii) $|\mathfrak{G}/\mathbf{O}^p(\mathfrak{G})\mathfrak{G}'| = |\mathfrak{H}/\mathbf{O}^p(\mathfrak{H})\mathfrak{H}'|$.

(iii) $\mathfrak{H} \cap \mathbf{O}^p(\mathfrak{G})\mathfrak{G}' = \mathbf{O}^p(\mathfrak{H})\mathfrak{H}'$.

(iv) $\mathfrak{S} \cap \mathfrak{G}' = \mathfrak{S} \cap \mathfrak{H}'$.

(v) $\mathfrak{S}\mathfrak{G}'/\mathfrak{G}' \cong \mathfrak{S}\mathfrak{H}'/\mathfrak{H}'$.

Proof. a) Since $\mathfrak{G}/\mathbf{O}^p(\mathfrak{G})$ is a p-group, $\mathfrak{S}\mathbf{O}^p(\mathfrak{G}) = \mathfrak{G}$. Since $\mathfrak{S} \leq \mathfrak{H}$, $\mathfrak{H}\mathbf{O}^p(\mathfrak{G}) = \mathfrak{G}$. Thus $\mathfrak{G}/\mathbf{O}^p(\mathfrak{G}) = \mathfrak{H}\mathbf{O}^p(\mathfrak{G})/\mathbf{O}^p(\mathfrak{G}) \cong \mathfrak{H}/(\mathfrak{H} \cap \mathbf{O}^p(\mathfrak{G}))$. Thus $\mathfrak{H}/(\mathfrak{H} \cap \mathbf{O}^p(\mathfrak{G}))$ is a p-group and $\mathfrak{H} \cap \mathbf{O}^p(\mathfrak{G}) \geq \mathbf{O}^p(\mathfrak{H})$.

b) (i) implies (ii) trivially. If (ii) holds, then $|\mathfrak{H} \cap \mathbf{O}^p(\mathfrak{G})| = |\mathbf{O}^p(\mathfrak{H})|$ by a). Since $\mathfrak{H} \cap \mathbf{O}^p(\mathfrak{G}) \geq \mathbf{O}^p(\mathfrak{H})$, (iii) holds. Finally (iii) implies (i), by a).

c) As in a), $\mathfrak{G} = \mathfrak{H}\mathbf{O}^p(\mathfrak{G})$, so $\mathfrak{G} = \mathfrak{H}(\mathbf{O}^p(\mathfrak{G})\mathfrak{G}')$. Thus $\mathfrak{G}/\mathbf{O}^p(\mathfrak{G})\mathfrak{G}' \cong \mathfrak{H}/(\mathfrak{H} \cap \mathbf{O}^p(\mathfrak{G})\mathfrak{G}')$, $\mathfrak{H}/(\mathfrak{H} \cap \mathbf{O}^p(\mathfrak{G})\mathfrak{G}')$ is an Abelian p-group and $\mathfrak{H} \cap \mathbf{O}^p(\mathfrak{G})\mathfrak{G}' \geq \mathbf{O}^p(\mathfrak{H})\mathfrak{H}'$.

Since $\mathfrak{S}\mathfrak{G}'/\mathfrak{G}' \in S_p(\mathfrak{G}/\mathfrak{G}')$ and $\mathfrak{G}/\mathfrak{G}'$ is Abelian, $\mathfrak{G}/\mathfrak{G}' = \mathfrak{S}\mathfrak{G}'/\mathfrak{G}' \times \mathfrak{T}/\mathfrak{G}'$ for some $\mathfrak{T} \trianglelefteq \mathfrak{G}$. Thus $\mathfrak{G}/\mathfrak{T}$ is an Abelian p-group. Hence $\mathfrak{T} \geq \mathbf{O}^p(\mathfrak{G})\mathfrak{G}'$, and $(\mathfrak{S}\mathfrak{G}'/\mathfrak{G}') \cap (\mathbf{O}^p(\mathfrak{G})\mathfrak{G}'/\mathfrak{G}') = 1$. But $\mathfrak{G} = \mathfrak{G}\mathbf{O}^p(\mathfrak{G})$, so $\mathfrak{G}/\mathfrak{G}' = (\mathfrak{S}\mathfrak{G}'/\mathfrak{G}')(\mathbf{O}^p(\mathfrak{G})\mathfrak{G}'/\mathfrak{G}') = (\mathfrak{S}\mathfrak{G}'/\mathfrak{G}') \times (\mathbf{O}^p(\mathfrak{G})\mathfrak{G}'/\mathfrak{G}')$.

d) The equivalence of (i), (ii), (iii) follows from c) just as b) follows from a). By c), $\mathfrak{G}/\mathbf{O}^p(\mathfrak{G})\mathfrak{G}' \cong \mathfrak{S}\mathfrak{G}'/\mathfrak{G}'$. Thus (i) and (v) are equivalent. Clearly (iv) and (v) are equivalent. **q.e.d.**

In this lemma, d) is of course an "Abelian version" of b). In 6.18, we prove that the conditions in b) are equivalent to those in d). To do this, we prove the following, which is the essential part of the proof of IV, 4.6.

6.16 Theorem. *Let $\mathfrak{G}$ be a group, $\mathfrak{H}$ a subgroup and $\mathfrak{N}$ a normal subgroup of $\mathfrak{G}$ such that $\mathfrak{G} = \mathfrak{N}\mathfrak{H}$. We put $\mathfrak{Q} = \mathfrak{N} \cap \mathfrak{H}$. Let $\mathfrak{T}$ be a normal subgroup of $\mathfrak{H}$ such that $\mathfrak{T} \leq \mathfrak{Q}$ and $\mathfrak{Q}/\mathfrak{T}$ is an Abelian p-group for some prime p.*

a) *If $|\mathfrak{G} : \mathfrak{H}|$ and $|\mathfrak{Q} : \mathfrak{T}|$ are coprime, there is a homomorphism ϕ of $\mathfrak{G}$ into $\mathfrak{H}/\mathfrak{T}$ such that*

$$\mathfrak{H}/\mathfrak{T} = (\mathfrak{Q}/\mathfrak{T})(\operatorname{im} \phi).$$

b) *If in addition $\mathbf{O}^p(\mathfrak{N}) = \mathfrak{N}$, then*

$$(\mathfrak{Q}/\mathfrak{T}) \cap (\operatorname{im} \phi) = 1.$$

Proof. a) In the direct product $\mathfrak{G} \times (\mathfrak{H}/\mathfrak{T})$, let

$$\mathfrak{F} = \{(g, \mathfrak{T}h) | g \in \mathfrak{G}, h \in \mathfrak{H}, \mathfrak{N}g = \mathfrak{N}\mathfrak{T}h = \mathfrak{N}h\}$$

(cf. I, 9.11). Since $\mathfrak{G} = \mathfrak{N}\mathfrak{H}$, there is an epimorphism π_1 of $\mathfrak{F}$ onto $\mathfrak{G}$ given by

$$(g, \mathfrak{T}h)\pi_1 = g.$$

Let $\mathfrak{M} = \ker \pi_1$; thus

$$\mathfrak{M} = \{(1, \mathfrak{T}h) | h \in \mathfrak{Q}\}$$

and $\mathfrak{M} \cong \mathfrak{Q}/\mathfrak{T}$. Let $\mathfrak{H}^*$ be the inverse image of $\mathfrak{H}$ under π_1, so that $|\mathfrak{F} : \mathfrak{H}^*| = |\mathfrak{G} : \mathfrak{H}|$ and

$$\mathfrak{H}^* = \{(h', \mathfrak{T}h) | h \in \mathfrak{H}, h' \in \mathfrak{Q}h\}.$$

And if

$$\mathfrak{D} = \{(h, \mathfrak{T}h) | h \in \mathfrak{H}\},$$

then $\mathfrak{H}^* = \mathfrak{M}\mathfrak{D}$ and $\mathfrak{M} \cap \mathfrak{D} = 1$. If $\mathfrak{S} \in S_p(\mathfrak{H}^*)$, then $\mathfrak{S} \in S_p(\mathfrak{F})$, since $|\mathfrak{F} : \mathfrak{H}^*| = |\mathfrak{G} : \mathfrak{H}|$ is prime to p. Also $\mathfrak{M}$ is an Abelian p-group and $\mathfrak{S}$ splits over $\mathfrak{M}$, so by I, 17.4, there exists a complement $\overline{\mathfrak{G}}$ for $\mathfrak{M}$ in $\mathfrak{F}$. We have

$$\overline{\mathfrak{G}} \cong \mathfrak{F}/\mathfrak{M} \cong \operatorname{im} \pi_1 = \mathfrak{G},$$

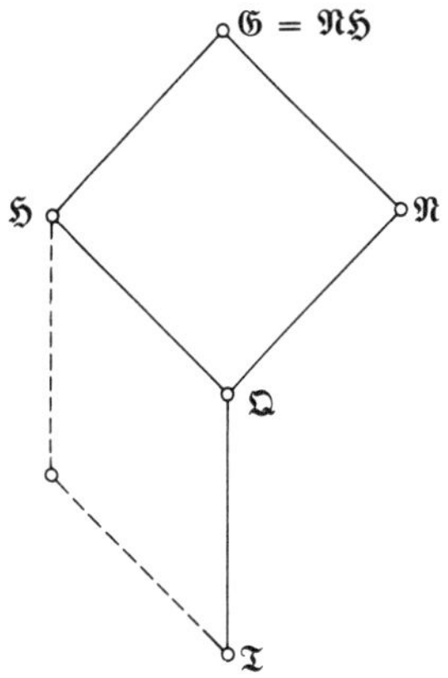

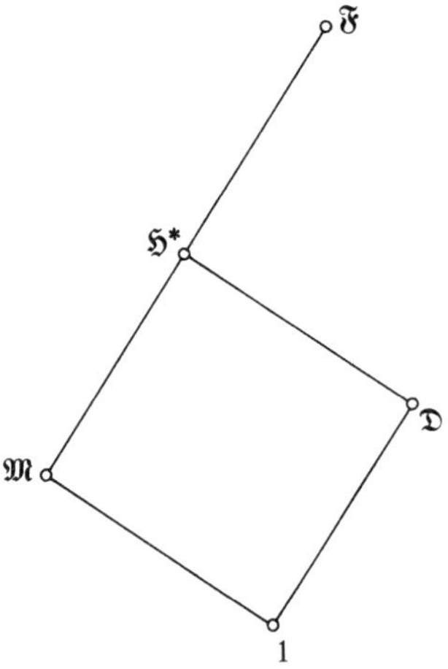

and the restriction π of π_1 to $\overline{\mathfrak{G}}$ is an isomorphism of $\overline{\mathfrak{G}}$ onto $\mathfrak{G}$. Let π_2 be the projection of $\mathfrak{F}$ onto $\mathfrak{H}/\mathfrak{T}$ given by

$$(g, \mathfrak{T}h)\pi_2 = \mathfrak{T}h,$$

and let $\phi = \pi^{-1}\pi_2$. Then

$$\mathfrak{H}/\mathfrak{T} = \mathfrak{F}\pi_2 = (\mathfrak{M}\overline{\mathfrak{G}})\pi_2 = (\mathfrak{M}\pi_2)(\overline{\mathfrak{G}}\pi\phi) = (\mathfrak{Q}/\mathfrak{T})(\mathfrak{G}\phi) = (\mathfrak{Q}/\mathfrak{T})(\operatorname{im}\phi).$$

b) Let $\mathfrak{N}^*$ be the inverse image of $\mathfrak{N}$ under π_1. Thus

$$\mathfrak{N}^* = \{(x, \mathfrak{T}h) \mid x \in \mathfrak{N}, h \in \mathfrak{Q}\}$$

and $\mathfrak{N}\pi^{-1} = \mathfrak{N}^* \cap \overline{\mathfrak{G}}$. Thus $\mathfrak{N}\phi \leq \mathfrak{Q}/\mathfrak{T}$. But $\mathfrak{Q}/\mathfrak{T}$ is a p-group and $\mathbf{O}^p(\mathfrak{N}) = \mathfrak{N}$, so $\mathfrak{N}\phi = 1$. Hence $|\operatorname{im}\phi|$ divides $|\mathfrak{G}/\mathfrak{N}| = |\mathfrak{H}/\mathfrak{Q}|$. But since $\mathfrak{H}/\mathfrak{T} = (\mathfrak{Q}/\mathfrak{T})(\operatorname{im}\phi)$,

$$|\mathfrak{H}/\mathfrak{Q}| = \frac{|\operatorname{im}\phi|}{|(\mathfrak{Q}/\mathfrak{T}) \cap (\operatorname{im}\phi)|}.$$

This implies that $(\mathfrak{Q}/\mathfrak{T}) \cap (\operatorname{im}\phi) = 1$. **q.e.d.**

6.17 Corollary. *If $\mathfrak{S} \in S_p(\mathfrak{G})$ and $\mathfrak{Q} = \mathfrak{S} \cap \mathbf{O}^p(\mathfrak{G})$, $\mathfrak{S}/\mathfrak{Q}'$ splits over $\mathfrak{Q}/\mathfrak{Q}'$.*

Proof. In 6.16, take $\mathfrak{H} = \mathfrak{S}$, $\mathfrak{N} = \mathbf{O}^p(\mathfrak{G})$ and $\mathfrak{T} = \mathfrak{Q}'$. **q.e.d.**

6.18 Theorem (TATE). *Suppose that $\mathfrak{H} \leq \mathfrak{G}$ and $|\mathfrak{G} : \mathfrak{H}|$ is prime to p. Then $\mathfrak{G}/\mathbf{O}^p(\mathfrak{G}) \cong \mathfrak{H}/\mathbf{O}^p(\mathfrak{H})$ if and only if $\mathfrak{G}/\mathbf{O}^p(\mathfrak{G})\mathfrak{G}' \cong \mathfrak{H}/\mathbf{O}^p(\mathfrak{H})\mathfrak{H}'$.*

Proof. If $\mathfrak{G}/\mathbf{O}^p(\mathfrak{G}) \cong \mathfrak{H}/\mathbf{O}^p(\mathfrak{H})$,

$$(\mathfrak{G}/\mathbf{O}^p(\mathfrak{G}))/(\mathfrak{G}/\mathbf{O}^p(\mathfrak{G}))' \cong (\mathfrak{H}/\mathbf{O}^p(\mathfrak{H}))/(\mathfrak{H}/\mathbf{O}^p(\mathfrak{H}))',$$

which implies that $\mathfrak{G}/\mathbf{O}^p(\mathfrak{G})\mathfrak{G}' \cong \mathfrak{H}/\mathbf{O}^p(\mathfrak{H})\mathfrak{H}'$.

If $\mathfrak{G}/\mathbf{O}^p(\mathfrak{G})$ and $\mathfrak{H}/\mathbf{O}^p(\mathfrak{H})$ are not isomorphic, then $\mathfrak{Q} = \mathfrak{H} \cap \mathbf{O}^p(\mathfrak{G}) > \mathbf{O}^p(\mathfrak{H})$, by 6.15b). Now $\mathfrak{Q} \trianglelefteq \mathfrak{H}$, so there exists a maximal normal subgroup $\mathfrak{T}$ of $\mathfrak{H}$ for which $\mathbf{O}^p(\mathfrak{H}) \leq \mathfrak{T} < \mathfrak{Q}$. Thus $\mathfrak{Q}/\mathfrak{T}$ is an Abelian p-group. Hence by 6.16, there exists $\mathfrak{K} \leq \mathfrak{H}$ such that $\mathfrak{H} = \mathfrak{K}\mathfrak{Q}$ and $\mathfrak{K} \cap \mathfrak{Q} = \mathfrak{T}$. We show that $\mathfrak{K}$ is a maximal subgroup of $\mathfrak{H}$. For if $\mathfrak{K} \leq \mathfrak{L} \leq \mathfrak{H}$, then $\mathfrak{T} \leq \mathfrak{L} \cap \mathfrak{Q} \leq \mathfrak{Q}$ and $\mathfrak{L} \cap \mathfrak{Q} \trianglelefteq \mathfrak{L}\mathfrak{Q} = \mathfrak{H}$. Thus by definition of $\mathfrak{T}$, either $\mathfrak{L} \cap \mathfrak{Q} = \mathfrak{T}$ or $\mathfrak{Q} \leq \mathfrak{L}$. But $\mathfrak{L} = \mathfrak{K}(\mathfrak{L} \cap \mathfrak{Q})$, so either

$\mathfrak{L} = \mathfrak{K}$ or $\mathfrak{L} = \mathfrak{H}$. Hence $\mathfrak{K}/\mathbf{O}^p(\mathfrak{H})$ is a maximal subgroup of the p-group $\mathfrak{H}/\mathbf{O}^p(\mathfrak{H})$. Thus $\mathfrak{K} \trianglelefteq \mathfrak{H}$ and $\mathfrak{K} \geq \mathbf{O}^p(\mathfrak{H})\mathfrak{H}'$. But $\mathfrak{H} \cap \mathbf{O}^p(\mathfrak{G})\mathfrak{G}' \geq \mathfrak{Q}$ and, since $\mathfrak{Q} > \mathfrak{T}$, $\mathfrak{Q} \not\leq \mathfrak{K}$, so $\mathfrak{H} \cap \mathbf{O}^p(\mathfrak{G})\mathfrak{G}' \not\leq \mathfrak{K}$. Thus $\mathfrak{H} \cap \mathbf{O}^p(\mathfrak{G})\mathfrak{G}' \neq \mathbf{O}^p(\mathfrak{H})\mathfrak{H}'$, and by 6.15, $\mathfrak{G}/\mathbf{O}^p(\mathfrak{G})\mathfrak{G}'$ is not isomorphic to $\mathfrak{H}/\mathbf{O}^p(\mathfrak{H})\mathfrak{H}'$.

q.e.d.

§ 7. Maximal p-Factor Groups

We now combine the results of the last two sections.

7.1 Definition. Suppose that $\mathbf{W}$ is a characteristic p-functor and $\mathfrak{G}$ is a finite group. We say that $\mathbf{W}$ *controls transfer* in $\mathfrak{G}$ if $\mathfrak{G}/\mathbf{O}^p(\mathfrak{G}) \cong \mathfrak{N}/\mathbf{O}^p(\mathfrak{N})$, where $\mathfrak{N} = \mathbf{N}_{\mathfrak{G}}(\mathbf{W}(\mathfrak{S}))$ for some $\mathfrak{S} \in S_p(\mathfrak{G})$ (and hence for any $\mathfrak{S} \in S_p(\mathfrak{G})$).

Suppose that $\mathfrak{S} \in S_p(\mathfrak{G})$ and $\mathfrak{N} = \mathbf{N}_{\mathfrak{G}}(\mathbf{W}(\mathfrak{S}))$. By 6.15 and 6.18, the following conditions are equivalent.

(i) $\mathbf{W}$ controls transfer in $\mathfrak{G}$.
(ii) $\mathfrak{G}/\mathbf{O}^p(\mathfrak{G}) \cong \mathfrak{N}/\mathbf{O}^p(\mathfrak{N})$.
(iii) $|\mathfrak{G}/\mathbf{O}^p(\mathfrak{G})| = |\mathfrak{N}/\mathbf{O}^p(\mathfrak{N})|$.
(iv) $\mathfrak{N} \cap \mathbf{O}^p(\mathfrak{G}) = \mathbf{O}^p(\mathfrak{N})$.
(v) $\mathfrak{G}/\mathbf{O}^p(\mathfrak{G})\mathfrak{G}' \cong \mathfrak{N}/\mathbf{O}^p(\mathfrak{N})\mathfrak{N}'$.
(vi) $|\mathfrak{G}/\mathbf{O}^p(\mathfrak{G})\mathfrak{G}'| = |\mathfrak{N}/\mathbf{O}^p(\mathfrak{N})\mathfrak{N}'|$.
(vii) $\mathfrak{N} \cap \mathbf{O}^p(\mathfrak{G})\mathfrak{G}' = \mathbf{O}^p(\mathfrak{N})\mathfrak{N}'$.
(viii) $\mathfrak{S} \cap \mathfrak{G}' = \mathfrak{S} \cap \mathfrak{N}'$.
(ix) $\mathfrak{S}\mathfrak{G}'/\mathfrak{G}' \cong \mathfrak{S}\mathfrak{N}'/\mathfrak{N}'$.

It will be noted that in this terminology, Grün's second theorem (IV, 3.7) states that if $\mathfrak{G}$ is p-normal, $\mathbf{Z}$ controls transfer in $\mathfrak{G}$.

In order to prove the first reduction theorem on controlling transfer, we restate 6.2 in the terminology of 5.7.

7.2 Theorem. *Suppose that* $\mathfrak{S} \in S_p(\mathfrak{G})$, $\mathfrak{K} \leq \mathfrak{G}$ *and* $\mathfrak{S} \cap \mathfrak{K} \in S_p(\mathfrak{K})$. *Suppose that* $\mathfrak{H} \leq \mathfrak{G}$ *and the relation* $\sim$ *on* $\mathscr{S}(\mathfrak{S})$ *is defined as follows:* $\mathbf{x} \sim \mathbf{y}$ *if either* (i) $\mathbf{x}$ *and* $\mathbf{y}$ *are both sequences of length greater than* 1, *or* (ii) $\mathbf{x} = (x)$, $\mathbf{y} = (y)$ *and* $xy^{-1} \in \mathfrak{H}$. *Then* $\sim$ *is an equivalence relation, and* $\sim$ *contains fusion in* $\mathfrak{K}$ *if and only if* $\mathfrak{S} \cap \mathfrak{K}' \leq \mathfrak{H}$.

Proof. Obviously, $\sim$ is an equivalence relation on $\mathscr{S}(\mathfrak{S})$.

First suppose that $\mathfrak{S} \cap \mathfrak{K}' \leq \mathfrak{H}$. It is to be proved that if $\mathbf{x} \in \mathscr{S}(\mathfrak{S} \cap \mathfrak{K})$, $\mathbf{y} \in \mathscr{S}(\mathfrak{S} \cap \mathfrak{K})$ and $\mathbf{y} = \mathbf{x}^g$ for some $g \in \mathfrak{K}$, then either (i) $\mathbf{x}$ and $\mathbf{y}$ are both of length greater than 1, or (ii) $\mathbf{x} = (x)$, $\mathbf{y} = (y)$ and

$xy^{-1} \in \mathfrak{H}$. If $\mathbf{x}$ is of length greater than 1, so is $\mathbf{y} = \mathbf{x}^g$ and (i) holds. If $\mathbf{x} = (x)$, where $x \in \mathfrak{S} \cap \mathfrak{K}$, then $\mathbf{y} = (x^g)$ and $x^g \in \mathfrak{S}$, so $xx^{-g} = [x^{-1}, g] \in \mathfrak{S} \cap \mathfrak{K}'$; since $\mathfrak{S} \cap \mathfrak{K}' \leq \mathfrak{H}$, (ii) holds.

Conversely, suppose that $\sim$ contains fusion in $\mathfrak{K}$. Thus, if $x \in \mathfrak{S} \cap \mathfrak{K}$, $g \in \mathfrak{K}$ and $x^g \in \mathfrak{S} \cap \mathfrak{K}$, then $(x^{-1}) \sim (x^{-g})$ and $x^{-1}x^g \in \mathfrak{H}$. Hence, by 6.2 (applied to the Sylow p-subgroup $\mathfrak{S} \cap \mathfrak{K}$ of $\mathfrak{K}$), $\mathfrak{S} \cap \mathfrak{K}' \leq \mathfrak{H}$. **q.e.d.**

7.3 Theorem (ALPERIN and GORENSTEIN [2]). *Suppose that* $\mathbf{W}$ *is a positive characteristic p-functor and that* $\mathfrak{G}$ *is a finite group. If* $\mathbf{W}$ *controls transfer in* $\mathbf{N}_{\mathfrak{G}}(\mathfrak{P})$ *for every non-identity p-subgroup* $\mathfrak{P}$ *of* $\mathfrak{G}$, *then* $\mathbf{W}$ *controls transfer in* $\mathfrak{G}$.

Proof. Suppose that $\mathfrak{S} \in S_p(\mathfrak{G})$ and $\mathfrak{N} = \mathbf{N}_{\mathfrak{G}}(\mathbf{W}(\mathfrak{S}))$. Define the relation $\sim$ on $\mathscr{S}(\mathfrak{S})$ as follows: $\mathbf{x} \sim \mathbf{y}$ if either (i) $\mathbf{x}$, $\mathbf{y}$ are both sequences of length greater than 1, or (ii) $\mathbf{x} = (x)$, $\mathbf{y} = (y)$ and $xy^{-1} \in \mathfrak{N}'$. It follows from 7.2 that if $\mathfrak{K} \leq \mathfrak{G}$ and $\mathfrak{S} \cap \mathfrak{K} \in S_p(\mathfrak{K})$, then $\sim$ contains fusion in $\mathfrak{K}$ if and only if $\mathfrak{S} \cap \mathfrak{K}' \leq \mathfrak{N}'$.

Suppose that $\mathbf{W}$ does not control transfer in $\mathfrak{G}$. Then $\mathfrak{S} \cap \mathfrak{G}' \nleq \mathfrak{N}'$, so $\sim$ does not contain fusion in $\mathfrak{G}$. However, $\sim$ contains fusion in $\mathfrak{N}$, so by 5.9, there exists a non-identity $\mathbf{W}$-extremal subgroup $\mathfrak{P}$ of $\mathfrak{S}$ such that

a) $\sim$ does not contain fusion in $\mathbf{N}_{\mathfrak{G}}(\mathfrak{P})$, and

b) $\sim$ contains fusion in $\mathbf{N}_{\mathfrak{G}}(\mathbf{W}(\mathbf{N}_{\mathfrak{S}}(\mathfrak{P})))$.

Now since $\mathfrak{P}$ is $\mathbf{W}$-extremal, $\mathfrak{S} \cap \mathbf{N}_{\mathfrak{G}}(\mathfrak{P}) \in S_p(\mathbf{N}_{\mathfrak{G}}(\mathfrak{P}))$ and $\mathfrak{S} \cap \mathbf{N}_{\mathfrak{G}}(\mathbf{W}(\mathbf{N}_{\mathfrak{S}}(\mathfrak{P})))$ is a Sylow p-subgroup of $\mathbf{N}_{\mathfrak{G}}(\mathbf{W}(\mathbf{N}_{\mathfrak{S}}(\mathfrak{P})))$. Hence a), b) are respectively equivalent to the following.

c) $\mathfrak{S} \cap \mathbf{N}_{\mathfrak{G}}(\mathfrak{P})' \nleq \mathfrak{N}'$.

d) $\mathfrak{S} \cap \mathbf{N}_{\mathfrak{G}}(\mathbf{W}(\mathbf{N}_{\mathfrak{S}}(\mathfrak{P})))' \leq \mathfrak{N}'$.

Thus $\mathfrak{S} \cap \mathbf{N}_{\mathfrak{G}}(\mathfrak{P})' \nleq \mathbf{N}_{\mathfrak{G}}(\mathbf{W}(\mathbf{N}_{\mathfrak{S}}(\mathfrak{P})))'$. Since $\mathbf{N}_{\mathfrak{S}}(\mathfrak{P}) \in S_p(\mathbf{N}_{\mathfrak{G}}(\mathfrak{P}))$, this shows that $\mathbf{W}$ does not control transfer in $\mathbf{N}_{\mathfrak{G}}(\mathfrak{P})$, by 7.1 (viii). **q.e.d.**

7.4 Lemma. *Suppose that* $\mathfrak{H}$, $\mathfrak{K}$ *are subgroups of the finite group* $\mathfrak{G}$ *such that* $\mathfrak{G} = \mathfrak{H}\mathfrak{K}$ *and* $\mathfrak{H} \cap \mathfrak{K} \geq \mathfrak{S} \in S_p(\mathfrak{G})$.

a) *If* $A \subseteq \mathfrak{S}$, $g \in \mathfrak{G}$ *and* $A^g \subseteq \mathfrak{S}$, *there exist* $u \in \mathfrak{H}$, $v \in \mathfrak{K}$ *such that* $g = uv$ *and* $A^u \subseteq \mathfrak{S}$.

b) *If* $\sim$ *is an equivalence relation on* $\mathscr{S}(\mathfrak{S})$ *and* $\sim$ *does not contain fusion in* $\mathfrak{G}$, *then either* $\sim$ *does not contain fusion in* $\mathfrak{H}$ *or* $\sim$ *does not contain fusion in* $\mathfrak{K}$.

Proof. a) Since $\mathfrak{G} = \mathfrak{H}\mathfrak{K}$, there exist $x \in \mathfrak{H}$, $y \in \mathfrak{K}$ such that $g = xy$. It follows from $A^g \subseteq \mathfrak{S}$ that $A^x \subseteq \mathfrak{S}^{y^{-1}}$. Thus $A^x \subseteq \mathfrak{H}^x \cap \mathfrak{K}^{y^{-1}} = \mathfrak{H} \cap \mathfrak{K}$. Since $A \subseteq \mathfrak{S}$, $\langle A \rangle$ is a p-group. Hence $\langle A^x \rangle$ is contained in some Sylow p-subgroup $\mathfrak{S}_1$ of $\mathfrak{H} \cap \mathfrak{K}$. Since $\mathfrak{S} \leq \mathfrak{H} \cap \mathfrak{K}$, we have $\mathfrak{S}_1 = \mathfrak{S}^z$ for some $z \in \mathfrak{H} \cap \mathfrak{K}$. Thus $A^x \subseteq \mathfrak{S}_1 = \mathfrak{S}^z$ and $A^{xz^{-1}} \subseteq \mathfrak{S}$. If $u = xz^{-1}$ and $v = zy$, then $uv = xy = g$, $u \in \mathfrak{H}$, $v \in \mathfrak{K}$ and $A^u \subseteq \mathfrak{S}$.

b) Since $\sim$ does not contain fusion in $\mathfrak{G}$, there exist elements $\mathbf{x}$, $\mathbf{y}$ of $\mathscr{S}(\mathfrak{S})$ such that $\mathbf{y} = \mathbf{x}^g$ for some $g \in \mathfrak{G}$ but $\mathbf{x} \not\sim \mathbf{y}$. If A is the set of elements of $\mathfrak{S}$ in $\mathbf{x}$, then $A \subseteq \mathfrak{S}$ and $A^g \subseteq \mathfrak{S}$. By a), there exist $u \in \mathfrak{H}$, $v \in \mathfrak{K}$ such that $g = uv$ and $A^u \subseteq \mathfrak{S}$. Thus $\mathbf{x}^u \in \mathscr{S}(\mathfrak{S})$ and $\mathbf{x}^{uv} = \mathbf{x}^g = \mathbf{y} \in \mathscr{S}(\mathfrak{S})$. Since $\mathbf{x} \not\sim \mathbf{x}^{uv}$, it follows from the transitivity of $\sim$ that either $\mathbf{x} \not\sim \mathbf{x}^u$ or $\mathbf{x}^u \not\sim \mathbf{x}^{uv}$. In the former case, $\sim$ does not contain fusion in $\mathfrak{H}$, and in the latter, $\sim$ does not contain fusion in $\mathfrak{K}$. **q.e.d.**

7.5 Lemma. a) *Suppose that π is a set of primes and $\mathfrak{P} = \mathbf{O}_\pi(\mathfrak{G})$. If $\mathfrak{P}\mathbf{C}_\mathfrak{G}(\mathfrak{P})/\mathfrak{P}$ is a π'-group, then $\mathbf{C}_\mathfrak{G}(\mathfrak{P}) = \mathbf{O}_{\pi'}(\mathfrak{G}) \times \mathbf{Z}(\mathfrak{P})$ and $\mathbf{O}_\pi(\overline{\mathfrak{G}}) \geq \mathbf{C}_{\overline{\mathfrak{G}}}(\mathbf{O}_\pi(\overline{\mathfrak{G}}))$, where $\overline{\mathfrak{G}} = \mathfrak{G}/\mathbf{O}_{\pi'}(\mathfrak{G})$.*

b) *Suppose that $\mathfrak{S} \in S_p(\mathfrak{G})$ and that $\mathfrak{G}$ cannot be expressed as the product of two proper subgroups of $\mathfrak{G}$ each of which contains $\mathfrak{S}$. Then either $\mathfrak{G} = \mathfrak{S}\mathbf{C}_\mathfrak{G}(\mathbf{O}_p(\mathfrak{G}))$, or $\mathbf{O}_p(\overline{\mathfrak{G}}) \geq \mathbf{C}_{\overline{\mathfrak{G}}}(\mathbf{O}_p(\overline{\mathfrak{G}}))$, where $\overline{\mathfrak{G}} = \mathfrak{G}/\mathbf{O}_{p'}(\mathfrak{G})$.*

Proof. a) Since $\mathfrak{P}\mathbf{C}_\mathfrak{G}(\mathfrak{P})/\mathfrak{P}$ is a π'-group, it follows from IX, 1.2 that

$$\mathbf{C}_\mathfrak{G}(\mathfrak{P}) = \mathfrak{L} \times (\mathbf{C}_\mathfrak{G}(\mathfrak{P}) \cap \mathfrak{P}) = \mathfrak{L} \times \mathbf{Z}(\mathfrak{P}),$$

where $\mathfrak{L}$ is a π'-subgroup of $\mathbf{C}_\mathfrak{G}(\mathfrak{P})$ and $\mathfrak{L} \trianglelefteq \mathfrak{G}$. Clearly $\mathfrak{L} \leq \mathbf{O}_{\pi'}(\mathfrak{G}) \leq \mathbf{C}_\mathfrak{G}(\mathfrak{P})$. Since $\mathbf{C}_\mathfrak{G}(\mathfrak{P})/\mathfrak{L} \cong \mathbf{Z}(\mathfrak{P})$ is a π-group, it follows that $\mathbf{O}_{\pi'}(\mathfrak{G}) = \mathfrak{L}$. Thus $\mathbf{C}_\mathfrak{G}(\mathfrak{P}) = \mathbf{O}_{\pi'}(\mathfrak{G}) \times \mathbf{Z}(\mathfrak{P})$.

Let $\mathfrak{B}/\mathfrak{L} = \mathbf{C}_{\overline{\mathfrak{G}}}(\mathbf{O}_\pi(\overline{\mathfrak{G}}))$, where $\overline{\mathfrak{G}} = \mathfrak{G}/\mathfrak{L}$. Clearly $\mathfrak{P}\mathfrak{L}/\mathfrak{L} \leq \mathbf{O}_\pi(\overline{\mathfrak{G}})$, so $[\mathfrak{B}, \mathfrak{P}] \leq \mathfrak{L}$. Since $\mathfrak{P} \trianglelefteq \mathfrak{G}$, $[\mathfrak{B}, \mathfrak{P}] \leq \mathfrak{P}$; thus $[\mathfrak{B}, \mathfrak{P}] \leq \mathfrak{P} \cap \mathfrak{L} = 1$ and $\mathfrak{B} \leq \mathbf{C}_\mathfrak{G}(\mathfrak{P}) = \mathfrak{L} \times \mathbf{Z}(\mathfrak{P})$. Thus $\mathfrak{B}/\mathfrak{L}$ is a π-group and $\mathfrak{B}/\mathfrak{L} \leq \mathbf{O}_\pi(\overline{\mathfrak{G}})$.

b) Let $\mathfrak{P} = \mathbf{O}_p(\mathfrak{G})$ and $\mathfrak{R} = \mathfrak{S} \cap \mathfrak{P}\mathbf{C}_\mathfrak{G}(\mathfrak{P})$. Since $\mathfrak{P}\mathbf{C}_\mathfrak{G}(\mathfrak{P}) \trianglelefteq \mathfrak{G}$, $\mathfrak{R} \trianglelefteq \mathfrak{S}$ and $\mathfrak{R} \in S_p(\mathfrak{P}\mathbf{C}_\mathfrak{G}(\mathfrak{P}))$. By the Frattini argument (I, 7.8),

$$\mathfrak{G} = (\mathfrak{P}\mathbf{C}_\mathfrak{G}(\mathfrak{P}))\mathbf{N}_\mathfrak{G}(\mathfrak{R}) = (\mathfrak{S}\mathbf{C}_\mathfrak{G}(\mathfrak{P}))\mathbf{N}_\mathfrak{G}(\mathfrak{R}).$$

Since $\mathfrak{R} \trianglelefteq \mathfrak{S}$, $\mathbf{N}_\mathfrak{G}(\mathfrak{R}) \geq \mathfrak{S}$. It follows from the hypothesis that either $\mathfrak{S}\mathbf{C}_\mathfrak{G}(\mathfrak{P}) = \mathfrak{G}$ or $\mathbf{N}_\mathfrak{G}(\mathfrak{R}) = \mathfrak{G}$. In the latter case, $\mathfrak{R} \trianglelefteq \mathfrak{G}$, so $\mathfrak{R} \leq \mathbf{O}_p(\mathfrak{G}) = \mathfrak{P}$. But it is obvious from the definition of $\mathfrak{R}$ that $\mathfrak{P} \leq \mathfrak{R}$, so $\mathfrak{P} = \mathfrak{R}$. Thus $\mathfrak{P} \in S_p(\mathfrak{P}\mathbf{C}_\mathfrak{G}(\mathfrak{P}))$ and $\mathfrak{P}\mathbf{C}_\mathfrak{G}(\mathfrak{P})/\mathfrak{P}$ is a p'-group. By a), $\mathbf{O}_p(\overline{\mathfrak{G}}) \geq \mathbf{C}_{\overline{\mathfrak{G}}}(\mathbf{O}_p(\overline{\mathfrak{G}}))$, where $\overline{\mathfrak{G}} = \mathfrak{G}/\mathbf{O}_{p'}(\mathfrak{G})$. **q.e.d.**

7.6 Theorem. *Suppose that $\mathbf{W}$ is a positive characteristic p-functor. Suppose that $\mathfrak{G}$ is a finite group and that for every section $\mathfrak{Q}$ of $\mathfrak{G}$ which* *satisfies* $\mathbf{O}_p(\mathfrak{Q}) \geq \mathbf{C}_{\mathfrak{Q}}(\mathbf{O}_p(\mathfrak{Q}))$, *then* $\mathbf{O}_p(\mathfrak{Q}) \cap \mathfrak{Q}' = \mathbf{O}_p(\mathfrak{Q}) \cap \mathbf{N}_{\mathfrak{Q}}(\mathbf{W}(\mathfrak{S}_0))'$ *for* $\mathfrak{S}_0 \in S_p(\mathfrak{Q})$. *Then* $\mathbf{W}$ *controls transfer in* $\mathfrak{G}$.

Proof. Suppose that this is false and that $\mathfrak{G}$ is a counterexample of minimal order. Let $\mathfrak{S} \in S_p(\mathfrak{G})$, $\mathfrak{N} = \mathbf{N}_{\mathfrak{G}}(\mathbf{W}(\mathfrak{S}))$. By 7.1, $\mathfrak{S} \cap \mathfrak{G}' \neq \mathfrak{S} \cap \mathfrak{N}'$, since $\mathbf{W}$ does not control transfer in $\mathfrak{G}$. We obtain a contradiction in a number of steps.

(1) $\mathbf{W}$ controls transfer in every proper section of $\mathfrak{G}$.

For every proper section of $\mathfrak{G}$ satisfies the hypothesis of the theorem and is of smaller order than $\mathfrak{G}$.

(2) $\mathfrak{P} \neq 1$, where $\mathfrak{P} = \mathbf{O}_p(\mathfrak{G})$.

$\mathbf{W}$ does not control transfer in $\mathfrak{G}$. By 7.3, there exists a non-identity p-subgroup $\mathfrak{T}$ of $\mathfrak{G}$ such that $\mathbf{W}$ does not control transfer in $\mathbf{N}_{\mathfrak{G}}(\mathfrak{T})$. By (1), $\mathbf{N}_{\mathfrak{G}}(\mathfrak{T}) = \mathfrak{G}$, or $\mathfrak{T} \trianglelefteq \mathfrak{G}$. Thus $1 \neq \mathfrak{T} \leq \mathfrak{P}$.

(3) If $1 \neq \mathfrak{K} \trianglelefteq \mathfrak{G}$ and $\mathfrak{F}/\mathfrak{K} = \mathbf{N}_{\mathfrak{G}/\mathfrak{K}}(\mathbf{W}(\mathfrak{S}\mathfrak{K}/\mathfrak{K}))$, then $\mathfrak{S} \cap \mathfrak{G}' \leq \mathfrak{S} \cap \mathfrak{F}'\mathfrak{K}$.

By (1), $\mathbf{W}$ controls transfer in $\mathfrak{G}/\mathfrak{K}$, so

$$(\mathfrak{S}\mathfrak{K}/\mathfrak{K}) \cap (\mathfrak{G}'\mathfrak{K}/\mathfrak{K}) = (\mathfrak{S}\mathfrak{K}/\mathfrak{K}) \cap (\mathfrak{F}'\mathfrak{K}/\mathfrak{K}),$$

or $\mathfrak{S}\mathfrak{K} \cap \mathfrak{G}'\mathfrak{K} = \mathfrak{S}\mathfrak{K} \cap \mathfrak{F}'\mathfrak{K}$. Thus $\mathfrak{S} \cap \mathfrak{G}' \leq \mathfrak{S} \cap \mathfrak{F}'\mathfrak{K}$.

(4) $\mathbf{O}_{p'}(\mathfrak{G}) = 1$.

Suppose false. By 5.2c), $\mathbf{N}_{\mathfrak{G}/\mathbf{O}_{p'}(\mathfrak{G})}(\mathbf{W}(\mathfrak{S}\mathbf{O}_{p'}(\mathfrak{G})/\mathbf{O}_{p'}(\mathfrak{G}))) = \mathfrak{N}\mathbf{O}_{p'}(\mathfrak{G})/\mathbf{O}_{p'}(\mathfrak{G})$, (where $\mathfrak{N} = \mathbf{N}_{\mathfrak{G}}(\mathbf{W}(\mathfrak{S}))$). By (1), $|\mathfrak{G} : \mathbf{O}^p(\mathfrak{G})\mathfrak{G}'| = |\mathfrak{N}\mathbf{O}_{p'}(\mathfrak{G}) : \mathbf{O}^p(\mathfrak{N})\mathfrak{N}'\mathbf{O}_{p'}(\mathfrak{G})|$ since $\mathbf{O}_{p'}(\mathfrak{G}) \leq \mathbf{O}^p(\mathfrak{G})$. Thus $|\mathfrak{G} : \mathbf{O}^p(\mathfrak{G})\mathfrak{G}'| = |\mathfrak{N} : \mathbf{O}^p(\mathfrak{N})\mathfrak{N}'(\mathfrak{N} \cap \mathbf{O}_{p'}(\mathfrak{G}))|$. But $\mathbf{O}^p(\mathfrak{N})\mathfrak{N}'(\mathfrak{N} \cap \mathbf{O}_{p'}(\mathfrak{G})) = \mathbf{O}^p(\mathfrak{N})\mathfrak{N}'$ since $\mathfrak{N} \cap \mathbf{O}_{p'}(\mathfrak{G}) \leq \mathbf{O}_{p'}(\mathfrak{N}) \leq \mathbf{O}^p(\mathfrak{N})$, so $|\mathfrak{G} : \mathbf{O}^p(\mathfrak{G})\mathfrak{G}'| = |\mathfrak{N} : \mathbf{O}^p(\mathfrak{N})\mathfrak{N}'|$ and $\mathbf{W}$ controls transfer in $\mathfrak{G}$.

(5) $\mathfrak{P} \cap \mathfrak{G}' \nleq \mathfrak{N}'$.

Since $\mathbf{W}$ does not control transfer in $\mathfrak{G}$, $\mathbf{W}(\mathfrak{S})$ is not normal in $\mathfrak{G}$. Therefore $\mathfrak{S} \neq \mathfrak{P}$. Put $\mathfrak{U}/\mathfrak{P} = \mathbf{W}(\mathfrak{S}/\mathfrak{P})$. Since $\mathbf{W}$ is positive, $\mathfrak{U} > \mathfrak{P}$. Thus if $\mathfrak{M} = \mathbf{N}_{\mathfrak{G}}(\mathfrak{U})$, $\mathfrak{M} < \mathfrak{G}$. By (1), $\mathbf{W}$ controls transfer in $\mathfrak{M}$. Since $\mathfrak{U} \trianglelefteq \mathfrak{S}$, $\mathfrak{S} \leq \mathfrak{M}$; hence

$$\mathfrak{S} \cap \mathfrak{M}' = \mathfrak{S} \cap \mathbf{N}_{\mathfrak{M}}(\mathbf{W}(\mathfrak{S}))' = \mathfrak{S} \cap (\mathfrak{N} \cap \mathfrak{M})' \leq \mathfrak{S} \cap \mathfrak{N}'.$$

But by (2), we may apply (3) with $\mathfrak{K} = \mathfrak{P}$; this gives

$$\mathfrak{S} \cap \mathfrak{G}' \leq \mathfrak{S} \cap \mathfrak{P}\mathfrak{M}',$$

since $\mathfrak{M}/\mathfrak{P} = \mathbf{N}_{\mathfrak{G}/\mathfrak{P}}(\mathfrak{U}/\mathfrak{P}) = \mathbf{N}_{\mathfrak{G}/\mathfrak{P}}(\mathbf{W}(\mathfrak{S}/\mathfrak{P}))$. Hence

$$\mathfrak{S} \cap \mathfrak{G}' \leq \mathfrak{S} \cap \mathfrak{P}\mathfrak{M}' = \mathfrak{P}(\mathfrak{S} \cap \mathfrak{M}') \leq \mathfrak{P}(\mathfrak{S} \cap \mathfrak{N}'),$$

and

$$\mathfrak{S} \cap \mathfrak{G}' = (\mathfrak{P} \cap \mathfrak{G}')(\mathfrak{S} \cap \mathfrak{N}').$$

Since $\mathfrak{S} \cap \mathfrak{N}' \neq \mathfrak{S} \cap \mathfrak{G}'$, it follows that $\mathfrak{P} \cap \mathfrak{G}' \nleq \mathfrak{N}'$.

(6) If $\mathfrak{G} = \mathfrak{H}\mathfrak{K}$, where $\mathfrak{S} \leq \mathfrak{H} \leq \mathfrak{G}$ and $\mathfrak{S} \leq \mathfrak{K} \leq \mathfrak{G}$, then either $\mathfrak{H} = \mathfrak{G}$ or $\mathfrak{K} = \mathfrak{G}$.

Let $\sim$ be the equivalence relation defined on $\mathscr{S}(\mathfrak{S})$ by putting $\mathbf{x} \sim \mathbf{y}$ if either (i) $\mathbf{x}$ and $\mathbf{y}$ are both sequences of length greater than 1, or (ii) $\mathbf{x} = (x)$, $\mathbf{y} = (y)$ and $xy^{-1} \in \mathfrak{N}'$. It follows from 7.2 that if $\mathfrak{S} \leq \mathfrak{L} \leq \mathfrak{G}$, then $\sim$ contains fusion in $\mathfrak{L}$ if and only if $\mathfrak{S} \cap \mathfrak{L}' \leq \mathfrak{N}'$. Hence if $\sim$ does not contain fusion in $\mathfrak{L}$, then $\mathfrak{S} \cap \mathfrak{L}' \nleq (\mathfrak{N} \cap \mathfrak{L})'$, so $\mathfrak{S} \cap \mathfrak{L}' \neq \mathfrak{S} \cap (\mathfrak{N} \cap \mathfrak{L})'$ and $\mathbf{W}$ does not control transfer in $\mathfrak{L}$.

Since $\mathfrak{S} \cap \mathfrak{G}' \nleq \mathfrak{N}'$, $\sim$ does not contain fusion in $\mathfrak{G}$. By 7.4, either $\sim$ does not contain fusion in $\mathfrak{H}$ or $\sim$ does not contain fusion in $\mathfrak{K}$. From above, it follows that $\mathbf{W}$ does not control transfer in $\mathfrak{H}$ or $\mathbf{W}$ does not control transfer in $\mathfrak{K}$. By (1), either $\mathfrak{H} = \mathfrak{G}$ or $\mathfrak{K} = \mathfrak{G}$.

(7) $\mathfrak{G} \neq \mathfrak{S}\mathbf{C}_{\mathfrak{G}}(\mathfrak{P})$.

For suppose $\mathfrak{G} = \mathfrak{S}\mathbf{C}_{\mathfrak{G}}(\mathfrak{P})$. Then by III, 1.10,

$$[\mathfrak{P}, \mathfrak{G}] = [\mathfrak{P}, \mathfrak{S}\mathbf{C}_{\mathfrak{G}}(\mathfrak{P})] = [\mathfrak{P}, \mathfrak{S}][\mathfrak{P}, \mathbf{C}_{\mathfrak{G}}(\mathfrak{P})] = [\mathfrak{P}, \mathfrak{S}].$$
$$\mathfrak{G}$$

By 6.1, $\mathfrak{P} \cap \mathfrak{G}' = \mathfrak{P} \cap \mathfrak{S}'$. But $\mathfrak{S} \leq \mathfrak{N}$, so $\mathfrak{P} \cap \mathfrak{G}' \leq \mathfrak{N}'$, contrary to (5).

(8) By (4), (6), (7) and 7.5b), $\mathbf{O}_p(\mathfrak{G}) \geq \mathbf{C}_{\mathfrak{G}}(\mathbf{O}_p(\mathfrak{G}))$. By the hypothesis of the theorem, $\mathfrak{P} \cap \mathfrak{G}' = \mathfrak{P} \cap \mathfrak{N}'$, contrary to (5). **q.e.d.**

7.7 Theorem. *Suppose* $\mathbf{W}$ *is a positive characteristic p-functor and that* $\mathfrak{G}$ *is a finite group. Suppose that for every section* $\mathfrak{Q}$ *of* $\mathfrak{G}$ *satisfying* $\mathbf{C}_{\mathfrak{Q}}(\mathbf{O}_p(\mathfrak{Q})) \leq \mathbf{O}_p(\mathfrak{Q})$, *and for every* $\mathfrak{S}_0 \in S_p(\mathfrak{Q})$, *either* $\mathbf{W}(\mathfrak{S}_0) \trianglelefteq \mathfrak{Q}$ *or there exists* $g \in \mathfrak{S}_0 - \mathbf{O}_p(\mathfrak{Q})$ *such that* $[\mathfrak{X}, \underbrace{g, \ldots, g}_{p-1}] \leq \mathfrak{Y}$ *for every chief factor* $\mathfrak{X}/\mathfrak{Y}$ *of* $\mathfrak{Q}$ *satisfying* $\mathfrak{X} \leq \mathbf{O}_p(\mathfrak{Q})$. *Then* $\mathbf{W}$ *controls transfer in* $\mathfrak{G}$.

Proof. This is proved by induction on $|\mathfrak{G}|$. If $\mathfrak{G} = 1$, there is nothing to prove. Suppose that $\mathfrak{G} \neq 1$. Note that by the inductive hypothesis, $\mathbf{W}$ controls transfer in every proper section of $\mathfrak{G}$.

We prove that the conditions of 7.6 are satisfied. Thus let $\mathfrak{Q}$ be a section of $\mathfrak{G}$ which satisfies $\mathbf{O}_p(\mathfrak{Q}) \geq \mathbf{C}_{\mathfrak{Q}}(\mathbf{O}_p(\mathfrak{Q}))$, and let $\mathfrak{S}_0 \in S_p(\mathfrak{Q})$. It is to be proved that $\mathbf{O}_p(\mathfrak{Q}) \cap \mathfrak{Q}' = \mathbf{O}_p(\mathfrak{Q}) \cap \mathbf{N}_{\mathfrak{Q}}(\mathbf{W}(\mathfrak{S}_0))'$. This is clear if $\mathbf{N}_{\mathfrak{Q}}(\mathbf{W}(\mathfrak{S}_0)) = \mathfrak{Q}$, so we may suppose that $\mathbf{W}(\mathfrak{S}_0)$ is not normal in $\mathfrak{Q}$. Let A be the set of all elements a of $\mathfrak{S}_0$ which have the property that $[\mathfrak{X}, \underbrace{a, \ldots, a}_{p-1}] \leq \mathfrak{Y}$ for every chief factor $\mathfrak{X}/\mathfrak{Y}$ of $\mathfrak{Q}$ satisfying $\mathfrak{X} \leq \mathbf{O}_p(\mathfrak{Q})$. By hypothesis $A \not\subseteq \mathbf{O}_p(\mathfrak{Q})$. If $a \in A$ and $g \in \mathfrak{Q}$, $[\mathfrak{X}, a^g, \ldots, a^g] \leq \mathfrak{Y}$ for any chief factor $\mathfrak{X}/\mathfrak{Y}$ of $\mathfrak{Q}$ satisfying $\mathfrak{X} \leq \mathbf{O}_p(\mathfrak{Q})$. Hence if $a \in A$, $g \in \mathfrak{Q}$ and $a^g \in \mathfrak{S}_0$, then $a^g \in A$. Thus if $g \in \mathfrak{Q}$ and $A^g \subseteq \mathfrak{S}_0$, $A^g = A$. It follows that if $\mathfrak{B} = \langle A \rangle$, $\mathfrak{B}$ is weakly closed in $\mathfrak{S}_0$ with respect to $\mathfrak{Q}$ (6.3). Hence by 6.14, $\mathbf{O}_p(\mathfrak{Q}) \cap \mathfrak{M}' = \mathbf{O}_p(\mathfrak{Q}) \cap \mathfrak{Q}'$, where $\mathfrak{M} = \mathbf{N}_{\mathfrak{Q}}(\mathfrak{B}) \geq \mathfrak{S}_0$. But since $A \not\subseteq \mathbf{O}_p(\mathfrak{Q})$, $\mathfrak{B} \not\leq \mathbf{O}_p(\mathfrak{Q})$, $\mathfrak{B}$ is not normal in $\mathfrak{Q}$ and $\mathfrak{M} < \mathfrak{Q}$. Thus $\mathfrak{M}$ is a proper section of $\mathfrak{G}$. By the inductive hypothesis, $\mathbf{W}$ controls transfer in $\mathfrak{M}$, that is, $\mathfrak{S}_0 \cap \mathfrak{M}' = \mathfrak{S}_0 \cap \mathbf{N}_{\mathfrak{M}}(\mathbf{W}(\mathfrak{S}_0))'$. Thus

$$\begin{aligned}\mathbf{O}_p(\mathfrak{Q}) \cap \mathfrak{Q}' &= \mathbf{O}_p(\mathfrak{Q}) \cap \mathfrak{M}' = \mathfrak{S}_0 \cap \mathbf{O}_p(\mathfrak{Q}) \cap \mathfrak{M}' \\ &= \mathbf{O}_p(\mathfrak{Q}) \cap \mathfrak{S}_0 \cap \mathbf{N}_{\mathfrak{M}}(\mathbf{W}(\mathfrak{S}_0))' \\ &\leq \mathbf{O}_p(\mathfrak{Q}) \cap \mathbf{N}_{\mathfrak{Q}}(\mathbf{W}(\mathfrak{S}_0))' \leq \mathbf{O}_p(\mathfrak{Q}) \cap \mathfrak{Q}'.\end{aligned}$$

Thus $\mathbf{O}_p(\mathfrak{Q}) \cap \mathfrak{Q}' = \mathbf{O}_p(\mathfrak{Q}) \cap \mathbf{N}_{\mathfrak{Q}}(\mathbf{W}(\mathfrak{S}_0))'$. Hence the conditions of 7.6 are satisfied, and $\mathbf{W}$ controls transfer in $\mathfrak{G}$. **q.e.d.**

§ 8. Glauberman's **K**-Subgroups

In 7.7 it was shown that if $\mathbf{W}$ does not control transfer in $\mathfrak{G}$, there exists a section $\mathfrak{Q}$ of $\mathfrak{G}$ such that

(1) $\mathbf{W}(\mathfrak{S}_0) \not\trianglelefteq \mathfrak{Q}$, where $\mathfrak{S}_0 \in S_p(\mathfrak{Q})$, and

(2) given $g \in \mathfrak{S}_0 - \mathbf{O}_p(\mathfrak{Q})$, there exists a chief factor $\mathfrak{X}/\mathfrak{Y}$ of $\mathfrak{Q}$ with $\mathfrak{X} \leq \mathbf{O}_p(\mathfrak{Q})$ such that $[\mathfrak{X}, \underbrace{g, \ldots, g}_{p-1}] \not\leq \mathfrak{Y}$.

Condition (1) immediately implies that $\mathbf{W}(\mathfrak{S}_0) \neq \mathbf{W}(\mathbf{O}_p(\mathfrak{Q}))$. We therefore seek a characteristic p-functor $\mathbf{K}$ with the property that whenever $\mathfrak{P}$ is a p-group, $\mathfrak{N} \trianglelefteq \mathfrak{P}$ and $\mathbf{K}(\mathfrak{P}) \neq \mathbf{K}(\mathfrak{N})$, then there exists $g \in \mathfrak{P} - \mathfrak{N}$ such that $[\mathfrak{X}, \underbrace{g, \ldots, g}_{p-1}] \leq \mathfrak{Y}$ whenever $\mathfrak{X}/\mathfrak{Y}$ is a minimal characteristic factor of $\mathfrak{N}$. The problem is thus reduced to one on p-groups. By 1.1, there is a characteristic subgroup $\mathfrak{K}$ of $\mathfrak{N}$ such that it is sufficient to consider the operation of g on $\mathfrak{K}$, so $\mathbf{K}$ has to be defined in such a way that $\mathbf{K}(\mathfrak{P}) \neq \mathbf{K}(\mathfrak{N})$ implies the existence of a subgroup $\mathfrak{R}$ such that $\mathfrak{R} \not\leq \mathfrak{N}$ and $[\mathfrak{K}, \underbrace{\mathfrak{R}, \ldots, \mathfrak{R}}_{k}] = 1$ for some k. Glauberman's definition of such a characteristic p-functor $\mathbf{K}$ is quite complicated.

8.1 Definitions. Let $\mathfrak{P}$ be a p-group and let $\mathscr{B}$ be the set of subgroups of $\mathfrak{P}$ of class at most 2. We define subsets $\mathscr{K}_n(\mathfrak{P})$ of $\mathscr{B}$ and subgroups $\mathbf{K}_n(\mathfrak{P})$ of $\mathfrak{P}$ for $n = -1, 0, 1, \ldots$ by induction on n.

(1) For $n = -1$, put $\mathscr{K}_{-1}(\mathfrak{P}) = \mathscr{B}$ and $\mathbf{K}_{-1}(\mathfrak{P}) = \mathfrak{P}$.

(2) For n even, define $\mathscr{K}_n(\mathfrak{P})$ to be the set of subgroups $\mathfrak{A} \in \mathscr{B}$ which satisfy the following conditions.

(2a) $$\mathbf{K}_{n-1}(\mathfrak{P}) \leq \mathbf{N}_{\mathfrak{P}}(\mathfrak{A}).$$

(2b) If $\mathfrak{B} \in \mathscr{B}$, $\mathfrak{B} \leq \mathbf{K}_{n-1}(\mathfrak{P})$, $\mathfrak{A} \leq \mathbf{N}_{\mathfrak{P}}(\mathfrak{B})$, $\mathfrak{A} \leq \mathbf{C}_{\mathfrak{P}}(\mathfrak{B}')$ and $[\mathbf{Z}(\mathfrak{A}), \mathfrak{B}, \mathfrak{B}] = 1$, then $[\mathfrak{A}, \mathfrak{B}] \leq \mathbf{Z}(\mathfrak{A})$.

(3) For n odd and $n > 0$, define $\mathscr{K}_n(\mathfrak{P})$ to be the set of subgroups $\mathfrak{A} \in \mathscr{B}$ for which $[\mathfrak{A}, \mathbf{K}_{n-1}(\mathfrak{P})] \leq \mathbf{Z}(\mathfrak{A})$.

(4) For $n \geq 0$, define $\mathbf{K}_n(\mathfrak{P}) = \langle \mathscr{K}_n(\mathfrak{P}) \rangle$. This holds also for $n = -1$.

Only rather formal use of the conditions in 8.1 will be made. In particular, only the simplest properties of commutators will be used.

8.2 Lemma. *Let $\mathfrak{P}$ be a p-group.*

a) $\mathbf{Z}_2(\mathfrak{P}) \in \mathscr{K}_n(\mathfrak{P})$ *and* $\mathbf{Z}_2(\mathfrak{P}) \leq \mathbf{K}_n(\mathfrak{P})$ *for all* $n \geq -1$.

b) *If $\mathfrak{A}$ is a normal Abelian subgroup of $\mathfrak{P}$, $\mathfrak{A} \in \mathscr{K}_n(\mathfrak{P})$ and $\mathfrak{A} \leq \mathbf{K}_n(\mathfrak{P})$ for all $n \geq -1$.*

c) *If α is an isomorphism of $\mathfrak{P}$ onto $\tilde{\mathfrak{P}}$, $\mathbf{K}_n(\mathfrak{P})\alpha = \mathbf{K}_n(\tilde{\mathfrak{P}})$ for all $n \geq -1$.*

d) *If $n \geq 0$ and $\mathfrak{A} \in \mathscr{K}_n(\mathfrak{P})$, then $\mathbf{K}_{n-1}(\mathfrak{P}) \leq \mathbf{N}_{\mathfrak{P}}(\mathfrak{A})$.*

Proof. It is a trivial consequence of 8.1 that $\mathscr{K}_n(\mathfrak{P})$ contains any normal subgroup $\mathfrak{N}$ of $\mathfrak{P}$ for which $[\mathfrak{N}, \mathfrak{P}] \leq \mathbf{Z}(\mathfrak{N})$. a) and b) are special cases of this. c) is proved by induction on n; it is only necessary to observe that if $\mathfrak{A} \in \mathscr{K}_n(\mathfrak{P})$, then $\mathfrak{A}\alpha \in \mathscr{K}_n(\tilde{\mathfrak{P}})$. d) is a trivial consequence of (2) and (3). **q.e.d.**

8.3 Lemma. *Let $\mathfrak{P}$ be a p-group.*

a) *Suppose $\mathfrak{Q}$, $\mathfrak{R}$ are subgroups of $\mathfrak{P}$, $m \geq -1$, $n \geq -1$ and $m \equiv n(2)$. If $\mathbf{K}_{n+1}(\mathfrak{Q}) \leq \mathfrak{R}$ and $\mathbf{K}_m(\mathfrak{R}) \leq \mathbf{K}_n(\mathfrak{Q})$, then $\mathscr{K}_{n+1}(\mathfrak{Q}) \subseteq \mathscr{K}_{m+1}(\mathfrak{R})$.*

b) $\mathscr{K}_{-1}(\mathfrak{P}) \supseteq \mathscr{K}_1(\mathfrak{P}) \supseteq \mathscr{K}_3(\mathfrak{P}) \supseteq \cdots,$
$\mathscr{K}_0(\mathfrak{P}) \subseteq \mathscr{K}_2(\mathfrak{P}) \subseteq \mathscr{K}_4(\mathfrak{P}) \subseteq \cdots.$

c) $\mathbf{K}_{-1}(\mathfrak{P}) \geq \mathbf{K}_1(\mathfrak{P}) \geq \mathbf{K}_3(\mathfrak{P}) \geq \cdots,$
$\mathbf{K}_0(\mathfrak{P}) \leq \mathbf{K}_2(\mathfrak{P}) \leq \mathbf{K}_4(\mathfrak{P}) \leq \cdots.$

Proof. a) Suppose that m, n are odd and $\mathfrak{A} \in \mathscr{K}_{n+1}(\mathfrak{Q})$. Since $\mathbf{K}_{n+1}(\mathfrak{Q}) \leq \mathfrak{R}$, $\mathfrak{A} \leq \mathfrak{R}$. By 8.2d), $\mathbf{K}_n(\mathfrak{Q}) \leq \mathbf{N}_{\mathfrak{P}}(\mathfrak{A})$. Since $\mathbf{K}_m(\mathfrak{R}) \leq \mathbf{K}_n(\mathfrak{Q})$, $\mathbf{K}_m(\mathfrak{R}) \leq \mathbf{N}_{\mathfrak{P}}(\mathfrak{A})$. Suppose that $\mathfrak{B} \in \mathscr{B}$, $\mathfrak{B} \leq \mathbf{K}_m(\mathfrak{R})$, $\mathfrak{A} \leq \mathbf{N}_{\mathfrak{R}}(\mathfrak{B}) \cap \mathbf{C}_{\mathfrak{R}}(\mathfrak{B}')$ and

$[\mathbf{Z}(\mathfrak{A}), \mathfrak{B}, \mathfrak{B}] = 1$. Then $\mathfrak{B} \leq \mathbf{K}_n(\mathfrak{Q})$, so by 2b), $[\mathfrak{A}, \mathfrak{B}] \leq \mathbf{Z}(\mathfrak{A})$. Hence by (2), $\mathfrak{A} \in \mathcal{K}_{m+1}(\mathfrak{R})$.

Next, suppose that m, n are even and $\mathfrak{A} \in \mathcal{K}_{n+1}(\mathfrak{Q})$. Again $\mathfrak{A} \leq \mathbf{K}_{n+1}(\mathfrak{Q}) \leq \mathfrak{R}$. Also $[\mathfrak{A}, \mathbf{K}_m(\mathfrak{R})] \leq [\mathfrak{A}, \mathbf{K}_n(\mathfrak{Q})] \leq \mathbf{Z}(\mathfrak{A})$ by (3); thus $\mathfrak{A} \in \mathcal{K}_{m+1}(\mathfrak{R})$.

b) By two applications of a) (with $\mathfrak{Q} = \mathfrak{R} = \mathfrak{P}$),

$$\mathcal{K}_l(\mathfrak{P}) \supseteq \mathcal{K}_{l+2}(\mathfrak{P}) \Rightarrow \mathcal{K}_{l+1}(\mathfrak{P}) \subseteq \mathcal{K}_{l+3}(\mathfrak{P}) \Rightarrow \mathcal{K}_{l+2}(\mathfrak{P}) \supseteq \mathcal{K}_{l+4}(\mathfrak{P})$$

for all $l \geq -1$. Since $\mathcal{K}_{-1}(\mathfrak{P}) = \mathscr{B} \supseteq \mathcal{K}_1(\mathfrak{P})$, the assertion follows at once.

c) is an immediate consequence of b). **q.e.d.**

8.4 Definition. For any p-group $\mathfrak{P}$, define

$$\underline{\mathbf{K}}(\mathfrak{P}) = \bigcap_{n \geq 0} \mathbf{K}_{2n-1}(\mathfrak{P}), \quad \overline{\mathbf{K}}(\mathfrak{P}) = \bigcup_{n \geq 0} \mathbf{K}_{2n}(\mathfrak{P}).$$

By 8.2c) and 8.3c), $\underline{\mathbf{K}}$ and $\overline{\mathbf{K}}$ are characteristic p-functors. By 8.2, $\underline{\mathbf{K}}(\mathfrak{P})$, $\overline{\mathbf{K}}(\mathfrak{P})$ both contain $\mathbf{Z}_2(\mathfrak{P})$ and all normal Abelian subgroups of $\mathfrak{P}$, so $\underline{\mathbf{K}}$, $\overline{\mathbf{K}}$ are positive characteristic p-functors.

8.5 Lemma. *If $\mathfrak{P}$ is any p-group,*

$$\overline{\mathbf{K}}(\mathfrak{P}) \geq \mathbf{C}_{\mathfrak{P}}(\overline{\mathbf{K}}(\mathfrak{P})), \quad \underline{\mathbf{K}}(\mathfrak{P}) \geq \mathbf{C}_{\mathfrak{P}}(\underline{\mathbf{K}}(\mathfrak{P})).$$

Proof. If $\mathfrak{A}$ is a maximal normal Abelian subgroup of $\mathfrak{P}$, $\mathfrak{A} \leq \overline{\mathbf{K}}(\mathfrak{P})$, so $\mathbf{C}_{\mathfrak{P}}(\mathfrak{A}) \geq \mathbf{C}_{\mathfrak{P}}(\overline{\mathbf{K}}(\mathfrak{P}))$. But by III, 7.3, $\mathbf{C}_{\mathfrak{P}}(\mathfrak{A}) = \mathfrak{A}$, so $\mathbf{C}_{\mathfrak{P}}(\overline{\mathbf{K}}(\mathfrak{P})) \leq \mathfrak{A} \leq \overline{\mathbf{K}}(\mathfrak{P})$. Similarly $\mathbf{C}_{\mathfrak{P}}(\underline{\mathbf{K}}(\mathfrak{P})) \leq \underline{\mathbf{K}}(\mathfrak{P})$. **q.e.d.**

8.6 Lemma. *Suppose that $\mathfrak{P}$ is a p-group and $\mathfrak{Q} \leq \mathfrak{P}$.*

a) *If $\overline{\mathbf{K}}(\mathfrak{P}) \leq \mathfrak{Q}$, then $\overline{\mathbf{K}}(\mathfrak{P}) \leq \overline{\mathbf{K}}(\mathfrak{Q})$ and $\underline{\mathbf{K}}(\mathfrak{P}) \geq \underline{\mathbf{K}}(\mathfrak{Q})$.*

b) *If $\underline{\mathbf{K}}(\mathfrak{P}) \leq \mathfrak{Q}$, then $\underline{\mathbf{K}}(\mathfrak{P}) \leq \underline{\mathbf{K}}(\mathfrak{Q})$ and $\overline{\mathbf{K}}(\mathfrak{P}) \geq \overline{\mathbf{K}}(\mathfrak{Q})$.*

c) *If $\overline{\mathbf{K}}(\mathfrak{P})\underline{\mathbf{K}}(\mathfrak{P}) \leq \mathfrak{Q}$, then $\overline{\mathbf{K}}(\mathfrak{P}) = \overline{\mathbf{K}}(\mathfrak{Q})$ and $\underline{\mathbf{K}}(\mathfrak{P}) = \underline{\mathbf{K}}(\mathfrak{Q})$.*

Proof. a) By hypothesis, $\mathbf{K}_{2n}(\mathfrak{P}) \leq \mathfrak{Q}$ for all n; and of course $\mathbf{K}_{2n+1}(\mathfrak{Q}) \leq \mathfrak{P}$. Hence by 8.3a),

$$\mathcal{K}_{2n-1}(\mathfrak{Q}) \subseteq \mathcal{K}_{2n-1}(\mathfrak{P}) \Rightarrow \mathcal{K}_{2n}(\mathfrak{P}) \subseteq \mathcal{K}_{2n}(\mathfrak{Q})$$
$$\Rightarrow \mathcal{K}_{2n+1}(\mathfrak{Q}) \subseteq \mathcal{K}_{2n+1}(\mathfrak{P}).$$

Since $\mathcal{K}_{-1}(\mathfrak{Q}) \subseteq \mathcal{K}_{-1}(\mathfrak{P})$, it follows that $\mathcal{K}_{2n}(\mathfrak{P}) \subseteq \mathcal{K}_{2n}(\mathfrak{Q})$ and

$\mathscr{K}_{2n-1}(\mathfrak{Q}) \subseteq \mathscr{K}_{2n-1}(\mathfrak{P})$ for all $n \geq 0$. Hence $\mathbf{K}_{2n}(\mathfrak{P}) \leq \mathbf{K}_{2n}(\mathfrak{Q})$ and $\overline{\mathbf{K}}(\mathfrak{P}) \leq \overline{\mathbf{K}}(\mathfrak{Q})$; also $\mathbf{K}_{2n-1}(\mathfrak{Q}) \leq \mathbf{K}_{2n-1}(\mathfrak{P})$ and $\underline{\mathbf{K}}(\mathfrak{Q}) \leq \underline{\mathbf{K}}(\mathfrak{P})$.

b) Since $\mathfrak{P}$ is finite, there is an integer r such that $\mathscr{K}_{2n+1}(\mathfrak{P}) = \mathscr{K}_{2r+1}(\mathfrak{P})$ for all $n \geq r$, by 8.3b). Then $\mathbf{K}_{2n+1}(\mathfrak{P}) = \underline{\mathbf{K}}(\mathfrak{P})$ for all $n \geq r$. It follows from 8.3a) that if $n \geq r$,

$$\mathscr{K}_{2n+1}(\mathfrak{P}) \subseteq \mathscr{K}_{2n-2r-1}(\mathfrak{Q}) \Rightarrow \mathscr{K}_{2n-2r}(\mathfrak{Q}) \subseteq \mathscr{K}_{2n+2}(\mathfrak{P})$$
$$\Rightarrow \mathscr{K}_{2n+3}(\mathfrak{P}) \subseteq \mathscr{K}_{2n-2r+1}(\mathfrak{Q}),$$

since $\mathbf{K}_{2n+3}(\mathfrak{P}) = \underline{\mathbf{K}}(\mathfrak{P}) \leq \mathfrak{Q}$. But $\mathscr{K}_{2r+1}(\mathfrak{P}) \subseteq \mathscr{K}_{-1}(\mathfrak{Q})$ since $\underline{\mathbf{K}}(\mathfrak{P}) \leq \mathfrak{Q}$; thus $\mathscr{K}_{2n+1}(\mathfrak{P}) \subseteq \mathscr{K}_{2n-2r-1}(\mathfrak{Q})$ and $\mathscr{K}_{2n-2r}(\mathfrak{Q}) \subseteq \mathscr{K}_{2n+2}(\mathfrak{P})$ for all $n \geq r$. Hence $\mathbf{K}_{2n+1}(\mathfrak{P}) \leq \mathbf{K}_{2n-2r-1}(\mathfrak{Q})$ and $\underline{\mathbf{K}}(\mathfrak{P}) \leq \underline{\mathbf{K}}(\mathfrak{Q})$; also $\mathbf{K}_{2n-2r}(\mathfrak{Q}) \leq \mathbf{K}_{2n+2}(\mathfrak{P})$ and $\overline{\mathbf{K}}(\mathfrak{P}) \geq \overline{\mathbf{K}}(\mathfrak{Q})$.

c) This follows from a) and b). **q.e.d.**

We shall now obtain the properties of $\underline{\mathbf{K}}$, $\overline{\mathbf{K}}$ mentioned in the introduction to this section. This is thus the last explicit use that we make of Definition 8.1.

8.7 Lemma. *Suppose that $\mathfrak{P}$ is a p-group, that $\mathfrak{N} \trianglelefteq \mathfrak{P}$ and that either $\underline{\mathbf{K}}(\mathfrak{N}) \neq \underline{\mathbf{K}}(\mathfrak{P})$ or $\overline{\mathbf{K}}(\mathfrak{N}) \neq \overline{\mathbf{K}}(\mathfrak{P})$. Suppose that $\mathfrak{K}$ is a characteristic subgroup of $\mathfrak{N}$, $\mathfrak{K} \geq \mathbf{C}_{\mathfrak{N}}(\mathfrak{K})$ and that $\mathfrak{K}/\mathbf{Z}(\mathfrak{K})$ is an elementary Abelian subgroup of $\mathbf{Z}(\mathfrak{N}/\mathbf{Z}(\mathfrak{K}))$. Then $\mathfrak{P}$ has a subgroup $\mathfrak{R}$ with the following properties.*

a) *The class of $\mathfrak{R}$ is at most 2.*

b) *$\mathfrak{K} \leq \mathbf{N}_{\mathfrak{P}}(\mathfrak{R})$.*

c) *$\mathfrak{R} \nleq \mathfrak{N}$.*

d) *There is a subgroup $\mathfrak{Z}$ of $\mathbf{Z}(\mathfrak{K}) \cap \mathbf{Z}(\mathfrak{R})$ such that if $\mathfrak{Q}/\mathfrak{Z} = \mathbf{Z}(\mathfrak{K}/\mathfrak{Z})$, then $[\mathfrak{K}, \mathfrak{R}, \mathfrak{R}] \leq \mathfrak{Q}$ and $[\mathfrak{Q}, \mathfrak{R}, \mathfrak{R}] = 1$.*

Proof. By 8.6c), $\overline{\mathbf{K}}(\mathfrak{P})\underline{\mathbf{K}}(\mathfrak{P}) \nleq \mathfrak{N}$. Hence $\mathbf{K}_i(\mathfrak{P}) \nleq \mathfrak{N}$ for some $i \geq 0$. Let n be the smallest integer such that $n \geq 0$ and $\mathbf{K}_n(\mathfrak{P}) \nleq \mathfrak{N}$.

First suppose that $\mathfrak{K} \notin \mathscr{K}_{n-1}(\mathfrak{P})$. Then $n > 0$, for $\mathfrak{K} \in \mathscr{K}_{-1}(\mathfrak{P})$. If n is even, then $n \geq 2$, so $\mathbf{K}_{n-2}(\mathfrak{P}) \leq \mathfrak{N}$ and

$$[\mathfrak{K}, \mathbf{K}_{n-2}(\mathfrak{P})] \leq [\mathfrak{K}, \mathfrak{N}] \leq \mathbf{Z}(\mathfrak{K});$$

thus by 8.1(3), $\mathfrak{K} \in \mathscr{K}_{n-1}(\mathfrak{P})$, a contradiction. Hence n is odd. Since $\mathfrak{K} \notin \mathscr{K}_{n-1}(\mathfrak{P})$, it follows from 8.1(2) that there exists a subgroup $\mathfrak{B}$ of class at most 2 such that

$$\mathfrak{K} \leq \mathbf{N}_{\mathfrak{P}}(\mathfrak{B}) \cap \mathbf{C}_{\mathfrak{P}}(\mathfrak{B}'), [\mathbf{Z}(\mathfrak{K}), \mathfrak{B}, \mathfrak{B}] = 1, \text{ but } [\mathfrak{K}, \mathfrak{B}] \nleq \mathbf{Z}(\mathfrak{K}).$$

Since $[\mathfrak{K}, \mathfrak{N}] \le \mathbf{Z}(\mathfrak{K})$, it follows from the last of these conditions that $\mathfrak{B} \not\le \mathfrak{N}$. Also

$$[\mathfrak{K}, \mathfrak{B}, \mathfrak{B}] \le \mathfrak{K} \cap \mathfrak{B}' \le \mathbf{C}_{\mathfrak{K}}(\mathfrak{K}) = \mathbf{Z}(\mathfrak{K}).$$

Hence we may take $\mathfrak{R} = \mathfrak{B}$, the condition d) being satisfied with $\mathfrak{Z} = 1$, $\mathfrak{Q} = \mathbf{Z}(\mathfrak{K})$.

Next suppose that $\mathfrak{K} \in \mathscr{K}_{n-1}(\mathfrak{P})$. Since $\mathbf{K}_n(\mathfrak{P}) \not\le \mathfrak{N}$, there exists $\mathfrak{C} \in \mathscr{K}_n(\mathfrak{P})$ such that $\mathfrak{C} \not\le \mathfrak{N}$. By 8.2d), $\mathbf{K}_{n-1}(\mathfrak{P}) \le \mathbf{N}_{\mathfrak{P}}(\mathfrak{C})$. Since $\mathfrak{K} \in \mathscr{K}_{n-1}(\mathfrak{P})$, $\mathfrak{K} \le \mathbf{N}_{\mathfrak{P}}(\mathfrak{C})$. Thus $[\mathfrak{K}, \mathbf{Z}(\mathfrak{C})] \le \mathbf{Z}(\mathfrak{C})$ and $[\mathfrak{K}, \mathbf{Z}(\mathfrak{C}), \mathbf{Z}(\mathfrak{C})] = 1$. If $\mathbf{Z}(\mathfrak{C}) \not\le \mathfrak{N}$, choose $\mathfrak{R} = \mathbf{Z}(\mathfrak{C})$; d) is satisfied with $\mathfrak{Z} = 1$, $\mathfrak{Q} = \mathbf{Z}(\mathfrak{K})$. Suppose, then, that $\mathbf{Z}(\mathfrak{C}) \le \mathfrak{N}$; we show that we may then take $\mathfrak{R} = \mathfrak{C}$. This is clear if $[\mathfrak{K}, \mathfrak{C}, \mathfrak{C}] = 1$, for then d) is again satisfied with $\mathfrak{Z} = 1$, $\mathfrak{Q} = \mathbf{Z}(\mathfrak{K})$. Suppose then that $[\mathfrak{K}, \mathfrak{C}, \mathfrak{C}] \ne 1$. Since $\mathfrak{K} \le \mathbf{K}_{n-1}(\mathfrak{P})$, $[\mathbf{K}_{n-1}(\mathfrak{P}), \mathfrak{C}, \mathfrak{C}] \ne 1$ and $[\mathbf{K}_{n-1}(\mathfrak{P}), \mathfrak{C}] \not\le \mathbf{Z}(\mathfrak{C})$. But $\mathfrak{C} \in \mathscr{K}_n(\mathfrak{P})$, so by 8.1(3), n is even. Let $\mathfrak{Z} = \mathbf{Z}(\mathfrak{K}) \cap \mathbf{Z}(\mathfrak{C})$, $\mathfrak{Q}/\mathfrak{Z} = \mathbf{Z}(\mathfrak{K}/\mathfrak{Z})$. Then

$$\mathfrak{Q} \le \mathfrak{K} \le \mathbf{K}_{n-1}(\mathfrak{P}), \quad \mathfrak{C} \le \mathbf{N}_{\mathfrak{P}}(\mathfrak{Q}), \quad [\mathfrak{C}, \mathfrak{Q}'] \le [\mathfrak{C}, \mathfrak{Z}] = 1,$$

and

$$[\mathbf{Z}(\mathfrak{C}), \mathfrak{Q}, \mathfrak{Q}] \le [\mathfrak{N}, \mathfrak{K}, \mathfrak{K}] \le [\mathbf{Z}(\mathfrak{K}), \mathfrak{K}] = 1.$$

Hence since $\mathfrak{C} \in \mathscr{K}_n(\mathfrak{P})$, 8.1(2) gives $[\mathfrak{C}, \mathfrak{Q}] \le \mathbf{Z}(\mathfrak{C})$. Thus $[\mathfrak{Q}, \mathfrak{C}, \mathfrak{C}] = 1$. And since $\mathfrak{K} \le \mathbf{N}_{\mathfrak{P}}(\mathfrak{C})$,

$$[\mathfrak{K}, \mathfrak{C}, \mathfrak{C}, \mathfrak{K}] \le \mathfrak{K}' \cap \mathfrak{C}' \le \mathbf{Z}(\mathfrak{K}) \cap \mathbf{Z}(\mathfrak{C}) = \mathfrak{Z}.$$

Thus $[\mathfrak{K}, \mathfrak{C}, \mathfrak{C}] \le \mathfrak{Q}$ and d) is satisfied. **q.e.d.**

8.8 Lemma. *Let* U, V *be* $\mathbb{Z}\mathfrak{G}$*-modules. If* $u \in$ U, $v \in$ V,

$$u \otimes v(g-1)^r = \sum_{j=0}^{r} (-1)^{r-j} \binom{r}{j} (ug^{-r}(g-1)^{r-j} \otimes v)(g-1)^j$$

for all $g \in \mathfrak{G}$ *and* $r \ge 0$.

Proof. This is a straight-forward calculation, using induction on r. It is trivial for $r = 0$. If $r > 0$, we have

$$u \otimes v(g-1) = (ug^{-1} \otimes v)(g-1) - ug^{-1}(g-1) \otimes v.$$

Replacing u by $ug^{-(r-1)}(g-1)^{r-1-j}$,

$$ug^{-(r-1)}(g-1)^{r-1-j}\otimes v(g-1)$$
$$=(ug^{-r}(g-1)^{r-1-j}\otimes v)(g-1)-ug^{-r}(g-1)^{r-j}\otimes v.$$

But by the inductive hypothesis,

$$u\otimes v(g-1)(g-1)^{r-1}$$
$$=\sum_{j=0}^{r-1}(-1)^{r-1-j}\binom{r-1}{j}(ug^{-(r-1)}(g-1)^{r-1-j}\otimes v(g-1))(g-1)^{j}.$$

Hence

$$\begin{aligned}u\otimes v(g-1)^{r}&=\sum_{j=0}^{r-1}(-1)^{r-1-j}\binom{r-1}{j}((ug^{-r}(g-1)^{r-1-j}\otimes v)(g-1)\\&\quad-ug^{-r}(g-1)^{r-j}\otimes v)(g-1)^{j}\\&=\sum_{j=1}^{r}(-1)^{r-j}\binom{r-1}{j-1}(ug^{-r}(g-1)^{r-j}\otimes v)(g-1)^{j}\\&\quad+\sum_{j=0}^{r-1}(-1)^{r-j}\binom{r-1}{j}(ug^{-r}(g-1)^{r-j}\otimes v)(g-1)^{j}\\&=\sum_{j=0}^{r}(-1)^{r-j}\binom{r}{j}(ug^{-r}(g-1)^{r-j}\otimes v)(g-1)^{j}.\end{aligned}$$ **q.e.d.**

8.9 Lemma. *Let $\mathfrak{X}$, $\mathfrak{U}$ be normal subgroups of the group $\mathfrak{G}$. Let α be an automorphism of $\mathfrak{G}$ for which $\mathfrak{X}$, $\mathfrak{U}$ are α-invariant and*

$$[\mathfrak{U},\underbrace{\alpha,\ldots,\alpha}_{m}]=[[\mathfrak{U},\mathfrak{X}],\underbrace{\alpha,\ldots,\alpha}_{n}]=1.$$

Then if $x\in\mathfrak{X}$, $u\in\mathfrak{U}$,

$$[u,[x,\underbrace{\alpha,\ldots,\alpha}_{m+n-1}]]\in[\mathfrak{U},\mathfrak{X},\mathfrak{G}][\mathfrak{G},\mathfrak{U},\mathfrak{X}][\mathfrak{X},\mathfrak{G},\mathfrak{U}].$$

Proof. Let $\mathfrak{H}=\langle\alpha\rangle$ and let U, V denote the $\mathbb{Z}\mathfrak{H}$-modules $\mathfrak{U}/[\mathfrak{U},\mathfrak{G}]$, $\mathfrak{X}/[\mathfrak{X},\mathfrak{G}]$ respectively, written additively. By VIII, 6.1, there is a $\mathbb{Z}$-bilinear mapping γ of $\mathsf{U}\times\mathsf{V}$ into $\mathsf{W}=[\mathfrak{U},\mathfrak{X}]/\mathfrak{N}$, where

$$\mathfrak{N}=[\mathfrak{U},\mathfrak{X},\mathfrak{G}][\mathfrak{G},\mathfrak{U},\mathfrak{X}][\mathfrak{X},\mathfrak{G},\mathfrak{U}],$$

given by

$$(y[\mathfrak{U}, \mathfrak{G}], x[\mathfrak{X}, \mathfrak{G}])\gamma = [y, x]\mathfrak{N} \quad (y \in \mathfrak{U}, x \in \mathfrak{X}).$$

This gives rise to a $\mathbb{Z}$-linear mapping β of $\mathsf{U} \otimes \mathsf{V}$ into W. Now W is also a $\mathbb{Z}\mathfrak{H}$-module and β is a $\mathbb{Z}\mathfrak{H}$-homomorphism, for

$$\begin{aligned}(((y[\mathfrak{U}, \mathfrak{G}]) \otimes (x[\mathfrak{X}, \mathfrak{G}]))\alpha)\beta &= ((y\alpha[\mathfrak{U}, \mathfrak{G}]), (x\alpha[\mathfrak{X}, \mathfrak{G}]))\gamma \\ &= [y\alpha, x\alpha]\mathfrak{N} = ([y, x]\mathfrak{N})\alpha \\ &= (((y[\mathfrak{U}, \mathfrak{G}]) \otimes (x[\mathfrak{X}, \mathfrak{G}]))\beta)\alpha.\end{aligned}$$

By hypothesis, $[\mathsf{U}, \underbrace{\alpha, \ldots, \alpha}_{m}] = [\mathsf{W}, \underbrace{\alpha, \ldots, \alpha}_{n}] = 0$. It follows from 8.8 that if $u \in \mathsf{U}$, $v \in \mathsf{V}$, then

$$\begin{aligned} &u \otimes v(\alpha - 1)^{m+n-1} \\ &= \sum_{j=0}^{m+n-1} (-1)^{m+n-j+1} \binom{m+n-1}{j} (u\alpha^{-(m+n-1)}(\alpha - 1)^{m+n-1-j} \otimes v)(\alpha - 1)^j \\ &= t(\alpha - 1)^n \end{aligned}$$

for some $t \in \mathsf{U} \otimes \mathsf{V}$. Thus

$$(u \otimes v(\alpha - 1)^{m+n-1})\beta = (t\beta)(\alpha - 1)^n = 0,$$

since $t\beta \in \mathsf{W}$. Hence if $y \in \mathfrak{U}$, $x \in \mathfrak{X}$,

$$(y[\mathfrak{U}, \mathfrak{G}], [x, \underbrace{\alpha, \ldots, \alpha}_{m+n-1}][\mathfrak{X}, \mathfrak{G}])\gamma = 1,$$

so

$$[y, [x, \underbrace{\alpha, \ldots, \alpha}_{m+n-1}]] \in \mathfrak{N}.$$

q.e.d.

8.10 Theorem. *Suppose that $\mathfrak{P}$ is a p-group and that $\mathfrak{N}$ is a normal subgroup of $\mathfrak{P}$ for which either $\underline{\mathbf{K}}(\mathfrak{N}) \neq \underline{\mathbf{K}}(\mathfrak{P})$ or $\overline{\mathbf{K}}(\mathfrak{N}) \neq \overline{\mathbf{K}}(\mathfrak{P})$. Then $\mathfrak{P}$ has an element c, not belonging to $\mathfrak{N}$, with the following properties.*

a) *$[\mathfrak{X}, c, c, c, c] \leq \mathfrak{Y}$, whenever $\mathfrak{X}$, $\mathfrak{Y}$ are characteristic subgroups of $\mathfrak{N}$, $\mathfrak{X} > \mathfrak{Y}$ and there is no characteristic subgroup of $\mathfrak{N}$ properly between $\mathfrak{X}$ and $\mathfrak{Y}$.*

b) $[\mathbf{Z}(\mathfrak{N}), c, c] = 1$.

Proof. By 1.1, there exists a characteristic subgroup $\mathfrak{K}$ of $\mathfrak{N}$ such that $\mathfrak{K} \geq \mathbf{C}_{\mathfrak{N}}(\mathfrak{K})$ and $\mathfrak{K}/\mathbf{Z}(\mathfrak{K})$ is an elementary Abelian subgroup of $\mathbf{Z}(\mathfrak{N}/\mathbf{Z}(\mathfrak{K}))$. By 8.7, there is a subgroup $\mathfrak{R}$ of class at most 2 such that $\mathfrak{K} \leq \mathbf{N}_{\mathfrak{P}}(\mathfrak{R})$, $\mathfrak{R} \nleq \mathfrak{N}$ and $[\mathfrak{Q}, \mathfrak{R}, \mathfrak{R}] = 1$, where $\mathfrak{Q}$ is a subgroup of $\mathfrak{K}$ containing $\mathbf{Z}(\mathfrak{K})$. Thus

$$[\mathbf{Z}(\mathfrak{K}), \mathfrak{R}, \mathfrak{R}] \leq [\mathfrak{Q}, \mathfrak{R}, \mathfrak{R}] = 1$$

and

$$[\mathfrak{K}, \mathfrak{R}, \mathfrak{R}, \mathfrak{R}] \leq [\mathfrak{R}, \mathfrak{R}, \mathfrak{R}] = 1.$$

Choose c to be an element of $\mathfrak{R}$ not lying in $\mathfrak{N}$. Since $\mathbf{Z}(\mathfrak{N}) \leq \mathbf{C}_{\mathfrak{N}}(\mathfrak{K}) \leq \mathbf{Z}(\mathfrak{K})$,

$$[\mathbf{Z}(\mathfrak{N}), c, c] \leq [\mathbf{Z}(\mathfrak{K}), \mathfrak{R}, \mathfrak{R}] = 1.$$

Thus b) is proved.

We prove a) by induction on $|\mathfrak{X}|$. Let $\mathfrak{G} = \mathbf{Aut}\,\mathfrak{N}$, and consider $\mathfrak{G}$ as a group of operators on $\mathfrak{N}$. Thus $\mathfrak{X}/\mathfrak{Y}$ is a $\mathfrak{G}$-chief factor of $\mathfrak{N}$. If $\mathfrak{X} \leq \mathfrak{K}$,

$$[\mathfrak{X}, c, c, c, c] \leq [\mathfrak{K}, \mathfrak{R}, \mathfrak{R}, \mathfrak{R}, \mathfrak{R}] = 1,$$

as required. Suppose that $\mathfrak{X} \nleq \mathfrak{K}$. Then $[\mathfrak{K}, \mathfrak{X}] \neq 1$ since $\mathfrak{K} \geq \mathbf{C}_{\mathfrak{N}}(\mathfrak{K})$. Let $\mathfrak{U}$ be a minimal characteristic subgroup of $\mathfrak{N}$ such that $\mathfrak{U} \leq \mathfrak{K}$ and $\mathfrak{V} = [\mathfrak{U}, \mathfrak{X}] \neq 1$. Since $[\mathfrak{N}, \mathfrak{U}] < \mathfrak{U}$, it follows that $[\mathfrak{N}, \mathfrak{U}, \mathfrak{X}] = 1$. Hence by Hall's three subgroup theorem (III, 1.10),

$$[\mathfrak{U}, \mathfrak{X}, \mathfrak{N}][\mathfrak{N}, \mathfrak{U}, \mathfrak{X}][\mathfrak{X}, \mathfrak{N}, \mathfrak{U}] \leq [\mathfrak{U}, \mathfrak{X}, \mathfrak{N}][\mathfrak{N}, \mathfrak{U}, \mathfrak{X}] = [\mathfrak{V}, \mathfrak{N}].$$

Let α be the automorphism of $\mathfrak{N}$ induced by c. Then

$$[\mathfrak{U}, \alpha; 3] = [\mathfrak{U}, c, c, c] \leq [\mathfrak{K}, \mathfrak{R}, \mathfrak{R}, \mathfrak{R}] = 1.$$

Also $\mathfrak{V} = [\mathfrak{U}, \mathfrak{X}] \leq [\mathfrak{K}, \mathfrak{N}] \leq \mathbf{Z}(\mathfrak{K})$, so

$$[\mathfrak{V}, \alpha; 2] = [\mathfrak{V}, c, c] \leq [\mathbf{Z}(\mathfrak{K}), \mathfrak{R}, \mathfrak{R}] = 1.$$

By 8.9,

$$[u, [x, \alpha, \alpha, \alpha, \alpha]] \in [\mathfrak{V}, \mathfrak{N}]$$

for any $u \in \mathfrak{U}$, $x \in \mathfrak{X}$. Let $\mathfrak{C}/[\mathfrak{V}, \mathfrak{N}] = \mathbf{C}_{\mathfrak{X}/[\mathfrak{V},\mathfrak{N}]}(\mathfrak{U})$. Thus $[x, \alpha, \alpha, \alpha, \alpha] \in \mathfrak{C}$ for all $x \in \mathfrak{X}$, and the assertion follows at once if $\mathfrak{C} \leq \mathfrak{Y}$. Suppose then that $\mathfrak{C} \not\leq \mathfrak{Y}$. Thus $\mathfrak{Y} < \mathfrak{Y}\mathfrak{C} \leq \mathfrak{X}$. But $\mathfrak{C}$ is a characteristic subgroup of $\mathfrak{N}$ and there is no characteristic subgroup properly between $\mathfrak{X}$ and $\mathfrak{Y}$. Hence $\mathfrak{Y}\mathfrak{C} = \mathfrak{X}$ and $\mathfrak{X}/\mathfrak{Y}$ is $\mathfrak{G}$-isomorphic to $\mathfrak{C}/(\mathfrak{C} \cap \mathfrak{Y})$. But $\mathfrak{C} < \mathfrak{X}$, since otherwise,

$$\mathfrak{V} = [\mathfrak{U}, \mathfrak{X}] \leq [\mathfrak{V}, \mathfrak{N}],$$

and this is impossible since $\mathfrak{V} \neq 1$. Hence by the inductive hypothesis, $[\mathfrak{C}, c, c, c, c] \leq \mathfrak{C} \cap \mathfrak{Y}$. Since c induces the automorphism α, $[\mathfrak{C}, \alpha, \alpha, \alpha, \alpha] \leq \mathfrak{C} \cap \mathfrak{Y}$ and

$$[\mathfrak{X}, c, c, c, c] = [\mathfrak{X}, \alpha, \alpha, \alpha, \alpha] \leq \mathfrak{Y}. \qquad \textbf{q.e.d.}$$

8.11 Theorem (GLAUBERMAN [5]). *If $p \geq 5$, $\underline{\mathbf{K}}$ and $\overline{\mathbf{K}}$ control transfer in any group $\mathfrak{G}$.*

Proof. Write $\mathbf{K}$ for $\underline{\mathbf{K}}$ or $\overline{\mathbf{K}}$. Let $\mathfrak{Q}$ be a section of $\mathfrak{G}$ for which $\mathbf{O}_p(\mathfrak{Q}) \geq \mathbf{C}_{\mathfrak{Q}}(\mathbf{O}_p(\mathfrak{Q}))$ and $\mathbf{K}(\mathfrak{S}_0) \not\trianglelefteq \mathfrak{Q}$, where $\mathfrak{S}_0 \in S_p(\mathfrak{Q})$. Since $\mathbf{K}(\mathbf{O}_p(\mathfrak{Q}))$ is a characteristic subgroup of $\mathbf{O}_p(\mathfrak{Q})$, $\mathbf{K}(\mathbf{O}_p(\mathfrak{Q})) \trianglelefteq \mathfrak{Q}$. Thus $\mathbf{K}(\mathfrak{S}_0) \neq \mathbf{K}(\mathbf{O}_p(\mathfrak{Q}))$. By 8.10, $\mathfrak{S}_0$ has an element g, not belonging to $\mathbf{O}_p(\mathfrak{Q})$, such that $[\mathfrak{X}, g, g, g, g] \leq \mathfrak{Y}$ for any pair $\mathfrak{X}$, $\mathfrak{Y}$ of characteristic subgroups of $\mathbf{O}_p(\mathfrak{Q})$ for which $\mathfrak{X} > \mathfrak{Y}$ and there is no characteristic subgroup properly between $\mathfrak{X}$ and $\mathfrak{Y}$. Since a characteristic series of $\mathbf{O}_p(\mathfrak{Q})$ may be refined to part of a chief series of $\mathfrak{Q}$, $[\mathfrak{X}, g, g, g, g] \leq \mathfrak{Y}$ for every chief factor $\mathfrak{X}/\mathfrak{Y}$ of $\mathfrak{Q}$ with $\mathfrak{X} \leq \mathbf{O}_p(\mathfrak{Q})$. Since $p \geq 5$, it follows that $[\mathfrak{X}, \underbrace{g, \ldots, g}_{p-1}] \leq \mathfrak{Y}$ for every such chief factor. Thus $\underline{\mathbf{K}}$ and $\overline{\mathbf{K}}$ control transfer in $\mathfrak{G}$, by 7.7. **q.e.d.**

8.12 Remarks. For $p = 2$, no positive characteristic 2-functor controls transfer in every finite group. For there exist simple groups $\mathfrak{G}$ in which the Sylow 2-subgroup $\mathfrak{S}$ is a maximal subgroup; (for example, $\mathfrak{G} = PSL(2, 17)$ (II, 8.27)). Then $\mathbf{N}_{\mathfrak{G}}(\mathbf{W}(\mathfrak{S})) = \mathfrak{S}$ for any positive characteristic 2-functor $\mathbf{W}$, and $\mathfrak{S} \cap \mathfrak{G}' = \mathfrak{S} > \mathfrak{S}' = \mathfrak{S} \cap \mathbf{N}_{\mathfrak{G}}(\mathbf{W}(\mathfrak{S}))'$.

For $p > 5$, $\mathbf{J}$ controls transfer in any group (GLAUBERMAN [2]).

As an application of Theorem 8.11, we have the following.

8.13 Theorem (THOMPSON). *Suppose that $p \geq 5$, $\mathfrak{S} \in S_p(\mathfrak{G})$ and $\mathfrak{S} \neq 1$. If $\mathbf{N}_{\mathfrak{G}}(\mathfrak{S})/\mathbf{C}_{\mathfrak{G}}(\mathfrak{S})$ is a p-group, then $\mathbf{O}^p(\mathfrak{G}) < \mathfrak{G}$.*

Proof. Suppose that this is false and that $\mathfrak{G}$ is a counterexample of minimal order. Thus $\mathfrak{G} = \mathbf{O}^p(\mathfrak{G})$. By 8.11, $\mathfrak{N} = \mathbf{O}^p(\mathfrak{N})$, where $\mathfrak{N} = \mathbf{N}_{\mathfrak{G}}(\underline{\mathbf{K}}(\mathfrak{S}))$.

Since $\mathfrak{G} = \mathbf{O}^p(\mathfrak{G})$ and $\mathfrak{S} \neq 1$, $\mathfrak{G}$ is not p-nilpotent. Hence by Burnside's transfer theorem (IV, 2.6), $\mathbf{C}_{\mathfrak{G}}(\mathfrak{S}) < \mathbf{N}_{\mathfrak{G}}(\mathfrak{S})$. But by hypothesis, $\mathbf{N}_{\mathfrak{G}}(\mathfrak{S})/\mathbf{C}_{\mathfrak{G}}(\mathfrak{S})$ is a p-group, so $\mathbf{C}_{\mathfrak{G}}(\mathfrak{S}) \geq \mathbf{O}^p(\mathbf{N}_{\mathfrak{G}}(\mathfrak{S}))$. Hence $\mathbf{N}_{\mathfrak{G}}(\mathfrak{S}) \neq \mathbf{O}^p(\mathbf{N}_{\mathfrak{G}}(\mathfrak{S}))$. Since $\mathfrak{N} = \mathbf{O}^p(\mathfrak{N})$, it follows that $\mathbf{N}_{\mathfrak{G}}(\mathfrak{S}) \neq \mathfrak{N}$. Thus $\mathfrak{S} \neq \underline{\mathbf{K}}(\mathfrak{S})$, so $\mathfrak{S}/\underline{\mathbf{K}}(\mathfrak{S})$ is a non-identity Sylow p-subgroup of $\mathfrak{N}/\underline{\mathbf{K}}(\mathfrak{S})$. Hence $\mathfrak{N}/\underline{\mathbf{K}}(\mathfrak{S})$ satisfies the hypothesis of the theorem. Since $\mathfrak{S} \neq 1$, $\underline{\mathbf{K}}(\mathfrak{S}) \neq 1$ and $|\mathfrak{N}/\underline{\mathbf{K}}(\mathfrak{S})| < |\mathfrak{G}|$. Since $\mathfrak{G}$ is a counterexample of minimal order, it follows that $\mathfrak{N}/\underline{\mathbf{K}}(\mathfrak{S}) > \mathbf{O}^p(\mathfrak{N}/\underline{\mathbf{K}}(\mathfrak{S}))$. But this implies that $\mathfrak{N} > \mathbf{O}^p(\mathfrak{N})$, a contradiction. **q.e.d.**

8.14 Remark. S. D. SMITH and A. P. TYRER [1] have studied groups in which $|\mathbf{N}_{\mathfrak{G}}(\mathfrak{S}) : \mathfrak{S}\mathbf{C}_{\mathfrak{G}}(\mathfrak{S})| = 2$. For p odd and $\mathbf{O}^p(\mathfrak{G}) = \mathfrak{G}$, they have shown that if $\mathfrak{S}$ is non-cyclic but of class at most 2, then $\mathfrak{G}$ is p-soluble of p-length 1. The proof is mainly by modular representation theory.

The following is the proof of a long-standing conjecture of Zassenhaus.

8.15 Theorem. *If $\mathfrak{G}$ is a finite group and $\mathbf{N}_{\mathfrak{G}}(\mathfrak{S}) = \mathfrak{S}$ for every non-identity Sylow subgroup $\mathfrak{S}$ of $\mathfrak{G}$, $|\mathfrak{G}|$ is a power of a prime.*

Proof. Suppose that $\mathfrak{G} \neq 1$; we show that $\mathbf{O}^p(\mathfrak{G}) < \mathfrak{G}$ for some prime p. This is clear if $\mathfrak{G}$ is soluble. But otherwise, $|\mathfrak{G}|$ is divisible by at least one prime $p \geq 5$, by 2.8. Let $\mathfrak{P} \in S_p(\mathfrak{G})$. By hypothesis, $\mathbf{N}_{\mathfrak{G}}(\mathfrak{P})/\mathbf{C}_{\mathfrak{G}}(\mathfrak{P})$ is a p-group, so by 8.13, $\mathbf{O}^p(\mathfrak{G}) < \mathfrak{G}$.

Now if $\mathfrak{G}$ is not a p-group, $\mathbf{O}^p(\mathfrak{G})$ contains a non-identity Sylow subgroup $\mathfrak{Q}$ of $\mathfrak{G}$. By the Frattini argument,

$$\mathfrak{G} = \mathbf{O}^p(\mathfrak{G})\mathbf{N}_{\mathfrak{G}}(\mathfrak{Q}).$$

But by hypothesis, $\mathbf{N}_{\mathfrak{G}}(\mathfrak{Q}) = \mathfrak{Q} \leq \mathbf{O}^p(\mathfrak{G})$, so $\mathfrak{G} = \mathbf{O}^p(\mathfrak{G})$, a contradiction. Thus $\mathfrak{G}$ is a p-group. **q.e.d.**

The results for $\overline{\mathbf{K}}$ and $\underline{\mathbf{K}}$ corresponding to the **ZJ**-theorem (3.9) hold also for $p = 2$. To prove them, we need the following lemma.

8.16 Lemma. *Suppose that $1 < \mathfrak{N} \leq \mathfrak{M} \leq \mathfrak{G}$, where $\mathfrak{N}$ is a minimal normal subgroup of $\mathfrak{G}$ and $\mathfrak{M}$ is a subnormal subgroup of $\mathfrak{G}$. Then $\mathfrak{N}$ is the direct product of minimal normal subgroups of $\mathfrak{M}$.*

Proof. We use induction on $|\mathfrak{G} : \mathfrak{M}|$. If $\mathfrak{M} = \mathfrak{G}$, the assertion is trivial. If $\mathfrak{M} < \mathfrak{G}$, there exists a normal subgroup $\mathfrak{L}$ of $\mathfrak{G}$ such that $\mathfrak{M} \leq \mathfrak{L} < \mathfrak{G}$. Let r be the greatest integer for which there exist minimal normal subgroups $\mathfrak{N}_1, \ldots, \mathfrak{N}_r$ of $\mathfrak{L}$ with $\mathfrak{N}_i \leq \mathfrak{N}$ and

$$\langle \mathfrak{N}_1, \ldots, \mathfrak{N}_r \rangle = \mathfrak{N}_1 \times \cdots \times \mathfrak{N}_r.$$

Since $\mathfrak{N} > 1$, $r \geq 1$. If $\overline{\mathfrak{N}} = \langle \mathfrak{N}_1, \ldots, \mathfrak{N}_r \rangle < \mathfrak{N}$, $\overline{\mathfrak{N}}$ is not normal in $\mathfrak{G}$ and there exists $g \in \mathfrak{G}$ such that $\overline{\mathfrak{N}}^g \neq \overline{\mathfrak{N}}$. Thus $\mathfrak{N}_i^g \not\leq \overline{\mathfrak{N}}$ for some i. If $\mathfrak{N}_{r+1} = \mathfrak{N}_i^g$, $\mathfrak{N}_{r+1} \trianglelefteq \mathfrak{L}^g = \mathfrak{L}$ and indeed, $\mathfrak{N}_{r+1}$ is a minimal normal subgroup of $\mathfrak{L}$. Since $\mathfrak{N}_{r+1} \cap \overline{\mathfrak{N}} \neq \mathfrak{N}_{r+1}$, we have $\mathfrak{N}_{r+1} \cap \overline{\mathfrak{N}} = 1$ and

$$\langle \mathfrak{N}_1, \ldots, \mathfrak{N}_{r+1} \rangle = \langle \overline{\mathfrak{N}}, \mathfrak{N}_{r+1} \rangle = \overline{\mathfrak{N}} \times \mathfrak{N}_{r+1} = \mathfrak{N}_1 \times \cdots \times \mathfrak{N}_{r+1},$$

contrary to the definition of r. Hence $\mathfrak{N} = \overline{\mathfrak{N}} = \mathfrak{N}_1 \times \cdots \times \mathfrak{N}_r$. The inductive hypothesis may now be applied to each $\mathfrak{N}_i$; thus each $\mathfrak{N}_i$ is the direct product of minimal normal subgroups of $\mathfrak{M}$. The same is therefore true of $\mathfrak{N}$. **q.e.d.**

8.17 Theorem (GLAUBERMAN). *Suppose that $\mathfrak{G}$ is a p-constrained group and that $\mathfrak{G}/\mathfrak{N}$ is p-stable for every $\mathfrak{N} \trianglelefteq \mathfrak{G}$. If $\mathfrak{S} \in S_p(\mathfrak{G})$, then* $\mathbf{O}_{p'}(\mathfrak{G})\underline{\mathbf{K}}(\mathfrak{S}) \trianglelefteq \mathfrak{G}$, $\mathbf{O}_{p'}(\mathfrak{G})\overline{\mathbf{K}}(\mathfrak{S}) \trianglelefteq \mathfrak{G}$ *and*

$$\mathfrak{G} = \mathbf{O}_{p'}(\mathfrak{G})\mathbf{N}_{\mathfrak{G}}(\overline{\mathbf{K}}(\mathfrak{S})) = \mathbf{O}_{p'}(\mathfrak{G})\mathbf{N}_{\mathfrak{G}}(\underline{\mathbf{K}}(\mathfrak{S})).$$

Proof. Suppose that $\mathfrak{G}$ is a minimal counterexample. Since the last assertion follows from the others by means of the Frattini argument, $\mathbf{O}_{p'}(\mathfrak{G})\mathbf{K}(\mathfrak{S})$ is not normal, where $\mathbf{K}$ is either $\underline{\mathbf{K}}$ or $\overline{\mathbf{K}}$. Since

$$\mathbf{K}(\mathfrak{S}\mathbf{O}_{p'}(\mathfrak{G})/\mathbf{O}_{p'}(\mathfrak{G})) = \mathbf{K}(\mathfrak{S})\mathbf{O}_{p'}(\mathfrak{G})/\mathbf{O}_{p'}(\mathfrak{G}),$$

it follows from the minimality of $\mathfrak{G}$ that $\mathbf{O}_{p'}(\mathfrak{G}) = 1$.

Write $\mathfrak{P} = \mathbf{O}_p(\mathfrak{G})$. Since $\mathfrak{G}$ is p-constrained, $\mathfrak{P} \geq \mathbf{C}_{\mathfrak{G}}(\mathfrak{P})$. Since $\mathfrak{G} \neq 1$, $\mathfrak{P} \neq 1$. By 1.1, $\mathfrak{P}$ has a characteristic subgroup $\mathfrak{A}$ such that $\mathfrak{A} \geq \mathbf{C}_{\mathfrak{P}}(\mathfrak{A})$ and $\mathfrak{A}/\mathbf{Z}(\mathfrak{A})$ is an elementary Abelian subgroup of $\mathbf{Z}(\mathfrak{P}/\mathbf{Z}(\mathfrak{A}))$. Thus $\mathfrak{A}$ and $\mathbf{Z}(\mathfrak{A})$ are non-identity normal subgroups of $\mathfrak{G}$.

(1) There is a subgroup $\mathfrak{R}$ of $\mathfrak{S}$ such that $\mathfrak{R} \not\leq \mathfrak{P}$ and

$$[\mathfrak{A}, \mathfrak{R}, \mathfrak{R}] \leq \mathfrak{B}, \quad [\mathfrak{B}, \mathfrak{R}, \mathfrak{R}] = 1,$$

where $\mathfrak{Z}$ is a subgroup of $\mathbf{Z}(\mathfrak{A}) \cap \mathbf{Z}(\mathfrak{R})$ and $\mathfrak{B}/\mathfrak{Z} = \mathbf{Z}(\mathfrak{A}/\mathfrak{Z})$.

For either $\underline{\mathbf{K}}(\mathfrak{S})$ or $\overline{\mathbf{K}}(\mathfrak{S})$ is not normal in $\mathfrak{G}$, but $\underline{\mathbf{K}}(\mathfrak{P})$ and $\overline{\mathbf{K}}(\mathfrak{P})$

are both normal. Hence either $\underline{\mathbf{K}}(\mathfrak{S}) \neq \underline{\mathbf{K}}(\mathfrak{P})$ or $\overline{\mathbf{K}}(\mathfrak{S}) \neq \overline{\mathbf{K}}(\mathfrak{P})$. By 8.7, $\mathfrak{P}$ has a subgroup $\mathfrak{R}$ with the stated properties.

(2) If $\mathfrak{C} = \mathbf{C}_{\mathfrak{G}}(\mathbf{Z}(\mathfrak{A}))$ and $\mathfrak{M} = \mathfrak{R}\mathfrak{C}$, then $\mathfrak{A} \leq \mathfrak{C} \trianglelefteq \mathfrak{G}$ and $\mathfrak{M}$ is a subnormal subgroup of $\mathfrak{G}$.

Clearly, $\mathfrak{A} \leq \mathfrak{C} \trianglelefteq \mathfrak{G}$. Since $\mathbf{Z}(\mathfrak{A}) \leq \mathfrak{B}$, $[\mathbf{Z}(\mathfrak{A}), \mathfrak{R}, \mathfrak{R}] = 1$, by (1). Since $\mathfrak{G}$ is p-stable, $\mathfrak{R}\mathfrak{C}/\mathfrak{C} \leq \mathbf{O}_p(\mathfrak{G}/\mathfrak{C})$. Since any subgroup of a p-group is subnormal, $\mathfrak{R}\mathfrak{C}/\mathfrak{C}$ is subnormal in $\mathbf{O}_p(\mathfrak{G}/\mathfrak{C})$ and hence in $\mathfrak{G}/\mathfrak{C}$. Thus $\mathfrak{M} = \mathfrak{R}\mathfrak{C}$ is subnormal in $\mathfrak{G}$.

(3) If $\mathfrak{X}_1/\mathfrak{Y}_1$ is a chief factor of $\mathfrak{M}$ and $\mathfrak{X}_1 \leq \mathfrak{A}$, then $[\mathfrak{X}_1, \mathfrak{R}, \mathfrak{R}] \leq \mathfrak{Y}_1$.

$\mathfrak{R}$ and $\mathfrak{C}$ both centralize $\mathfrak{Z}$, since $\mathfrak{Z} \leq \mathbf{Z}(\mathfrak{A}) \cap \mathbf{Z}(\mathfrak{R})$. Thus $\mathfrak{M} \leq \mathbf{N}_{\mathfrak{G}}(\mathfrak{Z})$. Also $\mathfrak{B} \trianglelefteq \mathbf{N}_{\mathfrak{G}}(\mathfrak{Z})$, since $\mathfrak{B}/\mathfrak{Z} = \mathbf{Z}(\mathfrak{A}/\mathfrak{Z}) \trianglelefteq \mathbf{N}_{\mathfrak{G}}(\mathfrak{Z})/\mathfrak{Z}$. Hence $\mathfrak{M} \leq \mathbf{N}_{\mathfrak{G}}(\mathfrak{B})$ and $\mathfrak{B} \trianglelefteq \mathfrak{M}$. Regarding $\mathfrak{A}$ as a group with $\mathfrak{M}$ as a group of operators, an $\mathfrak{M}$-chief series $\mathcal{C}$ of $\mathfrak{A}$ may be formed by refining the series $\mathfrak{A} \geq \mathfrak{B} > 1$. By (1), $[\mathfrak{X}_2, \mathfrak{R}, \mathfrak{R}] \leq \mathfrak{Y}_2$ for any factor $\mathfrak{X}_2/\mathfrak{Y}_2$ from $\mathcal{C}$. But $\mathfrak{X}_1/\mathfrak{Y}_1$ is an $\mathfrak{M}$-chief factor of $\mathfrak{A}$. By the Jordan-Hölder theorem, $\mathfrak{X}_1/\mathfrak{Y}_1$ is $\mathfrak{M}$-isomorphic to some factor $\mathfrak{X}_2/\mathfrak{Y}_2$ from $\mathcal{C}$. Since $\mathfrak{R} \leq \mathfrak{M}$, it follows that $[\mathfrak{X}_1, \mathfrak{R}, \mathfrak{R}] \leq \mathfrak{Y}_1$.

(4) If $\mathfrak{X}/\mathfrak{Y}$ is a chief factor of $\mathfrak{G}$ and $\mathfrak{X} \leq \mathfrak{A}$, then $[\mathfrak{X}, \mathfrak{R}] \leq \mathfrak{Y}$.

First, $\mathfrak{X}/\mathfrak{Y}$ is a direct product of chief factors $\mathfrak{X}_i/\mathfrak{Y}$ $(i = 1, 2, \dots)$ of $\mathfrak{M}$, by (2) and 8.16. By (3), $[\mathfrak{X}_i, \mathfrak{R}, \mathfrak{R}] \leq \mathfrak{Y}$. Thus $[\mathfrak{X}, \mathfrak{R}, \mathfrak{R}] \leq \mathfrak{Y}$. Since $\mathfrak{G}/\mathfrak{Y}$ is p-stable, it follows that $\mathfrak{R}\mathfrak{K}/\mathfrak{K} \leq \mathbf{O}_p(\mathfrak{G}/\mathfrak{K})$, where $\mathfrak{K}/\mathfrak{Y} = \mathbf{C}_{\mathfrak{G}/\mathfrak{Y}}(\mathfrak{X}/\mathfrak{Y})$. Now $\mathfrak{G}/\mathfrak{K}$ is isomorphic to a group of automorphisms of the chief factor $\mathfrak{X}/\mathfrak{Y}$, which is to say that $\mathfrak{G}/\mathfrak{K}$ possesses a faithful irreducible representation in $GF(p)$. By V, 5.17, $\mathbf{O}_p(\mathfrak{G}/\mathfrak{K}) = 1$. Thus $\mathfrak{R} \leq \mathfrak{K}$, and $[\mathfrak{X}, \mathfrak{R}] \leq \mathfrak{Y}$.

(5) Let $\mathfrak{X}_0, \mathfrak{X}_1, \dots, \mathfrak{X}_k$ be normal subgroups of $\mathfrak{G}$ such that

$$\mathfrak{A} = \mathfrak{X}_0 > \mathfrak{X}_1 > \cdots > \mathfrak{X}_k = 1,$$

and $\mathfrak{X}_{i-1}/\mathfrak{X}_i$ is a chief factor of $\mathfrak{G}$ $(i = 1, \dots, k)$. There is an isomorphism between $\mathfrak{G}/\mathbf{C}_{\mathfrak{G}}(\mathfrak{P})$ and a group $\mathfrak{H}$ of automorphisms of $\mathfrak{P}$; also, for each $\alpha \in \mathfrak{H}$, $\mathfrak{X}_i\alpha = \mathfrak{X}_i$ $(i = 0, \dots, k)$. If

$$\mathfrak{H}_0 = \{\beta \mid \beta \in \mathfrak{H}, (x\mathfrak{X}_i)\beta = x\mathfrak{X}_i \quad \text{for all} \quad x \in \mathfrak{X}_{i-1}\ (i = 1, \dots, k)\},$$

$\mathfrak{H}_0$ is a normal p-subgroup of $\mathfrak{H}$, by 1.3. Let $\mathfrak{L}$ be the group of automorphisms of $\mathfrak{P}$ induced by $\mathfrak{R}$. By (4), $[\mathfrak{X}_{i-1}, \mathfrak{R}] \leq \mathfrak{X}_i$ $(i = 1, \dots, k)$; hence $(x\mathfrak{X}_i)\gamma = x\mathfrak{X}_i$ for all $x \in \mathfrak{X}_{i-1}$, $\gamma \in \mathfrak{L}$. Thus $\mathfrak{L} \leq \mathfrak{H}_0$. Since $\mathfrak{H} \cong \mathfrak{G}/\mathbf{C}_{\mathfrak{G}}(\mathfrak{P})$, it follows that $\mathfrak{R}\mathbf{C}_{\mathfrak{G}}(\mathfrak{P})/\mathbf{C}_{\mathfrak{G}}(\mathfrak{P}) \leq \mathbf{O}_p(\mathfrak{G}/\mathbf{C}_{\mathfrak{G}}(\mathfrak{P}))$. By hypothesis, $\mathbf{C}_{\mathfrak{G}}(\mathfrak{P}) \leq \mathfrak{P}$; hence $\mathbf{O}_p(\mathfrak{G}/\mathbf{C}_{\mathfrak{G}}(\mathfrak{P})) = \mathbf{O}_p(\mathfrak{G})/\mathbf{C}_{\mathfrak{G}}(\mathfrak{P}) = \mathfrak{P}/\mathbf{C}_{\mathfrak{G}}(\mathfrak{P})$. Thus $\mathfrak{R} \leq \mathfrak{P}$, a contradiction. **q.e.d.**

§ 9. Further Properties of J, ZJ and K

9.1 Definition. The characteristic p-functor $\mathbf{W}$ is said to *control fusion strongly* in $\mathfrak{G}$ if whenever $\mathfrak{S} \in S_p(\mathfrak{G})$, $\mathfrak{K} \leq \mathfrak{S}$, $g \in \mathfrak{G}$ and $\mathfrak{K}^g \leq \mathfrak{S}$, there exists $h \in \mathbf{N}_{\mathfrak{G}}(\mathbf{W}(\mathfrak{S}))$ such that $gh^{-1} \in \mathbf{C}_{\mathfrak{G}}(\mathfrak{K})$.

We shall not need the corresponding notion of controlling fusion, (which means that $\mathfrak{K}^g \leq \mathfrak{S}$ implies that $gh^{-1} \in \mathbf{N}_{\mathfrak{G}}(\mathfrak{K})$ for some $h \in \mathbf{N}_{\mathfrak{G}}(\mathbf{W}(\mathfrak{S}))$).

9.2 Lemma. *Suppose that* $\mathfrak{S} \in S_p(\mathfrak{G})$ *and that* $\mathbf{W}$ *is a characteristic p-functor.*

a) $\mathbf{W}$ *strongly controls fusion in* $\mathfrak{G}$ *if and only if, whenever* $\mathbf{x}$, $\mathbf{y}$ *are elements of* $\mathscr{S}(\mathfrak{S})$ *(see 5.6) conjugate in* $\mathfrak{G}$, *then* $\mathbf{x}$, $\mathbf{y}$ *are conjugate in* $\mathbf{N}_{\mathfrak{G}}(\mathbf{W}(\mathfrak{S}))$.

b) *Suppose that* $\mathbf{W}$ *strongly controls fusion in* $\mathfrak{G}$. *Then* (i) $\mathbf{W}$ *controls transfer in* $\mathfrak{G}$, *and* (ii) *if* $\mathfrak{G}$ *is p-constrained,* $\mathbf{O}_{p'}(\mathfrak{G})\mathbf{W}(\mathfrak{S}) \trianglelefteq \mathfrak{G}$.

Proof. a) This is easily verified.

b) (i) If $x \in \mathfrak{S}$, $g \in \mathfrak{G}$ and $x^g \in \mathfrak{S}$, there exists $h \in \mathbf{N}_{\mathfrak{G}}(\mathbf{W}(\mathfrak{S}))$ such that $gh^{-1} \in \mathbf{C}_{\mathfrak{G}}(\langle x \rangle)$ and $x^g = x^h$. Hence

$$x^{-1}x^g = x^{-1}x^h = [x, h] \in \mathbf{N}_{\mathfrak{G}}(\mathbf{W}(\mathfrak{S}))'.$$

Hence by 6.2, $\mathfrak{S} \cap \mathfrak{G}' \leq \mathbf{N}_{\mathfrak{G}}(\mathbf{W}(\mathfrak{S}))'$ and $\mathfrak{S} \cap \mathfrak{G}' = \mathfrak{S} \cap \mathbf{N}_{\mathfrak{G}}(\mathbf{W}(\mathfrak{S}))'$. Thus $\mathbf{W}$ controls transfer in $\mathfrak{G}$.

(ii) Suppose that $\mathfrak{P} = \mathfrak{S} \cap \mathbf{O}_{p',p}(\mathfrak{G})$. Thus $\mathfrak{P} \in S_p(\mathbf{O}_{p',p}(\mathfrak{G}))$ and $\mathfrak{G} = \mathbf{O}_{p'}(\mathfrak{G})\mathbf{N}_{\mathfrak{G}}(\mathfrak{P})$. But if $g \in \mathbf{N}_{\mathfrak{G}}(\mathfrak{P})$, $\mathfrak{P}^g = \mathfrak{P} \leq \mathfrak{S}$, so $gh^{-1} \in \mathbf{C}_{\mathfrak{G}}(\mathfrak{P})$ for some $h \in \mathbf{N}_{\mathfrak{G}}(\mathbf{W}(\mathfrak{S}))$. Thus

$$\mathbf{N}_{\mathfrak{G}}(\mathfrak{P}) \subseteq \mathbf{C}_{\mathfrak{G}}(\mathfrak{P})\mathbf{N}_{\mathfrak{G}}(\mathbf{W}(\mathfrak{S})).$$

But since $\mathfrak{G}$ is p-constrained, $\mathbf{C}_{\mathfrak{G}}(\mathfrak{P}) \leq \mathbf{O}_{p'}(\mathfrak{G})\mathfrak{P}$, so

$$\mathfrak{G} = \mathbf{O}_{p'}(\mathfrak{G})\mathbf{N}_{\mathfrak{G}}(\mathfrak{P}) = \mathbf{O}_{p'}(\mathfrak{G})\mathbf{N}_{\mathfrak{G}}(\mathbf{W}(\mathfrak{S})).$$

Thus $\mathbf{O}_{p'}(\mathfrak{G})\mathbf{W}(\mathfrak{S}) \trianglelefteq \mathfrak{G}$. **q.e.d.**

9.3 Theorem (ALPERIN and GORENSTEIN). *Suppose that* $\mathbf{W}$ *is a positive characteristic p-functor and that* $\mathfrak{G}$ *is a finite group. If* $\mathbf{W}$ *strongly controls fusion in* $\mathbf{N}_{\mathfrak{G}}(\mathfrak{P})$ *for every non-identity p-subgroup* $\mathfrak{P}$ *of* $\mathfrak{G}$, *then* $\mathbf{W}$ *strongly controls fusion in* $\mathfrak{G}$.

Proof. Suppose that $\mathfrak{S} \in S_p(\mathfrak{G})$ and define the relation $\sim$ on $\mathscr{S}(\mathfrak{S})$ by putting $\mathbf{x} \sim \mathbf{y}$ if $\mathbf{x}, \mathbf{y}$ are conjugate in $\mathbf{N}_{\mathfrak{G}}(\mathbf{W}(\mathfrak{S}))$. Clearly, $\sim$ is an equivalence relation which contains fusion in $\mathbf{N}_{\mathfrak{G}}(\mathbf{W}(\mathfrak{S}))$. By 9.2a), $\mathbf{W}$ strongly controls fusion in $\mathfrak{G}$ if and only if $\sim$ contains fusion in $\mathfrak{G}$. Suppose that this is not the case. By 5.9, there exists a non-identity $\mathbf{W}$-extremal subgroup $\mathfrak{P}$ of $\mathfrak{S}$ such that $\sim$ does not contain fusion in $\mathbf{N}_{\mathfrak{G}}(\mathfrak{P})$ but $\sim$ contains fusion in $\mathbf{N}_{\mathfrak{G}}(\mathbf{W}(\mathbf{N}_{\mathfrak{S}}(\mathfrak{P})))$. Since $\mathbf{N}_{\mathfrak{S}}(\mathfrak{P}) \in S_p(\mathbf{N}_{\mathfrak{G}}(\mathfrak{P}))$, it follows that there exist $\mathbf{x}, \mathbf{y}$ in $\mathscr{S}(\mathbf{N}_{\mathfrak{G}}(\mathfrak{P}))$ such that $\mathbf{y} = \mathbf{x}^g$ for some $g \in \mathbf{N}_{\mathfrak{G}}(\mathfrak{P})$, but $\mathbf{x} \not\sim \mathbf{y}$. Since $\sim$ contains fusion in $\mathbf{N}_{\mathfrak{G}}(\mathbf{W}(\mathbf{N}_{\mathfrak{S}}(\mathfrak{P})))$ and $\mathfrak{S} \cap \mathbf{N}_{\mathfrak{G}}(\mathbf{W}(\mathbf{N}_{\mathfrak{G}}(\mathfrak{P})))$ $\geq \mathbf{N}_{\mathfrak{S}}(\mathfrak{P})$, it follows that $\mathbf{x}, \mathbf{y}$ are not conjugate in $\mathbf{N}_{\mathfrak{G}}(\mathbf{W}(\mathbf{N}_{\mathfrak{S}}(\mathfrak{P})))$. Thus $\mathbf{W}$ does not strongly control fusion in $\mathbf{N}_{\mathfrak{G}}(\mathfrak{P})$, contrary to hypothesis.

q.e.d.

9.4 Lemma. *Suppose that* $\mathbf{W}$ *is a positive characteristic p-functor. Let* $\mathfrak{P} = \mathbf{O}_p(\mathfrak{G})$. *Suppose that* $\mathfrak{P} \neq 1$ *and* $\mathfrak{G}/\mathbf{C}_{\mathfrak{G}}(\mathfrak{P})$ *is a p-group. If* $\mathbf{W}$ *strongly controls fusion in every proper section of* $\mathfrak{G}$, *then* $\mathbf{W}$ *strongly controls fusion in* $\mathfrak{G}$.

Proof. Suppose that $\mathfrak{S} \in S_p(\mathfrak{G})$. If $\mathfrak{S} = \mathfrak{P}$, $\mathbf{N}_{\mathfrak{G}}(\mathbf{W}(\mathfrak{S})) = \mathfrak{G}$ and there is nothing to prove. Suppose then that $\mathfrak{S} > \mathfrak{P}$. Since $\mathbf{W}$ is positive, $\mathbf{W}(\mathfrak{S}/\mathfrak{P}) \neq 1$. Thus if $\mathfrak{U}/\mathfrak{P} = \mathbf{W}(\mathfrak{S}/\mathfrak{P})$, $\mathfrak{U} > \mathfrak{P}$. Since $\mathfrak{P} = \mathbf{O}_p(\mathfrak{G})$, it follows that $\mathbf{N}_{\mathfrak{G}}(\mathfrak{U}) < \mathfrak{G}$. Clearly $\mathfrak{S} \leq \mathbf{N}_{\mathfrak{G}}(\mathfrak{U})$.

Suppose that $\mathfrak{K} \leq \mathfrak{S}$, $g \in \mathfrak{G}$ and $\mathfrak{K}^g \leq \mathfrak{S}$. Let $\mathfrak{Q} = \mathfrak{P}\mathfrak{K}$. Thus $\mathfrak{Q}^g \leq \mathfrak{S}$. By 4.3b), there exists $t \in \mathfrak{G}$ such that $\mathfrak{R} = \mathfrak{Q}^t$ is extremal in $\mathfrak{S}$. By hypothesis, $\mathfrak{P} \neq 1$ and $\mathbf{W}$ strongly controls fusion in $\mathfrak{G}/\mathfrak{P}$, so there exists $a' \in \mathbf{N}_{\mathfrak{G}}(\mathfrak{U})$ such that $\mathfrak{R} = \mathfrak{Q}^{a'}$. And since $\mathbf{N}_{\mathfrak{G}}(\mathfrak{U}) < \mathfrak{G}$, $\mathbf{W}$ strongly controls fusion in $\mathbf{N}_{\mathfrak{G}}(\mathfrak{U})$, so there exists $a \in \mathbf{N}_{\mathfrak{G}}(\mathbf{W}(\mathfrak{S}))$ such that $\mathfrak{R} = \mathfrak{Q}^a$. Thus $\mathfrak{R}^{a^{-1}g} \leq \mathfrak{S}$. Thus, again, there exists $c' \in \mathbf{N}_{\mathfrak{G}}(\mathfrak{U})$ such that $x^{a^{-1}g} \equiv x^{c'} \bmod \mathfrak{P}$ for all $x \in \mathfrak{R}$, and there exists $c \in \mathbf{N}_{\mathfrak{G}}(\mathbf{W}(\mathfrak{S}))$ such that $x^{c'} = x^c$ for all $x \in \mathfrak{R}$. Thus $x^{a^{-1}gc^{-1}} \equiv x \bmod \mathfrak{P}$, and $a^{-1}gc^{-1} \in \mathfrak{D}$, where $\mathfrak{D}/\mathfrak{P} = \mathbf{C}_{\mathfrak{G}/\mathfrak{P}}(\mathfrak{R}/\mathfrak{P})$.

Now

$$\mathfrak{D}/\mathfrak{P} = \mathbf{C}_{\mathfrak{G}/\mathfrak{P}}(\mathfrak{R}/\mathfrak{P}) \trianglelefteq \mathbf{N}_{\mathfrak{G}/\mathfrak{P}}(\mathfrak{R}/\mathfrak{P}) = \mathbf{N}_{\mathfrak{G}}(\mathfrak{R})/\mathfrak{P},$$

so $\mathfrak{D} \trianglelefteq \mathbf{N}_{\mathfrak{G}}(\mathfrak{R})$. Since $\mathfrak{R}$ is extremal in $\mathfrak{S}$, $\mathbf{N}_{\mathfrak{S}}(\mathfrak{R}) \in S_p(\mathbf{N}_{\mathfrak{G}}(\mathfrak{R}))$, and it follows that $\mathfrak{D} \cap \mathbf{N}_{\mathfrak{S}}(\mathfrak{R}) \in S_p(\mathfrak{D})$. Hence $\mathfrak{D} \cap \mathfrak{S} \in S_p(\mathfrak{D})$. On the other hand, $\mathfrak{D}/(\mathfrak{D} \cap \mathbf{C}_{\mathfrak{G}}(\mathfrak{P}))$ is a p-group, since $\mathfrak{G}/\mathbf{C}_{\mathfrak{G}}(\mathfrak{P})$ is a p-group. Further $\mathbf{C}_{\mathfrak{G}}(\mathfrak{R}) \leq \mathfrak{D} \cap \mathbf{C}_{\mathfrak{G}}(\mathfrak{P}) \leq \mathbf{N}_{\mathfrak{G}}(\mathfrak{R})$, and $(\mathfrak{D} \cap \mathbf{C}_{\mathfrak{G}}(\mathfrak{P}))/\mathbf{C}_{\mathfrak{G}}(\mathfrak{R})$ is isomorphic to a group of automorphisms of $\mathfrak{R}$ which leave fixed every element of $\mathfrak{R}/\mathfrak{P}$ and of $\mathfrak{P}$. By I, 4.4, $(\mathfrak{D} \cap \mathbf{C}_{\mathfrak{G}}(\mathfrak{P}))/\mathbf{C}_{\mathfrak{G}}(\mathfrak{R})$ is a p-group. Hence $\mathfrak{D}/\mathbf{C}_{\mathfrak{G}}(\mathfrak{R})$ is a p-group and $\mathfrak{D} = \mathbf{C}_{\mathfrak{G}}(\mathfrak{R})(\mathfrak{D} \cap \mathfrak{S})$.

Thus $a^{-1}gc^{-1} = bd$ for some $b \in \mathbf{C}_{\mathfrak{G}}(\mathfrak{R})$, $d \in \mathfrak{D} \cap \mathfrak{S}$. Thus $x^{a^{-1}gc^{-1}} = x^{bd} = x^d$ for all $x \in \mathfrak{R}$. Since $\mathfrak{R} = \mathfrak{Q}^a$, it follows that $y^{gc^{-1}} = y^{ad}$ for all $y \in \mathfrak{Q}$; hence $g(adc)^{-1} \in \mathbf{C}_{\mathfrak{G}}(\mathfrak{K})$. But $adc \in \mathbf{N}_{\mathfrak{G}}(\mathbf{W}(\mathfrak{S}))$. Thus **W** strongly controls fusion in $\mathfrak{G}$. **q.e.d.**

9.5 Theorem. *Let* **W** *be a positive characteristic p-functor. Suppose that* $\mathfrak{G}$ *is a finite group and that for any section* $\mathfrak{Q}$ *of* $\mathfrak{G}$ *which satisfies* $\mathbf{O}_p(\mathfrak{Q}) \geq \mathbf{C}_{\mathfrak{Q}}(\mathbf{O}_p(\mathfrak{Q}))$, $\mathbf{W}(\mathfrak{S}_0) \trianglelefteq \mathfrak{Q}$ *for* $\mathfrak{S}_0 \in S_p(\mathfrak{Q})$. *Then* **W** *strongly controls fusion in* $\mathfrak{G}$.

Proof. Suppose that this is false and that $\mathfrak{G}$ is a counterexample of minimal order. Let $\mathfrak{S} \in S_p(\mathfrak{G})$. We obtain a contradiction in a number of steps.

(1) **W** strongly controls fusion in every proper section of $\mathfrak{G}$.

For every proper section of $\mathfrak{G}$ satisfies the hypotheses of the theorem and is of smaller order than $\mathfrak{G}$.

(2) $\mathfrak{P} \neq 1$, where $\mathfrak{P} = \mathbf{O}_p(\mathfrak{G})$.

W does not strongly control fusion in $\mathfrak{G}$. By 9.3, there exists a non-identity p-subgroup $\mathfrak{T}$ of $\mathfrak{G}$ such that **W** does not strongly control fusion in $\mathbf{N}_{\mathfrak{G}}(\mathfrak{T})$. By (1), $\mathbf{N}_{\mathfrak{G}}(\mathfrak{T}) = \mathfrak{G}$, or $\mathfrak{T} \trianglelefteq \mathfrak{G}$. Thus $\mathfrak{T} \leq \mathfrak{P}$. Since $\mathfrak{T} \neq 1$, $\mathfrak{P} \neq 1$.

(3) $\mathfrak{G} \neq \mathfrak{S}\mathbf{C}_{\mathfrak{G}}(\mathfrak{P})$.

Otherwise $\mathfrak{G}/\mathbf{C}_{\mathfrak{G}}(\mathfrak{P}) \cong \mathfrak{S}/\mathbf{C}_{\mathfrak{S}}(\mathfrak{P})$ is a p-group. But then by 9.4, **W** strongly controls fusion in $\mathfrak{G}$, which is a contradiction.

(4) If $\mathfrak{G} = \mathfrak{H}\mathfrak{K}$, where $\mathfrak{S} \leq \mathfrak{H} \leq \mathfrak{G}$ and $\mathfrak{S} \leq \mathfrak{K} \leq \mathfrak{G}$, then either $\mathfrak{H} = \mathfrak{G}$ or $\mathfrak{K} = \mathfrak{G}$.

Let $\sim$ be the relation defined on $\mathscr{S}(\mathfrak{S})$ by putting $\mathbf{x} \sim \mathbf{y}$ if $\mathbf{x}, \mathbf{y}$ are conjugate in $\mathbf{N}_{\mathfrak{G}}(\mathbf{W}(\mathfrak{S}))$. Since **W** does not strongly control fusion in $\mathfrak{G}$, $\sim$ does not contain fusion in $\mathfrak{G}$. By 7.4, either $\sim$ does not contain fusion in $\mathfrak{H}$ or $\sim$ does not contain fusion in $\mathfrak{K}$. Thus **W** does not strongly control fusion in $\mathfrak{H}$ or **W** does not strongly control fusion in $\mathfrak{K}$. By (1), either $\mathfrak{H} = \mathfrak{G}$ or $\mathfrak{K} = \mathfrak{G}$.

(5) Write $\mathfrak{L} = \mathbf{O}_{p'}(\mathfrak{G})$ and $\mathfrak{Q} = \mathfrak{G}/\mathfrak{L}$. By (3), (4) and 7.5b), $\mathbf{O}_p(\mathfrak{Q}) \geq \mathbf{C}_{\mathfrak{Q}}(\mathbf{O}_p(\mathfrak{Q}))$. By the hypothesis of the theorem, $\mathbf{W}(\mathfrak{S}\mathfrak{L}/\mathfrak{L}) \trianglelefteq \mathfrak{Q}$. By 5.2c),

$$\mathfrak{G} = \mathbf{N}_{\mathfrak{G}}(\mathbf{W}(\mathfrak{S}))\mathfrak{L}.$$

Since $\mathfrak{L} = \mathbf{O}_{p'}(\mathfrak{G}) \leq \mathbf{C}_{\mathfrak{G}}(\mathfrak{P})$, it follows that

$$\mathfrak{G} = \mathbf{N}_{\mathfrak{G}}(\mathbf{W}(\mathfrak{S}))(\mathfrak{S}\mathbf{C}_{\mathfrak{G}}(\mathfrak{P})).$$

Since $\mathbf{N}_{\mathfrak{G}}(\mathbf{W}(\mathfrak{S})) \geq \mathfrak{S}$, it follows from (3) and (4) that $\mathbf{N}_{\mathfrak{G}}(\mathbf{W}(\mathfrak{S})) = \mathfrak{G}$. But then, obviously, $\mathbf{W}$ strongly controls fusion in $\mathfrak{G}$, a contradiction.

q.e.d.

9.6 Theorem. *Suppose that every section of $\mathfrak{G}$ is p-stable.*

a) *If p is odd, **ZJ** strongly controls fusion in $\mathfrak{G}$.*

b) *For any p, $\underline{\mathbf{K}}$ and $\overline{\mathbf{K}}$ strongly control fusion in $\mathfrak{G}$.*

Proof. Let $\mathfrak{Q}$ be a section of $\mathfrak{G}$ for which $\mathbf{O}_p(\mathfrak{Q}) \geq \mathbf{C}_{\mathfrak{Q}}(\mathbf{O}_p(\mathfrak{Q}))$. By hypothesis, every homomorphic image of $\mathfrak{Q}$ is p-stable. Suppose $\mathfrak{S}_0 \in S_p(\mathfrak{Q})$. By 3.8, $\mathbf{ZJ}(\mathfrak{S}_0) \trianglelefteq \mathfrak{Q}$ for p odd, and by 8.17, $\underline{\mathbf{K}}(\mathfrak{S}_0) \trianglelefteq \mathfrak{Q}$, $\overline{\mathbf{K}}(\mathfrak{S}_0) \trianglelefteq \mathfrak{Q}$. The assertions thus follow from 9.5. **q.e.d.**

9.7 Remarks. a) Using IX, 7.11, GLAUBERMAN [6] has proved that if p is odd and **ZJ** does not strongly control fusion in $\mathfrak{G}$, then $\mathbf{N}_{\mathfrak{G}}(\mathbf{ZJ}(\mathfrak{S}))$ has a section isomorphic to the normaliser $F(p)$ of a Sylow p-subgroup of $SA(2, p)$. Also $\mathfrak{G}$ has a section isomorphic to $F(p)$ if and only if $\mathbf{N}_{\mathfrak{G}}(\mathbf{ZJ}(\mathfrak{S}))$ has such a section, so by IX, 7.10, if $\mathbf{N}_{\mathfrak{G}}(\mathbf{ZJ}(\mathfrak{S}))$ has no section isomorphic to $F(p)$, $\mathfrak{G}$ is p-stable.

b) No positive characteristic p-functor strongly controls fusion in $SA(2, p)$.

We prove this using the notation of IX, 7.5. Let

$$\mathfrak{S} = \left\{ \begin{pmatrix} 1 & b & t \\ 0 & 1 & u \\ 0 & 0 & 1 \end{pmatrix} \middle| \, b, t, u \in GF(p) \right\},$$

so that $\mathfrak{S} \in S_p(SA(2, p))$. The only characteristic subgroups of $\mathfrak{S}$ are 1, $\mathbf{Z}(\mathfrak{S})$, $\mathfrak{S}$ and, for $p = 2$, a cyclic group of order 4. Thus if $\mathbf{W}$ is a positive characteristic p-functor, $\mathbf{Z}(\mathfrak{S})$ is a characteristic subgroup of $\mathbf{W}(\mathfrak{S})$, and $\mathbf{N}_{SA(2,p)}(\mathbf{W}(\mathfrak{S})) \leq \mathbf{N}_{SA(2,p)}(\mathbf{Z}(\mathfrak{S})) = \mathfrak{N}$, say. But

$$\mathbf{Z}(\mathfrak{S}) = \left\{ \begin{pmatrix} 1 & 0 & t \\ 0 & 1 & 0 \\ 0 & 0 & 1 \end{pmatrix} \middle| \, t \in GF(p) \right\}, \quad \mathfrak{N} = \left\{ \begin{pmatrix} a & b & t \\ 0 & a^{-1} & u \\ 0 & 0 & 1 \end{pmatrix} \middle| \, \begin{matrix} a, b, t, u \in GF(p), \\ a \neq 0 \end{matrix} \right\}.$$

However $\begin{pmatrix} 1 & 0 & 1 \\ 0 & 1 & 0 \\ 0 & 0 & 1 \end{pmatrix}, \begin{pmatrix} 1 & 0 & 0 \\ 0 & 1 & 1 \\ 0 & 0 & 1 \end{pmatrix}$ are conjugate in $SA(2, p)$ but not in $\mathfrak{N}$.

We turn next to the use of characteristic p-functors in establishing conditions for a group to be p-nilpotent. First we prove a lemma which is essentially the first half of the proof of IV, 6.2.

9.8 Lemma. *If $\mathfrak{G}$ is not p-nilpotent, $\mathfrak{G}$ has a section $\mathfrak{Q}$ with the following properties.*

a) *$\mathfrak{Q}$ is not p-nilpotent but $\mathfrak{Q}/\mathbf{O}_p(\mathfrak{Q})$ is p-nilpotent.*

b) $\mathbf{O}_p(\mathfrak{Q}) \geq \mathbf{C}_{\mathfrak{Q}}(\mathbf{O}_p(\mathfrak{Q}))$.

c) *The Sylow p-subgroups of $\mathfrak{Q}$ are maximal subgroups of $\mathfrak{Q}$.*

d) *Suppose $\mathfrak{S} \in S_p(\mathfrak{G})$, $\mathfrak{S}_0 \in S_p(\mathfrak{Q})$. If $\mathbf{W}$ is a positive characteristic p-functor and $\mathbf{N}_{\mathfrak{G}}(\mathbf{W}(\mathfrak{S}))$ is p-nilpotent, then $\mathbf{N}_{\mathfrak{Q}}(\mathbf{W}(\mathfrak{S}_0)) = \mathfrak{S}_0$.*

Proof. Suppose that this is false and that $\mathfrak{G}$ is a counterexample of minimal order.

(1) If $\mathfrak{Q}$ is a proper section of $\mathfrak{G}$ and $\mathfrak{Q}$ is not p-nilpotent, there exists a positive characteristic p-functor $\mathbf{W}$ such that $\mathbf{N}_{\mathfrak{G}}(\mathbf{W}(\mathfrak{S}))$ is p-nilpotent for $\mathfrak{S} \in S_p(\mathfrak{G})$ and $\mathbf{N}_{\mathfrak{Q}}(\mathbf{W}(\mathfrak{S}_0))$ is not p-nilpotent for $\mathfrak{S}_0 \in S_p(\mathfrak{Q})$.

Since $|\mathfrak{Q}| < |\mathfrak{G}|$, $\mathfrak{Q}$ has a section $\mathfrak{Q}^*$ which satisfies a), b), c) and which has the property that whenever $\mathbf{W}$ is a positive characteristic p-functor for which $\mathbf{N}_{\mathfrak{Q}}(\mathbf{W}(\mathfrak{S}_0))$ is p-nilpotent, $\mathbf{N}_{\mathfrak{Q}^*}(\mathbf{W}(\mathfrak{S}^*)) = \mathfrak{S}^*$ for $\mathfrak{S}^* \in S_p(\mathfrak{Q}^*)$. Since the theorem is false, $\mathfrak{Q}^*$ does not satisfy d). Thus for some $\mathbf{W}$, $\mathbf{N}_{\mathfrak{G}}(\mathbf{W}(\mathfrak{S}))$ is p-nilpotent and $\mathbf{N}_{\mathfrak{Q}^*}(\mathbf{W}(\mathfrak{S}^*)) \neq \mathfrak{S}^*$, whence $\mathbf{N}_{\mathfrak{Q}}(\mathbf{W}(\mathfrak{S}_0))$ is not p-nilpotent. This proves (1).

Let $\mathscr{X}$ be the set of non-identity p-subgroups $\mathfrak{X}$ of $\mathfrak{G}$ for which $\mathbf{N}_{\mathfrak{G}}(\mathfrak{X})$ is not p-nilpotent. Since $\mathfrak{G}$ is not p-nilpotent, $\mathscr{X}$ is non-empty, by IV, 5.8b). Let n be the greatest integer for which there exists $\mathfrak{X} \in \mathscr{X}$ such that p^n divides $|\mathbf{N}_{\mathfrak{G}}(\mathfrak{X})|$, and let

$$\mathscr{X}^* = \{\mathfrak{X} \mid \mathfrak{X} \in \mathscr{X},\ p^n \text{ divides } |\mathbf{N}_{\mathfrak{G}}(\mathfrak{X})|\}.$$

Let $\mathfrak{P}$ be an element of $\mathscr{X}^*$ of maximal order. If $\mathfrak{T} \in S_p(\mathbf{N}_{\mathfrak{G}}(\mathfrak{P}))$, then $\mathfrak{P} \leq \mathfrak{T}$ and $|\mathfrak{T}| = p^n$. Suppose that $\mathfrak{T} \leq \mathfrak{S} \in S_p(\mathfrak{G})$. Then $\mathfrak{T} \leq \mathbf{N}_{\mathfrak{S}}(\mathfrak{P})$ and, since $\mathbf{N}_{\mathfrak{S}}(\mathfrak{P})$ is a p-subgroup of $\mathbf{N}_{\mathfrak{G}}(\mathfrak{P})$, $\mathfrak{T} = \mathbf{N}_{\mathfrak{S}}(\mathfrak{P})$.

(2) $\mathfrak{P} \trianglelefteq \mathfrak{G}$.

Since $\mathfrak{P} \in \mathscr{X}$, $\mathbf{N}_{\mathfrak{G}}(\mathfrak{P})$ is not p-nilpotent. If $\mathbf{N}_{\mathfrak{G}}(\mathfrak{P}) < \mathfrak{G}$, then by (1), there exists a positive characteristic p-functor $\mathbf{W}$ such that $\mathbf{N}_{\mathfrak{G}}(\mathbf{W}(\mathfrak{S}))$ is p-nilpotent but $\mathbf{N}_{\mathfrak{G}}(\mathbf{W}(\mathfrak{T}))$ is not p-nilpotent. Thus $\mathfrak{T} \neq \mathfrak{S}$. And since $\mathbf{W}$ is positive, $\mathbf{W}(\mathfrak{T}) \neq 1$, so $\mathbf{W}(\mathfrak{T}) \in \mathscr{X}$. Since $\mathbf{W}(\mathfrak{T})$ is a characteristic subgroup of $\mathfrak{T}$, $\mathbf{W}(\mathfrak{T}) \trianglelefteq \mathbf{N}_{\mathfrak{S}}(\mathfrak{T})$. Hence by I, 8.8, $|\mathbf{N}_{\mathfrak{G}}(\mathbf{W}(\mathfrak{T}))|$ is divisible by p^{n+1}, contrary to the definition of n. Thus $\mathbf{N}_{\mathfrak{G}}(\mathfrak{P}) = \mathfrak{G}$ and $\mathfrak{P} \trianglelefteq \mathfrak{G}$.

(3) If $\mathfrak{P} < \mathfrak{R} \trianglelefteq \mathfrak{S}$, $\mathbf{N}_{\mathfrak{G}}(\mathfrak{R})$ is p-nilpotent. In particular, $\mathfrak{P} = \mathbf{O}_p(\mathfrak{G})$.

Since $|\mathfrak{R}| > |\mathfrak{P}|$, $\mathfrak{R} \notin \mathscr{X}^*$. Since $\mathfrak{R} \trianglelefteq \mathfrak{S}$, p^n divides $|\mathbf{N}_{\mathfrak{G}}(\mathfrak{R})|$, so $\mathfrak{R} \notin \mathscr{X}$. Thus $\mathbf{N}_{\mathfrak{G}}(\mathfrak{R})$ is p-nilpotent.

By (2), $\mathfrak{P} \leq \mathbf{O}_p(\mathfrak{G})$. If $\mathfrak{P} < \mathbf{O}_p(\mathfrak{G})$, $\mathfrak{G} = \mathbf{N}_{\mathfrak{G}}(\mathbf{O}_p(\mathfrak{G}))$ is p-nilpotent, a contradiction.

(4) $\mathfrak{G}/\mathfrak{P}$ is p-nilpotent.

Since $\mathfrak{P} \in \mathscr{X}$, $\mathfrak{P} \neq 1$, so $|\mathfrak{G}/\mathfrak{P}| < |\mathfrak{G}|$. It follows from (1) that if $\mathfrak{G}/\mathfrak{P}$ is not p-nilpotent, there exists a positive characteristic p-functor $\mathbf{W}$ such that $\mathbf{N}_{\mathfrak{G}/\mathfrak{P}}(\mathbf{W}(\mathfrak{S}/\mathfrak{P}))$ is not p-nilpotent. Let $\mathfrak{B}/\mathfrak{P} = \mathbf{W}(\mathfrak{S}/\mathfrak{P})$; thus $\mathbf{N}_{\mathfrak{G}}(\mathfrak{B})$ is not p-nilpotent. Since $\mathfrak{P} \leq \mathfrak{B} \trianglelefteq \mathfrak{S}$, it follows from (3) that $\mathfrak{B} = \mathfrak{P}$. But then $\mathfrak{S} = \mathfrak{P}$, since $\mathbf{W}$ is positive. Hence $\mathfrak{G}/\mathfrak{P}$ is a p'-group and is therefore p-nilpotent.

(5) $\mathbf{O}_{p'}(\mathfrak{G}) = 1$ and $\mathfrak{P} \geq \mathbf{C}_{\mathfrak{G}}(\mathfrak{P})$.

Suppose that $\mathfrak{N} = \mathbf{O}_{p'}(\mathfrak{G}) \neq 1$. Since $\mathfrak{G}$ is not p-nilpotent, $\mathfrak{G}/\mathfrak{N}$ is not p-nilpotent. By (1), there exists a positive characteristic p-functor $\mathbf{W}$ such that $\mathbf{N}_{\mathfrak{G}}(\mathbf{W}(\mathfrak{S}))$ is p-nilpotent but $\mathbf{N}_{\mathfrak{G}/\mathfrak{N}}(\mathbf{W}(\mathfrak{S}\mathfrak{N}/\mathfrak{N}))$ is not. This is impossible, since by 5.2c), the latter group is $\mathbf{N}_{\mathfrak{G}}(\mathbf{W}(\mathfrak{S}))\mathfrak{N}/\mathfrak{N}$. Hence $\mathbf{O}_{p'}(\mathfrak{G}) = 1$. By IX, 1.4 and (4), $\mathfrak{P} \geq \mathbf{C}_{\mathfrak{G}}(\mathfrak{P})$.

(6) $\mathfrak{S}$ is a maximal subgroup of $\mathfrak{G}$.

Suppose that $\mathfrak{S} \leq \mathfrak{M} < \mathfrak{G}$. By (1), $\mathfrak{M}$ is p-nilpotent. Thus $\mathbf{O}^p(\mathfrak{M})$ and $\mathfrak{P}$ are normal subgroups of $\mathfrak{M}$ of coprime orders and $\mathbf{O}^p(\mathfrak{M}) \leq \mathbf{C}_{\mathfrak{G}}(\mathfrak{P})$. Using (5), $\mathbf{O}^p(\mathfrak{M}) = 1$, whence $\mathfrak{M} = \mathfrak{S}$. Thus $\mathfrak{S}$ is a maximal subgroup of $\mathfrak{G}$.

But now, $\mathfrak{G}$ itself satisfies the conditions a)–d). For if $\mathbf{W}$ is a positive characteristic p-functor and $\mathbf{N}_{\mathfrak{G}}(\mathbf{W}(\mathfrak{S}))$ is p-nilpotent, then $\mathbf{N}_{\mathfrak{G}}(\mathbf{W}(\mathfrak{S})) = \mathfrak{S}$ by (6). Thus we have a contradiction. **q.e.d.**

The following lemma gives more information about groups satisfying the conditions a)–c) of 9.8.

9.9 Lemma. *Suppose that $\mathfrak{G}$ is a p-soluble group and $\mathfrak{S} \in S_p(\mathfrak{G})$.*

a) *If $\mathbf{O}_{p,p'}(\mathfrak{G}) \neq \mathfrak{G}$, there exists a maximal subgroup $\mathfrak{L}$ of $\mathfrak{G}$ such that $\mathfrak{S} \leq \mathfrak{L}$ but $\mathbf{O}_{p,p'}(\mathfrak{G}) \nleq \mathfrak{L}$.*

b) *Suppose that $\mathfrak{G}$ has precisely one maximal subgroup $\mathfrak{M}$ which contains $\mathfrak{S}$. Then $\mathfrak{G}$ is a $\{p, q\}$-group for some prime q and $\mathbf{O}_{p,q,p}(\mathfrak{G}) = \mathfrak{G}$. If $\mathfrak{R}/\mathbf{O}_p(\mathfrak{G}) = \Phi(\mathbf{O}_{p,q}(\mathfrak{G})/\mathbf{O}_p(\mathfrak{G}))$, then $\mathbf{O}_{p,q}(\mathfrak{G})/\mathfrak{R}$ is the only minimal normal subgroup of $\mathfrak{G}/\mathfrak{R}$ and $\mathfrak{M} = \mathfrak{R}\mathfrak{S}$.*

c) *Suppose that $\mathfrak{S}$ is a maximal subgroup of $\mathfrak{G}$. Then $\mathfrak{G}$ is a $\{p, q\}$-group for some prime q, $\mathbf{O}_{p,q,p}(\mathfrak{G}) = \mathfrak{G}$ and $\mathbf{O}_{p,q}(\mathfrak{G})/\mathbf{O}_p(\mathfrak{G})$ is the only minimal normal subgroup of $\mathfrak{G}/\mathbf{O}_p(\mathfrak{G})$. If p is odd, every section of $\mathfrak{G}$ is p-stable.*

Proof. Let $\mathfrak{Q} = \mathbf{O}_{p,p'}(\mathfrak{G})$, $\mathfrak{H} = \mathbf{O}_{p,p',p}(\mathfrak{G})$.

a) Since $\mathfrak{Q} \neq \mathfrak{G}$, $\mathfrak{H} > \mathfrak{Q}$. Since $\mathfrak{H} \trianglelefteq \mathfrak{G}$, $\mathfrak{H} \cap \mathfrak{S} \in S_p(\mathfrak{H})$ and $\mathfrak{H} \cap \mathfrak{S} \trianglelefteq \mathfrak{S}$. Since $\mathfrak{H}/\mathfrak{Q}$ is a p-group, $\mathfrak{H} = (\mathfrak{H} \cap \mathfrak{S})\mathfrak{Q}$. Also, by the

Frattini argument, $\mathfrak{G} = \mathbf{N}_{\mathfrak{G}}(\mathfrak{H} \cap \mathfrak{S})\mathfrak{H}$. Thus

$$\mathfrak{G} = \mathbf{N}_{\mathfrak{G}}(\mathfrak{H} \cap \mathfrak{S})(\mathfrak{H} \cap \mathfrak{S})\mathfrak{Q} = \mathbf{N}_{\mathfrak{G}}(\mathfrak{H} \cap \mathfrak{S})\mathfrak{Q}.$$

Since $\mathfrak{H} \neq \mathfrak{Q}$, $\mathfrak{H} \cap \mathfrak{S} \not\leq \mathfrak{Q}$. Thus $\mathfrak{H} \cap \mathfrak{S} \not\leq \mathbf{O}_p(\mathfrak{G})$ and $\mathfrak{H} \cap \mathfrak{S}$ is not normal in $\mathfrak{G}$. Hence $\mathbf{N}_{\mathfrak{G}}(\mathfrak{H} \cap \mathfrak{S}) \neq \mathfrak{G}$. Hence there exists a maximal subgroup $\mathfrak{L}$ of $\mathfrak{G}$ such that $\mathfrak{L} \geq \mathbf{N}_{\mathfrak{G}}(\mathfrak{H} \cap \mathfrak{S})$. Since $\mathfrak{H} \cap \mathfrak{S} \trianglelefteq \mathfrak{S}$, $\mathfrak{S} \leq \mathfrak{L}$. Also $\mathfrak{G} = \mathfrak{Q}\mathfrak{L}$, so $\mathfrak{Q} \not\leq \mathfrak{L}$.

b) By a), $\mathfrak{Q} \not\leq \mathfrak{M}$. Hence $\mathfrak{S}\mathfrak{Q}$ is contained in no maximal subgroup of $\mathfrak{G}$ and $\mathfrak{S}\mathfrak{Q} = \mathfrak{G}$. Thus $\mathbf{O}_{p,p',p}(\mathfrak{G}) = \mathfrak{G}$.

Let $\mathfrak{P} = \mathbf{O}_p(\mathfrak{G})$. Note that if $\mathfrak{P} \leq \mathfrak{X} \leq \mathfrak{Q}$ and $\mathfrak{S}\mathfrak{X} = \mathfrak{G}$, then $\mathfrak{X} = \mathfrak{Q}$. For then $\mathfrak{Q} = (\mathfrak{S} \cap \mathfrak{Q})\mathfrak{X}$, and since $\mathfrak{S} \cap \mathfrak{Q} \in S_p(\mathfrak{Q})$, $\mathfrak{S} \cap \mathfrak{Q} = \mathfrak{P} \leq \mathfrak{X}$.

Suppose that $\mathfrak{Q}/\mathfrak{P}$ is not of prime-power order. For each prime divisor r of $|\mathfrak{Q}/\mathfrak{P}|$, $\mathfrak{S}/\mathfrak{P}$ normalises some Sylow r-subgroup $\mathfrak{Q}_r/\mathfrak{P}$ of $\mathfrak{Q}/\mathfrak{P}$, by IX, 1.11. Since $\mathfrak{Q}_r < \mathfrak{Q}$, $\mathfrak{S}\mathfrak{Q}_r < \mathfrak{G}$. Thus $\mathfrak{Q}_r \leq \mathfrak{S}\mathfrak{Q}_r \leq \mathfrak{M}$ for all r. Thus $\mathfrak{Q}/\mathfrak{P} \leq \mathfrak{M}/\mathfrak{P}$ and $\mathfrak{M} \geq \mathfrak{S}\mathfrak{Q} = \mathfrak{G}$, a contradiction. Hence $\mathfrak{Q}/\mathfrak{P}$ is a q-group for some prime q.

Since $\mathfrak{R} < \mathfrak{Q}$, $\mathfrak{R}\mathfrak{S} < \mathfrak{G}$. Thus $\mathfrak{R}\mathfrak{S} \leq \mathfrak{M}$ and $\mathfrak{R} \leq \mathfrak{M} \cap \mathfrak{Q}$. If $\mathsf{K} = GF(q)$, $\mathfrak{Q}/\mathfrak{R}$ has the structure of a $\mathsf{K}(\mathfrak{G}/\mathfrak{Q})$-module in which the submodules are the normal subgroups of $\mathfrak{G}/\mathfrak{R}$ contained in $\mathfrak{Q}/\mathfrak{R}$. Now since $\mathfrak{Q} \trianglelefteq \mathfrak{G}$, $\mathfrak{R} \leq \mathfrak{M} \cap \mathfrak{Q}$ and $\mathfrak{Q}/\mathfrak{R}$ is Abelian, $\mathbf{N}_{\mathfrak{G}}(\mathfrak{M} \cap \mathfrak{Q}) \geq \mathfrak{M}\mathfrak{Q} = \mathfrak{G}$. Thus by the Maschke-Schur theorem, there exists $\mathfrak{N} \trianglelefteq \mathfrak{G}$ such that $\mathfrak{Q}/\mathfrak{R} = ((\mathfrak{M} \cap \mathfrak{Q})/\mathfrak{R}) \times (\mathfrak{N}/\mathfrak{R})$. If $\mathfrak{N} \leq \mathfrak{M}$, then $\mathfrak{Q} \leq \mathfrak{M}$ and $\mathfrak{M} = \mathfrak{G}$, a contradiction. Thus $\mathfrak{N} \not\leq \mathfrak{M}$ and $\mathfrak{S}\mathfrak{N} \not\leq \mathfrak{M}$. Hence $\mathfrak{S}\mathfrak{N} = \mathfrak{G}$ and $\mathfrak{N} = \mathfrak{Q}$. Thus $\mathfrak{M} \cap \mathfrak{Q} = \mathfrak{R}$, and $\mathfrak{M} = \mathfrak{S}(\mathfrak{M} \cap \mathfrak{Q}) = \mathfrak{R}\mathfrak{S}$.

If $\mathfrak{R} < \mathfrak{L} \leq \mathfrak{Q}$ and $\mathfrak{L} \trianglelefteq \mathfrak{G}$, then $\mathfrak{L} \not\leq \mathfrak{M}$, $\mathfrak{S}\mathfrak{L} \not\leq \mathfrak{M}$, $\mathfrak{S}\mathfrak{L} = \mathfrak{G}$ and $\mathfrak{L} = \mathfrak{Q}$. Thus $\mathfrak{Q}/\mathfrak{R}$ is a minimal normal subgroup of $\mathfrak{G}/\mathfrak{R}$. Suppose that $\mathfrak{K}/\mathfrak{R}$ is a minimal normal subgroup of $\mathfrak{G}/\mathfrak{R}$ and $\mathfrak{K}/\mathfrak{R} \neq \mathfrak{Q}/\mathfrak{R}$. Then $\mathfrak{K}/\mathfrak{R}$ centralizes $\mathfrak{Q}/\mathfrak{R}$. But if $\overline{\mathfrak{G}} = \mathfrak{G}/\mathfrak{P}$ and $\overline{\mathfrak{Q}} = \mathfrak{Q}/\mathfrak{P}$, then $\mathbf{O}_{q'}(\overline{\mathfrak{G}}) = 1$ and $\overline{\mathfrak{Q}} = \mathbf{O}_q(\overline{\mathfrak{G}})$. Since $\mathfrak{R}/\mathfrak{P} = \Phi(\overline{\mathfrak{Q}})$, it follows from IX, 1.6 that $\mathfrak{K}/\mathfrak{P} \leq \overline{\mathfrak{Q}}$, or $\mathfrak{K} \leq \mathfrak{Q}$. Thus $\mathfrak{K}/\mathfrak{R} = \mathfrak{Q}/\mathfrak{R}$, a contradiction, and b) is proved.

c) In the notation of b), we have $\mathfrak{M} = \mathfrak{S}$, so $\mathfrak{R} \leq \mathfrak{S} \cap \mathfrak{Q} = \mathfrak{P} \leq \mathfrak{R}$. Thus $\mathfrak{R} = \mathfrak{P}$ and $\mathfrak{Q}/\mathfrak{P}$ is the only minimal normal subgroup of $\mathfrak{G}/\mathfrak{P}$. If p is odd and $\mathfrak{G}$ has a section which is not p-stable, then $\mathfrak{G}$ has a section isomorphic to $SA(2, p)$, by IX, 7.10. By II, 8.10, the Sylow 2-subgroup of $SL(2, p)$ is non-Abelian. Thus the Sylow 2-subgroup of $\mathfrak{G}$ is non-Abelian. Since $\mathfrak{G}$ is a $\{p, q\}$-group and p is odd, $q = 2$. But the Sylow q-subgroup of $\mathfrak{G}$ is isomorphic to $\mathfrak{Q}/\mathfrak{P}$, and this is elementary Abelian. Thus we have a contradiction. **q.e.d.**

Remark. The p-stability can also be proved by using even the elementary form of Theorem B (IX, 1.10).

9.10 Theorem. *Suppose that $\mathfrak{G}$ is a group, p is an odd prime and $\mathfrak{S} \in S_p(\mathfrak{G})$.*

a) *If $\mathbf{N}_{\mathfrak{G}}(\mathbf{ZJ}(\mathfrak{S}))$ is p-nilpotent, $\mathfrak{G}$ is p-nilpotent.*

b) *If $\mathbf{N}_{\mathfrak{G}}(\underline{\mathbf{K}}(\mathfrak{S}))$ or $\mathbf{N}_{\mathfrak{G}}(\overline{\mathbf{K}}(\mathfrak{S}))$ is p-nilpotent, $\mathfrak{G}$ is p-nilpotent.*

c) *If $\mathbf{N}_{\mathfrak{G}}(\underline{\mathbf{K}}(\mathfrak{S}))/\mathbf{C}_{\mathfrak{G}}(\underline{\mathbf{K}}(\mathfrak{S}))$ or $\mathbf{N}_{\mathfrak{G}}(\overline{\mathbf{K}}(\mathfrak{S}))/\mathbf{C}_{\mathfrak{G}}(\overline{\mathbf{K}}(\mathfrak{S}))$ is a p-group, then $\mathfrak{G}$ is p-nilpotent.*

Proof. a, b) $\mathbf{N}_{\mathfrak{G}}(\mathbf{W}(\mathfrak{S}))$ is p-nilpotent, where $\mathbf{W}$ is $\mathbf{ZJ}$ in a) and $\mathbf{W}$ is either $\underline{\mathbf{K}}$ or $\overline{\mathbf{K}}$ in b). If $\mathfrak{G}$ is not p-nilpotent, then $\mathfrak{G}$ has a p-soluble section $\mathfrak{Q}$ such that $\mathfrak{Q}$ is not a p-group, $\mathbf{O}_p(\mathfrak{Q}) \geq \mathbf{C}_{\mathfrak{Q}}(\mathbf{O}_p(\mathfrak{Q}))$, the Sylow p-subgroup $\mathfrak{P}$ of $\mathfrak{Q}$ is a maximal subgroup of $\mathfrak{Q}$ and $\mathbf{N}_{\mathfrak{Q}}(\mathbf{W}(\mathfrak{P})) = \mathfrak{P}$, by 9.8. By 9.9, every section of $\mathfrak{Q}$ is p-stable. Hence by 3.8 and 8.17, $\mathbf{W}(\mathfrak{P}) \trianglelefteq \mathfrak{Q}$. This is a contradiction.

c) $\mathfrak{N}/\mathfrak{C}$ is a p-group, where $\mathfrak{N} = \mathbf{N}_{\mathfrak{G}}(\mathbf{K}(\mathfrak{S}))$, $\mathfrak{C} = \mathbf{C}_{\mathfrak{G}}(\mathbf{K}(\mathfrak{S})) \trianglelefteq \mathfrak{N}$ and $\mathbf{K}$ is either $\underline{\mathbf{K}}$ or $\overline{\mathbf{K}}$. Now $\mathfrak{S} \in S_p(\mathfrak{N})$, so $\mathfrak{S}_0 = \mathfrak{S} \cap \mathfrak{C} \in S_p(\mathfrak{C})$. By 8.5, $\mathfrak{S}_0 = \mathbf{C}_{\mathfrak{S}}(\mathbf{K}(\mathfrak{S})) \leq \mathbf{K}(\mathfrak{S})$, so $\mathfrak{S}_0 \leq \mathbf{Z}(\mathfrak{C})$. By IV, 2.6, $\mathfrak{C}$ is p-nilpotent. Since $\mathfrak{N}/\mathfrak{C}$ is a p-group, $\mathfrak{N}$ is p-nilpotent. By b), $\mathfrak{G}$ is p-nilpotent. **q.e.d.**

Note that for $p > 3$, 9.10b) follows immediately from 8.11.

9.11 Remark. Glauberman has shown that $\mathbf{J}, \underline{\mathbf{K}}, \overline{\mathbf{K}}$ also have the following properties.

a) If p is odd, $\mathbf{K}$ is $\underline{\mathbf{K}}$ or $\overline{\mathbf{K}}$ and $\mathfrak{S} \in S_p(\mathfrak{G})$, every element of $\mathfrak{S} \cap \mathbf{Z}(\mathbf{N}_{\mathfrak{G}}(\mathbf{K}(\mathfrak{S})))$ is weakly closed in $\mathfrak{S}$ with respect to $\mathfrak{G}$.

b) Suppose that either p is odd or that $\mathfrak{G}$ has no section isomorphic to $\mathfrak{S}_4$. Then, if $\mathfrak{S} \in S_p(\mathfrak{G})$, every element of $\mathfrak{S} \cap \mathbf{Z}(\mathbf{N}_{\mathfrak{G}}(\mathbf{J}(\mathfrak{S})))$ is weakly closed in $\mathfrak{S}$ with respect to $\mathfrak{G}$.

However these results are not true for $p = 2$ in general, as is shown by *PSL*(2, 17).

§ 10. The Product Theorem for J

For $p \geq 5$, if $\mathfrak{G}$ is a p-soluble group and $\mathbf{O}_{p'}(\mathfrak{G}) = 1$,

(1) $$\mathfrak{G} = \mathbf{C}_{\mathfrak{G}}(\mathbf{Z}(\mathfrak{S}))\mathbf{N}_{\mathfrak{G}}(\mathbf{J}(\mathfrak{S}))$$

for $\mathfrak{S} \in S_p(\mathfrak{G})$. (Actually $\mathbf{J}(\mathfrak{S}) \trianglelefteq \mathfrak{G}$ for $p > 5$, but we shall not prove this.) (1) holds frequently for $p \leq 3$ as well, and we shall therefore approach this equation by investigating what happens when it is false.

10.1 Lemma. *Suppose that* $\mathfrak{S} \in S_p(\mathfrak{G})$, $\mathfrak{N} \trianglelefteq \mathfrak{G}$ *and* $\mathbf{J}(\mathfrak{S}) \leq \mathfrak{N}$. *Then* $\mathfrak{G} = \mathfrak{N}\mathbf{N}_\mathfrak{G}(\mathbf{J}(\mathfrak{S}))$.

Proof. We have $\mathbf{J}(\mathfrak{S}) \leq \mathfrak{S} \cap \mathfrak{N}$, so $\mathbf{J}(\mathfrak{S}) = \mathbf{J}(\mathfrak{S} \cap \mathfrak{N})$ by definition of $\mathbf{J}(\mathfrak{S})$ (3.6). Thus $\mathbf{J}(\mathfrak{S})$ is a characteristic subgroup of $\mathfrak{S} \cap \mathfrak{N}$ and $\mathbf{J}(\mathfrak{S}) \trianglelefteq \mathbf{N}_\mathfrak{G}(\mathfrak{S} \cap \mathfrak{N})$. Thus $\mathbf{N}_\mathfrak{G}(\mathfrak{S} \cap \mathfrak{N}) \leq \mathbf{N}_\mathfrak{G}(\mathbf{J}(\mathfrak{S}))$. But $\mathfrak{S} \cap \mathfrak{N} \in S_p(\mathfrak{N})$, so using the Frattini argument,

$$\mathfrak{G} = \mathfrak{N}\mathbf{N}_\mathfrak{G}(\mathfrak{S} \cap \mathfrak{N}) = \mathfrak{N}\mathbf{N}_\mathfrak{G}(\mathbf{J}(\mathfrak{S})). \qquad \textbf{q.e.d.}$$

In view of 10.1, (1) can be approached by seeking $\mathfrak{N} \trianglelefteq \mathfrak{G}$ such that $\mathbf{J}(\mathfrak{S}) \leq \mathfrak{N} \leq \mathbf{C}_\mathfrak{G}(\mathbf{Z}(\mathfrak{S}))$. The most obvious choice for $\mathfrak{N}$ is the centralizer of $\langle \mathbf{Z}(\mathfrak{S}_0) | \mathfrak{S}_0 \in S_p(\mathfrak{G})\rangle$, but we shall in fact use the centralizer of the slightly larger normal subgroup $\mathbf{R}_p(\mathfrak{G})$ defined in the following lemma.

10.2 Lemma. *Let* $\mathfrak{G}$ *be a group and let p be a prime. Let* $\mathscr{R}_p(\mathfrak{G})$ *be the set of all normal p-subgroups* $\mathfrak{X}$ *of* $\mathfrak{G}$ *for which* $\mathbf{O}_p(\mathfrak{G}/\mathbf{C}_\mathfrak{G}(\mathfrak{X})) = 1$ *and let* $\mathbf{R}_p(\mathfrak{G}) = \langle \mathfrak{X} | \mathfrak{X} \in \mathscr{R}_p(\mathfrak{G})\rangle$.

a) $\mathbf{R}_p(\mathfrak{G}) \in \mathscr{R}_p(\mathfrak{G})$ *and* $\mathbf{R}_p(\mathfrak{G}) \leq \mathbf{Z}(\mathbf{O}_p(\mathfrak{G}))$.

b) $\mathbf{C}_\mathfrak{G}(\mathbf{R}_p(\mathfrak{G})) = \mathbf{C}_\mathfrak{G}(\Omega_1(\mathbf{R}_p(\mathfrak{G})))$.

c) *Suppose that* $\mathbf{O}_p(\mathfrak{G}) \geq \mathbf{C}_\mathfrak{G}(\mathbf{O}_p(\mathfrak{G}))$. *If* $\mathfrak{S} \in S_p(\mathfrak{G})$ *and* $\mathfrak{Y}$ *is the normal closure of* $\mathbf{Z}(\mathfrak{S})$ *in* $\mathfrak{G}$, *then* $\mathfrak{Y} \in \mathscr{R}_p(\mathfrak{G})$. *Hence* $\mathbf{Z}(\mathfrak{S}) \leq \mathfrak{Y} \leq \mathbf{R}_p(\mathfrak{G})$ *and* $\mathfrak{Y} \leq \mathbf{Z}(\mathbf{O}_p(\mathfrak{G}))$.

Proof. a) Clearly, $\mathbf{R}_p(\mathfrak{G})$ is a normal p-subgroup of $\mathfrak{G}$. Let $\mathfrak{P}/\mathbf{C}_\mathfrak{G}(\mathbf{R}_p(\mathfrak{G})) = \mathbf{O}_p(\mathfrak{G}/\mathbf{C}_\mathfrak{G}(\mathbf{R}_p(\mathfrak{G})))$. If $\mathfrak{X} \in \mathscr{R}_p(\mathfrak{G})$, then $\mathfrak{X} \leq \mathbf{R}_p(\mathfrak{G})$, $\mathbf{C}_\mathfrak{G}(\mathbf{R}_p(\mathfrak{G})) \leq \mathbf{C}_\mathfrak{G}(\mathfrak{X})$, $\mathfrak{P}\mathbf{C}_\mathfrak{G}(\mathfrak{X})/\mathbf{C}_\mathfrak{G}(\mathfrak{X})$ is a normal p-subgroup of $\mathfrak{G}/\mathbf{C}_\mathfrak{G}(\mathfrak{X})$, $\mathfrak{P}\mathbf{C}_\mathfrak{G}(\mathfrak{X})/\mathbf{C}_\mathfrak{G}(\mathfrak{X}) \leq \mathbf{O}_p(\mathfrak{G}/\mathbf{C}_\mathfrak{G}(\mathfrak{X})) = 1$ and $\mathfrak{P} \leq \mathbf{C}_\mathfrak{G}(\mathfrak{X})$. Thus $\mathfrak{P} \leq \mathbf{C}_\mathfrak{G}(\mathbf{R}_p(\mathfrak{G}))$.

If $\mathfrak{X} \in \mathscr{R}_p(\mathfrak{G})$, then $\mathbf{O}_p(\mathfrak{G}/\mathbf{C}_\mathfrak{G}(\mathfrak{X})) = 1$, so $\mathbf{O}_p(\mathfrak{G}) \leq \mathbf{C}_\mathfrak{G}(\mathfrak{X})$. Thus

$$\mathfrak{X} \leq \mathbf{C}_\mathfrak{G}(\mathbf{O}_p(\mathfrak{G})) \cap \mathbf{O}_p(\mathfrak{G}) = \mathbf{Z}(\mathbf{O}_p(\mathfrak{G})).$$

Hence $\mathbf{R}_p(\mathfrak{G}) \leq \mathbf{Z}(\mathbf{O}_p(\mathfrak{G}))$.

b) Let $\mathfrak{C} = \mathbf{C}_\mathfrak{G}(\mathbf{R}_p(\mathfrak{G}))$, $\mathfrak{C}_1 = \mathbf{C}_\mathfrak{G}(\Omega_1(\mathbf{R}_p(\mathfrak{G})))$; clearly $\mathfrak{C} \leq \mathfrak{C}_1$. If $x \in \mathfrak{C}_1$ and ξ is the automorphism of $\mathbf{R}_p(\mathfrak{G})$ induced by x, ξ induces the identity automorphism on $\Omega_1(\mathbf{R}_p(\mathfrak{G}))$. Since $\mathbf{R}_p(\mathfrak{G})$ is Abelian, it follows from 1.4 that the order of ξ is p^k for some k. Thus $x^{p^k} \in \mathfrak{C}$. Hence $\mathfrak{C}_1/\mathfrak{C}$ is a normal p-subgroup of $\mathfrak{G}/\mathfrak{C}$. By a), $\mathbf{O}_p(\mathfrak{G}/\mathfrak{C}) = 1$, so $\mathfrak{C}_1 = \mathfrak{C}$.

c) Since $\mathbf{O}_p(\mathfrak{G}) \geq \mathbf{C}_\mathfrak{G}(\mathbf{O}_p(\mathfrak{G}))$, $\mathbf{Z}(\mathfrak{S}) \leq \mathbf{O}_p(\mathfrak{G})$. Thus $\mathfrak{Y}$ is a normal p-subgroup of $\mathfrak{G}$. Let $\mathfrak{B} = \mathbf{C}_\mathfrak{G}(\mathfrak{Y})$, $\mathfrak{K}/\mathfrak{B} = \mathbf{O}_p(\mathfrak{G}/\mathfrak{B})$. Since $\mathfrak{K} \trianglelefteq \mathfrak{G}$, $\mathfrak{S} \cap \mathfrak{K} \in S_p(\mathfrak{K})$. But $\mathfrak{K}/\mathfrak{B}$ is a p-group, so it follows that $\mathfrak{K} = \mathfrak{B}(\mathfrak{S} \cap \mathfrak{K}) \leq \mathbf{C}_\mathfrak{G}(\mathbf{Z}(\mathfrak{S}))$. Since $\mathfrak{K} \trianglelefteq \mathfrak{G}$, it follows that $\mathfrak{K}$ centralizes every conjugate of $\mathbf{Z}(\mathfrak{S})$ in $\mathfrak{G}$, so $\mathfrak{K} \leq \mathbf{C}_{\mathfrak{G}}(\mathfrak{Y}) = \mathfrak{B}$. Thus $\mathbf{O}_p(\mathfrak{G}/\mathfrak{B}) = 1$ and $\mathfrak{Y} \in \mathscr{R}_p(\mathfrak{G})$. Thus $\mathbf{Z}(\mathfrak{S}) \leq \mathfrak{Y} \leq \mathbf{R}_p(\mathfrak{G})$ and, by a), $\mathfrak{Y} \leq \mathbf{R}_p(\mathfrak{G}) \leq \mathbf{Z}(\mathbf{O}_p(\mathfrak{G}))$. **q.e.d.**

10.3 Corollary. *Suppose that* $\mathbf{O}_p(\mathfrak{G}) \geq \mathbf{C}_{\mathfrak{G}}(\mathbf{O}_p(\mathfrak{G}))$ *and* $\mathfrak{S} \in S_p(\mathfrak{G})$. *If* $\mathfrak{G} \neq \mathbf{C}_{\mathfrak{G}}(\mathbf{Z}(\mathfrak{S}))\mathbf{N}_{\mathfrak{G}}(\mathbf{J}(\mathfrak{S}))$, *then* $\mathbf{J}(\mathfrak{S}) \nleq \mathbf{C}_{\mathfrak{G}}(\mathbf{R}_p(\mathfrak{G}))$.

Proof. Let $\mathfrak{N} = \mathbf{C}_{\mathfrak{G}}(\mathbf{R}_p(\mathfrak{G}))$; thus $\mathfrak{N} \trianglelefteq \mathfrak{G}$ and, by 10.2c), $\mathfrak{N} \leq \mathbf{C}_{\mathfrak{G}}(\mathbf{Z}(\mathfrak{S}))$. Hence $\mathfrak{G} \neq \mathfrak{N}\mathbf{N}_{\mathfrak{G}}(\mathbf{J}(\mathfrak{S}))$ and, by 10.1, $\mathbf{J}(\mathfrak{S}) \nleq \mathfrak{N}$. **q.e.d.**

10.4 Lemma. *Suppose that* $\mathfrak{G}$ *is p-soluble and* $\mathbf{O}_{p'}(\mathfrak{G}) = 1$. *Let* $\mathfrak{S} \in S_p(\mathfrak{G})$. *Suppose that* $\mathbf{C}_{\mathfrak{G}}(\mathbf{R}_p(\mathfrak{G})) \leq \mathfrak{N} \trianglelefteq \mathfrak{G}$ *and* $\mathbf{J}(\mathfrak{S}) \nleq \mathfrak{N}$. *Then there exists* $\mathfrak{A} \in \mathscr{A}(\mathfrak{S})$ *(see 3.6) such that* $|\mathfrak{A} : \mathfrak{A} \cap \mathfrak{N}| = p$.

Proof. Write $\mathfrak{Z} = \mathbf{R}_p(\mathfrak{G})$, $\mathfrak{C} = \mathbf{C}_{\mathfrak{G}}(\mathfrak{Z})$, and let

$$\mathscr{A} = \{\mathfrak{A} | \mathfrak{A} \in \mathscr{A}(\mathfrak{S}), \mathfrak{A} \nleq \mathfrak{N}\}.$$

Since $\mathbf{J}(\mathfrak{S}) \nleq \mathfrak{N}$, $\mathscr{A}$ is non-empty. Let $\mathfrak{A}$ be an element of $\mathscr{A}$ for which $|\mathfrak{A} \cap \mathfrak{Z}|$ is maximal.

Let $\overline{\mathfrak{G}} = \mathfrak{G}/\mathfrak{C}$, $\overline{\mathfrak{A}} = \mathfrak{A}\mathfrak{C}/\mathfrak{C}$. Since $\mathfrak{A} \in \mathscr{A}$ and $\mathfrak{C} \leq \mathfrak{N}$, $\overline{\mathfrak{A}} \neq 1$. But by 10.2, $\mathbf{O}_p(\overline{\mathfrak{G}}) = 1$, so by IX, 1.4, $\mathbf{O}_{p'}(\overline{\mathfrak{G}}) \geq \mathbf{C}_{\overline{\mathfrak{G}}}(\mathbf{O}_{p'}(\overline{\mathfrak{G}}))$. Since $\overline{\mathfrak{A}}$ is a p-group, it follows that $\mathbf{O}_{p'}(\overline{\mathfrak{G}})$ is normalised but not centralized by $\overline{\mathfrak{A}}$. Let $\overline{\mathfrak{Q}}$ be a minimal subgroup of $\mathbf{O}_{p'}(\overline{\mathfrak{G}})$ which is normalised but not centralized by $\overline{\mathfrak{A}}$. By IX, 1.12, $\overline{\mathfrak{Q}}$ is a q-group for some prime $q \neq p$. Write $\overline{\mathfrak{Q}} = \mathfrak{Q}/\mathfrak{C}$, $\Phi(\overline{\mathfrak{Q}}) = \mathfrak{R}/\mathfrak{C}$. Thus $\mathfrak{A} \leq \mathbf{N}_{\mathfrak{G}}(\mathfrak{Q})$.

(1) $[\mathfrak{R}, \mathfrak{A}] \leq \mathfrak{C}$, $\mathfrak{Q} = [\mathfrak{Q}, \mathfrak{A}]\mathfrak{C}$.

This follows at once from III, 13.5.

Let $\mathfrak{A}_0 = \ker(\mathfrak{A}$ on $\mathfrak{Q}/\mathfrak{R})$. Clearly $\mathfrak{A} \cap \mathfrak{C} \leq \mathfrak{A}_0$.

(2) $\mathfrak{A}/\mathfrak{A}_0$ is cyclic.

For by III, 13.5, $\mathfrak{Q}/\mathfrak{R}$ is an irreducible $\mathfrak{A}$-module. Hence by II, 3.10, $\mathfrak{A}/\mathfrak{A}_0$ is cyclic.

(3) $[\mathfrak{Q}, \mathfrak{A}_0] \leq \mathfrak{C}$.

If $a \in \mathfrak{A}_0$ and α is the automorphism of $\mathfrak{Q}/\mathfrak{C}$ induced by a, α induces the identity automorphism on $\mathfrak{Q}/\mathfrak{R}$ and on $\overline{\mathfrak{Q}}/\Phi(\overline{\mathfrak{Q}})$. By III, 3.18, $\alpha = 1$, since the order of α is a power of p. Hence a centralizes $\mathfrak{Q}/\mathfrak{C}$.

By 10.2a), $\mathfrak{Z}$ is an Abelian p-subgroup of $\mathfrak{G}$. Let $\mathfrak{Z}_1 = \Omega_1(\mathfrak{Z})$. By 10.2b), $\mathbf{C}_{\mathfrak{G}}(\mathfrak{Z}_1) = \mathfrak{C}$. Since $\mathfrak{Q} > \mathfrak{C}$, it follows that if $\mathfrak{Z}_2 = [\mathfrak{Z}_1, \mathfrak{Q}]$, $\mathfrak{Z}_2 \neq 1$. But by III, 13.4, $\mathfrak{Z}_1 = \mathfrak{Z}_2 \times \mathbf{C}_{\mathfrak{Z}_1}(\mathfrak{Q})$, since $\mathfrak{Q}/\mathfrak{C}$ is a q-group of operators on $\mathfrak{Z}_1$. Thus $\mathbf{C}_{\mathfrak{Z}_2}(\mathfrak{Q}) = 1$. Let $\mathfrak{Y} = \mathbf{C}_{\mathfrak{Z}_2}(\mathfrak{A}_0)$. Since $\mathfrak{A} \leq \mathbf{N}_{\mathfrak{G}}(\mathfrak{Z}_2)$, $\mathfrak{A}_0\mathfrak{Z}_2$ is a p-group and $\mathfrak{Y} \neq 1$. Thus $\mathfrak{Q} \nleq \mathbf{C}_{\mathfrak{G}}(\mathfrak{Y})$. But $\mathfrak{Q} \leq \mathbf{N}_{\mathfrak{G}}(\mathfrak{A}_0\mathfrak{C})$ by (3) and $\mathfrak{Q} \leq \mathbf{N}_{\mathfrak{G}}(\mathfrak{Z}_2)$ by III, 1.6b). Since $\mathfrak{Y} = \mathbf{C}_{\mathfrak{Z}_2}(\mathfrak{A}_0\mathfrak{C})$, it follows that $\mathfrak{Q} \leq \mathbf{N}_{\mathfrak{G}}(\mathfrak{Y})$. Since $\mathfrak{Q} \not\leq \mathbf{C}_{\mathfrak{G}}(\mathfrak{Y})$, $[\mathfrak{Q}, \mathfrak{A}] \not\leq \mathbf{C}_{\mathfrak{G}}(\mathfrak{Y})$, by (1). But $\mathfrak{A}$ normalises $\mathfrak{Z}_1$, $\mathfrak{Q}$, $\mathfrak{Z}_2$, $\mathfrak{A}_0$ and $\mathfrak{Y}$; since $\mathfrak{Q} \leq \mathbf{N}_{\mathfrak{G}}(\mathfrak{Y})$, it follows that $\mathfrak{A} \not\leq \mathbf{C}_{\mathfrak{G}}(\mathfrak{Y})$. Let $\mathfrak{Y}_1 = \mathbf{C}_{\mathfrak{Y}}(\mathfrak{A})$, so $\mathfrak{Y}_1 < \mathfrak{Y}$. Let $\mathfrak{Y}_2/\mathfrak{Y}_1 = \mathbf{C}_{\mathfrak{Y}/\mathfrak{Y}_1}(\mathfrak{A})$; by III, 2.6, $\mathfrak{Y}_2 > \mathfrak{Y}_1$. Choose $y \in \mathfrak{Y}_2 - \mathfrak{Y}_1$. For any $a \in \mathfrak{A}$, $y^a = yy_1$ for some $y_1 \in \mathfrak{Y}_1$ and $y^{a^p} = yy_1^p = y$. Thus $\Phi(\mathfrak{A}) \leq \mathbf{C}_{\mathfrak{G}}(y)$. By definition of $\mathfrak{Y}$, $\mathfrak{A}_0 \leq \mathbf{C}_{\mathfrak{G}}(y)$, so if $\mathfrak{A}_1 = \mathfrak{A}_0\Phi(\mathfrak{A})$, $y \in \mathbf{C}_{\mathfrak{Y}}(\mathfrak{A}_1)$. But $y \notin \mathfrak{Y}_1$, so $y \notin \mathfrak{A}$; hence $\mathbf{C}_{\mathfrak{Y}}(\mathfrak{A}_1) \not\leq \mathfrak{A}$ and $\mathfrak{A}_1\mathbf{C}_{\mathfrak{Y}}(\mathfrak{A}_1) > \mathfrak{A}_1$. But by (2), $|\mathfrak{A} : \mathfrak{A}_1| = p$, so $|\mathfrak{A}_1\mathbf{C}_{\mathfrak{Y}}(\mathfrak{A}_1)| \geq |\mathfrak{A}|$. Since $\mathfrak{A}_1\mathbf{C}_{\mathfrak{Y}}(\mathfrak{A}_1)$ is Abelian, $\mathfrak{A}_1\mathbf{C}_{\mathfrak{Y}}(\mathfrak{A}_1) \in \mathscr{A}(\mathfrak{S})$. But

$$\mathfrak{A} \cap \mathfrak{Z} = \mathfrak{A} \cap \mathfrak{C} \cap \mathfrak{Z} \leq \mathfrak{A}_0 \cap \mathfrak{Z} \leq \mathfrak{A}_1 \cap \mathfrak{Z} \leq \mathfrak{A} \cap \mathfrak{Z},$$

so $\mathfrak{A}_1 \cap \mathfrak{Z} = \mathfrak{A} \cap \mathfrak{Z}$ and

$$(\mathfrak{A}_1\mathbf{C}_{\mathfrak{Y}}(\mathfrak{A}_1)) \cap \mathfrak{Z} = (\mathfrak{A}_1 \cap \mathfrak{Z})\mathbf{C}_{\mathfrak{Y}}(\mathfrak{A}_1) = (\mathfrak{A} \cap \mathfrak{Z})\mathbf{C}_{\mathfrak{Y}}(\mathfrak{A}_1) > \mathfrak{A} \cap \mathfrak{Z}.$$

Since $\mathfrak{A}$ is an element of $\mathscr{A}$ for which $|\mathfrak{A} \cap \mathfrak{Z}|$ is maximal, $\mathfrak{A}_1\mathbf{C}_{\mathfrak{Y}}(\mathfrak{A}_1) \notin \mathscr{A}$. Thus $\mathfrak{A}_1 \leq \mathfrak{N}$ and $\mathfrak{A}_1 \leq \mathfrak{A} \cap \mathfrak{N} < \mathfrak{A}$. Since $|\mathfrak{A} : \mathfrak{A}_1| = p$, $|\mathfrak{A} : \mathfrak{A} \cap \mathfrak{N}| = p$. **q.e.d.**

10.5 Theorem. *Suppose that $\mathfrak{G}$ is p-soluble and $\mathbf{O}_{p'}(\mathfrak{G}) = 1$. Put $\mathfrak{Z} = \mathbf{R}_p(\mathfrak{G})$. Suppose that $\mathfrak{S} \in S_p(\mathfrak{G})$ and $\mathbf{J}(\mathfrak{S}) \not\leq \mathbf{C}_{\mathfrak{G}}(\mathfrak{Z})$. Then $p \leq 3$ and there exists a p-element x of $\mathfrak{G}$ such that $|\mathfrak{Z} : \mathbf{C}_{\mathfrak{Z}}(x)| = p$.*

Proof. Put $\mathfrak{C} = \mathbf{C}_{\mathfrak{G}}(\mathfrak{Z})$. By 10.4 there exists $\mathfrak{A} \in \mathscr{A}(\mathfrak{S})$ such that $|\mathfrak{A} : \mathfrak{A}_0| = p$, where $\mathfrak{A}_0 = \mathfrak{A} \cap \mathfrak{C}$. Suppose that $\mathfrak{A} = \langle x, \mathfrak{A}_0 \rangle$. Since $x \notin \mathfrak{C}$, $\mathbf{C}_{\mathfrak{Z}}(x) < \mathfrak{Z}$. But since $\mathfrak{Z}$ centralizes $\mathfrak{A}_0$,

$$\mathbf{C}_{\mathfrak{Z}}(x) \leq \mathbf{C}_{\mathfrak{S}}(\mathfrak{A}) \cap \mathfrak{Z} = \mathfrak{A} \cap \mathfrak{Z} \leq \mathbf{C}_{\mathfrak{Z}}(x)$$

and $\mathbf{C}_{\mathfrak{Z}}(x) = \mathfrak{A} \cap \mathfrak{Z} = \mathfrak{A} \cap \mathfrak{C} \cap \mathfrak{Z} = \mathfrak{A}_0 \cap \mathfrak{Z}$. Thus

$$|\mathfrak{A}_0\mathfrak{Z} : \mathfrak{A}_0| = |\mathfrak{Z} : \mathbf{C}_{\mathfrak{Z}}(x)| \geq p.$$

But since $\mathfrak{A}_0\mathfrak{Z}$ is Abelian, $|\mathfrak{A}_0\mathfrak{Z} : \mathfrak{A}_0| \leq |\mathfrak{A} : \mathfrak{A}_0| = p$, so $|\mathfrak{Z} : \mathbf{C}_{\mathfrak{Z}}(x)| = p$. Hence x centralizes $\mathfrak{Z}/\mathbf{C}_{\mathfrak{Z}}(x)$ and $[\mathfrak{Z}, x] \leq \mathbf{C}_{\mathfrak{Z}}(x)$. Thus if $\mathfrak{Z}_1 = \Omega_1(\mathfrak{Z})$, $[\mathfrak{Z}_1, x] \leq \mathbf{C}_{\mathfrak{Z}_1}(x)$ and $[\mathfrak{Z}_1, x, x] = 1$. But by 10.2b), $\mathbf{C}_{\mathfrak{G}}(\mathfrak{Z}_1) = \mathfrak{C}$, so there is a faithful representation of $\mathfrak{G}/\mathfrak{C}$ on $\mathfrak{Z}_1$. Since $\mathbf{O}_p(\mathfrak{G}/\mathfrak{C}) = 1$, it follows from IX, 2.19 that the minimum polynomial of the linear transformation of $\mathfrak{Z}_1$ induced by x is $(t - 1)^r$, where $r \geq p - 1$. Since $r \leq 2$, $p \leq 3$. **q.e.d.**

In the remainder of this section, the groups $\mathfrak{G} = SL(2, p)$ for $p = 2$ or $p = 3$ will occur quite often, and we remind the reader of the following facts about them. $\mathfrak{G}$ is in both cases a $\{2, 3\}$-group and has a normal p-complement $\mathfrak{Q}$. In fact $\mathfrak{Q} = \mathfrak{G}'$. For $p = 2$, $\mathfrak{Q}$ is of order 3, and for $p = 3$, $\mathfrak{Q}$ is the quaternion group of order 8. In either case, if $\mathfrak{S} \in S_p(\mathfrak{G})$, $\mathfrak{S}$ operates fixed point freely on $\mathfrak{Q}/\mathfrak{Q}'$ and $\mathfrak{S}$ centralizes $\mathfrak{Q}'$. Thus $\mathfrak{G}$ has no proper subgroup $\mathfrak{H}$ which is generated by p-elements but is not a p-group. For $p = 2$, $\mathbf{Z}(\mathfrak{G}) = 1$ and for $p = 3$, $|\mathbf{Z}(\mathfrak{G})| = 2$.

In 10.5, the element x induces a transvection (II, 6.3) on $\Omega_1(\mathfrak{Z})$ (by 10.2b)). To apply this, we use the following, which is a very special case of results of MCLAUGHLIN [1, 2].

10.6 Theorem. *Let V be a vector space of dimension $n \geq 2$ over $GF(p)$. Let $\mathfrak{G}$ be a non-identity p-soluble, irreducible group of linear transformations of V generated by transvections. Then $p \leq 3$, $n = 2$ and $\mathfrak{G} = SL(2, p)$.*

Proof (Hering). (1) $\mathbf{O}_{p'}(\mathfrak{G}) \geq \mathbf{C}_{\mathfrak{G}}(\mathbf{O}_{p'}(\mathfrak{G}))$ and $p \leq 3$.

Since $\mathfrak{G}$ is irreducible, $\mathbf{O}_p(\mathfrak{G}) = 1$. By IX, 1.4, $\mathbf{O}_{p'}(\mathfrak{G}) \geq \mathbf{C}_{\mathfrak{G}}(\mathbf{O}_{p'}(\mathfrak{G}))$. And by IX, 2.19, the minimum polynomial of a transvection x in $\mathfrak{G}$ is $(t - 1)^r$, where $r \geq p - 1$. Since $(x - 1)^2 = 0$, $r \leq 2$ and $p \leq 3$.

(2) There exists a transvection x in $\mathfrak{G}$ such that x does not centralize $\mathbf{O}_{p'}(\mathfrak{G})$.

Since $\mathfrak{G}$ is generated by transvections, this follows from (1).

(3) There exists a conjugate y of x in $\mathbf{O}_{p'}(\mathfrak{G})\langle x \rangle$ such that $\mathfrak{K} = \langle x, y \rangle \cong SL(2, p)$. Also $\mathfrak{K}$ leaves fixed and is faithfully represented on a two-dimensional subspace U of V by the linear transformations of U of determinant 1.

If $\mathfrak{G}_1 = \mathbf{O}_{p'}(\mathfrak{G})\langle x \rangle$, then $\mathbf{O}_p(\mathfrak{G}_1) = 1$, by (1). Thus $x \notin \mathbf{O}_p(\mathfrak{G}_1)$. By IX, 7.8, there is a conjugate y of x in $\mathfrak{G}_1$ such that $\mathfrak{K} = \langle x, y \rangle$ is not a p-group. Let $\mathsf{U}_1 = \ker(x - 1) \cap \ker(y - 1)$. Then U_1 is $\mathfrak{K}$-invariant and $\mathfrak{K}$ is represented trivially on it. By I, 4.4, the group $\tilde{\mathfrak{K}}$ of linear transformations of V/U_1 induced by $\mathfrak{K}$ is generated by elements of order p but is not a p-group. Thus $|\mathsf{V}/\mathsf{U}_1| \geq p^2$. Since x, y are transvections, $\ker(x - 1)$ and $\ker(y - 1)$ are of codimension 1. Hence $|\mathsf{V}/\mathsf{U}_1| = p^2$. Thus $\tilde{\mathfrak{K}} \leq GL(2, p)$, and indeed, since transvections are of determinant 1, $\tilde{\mathfrak{K}} \leq SL(2, p)$. But since $p \leq 3$, the only subgroup of $SL(2, p)$ which is generated by elements of order p but is not a p-group is $SL(2, p)$ itself. Thus $\tilde{\mathfrak{K}} = SL(2, p)$. Hence neither x nor y induces the identity mapping on V/U_1, so $\operatorname{im}(x - 1) \nsubseteq \mathsf{U}_1$ and $\operatorname{im}(y - 1) \nsubseteq \mathsf{U}_1$. Thus if $\mathsf{U} = \operatorname{im}(x - 1) + \operatorname{im}(y - 1)$, $\mathsf{U} \nsubseteq \mathsf{U}_1$. But U is $\mathfrak{K}$-invariant, so $(\mathsf{U} + \mathsf{U}_1)/\mathsf{U}_1$ is a non-zero $\mathfrak{K}$-invariant subspace of V/U_1. Since

$\bar{\mathfrak{K}} = SL(2, p)$, $\mathfrak{K}$ is irreducible on V/U_1, so $\mathsf{U} + \mathsf{U}_1 = \mathsf{V}$. Since x, y are transvections, the dimension of U is 2 and $\mathsf{U} \oplus \mathsf{U}_1 = \mathsf{V}$. Since $\mathfrak{K}$ is trivial on U_1, $\mathfrak{K}$ is faithfully represented on U and on V/U_1. Thus $\mathfrak{K} \cong SL(2, p)$.

Let $\mathfrak{N} = \mathfrak{K} \cap \mathbf{O}_{p'}(\mathfrak{G})$.

(4) For $p = 2$, $\mathfrak{N}$ is of order 3; for $p = 3$, $\mathfrak{N}$ is the quaternion group of order 8.

By (3), $\mathfrak{K} \leq \mathbf{O}_{p'}(\mathfrak{G})\langle x\rangle$, so $\mathfrak{K} = \mathfrak{N}\langle x\rangle$. Thus $\mathfrak{N} = \mathbf{O}^p(\mathfrak{K})$. Since $\mathfrak{K} \cong SL(2, p)$, $\mathfrak{N}$ has the stated structure.

If $\mathsf{V} = \mathsf{U}$, there is nothing further to prove. Suppose, then, that $\mathsf{V} \supset \mathsf{U}$. Since $\mathfrak{G}$ is irreducible, U is not $\mathfrak{G}$-invariant. Thus there exists a transvection z in $\mathfrak{G}$ such that $\mathsf{U}z \neq \mathsf{U}$. Thus $\operatorname{im}(z - 1) \nsubseteq \mathsf{U}$. Let $\mathsf{W} = \mathsf{U} + \operatorname{im}(z - 1)$. Then W is of dimension 3 and is $\mathfrak{H}$-invariant, where $\mathfrak{H} = \langle x, y, z\rangle$. Let $\bar{\mathfrak{H}}$, $\bar{\mathfrak{N}}$ be the groups of linear transformations of W induced by $\mathfrak{H}$, $\mathfrak{N}$ respectively.

(5) $\bar{\mathfrak{H}}$ is a p-soluble subgroup of $SL(3, p)$, $\bar{\mathfrak{N}} \cong \mathfrak{N}$ and $\bar{\mathfrak{N}} \leq \mathbf{O}_{p'}(\bar{\mathfrak{H}}) < \bar{\mathfrak{H}}$.

Since $\bar{\mathfrak{H}}$ is generated by transvections, $\bar{\mathfrak{H}} \leq SL(3, p)$. By (3), $\mathfrak{K}$ is represented faithfully on U and hence on W. Thus $\bar{\mathfrak{N}} \cong \mathfrak{N}$ and $\bar{\mathfrak{H}}$ is not a p'-group. Since $\mathfrak{N} \leq \mathfrak{H} \cap \mathbf{O}_{p'}(\mathfrak{G}) \leq \mathbf{O}_{p'}(\mathfrak{H})$, $\bar{\mathfrak{N}} \leq \mathbf{O}_{p'}(\bar{\mathfrak{H}}) < \bar{\mathfrak{H}}$.

Now suppose that $p = 2$. By (4) and (5), 3 divides $|\mathbf{O}_{2'}(\bar{\mathfrak{H}})|$. Now $SL(3, 2) \cong PSL(2, 7)$ (II, 6.14), so we can read off the possibilities for $\bar{\mathfrak{H}}$ from II, 8.27. The only one is the dihedral group of order 6. But if $\bar{\mathfrak{H}}$ is dihedral of order 6, $\mathbf{O}_{2'}(\bar{\mathfrak{H}}) = \bar{\mathfrak{N}}$ leaves U fixed. Hence $\mathbf{O}_{2'}(\bar{\mathfrak{H}})$ leaves $\mathsf{U}z$ and $\mathsf{U} \cap \mathsf{U}z$ fixed. Since $\mathsf{U}z \neq \mathsf{U}$, $|\mathsf{U} \cap \mathsf{U}z| = 2$, so by I, 4.4, $\mathbf{O}_{2'}(\bar{\mathfrak{H}})$ is a 2-group. This is a contradiction.

In the case $p = 3$, (4) and (5) show that 8 divides $|\mathbf{O}_{3'}(\bar{\mathfrak{H}})|$. Since $|SL(3, 3)| = 2^4 \cdot 3^3 \cdot 13$, $|\mathbf{O}_{3'}(\bar{\mathfrak{H}})|$ is 2^3, 2^4, $2^3 \cdot 13$ or $2^4 \cdot 13$. In the latter two cases, the Sylow 13-subgroup of $\mathbf{O}_{3'}(\bar{\mathfrak{H}})$ is unique and is centralized by some non-identity 2-element. But it is easy to verify that the Sylow 13-subgroup of $SL(3, 3)$ is its own centralizer (cf. II, 7.3), so this is impossible. If $|\mathbf{O}_{3'}(\bar{\mathfrak{H}})| = 8$, we again have $\mathbf{O}_{3'}(\bar{\mathfrak{H}}) = \bar{\mathfrak{N}}$ and this leaves fixed U and $\mathsf{U} \cap \mathsf{U}z$. Hence if $a \in \mathbf{O}_{3'}(\bar{\mathfrak{H}})$, a^2 induces the identity mapping on W/U, $\mathsf{U}/(\mathsf{U} \cap \mathsf{U}z)$ and $\mathsf{U} \cap \mathsf{U}z$. Thus $a^2 = 1$. But $\mathbf{O}_{3'}(\bar{\mathfrak{H}})$ is a quaternion group of order 8, so this case is also impossible.

There remains the case $|\mathbf{O}_{3'}(\bar{\mathfrak{H}})| = 16$. Since $\mathbf{O}_{3'}(\bar{\mathfrak{H}})$ is non-Abelian, it is easy to see from III, 11.9 that $\mathbf{O}_{3'}(\bar{\mathfrak{H}})$ has a characteristic subgroup $\langle c\rangle$ of order 2 contained in $\mathbf{O}_{3'}(\bar{\mathfrak{H}})'$. Thus $\langle c\rangle \trianglelefteq \bar{\mathfrak{H}}$ and c is in the centre of $\bar{\mathfrak{H}}$. Also $c \in \mathbf{O}_{3'}(\bar{\mathfrak{H}})' \leq \Phi(\mathbf{O}_{3'}(\bar{\mathfrak{H}})) \leq \bar{\mathfrak{N}}$, since $\bar{\mathfrak{N}}$ is a maximal subgroup of $\mathbf{O}_{3'}(\bar{\mathfrak{H}})$. Thus $\langle c\rangle = \mathbf{Z}(\bar{\mathfrak{N}})$, so $uc = -u$ for all $u \in \mathsf{U}$. Hence $(uz)c = ucz = -uz$. Since $\mathsf{W} = \mathsf{U} + \mathsf{U}z$, it follows that $wc = -w$ for all $w \in \mathsf{W}$. Thus $\det c = -1$, a contradiction. **q.e.d.**

10.7 Theorem. *Suppose that* $\mathfrak{G}$ *is a p-soluble group and that* $\mathbf{O}_{p'}(\mathfrak{G}) = 1$. *Let* $\mathfrak{Z} = \mathbf{R}_p(\mathfrak{G})$, $\mathfrak{C} = \mathbf{C}_{\mathfrak{G}}(\mathfrak{Z})$. *Suppose that the set T of p-elements x of* $\mathfrak{G}$ *for which* $|\mathfrak{Z} : \mathbf{C}_{\mathfrak{Z}}(x)| = p$ *is non-empty and let* $\mathfrak{H} = \langle T, \mathfrak{C} \rangle$. *Thus* $\mathfrak{H}$ *is a characteristic subgroup of* $\mathfrak{G}$. *Let* $\mathfrak{V} = [\mathfrak{Z}, \mathfrak{H}]$.

a) $p \leq 3$.

b) *If* $\mathfrak{V}_1, \ldots, \mathfrak{V}_n$ *are all the minimal normal subgroups of* $\mathfrak{H}$ *contained in* $\mathfrak{V}$, *then* $\mathfrak{V} = \mathfrak{V}_1 \times \cdots \times \mathfrak{V}_n$ *and* $|\mathfrak{V}_i| = p^2$ $(i = 1, \ldots, n)$. *If* $\mathfrak{H}_i = \mathbf{C}_{\mathfrak{H}}(\mathfrak{Z}/\mathfrak{V}_i)$, *then*

$$\mathfrak{H}/\mathfrak{C} = (\mathfrak{H}_1/\mathfrak{C}) \times \cdots \times (\mathfrak{H}_n/\mathfrak{C}),$$

$\mathfrak{H}_i$ *centralizes* $\mathfrak{V}_j$ *for* $j \neq i$ *and* $\mathfrak{H}_i/\mathfrak{C}$ *is faithfully represented on* $\mathfrak{V}_i$ *by the set of all elements of* $SL(2, p)$.

c) $\mathfrak{Z} = \mathbf{C}_{\mathfrak{Z}}(\mathfrak{H}) \times \mathfrak{V}$, $\mathfrak{V} = [\mathfrak{V}, \mathfrak{H}]$ *and* $\mathfrak{V}_i = [\mathfrak{Z}, \mathfrak{H}_i]$ $(i = 1, \ldots, n)$.

Proof. It is clear that $\mathfrak{H}$ is a characteristic subgroup of $\mathfrak{G}$. Let

$$1 = \mathfrak{W}_0 < \mathfrak{W}_1 < \cdots < \mathfrak{W}_r = \mathfrak{Z}$$

be an $\mathfrak{H}$-composition series of $\mathfrak{Z}$, and let $\mathfrak{U}_i = \mathfrak{W}_i/\mathfrak{W}_{i-1}$ $(i = 1, \ldots, r)$. Thus $\mathfrak{U}_1, \ldots, \mathfrak{U}_r$ are irreducible $\mathsf{K}\mathfrak{H}$-modules, where $\mathsf{K} = GF(p)$.

(1) $\bigcap_{i=1}^r \mathbf{C}_{\mathfrak{H}}(\mathfrak{U}_i) = \mathfrak{C}$.

Let $\mathfrak{D} = \bigcap_{i=1}^r \mathbf{C}_{\mathfrak{H}}(\mathfrak{U}_i)$. $\mathfrak{H}/\mathfrak{C}$ is isomorphic to a group of automorphisms of $\mathfrak{Z}$ which leave each $\mathfrak{W}_i$ fixed, and $\mathfrak{D}/\mathfrak{C}$ is precisely the set of elements corresponding to the automorphisms which leave fixed each coset of $\mathfrak{W}_{i-1}$ in $\mathfrak{W}_i$. By 1.3, $\mathfrak{D}/\mathfrak{C}$ is a normal p-subgroup of $\mathfrak{H}/\mathfrak{C}$, so $\mathfrak{D}/\mathfrak{C} \leq \mathbf{O}_p(\mathfrak{H}/\mathfrak{C})$. Since $\mathfrak{C}$ and $\mathfrak{H}$ are characteristic subgroups of $\mathfrak{G}$, $\mathbf{O}_p(\mathfrak{H}/\mathfrak{C}) \trianglelefteq \mathfrak{G}/\mathfrak{C}$ and so $\mathbf{O}_p(\mathfrak{H}/\mathfrak{C}) \leq \mathbf{O}_p(\mathfrak{G}/\mathfrak{C}) = 1$. Hence $\mathfrak{D} \leq \mathfrak{C}$.

(2) If $\mathbf{C}_{\mathfrak{H}}(\mathfrak{U}_i) \neq \mathfrak{H}$, then $|\mathfrak{U}_i| = p^2$ and $\mathfrak{H}/\mathbf{C}_{\mathfrak{H}}(\mathfrak{U}_i) \cong SL(2, p)$. Also $p \leq 3$.

$\mathfrak{H}/\mathbf{C}_{\mathfrak{H}}(\mathfrak{U}_i)$ is isomorphic to an irreducible group $\mathfrak{H}_i^*$ of linear transformations of $\mathfrak{U}_i$. Since $\mathfrak{H} = \langle T, \mathfrak{C} \rangle$ and $\mathfrak{C}$ centralizes $\mathfrak{U}_i$, $\mathfrak{H}_i^*$ is generated by the set T^* of linear transformations induced by elements of T. But if $t \in T$, $|\mathfrak{Z} : \mathbf{C}_{\mathfrak{Z}}(t)| = p$, so $|\mathfrak{U}_i : \mathbf{C}_{\mathfrak{U}_i}(t)| \leq p$. Thus $\mathfrak{H}_i^*$ is generated by transvections. Since $\mathfrak{G}$ is p-soluble, it follows from 10.6 that either $\mathfrak{H}_i^* = 1$ or $p \leq 3$, $|\mathfrak{U}_i| = p^2$ and $\mathfrak{H}_i^* \cong SL(2, p)$. If $\mathfrak{H}_i^* = 1$ for all $i = 1, \ldots, r$, then by (1) $\mathfrak{H} = \mathfrak{C}$, which implies that T is empty, contrary to hypothesis. Thus $p \leq 3$.

(3) $\mathfrak{Z} = \mathbf{C}_{\mathfrak{Z}}(\mathfrak{H}) \times \mathfrak{V}$ and $\mathfrak{V} = [\mathfrak{V}, \mathfrak{H}]$.

For $p \leq 3$, $SL(2, p)$ is p-nilpotent. Hence by (2), $\mathfrak{H}/\mathbf{C}_{\mathfrak{H}}(\mathfrak{U}_i)$ is p-nilpotent. Hence by (1), $\mathfrak{H}/\mathfrak{C}$ is p-nilpotent. Let $\mathfrak{K}/\mathfrak{C} = \mathbf{O}_{p'}(\mathfrak{H}/\mathfrak{C})$; thus $\mathfrak{H}/\mathfrak{K}$ is a p-group. Let $\mathfrak{X} = [\mathfrak{Z}, \mathfrak{K}]$. By III, 13.4, $\mathfrak{Z} = \mathfrak{X} \times \mathbf{C}_{\mathfrak{Z}}(\mathfrak{K})$. Also, since $\mathbf{O}_p(\mathfrak{H}/\mathfrak{C}) = 1$, $\mathfrak{K}/\mathfrak{C} \geq \mathbf{C}_{\mathfrak{H}/\mathfrak{C}}(\mathfrak{K}/\mathfrak{C})$ by IX, 1.4. Now if $g \in T$, $g \notin \mathbf{C}_{\mathfrak{G}}(\mathfrak{Z})$ and so $g \notin \mathfrak{K}$, since $\mathfrak{K}/\mathfrak{C}$ is a p'-group. Hence $[\mathfrak{K}, g] \not\leq \mathfrak{C}$. Thus $[\mathfrak{K}, g]$ centralizes $\mathbf{C}_{\mathfrak{Z}}(\mathfrak{K})$ but not $\mathfrak{Z} = \mathfrak{X} \times \mathbf{C}_{\mathfrak{Z}}(\mathfrak{K})$. Thus $[\mathfrak{K}, g]$ does not centralize $\mathfrak{X}$. Since $\mathfrak{X} \trianglelefteq \mathfrak{G}$, it follows that g does not centralize $\mathfrak{X}$. Since also $\mathbf{C}_{\mathfrak{Z}}(\mathfrak{K}) \trianglelefteq \mathfrak{G}$, $\mathbf{C}_{\mathfrak{Z}}(g) = \mathbf{C}_{\mathfrak{X}}(g) \times (\mathbf{C}_{\mathfrak{Z}}(\mathfrak{K}) \cap \mathbf{C}_{\mathfrak{Z}}(g))$. Since $|\mathfrak{Z} : \mathbf{C}_{\mathfrak{Z}}(g)| = p$ and $|\mathfrak{X} : \mathbf{C}_{\mathfrak{X}}(g)| \geq p$, it follows that $\mathbf{C}_{\mathfrak{Z}}(\mathfrak{K}) \leq \mathbf{C}_{\mathfrak{Z}}(g)$. Thus $\mathbf{C}_{\mathfrak{Z}}(\mathfrak{K})$ is centralized by every element of T and by $\mathfrak{C}$. Since $\mathfrak{H} = \langle T, \mathfrak{C} \rangle$, $\mathbf{C}_{\mathfrak{Z}}(\mathfrak{K}) = \mathbf{C}_{\mathfrak{Z}}(\mathfrak{H})$. Thus

$$\begin{aligned} \mathfrak{V} &= [\mathfrak{Z}, \mathfrak{H}] = [\mathfrak{X} \times \mathbf{C}_{\mathfrak{Z}}(\mathfrak{H}), \mathfrak{H}] = [\mathfrak{X}, \mathfrak{H}] \leq \mathfrak{X} \\ &= [\mathfrak{Z}, \mathfrak{K}] \leq [\mathfrak{Z}, \mathfrak{H}] = \mathfrak{V}. \end{aligned}$$

Hence $\mathfrak{V} = \mathfrak{X}$, $\mathfrak{Z} = \mathfrak{X} \times \mathbf{C}_{\mathfrak{Z}}(\mathfrak{K}) = \mathfrak{V} \times \mathbf{C}_{\mathfrak{Z}}(\mathfrak{H})$, and $\mathfrak{V} = [\mathfrak{X}, \mathfrak{H}] = [\mathfrak{V}, \mathfrak{H}]$.

Let $\mathfrak{V}_1, \ldots, \mathfrak{V}_n$ be the minimal normal subgroups of $\mathfrak{H}$ contained in $\mathfrak{V}$.

(4) $|\mathfrak{V}_i| = p^2$ and $\mathfrak{H}/\mathbf{C}_{\mathfrak{H}}(\mathfrak{V}_i) \cong SL(2, p)$ $(i = 1, \ldots, n)$.

Each $\mathfrak{V}_i$ is a term of an $\mathfrak{H}$-composition series of $\mathfrak{Z}$, so by (2), either (4) holds, or $\mathbf{C}_{\mathfrak{H}}(\mathfrak{V}_i) = \mathfrak{H}$. But by (3), $\mathbf{C}_{\mathfrak{V}}(\mathfrak{H}) = 1$, so $\mathbf{C}_{\mathfrak{V}_i}(\mathfrak{H}) = 1$ and $\mathbf{C}_{\mathfrak{H}}(\mathfrak{V}_i) \neq \mathfrak{H}$.

By (4), there exist elements t_i, t_i' in T such that $t_i \notin \mathbf{C}_{\mathfrak{H}}(\mathfrak{V}_i)$ and $t_i' \notin \langle \mathbf{C}_{\mathfrak{H}}(\mathfrak{V}_i), t_i \rangle$. Let $\mathfrak{H}_i = \langle t_i, t_i', \mathfrak{C} \rangle$.

(5) $\mathfrak{H} = \mathfrak{H}_i \mathbf{C}_{\mathfrak{H}}(\mathfrak{V}_i)$ $(i = 1, \ldots, n)$.

$SL(2, p)$ is generated by any two of its Sylow p-subgroups, as is easily verified for $p < 3$. Thus $\langle t_i \mathbf{C}_{\mathfrak{H}}(\mathfrak{V}_i), t_i' \mathbf{C}_{\mathfrak{H}}(\mathfrak{V}_i) \rangle = \mathfrak{H}/\mathbf{C}_{\mathfrak{H}}(\mathfrak{V}_i)$ by (4). Hence $\mathfrak{H} = \langle t_i, t_i', \mathbf{C}_{\mathfrak{H}}(\mathfrak{V}_i) \rangle = \mathfrak{H}_i \mathbf{C}_{\mathfrak{H}}(\mathfrak{V}_i)$.

(6) $\mathfrak{Z} = \mathfrak{V}_i \times \mathbf{C}_{\mathfrak{Z}}(\mathfrak{H}_i)$, $\mathbf{C}_{\mathfrak{H}_i}(\mathfrak{V}_i) = \mathfrak{C}$ and $[\mathfrak{Z}, \mathfrak{H}_i] = \mathfrak{V}_i$.

For $\mathbf{C}_{\mathfrak{Z}}(\mathfrak{H}_i) = \mathbf{C}_{\mathfrak{Z}}(t_i) \cap \mathbf{C}_{\mathfrak{Z}}(t_i')$ and $|\mathfrak{Z} : \mathbf{C}_{\mathfrak{Z}}(t_i)| = |\mathfrak{Z} : \mathbf{C}_{\mathfrak{Z}}(t_i')| = p$. Thus $|\mathfrak{Z} : \mathbf{C}_{\mathfrak{Z}}(\mathfrak{H}_i)| \leq p^2$. By (4) and (5), $\mathbf{C}_{\mathfrak{V}_i}(\mathfrak{H}_i) = 1$, so $|\mathfrak{V}_i \mathbf{C}_{\mathfrak{Z}}(\mathfrak{H}_i)| = p^2 |\mathbf{C}_{\mathfrak{Z}}(\mathfrak{H}_i)|$. Thus $\mathfrak{Z} = \mathfrak{V}_i \mathbf{C}_{\mathfrak{Z}}(\mathfrak{H}_i) = \mathfrak{V}_i \times \mathbf{C}_{\mathfrak{Z}}(\mathfrak{H}_i)$. If $a \in \mathbf{C}_{\mathfrak{H}_i}(\mathfrak{V}_i)$, then a centralizes $\mathfrak{V}_i$ and $\mathbf{C}_{\mathfrak{Z}}(\mathfrak{H}_i)$, whence $a \in \mathbf{C}_{\mathfrak{G}}(\mathfrak{Z}) = \mathfrak{C}$. Thus $\mathbf{C}_{\mathfrak{H}_i}(\mathfrak{V}_i) = \mathfrak{C}$. Finally, $[\mathfrak{Z}, \mathfrak{H}_i] = [\mathfrak{V}_i, \mathfrak{H}_i] = [\mathfrak{V}_i, \mathfrak{H}] = \mathfrak{V}_i$.

(7) $\mathfrak{H}_i = \mathbf{C}_{\mathfrak{H}}(\mathfrak{Z}/\mathfrak{V}_i)$ $(i = 1, \ldots, n)$. Thus $\mathfrak{H}_i \trianglelefteq \mathfrak{H}$.

Since $\mathfrak{H} = \mathfrak{H}_i \mathbf{C}_{\mathfrak{H}}(\mathfrak{V}_i)$ and by (6), $\mathfrak{H}_i \leq \mathbf{C}_{\mathfrak{H}}(\mathfrak{Z}/\mathfrak{V}_i)$,

$$\mathbf{C}_{\mathfrak{H}}(\mathfrak{Z}/\mathfrak{V}_i) = \mathfrak{H}_i \mathfrak{B},$$

where $\mathfrak{B} = \mathbf{C}_{\mathfrak{H}}(\mathfrak{Z}/\mathfrak{V}_i) \cap \mathbf{C}_{\mathfrak{H}}(\mathfrak{V}_i)$. Now $\mathfrak{H}/\mathfrak{C}$ is isomorphic to a group of automorphisms of $\mathfrak{Z}$ which leave $\mathfrak{V}_i$ fixed, and $\mathfrak{B}/\mathfrak{C}$ is precisely the set of elements corresponding to the automorphisms which leave fixed each element of $\mathfrak{V}_i$ and each coset of $\mathfrak{V}_i$ in $\mathfrak{Z}$. By 1.3, $\mathfrak{B}/\mathfrak{C}$ is a normal

p-subgroup of $\mathfrak{H}/\mathfrak{C}$. Since $\mathbf{O}_p(\mathfrak{H}/\mathfrak{C}) = 1$, $\mathfrak{B} = \mathfrak{C}$. Thus $\mathbf{C}_{\mathfrak{H}}(\mathfrak{Z}/\mathfrak{V}_i) = \mathfrak{H}_i\mathfrak{C} = \mathfrak{H}_i$.

By (5) and (6), $\mathfrak{H}/\mathbf{C}_{\mathfrak{H}}(\mathfrak{V}_i) \cong \mathfrak{H}_i/\mathbf{C}_{\mathfrak{H}_i}(\mathfrak{V}_i) = \mathfrak{H}_i/\mathfrak{C}$, so by (4), $\mathfrak{H}_i/\mathfrak{C}$ is faithfully represented on $\mathfrak{V}_i$ by the whole of $SL(2, p)$. If $i \neq j$, then by (6),

$$[\mathfrak{V}_j, \mathfrak{H}_i] \leq \mathfrak{V}_j \cap [\mathfrak{Z}, \mathfrak{H}_i] = \mathfrak{V}_j \cap \mathfrak{V}_i = 1,$$

so $\mathfrak{H}_i$ centralizes $\mathfrak{V}_j$. Hence $[\mathfrak{V}_1 \cdots \mathfrak{V}_{i-1}\mathfrak{V}_{i+1} \cdots \mathfrak{V}_n, \mathfrak{H}_i] = 1$. By (6), $\mathbf{C}_{\mathfrak{V}_i}(\mathfrak{H}_i) = 1$, so $\mathfrak{V}_i \cap (\mathfrak{V}_1 \cdots \mathfrak{V}_{i-1}\mathfrak{V}_{i+1} \cdots \mathfrak{V}_n) = 1$. Hence if $\mathfrak{V}^* = \mathfrak{V}_1 \cdots \mathfrak{V}_n$, then $\mathfrak{V}^* = \mathfrak{V}_1 \times \cdots \times \mathfrak{V}_n$ and $|\mathfrak{V}^*| = p^{2n}$.

By (6), $\mathfrak{V} = \mathfrak{V}_i \times \mathbf{C}_{\mathfrak{V}}(\mathfrak{H}_i)$ and $|\mathfrak{V} : \mathbf{C}_{\mathfrak{V}}(\mathfrak{H}_i)| = p^2$. Thus if $\mathfrak{H}^* = \mathfrak{H}_1 \cdots \mathfrak{H}_n$, $|\mathfrak{V} : \mathbf{C}_{\mathfrak{V}}(\mathfrak{H}^*)| = |\mathfrak{V} : \bigcap_{i=1}^n \mathbf{C}_{\mathfrak{V}}(\mathfrak{H}_i)| \leq p^{2n}$. But

$$\begin{aligned}\mathbf{C}_{\mathfrak{V}^*}(\mathfrak{H}^*) &= \bigcap_{i=1}^{n} \mathbf{C}_{\mathfrak{V}^*}(\mathfrak{H}_i)\\ &= \bigcap_{i=1}^{n} (\mathfrak{V}_1 \times \cdots \times \mathfrak{V}_{i-1} \times \mathfrak{V}_{i+1} \times \cdots \times \mathfrak{V}_n) = 1.\end{aligned}$$

Thus $|\mathfrak{V}^*\mathbf{C}_{\mathfrak{V}}(\mathfrak{H}^*)| \geq |\mathfrak{V}|$ and $\mathfrak{V} = \mathfrak{V}^* \times \mathbf{C}_{\mathfrak{V}}(\mathfrak{H}^*)$. Since $\mathbf{C}_{\mathfrak{V}}(\mathfrak{H}^*) \cap \mathfrak{V}_i = 1$ and $\mathbf{C}_{\mathfrak{V}}(\mathfrak{H}^*) \trianglelefteq \mathfrak{H}$, it follows from the definition of the $\mathfrak{V}_i$ that $\mathbf{C}_{\mathfrak{V}}(\mathfrak{H}^*) = 1$. Thus $\mathfrak{V} = \mathfrak{V}^* = \mathfrak{V}_1 \times \cdots \times \mathfrak{V}_n$.

By (6), $\mathbf{C}_{\mathfrak{H}_i}(\mathfrak{V}_i) = \mathfrak{C}$ $(i = 1, \ldots, n)$. Since $[\mathfrak{V}_j, \mathfrak{H}_i] = 1$ $(j \neq i)$, $\mathfrak{H}_1 \cdots \mathfrak{H}_{i-1}\mathfrak{H}_{i+1} \cdots \mathfrak{H}_n$ centralizes $\mathfrak{V}_i$. Thus $(\mathfrak{H}_1 \cdots \mathfrak{H}_{i-1}\mathfrak{H}_{i+1} \cdots \mathfrak{H}_n) \cap \mathfrak{H}_i = \mathfrak{C}$, and $\mathfrak{H}^*/\mathfrak{C}$ is the direct product of $\mathfrak{H}_1/\mathfrak{C}, \ldots, \mathfrak{H}_n/\mathfrak{C}$. Hence $|\mathfrak{H}^*/\mathfrak{C}| = |SL(2, p)|^n$. But since $\mathfrak{V} = \mathfrak{V}_1 \cdots \mathfrak{V}_n$,

$$\bigcap_{i=1}^{n} \mathbf{C}_{\mathfrak{H}}(\mathfrak{V}_i) = \mathbf{C}_{\mathfrak{H}}(\mathfrak{V}) = \mathbf{C}_{\mathfrak{H}}(\mathfrak{Z}) = \mathfrak{C},$$

by (3). Thus, by (4),

$$|\mathfrak{H}/\mathfrak{C}| = \left|\mathfrak{H}\Big/\left(\bigcap_{i=1}^{n} \mathbf{C}_{\mathfrak{H}}(\mathfrak{V}_i)\right)\right| \leq \prod_{i=1}^{n} |\mathfrak{H}/\mathbf{C}_{\mathfrak{H}}(\mathfrak{V}_i)| = |SL(2, p)|^n.$$

Thus $\mathfrak{H} = \mathfrak{H}^*$, and

$$\mathfrak{H}/\mathfrak{C} = (\mathfrak{H}_1/\mathfrak{C}) \times \cdots \times (\mathfrak{H}_n/\mathfrak{C}). \qquad \textbf{q.e.d.}$$

10.8 Theorem. *Suppose that $\mathfrak{G}$ is a p-soluble group and $\mathbf{O}_{p'}(\mathfrak{G}) = 1$. Write $\mathfrak{Z} = \mathbf{R}_p(\mathfrak{G})$, $\mathfrak{C} = \mathbf{C}_{\mathfrak{G}}(\mathfrak{Z})$. Suppose $\mathfrak{S} \in S_p(\mathfrak{G})$.*

a) *If* $p \geq 5$, *or if no section of* $\mathfrak{G}$ *is isomorphic to* $SL(2, p)$, $\mathbf{J}(\mathfrak{S}) \leq \mathfrak{C}$ *and*

$$\mathfrak{G} = \mathfrak{C}\mathbf{N}_{\mathfrak{G}}(\mathbf{J}(\mathfrak{S})) = \mathbf{C}_{\mathfrak{G}}(\mathbf{Z}(\mathfrak{S}))\mathbf{N}_{\mathfrak{G}}(\mathbf{J}(\mathfrak{S})).$$

b) $\mathbf{J}(\mathfrak{S}) \leq \mathfrak{H}$, *where* $\mathfrak{H}$ *is the group generated by* $\mathfrak{C}$ *and all p-elements* x *for which* $|\mathfrak{Z} : \mathbf{C}_{\mathfrak{Z}}(x)| = p$. *Further*

$$\mathfrak{G} = \mathfrak{H}\mathbf{N}_{\mathfrak{G}}(\mathbf{J}(\mathfrak{S})).$$

Proof. a) By 10.7, $\mathfrak{G}$ has no p-elements x for which $|\mathfrak{Z} : \mathbf{C}_{\mathfrak{Z}}(x)| = p$. Hence by 10.5, $\mathbf{J}(\mathfrak{S}) \leq \mathfrak{C}$. By IX, 1.4, $\mathbf{O}_p(\mathfrak{G}) \geq \mathbf{C}_{\mathfrak{G}}(\mathbf{O}_p(\mathfrak{G}))$, so by 10.3, $\mathfrak{G} = \mathbf{N}_{\mathfrak{G}}(\mathbf{J}(\mathfrak{S}))\mathbf{C}_{\mathfrak{G}}(\mathbf{Z}(\mathfrak{S}))$. By 10.1, $\mathfrak{G} = \mathfrak{C}\mathbf{N}_{\mathfrak{G}}(\mathbf{J}(\mathfrak{S}))$.

b) Suppose that $\mathbf{J}(\mathfrak{S}) \not\leq \mathfrak{H}$. Then certainly $\mathbf{J}(\mathfrak{S}) \not\leq \mathfrak{C}$, so by 10.5 the set T of p-elements x of $\mathfrak{G}$ for which $|\mathfrak{Z} : \mathbf{C}_{\mathfrak{Z}}(x)| = p$ is non-empty. Let $\mathfrak{B}_1, \ldots, \mathfrak{B}_n$ be the minimal normal subgroups of $\mathfrak{H}$ contained in $\mathfrak{B} = [\mathfrak{Z}, \mathfrak{H}]$. By 10.7, $|\mathfrak{B}_i| = p^2$ and

$$\mathfrak{B} = \mathfrak{B}_1 \times \cdots \times \mathfrak{B}_n.$$

By 10.4, there exists $\mathfrak{A} \in \mathscr{A}(\mathfrak{S})$ such that $|\mathfrak{A} : \mathfrak{A}_1| = p$, where $\mathfrak{A}_1 = \mathfrak{A} \cap \mathfrak{H}$. Write $\mathfrak{A} = \langle \mathfrak{A}_1, a \rangle$.

(1) $\mathfrak{A} \leq \mathbf{N}_{\mathfrak{G}}(\mathfrak{B}_i)$ $(i = 1, \ldots, n)$.

If $x \in \mathfrak{A}$, $\mathfrak{B}_i^x$ is a minimal normal subgroup of $\mathfrak{H}$, and $\mathfrak{B}_i^x = \mathfrak{B}_j$ for some j. Thus there is a permutation representation ρ of $\mathfrak{A}$ on $\{\mathfrak{B}_1, \ldots, \mathfrak{B}_n\}$ such that $\mathfrak{B}_i\rho(x) = \mathfrak{B}_i^x$. Since $\mathfrak{A}_1 \leq \mathfrak{H}$, $\mathfrak{A}_1 \leq \mathbf{N}_{\mathfrak{G}}(\mathfrak{B}_i)$ $(i = 1, \ldots, n)$, so $\rho(x) = 1$ for all $x \in \mathfrak{A}_1$. Hence $\rho(a)^p = \rho(a^p) = 1$. Suppose that $\rho(a) \neq 1$. Then $\rho(a)$ has a cycle $(\mathfrak{B}_{i_1}, \mathfrak{B}_{i_2}, \ldots, \mathfrak{B}_{i_p})$ of length p; let

$$\mathfrak{B}^* = \mathfrak{B}_{i_1} \times \cdots \times \mathfrak{B}_{i_p}.$$

Thus $\mathfrak{B}^{*a} = \mathfrak{B}_{i_1}\rho(a) \times \cdots \times \mathfrak{B}_{i_p}\rho(a) = \mathfrak{B}^*$ and $\mathfrak{A} \leq \mathbf{N}_{\mathfrak{G}}(\mathfrak{B}^*)$. Also, if $\mathfrak{B}_{i_1} = \langle u, v \rangle$, then

$$\mathfrak{B}^* = \langle u \rangle \times \langle u^a \rangle \times \cdots \times \langle u^{a^{p-1}} \rangle \times \langle v \rangle \times \langle v^a \rangle \times \cdots \times \langle v^{a^{p-1}} \rangle,$$

and it is easy to see that

$$\mathbf{C}_{\mathfrak{B}^*}(a) = \langle uu^a \cdots u^{a^{p-1}} \rangle \times \langle vv^a \cdots v^{a^{p-1}} \rangle.$$

Let $\mathfrak{A}_2 = \mathbf{C}_{\mathfrak{A}_1}(\mathfrak{B}^*)$. We show that

(1a) $$|\mathfrak{A}_1 : \mathfrak{A}_2||\mathfrak{A} \cap \mathfrak{B}^*| = p^2.$$

Indeed, if $\mathfrak{A}_1 = \mathfrak{A}_2$, then $\mathfrak{A}_1$ centralizes $\mathfrak{B}^*$ and

$$\mathfrak{A} \cap \mathfrak{B}^* = \mathbf{C}_{\mathfrak{B}^*}(\mathfrak{A}) = \mathbf{C}_{\mathfrak{B}^*}(a),$$

so from above, $|\mathfrak{A} \cap \mathfrak{B}^*| = p^2$ and (1a) holds. Now suppose that $\mathfrak{A}_1 > \mathfrak{A}_2$. Then $\mathfrak{A}_1$ does not centralize $\mathfrak{B}_{i_j}$ for some j. If $\tilde{\mathfrak{A}} = \mathbf{C}_{\mathfrak{A}_1}(\mathfrak{B}_{i_j})$, then $|\mathfrak{A}_1 : \tilde{\mathfrak{A}}| = p$, since $\mathfrak{A}_1/\tilde{\mathfrak{A}}$ is isomorphic to a non-identity p-subgroup of $\mathbf{Aut}\,\mathfrak{B}_{i_j}$ and $|\mathbf{Aut}\,\mathfrak{B}_{i_j}| = (p^2 - 1)(p^2 - p)$. Also

$$\tilde{\mathfrak{A}} = \tilde{\mathfrak{A}}^{a^k} = \mathbf{C}_{\mathfrak{A}_1}(\mathfrak{B}_{i_j}\rho(a^k))$$

for all k, so $\tilde{\mathfrak{A}} = \mathbf{C}_{\mathfrak{A}_1}(\mathfrak{B}^*) = \mathfrak{A}_2$ and $|\mathfrak{A}_1 : \mathfrak{A}_2| = p$. Also $\mathfrak{A}_2 = \mathbf{C}_{\mathfrak{A}_1}(\mathfrak{B}_{i_1})$. Thus if $\mathfrak{A}_1 = \langle \mathfrak{A}_2, b \rangle$, b does not centralize $\mathfrak{B}_{i_1}$. It follows from above that b does not centralize $\mathbf{C}_{\mathfrak{B}^*}(a)$, for u^{a^j} and $(u^{a^j})^b$ both lie in $\mathfrak{B}_{i_{j+1}}$. Hence $|\mathbf{C}_{\mathfrak{B}^*}(a, b)| = p$. But

$$\mathbf{C}_{\mathfrak{B}^*}(a, b) = \mathbf{C}_{\mathfrak{B}^*}(\mathfrak{A}) = \mathfrak{A} \cap \mathfrak{B}^*,$$

so again (1a) holds.

Since $\mathfrak{A}_2\mathfrak{B}^*$ is Abelian and $\mathfrak{A} \in \mathscr{A}(\mathfrak{S})$, $|\mathfrak{A}_2\mathfrak{B}^*| \leq |\mathfrak{A}| = p|\mathfrak{A}_1|$, so

$$|\mathfrak{B}^* : \mathfrak{A}_2 \cap \mathfrak{B}^*| = |\mathfrak{A}_2\mathfrak{B}^* : \mathfrak{A}_2| \leq p|\mathfrak{A}_1 : \mathfrak{A}_2|.$$

But $\mathfrak{A}_2 \cap \mathfrak{B}^* = \mathfrak{A} \cap \mathfrak{B}^*$, since $\mathfrak{A} \cap \mathfrak{B}^* \leq \mathfrak{A} \cap \mathfrak{H} \cap \mathbf{C}_{\mathfrak{G}}(\mathfrak{B}^*) = \mathbf{C}_{\mathfrak{A}_1}(\mathfrak{B}^*) = \mathfrak{A}_2$. Hence

$$|\mathfrak{B}^*| \leq p|\mathfrak{A}_1 : \mathfrak{A}_2||\mathfrak{A} \cap \mathfrak{B}^*| = p^3,$$

by (1a). But $|\mathfrak{B}^*| = p^{2p}$, so this gives a contradiction. Thus $\rho(a) = 1$ and $a \in \mathbf{N}_{\mathfrak{G}}(\mathfrak{B}_i)$ $(i = 1, \ldots, n)$. Therefore $\mathfrak{A} \leq \mathbf{N}_{\mathfrak{G}}(\mathfrak{B}_i)$.

Let $\mathfrak{H}^* = \mathfrak{H}\mathbf{C}_{\mathfrak{G}}(\mathfrak{B})$.

(2) $\mathfrak{A} \leq \mathfrak{H}^*$.

By 10.7, $\mathfrak{H} = \mathfrak{H}_1 \cdots \mathfrak{H}_n$, where $\mathfrak{H}_i$ centralizes $\mathfrak{B}_j$ for $j \neq i$ and is represented on $\mathfrak{B}_i$ by the whole of $SL(2, p)$. Now if $x \in \mathfrak{A}$, then $x \in \mathbf{N}_{\mathfrak{G}}(\mathfrak{B}_i)$ by (1). Let ξ_i be the automorphism $v \to v^x$ of $\mathfrak{B}_i$. Since the order of x is a power of p, $\det \xi_i = 1$, so there exists $g_i \in \mathfrak{H}_i$ such that $v^{g_i} = v\xi_i$ for all $v \in \mathfrak{B}_i$. Hence if $g = g_1 \cdots g_n$, $g \in \mathfrak{H}$ and

$$v^g = v^{g_1 \cdots g_n} = v^{g_i \cdots g_n} = (v\xi_i)^{g_{i+1} \cdots g_n} = v\xi_i = v^x$$

for all $v \in \mathfrak{B}_i$. Thus $xg^{-1} \in \mathbf{C}_{\mathfrak{G}}(\mathfrak{B})$ and $x \in \mathbf{C}_{\mathfrak{G}}(\mathfrak{B})\mathfrak{H} = \mathfrak{H}\mathbf{C}_{\mathfrak{G}}(\mathfrak{B}) = \mathfrak{H}^*$. Hence $\mathfrak{A} \leq \mathfrak{H}^*$.

To obtain a contradiction, let $\mathfrak{U} = \mathbf{C}_{\mathfrak{Z}}(\mathfrak{H})$. By 10.7, $\mathfrak{Z} = \mathfrak{U} \times \mathfrak{V}$. Since $\mathfrak{H}^* = \mathfrak{H}\mathbf{C}_{\mathfrak{G}}(\mathfrak{V})$ and $\mathbf{C}_{\mathfrak{H}^*}(\mathfrak{U}) \geq \mathfrak{H}$,

$$\mathbf{C}_{\mathfrak{H}^*}(\mathfrak{U}) = \mathfrak{H}(\mathbf{C}_{\mathfrak{G}}(\mathfrak{V}) \cap \mathbf{C}_{\mathfrak{H}^*}(\mathfrak{U})) = \mathfrak{H}\mathbf{C}_{\mathfrak{H}^*}(\mathfrak{Z}) = \mathfrak{H}\mathfrak{C} = \mathfrak{H}.$$

But $\mathfrak{A} \not\leq \mathfrak{H}$ and, by (2), $\mathfrak{A} \leq \mathfrak{H}^*$; thus $\mathfrak{A} \not\leq \mathbf{C}_{\mathfrak{G}}(\mathfrak{U})$. Since $\mathfrak{A}_1 \leq \mathfrak{H} \leq \mathbf{C}_{\mathfrak{G}}(\mathfrak{U})$, it follows that $a \notin \mathbf{C}_{\mathfrak{G}}(\mathfrak{U})$.

Now $\mathfrak{H}\mathfrak{A}/\mathfrak{H}$ is contained in a Sylow p-subgroup of $\mathfrak{H}^*/\mathfrak{H}$, so $\mathfrak{H}\mathfrak{A} \leq \mathfrak{H}\mathfrak{P}$ for some $\mathfrak{P} \in S_p(\mathbf{C}_{\mathfrak{G}}(\mathfrak{V}))$. Thus $a = bc$ for some $b \in \mathfrak{H}$, $c \in \mathfrak{P}$. Since $b \in \mathfrak{H} \leq \mathbf{C}_{\mathfrak{G}}(\mathfrak{U})$, $\mathbf{C}_{\mathfrak{U}}(a) = \mathbf{C}_{\mathfrak{U}}(c) = \mathfrak{W}$, say, so c does not centralize $\mathfrak{U}$. But $c \in \mathbf{C}_{\mathfrak{G}}(\mathfrak{V})$, so $\mathfrak{Z} > \mathbf{C}_{\mathfrak{Z}}(c) \geq \mathfrak{V}\mathfrak{W}$. But since $a \notin \mathfrak{H}$, we have $c \notin \mathfrak{H}$ and $c \notin T$. Thus $|\mathfrak{Z} : \mathfrak{V}\mathfrak{W}| \geq p^2$. Hence $|\mathfrak{U} : \mathfrak{W}| \geq p^2$, since $\mathfrak{Z} = \mathfrak{V}\mathfrak{U}$. Now $\mathfrak{U} \leq \mathbf{C}_{\mathfrak{G}}(\mathfrak{A}_1)$. Thus

$$\mathfrak{W} = \mathfrak{U} \cap \mathbf{C}_{\mathfrak{S}}(a) \leq \mathbf{C}_{\mathfrak{G}}(\mathfrak{A}_1) \cap \mathbf{C}_{\mathfrak{S}}(a) \leq \mathbf{C}_{\mathfrak{S}}(\mathfrak{A}) = \mathfrak{A},$$

so $\mathfrak{W} = \mathfrak{U} \cap \mathfrak{A} = \mathfrak{U} \cap \mathfrak{H} \cap \mathfrak{A} = \mathfrak{U} \cap \mathfrak{A}_1$. Hence

$$|\mathfrak{U}\mathfrak{A}_1 : \mathfrak{A}_1| = |\mathfrak{U} : \mathfrak{U} \cap \mathfrak{A}_1| = |\mathfrak{U} : \mathfrak{W}| \geq p^2$$

and $|\mathfrak{U}\mathfrak{A}_1| \geq p^2|\mathfrak{A}_1| > |\mathfrak{A}|$. But since $\mathfrak{U} \leq \mathbf{C}_{\mathfrak{G}}(\mathfrak{A}_1)$, $\mathfrak{U}\mathfrak{A}_1$ is an Abelian p-group. This contradicts $\mathfrak{A} \in \mathscr{A}(\mathfrak{S})$.

Hence $\mathbf{J}(\mathfrak{S}) \leq \mathfrak{H}$. By 10.1, $\mathfrak{G} = \mathfrak{H}\mathbf{N}_{\mathfrak{G}}(\mathbf{J}(\mathfrak{S}))$. **q.e.d.**

As an application, we mention the following analogue of IV, 6.2. (We remind the reader that **J** has a meaning here different from that in IV, 6.2).

10.9 Theorem. *Suppose that p is odd and* $\mathfrak{S} \in S_p(\mathfrak{G})$. *If* $\mathbf{N}_{\mathfrak{G}}(\mathbf{J}(\mathfrak{S}))$ *and* $\mathbf{C}_{\mathfrak{G}}(\mathbf{Z}(\mathfrak{S}))$ *are p-nilpotent,* $\mathfrak{G}$ *is p-nilpotent.*

Proof. First observe $\mathbf{N}_{\mathfrak{G}}(\mathfrak{S}) \leq \mathbf{N}_{\mathfrak{G}}(\mathbf{J}(\mathfrak{S}))$, so $\mathbf{N}_{\mathfrak{G}}(\mathfrak{S})$ is p-nilpotent. Thus $\mathbf{N}_{\mathfrak{G}}(\mathfrak{S}) = \mathfrak{S}\mathbf{O}_{p'}(\mathbf{N}_{\mathfrak{G}}(\mathfrak{S})) \leq \mathbf{C}_{\mathfrak{G}}(\mathbf{Z}(\mathfrak{S}))$. Since $\mathbf{C}_{\mathfrak{G}}(\mathbf{Z}(\mathfrak{S})) \trianglelefteq \mathbf{N}_{\mathfrak{G}}(\mathbf{Z}(\mathfrak{S}))$ and $\mathfrak{S} \in S_p(\mathbf{C}_{\mathfrak{G}}(\mathbf{Z}(\mathfrak{S})))$, it follows from the Frattini argument that $\mathbf{N}_{\mathfrak{G}}(\mathbf{Z}(\mathfrak{S})) = \mathbf{N}_{\mathfrak{G}}(\mathfrak{S})\mathbf{C}_{\mathfrak{G}}(\mathbf{Z}(\mathfrak{S})) = \mathbf{C}_{\mathfrak{G}}(\mathbf{Z}(\mathfrak{S}))$. Hence $\mathbf{N}_{\mathfrak{G}}(\mathbf{Z}(\mathfrak{S}))$ is p-nilpotent.

Suppose that $\mathfrak{G}$ is not p-nilpotent. By 9.8 and 9.9, $\mathfrak{G}$ has a $\{p, q\}$-section $\mathfrak{Q}$ for some prime q such that the Sylow q-subgroups of $\mathfrak{Q}$ are Abelian, $\mathbf{O}_{p'}(\mathfrak{Q}) = 1$ and $\mathbf{N}_{\mathfrak{Q}}(\mathbf{J}(\mathfrak{S}_0)) = \mathbf{N}_{\mathfrak{Q}}(\mathbf{Z}(\mathfrak{S}_0)) = \mathfrak{S}_0$, where $\mathfrak{S}_0 \in S_p(\mathfrak{Q})$. It follows from 10.8a) that $p < 5$ and that $\mathfrak{Q}$ has a section isomorphic to $SL(2, p)$. Thus $p = 3$ and $q = 2$. But the Sylow 2-subgroup of $SL(2, 3)$ is a quaternion group of order 8 and hence non-Abelian. This is impossible, so $\mathfrak{G}$ is p-nilpotent. **q.e.d.**

The conclusions of 10.7 allow the construction of fixed points of automorphisms.

10.10 Lemma. *Suppose that* $\mathfrak{B}_1$ *is a minimal normal subgroup of a group* $\mathfrak{H}$. *Let* $\mathfrak{A}$ *be a group of automorphisms of* $\mathfrak{H}$ *and let* $\mathfrak{B}_1, \ldots, \mathfrak{B}_m$ *be the distinct images of* $\mathfrak{B}_1$ *under all elements of* $\mathfrak{A}$. *Suppose that the group* $\mathfrak{B}$ *generated by* $\mathfrak{B}_1, \ldots, \mathfrak{B}_m$ *is their direct product:*

$$\mathfrak{B} = \mathfrak{B}_1 \times \cdots \times \mathfrak{B}_m.$$

Suppose that $x \in \mathfrak{B}_1$ *and* $x\alpha = x$ *for all* $\alpha \in \mathfrak{A}$ *which satisfy* $\mathfrak{B}_1\alpha = \mathfrak{B}_1$. *Then*

$$(x\alpha_1) \cdots (x\alpha_m) \in \mathbf{C}_{\mathfrak{H}}(\mathfrak{A})$$

for any elements $\alpha_1, \ldots, \alpha_m$ *of* $\mathfrak{A}$ *for which* $\mathfrak{B}_1\alpha_i = \mathfrak{B}_i$.

Proof. Let $y = (x\alpha_1) \cdots (x\alpha_m)$ and suppose $\alpha \in \mathfrak{A}$. For $i = 1, \ldots, m$, $\mathfrak{B}_1\alpha_i\alpha = \mathfrak{B}_{\pi(i)} = \mathfrak{B}_1\alpha_{\pi(i)}$ for some $\pi(i)$, and π is a permutation of $\{1, \ldots, m\}$. Thus $\mathfrak{B}_1\alpha_i\alpha\alpha_{\pi(i)}^{-1} = \mathfrak{B}_1$ and $x\alpha_i\alpha = x\alpha_{\pi(i)}$. Thus

$$y\alpha = (x\alpha_{\pi(1)}) \cdots (x\alpha_{\pi(m)}) = (x\alpha_1) \cdots (x\alpha_m) = y.$$

Hence $y \in \mathbf{C}_{\mathfrak{H}}(\mathfrak{A})$. **q.e.d.**

10.11 Lemma. *Suppose that* $\mathfrak{H}$, $\mathfrak{B}$ *are characteristic subgroups of a group* $\mathfrak{G}$. *Suppose also that* $\mathfrak{B}$ *is a non-identity p-subgroup of* $\mathfrak{H}$ *and that* $\mathfrak{B} = \mathfrak{B}_1 \times \cdots \times \mathfrak{B}_n$, *where* $\mathfrak{B}_1, \ldots, \mathfrak{B}_n$ *are all the minimal normal subgroups of* $\mathfrak{H}$ *contained in* $\mathfrak{B}$. *Let* $\mathfrak{A}$ *be a soluble group of automorphisms of* $\mathfrak{G}$ *such that*

$$(|\mathfrak{A}|, |\mathfrak{G}||\mathbf{Aut}\,\mathfrak{B}_1|) = 1.$$

Then $\mathbf{C}_{\mathfrak{G}}(\mathfrak{A}) \cap \mathfrak{B} \cap \mathbf{Z}(\mathfrak{S}) \neq 1$ *for some* $\mathfrak{S} \in S_p(\mathfrak{G})$.

Proof. Let $\mathfrak{G}^*$ be the semidirect product of $\mathfrak{G}$ and $\mathfrak{A}$. By IX, 1.11, there exists $\mathfrak{S} \in S_p(\mathfrak{G})$ such that $\mathfrak{A} \leq \mathbf{N}_{\mathfrak{G}^*}(\mathfrak{S})$. Let $\overline{\mathfrak{A}}$ be the group of automorphisms of $\mathfrak{H}$ induced by $\mathfrak{A}\mathfrak{S}$. Since $\mathfrak{A}\mathfrak{S} \leq \mathbf{N}_{\mathfrak{G}^*}(\mathfrak{B})$, $\overline{\mathfrak{A}}$ permutes $\mathfrak{B}_1, \ldots, \mathfrak{B}_n$ among themselves. Let

$$\overline{\mathfrak{B}} = \{\alpha | \alpha \in \overline{\mathfrak{A}}, \mathfrak{B}_1\alpha = \mathfrak{B}_1\}.$$

Then $|\overline{\mathfrak{B}}|$ divides $|\mathfrak{A}\mathfrak{S}|$. Also $\overline{\mathfrak{B}}$ induces a group $\tilde{\mathfrak{B}}$ of automorphisms on $\mathfrak{B}_1$. Thus $|\tilde{\mathfrak{B}}|$ divides both $|\mathbf{Aut}\,\mathfrak{B}_1|$ and $|\mathfrak{A}\mathfrak{S}|$. Since $(|\mathbf{Aut}\,\mathfrak{B}_1|, |\mathfrak{A}|) = 1$, $\tilde{\mathfrak{B}}$ is a p-group. Hence $\tilde{\mathfrak{B}}$ has a non-identity fixed point x in $\mathfrak{B}_1$ and $x\alpha = x$ for all $\alpha \in \overline{\mathfrak{B}}$. By 10.10, $\mathbf{C}_{\mathfrak{B}}(\overline{\mathfrak{A}}) \neq 1$. Since $\mathfrak{B} \leq \mathfrak{S}$, $\mathbf{C}_{\mathfrak{B}}(\overline{\mathfrak{A}}) \leq \mathbf{Z}(\mathfrak{S})$, so $\mathbf{C}_{\mathfrak{G}}(\mathfrak{A}) \cap \mathfrak{B} \cap \mathbf{Z}(\mathfrak{S}) \neq 1$. **q.e.d.**

10.12 Theorem. *Suppose that $\mathfrak{G}$ is a p-soluble group, $\mathbf{O}_{p'}(\mathfrak{G}) = 1$ and $\mathfrak{A}$ is a soluble group of automorphisms of $\mathfrak{G}$ for which $(|\mathfrak{A}|, |\mathfrak{G}|) = 1$ and $\mathbf{C}_{\mathfrak{G}}(\mathfrak{A})$ is a p′-group. Then if $\mathfrak{S} \in S_p(\mathfrak{G})$,*

$$\mathfrak{G} = \mathbf{C}_{\mathfrak{G}}(\mathbf{Z}(\mathfrak{S}))\mathbf{N}_{\mathfrak{G}}(\mathbf{J}(\mathfrak{S})).$$

Proof. Suppose that this is false. By 10.3, $\mathbf{J}(\mathfrak{S}) \nleq \mathbf{C}_{\mathfrak{G}}(\mathbf{R}_p(\mathfrak{G}))$. By 10.5 and 10.7, $\mathfrak{G}$ has characteristic subgroups $\mathfrak{H}$, $\mathfrak{B}$ such that $\mathfrak{B}$ is a non-identity p-subgroup of $\mathfrak{H}$ and that $\mathfrak{B}$ is the direct product of all the minimal normal subgroups $\mathfrak{B}_1, \mathfrak{B}_2, \ldots$ of $\mathfrak{H}$ contained in $\mathfrak{B}$. Further, $|\mathfrak{B}_1| = p^2$ and $|SL(2, p)|$ divides $|\mathfrak{G}|$. Hence

$$(|\mathfrak{A}|, |\mathfrak{G}||\mathbf{Aut}\,\mathfrak{B}_1|) = 1.$$

By 10.11, $|\mathbf{C}_{\mathfrak{G}}(\mathfrak{A})|$ is divisible by p, a contradiction. **q.e.d.**

10.13 Remark. GLAUBERMAN [7] showed that if $\mathfrak{G}$ is p-soluble, $\mathbf{O}_{p'}(\mathfrak{G}) = 1$ and $\mathfrak{A}$ is a soluble group of automorphisms of $\mathfrak{G}$ for which $(|\mathfrak{A}|, |\mathfrak{G}|) = 1$, then

$$\mathfrak{G} = \langle \mathbf{N}_{\mathfrak{G}}(\mathbf{J}(\mathfrak{S})), \mathbf{C}_{\mathfrak{G}}(\mathbf{Z}(\mathfrak{S})), \mathbf{C}_{\mathfrak{G}}(\mathfrak{A})\rangle,$$

where $\mathfrak{S} \in S_p(\mathfrak{G})$. The proof makes use of 10.8b).

10.14 Remarks. a) In his determination of the N-groups (that is, groups in which the normaliser of every non-identity soluble subgroup is soluble), THOMPSON made fairly extensive use of the following factorization theorems [3, 6].

For any p-group $\mathfrak{P}$, let $m = m(\mathfrak{P})$ be the greatest integer for which $\mathfrak{P}$ has an Abelian subgroup $\mathfrak{A}$ such that $|\mathfrak{A} : \Phi(\mathfrak{A})| = p^m$. Let $\mathbf{J}_n(\mathfrak{P})$ $(n \geq 0)$ denote the subgroup of $\mathfrak{P}$ generated by all Abelian subgroups $\mathfrak{A}$ for which $|\mathfrak{A} : \Phi(\mathfrak{A})| \geq p^{m-n}$. Thus the $\mathbf{J}_n(\mathfrak{P})$ form an increasing series of characteristic subgroups of $\mathfrak{P}$.

Now let $\mathfrak{G}$ be a p-soluble group satisfying $\mathbf{O}_{p'}(\mathfrak{G}) = 1$, let $\mathfrak{S} \in S_p(\mathfrak{G})$ and let $\mathfrak{C} = \mathbf{C}_{\mathfrak{G}}(\langle \mathbf{Z}(\mathfrak{S})^g | g \in \mathfrak{G}\rangle)$.

(1) If $p \geq 5$, $\mathbf{J}_0(\mathfrak{S}) \leq \cdots \leq \mathbf{J}_{p-4}(\mathfrak{S}) \leq \mathfrak{C}$.

(2) If $p = 3$ and no section of $\mathfrak{G}$ is isomorphic to $SA(2, p)$, $\mathbf{J}_0(\mathfrak{S}) \leq \mathfrak{C}$.

(3) If $p = 2$ and no section of $\mathfrak{G}$ is isomorphic to $\mathfrak{S}_4$, $\mathbf{J}_0(\mathfrak{S}) \leq \mathfrak{C}$. In all these circumstances $\mathfrak{G} = \mathbf{N}_{\mathfrak{G}}(\mathbf{J}_0(\mathfrak{S}))\mathbf{C}_{\mathfrak{G}}(\mathbf{Z}(\mathfrak{S}))$.

Next, suppose that $\mathfrak{S} \in S_p(\mathfrak{G})$ and that $\{1\}$ is the only p'-subgroup of $\mathfrak{G}$ normalised by $\mathfrak{S}$. Write $\mathfrak{H}_1 = \mathbf{C}_{\mathfrak{G}}(\mathbf{Z}(\mathfrak{S}))$, $\mathfrak{H}_2 = \mathbf{N}_{\mathfrak{G}}(\mathbf{J}_0(\mathfrak{S}))$, $\mathfrak{H}_3 = \mathbf{N}_{\mathfrak{G}}(\mathbf{Z}(\mathbf{J}_1(\mathfrak{S})))$, and suppose that $\mathfrak{H}_1, \mathfrak{H}_2, \mathfrak{H}_3$ are all p-soluble.

(4) If $p \geq 5$ or if $p = 2$ and $\mathfrak{G}$ is a $3'$-group, $\mathfrak{H}_1 \subseteq \mathfrak{H}_2\mathfrak{H}_3$.

(5) If $p \geq 5$, then $\mathfrak{H}_1\mathfrak{H}_2$ is a subgroup of $\mathfrak{G}$ and $\mathfrak{H}_1\mathfrak{H}_2$ contains every p-soluble subgroup of $\mathfrak{G}$ containing $\mathfrak{S}$.

For $p = 2$, this culminates in the 3-against-2 theorem:

$$\begin{aligned}\mathfrak{G} &= \mathbf{N}_{\mathfrak{G}}(\mathbf{J}_0(\mathfrak{S}))\mathbf{N}_{\mathfrak{G}}(\mathbf{Z}(\mathbf{J}_1(\mathfrak{S})))\\ &= \mathbf{N}_{\mathfrak{G}}(\mathbf{Z}(\mathbf{J}_1(\mathfrak{S})))\mathbf{C}_{\mathfrak{G}}(\mathbf{Z}(\mathfrak{S}))\\ &= \mathbf{C}_{\mathfrak{G}}(\mathbf{Z}(\mathfrak{S}))\mathbf{N}_{\mathfrak{G}}(\mathbf{J}_0(\mathfrak{S})),\end{aligned}$$

where $\mathfrak{S} \in S_2(\mathfrak{G})$, provided that $\mathfrak{G}$ is a soluble $\{3, 5\}'$-group in which $\mathbf{O}_{2'}(\mathfrak{G}) = 1$. Of these, the last one is from (3) and the first then follows from (4).

b) Another factorization theorem was used by GLAUBERMAN [10] to determine the simple groups which have no section isomorphic to $\mathfrak{S}_4$. They are the following.

(i) $PSL(2, 2^n)$ $(n \geq 3)$.

(ii) $Sz(2^{2n+1})$ $(n \geq 1)$.

(iii) $PSU(3, 2^n)$ $(n \geq 2)$.

(iv) $PSL(2, q)$ $(q \equiv 3\ (8)$ or $q \equiv 5\ (8), q \geq 5)$.

(v) Janko's first simple group $\mathfrak{J}_1$.

In particular, the only simple $3'$-groups are the Suzuki groups.

This is proved by combining with an important theorem of GOLDSCHMIDT [4] the following factorization theorem.

$$\begin{aligned}\mathfrak{G} &= \mathbf{C}_{\mathfrak{G}}(\mathbf{Z}(\mathfrak{S}))\mathbf{N}_{\mathfrak{G}}(\mathbf{J}_e(\mathfrak{S}))\\ &= \mathbf{C}_{\mathfrak{G}}(\mathbf{Z}(\mathfrak{S}))\mathbf{N}_{\mathfrak{G}}(\hat{\mathbf{J}}(\mathfrak{S}))\\ &= \mathbf{C}_{\mathfrak{G}}(\Omega_1\mathbf{Z}\hat{\mathbf{J}}(\mathfrak{S}))\mathbf{N}_{\mathfrak{G}}(\mathbf{J}_e(\mathfrak{S})).\end{aligned}$$

Here $\mathfrak{S} \in S_2(\mathfrak{G})$ and it is supposed that $\mathfrak{G}$ satisfies the following conditions.

(1) $\mathbf{O}_2(\mathfrak{G}) \geq \mathbf{C}_{\mathfrak{G}}(\mathbf{O}_2(\mathfrak{G}))$.

(2) $\mathfrak{G}$ has no section isomorphic to $\mathfrak{S}_4$.

(3) Every non-Abelian composition factor of $\mathfrak{G}$ is isomorphic to either $Sz(2^{2n+1})$ or to $PSL(2, 3^{2n+1})$.

The characteristic p-functor $\mathbf{J}_e$ is the same as the one used in § 2. But the definition of $\hat{\mathbf{J}}$ is more complicated. In the p-group $\mathfrak{P}$, let $\mathscr{E}$ denote the set of subgroups $\mathfrak{Q}$ of $\mathfrak{P}$ for which $\mathbf{Z}(\mathfrak{Q})$ contains every normal elementary Abelian subgroup $\mathfrak{B}$ of $\mathfrak{Q}$ having the following property: whenever $\mathfrak{R}$ is a non-identity elementary Abelian subgroup of $\mathfrak{Q}/\mathbf{C}_{\mathfrak{Q}}(\mathfrak{B})$, then

$$|\mathfrak{B}/\mathbf{C}_{\mathfrak{B}}(\mathfrak{R})| > |\mathfrak{R}|^{3/2} \text{ and } |[\mathfrak{B}, \mathfrak{R}]| > |\mathfrak{R}|.$$

Then $\hat{\mathbf{J}}(\mathfrak{P})$ is the group generated by all the subgroups in $\mathscr{E}$ which contain $\mathbf{J}_e(\mathfrak{P})$. (In fact $\mathbf{J}_e(\mathfrak{P}) \in \mathscr{E}$ and $\mathbf{J}_e(\mathfrak{P}) \leq \hat{\mathbf{J}}(\mathfrak{P})$.)

§ 11. Fixed Point Free Automorphism Groups

Let $\mathfrak{A}$ be a group of automorphisms of a group $\mathfrak{G}$ such that $\mathbf{C}_{\mathfrak{G}}(\mathfrak{A}) = 1$. If $|\mathfrak{A}|$ is prime, then $\mathfrak{G}$ is nilpotent (V, 8.14). It is conjectured that if $\mathfrak{A}$ is cyclic or if $(|\mathfrak{G}|, |\mathfrak{A}|) = 1$, then $\mathfrak{G}$ is soluble. For instance it has been proved that this is the case if $\mathfrak{A}$ is cyclic of order the product of two primes by RALSTON [1] and RICKMAN [1]. In this section we shall obtain some information about a minimal counterexample in the case $(|\mathfrak{G}|, |\mathfrak{A}|) = 1$, and we use this to prove that the conjecture is correct if $\mathfrak{A}$ is elementary Abelian.

First we prove some elementary facts.

11.1 Lemma. *Suppose that $\mathfrak{A}$ is a soluble group of automorphisms of $\mathfrak{G}$, $\mathbf{C}_{\mathfrak{G}}(\mathfrak{A}) = 1$ and $(|\mathfrak{A}|, |\mathfrak{G}|) = 1$.*

a) *If $\mathfrak{N}$ is a normal $\mathfrak{A}$-invariant subgroup of $\mathfrak{G}$, $\mathbf{C}_{\mathfrak{G}/\mathfrak{N}}(\mathfrak{A}) = 1$.*

b) *For any prime p, $\mathfrak{G}$ has a unique $\mathfrak{A}$-invariant Sylow p-subgroup $\mathfrak{S}$. If $\mathfrak{H}$ is an $\mathfrak{A}$-invariant subgroup of $\mathfrak{G}$, $\mathfrak{H} \cap \mathfrak{S} \in S_p(\mathfrak{H})$. Also any $\mathfrak{A}$-invariant p-subgroup of $\mathfrak{G}$ is contained in $\mathfrak{S}$.*

c) *Suppose that $\mathfrak{G}$ is soluble. If π is a set of primes, $\mathfrak{G}$ has a unique $\mathfrak{A}$-invariant Hall π-subgroup $\mathfrak{P}$. If $\mathfrak{H}$ is an $\mathfrak{A}$-invariant subgroup of $\mathfrak{G}$, $\mathfrak{H} \cap \mathfrak{P}$ is a Hall π-subgroup of $\mathfrak{H}$, and any $\mathfrak{A}$-invariant π-subgroup of $\mathfrak{G}$ is contained in $\mathfrak{P}$.*

Proof. a) This follows from I, 18.6.

b) The existence of an $\mathfrak{A}$-invariant Sylow p-subgroup $\mathfrak{S}$ of $\mathfrak{G}$ follows from IX, 1.11a) and the uniqueness from IX, 1.11b), since $\mathbf{C}_{\mathfrak{G}}(\mathfrak{A}) = 1$.

If $\mathfrak{H}$ is an $\mathfrak{A}$-invariant subgroup of $\mathfrak{G}$, then $\mathfrak{A}$ is a group of operators on $\mathfrak{H}$, so $\mathfrak{H}$ possesses an $\mathfrak{A}$-invariant Sylow p-subgroup $\mathfrak{S}_1$. By IX, 1.11c), $\mathfrak{S}_1 \leq \mathfrak{S}$. Thus $\mathfrak{S}_1 \leq \mathfrak{S} \cap \mathfrak{H}$. Since $\mathfrak{S}_1 \in S_p(\mathfrak{H})$ and $\mathfrak{S} \cap \mathfrak{H}$ is a

p-subgroup of $\mathfrak{H}$, it follows that $\mathfrak{S}_1 = \mathfrak{S} \cap \mathfrak{H}$. If, in particular, $\mathfrak{H}$ is a p-group, $\mathfrak{S} \cap \mathfrak{H} = \mathfrak{H}$ and $\mathfrak{H} \leq \mathfrak{S}$.

c) This also follows from IX, 1.11 as in b). **q.e.d.**

The following lemma is very similar to 2.10; the idea is due to Bender.

11.2 Lemma. *Suppose that $\mathfrak{G}$ is a group and that $\mathfrak{A}$ is a soluble group of operators on $\mathfrak{G}$, where $(|\mathfrak{A}|, |\mathfrak{G}|) = 1$. Let p, q be distinct primes. Suppose that for any non-identity $\mathfrak{A}$-invariant $\{p, q\}$-subgroup $\mathfrak{B}$ of $\mathfrak{G}$, $\mathbf{N}_{\mathfrak{G}}(\mathfrak{B})$ is soluble and possesses a unique $\mathfrak{A}$-invariant Hall $\{p, q\}$-subgroup $\mathbf{H}(\mathfrak{B})$.*

Let $\mathfrak{H}$ be a maximal $\mathfrak{A}$-invariant $\{p, q\}$-subgroup of $\mathfrak{G}$ and let $\mathfrak{L}$ be an $\mathfrak{A}$-invariant subgroup of $\mathbf{F}(\mathfrak{H})$ for which $\mathbf{O}_p(\mathfrak{L}) \neq 1$, $\mathbf{O}_q(\mathfrak{L}) \neq 1$. Then any $\mathfrak{A}$-invariant $\{p, q\}$-subgroup of $\mathfrak{G}$ containing $\mathfrak{L}$ is contained in $\mathfrak{H}$.

Proof. (1) Suppose that $\mathfrak{M}$ is a maximal $\mathfrak{A}$-invariant $\{p, q\}$-subgroup of $\mathfrak{G}$ and $\mathbf{O}_p(\mathfrak{M}) \neq 1$, $\mathbf{O}_q(\mathfrak{M}) \neq 1$. Let $\mathfrak{Z} = \mathbf{Z}(\mathbf{F}(\mathfrak{M}))$ and suppose that $\mathfrak{K}$ is an $\mathfrak{A}$-invariant $\{p, q\}$-subgroup of $\mathfrak{G}$ containing $\mathfrak{Z}$. Then $\mathfrak{Z} \leq \mathbf{F}(\mathfrak{K}) \leq \mathfrak{M}$.

Write $\mathfrak{Z} = \mathfrak{Z}_p \times \mathfrak{Z}_q$, where $\mathfrak{Z}_p \in S_p(\mathfrak{Z})$, $\mathfrak{Z}_q \in S_q(\mathfrak{Z})$. Since $\mathfrak{Z}_p = \mathbf{Z}(\mathbf{O}_p(\mathfrak{M}))$, $\mathfrak{Z}_p \neq 1$. Thus $\mathbf{N}_{\mathfrak{G}}(\mathfrak{Z}_p)$ is soluble and $\mathfrak{A}$-invariant. Also $\mathfrak{M}$ is an $\mathfrak{A}$-invariant $\{p, q\}$-subgroup of $\mathbf{N}_{\mathfrak{G}}(\mathfrak{Z}_p)$. By IX, 1.11c), $\mathfrak{M}$ is contained in an $\mathfrak{A}$-invariant Hall $\{p, q\}$-subgroup of $\mathbf{N}_{\mathfrak{G}}(\mathfrak{Z}_p)$. Thus $\mathfrak{M} \leq \mathbf{H}(\mathfrak{Z}_p)$. By maximality of $\mathfrak{M}$, $\mathfrak{M} = \mathbf{H}(\mathfrak{Z}_p)$. But similarly, $\mathbf{N}_{\mathfrak{K}}(\mathfrak{Z}_p)$, being an $\mathfrak{A}$-invariant $\{p, q\}$-subgroup of $\mathbf{N}_{\mathfrak{G}}(\mathfrak{Z}_p)$, is contained in $\mathbf{H}(\mathfrak{Z}_p)$. Thus $\mathbf{N}_{\mathfrak{K}}(\mathfrak{Z}_p) \leq \mathfrak{M}$. Thus $\mathbf{O}_q(\mathfrak{M}) \cap \mathbf{N}_{\mathfrak{K}}(\mathfrak{Z}_p) \leq \mathbf{O}_q(\mathbf{N}_{\mathfrak{K}}(\mathfrak{Z}_p))$, and by 1.6, $\mathbf{O}_q(\mathbf{N}_{\mathfrak{K}}(\mathfrak{Z}_p)) = \mathbf{O}_{p'}(\mathbf{N}_{\mathfrak{K}}(\mathfrak{Z}_p)) \leq \mathbf{O}_{p'}(\mathfrak{K}) = \mathbf{O}_q(\mathfrak{K})$. Thus $\mathbf{O}_q(\mathfrak{M}) \cap \mathbf{N}_{\mathfrak{K}}(\mathfrak{Z}_p) \leq \mathbf{O}_q(\mathfrak{K})$. Since $\mathfrak{Z} \leq \mathfrak{K}$, $\mathfrak{Z}_q \leq \mathbf{N}_{\mathfrak{K}}(\mathfrak{Z}_p)$. Thus $\mathfrak{Z}_q \leq \mathbf{O}_q(\mathfrak{M}) \cap \mathbf{N}_{\mathfrak{K}}(\mathfrak{Z}_p) \leq \mathbf{O}_q(\mathfrak{K})$. Similarly, $\mathfrak{Z}_p \leq \mathbf{O}_p(\mathfrak{K})$, so $\mathfrak{Z} \leq \mathbf{F}(\mathfrak{K})$.

Also $\mathbf{O}_q(\mathfrak{K}) \leq \mathbf{C}_{\mathfrak{G}}(\mathbf{O}_p(\mathfrak{K})) \leq \mathbf{C}_{\mathfrak{G}}(\mathfrak{Z}_p) \leq \mathbf{N}_{\mathfrak{G}}(\mathfrak{Z}_p)$. Since $\mathbf{O}_q(\mathfrak{K})$ is $\mathfrak{A}$-invariant, it follows that $\mathbf{O}_q(\mathfrak{K}) \leq \mathbf{H}(\mathfrak{Z}_p) = \mathfrak{M}$. Similarly $\mathbf{O}_p(\mathfrak{K}) \leq \mathfrak{M}$ and $\mathbf{F}(\mathfrak{K}) \leq \mathfrak{M}$.

(2) Suppose that $\mathfrak{M}$ is a maximal $\mathfrak{A}$-invariant $\{p, q\}$-subgroup of $\mathfrak{G}$ and $\mathbf{O}_p(\mathfrak{M}) \neq 1$, $\mathbf{O}_q(\mathfrak{M}) \neq 1$. Then $\mathfrak{M}$ is the only maximal $\mathfrak{A}$-invariant $\{p, q\}$-subgroup of $\mathfrak{G}$ which contains $\mathfrak{Z} = \mathbf{Z}(\mathbf{F}(\mathfrak{M}))$.

Let $\mathfrak{K}$ be any maximal $\mathfrak{A}$-invariant $\{p, q\}$-subgroup of $\mathfrak{G}$ which contains $\mathfrak{Z}$. By (1), $\mathfrak{Z} \leq \mathbf{F}(\mathfrak{K}) \leq \mathfrak{M}$. Since $\mathfrak{Z}$ is not of prime-power order, neither is $\mathbf{F}(\mathfrak{K})$; thus $\mathbf{O}_p(\mathfrak{K}) \neq 1$ and $\mathbf{O}_q(\mathfrak{K}) \neq 1$. Also $\mathfrak{M} \geq \mathbf{Z}(\mathbf{F}(\mathfrak{K}))$, so it follows from (1) (with $\mathfrak{K}$, $\mathfrak{M}$ the other way round) that $\mathbf{F}(\mathfrak{M}) \leq \mathfrak{K}$.

We now use the same argument as in (1) on $\mathbf{O}_p(\mathfrak{M})$. Thus if $\mathfrak{X} = \mathbf{N}_{\mathfrak{G}}(\mathbf{O}_p(\mathfrak{M}))$, $\mathfrak{X}$ is soluble and $\mathfrak{A}$-invariant, and $\mathfrak{M}$ is an $\mathfrak{A}$-invariant $\{p, q\}$-subgroup of $\mathfrak{X}$. Thus $\mathfrak{M} \leq \mathbf{H}(\mathbf{O}_p(\mathfrak{M}))$, and by maximality, $\mathfrak{M} = \mathbf{H}(\mathbf{O}_p(\mathfrak{M}))$. Hence $\mathbf{N}_{\mathfrak{K}}(\mathbf{O}_p(\mathfrak{M})) \leq \mathbf{H}(\mathbf{O}_p(\mathfrak{M})) = \mathfrak{M}$. Thus, since $\mathbf{O}_q(\mathfrak{M})$

$\leq \mathfrak{K}$, $\mathbf{O}_q(\mathfrak{M}) \trianglelefteq \mathbf{N}_{\mathfrak{K}}(\mathbf{O}_p(\mathfrak{M}))$ and $\mathbf{O}_q(\mathfrak{M}) \leq \mathbf{O}_{p'}(\mathbf{N}_{\mathfrak{K}}(\mathbf{O}_p(\mathfrak{M})))$. Using 1.6, $\mathbf{O}_q(\mathfrak{M}) \leq \mathbf{O}_{p'}(\mathfrak{K}) = \mathbf{O}_q(\mathfrak{K})$. Similarly, $\mathbf{O}_q(\mathfrak{K}) \leq \mathbf{O}_q(\mathfrak{M})$, so $\mathbf{O}_q(\mathfrak{K}) = \mathbf{O}_q(\mathfrak{M})$. Similarly, $\mathbf{O}_p(\mathfrak{K}) = \mathbf{O}_p(\mathfrak{M})$, so $\mathbf{F}(\mathfrak{K}) = \mathbf{F}(\mathfrak{M}) = \mathfrak{F}$, say. Thus $\mathfrak{K}$, $\mathfrak{M}$ are $\mathfrak{A}$-invariant $\{p, q\}$-subgroups of $\mathbf{N}_{\mathfrak{G}}(\mathfrak{F})$. Hence $\mathfrak{K} \leq \mathbf{H}(\mathfrak{F})$, $\mathfrak{M} \leq \mathbf{H}(\mathfrak{F})$. By maximility of $\mathfrak{M}$ and $\mathfrak{K}$, $\mathfrak{K} = \mathfrak{M} = \mathbf{H}(\mathfrak{F})$.

(3) To prove the assertion, let $\mathfrak{K}$ be a maximal $\mathfrak{A}$-invariant $\{p, q\}$-subgroup of $\mathfrak{G}$ such that $\mathfrak{K} \geq \mathfrak{L}$; it suffices to prove that $\mathfrak{K} \leq \mathfrak{H}$. Write $\mathfrak{L} = \mathfrak{L}_p \times \mathfrak{L}_q$, where $\mathfrak{L}_p \in S_p(\mathfrak{L})$, $\mathfrak{L}_q \in S_q(\mathfrak{L})$. By hypothesis, $\mathfrak{L}_p \neq 1$, $\mathfrak{L}_q \neq 1$. Let $\mathfrak{C} = \mathbf{H}(\mathfrak{L}_q) \cap \mathbf{C}_{\mathfrak{G}}(\mathfrak{L}_q)$. Since $\mathbf{C}_{\mathfrak{G}}(\mathfrak{L}_q) \trianglelefteq \mathbf{N}_{\mathfrak{G}}(\mathfrak{L}_q)$, $\mathfrak{C}$ is an $\mathfrak{A}$-invariant Hall $\{p, q\}$-subgroup of $\mathbf{C}_{\mathfrak{G}}(\mathfrak{L}_q)$. Note that by IX, 1.11c), any $\mathfrak{A}$-invariant $\{p, q\}$-subgroup of $\mathbf{C}_{\mathfrak{G}}(\mathfrak{L}_q)$ lies in $\mathbf{H}(\mathfrak{L}_q)$ and hence in $\mathfrak{C}$.

We apply (2) to $\mathfrak{H}$; we have $\mathbf{O}_p(\mathfrak{H}) \neq 1$ and $\mathbf{O}_q(\mathfrak{H}) \neq 1$ since $\mathfrak{L} \leq \mathbf{F}(\mathfrak{H})$ and $\mathfrak{L}_p \neq 1$, $\mathfrak{L}_q \neq 1$. Thus $\mathfrak{H}$ contains any $\mathfrak{A}$-invariant $\{p, q\}$-subgroup which contains $\mathbf{Z}(\mathbf{F}(\mathfrak{H}))$. But $\mathbf{Z}(\mathbf{F}(\mathfrak{H}))$ is an $\mathfrak{A}$-invariant $\{p, q\}$-subgroup of $\mathbf{C}_{\mathfrak{G}}(\mathfrak{L}_q)$, so $\mathbf{Z}(\mathbf{F}(\mathfrak{H})) \leq \mathfrak{C}$. Thus $\mathfrak{H} \geq \mathfrak{C}$. But also $\mathbf{C}_{\mathfrak{K}}(\mathfrak{L}_q) \leq \mathfrak{C}$, since $\mathbf{C}_{\mathfrak{K}}(\mathfrak{L}_q)$ is an $\mathfrak{A}$-invariant $\{p, q\}$-subgroup of $\mathbf{C}_{\mathfrak{G}}(\mathfrak{L}_q)$. Thus $\mathbf{C}_{\mathfrak{K}}(\mathfrak{L}_q) \leq \mathfrak{H}$ and

$$\mathbf{O}_p(\mathfrak{H}) \cap \mathbf{C}_{\mathfrak{K}}(\mathfrak{L}_q) \leq \mathbf{O}_p(\mathbf{C}_{\mathfrak{K}}(\mathfrak{L}_q)) = \mathbf{O}_{q'}(\mathbf{C}_{\mathfrak{K}}(\mathfrak{L}_q)) \leq \mathbf{O}_{q'}(\mathbf{N}_{\mathfrak{K}}(\mathfrak{L}_q)).$$

Using 1.6, $\mathbf{O}_p(\mathfrak{H}) \cap \mathbf{C}_{\mathfrak{K}}(\mathfrak{L}_q) \leq \mathbf{O}_{q'}(\mathfrak{K}) = \mathbf{O}_p(\mathfrak{K})$. Hence $\mathfrak{L}_p \leq \mathbf{O}_p(\mathfrak{K})$. Similarly, $\mathfrak{L}_q \leq \mathbf{O}_q(\mathfrak{K})$, so $\mathbf{O}_p(\mathfrak{K}) \neq 1$ and $\mathbf{O}_q(\mathfrak{K}) \neq 1$. Also $\mathbf{O}_p(\mathfrak{K}) \leq \mathbf{C}_{\mathfrak{G}}(\mathbf{O}_q(\mathfrak{K})) \leq \mathbf{C}_{\mathfrak{G}}(\mathfrak{L}_q)$. Since $\mathbf{O}_p(\mathfrak{K})$ is $\mathfrak{A}$-invariant, it follows that $\mathbf{O}_p(\mathfrak{K}) \leq \mathfrak{C}$. Thus $\mathbf{O}_p(\mathfrak{K}) \leq \mathfrak{H}$. Similarly, $\mathbf{O}_q(\mathfrak{K}) \leq \mathfrak{H}$ and $\mathbf{F}(\mathfrak{K}) \leq \mathfrak{H}$. It follows from (2) (applied to $\mathfrak{K}$) that $\mathfrak{H} = \mathfrak{K}$, on account of the maximality of $\mathfrak{H}$. **q.e.d.**

In the main theorems of this section, solubility will be established by showing that any two $\mathfrak{A}$-invariant Sylow subgroups of $\mathfrak{G}$ commute. The study of a minimal counterexample thus leads to the following hypothesis.

11.3 Hypothesis. a) $\mathfrak{A}$ is a soluble group of automorphisms of $\mathfrak{G}$, $\mathbf{C}_{\mathfrak{G}}(\mathfrak{A}) = 1$ and $(|\mathfrak{A}|, |\mathfrak{G}|) = 1$.

b) $\mathfrak{G}$ is insoluble and $\mathfrak{G}$ has no non-identity $\mathfrak{A}$-invariant proper normal subgroup.

c) Any proper $\mathfrak{A}$-invariant subgroup of $\mathfrak{G}$ is soluble.

d) p, q are prime divisors of $|\mathfrak{G}|$, $\mathfrak{P}$ is the $\mathfrak{A}$-invariant Sylow p-subgroup of $\mathfrak{G}$ and $\mathfrak{Q}$ is the $\mathfrak{A}$-invariant Sylow q-subgroup of $\mathfrak{G}$ (see 11.1).

e) $\mathscr{P} = \{\mathfrak{X} \mid \mathfrak{X} \leq \mathfrak{P}, \mathfrak{X}$ is $\mathfrak{A}$-invariant, $\mathfrak{X}\mathfrak{Q} = \mathfrak{Q}\mathfrak{X}\}$, $\mathscr{Q} = \{\mathfrak{Y} \mid \mathfrak{Y} \leq \mathfrak{Q}$, $\mathfrak{Y}$ is $\mathfrak{A}$-invariant, $\mathfrak{P}\mathfrak{Y} = \mathfrak{Y}\mathfrak{P}\}$. Also $\tilde{\mathfrak{P}} = \langle \mathfrak{X} \mid \mathfrak{X} \in \mathscr{P} \rangle$, $\tilde{\mathfrak{Q}} = \langle \mathfrak{Y} \mid \mathfrak{Y} \in \mathscr{Q} \rangle$.

f) $\mathscr{K}$ is the set of maximal $\mathfrak{A}$-invariant $\{p, q\}$-subgroups of $\mathfrak{G}$.

11.4 Lemma. *Suppose that Hypothesis 11.3 holds.*

a) $\tilde{\mathfrak{P}} \in \mathscr{P}$ *and* $\tilde{\mathfrak{P}}\mathfrak{Q} = \mathfrak{Q}\tilde{\mathfrak{P}}$; *also* $\tilde{\mathfrak{Q}} \in \mathscr{Q}$ *and* $\mathfrak{P}\tilde{\mathfrak{Q}} = \tilde{\mathfrak{Q}}\mathfrak{P}$.

b) *If* $\mathfrak{P}_0 \leq \mathfrak{P}$ *and* $\mathfrak{P}_0$ *normalises a non-identity normal* $\mathfrak{A}$*-invariant subgroup of* $\mathfrak{Q}$, $\mathfrak{P}_0 \leq \tilde{\mathfrak{P}}$.

c) $\tilde{\mathfrak{P}}\mathfrak{Q}$ *and* $\mathfrak{P}\tilde{\mathfrak{Q}}$ *both lie in* $\mathscr{K}$.

Proof. a) If $\mathscr{P} = \{\mathfrak{X}_1, \ldots, \mathfrak{X}_r\}$,

$$\tilde{\mathfrak{P}} = \mathfrak{X}_1 \cdots \mathfrak{X}_r \mathfrak{X}_1 \cdots \mathfrak{X}_r \cdots \mathfrak{X}_1 \cdots \mathfrak{X}_r.$$

Since $\mathfrak{Q}\mathfrak{X}_i = \mathfrak{X}_i\mathfrak{Q}$, $\tilde{\mathfrak{P}}\mathfrak{Q} = \mathfrak{Q}\tilde{\mathfrak{P}}$. Similarly, $\mathfrak{P}\tilde{\mathfrak{Q}} = \tilde{\mathfrak{Q}}\mathfrak{P}$. Clearly $\tilde{\mathfrak{P}}$ and $\tilde{\mathfrak{Q}}$ are $\mathfrak{A}$-invariant.

b) We have $\mathfrak{P}_0 \leq \mathbf{N}_{\mathfrak{G}}(\hat{\mathfrak{Q}})$ for some $\mathfrak{A}$-invariant subgroup $\hat{\mathfrak{Q}}$ satisfying $1 \neq \hat{\mathfrak{Q}} \trianglelefteq \mathfrak{Q}$. Thus $\mathfrak{Q} \leq \mathbf{N}_{\mathfrak{G}}(\hat{\mathfrak{Q}}) < \mathfrak{G}$, by 11.3b). But $\mathbf{N}_{\mathfrak{G}}(\hat{\mathfrak{Q}})$ is $\mathfrak{A}$-invariant, so by 11.3c), $\mathbf{N}_{\mathfrak{G}}(\hat{\mathfrak{Q}})$ is soluble. By 11.1c), $\mathbf{N}_{\mathfrak{G}}(\hat{\mathfrak{Q}})$ has a unique $\mathfrak{A}$-invariant Hall $\{p, q\}$-subgroup $\mathfrak{H}$, and $\mathfrak{H} = \mathfrak{P}_1\mathfrak{Q}_1$ for the $\mathfrak{A}$-invariant Sylow p-subgroup $\mathfrak{P}_1$ and the $\mathfrak{A}$-invariant Sylow q-subgroup $\mathfrak{Q}_1$. Since $\mathfrak{Q} \leq \mathbf{N}_{\mathfrak{G}}(\hat{\mathfrak{Q}})$, $\mathfrak{Q}_1 = \mathfrak{Q}$, by 11.1b); also $\mathfrak{P}_0 \leq \mathfrak{P} \cap \mathbf{N}_{\mathfrak{G}}(\hat{\mathfrak{Q}}) \leq \mathfrak{P}_1$. But then $\mathfrak{P}_1\mathfrak{Q} = \mathfrak{Q}\mathfrak{P}_1 = \mathfrak{H}$, so $\mathfrak{P}_1 \in \mathscr{P}$ and $\mathfrak{P}_0 \leq \mathfrak{P}_1 \leq \tilde{\mathfrak{P}}$.

c) Suppose that $\tilde{\mathfrak{P}}\mathfrak{Q} \leq \mathfrak{K} \in \mathscr{K}$. Then $\mathfrak{Q} \in S_q(\mathfrak{K})$. If $\mathfrak{P}^*$ is the $\mathfrak{A}$-invariant Sylow p-subgroup of $\mathfrak{K}$, then $\tilde{\mathfrak{P}} \leq \mathfrak{P}^*$ by 11.1b). But $\mathfrak{P}^*\mathfrak{Q} = \mathfrak{Q}\mathfrak{P}^* = \mathfrak{K}$, so $\mathfrak{P}^* \in \mathscr{P}$ and $\mathfrak{P}^* \leq \tilde{\mathfrak{P}}$. Hence $\mathfrak{P}^* = \tilde{\mathfrak{P}}$ and $\mathfrak{K} = \tilde{\mathfrak{P}}\mathfrak{Q} \in \mathscr{K}$.

q.e.d.

11.5 Lemma. *Suppose that Hypothesis 11.3 holds and that* $\mathfrak{H}$ *is an* $\mathfrak{A}$*-invariant* $\{p, q\}$*-subgroup of* $\mathfrak{G}$. *Then* $\mathfrak{H} = (\mathfrak{H} \cap \mathfrak{P})(\mathfrak{H} \cap \mathfrak{Q})$ *and*

$$\mathfrak{H} \cap \mathfrak{P} = \mathbf{O}_p(\mathfrak{H})\mathbf{C}_{\mathfrak{H} \cap \mathfrak{P}}(\mathbf{Z}(\mathfrak{H} \cap \mathfrak{Q}))\mathbf{N}_{\mathfrak{H} \cap \mathfrak{P}}(\mathbf{J}(\mathfrak{H} \cap \mathfrak{Q})).$$

Proof. It follows from 11.1 that $\mathfrak{H} \cap \mathfrak{P}$, $\mathfrak{H} \cap \mathfrak{Q}$ are Sylow subgroups of $\mathfrak{H}$, so $\mathfrak{H} = (\mathfrak{H} \cap \mathfrak{P})(\mathfrak{H} \cap \mathfrak{Q})$. Let $\overline{\mathfrak{H}} = \mathfrak{H}/\mathbf{O}_p(\mathfrak{H})$, $\overline{\mathfrak{Q}} = (\mathfrak{H} \cap \mathfrak{Q})\mathbf{O}_p(\mathfrak{H})/\mathbf{O}_p(\mathfrak{H})$. Since $\mathfrak{H} \cap \mathfrak{Q} \in S_q(\mathfrak{H})$, $\overline{\mathfrak{Q}} \in S_q(\overline{\mathfrak{H}})$. Now $\overline{\mathfrak{H}}$ is soluble (by 2.8) and $\mathbf{O}_{q'}(\overline{\mathfrak{H}}) = \mathbf{O}_p(\overline{\mathfrak{H}}) = 1$; also $\mathfrak{A}$ operates without fixed points on $\overline{\mathfrak{H}}$ by 11.1a). Hence by 10.12,

$$\overline{\mathfrak{H}} = \mathbf{C}_{\overline{\mathfrak{H}}}(\mathbf{Z}(\overline{\mathfrak{Q}}))\mathbf{N}_{\overline{\mathfrak{H}}}(\mathbf{J}(\overline{\mathfrak{Q}})).$$

By 5.2, $\mathbf{N}_{\overline{\mathfrak{H}}}(\mathbf{J}(\overline{\mathfrak{Q}})) = \mathbf{N}_{\mathfrak{H}}(\mathbf{J}(\mathfrak{H} \cap \mathfrak{Q}))\mathbf{O}_p(\mathfrak{H})/\mathbf{O}_p(\mathfrak{H})$ and $\mathbf{N}_{\overline{\mathfrak{H}}}(\mathbf{Z}(\overline{\mathfrak{Q}})) = \mathbf{N}_{\mathfrak{H}}(\mathbf{Z}(\mathfrak{H} \cap \mathfrak{Q}))\mathbf{O}_p(\mathfrak{H})/\mathbf{O}_p(\mathfrak{H})$. Thus if $\mathbf{C}_{\overline{\mathfrak{H}}}(\mathbf{Z}(\overline{\mathfrak{Q}})) = \mathfrak{C}/\mathbf{O}_p(\mathfrak{H})$, $\mathfrak{C} = \mathfrak{C}_1\mathbf{O}_p(\mathfrak{H})$, where $\mathfrak{C}_1 = \mathfrak{C} \cap \mathbf{N}_{\mathfrak{H}}(\mathbf{Z}(\mathfrak{H} \cap \mathfrak{Q}))$. Thus $[\mathfrak{C}_1, \mathbf{Z}(\mathfrak{H} \cap \mathfrak{Q})] \leq \mathbf{Z}(\mathfrak{H} \cap \mathfrak{Q})$. But $\mathbf{Z}(\mathfrak{H} \cap \mathfrak{Q})\mathbf{O}_p(\mathfrak{H})/\mathbf{O}_p(\mathfrak{H}) \leq \mathbf{Z}(\overline{\mathfrak{Q}})$, so by definition of $\mathfrak{C}$,

$$[\mathfrak{C}_1, \mathbf{Z}(\mathfrak{H} \cap \mathfrak{Q})] \leq \mathbf{O}_p(\mathfrak{H}) \cap \mathbf{Z}(\mathfrak{H} \cap \mathfrak{Q}) = 1.$$

Thus $\mathfrak{C}_1 = \mathbf{C}_{\mathfrak{H}}(\mathbf{Z}(\mathfrak{H} \cap \mathfrak{Q}))$ and $\mathbf{C}_{\bar{\mathfrak{H}}}(\mathbf{Z}(\bar{\mathfrak{Q}})) = \mathbf{C}_{\mathfrak{H}}(\mathbf{Z}(\mathfrak{H} \cap \mathfrak{Q}))\mathbf{O}_p(\mathfrak{H})/\mathbf{O}_p(\mathfrak{H})$. Thus

$$\mathfrak{H} = \mathbf{O}_p(\mathfrak{H})\mathbf{C}_{\mathfrak{H}}(\mathbf{Z}(\mathfrak{H} \cap \mathfrak{Q}))\mathbf{N}_{\mathfrak{H}}(\mathbf{J}(\mathfrak{H} \cap \mathfrak{Q})).$$

Now $\mathfrak{H} \cap \mathfrak{Q}$ is contained in both $\mathbf{C}_{\mathfrak{H}}(\mathbf{Z}(\mathfrak{H} \cap \mathfrak{Q}))$ and $\mathbf{N}_{\mathfrak{H}}(\mathbf{J}(\mathfrak{H} \cap \mathfrak{Q}))$; also $\mathfrak{H} = (\mathfrak{H} \cap \mathfrak{P})(\mathfrak{H} \cap \mathfrak{Q})$. Hence

$$\begin{aligned} \mathfrak{H} &= \mathbf{O}_p(\mathfrak{H})\mathbf{C}_{\mathfrak{H} \cap \mathfrak{P}}(\mathbf{Z}(\mathfrak{H} \cap \mathfrak{Q}))(\mathfrak{H} \cap \mathfrak{Q})\mathbf{N}_{\mathfrak{H}}(\mathbf{J}(\mathfrak{H} \cap \mathfrak{Q})) \\ &= \mathbf{O}_p(\mathfrak{H})\mathbf{C}_{\mathfrak{H} \cap \mathfrak{P}}(\mathbf{Z}(\mathfrak{H} \cap \mathfrak{Q}))\mathbf{N}_{\mathfrak{H} \cap \mathfrak{P}}(\mathbf{J}(\mathfrak{H} \cap \mathfrak{Q}))(\mathfrak{H} \cap \mathfrak{Q}) \end{aligned}$$

and

$$\mathfrak{H} \cap \mathfrak{P} = \mathbf{O}_p(\mathfrak{H})\mathbf{C}_{\mathfrak{H} \cap \mathfrak{P}}(\mathbf{Z}(\mathfrak{H} \cap \mathfrak{Q}))\mathbf{N}_{\mathfrak{H} \cap \mathfrak{P}}(\mathbf{J}(\mathfrak{H} \cap \mathfrak{Q})). \qquad \textbf{q.e.d.}$$

11.6 Lemma. *Suppose that Hypothesis 11.3 holds and that* $\mathbf{Z}(\mathfrak{Q}) \leq \mathfrak{H} \in \mathscr{K}$. *Then* $\mathfrak{H} \cap \mathfrak{P} = \mathbf{O}_p(\mathfrak{H})(\mathfrak{H} \cap \tilde{\mathfrak{P}})$.

Proof. Suppose that this is false. By 11.5,

$$\mathfrak{H} \cap \mathfrak{P} = \mathbf{O}_p(\mathfrak{H})\mathbf{C}_{\mathfrak{H} \cap \mathfrak{P}}(\mathbf{Z}(\mathfrak{H} \cap \mathfrak{Q}))\mathbf{N}_{\mathfrak{H} \cap \mathfrak{P}}(\mathbf{J}(\mathfrak{H} \cap \mathfrak{Q})).$$

Since $\mathbf{Z}(\mathfrak{Q}) \leq \mathfrak{H}$, we have $\mathbf{Z}(\mathfrak{Q}) \leq \mathbf{Z}(\mathfrak{H} \cap \mathfrak{Q})$, whence $\mathbf{C}_{\mathfrak{H} \cap \mathfrak{P}}(\mathbf{Z}(\mathfrak{H} \cap \mathfrak{Q}))$ centralizes and normalises $\mathbf{Z}(\mathfrak{Q})$, so by 11.4b), $\mathbf{C}_{\mathfrak{H} \cap \mathfrak{P}}(\mathbf{Z}(\mathfrak{H} \cap \mathfrak{Q})) \leq \mathfrak{H} \cap \tilde{\mathfrak{P}}$. Since 11.6 is false, it follows that $\mathbf{N}_{\mathfrak{H} \cap \mathfrak{P}}(\mathbf{J}(\mathfrak{H} \cap \mathfrak{Q})) \not\leq \mathfrak{H} \cap \tilde{\mathfrak{P}}$, or $\mathbf{N}_{\mathfrak{P}}(\mathbf{J}(\mathfrak{H} \cap \mathfrak{Q})) \not\leq \tilde{\mathfrak{P}}$. Thus $\mathfrak{H} \cap \mathfrak{Q}$ lies in the set $\mathscr{R}$ of $\mathfrak{A}$ invariant subgroups $\hat{\mathfrak{Q}}$ of $\mathfrak{Q}$ for which $\mathbf{Z}(\mathfrak{Q}) \leq \hat{\mathfrak{Q}}$ and $\mathbf{N}_{\mathfrak{P}}(\mathbf{J}(\hat{\mathfrak{Q}})) \not\leq \tilde{\mathfrak{P}}$. Let $\mathfrak{Q}_1$ be a maximal element of $\mathscr{R}$. Since $\mathfrak{Q}_1 \neq 1$, $\mathbf{N}_{\mathfrak{G}}(\mathbf{J}(\mathfrak{Q}_1))$ is soluble. By 11.1c), $\mathbf{N}_{\mathfrak{G}}(\mathbf{J}(\mathfrak{Q}_1))$ has an $\mathfrak{A}$-invariant Hall $\{p, q\}$-subgroup $\mathfrak{R}$ and $\mathfrak{R} = \mathfrak{P}^*\mathfrak{Q}^*$, where $\mathfrak{P}^*$, $\mathfrak{Q}^*$ are respectively $\mathfrak{A}$-invariant Sylow p-, q-subgroups of $\mathfrak{R}$ and of $\mathbf{N}_{\mathfrak{G}}(\mathbf{J}(\mathfrak{Q}_1))$. By 11.5,

$$\mathfrak{P}^* = \mathbf{O}_p(\mathfrak{R})\mathbf{C}_{\mathfrak{P}^*}(\mathbf{Z}(\mathfrak{Q}^*))\mathbf{N}_{\mathfrak{P}^*}(\mathbf{J}(\mathfrak{Q}^*)),$$

since $\mathfrak{P}^* = \mathfrak{R} \cap \mathfrak{P}$, $\mathfrak{Q}^* = \mathfrak{R} \cap \mathfrak{Q}$. Now $\mathbf{Z}(\mathfrak{Q}) \leq \mathfrak{Q}_1$, so $\mathbf{Z}(\mathfrak{Q}) \leq \mathbf{Z}(\mathfrak{Q}_1) \leq \mathbf{J}(\mathfrak{Q}_1) \leq \mathbf{O}_q(\mathfrak{R})$ (since $\mathfrak{R} \leq \mathbf{N}_{\mathfrak{G}}(\mathbf{J}(\mathfrak{Q}_1))$). Thus $\mathbf{O}_p(\mathfrak{R}) \leq \mathbf{C}_{\mathfrak{R}}(\mathbf{Z}(\mathfrak{Q}))$ and $\mathbf{O}_p(\mathfrak{R}) \leq \tilde{\mathfrak{P}}$ by 11.4b). Again, $\mathfrak{Q}_1 \leq \mathfrak{Q}^*$ by 11.1, so $\mathbf{Z}(\mathfrak{Q}) \leq \mathbf{Z}(\mathfrak{Q}^*)$; thus $\mathbf{C}_{\mathfrak{P}^*}(\mathbf{Z}(\mathfrak{Q}^*)) \leq \mathbf{C}_{\mathfrak{R}}(\mathbf{Z}(\mathfrak{Q}))$ and again, $\mathbf{C}_{\mathfrak{P}^*}(\mathbf{Z}(\mathfrak{Q}^*)) \leq \tilde{\mathfrak{P}}$. But $\mathfrak{P}^* = \mathbf{N}_{\mathfrak{P}}(\mathbf{J}(\mathfrak{Q}_1)) \not\leq \tilde{\mathfrak{P}}$. Thus $\mathbf{N}_{\mathfrak{P}^*}(\mathbf{J}(\mathfrak{Q}^*)) \not\leq \tilde{\mathfrak{P}}$ and $\mathfrak{Q}^* \in \mathscr{R}$. By maximality of $\mathfrak{Q}_1$, $\mathfrak{Q}^* = \mathfrak{Q}_1 \in S_q(\mathbf{N}_{\mathfrak{G}}(\mathbf{J}(\mathfrak{Q}_1)))$. By 2.4, $\mathfrak{Q}_1 \in S_q(\mathfrak{G})$, so $\mathfrak{Q}_1 = \mathfrak{Q}$. This gives a contradiction, since by 11.4b), $\mathbf{N}_{\mathfrak{P}}(\mathbf{J}(\mathfrak{Q})) \leq \tilde{\mathfrak{P}}$. **q.e.d.**

By 11.4c), $\tilde{\mathfrak{P}}\mathfrak{Q}$ and $\mathfrak{P}\tilde{\mathfrak{Q}}$ lie in $\mathscr{K}$. Let $\overline{\mathscr{K}} = \mathscr{K} - \{\tilde{\mathfrak{P}}\mathfrak{Q}, \mathfrak{P}\tilde{\mathfrak{Q}}\}$.

11.7 Corollary. *Under Hypothesis 11.3, if* $\mathfrak{H} \in \overline{\mathscr{K}}$, *then* $\mathbf{O}_p(\mathfrak{H}) \neq 1$ *and* $\mathbf{O}_q(\mathfrak{H}) \neq 1$.

Proof. Suppose, for instance, that $\mathbf{O}_p(\mathfrak{H}) = 1$. Then $\mathbf{O}_q(\mathfrak{H}) \neq 1$. Thus $\mathbf{N}_{\mathfrak{G}}(\mathbf{O}_q(\mathfrak{H}))$ is soluble and has an $\mathfrak{A}$-invariant Hall $\{p, q\}$-subgroup $\mathfrak{H}^*$ containing $\mathfrak{H}$, by 11.1. Since $\mathfrak{H} \in \mathscr{K}$, $\mathfrak{H} = \mathfrak{H}^*$. But $\mathbf{O}_q(\mathfrak{H}) \leq \mathfrak{Q}$, so $\mathbf{Z}(\mathfrak{Q}) \leq \mathbf{N}_{\mathfrak{G}}(\mathbf{O}_q(\mathfrak{H}))$; by 11.1, $\mathbf{Z}(\mathfrak{Q}) \leq \mathfrak{H}^* = \mathfrak{H}$. By 11.6, $\mathfrak{H} \cap \mathfrak{P} = \mathfrak{H} \cap \tilde{\mathfrak{P}}$, so $\mathfrak{H} = (\mathfrak{H} \cap \mathfrak{P})(\mathfrak{H} \cap \mathfrak{Q}) \leq \tilde{\mathfrak{P}}\mathfrak{Q}$. Thus $\mathfrak{H} = \tilde{\mathfrak{P}}\mathfrak{Q}$ and $\mathfrak{H} \notin \overline{\mathscr{K}}$. **q.e.d.**

11.8 Lemma. *Under Hypothesis 11.3, if* $\mathfrak{H} \in \overline{\mathscr{K}}$, *then* $\mathbf{Z}(\mathfrak{P}) \leq \mathfrak{H}$, $\mathbf{Z}(\mathfrak{Q}) \leq \mathfrak{H}$ *and*

$$\mathfrak{H} = \mathbf{F}(\mathfrak{H})(\mathfrak{H} \cap \tilde{\mathfrak{P}})(\mathfrak{H} \cap \tilde{\mathfrak{Q}}).$$

Proof. By 11.7, $\mathbf{O}_p(\mathfrak{H}) \neq 1$. As in 11.7, $\mathbf{N}_{\mathfrak{G}}(\mathbf{O}_p(\mathfrak{H}))$ has a unique $\mathfrak{A}$-invariant Hall $\{p, q\}$-subgroup which must be $\mathfrak{H}$ itself. But since $\mathbf{O}_p(\mathfrak{H}) \leq \mathfrak{P}$, $\mathbf{Z}(\mathfrak{P}) \leq \mathbf{N}_{\mathfrak{G}}(\mathbf{O}_p(\mathfrak{H}))$, so $\mathbf{Z}(\mathfrak{P}) \leq \mathfrak{H}$. Similarly $\mathbf{Z}(\mathfrak{Q}) \leq \mathfrak{H}$. By 11.6, $\mathfrak{H} \cap \mathfrak{P} = \mathbf{O}_p(\mathfrak{H})(\mathfrak{H} \cap \tilde{\mathfrak{P}})$ and $\mathfrak{H} \cap \mathfrak{Q} = \mathbf{O}_q(\mathfrak{H})(\mathfrak{H} \cap \tilde{\mathfrak{Q}})$. Thus

$$\begin{aligned}\mathfrak{H} &= (\mathfrak{H} \cap \mathfrak{P})(\mathfrak{H} \cap \mathfrak{Q}) = \mathbf{O}_p(\mathfrak{H})(\mathfrak{H} \cap \tilde{\mathfrak{P}})\mathbf{O}_q(\mathfrak{H})(\mathfrak{H} \cap \tilde{\mathfrak{Q}}) \\ &= \mathbf{O}_p(\mathfrak{H})\mathbf{O}_q(\mathfrak{H})(\mathfrak{H} \cap \tilde{\mathfrak{P}})(\mathfrak{H} \cap \tilde{\mathfrak{Q}}) = \mathbf{F}(\mathfrak{H})(\mathfrak{H} \cap \tilde{\mathfrak{P}})(\mathfrak{H} \cap \tilde{\mathfrak{Q}}).\end{aligned}$$

q.e.d.

11.9 Lemma. *Under Hypothesis 11.3, if* $\mathfrak{H} \in \overline{\mathscr{K}}$, *then* $\tilde{\mathfrak{P}} \cap \mathbf{O}_p(\mathfrak{H}) = \tilde{\mathfrak{Q}} \cap \mathbf{O}_q(\mathfrak{H}) = 1$.

Proof. Suppose that $\mathfrak{X} = \tilde{\mathfrak{P}} \cap \mathbf{O}_p(\mathfrak{H}) \neq 1$. Then $\mathfrak{X}\mathbf{O}_q(\mathfrak{H})$ is a subgroup of $\mathbf{F}(\mathfrak{H})$ of order divisible by pq. Also $\mathfrak{X}\mathbf{O}_q(\mathfrak{H}) \leq \tilde{\mathfrak{P}}\mathfrak{Q}$, so by 11.2, $\tilde{\mathfrak{P}}\mathfrak{Q} \leq \mathfrak{H}$. By 11.4c), $\mathfrak{H} = \tilde{\mathfrak{P}}\mathfrak{Q} \notin \overline{\mathscr{K}}$, a contradiction. Thus $\tilde{\mathfrak{P}} \cap \mathbf{O}_p(\mathfrak{H}) = 1$. Similarly $\tilde{\mathfrak{Q}} \cap \mathbf{O}_q(\mathfrak{H}) = 1$. **q.e.d.**

11.10 Lemma. *Suppose that Hypothesis 11.3 holds. If* $\mathfrak{H} \in \overline{\mathscr{K}}$, $\mathfrak{M}$ *is a non-identity* $\mathfrak{A}$-*invariant subgroup of* $\mathbf{F}(\mathfrak{H})$ *and* $\mathfrak{M} \leq \mathfrak{K} \in \overline{\mathscr{K}}$, *then* $\mathfrak{K} = \mathfrak{H}$.

Proof. Since $\mathfrak{M} \neq 1$, either $\mathbf{O}_p(\mathfrak{M}) \neq 1$ or $\mathbf{O}_q(\mathfrak{M}) \neq 1$. Replacing $\mathfrak{M}$ by $\mathbf{O}_p(\mathfrak{M})$ if $\mathbf{O}_p(\mathfrak{M}) \neq 1$ and by $\mathbf{O}_q(\mathfrak{M})$ otherwise, we may suppose that $\mathfrak{M}$ is of prime-power order. Indeed, we may suppose that $\mathfrak{M}$ is a q-group. Let $\mathfrak{Q}_1 = [\mathbf{Z}(\mathfrak{P}), \mathfrak{M}]$. By 11.8, $\mathbf{Z}(\mathfrak{P}) \leq \mathfrak{H}$, so

$$\mathfrak{Q}_1 \leq [\mathfrak{H}, \mathbf{O}_q(\mathfrak{H})] \leq \mathbf{O}_q(\mathfrak{H}).$$

But also by 11.8, $\mathbf{Z}(\mathfrak{P}) \leq \mathfrak{K}$, and by 11.1, $\mathbf{O}_{p',p}(\mathfrak{K}) = (\mathbf{O}_{p',p}(\mathfrak{K}) \cap \mathfrak{P})\mathbf{O}_{p'}(\mathfrak{K})$. Thus $\mathbf{Z}(\mathfrak{P})$ centralizes $\mathbf{O}_{p',p}(\mathfrak{K})/\mathbf{O}_{p'}(\mathfrak{K})$. By IX, 1.4, $\mathbf{Z}(\mathfrak{P}) \leq \mathbf{O}_{p',p}(\mathfrak{K})$, so

$$\mathfrak{Q}_1 \leq [\mathbf{O}_{p',p}(\mathfrak{K}), \mathfrak{K}] \leq \mathbf{O}_{p',p}(\mathfrak{K}).$$

Since $\mathfrak{Q}_1$ is a q-group, $\mathfrak{Q}_1 \leq \mathbf{O}_{p'}(\mathfrak{K}) = \mathbf{O}_q(\mathfrak{K})$. Thus $\mathbf{O}_p(\mathfrak{K}) \leq \mathbf{C}_{\mathfrak{P}}(\mathfrak{Q}_1)$.

If $\mathfrak{Q}_1 = 1$, $\mathfrak{M} \leq \mathbf{C}_{\mathfrak{G}}(\mathbf{Z}(\mathfrak{P}))$, and by 11.4b), $\mathfrak{M} \leq \tilde{\mathfrak{Q}}$. But this is impossible by 11.9, since $1 \neq \mathfrak{M} \leq \mathbf{O}_q(\mathfrak{H})$. Thus $\mathfrak{Q}_1 \neq 1$. Hence $\mathbf{C}_{\mathfrak{G}}(\mathfrak{Q}_1)$ is soluble and $\mathfrak{A}$-invariant. Hence the $\mathfrak{A}$-invariant Hall $\{p, q\}$-subgroup of $\mathbf{C}_{\mathfrak{G}}(\mathfrak{Q}_1)$ is $\mathbf{C}_{\mathfrak{P}}(\mathfrak{Q}_1)\mathbf{C}_{\mathfrak{Q}}(\mathfrak{Q}_1)$. This contains $\mathbf{C}_{\mathbf{F}(\mathfrak{H})}(\mathfrak{Q}_1)$, the order of which is divisible by pq. By 11.2, $\mathbf{C}_{\mathfrak{P}}(\mathfrak{Q}_1)\mathbf{C}_{\mathfrak{Q}}(\mathfrak{Q}_1) \leq \mathfrak{H}$. Hence $\mathbf{O}_p(\mathfrak{K}) \leq \mathfrak{H}$ and $\mathbf{O}_p(\mathfrak{K})\mathfrak{Q}_1 \leq \mathfrak{H}$; further $\mathbf{O}_p(\mathfrak{K})\mathfrak{Q}_1$ is an $\mathfrak{A}$-invariant subgroup of $\mathbf{F}(\mathfrak{K})$ of order divisible by pq. Again by 11.2 (applied to $\mathfrak{K}$), $\mathfrak{H} = \mathfrak{K}$. **q.e.d.**

11.11 Lemma. *Suppose that Hypothesis 11.3 holds and that* $\mathbf{Z}(\mathfrak{A})$ *contains an elementary Abelian subgroup* $\mathfrak{A}_0$ *of order* r^3 *for some prime* r. *Then* $|\overline{\mathscr{K}}| \leq 1$.

Proof. Suppose that $\mathfrak{H}_1 \in \overline{\mathscr{K}}$, $\mathfrak{H}_2 \in \overline{\mathscr{K}}$. By 11.7, $\mathbf{O}_p(\mathfrak{H}_1) \neq 1$. By 1.9,

$$\mathbf{O}_p(\mathfrak{H}_1) = \langle \mathbf{C}_{\mathfrak{G}}(\mathfrak{B}) \cap \mathbf{O}_p(\mathfrak{H}_1) | \mathfrak{B} \leq \mathfrak{A}_0, \mathfrak{A}_0/\mathfrak{B} \text{ is cyclic}\rangle.$$

Thus there exists a subgroup $\mathfrak{A}_1$ of $\mathfrak{A}_0$ of order r^2 for which $\mathbf{C}_{\mathfrak{G}}(\mathfrak{A}_1) \cap \mathbf{O}_p(\mathfrak{H}_1) \neq 1$. Similarly, there exists a subgroup $\mathfrak{A}_2$ of $\mathfrak{A}_0$ of order r^2 for which $\mathbf{C}_{\mathfrak{G}}(\mathfrak{A}_2) \cap \mathbf{O}_q(\mathfrak{H}_2) \neq 1$. Hence $\mathfrak{A}_1 \cap \mathfrak{A}_2 \neq 1$; choose $\alpha \in \mathfrak{A}_1 \cap \mathfrak{A}_2$, $\alpha \neq 1$. Since $\alpha \in \mathbf{Z}(\mathfrak{A})$, $\mathbf{C}_{\mathfrak{G}}(\alpha)$ is $\mathfrak{A}$-invariant. Since $\alpha \neq 1$, $\mathbf{C}_{\mathfrak{G}}(\alpha) < \mathfrak{G}$ and so $\mathbf{C}_{\mathfrak{G}}(\alpha)$ is soluble; the $\mathfrak{A}$-invariant Hall $\{p, q\}$-subgroup of $\mathbf{C}_{\mathfrak{G}}(\alpha)$ is $\mathbf{C}_{\mathfrak{P}}(\alpha)\mathbf{C}_{\mathfrak{Q}}(\alpha)$. Let $\mathfrak{K}$ be an element of $\mathscr{K}$ for which $\mathfrak{K} \geq \mathbf{C}_{\mathfrak{P}}(\alpha)\mathbf{C}_{\mathfrak{Q}}(\alpha)$. If $\mathfrak{K} = \tilde{\mathfrak{P}}\mathfrak{Q}$, then $\mathbf{C}_{\mathfrak{P}}(\alpha) \leq \tilde{\mathfrak{P}}$ and $1 \neq \mathbf{C}_{\mathfrak{G}}(\mathfrak{A}_1) \cap \mathbf{O}_p(\mathfrak{H}_1) \leq \mathbf{C}_{\mathfrak{G}}(\alpha) \cap \mathbf{O}_p(\mathfrak{H}_1) \leq \tilde{\mathfrak{P}} \cap \mathbf{O}_p(\mathfrak{H}_1)$, contrary to 11.9. Thus $\mathfrak{K} \neq \tilde{\mathfrak{P}}\mathfrak{Q}$. Similarly, $\mathfrak{K} \neq \mathfrak{P}\tilde{\mathfrak{Q}}$. Hence $\mathfrak{K} \in \overline{\mathscr{K}}$. Now $\mathbf{C}_{\mathfrak{P}}(\mathfrak{A}_1) \cap \mathbf{O}_p(\mathfrak{H}_1)$ is a non-identity $\mathfrak{A}$-invariant subgroup of $\mathbf{F}(\mathfrak{H}_1)$ and of $\mathfrak{K}$, so by 11.10, $\mathfrak{K} = \mathfrak{H}_1$. Similarly, $\mathfrak{K} = \mathfrak{H}_2$. Thus $\mathfrak{H}_1 = \mathfrak{H}_2$. **q.e.d.**

To apply this, we need a slight modification of 1.9.

11.12 Lemma (GORENSTEIN [1, Theorem 5.3.16]). *Suppose that* $\mathfrak{A}$ *is an elementary Abelian p-group of operators on a q-group* $\mathfrak{G}$, *where* p, q *are distinct primes. If* $\mathfrak{A}$ *is not cyclic and* $\alpha_1, \ldots, \alpha_n$ *are generators of all the non-identity cyclic subgroups of* $\mathfrak{A}$,

$$\mathfrak{G} = \mathbf{C}_{\mathfrak{G}}(\alpha_1) \cdots \mathbf{C}_{\mathfrak{G}}(\alpha_n).$$

Proof. This is proved by induction on $|\mathfrak{G}|$; suppose $\mathfrak{G} \neq 1$ and $\mathfrak{Z} = \mathbf{Z}(\mathfrak{G})$. By IX, 6.11, $\mathbf{C}_{\mathfrak{G}/\mathfrak{Z}}(\alpha_i) = \mathbf{C}_{\mathfrak{G}}(\alpha_i)\mathfrak{Z}/\mathfrak{Z}$, so by the inductive hypothesis,

$$\mathfrak{G} = \mathfrak{Z}\mathbf{C}_{\mathfrak{G}}(\alpha_1) \cdots \mathbf{C}_{\mathfrak{G}}(\alpha_n).$$

By 1.9,

$$\mathfrak{Z} = \langle \mathbf{C}_{\mathfrak{Z}}(\alpha) | \alpha \in \mathfrak{A}, \alpha \neq 1 \rangle = \mathbf{C}_{\mathfrak{Z}}(\alpha_1) \cdots \mathbf{C}_{\mathfrak{Z}}(\alpha_n).$$

Since $\mathfrak{Z} = \mathbf{Z}(\mathfrak{G})$ and $\mathbf{C}_{\mathfrak{Z}}(\alpha_i) \leq \mathbf{C}_{\mathfrak{G}}(\alpha_i)$, it follows at once that

$$\mathfrak{G} = \mathbf{C}_{\mathfrak{G}}(\alpha_1) \cdots \mathbf{C}_{\mathfrak{G}}(\alpha_n).$$ **q.e.d.**

11.13 Theorem (MARTINEAU). *Let $\mathfrak{A}$ be a soluble group of automorphisms of $\mathfrak{G}$ such that $\mathbf{C}_{\mathfrak{G}}(\mathfrak{A}) = 1$ and $(|\mathfrak{A}|, |\mathfrak{G}|) = 1$. Suppose that $\mathbf{Z}(\mathfrak{A})$ contains an elementary Abelian subgroup $\mathfrak{A}_0$ of order r^3 for some prime r. If $\mathbf{C}_{\mathfrak{G}}(\alpha)$ is soluble for each non-identity element α of $\mathfrak{A}$, $\mathfrak{G}$ is soluble.*

Proof. Suppose that this is false and let $\mathfrak{G}$ be a counterexample of minimal order. If $\mathfrak{S}_1, \ldots, \mathfrak{S}_n$ are the $\mathfrak{A}$-invariant Sylow subgroups of $\mathfrak{G}$ and $\mathfrak{S}_i\mathfrak{S}_j = \mathfrak{S}_j\mathfrak{S}_i$ for all i, j then $\prod_{j \neq i} \mathfrak{S}_j$ is a Sylow complement of $\mathfrak{G}$ $(i = 1, \ldots, n)$. But then $\mathfrak{G}$ is soluble, by VI, 1.10. This is not so, so there exist distinct primes p, q such that $\mathfrak{P}\mathfrak{Q} \neq \mathfrak{Q}\mathfrak{P}$, where $\mathfrak{P}$, $\mathfrak{Q}$ are respectively the $\mathfrak{A}$-invariant Sylow p-, q-subgroups of $\mathfrak{G}$.

If $\mathfrak{H}$ is a proper $\mathfrak{A}$-invariant subgroup of $\mathfrak{G}$, then $\mathfrak{H}$ is soluble. For if $\mathfrak{A}$ operates faithfully on $\mathfrak{H}$, then $\mathfrak{H}$ is soluble by minimality of $\mathfrak{G}$. And otherwise, there exists $\alpha \in \mathfrak{A}$ such that $\mathfrak{H} \leq \mathbf{C}_{\mathfrak{G}}(\alpha)$; but $\mathbf{C}_{\mathfrak{G}}(\alpha)$ is soluble by hypothesis.

If $\mathfrak{N}$ is a non-identity $\mathfrak{A}$-invariant proper normal subgroup of $\mathfrak{G}$, then $\mathfrak{N}$ is soluble, so $\mathfrak{G}/\mathfrak{N}$ is not. By minimality of $\mathfrak{G}$, $\mathfrak{A}$ does not operate faithfully on $\mathfrak{G}/\mathfrak{N}$. Thus there exists $\alpha \in \mathfrak{A}$ such that $\mathbf{C}_{\mathfrak{G}/\mathfrak{N}}(\alpha) = \mathfrak{G}/\mathfrak{N}$. But by I, 18.6, $\mathbf{C}_{\mathfrak{G}/\mathfrak{N}}(\alpha) = \mathbf{C}_{\mathfrak{G}}(\alpha)\,\mathfrak{N}/\mathfrak{N}$, so $\mathbf{C}_{\mathfrak{G}/\mathfrak{N}}(\alpha)$ is soluble. This is a contradiction, so $\mathfrak{G}$ has no non-identity $\mathfrak{A}$-invariant proper normal subgroup. Thus Hypothesis 11.3 holds in $\mathfrak{G}$. Since $\mathfrak{P}\mathfrak{Q} \neq \mathfrak{Q}\mathfrak{P}$, $\tilde{\mathfrak{Q}} < \mathfrak{Q}$ and $\tilde{\mathfrak{P}} < \mathfrak{P}$.

By 1.9, $\mathfrak{Q} = \langle \mathbf{C}_{\mathfrak{Q}}(\mathfrak{B}) | \mathfrak{B} \leq \mathfrak{A}_0, |\mathfrak{B}| = r^2 \rangle$. Hence there exists a subgroup $\mathfrak{C}$ of $\mathfrak{A}_0$ of order r^2 such that $\mathbf{C}_{\mathfrak{Q}}(\mathfrak{C}) \not\leq \tilde{\mathfrak{Q}}$. Thus if $\gamma \in \mathfrak{C} - \{1\}$, $\mathbf{C}_{\mathfrak{Q}}(\gamma) \not\leq \tilde{\mathfrak{Q}}$. Hence the $\mathfrak{A}$-invariant Hall $\{p, q\}$-subgroup $\mathbf{C}_{\mathfrak{P}}(\gamma)\mathbf{C}_{\mathfrak{Q}}(\gamma)$ of $\mathbf{C}_{\mathfrak{G}}(\gamma)$ is not contained in $\mathfrak{P}\tilde{\mathfrak{Q}}$. Now by 11.11, $\mathscr{K}$ contains at most one element other than $\mathfrak{P}\tilde{\mathfrak{Q}}$, $\tilde{\mathfrak{P}}\mathfrak{Q}$; denote it by $\mathfrak{H}$ if it exists. Thus either $\mathbf{C}_{\mathfrak{P}}(\gamma)\mathbf{C}_{\mathfrak{Q}}(\gamma) \leq \tilde{\mathfrak{P}}\mathfrak{Q}$ or $\mathbf{C}_{\mathfrak{P}}(\gamma)\mathbf{C}_{\mathfrak{Q}}(\gamma) \leq \mathfrak{H}$. Hence, for each $\gamma \in \mathfrak{C} - \{1\}$, either $\mathbf{C}_{\mathfrak{P}}(\gamma) \leq \tilde{\mathfrak{P}}$ or $\mathbf{C}_{\mathfrak{P}}(\gamma) \leq \mathfrak{H} \cap \mathfrak{P}$. Let $\gamma_1, \ldots, \gamma_{r+1}$ be generators of all the non-identity cyclic subgroups of $\mathfrak{C}$, the order being so chosen that $\mathbf{C}_{\mathfrak{P}}(\gamma_i) \leq \tilde{\mathfrak{P}}$ $(i = 1, \ldots, n)$ and $\mathbf{C}_{\mathfrak{P}}(\gamma_i) \leq \mathfrak{H} \cap \mathfrak{P}$ $(i = n + 1, \ldots, r + 1)$. By 11.12,

$$\mathfrak{P} = \mathbf{C}_{\mathfrak{P}}(\gamma_1) \cdots \mathbf{C}_{\mathfrak{P}}(\gamma_{r+1}) \leq \tilde{\mathfrak{P}}(\mathfrak{H} \cap \mathfrak{P}).$$

Thus $\mathfrak{H}$ exists; using 11.8 and 11.6, it follows that $\mathfrak{P} = \tilde{\mathfrak{P}}\mathbf{O}_p(\mathfrak{H})$. Similarly $\mathfrak{Q} = \mathbf{O}_q(\mathfrak{H})\tilde{\mathfrak{Q}}$. Hence

$$\begin{aligned}\mathfrak{P}\mathfrak{Q} &= \mathbf{O}_p(\mathfrak{H})\tilde{\mathfrak{P}}\mathfrak{Q} = \mathbf{O}_p(\mathfrak{H})\mathfrak{Q}\tilde{\mathfrak{P}}\\ &= \mathbf{O}_p(\mathfrak{H})\mathbf{O}_q(\mathfrak{H})\tilde{\mathfrak{Q}}\tilde{\mathfrak{P}} = \mathbf{O}_q(\mathfrak{H})\mathbf{O}_p(\mathfrak{H})\tilde{\mathfrak{Q}}\tilde{\mathfrak{P}}\\ &\subseteq \mathbf{O}_q(\mathfrak{H})\mathfrak{P}\tilde{\mathfrak{Q}}\tilde{\mathfrak{P}} = \mathbf{O}_q(\mathfrak{H})\tilde{\mathfrak{Q}}\mathfrak{P} = \mathfrak{Q}\mathfrak{P},\end{aligned}$$

so $\mathfrak{P}\mathfrak{Q} = \mathfrak{Q}\mathfrak{P}$, a contradiction. **q.e.d.**

Theorem 11.13 reduces the problem in the case when $\mathfrak{A}$ is elementary Abelian to the case when $|\mathfrak{A}| = r^2$. To deal with this, we need several more lemmas. First we observe that the condition $(|\mathfrak{A}|, |\mathfrak{G}|) = 1$ can be dropped when $|\mathfrak{A}|$ is a prime-power.

11.14 Lemma. *Suppose that $\mathfrak{A}$ is an r-group of operators on a group $\mathfrak{G}$, where r is a prime and $\mathbf{C}_{\mathfrak{G}}(\mathfrak{A}) = 1$. Then $|\mathfrak{G}| \equiv 1\ (r)$ and $(|\mathfrak{A}|, |\mathfrak{G}|) = 1$.*

Proof. If $A_1, \ldots, A_s$ are the orbits of the permutation representation of $\mathfrak{A}$ on $\mathfrak{G}$ with $A_1 = \{1\}$, then $|\mathfrak{G}| = |A_1| + \cdots + |A_s|$. Since $\mathbf{C}_{\mathfrak{G}}(\mathfrak{A}) = 1$, $|A_i| > 1$ for $i = 2, \ldots, s$. But by I, 5.10, $|A_i|$ is a divisor of $|\mathfrak{A}|$ and is therefore a power of r. Thus for $i > 1$, $|A_i|$ is divisible by r. Hence $|\mathfrak{G}| \equiv 1\ (r)$ and $(|\mathfrak{A}|, |\mathfrak{G}|) = 1$. **q.e.d.**

11.15 Lemma. *Suppose that $\mathfrak{G}$ is a p-soluble group of p-length 1. Then $S_p(\mathbf{C}_{\mathfrak{G}}(\mathbf{O}_{p'}(\mathfrak{G}))) = \{\mathbf{O}_p(\mathfrak{G})\}$.*

Proof. Let $\mathfrak{C} = \mathbf{C}_{\mathfrak{G}}(\mathbf{O}_{p'}(\mathfrak{G}))$. Since $\mathfrak{C} \trianglelefteq \mathfrak{G}$, $\mathbf{O}_{p'}(\mathfrak{C}) \leq \mathbf{O}_{p'}(\mathfrak{G})$. Thus $\mathbf{O}_{p'}(\mathfrak{C}) \leq \mathbf{Z}(\mathbf{O}_{p'}(\mathfrak{G})) = \mathfrak{Z}$, say, and $\mathfrak{Z} = \mathbf{O}_{p'}(\mathfrak{C})$. But the p-length of $\mathfrak{C}$ is at most 1, so if $\mathfrak{P} \in S_p(\mathfrak{C})$, $\mathfrak{P}\mathfrak{Z} \trianglelefteq \mathfrak{C}$. And since $\mathfrak{P} \leq \mathbf{C}_{\mathfrak{G}}(\mathfrak{Z})$, $\mathfrak{P} \trianglelefteq \mathfrak{P}\mathfrak{Z}$. Thus $\mathfrak{P}$ is a characteristic subgroup of $\mathfrak{P}\mathfrak{Z}$ and of $\mathfrak{C}$. Hence $\mathfrak{P} \trianglelefteq \mathfrak{G}$ and $\mathfrak{P} \leq \mathbf{O}_p(\mathfrak{G})$. But $\mathbf{O}_p(\mathfrak{G}) \leq \mathfrak{C}$ and $\mathfrak{P} \in S_p(\mathfrak{C})$, so $\mathfrak{P} = \mathbf{O}_p(\mathfrak{G})$. **q.e.d.**

11.16 Lemma. *Suppose that $\mathfrak{A}$ is an elementary Abelian r-group of operators on a group $\mathfrak{G}$ and that $\mathbf{C}_{\mathfrak{G}}(\mathfrak{A}) = 1$.*

a) *Suppose that $\mathfrak{H}$, $\mathfrak{K}$ are $\mathfrak{A}$-invariant subgroups of $\mathfrak{G}$ of coprime orders and that $\mathfrak{H} \leq \mathbf{N}_{\mathfrak{G}}(\mathfrak{K})$. Then $\mathfrak{H} = \mathbf{C}_{\mathfrak{H}}(\mathfrak{K})\mathbf{C}_{\mathfrak{H}}(\mathfrak{C})$, where $\mathfrak{C} = \langle \alpha \mid \alpha \in \mathfrak{A}, \mathbf{C}_{\mathfrak{K}}(\alpha) = 1\rangle$.*

b) *Suppose that $\mathfrak{K}$ is an $\mathfrak{A}$-invariant π-subgroup of $\mathfrak{G}$. If $\mathfrak{A} = \langle \alpha_1, \ldots, \alpha_n\rangle$ and $\mathbf{C}_{\mathfrak{K}}(\alpha_i) = 1$ $(i = 1, \ldots, n)$, then $\mathbf{N}_{\mathfrak{G}}(\mathfrak{K})/\mathbf{C}_{\mathfrak{G}}(\mathfrak{K})$ is a π-group.*

c) *Suppose that q is odd and $\mathfrak{Q}$ is an $\mathfrak{A}$-invariant Sylow q-subgroup of $\mathfrak{G}$. If $\mathfrak{A} = \langle \alpha_1, \ldots, \alpha_n \rangle$ and $\mathbf{C}_{\mathfrak{Q}}(\alpha_i) = 1$ $(i = 1, \ldots, n)$, then $\mathfrak{G}$ is q-nilpotent.*

Proof. a) Let $\mathfrak{N} = \mathbf{C}_{\mathfrak{H}}(\mathfrak{K})$. Then $\mathfrak{N}$ is a normal $\mathfrak{A}$-invariant subgroup of $\mathfrak{H}\mathfrak{K}$. Suppose that $\alpha \in \mathfrak{A}$ and $\mathbf{C}_{\mathfrak{K}}(\alpha) = 1$.

We show that if $\mathfrak{B}$ is a maximal subgroup of $\mathfrak{A}$, either $\mathbf{C}_{\mathfrak{H}}(\mathfrak{B}) \le \mathbf{C}_{\mathfrak{H}}(\alpha)$ or $\mathbf{C}_{\mathfrak{H}}(\mathfrak{B}) \le \mathfrak{N}$. If $\alpha \in \mathfrak{B}$, $\mathbf{C}_{\mathfrak{H}}(\mathfrak{B}) \le \mathbf{C}_{\mathfrak{H}}(\alpha)$. If $\alpha \notin \mathfrak{B}$, $\mathfrak{A} = \mathfrak{B}\langle\alpha\rangle$. Since $\mathbf{C}_{\mathfrak{H}}(\mathfrak{A}) = 1$, it follows that α leaves fixed no non-identity element of $\mathbf{C}_{\mathfrak{H}}(\mathfrak{B})$. But $\mathbf{C}_{\mathfrak{H}}(\mathfrak{B})$ is $\langle\alpha\rangle$-invariant. It follows from 11.1a) that $\langle\alpha\rangle$ operates fixed point freely on $\mathbf{C}_{\mathfrak{H}}(\mathfrak{B})/(\mathbf{C}_{\mathfrak{H}}(\mathfrak{B}) \cap \mathfrak{K})$ and hence on $\mathbf{C}_{\mathfrak{H}}(\mathfrak{B})\mathfrak{K}/\mathfrak{K}$. But also $\mathbf{C}_{\mathfrak{K}}(\alpha) = 1$, so $\langle\alpha\rangle$ acts fixed-point-freely on $\mathbf{C}_{\mathfrak{H}}(\mathfrak{B})\mathfrak{K}$. By V, 8.14, $\mathbf{C}_{\mathfrak{H}}(\mathfrak{B})\mathfrak{K}$ is nilpotent. Since $\mathfrak{H}$, $\mathfrak{K}$ are of coprime orders, it follows that $\mathbf{C}_{\mathfrak{H}}(\mathfrak{B}) \le \mathbf{C}_{\mathfrak{H}}(\mathfrak{K}) = \mathfrak{N}$.

Now $\mathfrak{H} = \langle \mathbf{C}_{\mathfrak{H}}(\mathfrak{B}) \mid \mathfrak{B} \le \mathfrak{A}, |\mathfrak{A} : \mathfrak{B}| = r \rangle$, by 1.9. Hence $\mathfrak{H} = \mathfrak{N}\mathbf{C}_{\mathfrak{H}}(\alpha)$. Thus, if $\alpha \in \mathfrak{A}$ and $\mathbf{C}_{\mathfrak{K}}(\alpha) = 1$, $\langle\alpha\rangle$ operates trivially on $\mathfrak{H}/\mathfrak{N}$. Hence $\mathfrak{C}$ operates trivially on $\mathfrak{H}/\mathfrak{N}$. But by IX, 6.11, $\mathbf{C}_{\mathfrak{H}/\mathfrak{N}}(\mathfrak{C}) = \mathbf{C}_{\mathfrak{H}}(\mathfrak{C})\mathfrak{N}/\mathfrak{N}$, so $\mathbf{C}_{\mathfrak{H}}(\mathfrak{C})\mathfrak{N} = \mathfrak{H}$.

b) Suppose that $q \in \pi'$. Since $\mathbf{N}_{\mathfrak{G}}(\mathfrak{K})$ is $\mathfrak{A}$-invariant, $\mathbf{N}_{\mathfrak{G}}(\mathfrak{K})$ possesses an $\mathfrak{A}$-invariant Sylow q-subgroup $\mathfrak{Q}_1$, by 11.1b). By a) applied to $\mathfrak{Q}_1$ and $\mathfrak{K}$, $\mathfrak{Q}_1 = \mathbf{C}_{\mathfrak{Q}_1}(\mathfrak{K})\mathbf{C}_{\mathfrak{Q}_1}(\mathfrak{C})$, where $\mathfrak{C} = \langle \gamma \mid \gamma \in \mathfrak{A}, \mathbf{C}_{\mathfrak{K}}(\gamma) = 1 \rangle$. By hypothesis, $\mathfrak{C} = \mathfrak{A}$. Since $\mathbf{C}_{\mathfrak{Q}_1}(\mathfrak{A}) = 1$, it follows that $\mathfrak{Q}_1 = \mathbf{C}_{\mathfrak{Q}_1}(\mathfrak{K}) \le \mathbf{C}_{\mathfrak{G}}(\mathfrak{K})$. Thus q does not divide $|\mathbf{N}_{\mathfrak{G}}(\mathfrak{K}) : \mathbf{C}_{\mathfrak{G}}(\mathfrak{K})|$. Hence $\mathbf{N}_{\mathfrak{G}}(\mathfrak{K})/\mathbf{C}_{\mathfrak{G}}(\mathfrak{K})$ is a π-group.

c) If $\mathfrak{Q}_0$ is any characteristic subgroup of $\mathfrak{Q}$, $\mathfrak{Q}_0$ is $\mathfrak{A}$-invariant. By b), $\mathbf{N}_{\mathfrak{G}}(\mathfrak{Q}_0)/\mathbf{C}_{\mathfrak{G}}(\mathfrak{Q}_0)$ is a q-group. By 10.9, $\mathfrak{G}$ is q-nilpotent. **q.e.d.**

11.17 Lemma. *Suppose that $\mathfrak{Q}$ is a q-group and that $\mathfrak{A}$ is an elementary Abelian p-group of operators on $\mathfrak{Q}$, where p, q are distinct primes. Suppose that $\mathbf{C}_{\mathfrak{Q}}(\mathfrak{A}) = 1$. Suppose that $\alpha \in \mathfrak{A}$ and $\mathfrak{Q}^* = \langle \mathbf{C}_{\mathfrak{Q}}(\mathfrak{B}) \mid \mathfrak{B} \le \mathfrak{A}, \langle \mathfrak{B}, \alpha \rangle = \mathfrak{A} \rangle$. Then $\mathfrak{Q}^* = [\mathfrak{Q}, \langle\alpha\rangle]$. In particular, $\mathfrak{Q}^*$ is a normal $\mathfrak{A}$-invariant subgroup of $\mathfrak{Q}$.*

Proof. Suppose that $\langle \mathfrak{B}, \alpha \rangle = \mathfrak{A}$. Since $\mathbf{C}_{\mathfrak{Q}}(\mathfrak{A}) = 1$ and $\mathbf{C}_{\mathfrak{Q}}(\mathfrak{B})$ is $\langle\alpha\rangle$-invariant, $\langle\alpha\rangle$ operates fixed point freely on $\mathbf{C}_{\mathfrak{Q}}(\mathfrak{B})$. By V, 8.9b), $\mathbf{C}_{\mathfrak{Q}}(\mathfrak{B}) = [\mathbf{C}_{\mathfrak{Q}}(\mathfrak{B}), \langle\alpha\rangle] \le [\mathfrak{Q}, \langle\alpha\rangle] = \mathfrak{S}$, say. Hence $\mathfrak{Q}^* \le \mathfrak{S}$.

Since $\mathbf{C}_{\mathfrak{Q}}(\mathfrak{B})$ is $\mathfrak{A}$-invariant, $\mathfrak{Q}^*$ is $\mathfrak{A}$-invariant. Also $\mathfrak{S}$ is $\mathfrak{A}$-invariant, so the normal closure $\mathfrak{T}$ of $\mathfrak{Q}^*$ in $\mathfrak{S}$ is $\mathfrak{A}$-invariant. And by 1.9, $\mathfrak{S} = \langle \mathbf{C}_{\mathfrak{S}}(\mathfrak{C}) \mid |\mathfrak{A} : \mathfrak{C}| = p \rangle$. We deduce that $\mathfrak{S} = \mathfrak{T}\mathbf{C}_{\mathfrak{S}}(\alpha)$. To do this, it is to be shown that if $|\mathfrak{A} : \mathfrak{C}| = p$, then $\mathbf{C}_{\mathfrak{S}}(\mathfrak{C}) \le \mathfrak{T}\mathbf{C}_{\mathfrak{S}}(\alpha)$. But if $\alpha \in \mathfrak{C}$, $\mathbf{C}_{\mathfrak{S}}(\mathfrak{C}) \le \mathbf{C}_{\mathfrak{S}}(\alpha) \le \mathfrak{T}\mathbf{C}_{\mathfrak{S}}(\alpha)$. And if $\alpha \notin \mathfrak{C}$, then by definition of $\mathfrak{Q}^*$, $\mathbf{C}_{\mathfrak{Q}}(\mathfrak{C}) \le \mathfrak{Q}^*$, whence $\mathbf{C}_{\mathfrak{S}}(\mathfrak{C}) \le \mathbf{C}_{\mathfrak{Q}}(\mathfrak{C}) \le \mathfrak{Q}^* \le \mathfrak{T} \le \mathfrak{T}\mathbf{C}_{\mathfrak{S}}(\alpha)$. Hence $\mathfrak{S} = \mathfrak{T}\mathbf{C}_{\mathfrak{S}}(\alpha)$.

Thus $[\mathfrak{S}, \langle\alpha\rangle] \leq \mathfrak{T}$. But by III, 13.3b), $[\mathfrak{S}, \langle\alpha\rangle] = [\mathfrak{Q}, \langle\alpha\rangle, \langle\alpha\rangle] = [\mathfrak{Q}, \langle\alpha\rangle] = \mathfrak{S}$. Thus $\mathfrak{T} = \mathfrak{S}$. By I, 8.9, $\mathfrak{Q}^* = \mathfrak{S}$. **q.e.d.**

11.18 Theorem (MARTINEAU). *Suppose that $\mathfrak{G}$ is a group, $\mathfrak{A}$ is an elementary Abelian group of automorphisms of $\mathfrak{G}$ and $\mathbf{C}_\mathfrak{G}(\mathfrak{A}) = 1$. Then $\mathfrak{G}$ is soluble.*

Proof. Suppose that this is false and consider a counterexample for which $|\mathfrak{G}||\mathfrak{A}|$ is minimal. Let $|\mathfrak{A}| = r^n$, where r is prime. By V, 8.14, $n > 1$. By 11.14, $\mathfrak{G}$ is a r'-group.

(1) If $\mathfrak{H}$ is an $\mathfrak{A}$-invariant proper subgroup of $\mathfrak{G}$, $\mathfrak{H}$ is soluble.

For $\mathfrak{A}$ induces a group $\mathfrak{A}_1$ of automorphisms of $\mathfrak{H}$ and $|\mathfrak{H}||\mathfrak{A}_1| < |\mathfrak{G}||\mathfrak{A}|$.

(2) $\mathfrak{G}$ has no non-identity proper $\mathfrak{A}$-invariant normal subgroup.

If $1 < \mathfrak{N} \lhd \mathfrak{G}$ and $\mathfrak{N}$ is $\mathfrak{A}$-invariant, then $\mathfrak{A}$ induces a group $\mathfrak{A}_2$ of automorphisms of $\mathfrak{G}/\mathfrak{N}$, and by 11.1a), $\mathbf{C}_{\mathfrak{G}/\mathfrak{N}}(\mathfrak{A}_2) = 1$. Since $|\mathfrak{G}/\mathfrak{N}|\,|\mathfrak{A}_2| < |\mathfrak{G}|\,|\mathfrak{A}|$, $\mathfrak{G}/\mathfrak{N}$ is soluble. By (1), $\mathfrak{N}$ is soluble; hence so is $\mathfrak{G}$. This is a contradiction.

(3) $|\mathfrak{A}| = r^2$.

If $n \geq 3$, we can apply 11.13, for by (1) $\mathbf{C}_\mathfrak{G}(\alpha)$ is soluble for every non-identity element α of $\mathfrak{A}$. Thus $n \leq 2$. Since $n > 1$, we have $n = 2$.

(4) If $\alpha \in \mathfrak{A} - \{1\}$, $\mathbf{C}_\mathfrak{G}(\alpha)$ is nilpotent.

For by (3), $\mathfrak{A} = \langle\alpha, \beta\rangle$ for some β. Then $\mathbf{C}_\mathfrak{G}(\alpha)$ is $\mathfrak{A}$-invariant and β operates fixed point freely on $\mathbf{C}_\mathfrak{G}(\alpha)$. By V, 8.14, $\mathbf{C}_\mathfrak{G}(\alpha)$ is nilpotent.

(5) If $\mathfrak{H}$ is an $\mathfrak{A}$-invariant proper subgroup of $\mathfrak{G}$, the Fitting height of $\mathfrak{H}$ is at most 2.

This follows from (1), (3), (4) and IX, 6.13.

(6) Let p, q be distinct prime divisors of $|\mathfrak{G}|$ and let $\mathfrak{H}$ be a maximal $\mathfrak{A}$-invariant $\{p, q\}$-subgroup of $\mathfrak{G}$. Suppose that $\mathbf{O}_p(\mathfrak{H}) \neq 1$ and $\mathbf{O}_q(\mathfrak{H}) \neq 1$. If $\mathfrak{L}$ is a non-identity $\mathfrak{A}$-invariant subgroup of $\mathbf{F}(\mathfrak{H})$, $\mathfrak{H}$ contains every $\mathfrak{A}$-invariant $\{p, q\}$-subgroup of $\mathbf{N}_\mathfrak{G}(\mathfrak{L})$.

Let $\mathfrak{L}^* = \mathbf{C}_{\mathbf{F}(\mathfrak{H})}(\mathfrak{L})$. Then $\mathfrak{L}^* \geq \mathbf{Z}(\mathbf{F}(\mathfrak{H}))$, so $\mathbf{O}_p(\mathfrak{L}^*) \neq 1$, $\mathbf{O}_q(\mathfrak{L}^*) \neq 1$. By (1) and 11.1, the conditions of 11.2 are satisfied, so any $\mathfrak{A}$-invariant $\{p, q\}$-subgroup of $\mathfrak{G}$ containing $\mathfrak{L}^*$ is contained in $\mathfrak{H}$. By 11.1, $\mathfrak{L}^*$ is contained in an $\mathfrak{A}$-invariant Hall $\{p, q\}$-subgroup $\mathfrak{K}$ of $\mathbf{N}_\mathfrak{G}(\mathfrak{L})$. Thus $\mathfrak{K} \leq \mathfrak{H}$. But again by 11.1, $\mathfrak{K}$ contains every $\mathfrak{A}$-invariant $\{p, q\}$-subgroup of $\mathbf{N}_\mathfrak{G}(\mathfrak{L})$. Thus (6) is proved.

Now suppose that $|\mathfrak{G}| = p_1^{a_1} \cdots p_n^{a_n}$, where $p_1, \ldots, p_n$ are distinct primes. By 11.1, $\mathfrak{G}$ has a unique $\mathfrak{A}$-invariant Sylow p_i-subgroup $\mathfrak{S}_i$. If $\mathfrak{S}_i\mathfrak{S}_j = \mathfrak{S}_j\mathfrak{S}_i$ for all i, j, $\prod_{j \neq i} \mathfrak{S}_j$ is a Sylow p_i-complement of $\mathfrak{G}$ ($i = 1, \ldots, n$). But then $\mathfrak{G}$ is soluble, by VI, 1.10. This is not so, so there exist distinct primes p, q such that $\mathfrak{P}\mathfrak{Q} \neq \mathfrak{Q}\mathfrak{P}$, where $\mathfrak{P}$, $\mathfrak{Q}$ are respectively the $\mathfrak{A}$-invariant Sylow p-, q-subgroups of $\mathfrak{G}$. Let $\{p, q\}$ be any such pair of primes.

(7) If $\mathfrak{H}$ is a maximal $\mathfrak{A}$-invariant $\{p, q\}$-subgroup of $\mathfrak{G}$, either $\mathfrak{P} \leqq \mathfrak{H}$ or $\mathfrak{Q} \leqq \mathfrak{H}$.

Suppose that this is false. Then since $\mathfrak{H} = (\mathfrak{H} \cap \mathfrak{P})(\mathfrak{H} \cap \mathfrak{Q})$, $\mathfrak{H} \cap \mathfrak{P} \neq 1$ and $\mathfrak{H} \cap \mathfrak{Q} \neq 1$.

a) $\mathbf{O}_p(\mathfrak{H}) \notin S_p(\mathfrak{H})$ and $\mathbf{O}_q(\mathfrak{H}) \notin S_q(\mathfrak{H})$.

For suppose that $\mathbf{O}_p(\mathfrak{H}) \in S_p(\mathfrak{H})$. Then $\mathbf{O}_p(\mathfrak{H}) = \mathfrak{H} \cap \mathfrak{P}$. Thus $\mathbf{O}_p(\mathfrak{H}) \neq 1$ and by (1) and (2), $\mathbf{N}_{\mathfrak{G}}(\mathbf{O}_p(\mathfrak{H}))$ is soluble. Thus by 11.1c), $\mathbf{N}_{\mathfrak{G}}(\mathbf{O}_p(\mathfrak{H}))$ has a unique $\mathfrak{A}$-invariant Hall $\{p, q\}$-subgroup which contains both $\mathbf{N}_{\mathfrak{P}}(\mathbf{O}_p(\mathfrak{H}))$ and $\mathfrak{H}$. It follows from the maximality of $\mathfrak{H}$ that $\mathfrak{H} \geq \mathbf{N}_{\mathfrak{P}}(\mathbf{O}_p(\mathfrak{H}))$. Hence $\mathbf{N}_{\mathfrak{P}}(\mathbf{O}_p(\mathfrak{H})) \leq \mathfrak{P} \cap \mathfrak{H} = \mathbf{O}_p(\mathfrak{H})$. By I, 8.8, $\mathbf{O}_p(\mathfrak{H}) = \mathfrak{P}$. Thus $\mathfrak{P} \leqq \mathfrak{H}$, a contradiction. Thus $\mathbf{O}_p(\mathfrak{H}) \notin S_p(\mathfrak{H})$. Similarly, $\mathbf{O}_q(\mathfrak{H}) \notin S_q(\mathfrak{H})$.

b) $\mathbf{O}_p(\mathfrak{H}) \neq 1$ and $\mathbf{O}_q(\mathfrak{H}) \neq 1$.

If $\mathbf{O}_q(\mathfrak{H}) = 1$, then $\mathbf{F}(\mathfrak{H}) = \mathbf{O}_p(\mathfrak{H})$. Hence by (5), $\mathfrak{H}/\mathbf{O}_p(\mathfrak{H})$ is nilpotent. Thus $\mathfrak{H}/\mathbf{O}_p(\mathfrak{H})$ is a q-group, and $\mathbf{O}_p(\mathfrak{H}) \in S_p(\mathfrak{H})$, contrary to a). Thus $\mathbf{O}_q(\mathfrak{H}) \neq 1$. Similarly $\mathbf{O}_p(\mathfrak{H}) \neq 1$.

c) If $\mathfrak{A} = \langle \alpha, \beta \rangle$, either $\mathbf{C}_{\mathbf{O}_p(\mathfrak{H})}(\alpha) \neq 1$ or $\mathbf{C}_{\mathbf{O}_p(\mathfrak{H})}(\beta) \neq 1$. Also either $\mathbf{C}_{\mathbf{O}_q(\mathfrak{H})}(\alpha) \neq 1$ or $\mathbf{C}_{\mathbf{O}_q(\mathfrak{H})}(\beta) \neq 1$.

Put $\mathfrak{R} = \mathbf{O}_p(\mathfrak{H})$ and suppose $\mathbf{C}_{\mathfrak{R}}(\alpha) = \mathbf{C}_{\mathfrak{R}}(\beta) = 1$. By 11.16b) applied to the normal $\mathfrak{A}$-invariant p-subgroup $\mathfrak{R}$ of $\mathfrak{H}$, $\mathfrak{H}/\mathbf{C}_{\mathfrak{H}}(\mathfrak{R})$ is a p-group. By 11.15, $\mathbf{O}_q(\mathfrak{H}) \in S_q(\mathbf{C}_{\mathfrak{H}}(\mathfrak{R}))$, since the Fitting height of $\mathfrak{H}$ is at most 2. Hence $\mathbf{O}_q(\mathfrak{H}) \in S_q(\mathfrak{H})$, contrary to a).

d) Suppose that $\alpha \in \mathfrak{A} - \{1\}$ and $\mathbf{C}_{\mathfrak{R}}(\alpha) \neq 1$, where $\mathfrak{R} = \mathbf{O}_p(\mathfrak{H})$. Then $\mathfrak{H} \geq \mathbf{C}_{\mathfrak{Q}}(\alpha)$.

Choose $\beta \in \mathfrak{A}$ such that $\mathfrak{A} = \langle \alpha, \beta \rangle$. By (4), $\mathbf{C}_{\mathfrak{G}}(\alpha)$ is nilpotent. Hence $\mathbf{C}_{\mathfrak{Q}}(\alpha)$ centralizes $\mathbf{C}_{\mathfrak{R}}(\alpha)$ and $\mathbf{C}_{\mathfrak{Q}}(\alpha) \leq \mathbf{N}_{\mathfrak{G}}(\mathbf{C}_{\mathfrak{R}}(\alpha))$. Since $\mathbf{C}_{\mathfrak{R}}(\alpha)$ is a non-identity $\mathfrak{A}$-invariant subgroup of $\mathbf{F}(\mathfrak{H})$, $\mathfrak{H}$ contains every $\mathfrak{A}$-invariant $\{p, q\}$-subgroup of $\mathbf{N}_{\mathfrak{G}}(\mathbf{C}_{\mathfrak{R}}(\alpha))$, by (6). Thus $\mathbf{C}_{\mathfrak{Q}}(\alpha) \leq \mathfrak{H}$, and d) is proved.

Since $\mathfrak{P}\mathfrak{Q} \neq \mathfrak{Q}\mathfrak{P}$, $\mathfrak{P}\mathfrak{Q} \neq \mathfrak{H}$. Since $\mathfrak{H}$ is a $\{p, q\}$-group and $\mathfrak{P}$, $\mathfrak{Q}$ are Sylow subgroups of $\mathfrak{G}$, it follows that either $\mathfrak{P} \nleq \mathfrak{H}$ or $\mathfrak{Q} \nleq \mathfrak{H}$. We suppose that $\mathfrak{Q} \nleq \mathfrak{H}$. Now by 1.9, $\mathfrak{Q} = \langle \mathbf{C}_{\mathfrak{Q}}(\gamma) | \gamma \in \mathfrak{A} - \{1\} \rangle$. Hence there exists $\alpha \in \mathfrak{A} - \{1\}$ such that $\mathbf{C}_{\mathfrak{Q}}(\alpha) \nleq \mathfrak{H}$. By d), $\mathbf{C}_{\mathfrak{R}}(\alpha) = 1$, where $\mathfrak{R} = \mathbf{O}_p(\mathfrak{H})$. By c), $\mathbf{C}_{\mathfrak{R}}(\beta) \neq 1$ for all $\beta \in \mathfrak{A} - \langle \alpha \rangle$. By d), $\mathbf{C}_{\mathfrak{Q}}(\beta) \leq \mathfrak{H}$ and $\mathbf{C}_{\mathfrak{Q}}(\beta) \leq \mathbf{N}_{\mathfrak{G}}(\mathfrak{R})$. By 11.16a), $\mathbf{C}_{\mathfrak{Q}}(\beta) = \mathbf{C}_{\mathbf{C}_{\mathfrak{Q}}(\beta)}(\mathfrak{R})\mathbf{C}_{\mathbf{C}_{\mathfrak{Q}}(\beta)}(\alpha)$. Thus $\mathbf{C}_{\mathfrak{Q}}(\beta) \leq \mathbf{C}_{\mathfrak{H}}(\mathfrak{R})$, for $\mathbf{C}_{\mathfrak{G}}(\alpha, \beta) = \mathbf{C}_{\mathfrak{G}}(\mathfrak{A}) = 1$. By 11.15, $\mathbf{O}_q(\mathfrak{H})$ is the unique Sylow q-subgroup of $\mathbf{C}_{\mathfrak{H}}(\mathfrak{R})$, so $\mathbf{C}_{\mathfrak{Q}}(\beta) \leq \mathbf{O}_q(\mathfrak{H})$, for all $\beta \in \mathfrak{A} - \langle \alpha \rangle$. Hence if $\mathfrak{Q}^* = \langle \mathbf{C}_{\mathfrak{Q}}(\beta) | \beta \in \mathfrak{A} - \langle \alpha \rangle \rangle$, $\mathfrak{Q}^* \leq \mathbf{O}_q(\mathfrak{H})$. But by 11.17, $\mathfrak{Q}^*$ is a normal $\mathfrak{A}$-invariant subgroup of $\mathfrak{Q}$. Since $\mathfrak{Q} \nleq \mathfrak{H}$, $\mathfrak{Q}^* = 1$ by (6). Hence $\mathbf{C}_{\mathfrak{Q}}(\beta) = 1$ for all $\beta \in \mathfrak{A} - \langle \alpha \rangle$. Choose elements β, γ in $\mathfrak{A} - \langle \alpha \rangle$ such that $\langle \beta, \gamma \rangle = \mathfrak{A}$. Then $\mathbf{C}_{\mathfrak{Q}}(\beta) = \mathbf{C}_{\mathfrak{Q}}(\gamma) = 1$, contrary to c), since $\mathbf{O}_q(\mathfrak{H}) \leq \mathfrak{Q}$. Thus (7) is proved.

(8) If $\mathbf{C}_{\mathfrak{P}}(\mathfrak{Q}) = \mathbf{N}_{\mathfrak{P}}(\mathfrak{Q})$, then $\mathbf{C}_{\mathfrak{P}}(\mathfrak{Q}) = 1$. If $\mathbf{C}_{\mathfrak{Q}}(\mathfrak{P}) = \mathbf{N}_{\mathfrak{Q}}(\mathfrak{P})$, then $\mathbf{C}_{\mathfrak{Q}}(\mathfrak{P}) = 1$.

Suppose $\mathfrak{C} = \mathbf{C}_{\mathfrak{P}}(\mathfrak{Q}) = \mathbf{N}_{\mathfrak{P}}(\mathfrak{Q}) \neq 1$. Then $\mathbf{N}_{\mathfrak{G}}(\mathfrak{C})$ is soluble. By 11.1c), $\mathbf{N}_{\mathfrak{G}}(\mathfrak{C})$ has an $\mathfrak{A}$-invariant Hall $\{p, q\}$-subgroup $\mathfrak{M}$ which contains every $\mathfrak{A}$-invariant $\{p, q\}$-subgroup of $\mathbf{N}_{\mathfrak{G}}(\mathfrak{C})$. In particular, $\mathfrak{M} \geq \mathbf{N}_{\mathfrak{P}}(\mathfrak{C})$ and, since $\mathfrak{Q} \leq \mathbf{C}_{\mathfrak{G}}(\mathfrak{C})$, $\mathfrak{Q} \leq \mathfrak{M}$. Let $\mathfrak{N}$ be a maximal $\mathfrak{A}$-invariant $\{p, q\}$-subgroup of $\mathfrak{G}$ for which $\mathfrak{N} \geq \mathfrak{M}$. Thus $\mathfrak{Q} \leq \mathfrak{N}$ but $\mathfrak{N} \neq \mathfrak{P}\mathfrak{Q}$. Since $\mathfrak{P} \in S_p(\mathfrak{G})$, it follows that $\mathfrak{P} \not\leq \mathfrak{N}$. By (7), $\mathfrak{Q} \trianglelefteq \mathfrak{N}$. Thus $\mathbf{N}_{\mathfrak{P}}(\mathfrak{C}) \leq \mathfrak{M} \leq \mathfrak{N} \leq \mathbf{N}_{\mathfrak{G}}(\mathfrak{Q})$ and $\mathbf{N}_{\mathfrak{P}}(\mathfrak{C}) \leq \mathbf{N}_{\mathfrak{P}}(\mathfrak{Q}) = \mathfrak{C}$. By I, 8.8, $\mathfrak{C} = \mathfrak{P}$. But then $\mathfrak{P} \leq \mathbf{C}_{\mathfrak{G}}(\mathfrak{Q})$ and $\mathfrak{P}\mathfrak{Q} = \mathfrak{Q}\mathfrak{P}$, a contradiction. Thus $\mathfrak{C} = 1$. Similarly, if $\mathbf{C}_{\mathfrak{Q}}(\mathfrak{P}) = \mathbf{N}_{\mathfrak{Q}}(\mathfrak{P})$, $\mathbf{C}_{\mathfrak{Q}}(\mathfrak{P}) = 1$.

(9) If $\gamma \in \mathfrak{A} - \{1\}$, either $\mathbf{C}_{\mathfrak{Q}}(\gamma) \leq \mathbf{N}_{\mathfrak{G}}(\mathfrak{P})$ or $\mathbf{C}_{\mathfrak{P}}(\gamma) \leq \mathbf{N}_{\mathfrak{G}}(\mathfrak{Q})$.

For $\mathbf{C}_{\mathfrak{P}}(\gamma)\mathbf{C}_{\mathfrak{Q}}(\gamma)$ is the $\mathfrak{A}$-invariant Hall $\{p, q\}$-subgroup of $\mathbf{C}_{\mathfrak{G}}(\gamma)$. Let $\mathfrak{H}$ be a maximal $\mathfrak{A}$-invariant $\{p, q\}$-subgroup of $\mathfrak{G}$ such that $\mathfrak{H} \geq \mathbf{C}_{\mathfrak{P}}(\gamma)\mathbf{C}_{\mathfrak{Q}}(\gamma)$. By (7), either $\mathfrak{P} \trianglelefteq \mathfrak{H}$ or $\mathfrak{Q} \trianglelefteq \mathfrak{H}$. Thus $\mathbf{C}_{\mathfrak{P}}(\gamma)\mathbf{C}_{\mathfrak{Q}}(\gamma)$ lies in either $\mathbf{N}_{\mathfrak{G}}(\mathfrak{P})$ or $\mathbf{N}_{\mathfrak{G}}(\mathfrak{Q})$. The assertion follows at once.

(10) With p, q interchanged if necessary, there exists $\alpha \in \mathfrak{A} - \{1\}$ such that $\mathbf{C}_{\mathfrak{P}}(\alpha) \leq \mathbf{N}_{\mathfrak{G}}(\mathfrak{Q})$ and, for all $\beta \in \mathfrak{A} - \langle\alpha\rangle$, $\mathbf{C}_{\mathfrak{P}}(\beta) \not\leq \mathbf{N}_{\mathfrak{G}}(\mathfrak{Q})$, $\mathbf{C}_{\mathfrak{Q}}(\beta) \leq \mathbf{N}_{\mathfrak{G}}(\mathfrak{P})$.

Suppose that this is false. Without loss of generality, we may suppose that p is odd.

Since $\mathfrak{Q}\mathfrak{P} \neq \mathfrak{P}\mathfrak{Q}$, $\mathfrak{Q} \not\leq \mathbf{N}_{\mathfrak{G}}(\mathfrak{P})$. By 1.9, $\mathfrak{Q} = \langle \mathbf{C}_{\mathfrak{Q}}(\gamma) | \gamma \in \mathfrak{A} - \{1\}\rangle$. Thus there exists $\alpha \in \mathfrak{A} - \{1\}$ such that $\mathbf{C}_{\mathfrak{Q}}(\alpha) \not\leq \mathbf{N}_{\mathfrak{G}}(\mathfrak{P})$. It follows from (9) that $\mathbf{C}_{\mathfrak{P}}(\alpha) \leq \mathbf{N}_{\mathfrak{G}}(\mathfrak{Q})$. Since (10) is false, there exists $\beta \in \mathfrak{A} - \langle\alpha\rangle$ such that either $\mathbf{C}_{\mathfrak{P}}(\beta) \leq \mathbf{N}_{\mathfrak{G}}(\mathfrak{Q})$ or $\mathbf{C}_{\mathfrak{Q}}(\beta) \not\leq \mathbf{N}_{\mathfrak{G}}(\mathfrak{P})$. It follows from (9) that $\mathbf{C}_{\mathfrak{P}}(\beta) < \mathbf{N}_{\mathfrak{G}}(\mathfrak{Q})$. Thus $\langle \mathbf{C}_{\mathfrak{P}}(\alpha), \mathbf{C}_{\mathfrak{P}}(\beta)\rangle \leq \mathbf{N}_{\mathfrak{P}}(\mathfrak{Q}) - \mathfrak{K}$, say. Clearly, $\mathfrak{G}$ is not p-nilpotent, so by 11.16c), either $\mathbf{C}_{\mathfrak{P}}(\alpha) \neq 1$ or $\mathbf{C}_{\mathfrak{P}}(\beta) \neq 1$. Thus $\mathfrak{K} \neq 1$. Let $\mathscr{K}$ be the set of non-identity $\mathfrak{A}$-invariant subgroups $\mathfrak{X}$ of $\mathfrak{P}$ such that $\mathfrak{X} \geq \langle \mathbf{C}_{\mathfrak{P}}(\alpha), \mathbf{C}_{\mathfrak{P}}(\beta)\rangle$. Thus $\mathfrak{K} \in \mathscr{K}$.

Let $\mathfrak{S} = \mathbf{N}_{\mathfrak{Q}}(\mathfrak{P})$. Thus $[\mathfrak{S}, \mathfrak{K}] \leq [\mathfrak{Q}, \mathbf{N}_{\mathfrak{P}}(\mathfrak{Q})] \leq \mathfrak{Q}$. Similarly $[\mathfrak{S}, \mathfrak{K}] \leq \mathfrak{P}$, so $[\mathfrak{S}, \mathfrak{K}] = 1$. Let $\mathscr{H}$ be the set of elements $\mathfrak{X}$ of $\mathscr{K}$ for which $[\mathfrak{S}, \mathfrak{X}] = 1$. Thus $\mathfrak{K} \in \mathscr{H}$.

We prove that if $\mathfrak{X} \in \mathscr{H}$, then $\mathbf{N}_{\mathfrak{P}}(\mathfrak{X}) \in \mathscr{H}$. Clearly $\mathbf{N}_{\mathfrak{P}}(\mathfrak{X}) \in \mathscr{K}$. Since $\mathfrak{X} \neq 1$, $\mathbf{N}_{\mathfrak{G}}(\mathfrak{X})$ is soluble. Thus by 11.1c), $\mathbf{N}_{\mathfrak{G}}(\mathfrak{X})$ has a unique $\mathfrak{A}$-invariant Hall $\{p, q\}$-subgroup $\mathfrak{L}$ and $\mathfrak{L}$ contains any $\mathfrak{A}$-invariant $\{p, q\}$-subgroup of $\mathbf{N}_{\mathfrak{G}}(\mathfrak{X})$. Since $[\mathfrak{S}, \mathfrak{X}] = 1$, it follows that $\mathfrak{S} \leq \mathfrak{L}$. Indeed, if $\mathfrak{U} = \mathfrak{L} \cap \mathbf{C}_{\mathfrak{G}}(\mathfrak{X})$, $\mathfrak{S} \leq \mathfrak{U}$. Also $\mathbf{Z}(\mathfrak{X}) \leq \mathfrak{L} \cap \mathbf{C}_{\mathfrak{G}}(\mathfrak{X}) = \mathfrak{U}$. Now $\mathfrak{A}$ is a group of operators on the $\{p, q\}$-group $\overline{\mathfrak{G}} = \mathfrak{U}/\mathbf{Z}(\mathfrak{X})$. By 11.1a), $\mathbf{C}_{\overline{\mathfrak{G}}}(\mathfrak{A}) = 1$. By 11.1b), $\mathfrak{U} \cap \mathfrak{P} \in S_p(\mathfrak{U})$, so if $\overline{\mathfrak{P}} = (\mathfrak{U} \cap \mathfrak{P})/\mathbf{Z}(\mathfrak{X})$, $\overline{\mathfrak{P}}$ is the $\mathfrak{A}$-invariant Sylow p-subgroup of $\overline{\mathfrak{G}}$. By IX, 6.11, $\mathbf{C}_{\overline{\mathfrak{P}}}(\alpha) = \mathbf{C}_{\mathfrak{U} \cap \mathfrak{P}}(\alpha)\mathbf{Z}(\mathfrak{X})/\mathbf{Z}(\mathfrak{X})$. But $\mathbf{C}_{\mathfrak{P}}(\alpha) \leq \mathfrak{X}$ since $\mathfrak{X} \in \mathscr{H}$. Thus $\mathbf{C}_{\mathfrak{U} \cap \mathfrak{P}}(\alpha) \leq \mathfrak{X} \cap \mathfrak{U} = \mathbf{Z}(\mathfrak{X})$. Hence $\mathbf{C}_{\overline{\mathfrak{P}}}(\alpha) = 1$ and similarly $\mathbf{C}_{\overline{\mathfrak{P}}}(\beta) = 1$. By 11.16c), $\overline{\mathfrak{G}}$ is p-nilpotent. Let $\mathfrak{V}/\mathbf{Z}(\mathfrak{X})$ be the normal p-complement of $\overline{\mathfrak{G}} = \mathfrak{U}/\mathbf{Z}(\mathfrak{X})$. Then $\mathfrak{V}/\mathbf{Z}(\mathfrak{X})$ is a q-group and $\mathfrak{V} = \mathfrak{W}\mathbf{Z}(\mathfrak{X})$, where $\mathfrak{W} \in S_q(\mathfrak{V})$. Since $\mathfrak{W} \leq \mathfrak{U} \leq \mathbf{C}_{\mathfrak{G}}(\mathfrak{X})$, $\mathfrak{W} \trianglelefteq \mathfrak{V}$. Since $\mathfrak{U}/\mathfrak{V}$ is a p-group, it follows that $\mathfrak{W}$ is the unique Sylow q-subgroup of $\mathfrak{U}$. Since $\mathfrak{S}$ is a q-subgroup of $\mathfrak{U}$, $\mathfrak{S} \leq \mathfrak{W}$. Also, since $\mathbf{C}_{\mathfrak{G}}(\mathfrak{X}) \trianglelefteq \mathbf{N}_{\mathfrak{G}}(\mathfrak{X})$, $\mathfrak{U} \trianglelefteq \mathfrak{L}$ and $\mathfrak{W} \trianglelefteq \mathfrak{L}$. Since $\mathfrak{L} \geq \mathbf{N}_{\mathfrak{P}}(\mathfrak{X})$, we have

$$[\mathfrak{S}, \mathbf{N}_{\mathfrak{P}}(\mathfrak{X})] \leq [\mathfrak{S}, \mathfrak{P}] \cap [\mathfrak{W}, \mathfrak{L}] \leq \mathfrak{P} \cap \mathfrak{W} = 1.$$

Thus $\mathbf{N}_{\mathfrak{P}}(\mathfrak{X}) \in \mathscr{H}$.

It follows from this and I, 8.8 that $\mathfrak{P} \in \mathscr{H}$; hence $[\mathfrak{S}, \mathfrak{P}] = 1$. Thus $\mathfrak{S} = \mathbf{N}_{\mathfrak{Q}}(\mathfrak{P}) = \mathbf{C}_{\mathfrak{Q}}(\mathfrak{P})$. By (8), $\mathfrak{S} = \mathbf{N}_{\mathfrak{Q}}(\mathfrak{P}) = 1$.

Since $\mathfrak{Q}\mathfrak{P} \neq \mathfrak{P}\mathfrak{Q}$, $\mathfrak{P} \nleq \mathbf{N}_{\mathfrak{G}}(\mathfrak{Q})$. By 1.9, $\mathfrak{P} = \langle \mathbf{C}_{\mathfrak{P}}(\gamma) | \gamma \in \mathfrak{A} - \{1\}\rangle$. Thus there exists $\alpha' \in \mathfrak{A} - \{1\}$ such that $\mathbf{C}_{\mathfrak{P}}(\alpha') \nleq \mathbf{N}_{\mathfrak{G}}(\mathfrak{Q})$. By (9), $\mathbf{C}_{\mathfrak{Q}}(\alpha') \leq \mathbf{N}_{\mathfrak{Q}}(\mathfrak{P}) = 1$. But also, since (10) is false with p, q interchanged, there exists $\beta' \in \mathfrak{A} - \langle \alpha' \rangle$ such that either $\mathbf{C}_{\mathfrak{Q}}(\beta') \leq \mathbf{N}_{\mathfrak{G}}(\mathfrak{P})$ or $\mathbf{C}_{\mathfrak{P}}(\beta') \nleq \mathbf{N}_{\mathfrak{G}}(\mathfrak{Q})$. By (9), $\mathbf{C}_{\mathfrak{Q}}(\beta') \leq \mathbf{N}_{\mathfrak{Q}}(\mathfrak{P}) = 1$. Thus $\mathbf{C}_{\mathfrak{Q}}(\alpha') = \mathbf{C}_{\mathfrak{Q}}(\beta') = 1$. By 11.16b), $\mathbf{N}_{\mathfrak{G}}(\mathfrak{Q})/\mathbf{C}_{\mathfrak{G}}(\mathfrak{Q})$ is a q-group. Hence $\mathfrak{K} = \mathbf{N}_{\mathfrak{P}}(\mathfrak{Q})$ centralizes $\mathfrak{Q}$ and $\mathbf{N}_{\mathfrak{P}}(\mathfrak{Q}) = \mathbf{C}_{\mathfrak{P}}(\mathfrak{Q})$. By (8), $\mathfrak{K} = \mathbf{N}_{\mathfrak{P}}(\mathfrak{Q}) = 1$. But it has been shown earlier that $\mathfrak{K} \neq 1$. This contradiction shows that (10) holds.

We assume henceforth that p, q are so chosen that the assertion of (10) holds. Thus $\alpha \in \mathfrak{A} - \{1\}$, $\mathbf{C}_{\mathfrak{P}}(\alpha) \leq \mathbf{N}_{\mathfrak{G}}(\mathfrak{Q})$ and, for all $\beta \in \mathfrak{A} - \langle \alpha \rangle$, $\mathbf{C}_{\mathfrak{P}}(\beta) \nleq \mathbf{N}_{\mathfrak{G}}(\mathfrak{Q})$, $\mathbf{C}_{\mathfrak{Q}}(\beta) \leq \mathbf{N}_{\mathfrak{G}}(\mathfrak{P})$.

(11) $\mathbf{C}_{\mathfrak{Q}}(\alpha) = \mathfrak{Q}$.

Suppose that this is false and let $\mathfrak{Q}^* = \langle \mathbf{C}_{\mathfrak{Q}}(\beta) | \beta \in \mathfrak{A} - \langle \alpha \rangle\rangle$. By 11.17, $\mathfrak{Q}^*$ is a normal $\mathfrak{A}$-invariant subgroup of $\mathfrak{Q}$, and $\mathfrak{Q}^* = [\mathfrak{Q}, \langle \alpha \rangle]$. Thus $\mathbf{C}_{\mathfrak{Q}/\mathfrak{Q}^*}(\alpha) = \mathfrak{Q}/\mathfrak{Q}^*$. It follows from IX, 6.11 that $\mathfrak{Q}^*\mathbf{C}_{\mathfrak{Q}}(\alpha) = \mathfrak{Q}$. Since $\mathbf{C}_{\mathfrak{Q}}(\alpha) \neq \mathfrak{Q}$, $\mathfrak{Q}^* \neq 1$. Thus $\mathbf{N}_{\mathfrak{G}}(\mathfrak{Q}^*)$ is soluble, and by 11.1c), $\mathbf{N}_{\mathfrak{G}}(\mathfrak{Q}^*)$ has an $\mathfrak{A}$-invariant Hall $\{p, q\}$-subgroup $\mathfrak{M}$ which contains every $\mathfrak{A}$-invariant $\{p, q\}$-subgroup of $\mathbf{N}_{\mathfrak{G}}(\mathfrak{Q}^*)$. Thus $\mathfrak{Q} \leq \mathfrak{M}$, so $\mathfrak{P} \nleq \mathfrak{M}$ and $\mathfrak{P} \nleq \mathbf{N}_{\mathfrak{G}}(\mathfrak{Q}^*)$. Thus $\mathfrak{Q}^* \nleq \mathbf{C}_{\mathfrak{G}}(\mathfrak{P})$, so there exists $\beta \in \mathfrak{A} - \langle \alpha \rangle$ such that $\mathfrak{C} = \mathbf{C}_{\mathfrak{Q}}(\beta) \nleq \mathbf{C}_{\mathfrak{G}}(\mathfrak{P})$. But by (10), $\mathfrak{C} \leq \mathbf{N}_{\mathfrak{G}}(\mathfrak{P})$, so by 11.16a), $\mathfrak{C} = \mathbf{C}_{\mathfrak{C}}(\mathfrak{P})\mathbf{C}_{\mathfrak{C}}(\mathfrak{X})$, where $\mathfrak{X} = \langle \gamma | \gamma \in \mathfrak{A}, \mathbf{C}_{\mathfrak{P}}(\gamma) = 1\rangle$. Since $\mathfrak{C} \neq \mathbf{C}_{\mathfrak{C}}(\mathfrak{P})$, $\mathbf{C}_{\mathfrak{C}}(\mathfrak{X}) \neq 1$. Since $\mathfrak{C} \leq \mathbf{C}_{\mathfrak{G}}(\beta)$, $\mathbf{C}_{\mathfrak{G}}(\mathfrak{X}) \cap \mathbf{C}_{\mathfrak{G}}(\beta) \neq 1$. Since $\mathbf{C}_{\mathfrak{G}}(\mathfrak{A}) = 1$, it follows that $\alpha \notin \mathfrak{X}$. Thus $\mathbf{C}_{\mathfrak{P}}(\alpha) \neq 1$.

Since $\mathbf{C}_{\mathfrak{P}}(\alpha) \leq \mathbf{N}_{\mathfrak{G}}(\mathfrak{Q})$ and $\mathbf{C}_{\mathfrak{Q}}(\beta) \leq \mathbf{N}_{\mathfrak{G}}(\mathfrak{P})$ for all $\beta \in \mathfrak{A} - \langle \alpha \rangle$,

$$[\mathbf{C}_{\mathfrak{P}}(\alpha), \mathbf{C}_{\mathfrak{Q}}(\beta)] \leq \mathfrak{Q} \cap \mathfrak{P} = 1.$$

Also, since $\mathbf{C}_{\mathfrak{G}}(\alpha)$ is nilpotent, $[\mathbf{C}_{\mathfrak{P}}(\alpha), \mathbf{C}_{\mathfrak{Q}}(\alpha)] = 1$. Thus

$$\mathbf{C}_{\mathfrak{G}}(\mathbf{C}_{\mathfrak{P}}(\alpha)) \geq \langle \mathbf{C}_{\mathfrak{Q}}(\gamma) | \gamma \in \mathfrak{A} - \{1\}\rangle = \mathfrak{Q},$$

by 1.9. Now $\mathfrak{Q} \times \mathbf{C}_{\mathfrak{P}}(\alpha)$ is an $\mathfrak{A}$-invariant $\{p, q\}$-subgroup, so there exists a maximal $\mathfrak{A}$-invariant $\{p, q\}$-subgroup $\mathfrak{H}$ containing $\mathfrak{Q} \times \mathbf{C}_{\mathfrak{P}}(\alpha)$. By IX, 1.4, $\mathbf{C}_{\mathfrak{P}}(\alpha) \leq \mathbf{O}_p(\mathfrak{H})$, so $\mathbf{O}_p(\mathfrak{H}) \neq 1$. Since $\mathfrak{Q} \leq \mathfrak{H}$, $\mathfrak{P} \not\leq \mathfrak{H}$. By (7), $\mathfrak{Q} \trianglelefteq \mathfrak{H}$, so $\mathbf{O}_q(\mathfrak{H}) \neq 1$. By (6), $\mathfrak{H}$ contains every $\mathfrak{A}$-invariant $\{p, q\}$-subgroup of $\mathbf{N}_{\mathfrak{G}}(\mathfrak{L})$ for any non-identity $\mathfrak{A}$-invariant subgroup $\mathfrak{L}$ of $\mathbf{F}(\mathfrak{H})$. It follows that $\mathbf{Z}(\mathfrak{P}) \cap \mathbf{O}_p(\mathfrak{H}) = 1$, since $\mathfrak{P} \not\leq \mathfrak{H}$. It follows also by taking $\mathfrak{L} = \mathbf{O}_p(\mathfrak{H})$ that $\mathbf{Z}(\mathfrak{P}) \leq \mathfrak{H}$, for by 11.1b), $\mathbf{O}_p(\mathfrak{H}) \leq \mathfrak{P}$.

Since $\mathbf{Z}(\mathfrak{P}) \cap \mathbf{O}_p(\mathfrak{H}) = 1$ and $\mathbf{C}_{\mathfrak{P}}(\alpha) \leq \mathbf{O}_p(\mathfrak{H})$, $\mathbf{C}_{\mathbf{Z}(\mathfrak{P})}(\alpha) = 1$. Thus if $\mathfrak{Y} = \langle \gamma \mid \gamma \in \mathfrak{A}, \mathbf{C}_{\mathbf{Z}(\mathfrak{P})}(\gamma) = 1 \rangle$, $\alpha \in \mathfrak{Y}$. Thus if $\beta \in \mathfrak{A} - \langle \alpha \rangle$ and $\mathfrak{D} = \mathbf{C}_{\mathfrak{Q}}(\beta)$, $\mathbf{C}_{\mathfrak{D}}(\mathfrak{Y}) \leq \mathbf{C}_{\mathfrak{G}}(\alpha, \beta) = 1$. Since $\mathbf{C}_{\mathfrak{Q}}(\beta) \leq \mathbf{N}_{\mathfrak{G}}(\mathfrak{P}) \leq \mathbf{N}_{\mathfrak{G}}(\mathbf{Z}(\mathfrak{P}))$, it follows from 11.16a) that $\mathbf{C}_{\mathfrak{Q}}(\beta) = \mathfrak{D} = \mathbf{C}_{\mathfrak{D}}(\mathbf{Z}(\mathfrak{P}))\mathbf{C}_{\mathfrak{D}}(\mathfrak{Y}) = \mathbf{C}_{\mathfrak{D}}(\mathbf{Z}(\mathfrak{P}))$. Hence $[\mathbf{C}_{\mathfrak{Q}}(\beta), \mathbf{Z}(\mathfrak{P})] = 1$ for all $\beta \in \mathfrak{A} - \langle \alpha \rangle$ and $[\mathfrak{Q}^*, \mathbf{Z}(\mathfrak{P})] = 1$.

Now $\mathfrak{A}$ is generated by $\mathfrak{A} - \langle \alpha \rangle$, so there exists $\beta \in \mathfrak{A} - \langle \alpha \rangle$ such that β does not leave every element of $\mathbf{Z}(\mathfrak{P})$ fixed. Let $\mathfrak{R} = [\mathbf{Z}(\mathfrak{P}), \langle \beta \rangle]$, so $\mathfrak{R} \neq 1$. Since $\mathbf{C}_{\mathfrak{Q}}(\beta) \leq \mathfrak{Q}^*$, it follows from IX, 6.11 that $\mathbf{C}_{\mathfrak{Q}/\mathfrak{Q}^*}(\beta) = 1$. Also, by III, 13.4b), $\mathbf{C}_{\mathfrak{R}}(\beta) = 1$. Now $\mathfrak{R} \leq \mathbf{Z}(\mathfrak{P}) \leq \mathfrak{H} \leq \mathbf{N}_{\mathfrak{G}}(\mathfrak{Q})$ and $\mathfrak{R} \leq \mathbf{Z}(\mathfrak{P}) \leq \mathbf{C}_{\mathfrak{G}}(\mathfrak{Q}^*)$; thus $\mathfrak{Q}\mathfrak{R}/\mathfrak{Q}^*$ is a group with $\mathfrak{A}$ as group of operators. Now $\mathbf{C}_{\mathfrak{Q}\mathfrak{R}/\mathfrak{Q}}(\beta) = 1$ and $\mathbf{C}_{\mathfrak{Q}/\mathfrak{Q}^*}(\beta) = 1$, so $\mathbf{C}_{\mathfrak{Q}\mathfrak{R}/\mathfrak{Q}^*}(\beta) = 1$. By V, 8.14, $\mathfrak{Q}\mathfrak{R}/\mathfrak{Q}^*$ is nilpotent. Thus, regarding $\mathfrak{R}$ as a group of operators on $\mathfrak{Q}$, $\mathfrak{R}$ centralizes $\mathfrak{Q}/\mathfrak{Q}^*$. But also $\mathfrak{R}$ centralizes $\mathfrak{Q}^*$, so by I, 4.4, $\mathfrak{R}$ centralizes $\mathfrak{Q}$. Hence by IX, 1.4, $\mathfrak{R} \leq \mathbf{O}_p(\mathfrak{H})$. But $\mathbf{O}_p(\mathfrak{H}) \cap \mathbf{Z}(\mathfrak{P}) = 1$, so $\mathfrak{R} = 1$, a contradiction. Hence $\mathbf{C}_{\mathfrak{Q}}(\alpha) = \mathfrak{Q}$.

(12) $q = 2$.

Let β, γ be elements of $\mathfrak{A} - \langle \alpha \rangle$ such that $\langle \beta, \gamma \rangle = \mathfrak{A}$. Since $\mathfrak{A} = \langle \alpha, \beta \rangle$, $\mathbf{C}_{\mathfrak{G}}(\alpha) \cap \mathbf{C}_{\mathfrak{G}}(\beta) = 1$. By (11), $\mathfrak{Q} \leq \mathbf{C}_{\mathfrak{G}}(\alpha)$, so $\mathbf{C}_{\mathfrak{Q}}(\beta) = 1$. Similarly, $\mathbf{C}_{\mathfrak{Q}}(\gamma) = 1$. But $\mathfrak{G}$ is not q-nilpotent, so by 11.16c), $q = 2$.

To obtain a contradiction, write $|\mathfrak{G}| = 2^a p_1^{a_1} \cdots p_n^{a_n}$, where $p_1, \ldots, p_n$ are distinct odd primes. Let $\mathfrak{P}_i$ be the $\mathfrak{A}$-invariant Sylow p_i-subgroup of $\mathfrak{G}$, and let $\mathfrak{Q}$ be the $\mathfrak{A}$-invariant Sylow 2-subgroup of $\mathfrak{G}$. By (12), $\mathfrak{P}_i\mathfrak{P}_j = \mathfrak{P}_j\mathfrak{P}_i$ for all i, j. Thus, if $\mathfrak{T} = \mathfrak{P}_1 \cdots \mathfrak{P}_n$, $\mathfrak{T}$ is an $\mathfrak{A}$-invariant soluble subgroup of $\mathfrak{G}$ of order $p_1^{a_1} \cdots p_n^{a_n}$. Thus $\mathfrak{G} = \mathfrak{Q}\mathfrak{T}$ and $\mathfrak{T} \neq \mathfrak{G}$. By (11), $[\mathfrak{Q}, \langle \alpha \rangle] = 1$. Thus if $x \in \mathfrak{G}$, there exist $y \in \mathfrak{Q}$, $z \in \mathfrak{T}$ such that $x = yz$, and

$$x\alpha^i = (y\alpha^i)(z\alpha^i) = y(z\alpha^i) = xz^{-1}(z\alpha^i),$$

so $[x, \alpha^i] = x^{-1}(x\alpha^i) = z^{-1}(z\alpha^i) \in \mathfrak{T}$. Hence $[\mathfrak{G}, \langle \alpha \rangle] \leq \mathfrak{T} \neq \mathfrak{G}$. But by III, 1.6b), $[\mathfrak{G}, \langle \alpha \rangle]$ is a normal $\mathfrak{A}$-invariant subgroup of $\mathfrak{G}$. Since $[\mathfrak{G}, \langle \alpha \rangle] \neq \mathfrak{G}$, $[\mathfrak{G}, \langle \alpha \rangle] = 1$, by (2). Thus $\alpha = 1$, a contradiction.

q.e.d.

11.19 Remarks. Of the various generalizations of Martineau's work, we mention the following.

a) (PETTET [1]). Let $\mathfrak{A}$ be an elementary Abelian group of automorphisms of $\mathfrak{G}$ for which $(|\mathfrak{G}|, |\mathfrak{A}|) = 1$. Suppose that for each prime divisor p of $|\mathfrak{G}|$, $\mathfrak{G}$ has a unique $\mathfrak{A}$-invariant Sylow p-subgroup. If $|\mathbf{C}_{\mathfrak{G}}(\mathfrak{A})|$ is odd, $\mathfrak{G}$ is soluble.

b) (ROWLEY [1]). Let $\mathfrak{A}$ be an Abelian group of automorphisms of $\mathfrak{G}$ such that $(|\mathfrak{G}|, |\mathfrak{A}|) = 1$, $\mathbf{C}_{\mathfrak{G}}(\mathfrak{A}) = 1$ and $\mathfrak{A}^{pq} = 1$ for distinct primes p, q. Then $\mathfrak{G}$ is soluble.

§ 12. Local Methods and Cohomology

It will now be shown how local methods may be used to study the cohomology groups of a group and in particular the Schur multiplier. In discussing $\mathbf{H}^n(\mathfrak{G}, \mathsf{M})$, we shall use the definition

$$\mathbf{H}^n(\mathfrak{G}, \mathsf{M}) = \tilde{\mathbf{Z}}^n(\mathfrak{G}, \mathsf{M})/\tilde{\mathbf{B}}^n(\mathfrak{G}, \mathsf{M})$$

given in I, 16.14.

12.1 Lemma. *Suppose that $f \in \tilde{\mathbf{Z}}^n(\mathfrak{G}, \mathsf{M})$ $(n \geq 1)$ and $g \in \mathfrak{G}$. Let*

$$a(x_1, \ldots, x_n) = \sum_{i=1}^{n} (-1)^i f(gx_1, \ldots, gx_{i-1}, x_i, gx_i, x_{i+1}, \ldots, x_n)$$

for any $x_i \in \mathfrak{G}$. Then $a \in \tilde{\mathbf{C}}^{n-1}(\mathfrak{G}, \mathsf{M})$ and

$$f(x_0, \ldots, x_n) - f(gx_0, \ldots, gx_n) = (a\tilde{\delta}_{n-1})(x_0, \ldots, x_n).$$

Proof. It is clear that $a \in \tilde{\mathbf{C}}^{n-1}(\mathfrak{G}, \mathsf{M})$. Using the definitions of $\tilde{\delta}_{n-1}$ and a,

$$\begin{aligned}
&(a\tilde{\delta}_{n-1})(x_0, \ldots, x_n) \\
&= \sum_{j=0}^{n} (-1)^j a(x_0, \ldots, \hat{x}_j, \ldots, x_n) \\
&= \sum_{i=1}^{n} (-1)^i \Bigg\{ \sum_{j=0}^{i-1} (-1)^j f(gx_0, \ldots, g\hat{x}_j, \ldots, gx_{i-1}, x_i, gx_i, x_{i+1}, \ldots, x_n) \\
&\qquad + \sum_{j=i}^{n} (-1)^j f(gx_0, \ldots, gx_{i-2}, x_{i-1}, gx_{i-1}, x_i, \ldots, \hat{x}_j, \ldots, x_n) \Bigg\}.
\end{aligned}$$

We subtract $0 = (f\tilde{\delta}_n)(gx_0, \ldots, gx_{i-1}, x_i, gx_i, x_{i+1}, \ldots, x_n)$ from the first sum and, in the case when $i = 1$, we subtract $0 = (f\tilde{\delta}_n)(x_0, gx_0, x_1, \ldots, x_n)$ from the second. This gives

$$\begin{aligned}
&(a\tilde{\delta}_{n-1})(x_0, \ldots, x_n)\\
&= \sum_{i=1}^{n} (-f(gx_0, \ldots, gx_i, x_{i+1}, \ldots, x_n) + f(gx_0, \ldots, gx_{i-1}, x_i, \ldots, x_n))\\
&\quad + \sum_{i=1}^{n} \sum_{j=i+1}^{n} (-1)^{i+j} f(gx_0, \ldots, gx_{i-1}, x_i, gx_i, x_{i+1}, \ldots, \hat{x}_j, \ldots, x_n)\\
&\quad + \sum_{i=2}^{n} \sum_{j=i}^{n} (-1)^{i+j} f(gx_0, \ldots, gx_{i-2}, x_{i-1}, gx_{i-1}, x_i, \ldots, \hat{x}_j, \ldots, x_n)\\
&\quad + (-f(gx_0, x_1, \ldots, x_n) + f(x_0, \ldots, x_n))\\
&= f(x_0, \ldots, x_n) - f(gx_0, \ldots, gx_n).
\end{aligned}$$

q.e.d.

12.2 Definition. Suppose that M is a $\mathfrak{G}$-module. Given $g \in \mathfrak{G}$ and $\mathfrak{H} \leq \mathfrak{G}$, we define $\alpha_n \in \mathrm{Hom}(\tilde{\mathbf{C}}^n(\mathfrak{H}, \mathsf{M}), \tilde{\mathbf{C}}^n(\mathfrak{H}^g, \mathsf{M}))$ $(n \geq 0)$ by putting

$$(f\alpha_n)(h_0^g, \ldots, h_n^g) = f(h_0, \ldots, h_n)g,$$

where $f \in \tilde{\mathbf{C}}^n(\mathfrak{H}, \mathsf{M})$, $h_i \in \mathfrak{H}$. Then α_n is an isomorphism and $\alpha_n\tilde{\delta}_n = \tilde{\delta}_n\alpha_{n+1}$. Hence for $n \geq 1$, α_n induces an isomorphism $T^g_{\mathfrak{H}}$ of $\mathbf{H}^n(\mathfrak{H}, \mathsf{M})$ onto $\mathbf{H}^n(\mathfrak{H}^g, \mathsf{M})$. If, further, $g' \in \mathfrak{G}$, then $T^g_{\mathfrak{H}}T^{g'}_{\mathfrak{H}^g} = T^{gg'}_{\mathfrak{H}}$. Also $T^g_{\mathfrak{G}} = 1$, for if $f \in \tilde{\mathbf{Z}}^n(\mathfrak{G}, \mathsf{M})$,

$$(f\alpha_n)(g_0, \ldots, g_n) = f(gg_0g^{-1}, \ldots, gg_ng^{-1})g = f(gg_0, \ldots, gg_n),$$

so by 12.1, $f\alpha_n = f - a\tilde{\delta}_{n-1}$.

We denote the *restriction* mapping of $\mathbf{H}^n(\mathfrak{G}, \mathsf{M})$ into $\mathbf{H}^n(\mathfrak{H}, \mathsf{M})$ by $R(\mathfrak{G}, \mathfrak{H})$ (see I, 16.15). Thus if $\mathfrak{K} \leq \mathfrak{H} \leq \mathfrak{G}$,

$$R(\mathfrak{G}, \mathfrak{H})R(\mathfrak{H}, \mathfrak{K}) = R(\mathfrak{G}, \mathfrak{K}).$$

In the case when $|\mathfrak{G} : \mathfrak{H}|$ is finite, we also have the *corestriction* mapping of $\mathbf{H}^n(\mathfrak{H}, \mathsf{M})$ into $\mathbf{H}^n(\mathfrak{G}, \mathsf{M})$; this is denoted by $C(\mathfrak{H}, \mathfrak{G})$. By I, 16.18,

$$R(\mathfrak{G}, \mathfrak{H})C(\mathfrak{H}, \mathfrak{G}) = |\mathfrak{G} : \mathfrak{H}|1,$$

where 1 denotes the identity mapping on $\mathbf{H}^n(\mathfrak{G}, \mathsf{M})$.

12.3 Lemma. a) *If* $\mathfrak{H} \leq \mathfrak{G}$ *and* $g \in \mathfrak{G}$,

$$R(\mathfrak{G}, \mathfrak{H}) T_{\mathfrak{H}}^{g} = R(\mathfrak{G}, \mathfrak{H}^{g}).$$

b) *Suppose that* $\mathfrak{H}$, $\mathfrak{K}$ *are subgroups of finite index in* $\mathfrak{G}$ *and that* S *is a complete set of representatives of the double cosets* $\mathfrak{H} g \mathfrak{K}$. *Then*

$$C(\mathfrak{H}, \mathfrak{G}) R(\mathfrak{G}, \mathfrak{K}) = \sum_{s \in \mathsf{S}} T_{\mathfrak{H}}^{s} R(\mathfrak{H}^{s}, \mathfrak{H}^{s} \cap \mathfrak{K}) C(\mathfrak{H}^{s} \cap \mathfrak{K}, \mathfrak{K}).$$

Proof. Let M be a $\mathfrak{G}$-module.

a) Define $\xi_n \in \operatorname{Hom}(\tilde{\mathbf{C}}^n(\mathfrak{G}, \mathsf{M}), \tilde{\mathbf{C}}^n(\mathfrak{H}, \mathsf{M}))$ and $\zeta_n \in \operatorname{Hom}(\tilde{\mathbf{C}}^n(\mathfrak{G}, \mathsf{M}), \tilde{\mathbf{C}}^n(\mathfrak{H}^g, \mathsf{M}))$ to be the restriction mappings and define $\alpha_n \in \operatorname{Hom}(\tilde{\mathbf{C}}^n(\mathfrak{H}, \mathsf{M}), \tilde{\mathbf{C}}^n(\mathfrak{H}^g, \mathsf{M}))$ as in 12.2. Then if $f \in \tilde{\mathbf{Z}}^n(\mathfrak{G}, \mathsf{M})$,

$$\begin{aligned}(f(\xi_n \alpha_n - \zeta_n))(h_0^g, \ldots, h_n^g) &= f(h_0, \ldots, h_n)g - f(h_0^g, \ldots, h_n^g) \\ &= f(gh_0^g, \ldots, gh_n^g) - f(h_0^g, \ldots, h_n^g)\end{aligned}$$

for any $h_i \in \mathfrak{H}$. Hence by 12.1, $f(\xi_n \alpha_n - \zeta_n) \in \tilde{\mathbf{B}}^n(\mathfrak{H}^g, \mathsf{M})$, so $\xi_n \alpha_n$, ζ_n induce the same mappings on $\mathbf{H}^n(\mathfrak{G}, \mathsf{M})$. Thus

$$R(\mathfrak{G}, \mathfrak{H}) T_{\mathfrak{H}}^{g} = R(\mathfrak{G}, \mathfrak{H}^{g}).$$

(Our previous remark that $T_{\mathfrak{G}}^{g} = 1$ is a special case of this.)

b) For each $s \in \mathsf{S}$, let T_s be a transversal of $\mathfrak{H}^s \cap \mathfrak{K}$ in $\mathfrak{K}$:

$$\mathfrak{K} = \bigcup_{t \in \mathsf{T}_s} (\mathfrak{H}^s \cap \mathfrak{K}) t = \bigcup_{t \in \mathsf{T}_s} t^{-1} (\mathfrak{H}^s \cap \mathfrak{K}).$$

For $f \in \tilde{\mathbf{C}}^n(\mathfrak{H}^s \cap \mathfrak{K}, \mathsf{M})$, define $f\zeta \in \tilde{\mathbf{C}}^n(\mathfrak{K}, \mathsf{M})$ by putting

$$(f\zeta)(x_0, \ldots, x_n) = \sum_{t \in \mathsf{T}_s} f(v_0 x_0 t^{-1}, \ldots, v_n x_n t^{-1}) t$$

where $x_i \in \mathfrak{K}$ and $v_i \in \mathsf{T}_s$ is so chosen that $t x_i^{-1} v_i^{-1} \in \mathfrak{H}^s \cap \mathfrak{K}$. By Definition I, 16.15, ζ induces the mapping $C(\mathfrak{H}^s \cap \mathfrak{K}, \mathfrak{K})$ on $\mathbf{H}^n(\mathfrak{H}^s \cap \mathfrak{K}, \mathsf{M})$. Hence the mapping $U_s = T_{\mathfrak{H}}^{s} R(\mathfrak{H}^s, \mathfrak{H}^s \cap \mathfrak{K}) C(\mathfrak{H}^s \cap \mathfrak{K}, \mathfrak{K})$ on $\mathbf{H}^n(\mathfrak{H}, \mathsf{M})$ is induced by $\xi \in \operatorname{Hom}(\tilde{\mathbf{C}}^n(\mathfrak{H}, \mathsf{M}), \tilde{\mathbf{C}}^n(\mathfrak{K}, \mathsf{M}))$, where, if $f \in \tilde{\mathbf{C}}^n(\mathfrak{H}, \mathsf{M})$,

$$(f\xi)(x_0, \ldots, x_n) = \sum_{t \in \mathsf{T}_s} f((v_0 x_0 t^{-1})^{s^{-1}}, \ldots, (v_n x_n t^{-1})^{s^{-1}}) st.$$

But by 6.8a), $s\mathsf{T}_s$ is a set of representatives of the cosets of $\mathfrak{H}$ in $\mathfrak{H}s\mathfrak{K}$, so $\bigcup_{s\in\mathsf{S}} s\mathsf{T}_s$ is a transversal of $\mathfrak{H}$ in $\mathfrak{G}$. Since $sv_ix_it^{-1}s^{-1} \in \mathfrak{H}$, it follows once again from I, 16.15 that

$$\sum_{s\in\mathsf{S}} U_s = C(\mathfrak{H}, \mathfrak{G})R(\mathfrak{G}, \mathfrak{K}).$$ **q.e.d.**

12.4 Definition. Suppose that M is a $\mathfrak{G}$-module and that $\mathfrak{H}$ is a subgroup of $\mathfrak{G}$ of finite index. The element α of $\mathbf{H}^n(\mathfrak{H}, \mathsf{M})$ $(n \geq 1)$ is said to be *stable* in $\mathfrak{G}$ if, for all $g \in \mathfrak{G}$,

$$\alpha R(\mathfrak{H}, \mathfrak{H} \cap \mathfrak{H}^g) = \alpha T_{\mathfrak{H}}^g R(\mathfrak{H}^g, \mathfrak{H} \cap \mathfrak{H}^g).$$

Thus the stable elements form a subgroup of $\mathbf{H}^n(\mathfrak{H}, \mathsf{M})$.

12.5 Lemma. *Suppose that* M *is a* $\mathfrak{G}$*-module and that* $\mathfrak{H}$ *is a subgroup of* $\mathfrak{G}$ *of finite index.*

a) *If* $\beta \in \mathbf{H}^n(\mathfrak{G}, \mathsf{M})$, $\beta R(\mathfrak{G}, \mathfrak{H})$ *is a stable element of* $\mathbf{H}^n(\mathfrak{H}, \mathsf{M})$.

b) *If* α *is a stable element of* $\mathbf{H}^n(\mathfrak{H}, \mathsf{M})$,

$$\alpha C(\mathfrak{H}, \mathfrak{G})R(\mathfrak{G}, \mathfrak{H}) = |\mathfrak{G} : \mathfrak{H}|\alpha.$$

Proof. a) It is to be shown that for any $g \in \mathfrak{G}$,

$$\beta R(\mathfrak{G}, \mathfrak{H})R(\mathfrak{H}, \mathfrak{H} \cap \mathfrak{H}^g) = \beta R(\mathfrak{G}, \mathfrak{H}) T_{\mathfrak{H}}^g R(\mathfrak{H}^g, \mathfrak{H} \cap \mathfrak{H}^g).$$

But by 12.3a),

$$R(\mathfrak{G}, \mathfrak{H}) T_{\mathfrak{H}}^g = R(\mathfrak{G}, \mathfrak{H}^g),$$

so

$$R(\mathfrak{G}, \mathfrak{H}) T_{\mathfrak{H}}^g R(\mathfrak{H}^g, \mathfrak{H} \cap \mathfrak{H}^g) = R(\mathfrak{G}, \mathfrak{H} \cap \mathfrak{H}^g) = R(\mathfrak{G}, \mathfrak{H})R(\mathfrak{H}, \mathfrak{H} \cap \mathfrak{H}^g).$$

b) Let S be a complete set of representatives of the double cosets $\mathfrak{H}g\mathfrak{H}$. By 12.3b),

$$C(\mathfrak{H}, \mathfrak{G})R(\mathfrak{G}, \mathfrak{H}) = \sum_{s\in\mathsf{S}} T_{\mathfrak{H}}^s R(\mathfrak{H}^s, \mathfrak{H} \cap \mathfrak{H}^s)C(\mathfrak{H} \cap \mathfrak{H}^s, \mathfrak{H}).$$

Since α is stable, it follows that

$$\begin{aligned}
\alpha C(\mathfrak{H}, \mathfrak{G}) R(\mathfrak{G}, \mathfrak{H}) &= \sum_{s \in \mathsf{S}} \alpha R(\mathfrak{H}, \mathfrak{H} \cap \mathfrak{H}^s) C(\mathfrak{H} \cap \mathfrak{H}^s, \mathfrak{H}) \\
&= \sum_{s \in \mathsf{S}} |\mathfrak{H} : \mathfrak{H} \cap \mathfrak{H}^s| \alpha \qquad \text{(by I, 16.18)} \\
&= \sum_{s \in \mathsf{S}} \frac{|\mathfrak{H}^s \mathfrak{H}|}{|\mathfrak{H}^s|} \alpha \\
&= \sum_{s \in \mathsf{S}} \frac{|\mathfrak{H} s \mathfrak{H}|}{|\mathfrak{H}|} \alpha \\
&= |\mathfrak{G} : \mathfrak{H}| \alpha.
\end{aligned}$$

q.e.d.

12.6 Lemma. *Let p be a prime and let $\mathfrak{H}$ be a subgroup of $\mathfrak{G}$ of finite index r, where $(r, p) = 1$. Let M be a $\mathfrak{G}$-module and let $\mathfrak{P}_{\mathfrak{G}}$, $\mathfrak{P}_{\mathfrak{H}}$ be the subgroups consisting of the p-elements of* $\mathbf{H}^n(\mathfrak{G}, \mathsf{M})$, $\mathbf{H}^n(\mathfrak{H}, \mathsf{M})$ *respectively. Then*

$$\mathfrak{P}_{\mathfrak{H}} = \mathfrak{S} \oplus \mathfrak{T},$$

where $\mathfrak{S}$ is the group of stable elements of $\mathfrak{P}_{\mathfrak{H}}$, and $\mathfrak{S} \cong \mathfrak{P}_{\mathfrak{G}}$.

Proof. Let ρ, σ denote the restrictions of $R(\mathfrak{G}, \mathfrak{H})$ to $\mathfrak{P}_{\mathfrak{G}}$ and of $C(\mathfrak{H}, \mathfrak{G})$ to $\mathfrak{P}_{\mathfrak{H}}$ respectively. Then $\rho \in \operatorname{Hom}(\mathfrak{P}_{\mathfrak{G}}, \mathfrak{P}_{\mathfrak{H}})$, $\sigma \in \operatorname{Hom}(\mathfrak{P}_{\mathfrak{H}}, \mathfrak{P}_{\mathfrak{G}})$ and

$$\rho\sigma = r1,$$

by I, 16.18. Since $(r, p) = 1$, it follows that ρ is a monomorphism. Hence $\mathfrak{S} = \operatorname{im} \rho$ is isomorphic to $\mathfrak{P}_{\mathfrak{G}}$. Also

$$\mathfrak{P}_{\mathfrak{H}} = \mathfrak{S} \oplus (\ker \sigma).$$

By 12.5a), any element of $\mathfrak{S}$ is stable. Conversely, suppose that α is a stable element of $\mathfrak{P}_{\mathfrak{H}}$. By 12.5b), $\alpha\sigma\rho = r\alpha$. Thus $\alpha \in \mathfrak{S}$ and $\mathfrak{S}$ is the group of stable p-elements of $\mathfrak{P}_{\mathfrak{H}}$. **q.e.d.**

12.7 Definition. a) If $\mathfrak{H} \leq \mathfrak{G}$, we write $i_{\mathfrak{H},\mathfrak{G}}$ for the injection mapping of $\mathfrak{H}$ into $\mathfrak{G}$. If also $g \in \mathfrak{G}$, we denote the mapping $x \to x^g$ of $\mathfrak{H}$ into $\mathfrak{H}^g$ by $t_{\mathfrak{H},g}$.

b) Suppose that $\mathfrak{G}_1$, $\mathfrak{G}_2$ are groups and that ρ is a homomorphism of $\mathfrak{G}_1$ into $\mathfrak{G}_2$. If M is any (additively written) Abelian group, M may be regarded as a trivial $\mathfrak{G}_1$- or $\mathfrak{G}_2$-module and, for $n \geq 0$, we obtain a homomorphism $\tilde{\rho}_n$ of $\tilde{\mathbf{C}}^n(\mathfrak{G}_2, \mathsf{M})$ into $\tilde{\mathbf{C}}^n(\mathfrak{G}_1, \mathsf{M})$ if for $f \in \tilde{\mathbf{C}}^n(\mathfrak{G}_2, \mathsf{M})$, $f\tilde{\rho}_n$ is defined by

$$(f\tilde{\rho}_n)(x_0, \ldots, x_n) = f(x_0\rho, \ldots, x_n\rho),$$

where $x_i \in \mathfrak{G}_1$ $(i = 0, \ldots, n)$. Then $\tilde{\rho}_n\tilde{\tilde{\delta}}_n = \tilde{\tilde{\delta}}_n\tilde{\rho}_{n+1}$, so there is a homomorphism ρ^* of $\mathbf{H}^n(\mathfrak{G}_2, \mathsf{M})$ into $\mathbf{H}^n(\mathfrak{G}_1, \mathsf{M})$ given by

$$(f + \tilde{\mathbf{B}}^n(\mathfrak{G}_2, \mathsf{M}))\rho^* = f\tilde{\rho}_n + \tilde{\mathbf{B}}^n(\mathfrak{G}_1, \mathsf{M})$$

$(f \in \tilde{\mathbf{Z}}^n(\mathfrak{G}_2, \mathsf{M}))$. The notation ρ^* will be used throughout this section whenever we have a homomorphism ρ of groups. In particular, if M is a trivial $\mathfrak{G}$-module, $\mathfrak{H} \leq \mathfrak{G}$ and $g \in \mathfrak{G}$,

$$i^*_{\mathfrak{H},\mathfrak{G}} = R(\mathfrak{G}, \mathfrak{H}), \quad t^*_{\mathfrak{H},g} = T^{g^{-1}}_{\mathfrak{H}^g}.$$

Observe that if ρ, σ are homomorphisms of $\mathfrak{G}_1$ into $\mathfrak{G}_2$, $\mathfrak{G}_2$ into $\mathfrak{G}_3$ respectively, then $(\rho\sigma)^* = \sigma^*\rho^*$.

12.8 Theorem. *Suppose that $\mathfrak{S} \in S_p(\mathfrak{G})$ and that $\mathbf{W}$ is a characteristic p-functor which strongly controls fusion in $\mathfrak{G}$. If M is a trivial $\mathfrak{G}$-module, then $\mathbf{H}^n(\mathfrak{G}, \mathsf{M})$ and $\mathbf{H}^n(\mathbf{N}_{\mathfrak{G}}(\mathbf{W}(\mathfrak{S})), \mathsf{M})$ have isomorphic Sylow p-subgroups for all $n \geq 1$.*

Proof. Let $\mathfrak{W} = \mathbf{W}(\mathfrak{S})$. We apply 12.6 to the subgroup $\mathfrak{S}$ of $\mathfrak{G}$ and of $\mathbf{N}_{\mathfrak{G}}(\mathfrak{W})$. Since $\mathbf{H}^n(\mathfrak{S}, \mathsf{M})$ is a p-group (I, 16.19), the Sylow p-subgroups of $\mathbf{H}^n(\mathfrak{G}, \mathsf{M})$, $\mathbf{H}^n(\mathbf{N}_{\mathfrak{G}}(\mathfrak{W}), \mathsf{M})$ are isomorphic to the groups $\mathfrak{P}_1$, $\mathfrak{P}_2$ of elements of $\mathbf{H}^n(\mathfrak{S}, \mathsf{M})$ stable under $\mathfrak{G}$, $\mathbf{N}_{\mathfrak{G}}(\mathfrak{W})$ respectively. We show that $\mathfrak{P}_1 = \mathfrak{P}_2$. Clearly $\mathfrak{P}_1 \leq \mathfrak{P}_2$.

Suppose that $\alpha \in \mathfrak{P}_2$ and $g \in \mathfrak{G}$. Since $\mathbf{W}$ strongly controls fusion in $\mathfrak{G}$ and $(\mathfrak{S} \cap \mathfrak{S}^g)^{g^{-1}} \leq \mathfrak{S}$, there exists $h \in \mathbf{N}_{\mathfrak{G}}(\mathfrak{W})$ such that $a^{g^{-1}} = a^{h^{-1}}$ for all $a \in \mathfrak{S} \cap \mathfrak{S}^g$. Thus $g^{-1}h \in \mathbf{C}_{\mathfrak{G}}(\mathfrak{S} \cap \mathfrak{S}^g)$ and

$$\mathfrak{S} \cap \mathfrak{S}^g = (\mathfrak{S} \cap \mathfrak{S}^g)^{g^{-1}h} \leq \mathfrak{S} \cap \mathfrak{S}^h.$$

Since $\alpha \in \mathfrak{P}_2$,

$$\alpha R(\mathfrak{S}, \mathfrak{S} \cap \mathfrak{S}^h) = \alpha T^h_{\mathfrak{S}} R(\mathfrak{S}^h, \mathfrak{S} \cap \mathfrak{S}^h).$$

Hence

$$\begin{aligned}\alpha R(\mathfrak{S}, \mathfrak{S} \cap \mathfrak{S}^g) &= \alpha T^h_{\mathfrak{S}} R(\mathfrak{S}^h, \mathfrak{S} \cap \mathfrak{S}^g)\\ &= \alpha t^*_{\mathfrak{S}^h, h^{-1}} i^*_{\mathfrak{S} \cap \mathfrak{S}^g, \mathfrak{S}^h}\\ &= \alpha(i_{\mathfrak{S} \cap \mathfrak{S}^g, \mathfrak{S}^h} t_{\mathfrak{S}^h, h^{-1}})^*.\end{aligned}$$

But since $a^{g^{-1}} = a^{h^{-1}}$ for all $a \in \mathfrak{S} \cap \mathfrak{S}^g$,

$$i_{\mathfrak{S} \cap \mathfrak{S}^g, \mathfrak{S}^h} t_{\mathfrak{S}^h, h^{-1}} = i_{\mathfrak{S} \cap \mathfrak{S}^g, \mathfrak{S}^g} t_{\mathfrak{S}^g, g^{-1}},$$

so

$$\begin{aligned} \alpha R(\mathfrak{S}, \mathfrak{S} \cap \mathfrak{S}^g) &= \alpha t^*_{\mathfrak{S}^g, g^{-1}} i^*_{\mathfrak{S} \cap \mathfrak{S}^g, \mathfrak{S}^g} \\ &= \alpha T^g_{\mathfrak{S}} R(\mathfrak{S}^g, \mathfrak{S} \cap \mathfrak{S}^g), \end{aligned}$$

and $\alpha \in \mathfrak{P}_1$. Hence $\mathfrak{P}_1 = \mathfrak{P}_2$. **q.e.d.**

12.9 Corollary. *Suppose that $\mathfrak{G}$ is p-normal and $\mathfrak{S} \in S_p(\mathfrak{G})$. Then the Sylow p-subgroups of* $\mathbf{H}^n(\mathfrak{G}, \mathsf{M})$ *and* $\mathbf{H}^n(\mathbf{N}_{\mathfrak{G}}(\mathbf{Z}(\mathfrak{S})), \mathsf{M})$ *are isomorphic for all trivial $\mathfrak{G}$-modules* M *and all $n \geq 1$. In particular,*

a) *the maximal Abelian p-factor groups of $\mathfrak{G}$ and* $\mathbf{N}_{\mathfrak{G}}(\mathbf{Z}(\mathfrak{S}))$ *are isomorphic (Grün's second theorem, IV, 3.7), and*

b) *the Sylow p-subgroups of the Schur multipliers of $\mathfrak{G}$ and* $\mathbf{N}_{\mathfrak{G}}(\mathbf{Z}(\mathfrak{S}))$ *are isomorphic.*

Proof. The main assertion follows from 12.8 if we show that $\mathbf{Z}$ strongly controls fusion in $\mathfrak{G}$. To do this let $\mathfrak{A}$, $\mathfrak{B}$ be subgroups of $\mathfrak{S}$ for which there exists $g \in \mathfrak{G}$ such that $\mathfrak{A}^g = \mathfrak{B}$. Then $\mathbf{Z}(\mathfrak{S}^{g^{-1}}) \leq \mathbf{C}_{\mathfrak{G}}(\mathfrak{B}^{g^{-1}}) = \mathbf{C}_{\mathfrak{G}}(\mathfrak{A})$ and $\mathbf{Z}(\mathfrak{S}) \leq \mathbf{C}_{\mathfrak{G}}(\mathfrak{A})$. Hence there exists $\mathfrak{P} \in S_p(\mathbf{C}_{\mathfrak{G}}(\mathfrak{A}))$ such that $\mathbf{Z}(\mathfrak{S}) \leq \mathfrak{P}$ and $\mathbf{Z}(\mathfrak{S}^{g^{-1}}) \leq \mathfrak{P}^c$ for some $c \in \mathbf{C}_{\mathfrak{G}}(\mathfrak{A})$. But $\mathfrak{P} \leq \mathfrak{S}^x$ for some $x \in \mathfrak{G}$, so $\mathbf{Z}(\mathfrak{S})^{x^{-1}} \leq \mathfrak{S}$ and $\mathbf{Z}(\mathfrak{S})^{g^{-1}c^{-1}x^{-1}} \leq \mathfrak{S}$. Since $\mathfrak{G}$ is p-normal, it follows that $\mathbf{Z}(\mathfrak{S})^{x^{-1}} = \mathbf{Z}(\mathfrak{S}) = \mathbf{Z}(\mathfrak{S})^{g^{-1}c^{-1}x^{-1}}$, so $\mathbf{Z}(\mathfrak{S})^{cg} = \mathbf{Z}(\mathfrak{S})$. Hence $cg \in \mathbf{N}_{\mathfrak{G}}(\mathbf{Z}(\mathfrak{S}))$; also $a^{cg} = a^g$ for all $a \in \mathfrak{A}$. Thus $\mathbf{Z}$ strongly controls fusion in $\mathfrak{G}$.

Taking $n = 1$ and $\mathsf{M} = \mathbb{C}^\times$, we see that $\mathbf{H}^1(\mathfrak{G}, \mathbb{C}^\times)$ and $\mathbf{H}^1(\mathbf{N}_{\mathfrak{G}}(\mathbf{Z}(\mathfrak{S})), \mathbb{C}^\times)$ have isomorphic Sylow p-subgroups. But $\mathbf{H}^1(\mathfrak{G}, \mathbb{C}^\times) \cong \mathrm{Hom}(\mathfrak{G}, \mathbb{C}^\times) \cong \mathfrak{G}/\mathfrak{G}'$, so a) follows at once. b) follows by taking $n = 2$, $\mathsf{M} = \mathbb{C}^\times$. **q.e.d.**

We now start to use local methods.

12.10 Lemma. *Suppose that $\mathfrak{S} \in S_p(\mathfrak{G})$ and that* M *is a trivial $\mathfrak{G}$-module. Let $\mathscr{F}$ be a conjugation family for $\mathfrak{S}$ (see 4.1). Then $\alpha \in$* $\mathbf{H}^n(\mathfrak{S}, \mathsf{M})$ *is stable in $\mathfrak{G}$ if and only if $T^g_{\mathfrak{P}}$ fixes $\alpha R(\mathfrak{S}, \mathfrak{P})$ whenever $\mathfrak{P} \in \mathscr{F}$ and $g \in$* $\mathbf{N}_{\mathfrak{G}}(\mathfrak{P})$.

Proof. First suppose that α is stable, $\mathfrak{P} \leq \mathfrak{S}$ and $g \in \mathbf{N}_{\mathfrak{G}}(\mathfrak{P})$. Then

$$\alpha R(\mathfrak{S}, \mathfrak{S} \cap \mathfrak{S}^{g^{-1}}) = \alpha T^{g^{-1}}_{\mathfrak{S}} R(\mathfrak{S}^{g^{-1}}, \mathfrak{S} \cap \mathfrak{S}^{g^{-1}}).$$

Since $\mathfrak{P} \leq \mathfrak{S} \cap \mathfrak{S}^{g^{-1}}$, it follows by applying $R(\mathfrak{S} \cap \mathfrak{S}^{g^{-1}}, \mathfrak{P})$ that

$$\alpha R(\mathfrak{S}, \mathfrak{P}) = \alpha T_{\mathfrak{S}}^{g^{-1}} R(\mathfrak{S}^{g^{-1}}, \mathfrak{P}),$$

so by 12.7,

$$\begin{aligned} \alpha R(\mathfrak{S}, \mathfrak{P}) T_{\mathfrak{P}}^{g} &= \alpha t^{*}_{\mathfrak{S}^{g^{-1}}, g} i^{*}_{\mathfrak{P}, \mathfrak{S}^{g^{-1}}} t^{*}_{\mathfrak{P}^{g}, g^{-1}} \\ &= \alpha(t_{\mathfrak{P}^{g}, g^{-1}} i_{\mathfrak{P}, \mathfrak{S}^{g^{-1}}} t_{\mathfrak{S}^{g^{-1}}, g})^{*} \\ &= \alpha i^{*}_{\mathfrak{P}^{g}, \mathfrak{S}} \\ &= \alpha R(\mathfrak{S}, \mathfrak{P}^{g}) \\ &= \alpha R(\mathfrak{S}, \mathfrak{P}). \end{aligned}$$

Conversely, suppose that $T_{\mathfrak{P}}^{h}$ fixes $\alpha R(\mathfrak{S}, \mathfrak{P})$ whenever $\mathfrak{P} \in \mathscr{F}$ and $h \in \mathbf{N}_{\mathfrak{G}}(\mathfrak{P})$. Choose $g \in \mathfrak{G}$ and put $\mathfrak{D} = \mathfrak{S} \cap \mathfrak{S}^{g}$. Then $\mathfrak{D}^{g^{-1}} \leq \mathfrak{S}$, so there exist subgroups $\mathfrak{P}_1, \ldots, \mathfrak{P}_m$ in $\mathscr{F}$ and elements $g_j \in \mathbf{N}_{\mathfrak{G}}(\mathfrak{P}_j)$ $(j = 1, \ldots, m)$ such that $g^{-1} = g_1 \cdots g_m$ and $\mathfrak{D}^{g_1 \cdots g_{j-1}} \leq \mathfrak{P}_j$ $(j = 1, \ldots, m)$. Put $h_j = g_1 \cdots g_j$, $\mathfrak{D}_j = \mathfrak{D}^{h_j}$ $(j = 0, \ldots, m)$. Thus $h_0 = 1$, $h_m = g^{-1}$, $\mathfrak{D}_0 = \mathfrak{D}$, and for $j = 1, \ldots, m$, $\mathfrak{D}_{j-1} \leq \mathfrak{P}_j$ and $\mathfrak{D}_j = \mathfrak{D}_{j-1}^{g} \leq \mathfrak{P}_j$. Define

$$t_j = t_{\mathfrak{D}, h_j} i_{\mathfrak{D}_j, \mathfrak{S}} \quad (j = 0, \ldots, m).$$

By hypothesis, for $j = 1, \ldots, m$,

$$\alpha i^{*}_{\mathfrak{P}_j, \mathfrak{S}} t^{*}_{\mathfrak{P}_j, g_j} = \alpha R(\mathfrak{S}, \mathfrak{P}_j) T_{\mathfrak{P}_j}^{g_j^{-1}} = \alpha R(\mathfrak{S}, \mathfrak{P}_j) = \alpha i^{*}_{\mathfrak{P}_j, \mathfrak{S}}.$$

Hence

$$\alpha i^{*}_{\mathfrak{P}_j, \mathfrak{S}} t^{*}_{\mathfrak{P}_j, g_j} i^{*}_{\mathfrak{D}_{j-1}, \mathfrak{P}_j} t^{*}_{\mathfrak{D}, h_{j-1}} = \alpha i^{*}_{\mathfrak{P}_j, \mathfrak{S}} i^{*}_{\mathfrak{D}_{j-1}, \mathfrak{P}_j} t^{*}_{\mathfrak{D}, h_{j-1}},$$

so

$$\alpha(t_{\mathfrak{D}, h_{j-1}} i_{\mathfrak{D}_{j-1}, \mathfrak{P}_j} t_{\mathfrak{P}_j, g_j} i_{\mathfrak{P}_j, \mathfrak{S}})^{*} = \alpha(t_{\mathfrak{D}, h_{j-1}} i_{\mathfrak{D}_{j-1}, \mathfrak{S}})^{*} = \alpha t^{*}_{j-1}.$$

Hence

$$\alpha t^{*}_{j-1} = \alpha(t_{\mathfrak{D}, h_j} i_{\mathfrak{D}_j, \mathfrak{S}})^{*} = \alpha t^{*}_{j} \quad (j = 1, \ldots, m)$$

and $\alpha t^{*}_{0} = \alpha t^{*}_{m}$. But $t^{*}_{0} = i^{*}_{\mathfrak{D}, \mathfrak{S}} = R(\mathfrak{S}, \mathfrak{S} \cap \mathfrak{S}^{g})$, and since

$$t_m = t_{\mathfrak{D}, g^{-1}} i_{\mathfrak{D}_m, \mathfrak{S}} = i_{\mathfrak{D}, \mathfrak{S}^g} t_{\mathfrak{S}^g, g^{-1}},$$

$t_m^* = T_{\mathfrak{S}}^g R(\mathfrak{S}^g, \mathfrak{S} \cap \mathfrak{S}^g)$. Hence α is stable. **q.e.d.**

In order to apply this to the Schur multiplier, we have to show that the isomorphism between $(\mathfrak{R} \cap \mathfrak{F}')/[\mathfrak{R}, \mathfrak{F}]$ and $\mathbf{H}^2(\mathfrak{F}/\mathfrak{R}, \mathbb{C}^\times)$ established in V, 23.5 is operator invariant. We use the following elementary facts.

12.11 Lemma. a) *If* M_1, M_2 *are* $\mathfrak{G}$*-modules, the group* $\operatorname{Hom}(\mathsf{M}_1, \mathsf{M}_2)$ *of additive homomorphisms of* M_1 *into* M_2 *has the structure of a* $\mathfrak{G}$*-module in which*

$$x(fg) = ((xg^{-1})f)g$$

for all $x \in \mathsf{M}_1$, $f \in \operatorname{Hom}(\mathsf{M}_1, \mathsf{M}_2)$, $g \in \mathfrak{G}$.

b) *If* $\mathfrak{F}$ *is a free group,* $\mathfrak{R} \trianglelefteq \mathfrak{F}$ *and* α *is an endomorphism of* $\mathfrak{F}/\mathfrak{R}$*, there is an endomorphism* γ *of* $\mathfrak{F}$ *such that*

$$(f\gamma)\mathfrak{R} = (f\mathfrak{R})\alpha$$

for all $f \in \mathfrak{F}$.

Proof. a) For if g_1, g_2 are in $\mathfrak{G}$,

$$x(f(g_1 g_2)) = ((xg_2^{-1}g_1^{-1})f)g_1 g_2 = ((xg_2^{-1})(fg_1))g_2 = x((fg_1)g_2).$$

b) Let X be a group-basis of $\mathfrak{F}$. For each $x \in X$, there exists $f_x \in \mathfrak{F}$ such that

$$(x\mathfrak{R})\alpha = f_x \mathfrak{R},$$

and there exists an endomorphism γ of $\mathfrak{F}$ such that

$$x\gamma = f_x.$$

Hence $(x\mathfrak{R})\alpha = (x\gamma)\mathfrak{R}$ and $(f\mathfrak{R})\alpha = (f\gamma)\mathfrak{R}$ for all $f \in \mathfrak{F}$. **q.e.d.**

12.12 Theorem. *Let* $\mathfrak{G}$ *be a group of operators on the finite group* $\mathfrak{Q}$*. Suppose that* $\mathfrak{F}$ *is a finitely generated free group,* $\mathfrak{R} \trianglelefteq \mathfrak{F}$ *and that* ζ *is an isomorphism of* $\mathfrak{F}/\mathfrak{R}$ *onto* $\mathfrak{Q}$.

a) *There is a unique* $\mathfrak{G}$*-module structure on* $(\mathfrak{R} \cap \mathfrak{F}')/[\mathfrak{R}, \mathfrak{F}]$ *with the property that whenever* $g \in \mathfrak{G}$ *and* ξ *is an endomorphism of* $\mathfrak{F}$ *such that*

$$((f\xi)\mathfrak{R})\zeta = ((f\mathfrak{R})\zeta)g$$

for all $f \in \mathfrak{F}$, *then*

$$(r[\mathfrak{R}, \mathfrak{F}])g = (r\xi)[\mathfrak{R}, \mathfrak{F}]$$

for all $r \in \mathfrak{R} \cap \mathfrak{F}'$.

b) *If* M *is a trivial* $\mathfrak{Q}$*-module, there is a* $\mathfrak{G}$*-module structure on* $\mathbf{H}^2(\mathfrak{Q}, \mathsf{M})$ *such that if* $c \in \mathbf{Z}^2(\mathfrak{Q}, \mathsf{M})$ *and* $g \in \mathfrak{G}$,

$$(c + \mathbf{B}^2(\mathfrak{Q}, \mathsf{M}))g = c' + \mathbf{B}^2(\mathfrak{Q}, \mathsf{M}),$$

*where** c' *is the element of* $\mathbf{Z}^2(\mathfrak{Q}, \mathsf{M})$ *given by*

$$c'(x, y) = c(xg^{-1}, yg^{-1}) \quad (x \in \mathfrak{Q},\ y \in \mathfrak{Q}).$$

c) *If* K *is an algebraically closed field of characteristic 0,* $\mathrm{Hom}((\mathfrak{R} \cap \mathfrak{F}')/[\mathfrak{R}, \mathfrak{F}], \mathsf{K}^\times)$ *and* $\mathbf{H}^2(\mathfrak{Q}, \mathsf{K}^\times)$ *are isomorphic* $\mathfrak{G}$*-modules.*

Proof. a) Suppose that $g \in \mathfrak{G}$. By 12.11b), there is an endomorphism γ of $\mathfrak{F}$ such that

$$(f\gamma)\mathfrak{R} = (f\mathfrak{R})\zeta g\zeta^{-1}$$

for all $f \in \mathfrak{F}$; for $g = 1$ we choose γ to be 1. Then $\mathfrak{R}\gamma \subseteq \mathfrak{R}$, so γ induces an endomorphism on $(\mathfrak{R} \cap \mathfrak{F}')/[\mathfrak{R}, \mathfrak{F}]$ and we may put

$$(r[\mathfrak{R}, \mathfrak{F}])g = (r\gamma)[\mathfrak{R}, \mathfrak{F}]$$

for any $r \in \mathfrak{R} \cap \mathfrak{F}'$. Now suppose that ξ is any endomorphism of $\mathfrak{F}$ such that

$$((f\xi)\mathfrak{R})\zeta = ((f\mathfrak{R})\zeta)g$$

for any $f \in \mathfrak{F}$. Then $(f\xi)\mathfrak{R} = (f\gamma)\mathfrak{R}$ and $f\xi = (f\gamma)r_f$ for some $r_f \in \mathfrak{R}$. Hence for any f_1, f_2 in $\mathfrak{F}$,

$$\begin{aligned}[f_1, f_2]\xi = [f_1\xi, f_2\xi] &= [(f_1\gamma)r_{f_1}, (f_2\gamma)r_{f_2}] \\ &\equiv [f_1\gamma, f_2\gamma] = [f_1, f_2]\gamma \mod [\mathfrak{R}, \mathfrak{F}],\end{aligned}$$

so

$$y\xi \equiv y\gamma \mod [\mathfrak{R}, \mathfrak{F}]$$

---

*For the definitions of $\mathbf{Z}^2$, $\mathbf{B}^2$, see I, 16.12.

for all $y \in \mathfrak{F}'$. In particular, if $r \in \mathfrak{R} \cap \mathfrak{F}'$,

$$(r[\mathfrak{R}, \mathfrak{F}])g = (r\gamma)[\mathfrak{R}, \mathfrak{F}] = (r\xi)[\mathfrak{R}, \mathfrak{F}].$$

To check that this defines a $\mathfrak{G}$-module structure, observe first that

$$(r[\mathfrak{R}, \mathfrak{F}])1_{\mathfrak{G}} = r[\mathfrak{R}, \mathfrak{F}],$$

since we chose γ to be 1 in this case. If g_1, g_2 are elements of $\mathfrak{G}$ and γ_1, γ_2 are endomorphisms such that

$$((f\gamma_1)\mathfrak{R})\zeta = ((f\mathfrak{R})\zeta)g_1, \quad ((f\gamma_2)\mathfrak{R})\zeta = ((f\mathfrak{R})\zeta)g_2$$

for all $f \in \mathfrak{F}$, then

$$((f\gamma_1\gamma_2)\mathfrak{R})\zeta = (((f\gamma_1)\mathfrak{R})\zeta)g_2 = ((f\mathfrak{R})\zeta)g_1g_2.$$

Hence if $r \in \mathfrak{R} \cap \mathfrak{F}'$,

$$\begin{aligned}(r[\mathfrak{R}, \mathfrak{F}])(g_1g_2) &= (r\gamma_1\gamma_2)[\mathfrak{R}, \mathfrak{F}]\\ &= ((r\gamma_1)[\mathfrak{R}, \mathfrak{F}])g_2\\ &= ((r[\mathfrak{R}, \mathfrak{F}])g_1)g_2.\end{aligned}$$

Thus $(\mathfrak{R} \cap \mathfrak{F}')/[\mathfrak{R}, \mathfrak{F}]$ is a $\mathfrak{G}$-module, and this is clearly the only $\mathfrak{G}$-module structure on $(\mathfrak{R} \cap \mathfrak{F}')/[\mathfrak{R}, \mathfrak{F}]$ with the stated property.

b) Given $c \in \mathbf{Z}^2(\mathfrak{Q}, \mathsf{M})$ and $g \in \mathfrak{G}$, define $c' \in \mathbf{C}^2(\mathfrak{Q}, \mathsf{M})$ by

$$c'(x, y) = c(xg^{-1}, yg^{-1}) \quad (x \in \mathfrak{Q}, y \in \mathfrak{Q}).$$

It is easy to verify that $c' \in \mathbf{Z}^2(\mathfrak{Q}, \mathsf{M})$ and that if $c \in \mathbf{B}^2(\mathfrak{Q}, \mathsf{M})$, then $c' \in \mathbf{B}^2(\mathfrak{Q}, \mathsf{M})$. Hence we may put

$$(c + \mathbf{B}^2(\mathfrak{Q}, \mathsf{M}))g = c' + \mathbf{B}^2(\mathfrak{Q}, \mathsf{M})$$

and $\mathbf{H}^2(\mathfrak{Q}, \mathsf{M})$ becomes a $\mathfrak{G}$-module.

c) Since $\mathsf{K}^\times$ is a multiplicative group, we use the multiplicative notation in $\mathbf{H}^2(\mathfrak{G}, \mathsf{K}^\times)$.

For each $x \in \mathfrak{Q}$, choose $t_x \in \mathfrak{F}$ such that $x\zeta^{-1} = t_x\mathfrak{R}$. Write

$$t_xt_y = r(x, y)t_{xy},$$

so $r(x, y) \in \mathfrak{R}$. Let $\mathfrak{H} = \mathrm{Hom}(\mathfrak{R}/[\mathfrak{R}, \mathfrak{F}], \mathsf{K}^\times)$. By V, 23.3a), there is a homomorphism ε of $\mathfrak{H}$ into $\mathbf{H}^2(\mathfrak{Q}, \mathsf{K}^\times)$ given by

$$\alpha\varepsilon = c\mathbf{B}^2(\mathfrak{Q}, \mathsf{K}^\times) \quad (\alpha \in \mathfrak{H}),$$

where c is the element of $\mathbf{Z}^2(\mathfrak{Q}, \mathsf{K}^\times)$ given by

$$c(x, y) = (r(x, y)[\mathfrak{R}, \mathfrak{F}])\alpha \quad (x \in \mathfrak{Q}, y \in \mathfrak{Q});$$

further, $\ker \varepsilon = \mathfrak{T}^\perp$, where $\mathfrak{T} = (\mathfrak{R} \cap \mathfrak{F}')/[\mathfrak{R}, \mathfrak{F}]$ and

$$\mathfrak{T}^\perp = \{\phi \mid \phi \in \mathfrak{H}, \mathfrak{T}\phi = 1\}.$$

By V, 6.3a), the restriction mapping of $\mathfrak{H}$ into $\hat{\mathfrak{T}} = \mathrm{Hom}(\mathfrak{T}, \mathsf{K}^\times)$ is an epimorphism with kernel $\mathfrak{T}^\perp$, so this gives rise to an isomorphism ρ of $\mathfrak{H}/\mathfrak{T}^\perp$ onto $\hat{\mathfrak{T}}$. Hence

$$\mathfrak{H}\varepsilon \cong \mathfrak{H}/\mathfrak{T}^\perp \cong \hat{\mathfrak{T}}.$$

By V, 23.5, $\mathfrak{T}$ is finite and $\mathfrak{T} \cong \mathbf{H}^2(\mathfrak{Q}, \mathsf{K}^\times)$. Thus by V, 6.4a), $|\hat{\mathfrak{T}}| = |\mathfrak{T}|$. Hence $|\mathfrak{H}\varepsilon| = |\mathbf{H}^2(\mathfrak{Q}, \mathsf{K}^\times)|$ and so ε is an epimorphism of $\mathfrak{H}$ onto $\mathbf{H}^2(\mathfrak{Q}, \mathsf{K}^\times)$. Hence there is an isomorphism η of $\mathfrak{H}/\mathfrak{T}^\perp$ onto $\mathbf{H}^2(\mathfrak{Q}, \mathsf{K}^\times)$ given by

$$(\alpha\mathfrak{T}^\perp)\eta = \alpha\varepsilon \quad (\alpha \in \mathfrak{H}),$$

and $\rho^{-1}\eta$ is an isomorphism of $\hat{\mathfrak{T}}$ onto $\mathbf{H}^2(\mathfrak{Q}, \mathsf{K}^\times)$. It is to be shown that $\rho^{-1}\eta$ is a $\mathfrak{G}$-isomorphism. Suppose then that $\beta \in \hat{\mathfrak{T}}$ and $g \in \mathfrak{G}$; we must prove that

$$(\beta g)\rho^{-1}\eta = (\beta\rho^{-1}\eta)g.$$

By 12.11b), there is an endomorphism γ of $\mathfrak{F}$ such that

$$((f\gamma)\mathfrak{R})\zeta = ((f\mathfrak{R})\zeta)g^{-1}$$

for all $f \in \mathfrak{F}$, and by a),

$$(r[\mathfrak{R}, \mathfrak{F}])g^{-1} = (r\gamma)[\mathfrak{R}, \mathfrak{F}] \quad (r \in \mathfrak{R} \cap \mathfrak{F}').$$

Now $\beta\rho^{-1} = \alpha\mathfrak{T}^\perp$ for some $\alpha \in \mathfrak{H}$ and β is the restriction of α to $\mathfrak{T}$. Hence if $r \in \mathfrak{R} \cap \mathfrak{F}'$, it follows using 12.11a) that

$$\begin{aligned}(r[\mathfrak{R}, \mathfrak{F}])(\beta g) &= ((r[\mathfrak{R}, \mathfrak{F}])g^{-1})\beta \\ &= ((r\gamma)[\mathfrak{R}, \mathfrak{F}])\beta \\ &= ((r\gamma)[\mathfrak{R}, \mathfrak{F}])\alpha.\end{aligned}$$

Thus if $\alpha' \in \mathfrak{H}$ is defined by

$$(r[\mathfrak{R}, \mathfrak{F}])\alpha' = ((r\gamma)[\mathfrak{R}, \mathfrak{F}])\alpha \quad (r \in \mathfrak{R}),$$

βg is the restriction of α' to $\mathfrak{T}$ and $(\beta g)\rho^{-1} = \alpha'\mathfrak{T}^{\perp}$. Thus $(\beta g)\rho^{-1}\eta = \alpha'\varepsilon = c'\mathbf{B}^2(\mathfrak{Q}, \mathsf{K}^{\times})$, where $c' \in \mathbf{Z}^2(\mathfrak{Q}, \mathsf{K}^{\times})$ is defined by

$$c'(x, y) = (r(x, y)[\mathfrak{R}, \mathfrak{F}])\alpha' = ((r(x, y)\gamma)[\mathfrak{R}, \mathfrak{F}])\alpha.$$

Similarly $\beta\rho^{-1}\eta = c\mathbf{B}^2(\mathfrak{Q}, \mathsf{K}^{\times})$, where

$$c(x, y) = (r(x, y)[\mathfrak{R}, \mathfrak{F}])\alpha.$$

Thus by b), $(\beta\rho^{-1}\eta)g = c''\mathbf{B}^2(\mathfrak{Q}, \mathsf{M})$, where

$$c''(x, y) = c(xg^{-1}, yg^{-1}) = (r(xg^{-1}, yg^{-1})[\mathfrak{R}, \mathfrak{F}])\alpha.$$

We show that $c''\mathbf{B}^2(\mathfrak{Q}, \mathsf{M}) = c'\mathbf{B}^2(\mathfrak{Q}, \mathsf{M})$.

Since

$$((t_x\gamma)\mathfrak{R})\zeta = ((t_x\mathfrak{R})\zeta)g^{-1} = xg^{-1} = (t_{xg^{-1}}\mathfrak{R})\zeta,$$

$t_{xg^{-1}} = b(x)(t_x\gamma)$ for some $b(x) \in \mathfrak{R}$. It follows from

$$t_{xg^{-1}}t_{yg^{-1}} = r(xg^{-1}, yg^{-1})t_{(xy)g^{-1}}$$

that

$$\begin{aligned}b(x)(t_x\gamma)b(y)(t_y\gamma) &= r(xg^{-1}, yg^{-1})b(xy)(t_{xy}\gamma) \\ &= r(xg^{-1}, yg^{-1})b(xy)(r(x, y)^{-1}\gamma)(t_x\gamma)(t_y\gamma).\end{aligned}$$

Hence

$$b(x)b(y) \equiv r(xg^{-1}, yg^{-1})b(xy)(r(x, y)^{-1}\gamma) \mod [\mathfrak{R}, \mathfrak{F}].$$

Applying α,

$$\sigma(x)\sigma(y) = c''(x, y)\sigma(xy)c'(x, y)^{-1},$$

where $\sigma(x) = (b(x)[\mathfrak{R}, \mathfrak{F}])\alpha$. Hence

$$c''(x, y) = c'(x, y)\sigma(y)\sigma(xy)^{-1}\sigma(x),$$

as required. **q.e.d.**

The module structure of $\mathbf{H}^2(\mathfrak{Q}, \mathsf{M})$ given by b) can also be described as follows.

12.13 Lemma. *If $\mathfrak{Q}$ is a normal subgroup of $\mathfrak{G}$ and M is a trivial $\mathfrak{G}$-module,*

$$\alpha g = \alpha T^g_{\mathfrak{Q}} \quad (\alpha \in \mathbf{H}^2(\mathfrak{Q}, \mathsf{M}), g \in \mathfrak{G}).$$

Proof. The module structure of $\mathbf{H}^2(\mathfrak{Q}, \mathsf{M})$ given in 12.12b) is in terms of the description $\mathbf{H}^2(\mathfrak{Q}, \mathsf{M}) \cong \mathbf{Z}^2(\mathfrak{Q}, \mathsf{M})/\mathbf{B}^2(\mathfrak{Q}, \mathsf{M})$; we need to change this to the other description $\mathbf{H}^2(\mathfrak{Q}, \mathsf{M}) \cong \tilde{\mathbf{Z}}^2(\mathfrak{Q}, \mathsf{M})/\tilde{\mathbf{B}}^2(\mathfrak{Q}, \mathsf{M})$. For this we use the proof of I, 16.14, according to which the element $\tilde{c}$ of $\tilde{\mathbf{C}}^2(\mathfrak{Q}, \mathsf{M})$ corresponding to $c \in \mathbf{C}^2(\mathfrak{Q}, \mathsf{M})$ is given by

$$\tilde{c}(x, y, z) = c(xy^{-1}, yz^{-1})z = c(xy^{-1}, yz^{-1}).$$

Thus if $\alpha = \tilde{c} + \tilde{\mathbf{B}}^2(\mathfrak{Q}, \mathsf{M})$, then $\alpha g = \tilde{c}' + \tilde{\mathbf{B}}^2(\mathfrak{Q}, \mathsf{M})$, where

$$\tilde{c}'(x, y, z) = c'(xy^{-1}, yz^{-1}) = c((xy^{-1})^{g^{-1}}, (yz^{-1})^{g^{-1}}) = \tilde{c}(x^{g^{-1}}, y^{g^{-1}}, z^{g^{-1}}),$$

or

$$\tilde{c}'(x^g, y^g, z^g) = \tilde{c}(x, y, z).$$

Hence by 12.2, $\alpha g = \alpha T^g_{\mathfrak{Q}}$. **q.e.d.**

12.14 Lemma. *Suppose that $\mathfrak{M}$, $\mathfrak{N}$ are normal subgroups of a group $\mathfrak{G}$. Then for $n \geq 1$,*

$$[\mathfrak{M}', \mathfrak{N}; n] \leq \prod_{i=0}^{n} [[\mathfrak{M}, \mathfrak{N}; i], [\mathfrak{M}, \mathfrak{N}; n - i]],$$

where $[\mathfrak{M}, \mathfrak{N}; i] = [\mathfrak{M}, \mathfrak{N}, \underset{i}{\ldots} \mathfrak{N}]$ *(and* $[\mathfrak{M}, \mathfrak{N}; 0] = \mathfrak{M}$*).*

Proof. This is proved by induction on n. For $n > 1$, the inductive hypothesis gives

$$[\mathfrak{M}', \mathfrak{N}; n - 1] \leq \prod_{i=0}^{n-1} [[\mathfrak{M}, \mathfrak{N}; i], [\mathfrak{M}, \mathfrak{N}; n - i - 1]],$$

and this is also true for $n = 1$. By III, 1.10,

$$\begin{aligned} &[\mathfrak{M}', \mathfrak{N}; n] \\ &\leq \prod_{i=0}^{n-1} [[\mathfrak{M}, \mathfrak{N}; i], [\mathfrak{M}, \mathfrak{N}; n - i - 1], \mathfrak{N}] \\ &\leq \prod_{i=0}^{n-1} [[\mathfrak{M}, \mathfrak{N}; n - i], [\mathfrak{M}, \mathfrak{N}; i]][[\mathfrak{M}, \mathfrak{N}; i + 1], [\mathfrak{M}, \mathfrak{N}; n - i - 1]] \\ &\leq \prod_{i=1}^{n} [[\mathfrak{M}, \mathfrak{N}; i], [\mathfrak{M}, \mathfrak{N}; n - i]]. \end{aligned}$$

q.e.d.

12.15 Lemma. *Suppose that $\mathfrak{G}$ is finite, $\mathfrak{Q} \trianglelefteq \mathfrak{G}$ and $\mathfrak{Q} \leq \mathbf{Z}_c(\mathfrak{G})$. If K is an algebraically closed field of characteristic 0,*

$$[\mathbf{H}^2(\mathfrak{Q}, \mathsf{K}^\times), \mathfrak{G}; 2c - 1] = 1.$$

Proof. Let $\mathfrak{F}$ be a finitely generated free group having a normal subgroup $\mathfrak{R}$ for which there is an isomorphism ζ of $\mathfrak{F}/\mathfrak{R}$ onto $\mathfrak{G}$. Write $\mathfrak{Q}\zeta^{-1} = \mathfrak{F}_1/\mathfrak{R}$; thus $\mathfrak{F}_1 \trianglelefteq \mathfrak{F}$ and by IX, 1.14, $\mathfrak{F}_1$ is a finitely generated free group. By 12.14,

$$[\mathfrak{F}_1', \mathfrak{F}; 2c - 1] \leq \prod_{i=1}^{2c-1} [[\mathfrak{F}_1, \mathfrak{F}; i], [\mathfrak{F}_1, \mathfrak{F}; 2c - i - 1]].$$

Since $\mathfrak{Q} \leq \mathbf{Z}_c(\mathfrak{G})$, $[\mathfrak{F}_1, \mathfrak{F}; c] \leq \mathfrak{R}$. Hence

$$[\mathfrak{F}_1', \mathfrak{F}; 2c - 1] \leq [\mathfrak{R}, \mathfrak{F}_1].$$

Now if $\mathfrak{T} = (\mathfrak{R} \cap \mathfrak{F}_1')/[\mathfrak{R}, \mathfrak{F}_1]$, $\mathfrak{T}$ is an $\mathfrak{F}$-module and this gives

$$[\mathfrak{T}, \mathfrak{F}; 2c - 1] = 1.$$

But also, by 12.12a), $\mathfrak{T}$ has a $\mathfrak{G}$-module structure and if $f \in \mathfrak{F}$,

$$(r[\mathfrak{R}, \mathfrak{F}_1])((f\mathfrak{R})\zeta) = r^f[\mathfrak{R}, \mathfrak{F}_1] \quad (r \in \mathfrak{R} \cap \mathfrak{F}_1').$$

This means that if $x \in \mathfrak{T}$ and $f \in \mathfrak{F}$, $x((f\mathfrak{R})\zeta) = xf$. Hence

$$[\mathfrak{T}, \mathfrak{G}; 2c - 1] = 1.$$

Let $\mathfrak{T}_{i+1} = [\mathfrak{T}, \mathfrak{G}; i]$ $(i = 0, \ldots, 2c - 1)$. Thus

$$\mathfrak{T} = \mathfrak{T}_1 \geq \mathfrak{T}_2 \geq \cdots \geq \mathfrak{T}_{2c-1} \geq \mathfrak{T}_{2c} = 1.$$

Let $\hat{\mathfrak{T}} = \operatorname{Hom}(\mathfrak{T}, \mathsf{K}^\times)$ and let

$$\mathfrak{T}_i^\perp = \{f \mid f \in \hat{\mathfrak{T}}, xf = 1 \quad \text{for all} \quad x \in \mathfrak{T}_i\} \quad (1 \leq i \leq 2c).$$

Then

$$\hat{\mathfrak{T}} = \mathfrak{T}_{2c}^\perp \geq \mathfrak{T}_{2c-1}^\perp \geq \cdots \geq \mathfrak{T}_2^\perp \geq \mathfrak{T}_1^\perp = 0.$$

Suppose that $f \in \mathfrak{T}_{i+1}^\perp$ $(i = 1, \ldots, 2c - 1)$ and $g \in \mathfrak{G}$. Then for any $x \in \mathfrak{T}_i$,

$$(xg^{-1})x^{-1} \in [\mathfrak{T}_i, \mathfrak{G}] = \mathfrak{T}_{i+1},$$

so $(xg^{-1})f = xf$ and $x(fg - f) = 1$. Thus $fg - f \in \mathfrak{T}_i^\perp$; hence $[\mathfrak{T}_{i+1}^\perp, \mathfrak{G}] \leq \mathfrak{T}_i^\perp$. Thus $[\hat{\mathfrak{T}}, \mathfrak{G}; 2c - 1] = 1$. By 12.12c), $[\mathbf{H}^2(\mathfrak{Q}, \mathsf{K}^\times), \mathfrak{G}; 2c - 1] = 1$. **q.e.d.**

12.16 Lemma. *Suppose that $\mathfrak{S} \in S_p(\mathfrak{G})$ and that the class of $\mathfrak{S}$ is at most $\frac{1}{2}p$. Let β be an element of $\mathbf{H}^2(\mathfrak{P}, \mathsf{K}^\times)$, where $\mathfrak{P}$ is a normal p-subgroup of $\mathfrak{G}$ and K is an algebraically closed field of characteristic 0. If $\beta T_{\mathfrak{P}}^g = \beta$ for all $g \in \mathbf{N}_{\mathfrak{G}}(\mathfrak{S})$, then $\beta T_{\mathfrak{P}}^g = \beta$ for all $g \in \mathfrak{G}$.*

Proof. $\mathbf{H}^2(\mathfrak{P}, \mathsf{K}^\times)$ is a $\mathfrak{G}$-module, so we can form the split extension

$$\mathfrak{H} = \mathfrak{G}\mathbf{H}^2(\mathfrak{P}, \mathsf{K}^\times);$$

thus $\mathfrak{H} = \{(g, \alpha) \mid g \in \mathfrak{G}, \alpha \in \mathbf{H}^2(\mathfrak{P}, \mathsf{K}^\times)\}$ and, by 12.13,

$$(g_1, \alpha_1)(g_2, \alpha_2) = (g_1 g_2, (\alpha_1 T_{\mathfrak{P}}^{g_2})\alpha_2).$$

By 12.15,

$$[\mathbf{H}^2(\mathfrak{P}, \mathsf{K}^\times), \mathfrak{S}; p - 1] = 1.$$

Since $\mathbf{H}^2(\mathfrak{P}, \mathsf{K}^\times)$ is a p-group (I, 16.19), $\mathfrak{S}_1 = \mathfrak{S}\mathbf{H}^2(\mathfrak{P}, \mathsf{K}^\times) \in S_p(\mathfrak{H})$, and by III, 1.10,

$$[\mathbf{H}^2(\mathfrak{P}, \mathsf{K}^\times), \mathfrak{S}_1; p - 1] = 1.$$

We apply 6.14 (with $A = \mathfrak{S}_1$); this gives

$$\mathbf{Z}(\mathfrak{M}) \cap \mathbf{H}^2(\mathfrak{P}, \mathsf{K}^\times) = \mathbf{Z}(\mathfrak{H}) \cap \mathbf{H}^2(\mathfrak{P}, \mathsf{K}^\times),$$

where $\mathfrak{M} = \mathbf{N}_{\mathfrak{H}}(\mathfrak{S}_1) = \mathbf{N}_{\mathfrak{G}}(\mathfrak{S})\mathbf{H}^2(\mathfrak{P}, \mathsf{K}^\times)$. The hypothesis on β is that $(1, \beta) \in \mathbf{Z}(\mathfrak{M})$. Thus $(1, \beta) \in \mathbf{Z}(\mathfrak{H})$ and $\beta T^g_{\mathfrak{P}} = \beta$ for all $g \in \mathfrak{G}$. **q.e.d.**

12.17 Theorem (HOLT [1]). *Suppose that $\mathfrak{S} \in S_p(\mathfrak{G})$ and that the class of $\mathfrak{S}$ is at most $\frac{1}{2}p$. Then the Sylow p-subgroups of the Schur multipliers of $\mathfrak{G}$ and of $\mathbf{N}_{\mathfrak{G}}(\mathfrak{S})$ are isomorphic.*

Proof. By 12.6, the Sylow p-subgroups of $\mathbf{H}^2(\mathfrak{G}, \mathsf{K}^\times)$, $\mathbf{H}^2(\mathbf{N}_{\mathfrak{G}}(\mathfrak{S}), \mathsf{K}^\times)$ are isomorphic to the groups $\mathfrak{A}_1$, $\mathfrak{A}_2$ of elements of $\mathbf{H}^2(\mathfrak{S}, \mathsf{K}^\times)$ stable in $\mathfrak{G}$, $\mathbf{N}_{\mathfrak{G}}(\mathfrak{S})$ respectively. We prove that $\mathfrak{A}_1 = \mathfrak{A}_2$. Clearly $\mathfrak{A}_1 \leq \mathfrak{A}_2$. Suppose that $\alpha \in \mathfrak{A}_2$. Then $\alpha T^g_{\mathfrak{S}} = \alpha$ for all $g \in \mathbf{N}_{\mathfrak{G}}(\mathfrak{S})$.

Let $\mathscr{F}$ be the set of **I**-extremal subgroups of $\mathfrak{S}$ (5.3), where **I** is the identity characteristic p-functor. Thus, if $\mathfrak{P} \in \mathscr{F}$, $\mathfrak{P}$ is extremal in $\mathfrak{S}$ and $\mathbf{N}_{\mathfrak{S}}(\mathfrak{P}) \in \mathscr{F}$. By 5.5, $\mathscr{F}$ is a conjugation family for $\mathfrak{S}$. By 12.10, it is sufficient to prove that $T^g_{\mathfrak{P}}$ fixes $\alpha R(\mathfrak{S}, \mathfrak{P})$ whenever $\mathfrak{P} \in \mathscr{F}$ and $g \in \mathbf{N}_{\mathfrak{G}}(\mathfrak{P})$. This is done by induction on $|\mathfrak{S} : \mathbf{N}_{\mathfrak{S}}(\mathfrak{P})|$. Let $\mathfrak{P}_1 = \mathbf{N}_{\mathfrak{S}}(\mathfrak{P})$. Then $\alpha R(\mathfrak{S}, \mathfrak{P}_1)T^g_{\mathfrak{P}_1} = \alpha R(\mathfrak{S}, \mathfrak{P}_1)$ for all $g \in \mathbf{N}_{\mathfrak{G}}(\mathfrak{P}_1)$; this follows from the inductive hypothesis if $\mathfrak{P}_1 < \mathfrak{S}$ and from the previous paragraph if $\mathfrak{P}_1 = \mathfrak{S}$. Hence if $h \in \mathbf{N}_{\mathfrak{G}}(\mathfrak{P}_1) \cap \mathbf{N}_{\mathfrak{G}}(\mathfrak{P})$,

$$\begin{aligned}
\alpha R(\mathfrak{S}, \mathfrak{P})T^h_{\mathfrak{P}} &= \alpha R(\mathfrak{S}, \mathfrak{P}_1)R(\mathfrak{P}_1, \mathfrak{P})T^h_{\mathfrak{P}} \\
&= \alpha R(\mathfrak{S}, \mathfrak{P}_1)i^*_{\mathfrak{P}, \mathfrak{P}_1}t^*_{\mathfrak{P}, h^{-1}} \\
&= \alpha R(\mathfrak{S}, \mathfrak{P}_1)(t_{\mathfrak{P}, h^{-1}}i_{\mathfrak{P}, \mathfrak{P}_1})^* \\
&= \alpha R(\mathfrak{S}, \mathfrak{P}_1)(i_{\mathfrak{P}, \mathfrak{P}_1}t_{\mathfrak{P}_1, h^{-1}})^* \\
&= \alpha R(\mathfrak{S}, \mathfrak{P}_1)T^h_{\mathfrak{P}_1}R(\mathfrak{P}_1, \mathfrak{P}) \\
&= \alpha R(\mathfrak{S}, \mathfrak{P}_1)R(\mathfrak{P}_1, \mathfrak{P}) \\
&= \alpha R(\mathfrak{S}, \mathfrak{P}).
\end{aligned}$$

Hence by 12.16 applied to the Sylow p-subgroup $\mathfrak{P}_1$ of $\mathbf{N}_{\mathfrak{G}}(\mathfrak{P})$ and to $\alpha R(\mathfrak{S}, \mathfrak{P})$,

$$\alpha R(\mathfrak{S}, \mathfrak{P}) T^g_{\mathfrak{P}} = \alpha R(\mathfrak{S}, \mathfrak{P})$$

for all $g \in \mathbf{N}_{\mathfrak{G}}(\mathfrak{P})$. **q.e.d.**

12.18 Remark. In [1], HOLT gives examples which show that in the condition on the class of $\mathfrak{S}$ in 12.17, $\frac{1}{2}p$ cannot be replaced by $\frac{1}{2}p + 1$.

§ 13. The Generalized Fitting Subgroup

13.1 Lemma. *Suppose that*

(1) $$\mathfrak{G} = \mathfrak{G}_0 > \mathfrak{G}_1 > \cdots > \mathfrak{G}_n = 1,$$

where $\mathfrak{G}_i \trianglelefteq \mathfrak{G}$ and $\mathfrak{G}_{i-1}/\mathfrak{G}_i$ is either Abelian or the direct product of non-Abelian simple groups ($i = 1, \ldots, n$). If $x \in \mathfrak{G}$ and x induces an inner automorphism on $\mathfrak{G}_{i-1}/\mathfrak{G}_i$ for each $i = 1, \ldots, n$, then x induces an inner automorphism on any chief factor of $\mathfrak{G}$.

Proof. The series (1) may be refined to a chief series of $\mathfrak{G}$, and by the Jordan-Hölder theorem, any chief factor of $\mathfrak{G}$ is $\mathfrak{G}$-isomorphic to one of the factors in this refinement. It is therefore sufficient to show that if $\mathfrak{G}_{i-1} \geq \mathfrak{K} > \mathfrak{L} \geq \mathfrak{G}_i$ and $\mathfrak{K} \trianglelefteq \mathfrak{G}$, $\mathfrak{L} \trianglelefteq \mathfrak{G}$, then x induces an inner automorphism on $\mathfrak{K}/\mathfrak{L}$.

Write $\overline{\mathfrak{G}} = \mathfrak{G}_{i-1}/\mathfrak{G}_i$. By hypothesis, the automorphism of $\overline{\mathfrak{G}}$ induced by x is the inner automorphism induced by some element g of $\overline{\mathfrak{G}}$. If $\overline{\mathfrak{G}}$ is Abelian, this is the identity automorphism and x induces the identity automorphism on $\mathfrak{K}/\mathfrak{L}$. Suppose, then, that $\overline{\mathfrak{G}}$ is non-Abelian. By hypothesis,

$$\overline{\mathfrak{G}} = \mathfrak{H}_1 \times \cdots \times \mathfrak{H}_r,$$

where $\mathfrak{H}_1, \ldots, \mathfrak{H}_r$ are non-Abelian simple groups. Hence

$$\mathfrak{K}/\mathfrak{G}_i = \mathfrak{H}_{j_1} \times \cdots \times \mathfrak{H}_{j_s}$$

for certain indices $j_1, \ldots, j_s$ (I, 9.12). If $g = g_1 \cdots g_r$ ($g_j \in \mathfrak{H}_j$), the automorphism of $\mathfrak{K}/\mathfrak{G}_i$ induced by g is also induced by $g' = g_{j_1} \cdots g_{j_s}$. Hence the automorphisms induced by x on $\mathfrak{K}/\mathfrak{G}_i$ and $\mathfrak{K}/\mathfrak{L}$ are inner. **q.e.d.**

13.2 Definition. A group $\mathfrak{G}$ is called *quasinilpotent* if given any chief factor $\mathfrak{X}$ of $\mathfrak{G}$, every automorphism of $\mathfrak{X}$ induced by an element of $\mathfrak{G}$ is inner.

13.3 Lemma. a) *If $\mathfrak{G}$ is quasinilpotent and $\mathfrak{N} \trianglelefteq \mathfrak{G}$, $\mathfrak{N}$ and $\mathfrak{G}/\mathfrak{N}$ are quasinilpotent.*

b) *A subnormal subgroup of a quasinilpotent group is quasinilpotent.*

c) *If $\mathfrak{G}/\mathfrak{M}$ and $\mathfrak{G}/\mathfrak{N}$ are quasinilpotent, $\mathfrak{G}/(\mathfrak{M} \cap \mathfrak{N})$ is quasinilpotent.*

d) *The direct product of quasinilpotent groups is quasinilpotent.*

Proof. a) To prove that $\mathfrak{N}$ is quasinilpotent, consider a chief series of $\mathfrak{G}$ of the form

$$\mathfrak{G} = \mathfrak{G}_0 > \cdots > \mathfrak{G}_{m-1} > \mathfrak{G}_m = \mathfrak{N} > \mathfrak{G}_{m+1} > \cdots > \mathfrak{G}_n = 1.$$

Thus $\mathfrak{G}_{i-1}/\mathfrak{G}_i$ is the direct product of isomorphic simple groups. If $x \in \mathfrak{N}$, then x induces an inner automorphism on $\mathfrak{G}_{i-1}/\mathfrak{G}_i$ for each $i = m + 1, \ldots, n$. Hence by 13.1, x induces an inner automorphism on any chief factor of $\mathfrak{N}$. Thus $\mathfrak{N}$ is quasinilpotent.

It is clear that $\mathfrak{G}/\mathfrak{N}$ is quasinilpotent.

b) This follows from a) by an immediate induction.

c) Any chief factor of $\mathfrak{G}/(\mathfrak{M} \cap \mathfrak{N})$ is $\mathfrak{G}$-isomorphic to one of either $\mathfrak{G}/\mathfrak{M}$ or $\mathfrak{G}/\mathfrak{N}$. Thus $\mathfrak{G}/(\mathfrak{M} \cap \mathfrak{N})$ is quasinilpotent.

d) This follows at once from c). **q.e.d.**

13.4 Lemma. *Suppose that $\mathfrak{G}$ is quasinilpotent, $\mathfrak{N} \trianglelefteq \mathfrak{G}$ and every composition factor of $\mathfrak{N}$ is non-Abelian. Then*

$$\mathfrak{G} = \mathfrak{N} \times \mathbf{C}_{\mathfrak{G}}(\mathfrak{N}).$$

Proof. We use induction on $|\mathfrak{N}|$. Suppose that $\mathfrak{N} \neq 1$ and let $\mathfrak{M}$ be a minimal normal subgroup of $\mathfrak{G}$ contained in $\mathfrak{N}$. Given $g \in \mathfrak{G}$, the automorphism of $\mathfrak{M}$ induced by g is also induced by some element x of $\mathfrak{M}$, so $gx^{-1} \in \mathbf{C}_{\mathfrak{G}}(\mathfrak{M})$. Hence $\mathfrak{G} = \mathfrak{M}\mathbf{C}_{\mathfrak{G}}(\mathfrak{M})$. Therefore $\mathfrak{N} = \mathfrak{M}\mathfrak{C}$, where $\mathfrak{C} = \mathfrak{N} \cap \mathbf{C}_{\mathfrak{G}}(\mathfrak{M}) \trianglelefteq \mathfrak{G}$. Now $\mathfrak{C} < \mathfrak{N}$, for otherwise $\mathfrak{N} \leq \mathbf{C}_{\mathfrak{G}}(\mathfrak{M})$, $\mathfrak{M} \leq \mathbf{Z}(\mathfrak{N})$ and $\mathfrak{N}$ has an Abelian composition factor. Also the composition factors of $\mathfrak{C}$ are all non-Abelian, since they are isomorphic to composition factors of $\mathfrak{N}$. Hence by the inductive hypothesis, $\mathfrak{G} = \mathfrak{C}\mathbf{C}_{\mathfrak{G}}(\mathfrak{C})$. But $\mathbf{C}_{\mathfrak{G}}(\mathfrak{C}) \geq \mathfrak{M}$, so

$$\mathbf{C}_{\mathfrak{G}}(\mathfrak{C}) = \mathfrak{M}(\mathbf{C}_{\mathfrak{G}}(\mathfrak{C}) \cap \mathbf{C}_{\mathfrak{G}}(\mathfrak{M})) = \mathfrak{M}\mathbf{C}_{\mathfrak{G}}(\mathfrak{N}).$$

Thus $\mathfrak{G} = \mathfrak{C}\mathfrak{M}\mathbf{C}_{\mathfrak{G}}(\mathfrak{N}) = \mathfrak{N}\mathbf{C}_{\mathfrak{G}}(\mathfrak{N})$. Since $\mathfrak{N} \cap \mathbf{C}_{\mathfrak{G}}(\mathfrak{N}) = \mathbf{Z}(\mathfrak{N}) = 1$,

$$\mathfrak{G} = \mathfrak{N} \times \mathbf{C}_{\mathfrak{G}}(\mathfrak{N}). \qquad \textbf{q.e.d.}$$

13.5 Definition. A group $\mathfrak{G}$ is called *semisimple* if $\mathfrak{G}$ is the direct product of non-Abelian simple groups. In particular the unit group is semisimple.

Clearly, a semisimple group is quasinilpotent.

13.6 Theorem. *The group $\mathfrak{G}$ is quasinilpotent if and only if $\mathfrak{G}/\mathbf{Z}_{\infty}(\mathfrak{G})$ is semisimple.*

Proof. Write $\mathfrak{Z} = \mathbf{Z}_{\infty}(\mathfrak{G})$.

Suppose first that $\mathfrak{G}/\mathfrak{Z}$ is semisimple. By refining $1 = \mathbf{Z}_0(\mathfrak{G}) \leq \mathbf{Z}_1(\mathfrak{G}) \leq \cdots \leq \mathfrak{Z} \leq \mathfrak{G}$, we obtain a chief series

$$\mathfrak{G} = \mathfrak{G}_0 > \cdots > \mathfrak{G}_{m-1} > \mathfrak{G}_m = \mathfrak{Z} > \mathfrak{G}_{m+1} > \cdots > \mathfrak{G}_{n-1} > \mathfrak{G}_n = 1,$$

where $\mathfrak{G}_{i-1}/\mathfrak{G}_i \leq \mathbf{Z}(\mathfrak{G}/\mathfrak{G}_i)$ $(i = m+1, \ldots, n)$. Thus for $m + 1 \leq i \leq n$, the only automorphism of $\mathfrak{G}_{i-1}/\mathfrak{G}_i$ induced by elements of $\mathfrak{G}$ is the identity. Since $\mathfrak{G}/\mathfrak{Z}$ is semisimple, $\mathfrak{G}/\mathfrak{Z}$ is quasinilpotent. Thus for $1 \leq i \leq m$, any automorphism of $\mathfrak{G}_{i-1}/\mathfrak{G}_i$ induced by an element of $\mathfrak{G}$ is inner. Hence $\mathfrak{G}$ is quasinilpotent.

We prove the converse by induction on $|\mathfrak{G}|$. Thus we may suppose that $\mathfrak{G} \neq 1$. If $\mathfrak{Z} \neq 1$, the inductive hypothesis may be applied to $\mathfrak{G}/\mathfrak{Z}$ by 13.3a), and this gives the result at once. Suppose, then, that $\mathfrak{Z} = 1$, and let $\mathfrak{N}$ be a minimal normal subgroup of $\mathfrak{G}$. Then $\mathfrak{G} = \mathfrak{N}\mathfrak{C}$, where $\mathfrak{C} = \mathbf{C}_{\mathfrak{G}}(\mathfrak{N})$, since the automorphism of $\mathfrak{N}$ induced by an element of $\mathfrak{G}$ is already induced by an element of $\mathfrak{N}$. Since $\mathfrak{N} \not\leq \mathbf{Z}(\mathfrak{G})$, $\mathfrak{C} \lhd \mathfrak{G}$. By 13.3, $\mathfrak{C}$ is quasinilpotent, so by the inductive hypothesis, $\mathfrak{C}/\mathbf{Z}_{\infty}(\mathfrak{C})$ is semisimple. But since $\mathfrak{G} = \mathfrak{N}\mathfrak{C}$ and $\mathfrak{C} = \mathbf{C}_{\mathfrak{G}}(\mathfrak{N})$, $\mathbf{Z}(\mathfrak{C}) \leq \mathbf{Z}(\mathfrak{G}) = 1$. Thus $\mathfrak{C}$ is semisimple. Also $\mathfrak{N} \cap \mathfrak{C} \leq \mathbf{Z}(\mathfrak{C}) = 1$, so $\mathfrak{G} = \mathfrak{N} \times \mathfrak{C}$. Thus $\mathbf{Z}(\mathfrak{N}) = 1$, $\mathfrak{N}$ is semisimple and $\mathfrak{G}$ is semisimple. **q.e.d.**

13.7 Corollary. a) *If $\mathfrak{G}$ is perfect and quasinilpotent, $\mathfrak{G}/\mathbf{Z}(\mathfrak{G})$ is semisimple.*

b) *If $\mathfrak{G}$ is quasinilpotent, $\mathbf{F}(\mathfrak{G}) = \mathbf{Z}_{\infty}(\mathfrak{G})$.*

c) *If $\mathfrak{N}$ is a soluble normal subgroup of the quasinilpotent group $\mathfrak{G}$, $\mathfrak{N} \leq \mathbf{Z}_{\infty}(\mathfrak{G})$.*

d) *$\mathfrak{G}$ is nilpotent if and only if $\mathfrak{G}$ is soluble and quasinilpotent.*

Proof. a) Since $[\mathbf{Z}_2(\mathfrak{G}), \mathfrak{G}] = [\mathbf{Z}_2(\mathfrak{G}), \mathfrak{G}'] = 1$ (III, 2.11), $\mathbf{Z}_2(\mathfrak{G}) \leq \mathbf{Z}_1(\mathfrak{G})$ and $\mathbf{Z}_{\infty}(\mathfrak{G}) = \mathbf{Z}(\mathfrak{G})$. Thus the assertion follows from 13.6.

b) By 13.6, $\mathbf{F}(\mathfrak{G}/\mathbf{Z}_\infty(\mathfrak{G})) = 1$, so $\mathbf{F}(\mathfrak{G}) = \mathbf{Z}_\infty(\mathfrak{G})$.

c) Put $\mathfrak{Z} = \mathbf{Z}_\infty(\mathfrak{G})$. Then $\mathfrak{N}\mathfrak{Z}/\mathfrak{Z}$ is a soluble normal subgroup of the semisimple group $\mathfrak{G}/\mathfrak{Z}$. By I, 9.12, $\mathfrak{N}\mathfrak{Z}/\mathfrak{Z} = 1$ and $\mathfrak{N} \leq \mathfrak{Z}$.

d) This follows from c). **q.e.d.**

Since simple non-Abelian groups possess subgroups which are soluble but not nilpotent, it is not in general the case that a subgroup of a quasinilpotent group is quasinilpotent.

13.8 Theorem. *Suppose that* $\mathfrak{G}$ *is quasinilpotent and put*

$$\mathfrak{D} = \bigcap_{n \geq 1} \gamma_n(\mathfrak{G}).$$

Then $\mathfrak{G} = \mathfrak{D}\mathbf{F}(\mathfrak{G})$ *and* $[\mathfrak{D}, \mathbf{F}(\mathfrak{G})] = 1$. *Also* $\mathfrak{D} \cap \mathbf{F}(\mathfrak{G}) = \mathbf{Z}(\mathfrak{D})$, $\mathfrak{D}$ *is a perfect quasinilpotent characteristic subgroup of* $\mathfrak{G}$, *and* $\mathfrak{D}/\mathbf{Z}(\mathfrak{D})$ *is semisimple.*

Proof. Let $\mathfrak{N} = \mathbf{F}(\mathfrak{G}) = \mathbf{Z}_\infty(\mathfrak{G})$. By 13.6, $\mathfrak{G}/\mathfrak{N}$ is semisimple, so $\mathfrak{G} = \gamma_n(\mathfrak{G})\mathfrak{N}$ for all $n \geq 1$. Since $\mathfrak{G}$ is finite, $\mathfrak{D} = \gamma_k(\mathfrak{G}) = \gamma_{k+1}(\mathfrak{G}) = \cdots$ for some k. Thus $\mathfrak{G} = \gamma_k(\mathfrak{G})\mathfrak{N} = \mathfrak{D}\mathfrak{N}$. Also $\mathfrak{N} = \mathbf{Z}_l(\mathfrak{G})$ for some $l \geq k$, so $[\mathfrak{D}, \mathfrak{N}] = [\gamma_l(\mathfrak{G}), \mathbf{Z}_l(\mathfrak{G})] = 1$. Thus $\mathfrak{D} \cap \mathfrak{N} = \mathbf{Z}(\mathfrak{D})$. Also

$$\mathfrak{D} = \gamma_{k+1}(\mathfrak{G}) = [\gamma_k(\mathfrak{G}), \mathfrak{G}] = [\mathfrak{D}, \mathfrak{G}] = [\mathfrak{D}, \mathfrak{D}\mathfrak{N}] = [\mathfrak{D}, \mathfrak{D}] = \mathfrak{D}'.$$

Thus $\mathfrak{D}$ is perfect. By 13.3, $\mathfrak{D}$ is quasinilpotent. By 13.7a), $\mathfrak{D}/\mathbf{Z}(\mathfrak{D})$ is semisimple. **q.e.d.**

13.9 Definition. For any group $\mathfrak{G}$, the *generalized Fitting subgroup* $\mathbf{F}^*(\mathfrak{G})$ is the set of all elements x of $\mathfrak{G}$ which induce an inner automorphism on every chief factor of $\mathfrak{G}$.

Clearly $\mathbf{F}^*(\mathfrak{G})$ is a characteristic subgroup of $\mathfrak{G}$. By III, 4.3, $\mathbf{F}(\mathfrak{G}) \leq \mathbf{F}^*(\mathfrak{G})$.

13.10 Theorem. $\mathbf{F}^*(\mathfrak{G})$ *is quasinilpotent, and every subnormal quasinilpotent subgroup of* $\mathfrak{G}$ *is contained in* $\mathbf{F}^*(\mathfrak{G})$.

Proof. To prove that $\mathbf{F}^*(\mathfrak{G})$ is quasinilpotent, consider a chief series of $\mathfrak{G}$ of the form

$$\mathfrak{G} = \mathfrak{G}_0 > \cdots > \mathfrak{G}_{m-1} > \mathfrak{G}_m = \mathbf{F}^*(\mathfrak{G}) > \mathfrak{G}_{m+1} > \cdots > \mathfrak{G}_n = 1.$$

Thus $\mathfrak{G}_{i-1}/\mathfrak{G}_i$ is the direct product of isomorphic simple groups. If $x \in \mathbf{F}^*(\mathfrak{G})$, then x induces an inner automorphism on $\mathfrak{G}_{i-1}/\mathfrak{G}_i$ for each $i = m + 1, \ldots, n$. Hence by 13.1, x induces an inner automorphism on any chief factor of $\mathbf{F}^*(\mathfrak{G})$. Thus $\mathbf{F}^*(\mathfrak{G})$ is quasinilpotent.

Let $\mathfrak{H}$ be a subnormal quasinilpotent subgroup of $\mathfrak{G}$. We prove that $\mathfrak{H} \leq \mathbf{F}^*(\mathfrak{G})$ by induction on $|\mathfrak{G} : \mathfrak{H}|$. If $\mathfrak{H} = \mathfrak{G}$, $\mathfrak{G}$ is quasinilpotent, so $\mathbf{F}^*(\mathfrak{G}) = \mathfrak{G} \geq \mathfrak{H}$. If $\mathfrak{H} < \mathfrak{G}$, there exists $\mathfrak{M} \lhd \mathfrak{G}$ such that $\mathfrak{H}$ is a subnormal subgroup of $\mathfrak{M}$. By the inductive hypothesis, $\mathfrak{H} \leq \mathfrak{N}$, where $\mathfrak{N} = \mathbf{F}^*(\mathfrak{M})$. Since $\mathfrak{N}$ is a characteristic subgroup of $\mathfrak{M}$, $\mathfrak{N} \trianglelefteq \mathfrak{G}$. Also $\mathfrak{N}$ is quasinilpotent. It is sufficient to prove that $\mathfrak{N} \leq \mathbf{F}^*(\mathfrak{G})$. Let

$$\mathfrak{G} = \mathfrak{G}_0 > \mathfrak{G}_1 > \cdots > \mathfrak{G}_n = 1$$

be a chief series of $\mathfrak{G}$ in which $\mathfrak{G}_l = \mathfrak{N}$, and suppose that $x \in \mathfrak{N}$. Then x induces the identity automorphism on $\mathfrak{G}_{i-1}/\mathfrak{G}_i$ for $1 \leq i \leq l$. By 8.16, $\mathfrak{G}_{i-1}/\mathfrak{G}_i$ is the direct product of chief factors of $\mathfrak{M}$ ($i = l + 1, \ldots, n$). Since $x \in \mathbf{F}^*(\mathfrak{M})$, x induces inner automorphisms on each of these factors, so x induces an inner automorphism on any chief factor of $\mathfrak{G}$. Thus $x \in \mathbf{F}^*(\mathfrak{G})$ and $\mathfrak{N} \leq \mathbf{F}^*(\mathfrak{G})$. **q.e.d.**

13.11 Corollary. a) $\mathbf{F}^*(\mathbf{F}^*(\mathfrak{G})) = \mathbf{F}^*(\mathfrak{G})$.

b) *If* $\mathfrak{M}$, $\mathfrak{N}$ *are normal quasinilpotent subgroups of* $\mathfrak{G}$, $\mathfrak{M}\mathfrak{N}$ *is quasinilpotent.*

c) *If* $\mathfrak{N} \trianglelefteq \mathfrak{G}$, $\mathbf{F}^*(\mathfrak{N}) = \mathfrak{N} \cap \mathbf{F}^*(\mathfrak{G})$.

Proof. a) Since $\mathbf{F}^*(\mathfrak{G})$ is quasinilpotent, $\mathbf{F}^*(\mathbf{F}^*(\mathfrak{G})) = \mathbf{F}^*(\mathfrak{G})$.

b) By 13.10, $\mathfrak{M} \leq \mathbf{F}^*(\mathfrak{G})$ and $\mathfrak{N} \leq \mathbf{F}^*(\mathfrak{G})$. Thus $\mathfrak{M}\mathfrak{N} \leq \mathbf{F}^*(\mathfrak{G})$, and $\mathbf{F}^*(\mathfrak{G})$ is quasinilpotent. By 13.3, $\mathfrak{M}\mathfrak{N}$ is quasinilpotent.

c) $\mathbf{F}^*(\mathfrak{N})$ is a normal quasinilpotent subgroup of $\mathfrak{G}$, so $\mathbf{F}^*(\mathfrak{N}) \leq \mathfrak{N} \cap \mathbf{F}^*(\mathfrak{G})$. And $\mathfrak{N} \cap \mathbf{F}^*(\mathfrak{G})$ is a normal quasinilpotent subgroup of $\mathfrak{N}$, so $\mathfrak{N} \cap \mathbf{F}^*(\mathfrak{G}) = \mathbf{F}^*(\mathfrak{N})$. **q.e.d.**

13.12 Theorem *(cf. III, 4.2)*. $\mathbf{C}_{\mathfrak{G}}(\mathbf{F}^*(\mathfrak{G})) \leq \mathbf{F}(\mathfrak{G})$.

Proof. Let $\mathfrak{C} = \mathbf{C}_{\mathfrak{G}}(\mathbf{F}^*(\mathfrak{G}))$. Thus $\mathfrak{C} \trianglelefteq \mathfrak{G}$. Suppose that $\mathfrak{C} \nleq \mathbf{F}(\mathfrak{G})$. Then there exists a normal subgroup $\mathfrak{N}$ of $\mathfrak{G}$ minimal with respect to satisfying $\mathfrak{C} \cap \mathbf{F}(\mathfrak{G}) < \mathfrak{N} \leq \mathfrak{C}$. Since $\mathfrak{N} \leq \mathfrak{C}$ and $\mathbf{F}(\mathfrak{G}) \leq \mathbf{F}^*(\mathfrak{G})$, $\mathfrak{C} \cap \mathbf{F}(\mathfrak{G}) \leq \mathbf{Z}(\mathfrak{N})$. By 13.6, $\mathfrak{N}$ is quasinilpotent. By 13.10, $\mathfrak{N} \leq \mathbf{F}^*(\mathfrak{G})$. Since $\mathfrak{N} \leq \mathfrak{C}$, $\mathfrak{N} \leq \mathbf{Z}(\mathbf{F}^*(\mathfrak{G})) \leq \mathbf{F}(\mathfrak{G})$, a contradiction. **q.e.d.**

13.13 Theorem. $\mathbf{F}^*(\mathfrak{G})/\mathbf{F}(\mathfrak{G})$ *is the group generated by all minimal normal subgroups of* $\mathfrak{C}\mathbf{F}(\mathfrak{G})/\mathbf{F}(\mathfrak{G})$, *where* $\mathfrak{C} = \mathbf{C}_{\mathfrak{G}}(\mathbf{F}(\mathfrak{G}))$.

Proof. Let $\mathfrak{K}/\mathbf{F}(\mathfrak{G})$ be the group generated by all minimal normal subgroups of $\mathfrak{C}\mathbf{F}(\mathfrak{G})/\mathbf{F}(\mathfrak{G})$. First we show that $\mathfrak{K} \leq \mathbf{F}^*(\mathfrak{G})$. This is trivial if $\mathfrak{C} \leq \mathbf{F}(\mathfrak{G})$. Otherwise, let $\mathfrak{L}/\mathbf{F}(\mathfrak{G})$ be a minimal normal subgroup of $\mathfrak{C}\mathbf{F}(\mathfrak{G})/\mathbf{F}(\mathfrak{G})$. Thus $\mathfrak{L} = (\mathfrak{C} \cap \mathfrak{L})\mathbf{F}(\mathfrak{G})$ and $\mathfrak{L}/\mathbf{F}(\mathfrak{G}) \cong (\mathfrak{C} \cap \mathfrak{L})/(\mathfrak{C} \cap \mathbf{F}(\mathfrak{G}))$. Hence $(\mathfrak{C} \cap \mathfrak{L})/(\mathfrak{C} \cap \mathbf{F}(\mathfrak{G}))$ is the direct product of simple groups. But $\mathfrak{C} \cap \mathbf{F}(\mathfrak{G}) \leq \mathbf{Z}(\mathfrak{C} \cap \mathfrak{L})$, so $\mathfrak{C} \cap \mathfrak{L}$ is a quasinilpotent subnormal subgroup of $\mathfrak{G}$. By 13.10, $\mathfrak{C} \cap \mathfrak{L} \leq \mathbf{F}^*(\mathfrak{G})$. Hence $\mathfrak{L} \leq \mathbf{F}^*(\mathfrak{G})$. Thus $\mathfrak{K} \leq \mathbf{F}^*(\mathfrak{G})$.

By 13.8, $\mathbf{F}^*(\mathfrak{G}) \leq \mathfrak{C}\mathbf{F}(\mathfrak{G})$. By 13.6, $\mathbf{F}^*(\mathfrak{G})/\mathbf{F}(\mathfrak{G})$ is the direct product of non-Abelian simple groups $\mathfrak{N}_1, \dots, \mathfrak{N}_k$. We apply I, 9.12. Thus if $g \in \mathfrak{C}\mathbf{F}(\mathfrak{G})/\mathbf{F}(\mathfrak{G})$ and $1 \leq i \leq k$, $\mathfrak{N}_i^g = \mathfrak{N}_j$ for some j, and if $\mathfrak{M}_1, \dots, \mathfrak{M}_l$ are the products of the $\mathfrak{N}_i$ in the orbits of $\mathfrak{C}\mathbf{F}(\mathfrak{G})$, $\mathfrak{M}_1, \dots, \mathfrak{M}_l$ are minimal normal subgroups of $\mathfrak{C}\mathbf{F}(\mathfrak{G})/\mathbf{F}(\mathfrak{G})$. Hence $\mathbf{F}^*(\mathfrak{G})/\mathbf{F}(\mathfrak{G}) = \mathfrak{M}_1 \times \cdots \times \mathfrak{M}_i \leq \mathfrak{K}/\mathbf{F}(\mathfrak{G})$ and $\mathfrak{K} = \mathbf{F}^*(\mathfrak{G})$. **q.e.d.**

13.14 Definition. For any group $\mathfrak{G}$, write

$$\mathbf{E}(\mathfrak{G}) = \bigcap_{n \geq 1} \gamma_n(\mathbf{F}^*(\mathfrak{G})).$$

($\mathbf{E}(\mathfrak{G})$ is sometimes called the *layer* of $\mathfrak{G}$). By 13.8, $\mathbf{E}(\mathfrak{G})$ is a perfect quasinilpotent characteristic subgroup of $\mathfrak{G}$ and

$$\mathbf{F}^*(\mathfrak{G}) = \mathbf{E}(\mathfrak{G})\mathbf{F}(\mathfrak{G}), \quad [\mathbf{E}(\mathfrak{G}), \mathbf{F}(\mathfrak{G})] = 1.$$

Hence $\mathbf{Z}(\mathbf{E}(\mathfrak{G})) = \mathbf{E}(\mathfrak{G}) \cap \mathbf{F}(\mathfrak{G}) \leq \mathbf{Z}(\mathbf{F}(\mathfrak{G}))$. By 13.7, $\mathbf{E}(\mathfrak{G})/\mathbf{Z}(\mathbf{E}(\mathfrak{G}))$ is the direct product of simple non-Abelian groups.

13.15 Theorem. a) *If $\mathfrak{H}$ is a perfect quasinilpotent subnormal subgroup of $\mathfrak{G}$, $\mathfrak{H} \leq \mathbf{E}(\mathfrak{G})$.*

b) $\mathbf{E}(\mathbf{E}(\mathfrak{G})) = \mathbf{E}(\mathfrak{G})$.

c) $\mathbf{C}_{\mathbf{F}^*(\mathfrak{G})}(\mathbf{E}(\mathfrak{G})) = \mathbf{F}(\mathfrak{G})$.

Proof. a) By 13.10, $\mathfrak{H} \leq \mathbf{F}^*(\mathfrak{G})$. Since $\mathbf{F}^*(\mathfrak{G})/\mathbf{E}(\mathfrak{G})$ is nilpotent, $\mathfrak{H}/(\mathfrak{H} \cap \mathbf{E}(\mathfrak{G}))$ is nilpotent. Since $\mathfrak{H}$ is perfect, $\mathfrak{H} \leq \mathbf{E}(\mathfrak{G})$.

b) Since $\mathbf{E}(\mathfrak{G})$ is a perfect quasinilpotent normal subgroup of $\mathbf{E}(\mathfrak{G})$, $\mathbf{E}(\mathfrak{G}) \leq \mathbf{E}(\mathbf{E}(\mathfrak{G}))$, by a). Thus $\mathbf{E}(\mathbf{E}(\mathfrak{G})) = \mathbf{E}(\mathfrak{G})$.

c) By 13.7, $\mathbf{Z}(\mathbf{E}(\mathfrak{G})/(\mathbf{E}(\mathfrak{G}) \cap \mathbf{F}(\mathfrak{G}))) = 1$, since $\mathbf{E}(\mathfrak{G}) \cap \mathbf{F}(\mathfrak{G}) = \mathbf{Z}(\mathbf{E}(\mathfrak{G}))$. Thus $\mathbf{Z}(\mathbf{E}(\mathfrak{G})\mathbf{F}(\mathfrak{G})/\mathbf{F}(\mathfrak{G})) = 1$. But $\mathbf{E}(\mathfrak{G})\mathbf{F}(\mathfrak{G}) = \mathbf{F}^*(\mathfrak{G})$, so $\mathbf{C}_{\mathbf{F}^*(\mathfrak{G})}(\mathbf{E}(\mathfrak{G})) = \mathbf{F}(\mathfrak{G})$. **q.e.d.**

13.16 Lemma. a) *Suppose that $\mathfrak{G} = \mathfrak{G}_1 \cdots \mathfrak{G}_n$, where each $\mathfrak{G}_i$ is a simple non-Abelian normal subgroup of $\mathfrak{G}$ and $\mathfrak{G}_i \neq \mathfrak{G}_j$ for $i \neq j$. Then $\mathfrak{G} =$*

$\mathfrak{G}_1 \times \cdots \times \mathfrak{G}_n$ *and any subnormal subgroup* $\mathfrak{H}$ *of* $\mathfrak{G}$ *is the direct product of certain* $\mathfrak{G}_i$.

b) *If* $\mathfrak{X}$ *is a soluble subgroup of* $\mathfrak{G}$ *and* $\mathbf{E}(\mathfrak{G}) \leq \mathbf{N}_{\mathfrak{G}}(\mathfrak{X})$, *then* $[\mathbf{E}(\mathfrak{G}), \mathfrak{X}] = 1$.

Proof. a) If $i \neq j$, $\mathfrak{G}_i \cap \mathfrak{G}_j \lhd \mathfrak{G}_j$, so $\mathfrak{G}_i \cap \mathfrak{G}_j = 1$ and $[\mathfrak{G}_i, \mathfrak{G}_j] = 1$. Thus $\mathfrak{G}_i \cap (\prod_{j \neq i} \mathfrak{G}_j) \leq \mathbf{Z}(\mathfrak{G}_i) = 1$. By I, 9.3, $\mathfrak{G}$ is the direct product of $\mathfrak{G}_1, \ldots, \mathfrak{G}_n$.

We prove that $\mathfrak{H}$ is the product of certain $\mathfrak{G}_i$ by induction on $|\mathfrak{G} : \mathfrak{H}|$. If $\mathfrak{H} < \mathfrak{G}$, we have $\mathfrak{H} \lhd \mathfrak{K}$ for some subnormal subgroup $\mathfrak{K}$ of $\mathfrak{G}$. By the inductive hypothesis, $\mathfrak{K}$ is the direct product of certain $\mathfrak{G}_i$. By I, 9.12, so is $\mathfrak{H}$.

b) Since $\mathbf{E}(\mathfrak{G}) \leq \mathbf{N}_{\mathfrak{G}}(\mathfrak{X})$, $[\mathbf{E}(\mathfrak{G}), \mathfrak{X}] \leq \mathbf{E}(\mathfrak{G}) \cap \mathfrak{X}$ and $\mathbf{E}(\mathfrak{G}) \cap \mathfrak{X}$ is a normal subgroup of $\mathbf{E}(\mathfrak{G})$. Also $\mathbf{E}(\mathfrak{G}) \cap \mathfrak{X}$ is soluble, so $\mathbf{E}(\mathfrak{G}) \cap \mathfrak{X} \leq \mathbf{Z}(\mathbf{E}(\mathfrak{G}))$, since $\mathbf{E}(\mathfrak{G})/\mathbf{Z}(\mathbf{E}(\mathfrak{G}))$ is the direct product of simple non-Abelian groups. Hence $[\mathbf{E}(\mathfrak{G}), \mathfrak{X}, \mathbf{E}(\mathfrak{G})] = 1$, and, using III, 1.10, $[\mathbf{E}(\mathfrak{G}), \mathfrak{X}] = [\mathbf{E}(\mathfrak{G}), \mathbf{E}(\mathfrak{G}), \mathfrak{X}] = 1$. **q.e.d.**

13.17 Definition. A *component* of $\mathbf{E}(\mathfrak{G})$ is a perfect normal subgroup $\mathfrak{H}$ of $\mathbf{E}(\mathfrak{G})$ such that $\mathfrak{H}/\mathbf{Z}(\mathfrak{H})$ is simple.

13.18 Theorem. *Suppose that* $\mathfrak{G}$ *is a group and* $\mathfrak{Z} = \mathbf{Z}(\mathbf{E}(\mathfrak{G}))$.

a) $\mathbf{E}(\mathfrak{G})$ *is the product of its components but is not the product of any proper subset of them.*

b) *If* $\mathfrak{H}$ *is a component of* $\mathbf{E}(\mathfrak{G})$, $\mathfrak{H}\mathfrak{Z}/\mathfrak{Z}$ *is a simple direct factor of* $\mathbf{E}(\mathfrak{G})/\mathfrak{Z}$, *and* $\mathbf{Z}(\mathfrak{H}) = \mathfrak{H} \cap \mathfrak{Z}$.

c) *If* $\mathfrak{K}_1, \mathfrak{K}_2$ *are distinct components of* $\mathbf{E}(\mathfrak{G})$, $[\mathfrak{K}_1, \mathfrak{K}_2] = 1$.

d) *If* $\mathfrak{N}$ *is a subnormal subgroup of* $\mathbf{E}(\mathfrak{G})$, $\mathfrak{N}$ *is the product of* $\mathfrak{N} \cap \mathfrak{Z}$ *and certain components of* $\mathbf{E}(\mathfrak{G})$. *In particular,* $\mathfrak{N} \trianglelefteq \mathbf{E}(\mathfrak{G})$. *Also* $\mathbf{Z}(\mathbf{E}(\mathfrak{G})/\mathfrak{N}) = \mathfrak{N}\mathfrak{Z}/\mathfrak{N}$ *and* $\mathbf{E}(\mathfrak{G}) = \mathfrak{N}\mathbf{C}_{\mathbf{E}(\mathfrak{G})}(\mathfrak{N})$.

e) *If* $\mathfrak{H}$ *is a component of* $\mathbf{E}(\mathfrak{G})$ *and* $\mathfrak{X} \leq \mathfrak{G}$, *either* $\mathfrak{H} \leq [\mathfrak{H}, \mathfrak{X}]$ *or* $[\mathfrak{H}, \mathfrak{X}] = 1$. *If, further,* $\mathfrak{H} \leq \mathbf{N}_{\mathfrak{G}}(\mathfrak{X})$, *either* $\mathfrak{H} \leq \mathbf{E}(\mathfrak{X})$ *or* $[\mathfrak{H}, \mathfrak{X}] = 1$.

Proof. By 13.14, $\mathbf{E}(\mathfrak{G})/\mathfrak{Z} = \overline{\mathfrak{G}}_1 \times \cdots \times \overline{\mathfrak{G}}_n$, where $\overline{\mathfrak{G}}_i = \mathfrak{G}_i/\mathfrak{Z}$ is simple and non-Abelian. Thus $\mathbf{Z}(\mathfrak{G}_i) = \mathfrak{Z}$. Let $\mathfrak{H}_i = \mathfrak{G}_i'$. Since $\mathfrak{G}_i/\mathfrak{Z}$ is perfect, $\mathfrak{G}_i = \mathfrak{H}_i\mathfrak{Z}$. Thus $\mathfrak{H}_i = \mathfrak{G}_i' = \mathfrak{H}_i'$ and $\mathfrak{H}_i$ is perfect. Also $\mathfrak{H}_i/(\mathfrak{H}_i \cap \mathfrak{Z}) \cong \mathfrak{G}_i/\mathfrak{Z}$ is simple, so $\mathbf{Z}(\mathfrak{H}_i) = \mathfrak{H}_i \cap \mathfrak{Z}$ and $\mathfrak{H}_i/\mathbf{Z}(\mathfrak{H}_i)$ is simple. Hence $\mathfrak{H}_i$ is a component of $\mathbf{E}(\mathfrak{G})$. If $\mathfrak{H} = \mathfrak{H}_1 \cdots \mathfrak{H}_n$, then $\mathfrak{H}\mathfrak{Z} \geq \mathfrak{G}_i$, so $\mathbf{E}(\mathfrak{G}) = \mathfrak{H}\mathfrak{Z}$. Since $\mathbf{E}(\mathfrak{G})$ is perfect, $\mathbf{E}(\mathfrak{G}) = \mathfrak{H} = \mathfrak{H}_1 \cdots \mathfrak{H}_n$.

If $\mathfrak{H}$ is any component of $\mathbf{E}(\mathfrak{G})$, then by 13.16a), $\mathfrak{H}\mathfrak{Z}/\mathfrak{Z}$ is the direct product of certain $\overline{\mathfrak{G}}_i$. Thus $\mathbf{Z}(\mathfrak{H}/\mathfrak{H} \cap \mathfrak{Z}) = 1$ and $\mathbf{Z}(\mathfrak{H}) = \mathfrak{H} \cap \mathfrak{Z}$. Since

$\mathfrak{H}$ is a component, $\mathfrak{H}\mathfrak{Z}/\mathfrak{Z} \cong \mathfrak{H}/(\mathfrak{H} \cap \mathfrak{Z})$ is simple, so $\mathfrak{H}\mathfrak{Z}/\mathfrak{Z} = \overline{\mathfrak{G}}_i$ for some i and $\mathfrak{H}\mathfrak{Z} = \mathfrak{G}_i$. Thus

$$\mathfrak{H}_i = \mathfrak{G}_i' = \mathfrak{H}' = \mathfrak{H}.$$

Hence $\mathfrak{H}_1, \ldots, \mathfrak{H}_n$ are all the components of $\mathbf{E}(\mathfrak{G})$. Clearly $\mathbf{E}(\mathfrak{G})$ is not the product of any proper subset of them. Thus a) and b) are proved.

To prove c), observe that $\mathfrak{K}_1\mathfrak{Z}/\mathfrak{Z}$ and $\mathfrak{K}_2\mathfrak{Z}/\mathfrak{Z}$ are distinct direct factors of $\mathbf{E}(\mathfrak{G})/\mathfrak{Z}$, so $[\mathfrak{K}_1, \mathfrak{K}_2] \leq \mathfrak{Z}$. Thus $[\mathfrak{K}_1, \mathfrak{K}_2, \mathfrak{K}_1] = 1$, and by III, 1.10, $[\mathfrak{K}_1', \mathfrak{K}_2] = 1$. Hence $[\mathfrak{K}_1, \mathfrak{K}_2] = 1$.

To prove d), let $\mathfrak{N}$ be a subnormal subgroup of $\mathbf{E}(\mathfrak{G})$. Then $\mathfrak{N}\mathfrak{Z}/\mathfrak{Z}$ is a subnormal subgroup of $\mathbf{E}(\mathfrak{G})/\mathfrak{Z}$. By 13.16, $\mathfrak{N}\mathfrak{Z}$ is the product of certain $\mathfrak{G}_i$. If $\mathfrak{K}$ is the product of the corresponding set of $\mathfrak{H}_i$, $\mathfrak{N}\mathfrak{Z} = \mathfrak{K}\mathfrak{Z}$, since $\mathfrak{G}_i = \mathfrak{H}_i\mathfrak{Z}$. Hence $\mathfrak{K} = \mathfrak{K}' = \mathfrak{N}' \leq \mathfrak{N}$. Thus $\mathfrak{N} = \mathfrak{K}(\mathfrak{N} \cap \mathfrak{Z})$ and $\mathfrak{N} \trianglelefteq \mathbf{E}(\mathfrak{G})$. Since $\mathfrak{N}\mathfrak{Z}/\mathfrak{Z}$ is the direct product of certain $\overline{\mathfrak{G}}_i$, $\mathbf{E}(\mathfrak{G})/\mathfrak{N}\mathfrak{Z}$ is isomorphic to the direct product of the others and $\mathbf{Z}(\mathbf{E}(\mathfrak{G})/\mathfrak{N}\mathfrak{Z}) = 1$. Hence $\mathbf{Z}(\mathbf{E}(\mathfrak{G})/\mathfrak{N}) = \mathfrak{N}\mathfrak{Z}/\mathfrak{N}$. And if $\mathfrak{M}$ is the product of the components not in $\mathfrak{K}$, $\mathbf{E}(\mathfrak{G}) = \mathfrak{M}\mathfrak{N}$ by a) and $\mathfrak{M} \leq \mathbf{C}_{\mathbf{E}(\mathfrak{G})}(\mathfrak{N})$ by c).

To prove e), put $\mathfrak{Y} = [\mathfrak{H}, \mathfrak{X}]$. Thus $\mathfrak{Y} \leq \mathbf{E}(\mathfrak{G})$ and so $\mathfrak{Y} \leq \mathbf{N}_{\mathfrak{G}}(\mathfrak{H})$. But also, $\mathfrak{H} \leq \mathbf{N}_{\mathfrak{G}}(\mathfrak{Y})$, by III, 1.6b), so $[\mathfrak{Y}, \mathfrak{H}] \leq \mathfrak{Y} \cap \mathfrak{H}$. Since $\mathfrak{H}/\mathbf{Z}(\mathfrak{H})$ is simple and $\mathfrak{Y} \cap \mathfrak{H} \trianglelefteq \mathfrak{H}$, either $(\mathfrak{Y} \cap \mathfrak{H})\mathbf{Z}(\mathfrak{H}) = \mathfrak{H}$ or $\mathfrak{Y} \cap \mathfrak{H} \leq \mathbf{Z}(\mathfrak{H})$. In the first case $\mathfrak{H} = \mathfrak{H}' = (\mathfrak{Y} \cap \mathfrak{H})' \leq \mathfrak{Y} = [\mathfrak{H}, \mathfrak{X}]$. In the second, $[\mathfrak{Y}, \mathfrak{H}, \mathfrak{H}] = 1$ and, by III, 1.10, $[\mathfrak{Y}, \mathfrak{H}'] = 1$. Hence $[\mathfrak{H}, \mathfrak{X}, \mathfrak{H}] = 1$ and, by III, 1.10, $[\mathfrak{H}', \mathfrak{X}] = 1$. Thus $[\mathfrak{H}, \mathfrak{X}] = 1$.

Now suppose that $\mathfrak{H} \leq \mathbf{N}_{\mathfrak{G}}(\mathfrak{X})$ and $[\mathfrak{H}, \mathfrak{X}] \neq 1$. Then $\mathfrak{H} \leq [\mathfrak{H}, \mathfrak{X}] \leq \mathfrak{X}$. By 13.15, $\mathfrak{H} \leq \mathbf{E}(\mathfrak{X})$. **q.e.d.**

13.19 Theorem. *If $\mathfrak{N}$ is a normal perfect quasinilpotent subgroup of $\mathfrak{G}$ and $\mathfrak{A} \leq \mathfrak{G}$, $[\mathfrak{N}, \mathfrak{A}]$ is the product of those components of $\mathfrak{N}$ not centralized by $\mathfrak{A}$, and*

$$\mathfrak{N} = [\mathfrak{N}, \mathfrak{A}]\mathbf{C}_{\mathfrak{N}}(\mathfrak{A}).$$

Proof. Let $\mathscr{H}$ be the set of components of $\mathfrak{N}$. If $\mathfrak{H} \in \mathscr{H}$ and $a \in \mathfrak{A}$, then $\mathfrak{H}^a \in \mathscr{H}$. Thus $\mathfrak{A}$ operates on $\mathscr{H}$. Let $\mathscr{H}_1, \ldots, \mathscr{H}_r$ be the orbits, and put

$$\mathfrak{H}_i = \langle \mathfrak{H} | \mathfrak{H} \in \mathscr{H}_i \rangle.$$

Thus $\mathfrak{H}_i \trianglelefteq \mathfrak{N}\mathfrak{A}$ and, by 13.18a), $\mathfrak{N} = \mathfrak{H}_1 \cdots \mathfrak{H}_r$. Also $\mathfrak{H}_j$ is perfect and quasinilpotent, so again by 13.18a), $\mathscr{H}_i$ is the set of components of $\mathfrak{H}_i$. Since $[\mathfrak{H}_i, \mathfrak{A}] \trianglelefteq \mathfrak{H}_i$, it follows from 13.18d) that $[\mathfrak{H}_i, \mathfrak{A}]$ is the product of a subgroup of $\mathbf{Z}(\mathfrak{H}_i)$ and the elements of

$$\mathscr{H}_i' = \{\mathfrak{H} \mid \mathfrak{H} \in \mathscr{H}_i, \mathfrak{H} \le [\mathfrak{H}_i, \mathfrak{A}]\}.$$

Since $\mathfrak{A}$ normalises $[\mathfrak{H}_i, \mathfrak{A}]$, $\mathscr{H}_i'$ is invariant under $\mathfrak{A}$. Thus either $\mathscr{H}_i' = \emptyset$ or $\mathscr{H}_i' = \mathscr{H}_i$. If $\mathscr{H}_i' = \emptyset$, then $[\mathfrak{H}_i, \mathfrak{A}] \le \mathbf{Z}(\mathfrak{H}_i)$, so $\mathfrak{A} \le \mathbf{C}_{\mathfrak{G}}(\mathfrak{H}_i')$ $= \mathbf{C}_{\mathfrak{G}}(\mathfrak{H}_i)$ by III, 1.10. If $\mathscr{H}_i' = \mathscr{H}_i$, $[\mathfrak{H}_i, \mathfrak{A}] = \mathfrak{H}_i$ and $\mathfrak{A}$ centralizes no element of $\mathscr{H}_i$. Thus $[\mathfrak{N}, \mathfrak{A}] = \prod_{\mathscr{H}_i' = \mathscr{H}_i} \mathfrak{H}_i$ and $\mathfrak{N} = [\mathfrak{N}, \mathfrak{A}]\mathbf{C}_{\mathfrak{N}}(\mathfrak{A})$.

q.e.d.

13.20 Theorem (BENDER [4]). *Suppose that $\mathfrak{P}$ is a p-subgroup of $\mathfrak{G}$.*

a) $\mathbf{O}^p(\mathbf{F}^*(\mathbf{C}_{\mathfrak{G}}(\mathfrak{P}))) = \mathbf{O}^p(\mathbf{F}^*(\mathbf{N}_{\mathfrak{G}}(\mathfrak{P}))) \le \mathbf{C}_{\mathfrak{G}}(\mathbf{O}_p(\mathfrak{G}))$.

b) *If* $\mathbf{F}^*(\mathfrak{G})$ *is a p-group,* $\mathbf{F}^*(\mathbf{C}_{\mathfrak{G}}(\mathfrak{P}))$ *is a p-group.*

c) *If $\mathfrak{P} \le \mathfrak{Q}$, where $\mathfrak{Q}$ is a p-subgroup and* $\mathbf{F}^*(\mathbf{C}_{\mathfrak{G}}(\mathfrak{P}))$ *is a p-group, then* $\mathbf{F}^*(\mathbf{C}_{\mathfrak{G}}(\mathfrak{Q}))$ *is a p-group.*

Proof. a) We have $\mathbf{E}(\mathbf{N}_{\mathfrak{G}}(\mathfrak{P})) \le \mathbf{C}_{\mathfrak{G}}(\mathbf{F}(\mathbf{N}_{\mathfrak{G}}(\mathfrak{P}))) \le \mathbf{C}_{\mathfrak{G}}(\mathfrak{P})$, so $\mathbf{E}(\mathbf{N}_{\mathfrak{G}}(\mathfrak{P})) = \mathbf{E}(\mathbf{C}_{\mathfrak{G}}(\mathfrak{P}))$ by 13.15. Clearly, $\mathbf{O}^p(\mathbf{F}(\mathbf{N}_{\mathfrak{G}}(\mathfrak{P}))) = \mathbf{O}^p(\mathbf{F}(\mathbf{C}_{\mathfrak{G}}(\mathfrak{P})))$. For any group $\mathfrak{X}$, $\mathbf{F}^*(\mathfrak{X}) = \mathbf{E}(\mathfrak{X})\mathbf{F}(\mathfrak{X})$, so $\mathbf{O}^p(\mathbf{F}^*(\mathfrak{X})) = \mathbf{E}(\mathfrak{X})\mathbf{O}^p(\mathbf{F}(\mathfrak{X}))$. Hence $\mathbf{O}^p(\mathbf{F}^*(\mathbf{C}_{\mathfrak{G}}(\mathfrak{P}))) = \mathbf{O}^p(\mathbf{F}^*(\mathbf{N}_{\mathfrak{G}}(\mathfrak{P}))) = \mathfrak{K}$, say.

Now $\mathbf{O}^p(\mathfrak{K}) = \mathfrak{K}$ and $[\mathfrak{K}, \mathfrak{P}] = 1$. If $\mathfrak{C} = \mathbf{C}_{\mathbf{O}_p(\mathfrak{G})}(\mathfrak{P})$, $\mathfrak{C}$ is a normal p-subgroup of $\mathbf{C}_{\mathfrak{G}}(\mathfrak{P})$, so $\mathbf{E}(\mathbf{C}_{\mathfrak{G}}(\mathfrak{P})) \le \mathbf{C}_{\mathfrak{G}}(\mathbf{F}(\mathbf{C}_{\mathfrak{G}}(\mathfrak{P}))) \le \mathbf{C}_{\mathfrak{G}}(\mathfrak{C})$. Also $\mathbf{O}^p(\mathbf{F}(\mathbf{C}_{\mathfrak{G}}(\mathfrak{P})))$ is a normal p'-subgroup of $\mathbf{C}_{\mathfrak{G}}(\mathfrak{P})$ and hence lies in $\mathbf{C}_{\mathfrak{G}}(\mathfrak{C})$. Thus $\mathfrak{K} = \mathbf{O}^p(\mathbf{F}^*(\mathbf{C}_{\mathfrak{G}}(\mathfrak{P}))) \le \mathbf{C}_{\mathfrak{G}}(\mathfrak{C})$ and $[\mathfrak{K}, \mathbf{C}_{\mathbf{O}_p(\mathfrak{G})}(\mathfrak{P})] = 1$. By 1.5, $[\mathfrak{K}, \mathbf{O}_p(\mathfrak{G})] = 1$.

b) $\mathbf{O}^p(\mathbf{F}^*(\mathbf{C}_{\mathfrak{G}}(\mathfrak{P}))) \le \mathbf{C}_{\mathfrak{G}}(\mathbf{O}_p(\mathfrak{G})) = \mathbf{C}_{\mathfrak{G}}(\mathbf{F}^*(\mathfrak{G})) \le \mathbf{F}^*(\mathfrak{G})$, by 13.12. Thus $\mathbf{O}^p(\mathbf{F}^*(\mathbf{C}_{\mathfrak{G}}(\mathfrak{P})))$ is a p-group. Hence $\mathbf{F}^*(\mathbf{C}_{\mathfrak{G}}(\mathfrak{P}))$ is a p-group.

c) This is proved by induction on $|\mathfrak{Q} : \mathfrak{P}|$. If $\mathfrak{Q} = \mathfrak{P}$, there is nothing to prove. If $\mathfrak{Q} > \mathfrak{P}$, let $\mathfrak{P}_0 = \mathbf{N}_{\mathfrak{Q}}(\mathfrak{P})$. By I, 8.8, $\mathfrak{P}_0 > \mathfrak{P}$. Let $\mathfrak{G}_0 = \mathbf{N}_{\mathfrak{G}}(\mathfrak{P})$. By a), $\mathbf{O}^p(\mathbf{F}^*(\mathfrak{G}_0)) = \mathbf{O}^p(\mathbf{F}^*(\mathbf{C}_{\mathfrak{G}}(\mathfrak{P}))) = 1$, so $\mathbf{F}^*(\mathfrak{G}_0)$ is a p-group. Since $\mathfrak{P}_0 \le \mathfrak{G}_0$, $\mathbf{F}^*(\mathbf{C}_{\mathfrak{G}_0}(\mathfrak{P}_0))$ is a p-group, by b). But $\mathbf{C}_{\mathfrak{G}}(\mathfrak{P}_0) = \mathbf{N}_{\mathfrak{G}}(\mathfrak{P}) \cap \mathbf{C}_{\mathfrak{G}}(\mathfrak{P}_0) = \mathbf{C}_{\mathfrak{G}_0}(\mathfrak{P}_0)$, so $\mathbf{F}^*(\mathbf{C}_{\mathfrak{G}}(\mathfrak{P}_0))$ is a p-group. By the inductive hypothesis, $\mathbf{F}^*(\mathbf{C}_{\mathfrak{G}}(\mathfrak{Q}))$ is a p-group. **q.e.d.**

§ 14. The Generalized p'-Core

The generalized Fitting subgroup of a group $\mathfrak{G}$ is a subgroup having a property (Theorem 13.12) which is analogous to one which the Fitting subgroup has for soluble groups. In this section we discuss a subgroup $\mathbf{O}_{p^*}(\mathfrak{G})$, which has the property that for any p-subgroup $\mathfrak{P}$ of $\mathfrak{G}$,

$$\mathbf{O}_{p^*}(\mathbf{C}_{\mathfrak{G}}(\mathfrak{P})) \le \mathbf{O}_{p^*}(\mathfrak{G});$$

this is analogous to 1.6, which, however, was proved only for p-constrained groups.

Before giving the definition, we make the following trivial remark.

14.1 Lemma. *If* $\mathfrak{N} \trianglelefteq \mathfrak{G}$ *and* $\mathfrak{P} \in S_p(\mathfrak{N})$, $\mathfrak{N}\mathbf{C}_{\mathfrak{G}}(\mathfrak{P}) \geq \mathbf{O}_{p'}(\mathfrak{G})$.

Proof. Let $\mathfrak{K} = \mathfrak{N} \cap \mathbf{O}_{p'}(\mathfrak{G})$. Since $[\mathfrak{P}, \mathbf{O}_{p'}(\mathfrak{G})] \leq [\mathfrak{N}, \mathbf{O}_{p'}(\mathfrak{G})] \leq \mathfrak{K}$, $\mathbf{O}_{p'}(\mathfrak{G}) \leq \mathbf{C}_{\mathfrak{G}}(\mathfrak{P}\mathfrak{K}/\mathfrak{K})$. By IX, 6.11, $\mathbf{C}_{\mathfrak{G}}(\mathfrak{P}\mathfrak{K}/\mathfrak{K}) = \mathfrak{K}\mathbf{C}_{\mathfrak{G}}(\mathfrak{P})$, so $\mathbf{O}_{p'}(\mathfrak{G}) \leq \mathfrak{K}\mathbf{C}_{\mathfrak{G}}(\mathfrak{P}) \leq \mathfrak{N}\mathbf{C}_{\mathfrak{G}}(\mathfrak{P})$. **q.e.d.**

14.2 Definition. A group $\mathfrak{G}$ is called a *p^*-group* if (i) $\mathbf{O}^p(\mathfrak{G}) = \mathfrak{G}$, and (ii) whenever $\mathfrak{N} \trianglelefteq \mathfrak{G}$ and $\mathfrak{P} \in S_p(\mathfrak{N})$, $\mathfrak{G} = \mathfrak{N}\mathbf{C}_{\mathfrak{G}}(\mathfrak{P})$.

For example, a p'-group is a p^*-group. Also if $\mathfrak{G}$ is the direct product of non-Abelian simple groups, $\mathfrak{G}$ is a p^*-group, since by I, 9.12, $\mathfrak{G} = \mathfrak{N}\mathbf{C}_{\mathfrak{G}}(\mathfrak{N})$ for all normal subgroups $\mathfrak{N}$ of $\mathfrak{G}$. Thus any chief factor of a group is either a p-group or a p^*-group. Note that the symmetric group $\mathfrak{S}_5$ is not a 5*-group, but $\mathfrak{A}_5$ and $\mathfrak{S}_5/\mathfrak{A}_5$ are both 5*-groups.

14.3 Lemma. *Let* $\mathfrak{G}$ *be a* p^**-group.*

a) *If* $\mathfrak{N} \trianglelefteq \mathfrak{G}$ *and* $\mathfrak{P} \in S_p(\mathfrak{N})$, $\mathfrak{G} = \mathfrak{N}\mathbf{O}^p(\mathbf{C}_{\mathfrak{G}}(\mathfrak{P}))$.

b) *If* $\mathfrak{M} \trianglelefteq \mathfrak{G}$, $\mathfrak{G}/\mathfrak{M}$ *is a* p^**-group.*

c) *If* $\mathfrak{M}$ *is a normal p-subgroup of* $\mathfrak{G}$, $\mathfrak{M} \leq \mathbf{Z}(\mathfrak{G})$.

d) *If* $\mathfrak{G}$ *is p-constrained,* $\mathfrak{G}$ *is a* p'*-group.*

Proof. a) By 14.2, $\mathfrak{G} = \mathfrak{N}\mathbf{C}_{\mathfrak{G}}(\mathfrak{P})$ and $\mathbf{O}^p(\mathfrak{G}) = \mathfrak{G}$. Thus $\mathfrak{N}\mathbf{O}^p(\mathbf{C}_{\mathfrak{G}}(\mathfrak{P}))$ is a normal subgroup of $\mathfrak{G}$ of index a power of p and $\mathfrak{N}\mathbf{O}^p(\mathbf{C}_{\mathfrak{G}}(\mathfrak{P})) = \mathfrak{G}$.

b) If $\mathfrak{N}/\mathfrak{M} \trianglelefteq \mathfrak{G}/\mathfrak{M}$ and $\mathfrak{P} \in S_p(\mathfrak{N})$, then $\mathfrak{P}\mathfrak{M}/\mathfrak{M} \in S_p(\mathfrak{N}/\mathfrak{M})$ and

$$\mathfrak{G}/\mathfrak{M} = \mathfrak{N}\mathbf{C}_{\mathfrak{G}}(\mathfrak{P})/\mathfrak{M} = (\mathfrak{N}/\mathfrak{M})(\mathfrak{M}\mathbf{C}_{\mathfrak{G}}(\mathfrak{P})/\mathfrak{M}) = (\mathfrak{N}/\mathfrak{M})\mathbf{C}_{\mathfrak{G}/\mathfrak{M}}(\mathfrak{P}\mathfrak{M}/\mathfrak{M}).$$

Clearly $\mathbf{O}^p(\mathfrak{G}/\mathfrak{M}) = \mathfrak{G}/\mathfrak{M}$, so $\mathfrak{G}/\mathfrak{M}$ is a p^*-group.

c) By definition, $\mathfrak{G} = \mathfrak{M}\mathbf{C}_{\mathfrak{G}}(\mathfrak{M})$. Thus $\mathfrak{G}/\mathbf{C}_{\mathfrak{G}}(\mathfrak{M})$ is a p-group. Since $\mathbf{O}^p(\mathfrak{G}) = \mathfrak{G}$, $\mathbf{C}_{\mathfrak{G}}(\mathfrak{M}) = \mathfrak{G}$ and $\mathfrak{M} \leq \mathbf{Z}(\mathfrak{G})$.

d) Suppose that $\mathfrak{P} \in S_p(\mathbf{O}_{p',p}(\mathfrak{G}))$. Since $\mathfrak{G}$ is p-constrained, $\mathbf{C}_{\mathfrak{G}}(\mathfrak{P}) \leq \mathbf{C}_{\mathfrak{G}}(\mathbf{O}_{p',p}(\mathfrak{G})/\mathbf{O}_{p'}(\mathfrak{G})) \leq \mathbf{O}_{p',p}(\mathfrak{G})$. Thus $\mathfrak{G} = \mathbf{O}_{p',p}(\mathfrak{G})\mathbf{C}_{\mathfrak{G}}(\mathfrak{P}) = \mathbf{O}_{p',p}(\mathfrak{G})$, and since $\mathbf{O}^p(\mathfrak{G}) = \mathfrak{G}$, $\mathfrak{G} = \mathbf{O}_{p'}(\mathfrak{G})$ is a p'-group. **q.e.d.**

14.4 Lemma. a) *Suppose that* $\mathfrak{H} \leq \mathfrak{G}$. *If* $\mathfrak{P}$ *is a p-subgroup of* $\mathfrak{G}$ *such that* $[\mathfrak{H}, \mathfrak{P}]$ *is a p-group and* $[\mathfrak{H}, \mathfrak{P}] \leq \mathbf{Z}(\mathfrak{H})$, *then* $[\mathbf{O}^p(\mathfrak{H}), \mathfrak{P}] = 1$.

b) *Suppose that* $\mathfrak{N}$, $\mathfrak{P}$ *are normal subgroups of* $\mathfrak{G}$, $\mathfrak{N}$ *is a* p^**-group and* $\mathfrak{P}$ *is a p-group. Then* $[\mathfrak{N}, \mathfrak{P}] = 1$.

Proof. a) Let g be a p'-element of $\mathfrak{H}$. Then g centralizes $\mathfrak{P}[\mathfrak{H}, \mathfrak{P}]/[\mathfrak{H}, \mathfrak{P}]$ and $[\mathfrak{H}, \mathfrak{P}]$, so by I, 4.4, g centralizes $\mathfrak{P}[\mathfrak{H}, \mathfrak{P}]$. Thus $g \in \mathbf{C}_{\mathfrak{G}}(\mathfrak{P})$ and $\mathbf{C}_{\mathfrak{G}}(\mathfrak{P}) \geq \mathbf{O}^p(\mathfrak{H})$.

b) $[\mathfrak{N}, \mathfrak{P}]$ is a normal p-subgroup of the p^*-group $\mathfrak{N}$, so by 14.3, $[\mathfrak{N}, \mathfrak{P}] \leq \mathbf{Z}(\mathfrak{N})$. By a), $[\mathbf{O}^p(\mathfrak{N}), \mathfrak{P}] = 1$. But $\mathbf{O}^p(\mathfrak{N}) = \mathfrak{N}$, so $[\mathfrak{N}, \mathfrak{P}] = 1$. **q.e.d.**

14.5 Theorem. a) *If $\mathfrak{K}$ is a normal p'-subgroup of $\mathfrak{G}$ and $\mathfrak{G}/\mathfrak{K}$ is a p^*-group, $\mathfrak{G}$ is a p^*-group.*

b) *Suppose that $\mathfrak{Z}$ is a normal p-subgroup of $\mathfrak{G}$, $[\mathfrak{Z}, \mathbf{O}^p(\mathfrak{G})] = 1$ and $\mathbf{O}^p(\mathfrak{G}/\mathfrak{Z})$ is a p^*-group. Then $\mathbf{O}^p(\mathfrak{G})$ is a p^*-group.*

Proof. a) First, we have

$$\mathfrak{G}/\mathfrak{K} = \mathbf{O}^p(\mathfrak{G}/\mathfrak{K}) = \mathbf{O}^p(\mathfrak{G})\mathfrak{K}/\mathfrak{K},$$

and $\mathfrak{K} \leq \mathbf{O}^p(\mathfrak{G})$ since $\mathfrak{K}$ is a p'-group. Thus $\mathfrak{G} = \mathbf{O}^p(\mathfrak{G})$.

Suppose that $\mathfrak{N} \trianglelefteq \mathfrak{G}$ and $\mathfrak{P} \in S_p(\mathfrak{N})$. Then $\mathfrak{N}\mathfrak{K}/\mathfrak{K} \trianglelefteq \mathfrak{G}/\mathfrak{K}$ and $\mathfrak{P}\mathfrak{K}/\mathfrak{K} \in S_p(\mathfrak{N}\mathfrak{K}/\mathfrak{K})$, so

$$\mathfrak{G} = \mathfrak{N}\mathfrak{K}\mathbf{C}_{\mathfrak{G}}(\mathfrak{P}\mathfrak{K}/\mathfrak{K}) = \mathfrak{N}\mathbf{C}_{\mathfrak{G}}(\mathfrak{P})\mathfrak{K},$$

by IX, 6.11. By 14.1, $\mathfrak{N}\mathbf{C}_{\mathfrak{G}}(\mathfrak{P}) \geq \mathbf{O}_{p'}(\mathfrak{G}) \geq \mathfrak{K}$, so $\mathfrak{G} = \mathfrak{N}\mathbf{C}_{\mathfrak{G}}(\mathfrak{P})$.

b) Let $\mathfrak{H} = \mathbf{O}^p(\mathfrak{G})$, $\mathfrak{Y} = \mathfrak{Z} \cap \mathfrak{H}$. Thus $\mathfrak{Y} \leq \mathbf{Z}(\mathfrak{H})$. Also $\mathfrak{H}/\mathfrak{Y} \cong \mathfrak{H}\mathfrak{Z}/\mathfrak{Z} = \mathbf{O}^p(\mathfrak{G}/\mathfrak{Z})$ is a p^*-group. If $\mathfrak{N} \trianglelefteq \mathfrak{H}$ and $\mathfrak{P} \in S_p(\mathfrak{N})$, then $\mathfrak{N}\mathfrak{Y}/\mathfrak{Y} \trianglelefteq \mathfrak{H}/\mathfrak{Y}$ and $\mathfrak{P}\mathfrak{Y}/\mathfrak{Y} \in S_p(\mathfrak{N}\mathfrak{Y}/\mathfrak{Y})$. Thus if $\mathfrak{C} = \mathbf{C}_{\mathfrak{H}}(\mathfrak{P}\mathfrak{Y}/\mathfrak{Y})$, $\mathfrak{H} = \mathfrak{N}\mathfrak{Y}\mathbf{O}^p(\mathfrak{C})$, by 14.3a). Now $[\mathfrak{C}, \mathfrak{P}] \leq \mathfrak{Y} \leq \mathbf{Z}(\mathfrak{C})$, so by 14.4a), $[\mathbf{O}^p(\mathfrak{C}), \mathfrak{P}] = 1$. Thus $\mathfrak{H} \leq \mathfrak{N}\mathbf{C}_{\mathfrak{G}}(\mathfrak{P})$ and $\mathfrak{H} = \mathfrak{N}\mathbf{C}_{\mathfrak{H}}(\mathfrak{P})$. Thus $\mathfrak{H}$ is a p^*-group. **q.e.d.**

14.6 Corollary. a) *Suppose that $\mathfrak{N} \trianglelefteq \mathfrak{G}$, $\mathfrak{S} \in S_p(\mathfrak{N})$, $\mathfrak{D} \leq \mathbf{C}_{\mathfrak{G}}(\mathfrak{S})$ and $\mathfrak{N}\mathfrak{D}/\mathfrak{N}$ is a p^*-group. Then $\mathbf{O}^p(\mathfrak{D})$ is a p^*-group.*

b) *If $\mathfrak{G}/\mathbf{Z}(\mathfrak{G})$ is a p^*-group, $\mathbf{O}^p(\mathfrak{G})$ is a p^*-group.*

c) *A perfect quasinilpotent group is a p^*-group.*

Proof. a) Since $\mathfrak{D} \leq \mathbf{C}_{\mathfrak{G}}(\mathfrak{S})$, $\mathfrak{S} \cap \mathfrak{D}$ is a p-subgroup of $\mathbf{Z}(\mathfrak{D})$. Also $\mathfrak{S}$ is the normal Sylow p-subgroup of $\mathfrak{S}(\mathfrak{N} \cap \mathfrak{D})$, so $\mathfrak{S} \cap (\mathfrak{N} \cap \mathfrak{D}) = \mathfrak{S} \cap \mathfrak{D}$ is the Sylow p-subgroup of $\mathfrak{N} \cap \mathfrak{D}$. Hence $(\mathfrak{N} \cap \mathfrak{D})/(\mathfrak{S} \cap \mathfrak{D})$ is a p'-group. Now $\mathfrak{D}/(\mathfrak{N} \cap \mathfrak{D}) \cong \mathfrak{N}\mathfrak{D}/\mathfrak{N}$ is a p^*-group, so by 14.5a), $\mathfrak{D}/(\mathfrak{S} \cap \mathfrak{D})$ is a p^*-group. And since $\mathfrak{S} \cap \mathfrak{D} \leq \mathbf{Z}(\mathfrak{D})$, $\mathbf{O}^p(\mathfrak{D})$ is a p^*-group, by 14.5b).

b) This follows from a) by taking $\mathfrak{N} = \mathbf{Z}(\mathfrak{G})$ and $\mathfrak{D} = \mathfrak{G}$.

c) If $\mathfrak{G}$ is a perfect quasinilpotent group, $\mathfrak{G}/\mathbf{Z}(\mathfrak{G})$ is the direct product of non-Abelian simple groups, by 13.7. Thus $\mathfrak{G}/\mathbf{Z}(\mathfrak{G})$ is a

p^*-group, and by b), $\mathbf{O}^p(\mathfrak{G})$ is a p^*-group. Since $\mathfrak{G}$ is perfect, $\mathfrak{G} = \mathbf{O}^p(\mathfrak{G})$, so $\mathfrak{G}$ is a p^*-group. **q.e.d.**

Next we generalize the formula $\mathfrak{G} = [\mathfrak{G}, \mathfrak{P}]\mathbf{C}_{\mathfrak{G}}(\mathfrak{P})$, which holds when $\mathfrak{P}$ is a p-group of operators on the p'-group $\mathfrak{G}$.

14.7 Lemma. *Suppose that $\mathfrak{P}$ is a p-group of operators on a p^*-group $\mathfrak{G}$. Let $\mathfrak{S}$ be a $\mathfrak{P}$-invariant Sylow p-subgroup of $[\mathfrak{G}, \mathfrak{P}]$. Then $\mathfrak{G}_0 = \mathbf{O}^p(\mathbf{C}_{\mathfrak{G}}(\mathfrak{P}) \cap \mathbf{C}_{\mathfrak{G}}(\mathfrak{S}))$ is a p^*-group and $\mathfrak{G} = [\mathfrak{G}, \mathfrak{P}]\mathfrak{G}_0$. In particular,*

$$\mathfrak{G} = [\mathfrak{G}, \mathfrak{P}]\mathbf{C}_{\mathfrak{G}}(\mathfrak{P}).$$

Proof. This is proved by induction on $|\mathfrak{G}|$. Let $\mathfrak{K} = [\mathfrak{G}, \mathfrak{P}]$, $\mathfrak{C} = \mathbf{C}_{\mathfrak{G}}(\mathfrak{S})$. Thus $\mathfrak{K}$, $\mathfrak{C}$ are $\mathfrak{P}$-invariant and $\mathfrak{K} \trianglelefteq \mathfrak{G}$. Since $\mathfrak{G}$ is a p^*-group and $\mathfrak{S} \in S_p(\mathfrak{K})$, it follows from 14.3a) that $\mathfrak{G} = \mathfrak{K}\mathfrak{B}$, where $\mathfrak{B} = \mathbf{O}^p(\mathfrak{C})$. Thus $\mathfrak{K}\mathfrak{B}/\mathfrak{K}$ is a p^*-group, and by 14.6a), $\mathfrak{B}$ is a p^*-group. If $\mathfrak{B} < \mathfrak{G}$, we may apply the inductive hypothesis to $\mathfrak{B}$; this gives $\mathfrak{B} = [\mathfrak{B}, \mathfrak{P}]\mathfrak{B}_0$, where $\mathfrak{B}_0 \leq \mathfrak{G}_0$. Hence

$$\mathfrak{G} = \mathfrak{K}[\mathfrak{B}, \mathfrak{P}]\mathfrak{B}_0 = \mathfrak{K}\mathfrak{G}_0,$$

and by 14.6a), $\mathfrak{G}_0$ is a p^*-group.

If, on the other hand, $\mathfrak{B} = \mathfrak{G}$, then $\mathfrak{S} \leq \mathbf{Z}(\mathfrak{G})$. Thus $\mathfrak{K} = \mathfrak{S} \times \mathfrak{T}$ for a characteristic p'-subgroup $\mathfrak{T}$ of $\mathfrak{K}$. Hence $\mathfrak{P}$ is a group of operators on $\mathfrak{G}/\mathfrak{T}$ and $[\mathfrak{G}/\mathfrak{T}, \mathfrak{P}] = \mathfrak{K}/\mathfrak{T}$ is a p-subgroup of the centre of $\mathfrak{G}/\mathfrak{T}$. By 14.4, $[\mathbf{O}^p(\mathfrak{G}/\mathfrak{T}), \mathfrak{P}] = 1$, so $\mathfrak{P}$ operates trivially on $\mathfrak{G}/\mathfrak{T}$. Hence $\mathfrak{K} = [\mathfrak{G}, \mathfrak{P}] \leq \mathfrak{T}$ and $\mathfrak{S} = 1$. $\mathfrak{K}$ is therefore a p'-group. Given $g \in \mathfrak{G}$, $\mathfrak{P}$ leaves the coset $g\mathfrak{K}$ fixed, and $\mathfrak{P}$ induces a group of permutations of the elements of $g\mathfrak{K}$. Since $\mathfrak{P}$ is a p-group, some element x of $g\mathfrak{K}$ is left fixed by $\mathfrak{P}$; that is, $x \in \mathbf{C}_{\mathfrak{G}}(\mathfrak{P})$. Since $g \in x\mathfrak{K}$, it follows that $\mathfrak{G} = \mathfrak{K}\mathbf{C}_{\mathfrak{G}}(\mathfrak{P})$. Since $\mathfrak{S} = 1$, $\mathfrak{G}_0 = \mathbf{O}^p(\mathbf{C}_{\mathfrak{G}}(\mathfrak{P}))$. Thus $\mathfrak{K}\mathfrak{G}_0 \trianglelefteq \mathfrak{G}$ and $|\mathfrak{G} : \mathfrak{K}\mathfrak{G}_0|$ is a power of p, so $\mathfrak{G} = \mathfrak{K}\mathfrak{G}_0$. By 14.6a), $\mathfrak{G}_0$ is a p^*-group. **q.e.d.**

14.8 Lemma. *Suppose that $\mathfrak{G} = \mathfrak{A}\mathfrak{B}$, where $\mathfrak{A}$, $\mathfrak{B}$ are p^*-subgroups of $\mathfrak{G}$ for which $|\mathfrak{G} : \mathbf{N}_{\mathfrak{G}}(\mathfrak{A})|$, $|\mathfrak{G} : \mathbf{N}_{\mathfrak{G}}(\mathfrak{B})|$ are both prime to p. Then $\mathfrak{G}$ is a p^*-group.*

Proof. Suppose that $\mathfrak{N} \trianglelefteq \mathfrak{G}$ and $\mathfrak{P} \in S_p(\mathfrak{N})$. Then $\mathfrak{P} \leq \mathfrak{S}$ for some $\mathfrak{S} \in S_p(\mathfrak{G})$. Since $|\mathfrak{G} : \mathbf{N}_{\mathfrak{G}}(\mathfrak{A})|$, $|\mathfrak{G} : \mathbf{N}_{\mathfrak{G}}(\mathfrak{B})|$ are prime to p, there exist elements x, y such that $\mathfrak{S} \leq \mathbf{N}_{\mathfrak{G}}(\mathfrak{A}^x)$ and $\mathfrak{S} \leq \mathbf{N}_{\mathfrak{G}}(\mathfrak{B}^y)$. Thus $\mathfrak{P}$ is a p-group of operators on the p^*-groups $\mathfrak{A}^x$, $\mathfrak{B}^y$. By 14.7,

$$\mathfrak{A}^x = [\mathfrak{A}^x, \mathfrak{P}]\mathbf{C}_{\mathfrak{A}^x}(\mathfrak{P}) \leq \mathfrak{N}\mathbf{C}_{\mathfrak{G}}(\mathfrak{P}).$$

Similarly $\mathfrak{B}^y \leq \mathfrak{N}\mathbf{C}_{\mathfrak{G}}(\mathfrak{P})$. But by VI, 4.5, $\mathfrak{G} = \mathfrak{A}^x\mathfrak{B}^y$, so $\mathfrak{G} = \mathfrak{N}\mathbf{C}_{\mathfrak{G}}(\mathfrak{P})$. Also $\mathbf{O}^p(\mathfrak{G}) \geq \langle \mathbf{O}^p(\mathfrak{A}), \mathbf{O}^p(\mathfrak{B})\rangle = \langle \mathfrak{A}, \mathfrak{B}\rangle = \mathfrak{G}$, so $\mathfrak{G}$ is a p^*-group.

q.e.d.

14.9 Definition. For any group $\mathfrak{G}$, denote the subgroup generated by all normal p^*-subgroups of $\mathfrak{G}$ by $\mathbf{O}_{p^*}(\mathfrak{G})$. $\mathbf{O}_{p^*}(\mathfrak{G})$ is called the *generalized p'-core* of $\mathfrak{G}$.

$\mathbf{O}_{p^*}(\mathfrak{G})$ is a characteristic subgroup of $\mathfrak{G}$, and by 14.8, $\mathbf{O}_{p^*}(\mathfrak{G})$ is a p^*-group. Thus if $\mathfrak{N} \trianglelefteq \mathfrak{G}$, $\mathbf{O}_{p^*}(\mathfrak{N}) \leq \mathbf{O}_{p^*}(\mathfrak{G})$. Clearly $\mathbf{O}_{p^*}(\mathfrak{G}) \geq \mathbf{O}_{p'}(\mathfrak{G})$. By 14.6c), $\mathbf{O}_{p^*}(\mathfrak{G}) \geq \mathbf{E}(\mathfrak{G})$. If $\mathfrak{G}$ is p-soluble, $\mathbf{O}_{p^*}(\mathfrak{G}) = \mathbf{O}_{p'}(\mathfrak{G})$, by 14.3d) and IX, 1.4 (see 14.12b)).

Write $\mathbf{O}_p(\mathfrak{G}/\mathbf{O}_{p^*}(\mathfrak{G})) = \mathbf{O}_{p^*,p}(\mathfrak{G})/\mathbf{O}_{p^*}(\mathfrak{G})$.

Next we generalize IX, 1.4.

14.10 Theorem. *If $\mathfrak{P} \in S_p(\mathbf{O}_{p^*,p}(\mathfrak{G}))$, then $\mathbf{C}_{\mathfrak{G}}(\mathfrak{P}) \leq \mathbf{O}_{p^*,p}(\mathfrak{G})$.*

Proof. Suppose that this is false. Since $\mathfrak{G} = \mathbf{N}_{\mathfrak{G}}(\mathfrak{P})\mathbf{O}_{p^*,p}(\mathfrak{G})$, $\mathbf{C}_{\mathfrak{G}}(\mathfrak{P})\mathbf{O}_{p^*,p}(\mathfrak{G}) \trianglelefteq \mathfrak{G}$. Now $\mathbf{C}_{\mathfrak{G}}(\mathfrak{P})\mathbf{O}_{p^*,p}(\mathfrak{G})/\mathbf{O}_{p^*,p}(\mathfrak{G}) \neq 1$, so there exists a minimal normal subgroup $\mathfrak{M}/\mathbf{O}_{p^*,p}(\mathfrak{G})$ of $\mathfrak{G}/\mathbf{O}_{p^*,p}(\mathfrak{G})$ for which $\mathfrak{M} \leq \mathbf{C}_{\mathfrak{G}}(\mathfrak{P})\mathbf{O}_{p^*,p}(\mathfrak{G})$. Being a chief factor of $\mathfrak{G}$, $\mathfrak{M}/\mathbf{O}_{p^*,p}(\mathfrak{G})$ is either a p-group or a p^*-group. But obviously $\mathbf{O}_p(\mathfrak{G}/\mathbf{O}_{p^*,p}(\mathfrak{G})) = 1$, so $\mathfrak{M}/\mathbf{O}_{p^*,p}(\mathfrak{G})$ is a p^*-group. Now $\mathfrak{M} = \mathbf{C}_{\mathfrak{M}}(\mathfrak{P})\mathbf{O}_{p^*,p}(\mathfrak{G})$, so by 14.6a), $\mathbf{O}^p(\mathbf{C}_{\mathfrak{M}}(\mathfrak{P}))$ is a p^*-group. Now $\mathbf{O}^p(\mathbf{C}_{\mathfrak{M}}(\mathfrak{P})) \trianglelefteq \mathbf{N}_{\mathfrak{M}}(\mathfrak{P})$ and $|\mathfrak{M} : \mathbf{N}_{\mathfrak{M}}(\mathfrak{P})|$ is prime to p, since if $\mathfrak{P} \leq \mathfrak{P}_0 \in S_p(\mathfrak{M})$, then $\mathfrak{P} = \mathfrak{P}_0 \cap \mathbf{O}_{p^*,p}(\mathfrak{G}) \trianglelefteq \mathfrak{P}_0$ and $\mathfrak{P}_0 \leq \mathbf{N}_{\mathfrak{M}}(\mathfrak{P})$. Thus if $\mathfrak{L} = \mathbf{O}^p(\mathbf{C}_{\mathfrak{M}}(\mathfrak{P}))\mathbf{O}_{p^*}(\mathfrak{G})$, $\mathfrak{L}$ is a p^*-group by 14.8. Now $\mathfrak{L} \trianglelefteq \mathfrak{M} = \mathbf{C}_{\mathfrak{M}}(\mathfrak{P})\mathfrak{P}\mathbf{O}_{p^*}(\mathfrak{G})$ and $|\mathfrak{M} : \mathfrak{L}|$ is a power of p, so $\mathfrak{L} \geq \mathbf{O}^p(\mathfrak{M})$. Since $\mathfrak{L}$ is a p^*-group, $\mathfrak{L} = \mathbf{O}^p(\mathfrak{L}) = \mathbf{O}^p(\mathfrak{M}) \trianglelefteq \mathfrak{G}$ and $\mathfrak{L} \leq \mathbf{O}_{p^*}(\mathfrak{G})$. Thus $\mathfrak{M} = \mathbf{O}^p(\mathfrak{M})\mathbf{O}_{p^*,p}(\mathfrak{G}) \leq \mathbf{O}_{p^*,p}(\mathfrak{G})$, a contradiction. **q.e.d.**

14.11 Corollary. *If $\mathbf{O}^p(\mathfrak{G})$ is a p^*-group and $\mathfrak{N} \trianglelefteq \mathfrak{G}$, $\mathbf{O}^p(\mathfrak{N})$ is a p^*-group.*

Proof. Suppose that $\mathfrak{P} \in S_p(\mathbf{O}_{p^*,p}(\mathfrak{N}))$. By 14.10, $\mathbf{C}_{\mathfrak{N}}(\mathfrak{P}) \leq \mathbf{O}_{p^*,p}(\mathfrak{N})$. By 14.7,

$$\mathbf{O}^p(\mathfrak{G}) = [\mathbf{O}^p(\mathfrak{G}), \mathfrak{P}]\mathbf{O}^p(\mathbf{C}_{\mathfrak{G}}(\mathfrak{P})) \leq \mathbf{O}_{p^*,p}(\mathfrak{N})\mathbf{C}_{\mathfrak{G}}(\mathfrak{P}),$$

so

$$\mathbf{O}^p(\mathfrak{N}) \leq \mathfrak{N} \cap \mathbf{O}^p(\mathfrak{G}) \leq \mathbf{O}_{p^*,p}(\mathfrak{N})\mathbf{C}_{\mathfrak{N}}(\mathfrak{P}) = \mathbf{O}_{p^*,p}(\mathfrak{N})$$

and $\mathbf{O}^p(\mathfrak{N}) = \mathbf{O}_{p^*}(\mathfrak{N})$ is a p^*-group. **q.e.d.**

Theorem 14.10 can be slightly improved as follows.

14.12 Theorem. a) $\mathbf{O}_{p^*,p}(\mathfrak{G})$ *is the intersection of all subgroups in the set* $\mathscr{U}$, *where* $\mathfrak{U} \in \mathscr{U}$ *if and only if* $\mathfrak{U} \trianglelefteq \mathfrak{G}$, $\mathbf{O}_p(\mathfrak{G}/\mathfrak{U}) = 1$ *and the centralizer in* $\mathfrak{G}$ *of any Sylow p-subgroup of* $\mathfrak{U}$ *is contained in* $\mathfrak{U}$.

b) $\mathfrak{G}$ *is p-constrained if and only if* $\mathbf{O}_{p^*}(\mathfrak{G}) = \mathbf{O}_{p'}(\mathfrak{G})$.

Proof. a) Let $\mathfrak{B}$ be the intersection of all subgroups in $\mathscr{U}$. By 14.10, $\mathbf{O}_{p^*,p}(\mathfrak{G}) \in \mathscr{U}$, so $\mathfrak{B} \leq \mathbf{O}_{p^*,p}(\mathfrak{G})$. We must therefore show that $\mathbf{O}_{p^*,p}(\mathfrak{G}) \leq \mathfrak{U}$ for every $\mathfrak{U} \in \mathscr{U}$. Suppose that $\mathfrak{P} \in S_p(\mathfrak{U})$. By 14.7,

$$\mathbf{O}_{p^*}(\mathfrak{G}) = [\mathbf{O}_{p^*}(\mathfrak{G}), \mathfrak{P}](\mathbf{O}_{p^*}(\mathfrak{G}) \cap \mathbf{C}_{\mathfrak{G}}(\mathfrak{P})).$$

Since $\mathfrak{U} \trianglelefteq \mathfrak{G}$ and $\mathbf{C}_{\mathfrak{G}}(\mathfrak{P}) \leq \mathfrak{U}$, it follows that $\mathbf{O}_{p^*}(\mathfrak{G}) \leq \mathfrak{U}$. Thus

$$\mathbf{O}_{p^*,p}(\mathfrak{G})\mathfrak{U}/\mathfrak{U} \leq \mathbf{O}_p(\mathfrak{G}/\mathfrak{U}) = 1$$

and $\mathbf{O}_{p^*,p}(\mathfrak{G}) \leq \mathfrak{U}$.

b) If $\mathfrak{G}$ is p-constrained, $\mathbf{O}_{p',p}(\mathfrak{G}) \in \mathscr{U}$, so $\mathbf{O}_{p^*,p}(\mathfrak{G}) \leq \mathbf{O}_{p',p}(\mathfrak{G})$. Thus $\mathbf{O}_{p^*}(\mathfrak{G}) = \mathbf{O}^p(\mathbf{O}_{p^*,p}(\mathfrak{G})) \leq \mathbf{O}^p(\mathbf{O}_{p',p}(\mathfrak{G})) = \mathbf{O}_{p'}(\mathfrak{G})$.

If $\mathbf{O}_{p^*}(\mathfrak{G}) = \mathbf{O}_{p'}(\mathfrak{G})$ and $\mathfrak{P} \in S_p(\mathbf{O}_{p',p}(\mathfrak{G}))$, then $\mathfrak{P} \in S_p(\mathbf{O}_{p^*,p}(\mathfrak{G}))$, so $\mathbf{C}_{\mathfrak{G}}(\mathfrak{P}) \leq \mathbf{O}_{p^*,p}(\mathfrak{G}) = \mathbf{O}_{p',p}(\mathfrak{G})$, by 14.10. By IX, 6.11,

$$\mathbf{C}_{\mathfrak{G}}(\mathbf{O}_{p',p}(\mathfrak{G})/\mathbf{O}_{p'}(\mathfrak{G})) \leq \mathbf{C}_{\mathfrak{G}}(\mathfrak{P})\mathbf{O}_{p'}(\mathfrak{G}) \leq \mathbf{O}_{p',p}(\mathfrak{G}). \qquad \textbf{q.e.d.}$$

We now generalize 1.6.

14.13 Theorem. *If* $\mathfrak{P}$ *is a p-subgroup of the group* $\mathfrak{G}$,

$$\mathbf{O}_{p^*}(\mathbf{C}_{\mathfrak{G}}(\mathfrak{P})) \leq \mathbf{O}_{p^*}(\mathfrak{G}).$$

Proof. This is proved by induction on $|\mathfrak{G} : \mathfrak{P}|$. Let $\mathfrak{M} = \mathbf{O}_{p^*,p}(\mathfrak{G})$. First suppose that $\mathfrak{P} \in S_p(\mathfrak{P}\mathfrak{M})$. Then $\mathfrak{P} \cap \mathfrak{M} \in S_p(\mathfrak{M})$. By 14.10,

$$\mathbf{O}_{p^*,p}(\mathfrak{G}) \geq \mathbf{C}_{\mathfrak{G}}(\mathfrak{P} \cap \mathfrak{M}) \geq \mathbf{C}_{\mathfrak{G}}(\mathfrak{P}),$$

so

$$\mathbf{O}_{p^*}(\mathfrak{G}) = \mathbf{O}^p(\mathbf{O}_{p^*,p}(\mathfrak{G})) \geq \mathbf{O}^p(\mathbf{C}_\mathfrak{G}(\mathfrak{P})) \geq \mathbf{O}_{p^*}(\mathbf{C}_\mathfrak{G}(\mathfrak{P})).$$

If $\mathfrak{P} \notin S_p(\mathfrak{P}\mathfrak{M})$, then by I, 8.8, p divides

$$|\mathbf{N}_{\mathfrak{P}\mathfrak{M}}(\mathfrak{P}) : \mathfrak{P}| = |\mathbf{N}_\mathfrak{M}(\mathfrak{P}) : \mathfrak{P} \cap \mathfrak{M}|.$$

Thus if $\mathfrak{S} \in S_p(\mathbf{N}_\mathfrak{M}(\mathfrak{P}))$, $\mathfrak{S} > \mathfrak{P} \cap \mathfrak{M}$. Now $\mathbf{N}_\mathfrak{M}(\mathfrak{P})$ normalises $\mathbf{C}_\mathfrak{G}(\mathfrak{P})$, so if $\mathfrak{K} = \mathbf{O}_{p^*}(\mathbf{C}_\mathfrak{G}(\mathfrak{P}))$, $\mathfrak{S}$ normalises $\mathfrak{K}$. Let $\mathfrak{T}$ be a Sylow p-subgroup of $\mathfrak{S}[\mathfrak{K}, \mathfrak{S}]$ containing $\mathfrak{S}$. Thus $\mathfrak{T} \leq \mathbf{N}_\mathfrak{G}(\mathfrak{P})$ and $\mathfrak{T}\mathfrak{P} > \mathfrak{P}$. By the inductive hypothesis,

$$\mathbf{O}_{p^*}(\mathbf{C}_\mathfrak{G}(\mathfrak{T}\mathfrak{P})) \leq \mathbf{O}_{p^*}(\mathfrak{G}).$$

Now $\mathfrak{T} = \mathfrak{S}(\mathfrak{T} \cap [\mathfrak{K}, \mathfrak{S}])$ and $\mathfrak{T} \cap [\mathfrak{K}, \mathfrak{S}]$ is an $\mathfrak{S}$-invariant Sylow p-subgroup of $[\mathfrak{K}, \mathfrak{S}]$. Hence by 14.7, $\mathfrak{K} = [\mathfrak{K}, \mathfrak{S}]\mathfrak{K}_0$, where $\mathfrak{K}_0$ is a p^*-group and

$$\mathfrak{K}_0 = \mathbf{O}^p(\mathbf{C}_\mathfrak{K}(\mathfrak{S}) \cap \mathbf{C}_\mathfrak{K}(\mathfrak{T} \cap [\mathfrak{K}, \mathfrak{S}])) = \mathbf{O}^p(\mathbf{C}_\mathfrak{K}(\mathfrak{T})) = \mathbf{O}^p(\mathbf{C}_\mathfrak{K}(\mathfrak{T}\mathfrak{P})).$$

Now $\mathbf{C}_\mathfrak{G}(\mathfrak{T}\mathfrak{P})$ normalises $\mathfrak{K}$, so $\mathfrak{K}_0$ is a normal p^*-subgroup of $\mathbf{C}_\mathfrak{G}(\mathfrak{T}\mathfrak{P})$ and $\mathfrak{K}_0 \leq \mathbf{O}_{p^*}(\mathbf{C}_\mathfrak{G}(\mathfrak{T}\mathfrak{P})) \leq \mathbf{O}_{p^*}(\mathfrak{G}) \leq \mathfrak{M}$. Also $[\mathfrak{K}, \mathfrak{S}] \leq \mathfrak{M}$, so $\mathfrak{K} \leq \mathfrak{M}$. Hence

$$\mathbf{O}_{p^*}(\mathbf{C}_\mathfrak{G}(\mathfrak{P})) = \mathfrak{K} = \mathbf{O}^p(\mathfrak{K}) \leq \mathbf{O}^p(\mathfrak{M}) = \mathbf{O}_{p^*}(\mathfrak{G}). \qquad \textbf{q.e.d.}$$

14.14 Theorem. *Suppose that $\mathfrak{G}$ is a p^*-group and $\mathfrak{P}$ is a p-group of operators on $\mathfrak{G}$.*

a) $\mathfrak{G} = [\mathfrak{G}, \mathfrak{P}]\mathbf{O}_{p^*}(\mathbf{C}_\mathfrak{G}(\mathfrak{P}))$.

b) $[\mathfrak{G}, \mathfrak{P}]$ *is a p^*-group and* $[\mathfrak{G}, \mathfrak{P}, \mathfrak{P}] = [\mathfrak{G}, \mathfrak{P}]$.

c) *If $\mathfrak{N}$ is a normal $\mathfrak{P}$-invariant subgroup of $\mathfrak{G}$, $\mathbf{C}_\mathfrak{G}(\mathfrak{N}) \leq \mathfrak{N}$ and $[\mathbf{O}_{p^*}(\mathfrak{N}), \mathfrak{P}] = 1$, then $\mathfrak{P}$ operates trivially on $\mathfrak{G}$.*

Proof. a) Write $\mathfrak{K} = [\mathfrak{G}, \mathfrak{P}]$, $\mathfrak{U} = \mathbf{C}_\mathfrak{G}(\mathfrak{P})$. Let $\mathfrak{S}_1$ be a ($\mathfrak{P}$-invariant) Sylow p-subgroup of $\mathfrak{U} \cap \mathfrak{K}$ and let $\mathfrak{S}$ be a $\mathfrak{P}$-invariant Sylow p-subgroup of $\mathfrak{K}$ such that $\mathfrak{S} \geq \mathfrak{S}_1$. By 14.7, $\mathfrak{G} = \mathfrak{K}\mathfrak{G}_0$, where $\mathfrak{G}_0$ is a p^*-subgroup of $\mathfrak{U} \cap \mathbf{C}_\mathfrak{G}(\mathfrak{S})$. Thus $\mathfrak{U} = (\mathfrak{K} \cap \mathfrak{U})\mathfrak{G}_0$ and $\mathfrak{U}/(\mathfrak{K} \cap \mathfrak{U}) \cong \mathfrak{G}_0/(\mathfrak{G}_0 \cap \mathfrak{K})$ is a p^*-group. Also $\mathfrak{U} = (\mathfrak{K} \cap \mathfrak{U})\mathbf{C}_\mathfrak{U}(\mathfrak{S}_1)$, so by 14.6a), $\mathbf{O}^p(\mathbf{C}_\mathfrak{U}(\mathfrak{S}_1))$ is a p^*-group and

$$\mathfrak{G}_0 \leq \mathbf{O}^p(\mathbf{C}_\mathfrak{U}(\mathfrak{S}_1)) = \mathbf{O}_{p^*}(\mathbf{C}_\mathfrak{U}(\mathfrak{S}_1)).$$

But by 14.13,

$$\mathbf{O}_{p^*}(\mathbf{C}_{\mathfrak{U}}(\mathfrak{S}_1)) \le \mathbf{O}_{p^*}(\mathfrak{U}),$$

so $\mathfrak{G} = \mathfrak{K}\mathbf{O}_{p^*}(\mathfrak{U})$.

b) Let $\mathfrak{K} = [\mathfrak{G}, \mathfrak{P}]$ and $\overline{\mathfrak{K}} = \mathbf{O}^p(\mathfrak{K})$. By a), $\mathfrak{G}/\overline{\mathfrak{K}} = (\mathfrak{K}/\overline{\mathfrak{K}})(\mathbf{C}_{\mathfrak{G}}(\mathfrak{P})\overline{\mathfrak{K}}/\overline{\mathfrak{K}})$. By 14.3, $\mathfrak{K}/\overline{\mathfrak{K}} \le \mathbf{Z}(\mathfrak{G}/\overline{\mathfrak{K}})$, so

$$\mathfrak{G}/\overline{\mathfrak{K}} = \mathbf{O}^p(\mathfrak{G}/\overline{\mathfrak{K}}) = \mathbf{O}^p(\mathbf{C}_{\mathfrak{G}}(\mathfrak{P}))\overline{\mathfrak{K}}/\overline{\mathfrak{K}}$$

and $\mathfrak{G} = \overline{\mathfrak{K}}\mathbf{C}_{\mathfrak{G}}(\mathfrak{P})$. Hence $[\mathfrak{G}, \mathfrak{P}] \le \overline{\mathfrak{K}}$ and $\mathfrak{K} = \overline{\mathfrak{K}} = \mathbf{O}^p(\mathfrak{K})$. By 14.11, $\mathfrak{K}$ is a p^*-group. And

$$[\mathfrak{G}, \mathfrak{P}] = [\mathbf{C}_{\mathfrak{G}}(\mathfrak{P})[\mathfrak{G}, \mathfrak{P}], \mathfrak{P}] = [\mathfrak{G}, \mathfrak{P}, \mathfrak{P}].$$

c) Let $\mathfrak{R}$ be a $\mathfrak{P}$-invariant Sylow p-subgroup of $\mathfrak{N}$. By 14.11, $\mathbf{O}^p(\mathfrak{N})$ is a p^*-group, so $\mathbf{O}^p(\mathfrak{N}) = \mathbf{O}_{p^*}(\mathfrak{N})$ and $\mathfrak{N} = \mathfrak{R}\mathbf{O}_{p^*}(\mathfrak{N})$. Since $[\mathbf{O}_{p^*}(\mathfrak{N}), \mathfrak{P}] = 1$, we have $[\mathbf{O}_{p^*}(\mathfrak{N}), [\mathfrak{G}, \mathfrak{P}]] = 1$. Thus $[\mathbf{C}_{\mathfrak{G}}(\mathfrak{R}), \mathfrak{P}]$ centralizes both $\mathbf{O}_{p^*}(\mathfrak{N})$ and $\mathfrak{R}$. Hence

$$[\mathbf{C}_{\mathfrak{G}}(\mathfrak{R}), \mathfrak{P}] \le \mathbf{C}_{\mathfrak{G}}(\mathfrak{N}) \le \mathfrak{N}.$$

But by the definition of p^*-groups, $\mathfrak{G} = \mathfrak{N}\mathbf{C}_{\mathfrak{G}}(\mathfrak{R})$, so $[\mathfrak{G}, \mathfrak{P}] \le \mathfrak{N}$. By b), $[\mathfrak{G}, \mathfrak{P}] \le \mathbf{O}_{p^*}(\mathfrak{N})$, so

$$[\mathfrak{G}, \mathfrak{P}] = [\mathfrak{G}, \mathfrak{P}, \mathfrak{P}] \le [\mathbf{O}_{p^*}(\mathfrak{N}), \mathfrak{P}] = 1. \qquad \textbf{q.e.d.}$$

Finally, we generalize 1.12.

14.15 Theorem (BENDER). *Suppose that* $\mathfrak{S} \in S_p(\mathfrak{G})$ *and that* $\mathfrak{A}$ *is a normal subgroup of* $\mathfrak{S}$ *for which* $\mathfrak{A} \ge \mathbf{C}_{\mathfrak{S}}(\mathfrak{A})$. *If* $\mathfrak{H}$ *is a* p^**-subgroup of* $\mathfrak{G}$ *and* $\mathfrak{A} \le \mathbf{N}_{\mathfrak{G}}(\mathfrak{H})$, *then* $\mathfrak{H} \le \mathbf{O}_{p^*}(\mathfrak{G})$.

Proof. Suppose that this is false and let $\mathfrak{G}$ be a counterexample of minimal order. Let $\mathfrak{K}$ be a p^*-subgroup of minimal order for which $\mathfrak{A} \le \mathbf{N}_{\mathfrak{G}}(\mathfrak{K})$ but $\mathfrak{K} \not\le \mathbf{O}_{p^*}(\mathfrak{G})$. By 14.13, $\mathfrak{K} \not\le \mathbf{O}_{p^*}(\mathbf{C}_{\mathfrak{G}}(\mathfrak{A}))$. But $\mathfrak{S} \in S_p(\mathbf{N}_{\mathfrak{G}}(\mathfrak{A}))$, so $\mathfrak{S} \cap \mathbf{C}_{\mathfrak{G}}(\mathfrak{A}) = \mathbf{C}_{\mathfrak{S}}(\mathfrak{A}) = \mathbf{C}_{\mathfrak{S}}(\mathfrak{A}) \cap \mathfrak{A} = \mathbf{Z}(\mathfrak{A})$ is a Sylow p-subgroup of $\mathbf{C}_{\mathfrak{G}}(\mathfrak{A})$. Since $\mathbf{Z}(\mathfrak{A}) \le \mathbf{Z}(\mathbf{C}_{\mathfrak{G}}(\mathfrak{A}))$, it follows from 14.5b) that $\mathbf{O}^p(\mathbf{C}_{\mathfrak{G}}(\mathfrak{A}))$ is a p^*-group. Thus $\mathbf{O}_{p^*}(\mathbf{C}_{\mathfrak{G}}(\mathfrak{A})) = \mathbf{O}^p(\mathbf{C}_{\mathfrak{G}}(\mathfrak{A}))$ and $\mathfrak{K} \not\le \mathbf{O}^p(\mathbf{C}_{\mathfrak{G}}(\mathfrak{A}))$. Since $\mathfrak{K} = \mathbf{O}^p(\mathfrak{K})$, $\mathfrak{K} \not\le \mathbf{C}_{\mathfrak{G}}(\mathfrak{A})$. Therefore $\mathbf{O}_{p^*}(\mathbf{C}_{\mathfrak{K}}(\mathfrak{A}))$ is a proper $\mathfrak{A}$-invariant p^*-subgroup of $\mathfrak{K}$, and by minimality of $\mathfrak{K}$, $\mathbf{O}_{p^*}(\mathbf{C}_{\mathfrak{K}}(\mathfrak{A})) \le \mathbf{O}_{p^*}(\mathfrak{G})$. By 14.14, $[\mathfrak{K}, \mathfrak{A}]$ is a p^*-group and

$$\mathfrak{K} = [\mathfrak{K}, \mathfrak{A}]\mathbf{O}_{p*}(\mathbf{C}_{\mathfrak{K}}(\mathfrak{A})),$$

so $[\mathfrak{K}, \mathfrak{A}] \not\leq \mathbf{O}_{p*}(\mathfrak{G})$. By minimality of $\mathfrak{K}$, $[\mathfrak{K}, \mathfrak{A}] = \mathfrak{K}$.

Write $\mathfrak{M} = \mathbf{O}_{p*,p}(\mathfrak{G})$, $\mathfrak{P} = \mathfrak{S} \cap \mathfrak{M}$ and $\mathfrak{U} = [\mathfrak{A}, \mathfrak{P}]$. Thus $\mathfrak{P} \in S_p(\mathfrak{M})$, $\mathfrak{P} \trianglelefteq \mathfrak{S}$ and $\mathfrak{U} \leq \mathfrak{A} \cap \mathfrak{P}$. By 14.14a),

$$\mathfrak{K} = [\mathfrak{K}, \mathfrak{U}]\mathbf{O}_{p*}(\mathbf{C}_{\mathfrak{K}}(\mathfrak{U})).$$

Now $[\mathfrak{K}, \mathfrak{U}] \leq \mathfrak{M}$. If $\mathbf{O}_{p*}(\mathbf{C}_{\mathfrak{K}}(\mathfrak{U})) \leq \mathfrak{M}$, we have

$$\mathfrak{K} = \mathbf{O}^p(\mathfrak{K}) \leq \mathbf{O}^p(\mathfrak{M}) = \mathbf{O}_{p*}(\mathfrak{G}),$$

a contradiction. Thus $\mathbf{O}_{p*}(\mathbf{C}_{\mathfrak{K}}(\mathfrak{U})) \not\leq \mathfrak{M}$. But $\mathbf{C}_{\mathfrak{K}}(\mathfrak{U})$ is $\mathfrak{A}$-invariant, so by minimality of $\mathfrak{K}$, $\mathbf{C}_{\mathfrak{K}}(\mathfrak{U}) = \mathfrak{K}$. Thus $\mathfrak{K} \leq \mathbf{C}_{\mathfrak{G}}(\mathfrak{U})$ but $\mathfrak{K} \not\leq \mathbf{O}_{p*}(\mathbf{C}_{\mathfrak{G}}(\mathfrak{U}))$, since by 14.13, $\mathbf{O}_{p*}(\mathbf{C}_{\mathfrak{G}}(\mathfrak{U})) \leq \mathbf{O}_{p*}(\mathfrak{G})$. Thus $\mathfrak{K} \not\leq \mathbf{O}_{p*}(\mathfrak{G}_0)$, where $\mathfrak{G}_0 = \mathbf{C}_{\mathfrak{G}}(\mathfrak{U})\mathfrak{A}$. But since $\mathfrak{P} \trianglelefteq \mathfrak{S}$ and $\mathfrak{A} \trianglelefteq \mathfrak{S}$, we have $\mathfrak{U} \trianglelefteq \mathfrak{S}$ and $\mathfrak{S} \leq \mathbf{N}_{\mathfrak{G}}(\mathfrak{U})$. Thus $(\mathfrak{S} \cap \mathbf{C}_{\mathfrak{G}}(\mathfrak{U})) \in S_p(\mathbf{C}_{\mathfrak{G}}(\mathfrak{U}))$ and $\mathfrak{S}_0 = (\mathfrak{S} \cap \mathbf{C}_{\mathfrak{G}}(\mathfrak{U}))\mathfrak{A} \in S_p(\mathfrak{G}_0)$. Hence the theorem is not valid in $\mathfrak{G}_0$. By minimality of $\mathfrak{G}$, $\mathfrak{G} = \mathfrak{G}_0 = \mathbf{C}_{\mathfrak{G}}(\mathfrak{U})\mathfrak{A}$. In particular, $\mathfrak{U} \trianglelefteq \mathfrak{G}$ and $[\mathfrak{U}, \mathbf{O}^p(\mathfrak{G})] = 1$.

If $\mathfrak{M}^*/\mathfrak{U} = \mathbf{O}_{p*,p}(\mathfrak{G}/\mathfrak{U})$, $\mathbf{O}^p(\mathfrak{M}^*/\mathfrak{U})$ is a p^*-group. Hence by 14.5b), $\mathbf{O}^p(\mathfrak{M}^*)$ is a p^*-group and $\mathfrak{M}^* \leq \mathbf{O}_{p*,p}(\mathfrak{G}) = \mathfrak{M}$. But obviously $\mathfrak{M}/\mathfrak{U} \leq \mathbf{O}_{p*,p}(\mathfrak{G}/\mathfrak{U})$, so $\mathfrak{M}/\mathfrak{U} = \mathbf{O}_{p*,p}(\mathfrak{G}/\mathfrak{U})$. It follows from 14.10 that $\mathbf{C}_{\mathfrak{G}}(\mathfrak{P}/\mathfrak{U}) \leq \mathfrak{M}$. But

$$[\mathbf{N}_{\mathfrak{G}}(\mathfrak{P}), \mathfrak{P}, \mathfrak{A}] \leq [\mathfrak{P}, \mathfrak{A}] = \mathfrak{U}$$

and

$$[\mathfrak{P}, \mathfrak{A}, \mathbf{N}_{\mathfrak{G}}(\mathfrak{P})] \leq \mathfrak{U},$$

so $[\mathfrak{A}, \mathbf{N}_{\mathfrak{G}}(\mathfrak{P}), \mathfrak{P}] \leq \mathfrak{U}$ and $[\mathfrak{A}, \mathbf{N}_{\mathfrak{G}}(\mathfrak{P})] \leq \mathbf{C}_{\mathfrak{G}}(\mathfrak{P}/\mathfrak{U}) \leq \mathfrak{M}$. But by the Frattini argument, $\mathfrak{G} = \mathfrak{M}\mathbf{N}_{\mathfrak{G}}(\mathfrak{P})$, so $[\mathfrak{G}, \mathfrak{A}] \leq \mathfrak{M}$. Thus $\mathfrak{K} = [\mathfrak{K}, \mathfrak{A}] \leq \mathfrak{M}$ and $\mathfrak{K} = \mathbf{O}^p(\mathfrak{K}) \leq \mathbf{O}^p(\mathfrak{M}) = \mathbf{O}_{p*}(\mathfrak{G})$. **q.e.d.**

We now discuss the relationship with $\mathbf{E}(\mathfrak{G})$ and $\mathbf{F}^*(\mathfrak{G})$.

14.16 Definition. For any group $\mathfrak{G}$, write

$$\mathbf{O}_{p',\mathbf{E}}(\mathfrak{G})/\mathbf{O}_{p'}(\mathfrak{G}) = \mathbf{E}(\mathfrak{G}/\mathbf{O}_{p'}(\mathfrak{G})),$$

$$\mathbf{O}_{p',\mathbf{E},p}(\mathfrak{G})/\mathbf{O}_{p',\mathbf{E}}(\mathfrak{G}) = \mathbf{O}_p(\mathfrak{G}/\mathbf{O}_{p',\mathbf{E}}(\mathfrak{G})).$$

14.17 Theorem. a) *If* $\mathfrak{P}$ *is a p-subgroup of* $\mathbf{O}_{p',\mathbf{E},p}(\mathfrak{G})$, *then*

$$\mathbf{O}_{p^*}(\mathfrak{G}) = \mathbf{O}_{p',\mathbf{E}}(\mathfrak{G})\mathbf{O}_{p^*}(\mathbf{C}_{\mathfrak{G}}(\mathfrak{P})).$$

b) $\mathbf{O}_{p^*}(\mathfrak{G})$ *normalises every component of* $\mathbf{E}(\mathfrak{G}/\mathbf{O}_{p'}(\mathfrak{G}))$.

Proof. a) By 14.14a),

$$\mathbf{O}_{p^*}(\mathfrak{G}) = [\mathbf{O}_{p^*}(\mathfrak{G}), \mathfrak{P}]\mathbf{O}_{p^*}(\mathfrak{C}),$$

where $\mathfrak{C} = \mathbf{C}_{\mathfrak{G}}(\mathfrak{P}) \cap \mathbf{O}_{p^*}(\mathfrak{G})$. Thus $\mathfrak{C} \trianglelefteq \mathbf{C}_{\mathfrak{G}}(\mathfrak{P})$, so $\mathbf{O}_{p^*}(\mathfrak{C}) \leq \mathbf{O}_{p^*}(\mathbf{C}_{\mathfrak{G}}(\mathfrak{P}))$. By 14.13, $\mathbf{O}_{p^*}(\mathbf{C}_{\mathfrak{G}}(\mathfrak{P})) \leq \mathbf{O}_{p^*}(\mathfrak{G})$, so

$$\mathbf{O}_{p^*}(\mathfrak{G}) = [\mathbf{O}_{p^*}(\mathfrak{G}), \mathfrak{P}]\mathbf{O}_{p^*}(\mathbf{C}_{\mathfrak{G}}(\mathfrak{P})).$$

Also by 14.14b), $[\mathbf{O}_{p^*}(\mathfrak{G}), \mathfrak{P}]$ is a p^*-group, so since $\mathfrak{P} \leq \mathbf{O}_{p',\mathbf{E},p}(\mathfrak{G})$,

$$[\mathbf{O}_{p^*}(\mathfrak{G}), \mathfrak{P}] = \mathbf{O}^p([\mathbf{O}_{p^*}(\mathfrak{G}), \mathfrak{P}]) \leq \mathbf{O}^p(\mathbf{O}_{p',\mathbf{E},p}(\mathfrak{G})) = \mathbf{O}_{p',\mathbf{E}}(\mathfrak{G}).$$

By 14.6c), $\mathbf{E}(\mathfrak{G}/\mathbf{O}_{p'}(\mathfrak{G}))$ is a p^*-group, so by 14.5, $\mathbf{O}_{p',\mathbf{E}}(\mathfrak{G})$ is a p^*-group. Thus $\mathbf{O}_{p',\mathbf{E}}(\mathfrak{G}) \leq \mathbf{O}_{p^*}(\mathfrak{G})$ and

$$\mathbf{O}_{p^*}(\mathfrak{G}) = \mathbf{O}_{p',\mathbf{E}}(\mathfrak{G})\mathbf{O}_{p^*}(\mathbf{C}_{\mathfrak{G}}(\mathfrak{P})).$$

b) Suppose that $\mathfrak{S} \in S_p(\mathbf{O}_{p',\mathbf{E}}(\mathfrak{G}))$. By definition of a p^*-group, $\mathbf{O}_{p^*}(\mathfrak{G}) = \mathbf{O}_{p',\mathbf{E}}(\mathfrak{G})\mathfrak{C}$, where $\mathfrak{C} = \mathbf{C}_{\mathfrak{G}}(\mathfrak{S}) \cap \mathbf{O}_{p^*}(\mathfrak{G})$. Hence it is sufficient to show that $\mathbf{C}_{\mathfrak{G}}(\mathfrak{S})$ normalises every component of $\mathbf{E}(\mathfrak{G}/\mathbf{O}_{p'}(\mathfrak{G}))$. Suppose that this is not the case and that $\mathfrak{H}_1^x = \mathfrak{H}_2$, where $\mathfrak{H}_1/\mathbf{O}_{p'}(\mathfrak{G})$, $\mathfrak{H}_2/\mathbf{O}_{p'}(\mathfrak{G})$ are distinct components of $\mathbf{E}(\mathfrak{G}/\mathbf{O}_{p'}(\mathfrak{G}))$ and $x \in \mathbf{C}_{\mathfrak{G}}(\mathfrak{S})$. Then $\mathfrak{H}_1 \cap \mathfrak{S} = (\mathfrak{H}_1 \cap \mathfrak{S})^x = \mathfrak{H}_2 \cap \mathfrak{S} = (\mathfrak{H}_1 \cap \mathfrak{H}_2) \cap \mathfrak{S}$. But if $\mathfrak{Z}/\mathbf{O}_{p'}(\mathfrak{G}) = \mathbf{Z}(\mathbf{E}(\mathfrak{G}/\mathbf{O}_{p'}(\mathfrak{G})))$, $\mathfrak{H}_1 \cap \mathfrak{H}_2 \leq \mathfrak{Z}$, by 13.18. Thus $\mathfrak{H}_1 \cap \mathfrak{S} \leq \mathfrak{Z}$ and $\mathfrak{H}_1\mathfrak{Z}/\mathfrak{Z}$ is a p'-group. Let $\mathfrak{H}/\mathbf{O}_{p'}(\mathfrak{G})$ be the product of all components of $\mathbf{E}(\mathfrak{G}/\mathbf{O}_{p'}(\mathfrak{G}))$ isomorphic to $\mathfrak{H}_1$. Clearly $\mathfrak{H} \trianglelefteq \mathfrak{G}$, $\mathfrak{H}/\mathbf{O}_{p'}(\mathfrak{G})$ is perfect and $\mathfrak{H}/(\mathfrak{H} \cap \mathfrak{Z})$ is a p'-group. Since $(\mathfrak{H} \cap \mathfrak{Z})/\mathbf{O}_{p'}(\mathfrak{G}) \leq \mathbf{Z}(\mathfrak{H}/\mathbf{O}_{p'}(\mathfrak{G}))$, it follows from Burnside's transfer theorem (IV, 2.6) that $\mathfrak{H}/\mathbf{O}_{p'}(\mathfrak{G})$ is p-nilpotent. Since it is also perfect, $\mathfrak{H}/\mathbf{O}_{p'}(\mathfrak{G})$ is a p'-group and $\mathfrak{H} \leq \mathbf{O}_{p'}(\mathfrak{G})$. This is a contradiction. **q.e.d.**

14.18 Theorem. *Suppose that for each composition factor* $\mathfrak{X}$ *of* $\mathbf{E}(\mathfrak{G}/\mathbf{O}_{p'}(\mathfrak{G}))$, $\mathbf{C}_{\mathfrak{A}}(\mathfrak{T})$ *is p-soluble, where* $\mathfrak{A} = \mathbf{Aut}\ \mathfrak{X}$ *and* $\mathfrak{T} \in S_p(\mathfrak{X})$. *Let* $\mathfrak{P}$ *be a p-subgroup of* $\mathbf{O}_{p',\mathbf{E},p}(\mathfrak{G})$ *containing a Sylow p-subgroup of* $\mathbf{O}_{p',\mathbf{E}}(\mathfrak{G})$. *Then* $\mathbf{O}_{p^*}(\mathbf{C}_{\mathfrak{G}}(\mathfrak{P}))$ *is a p′-group and*

$$\mathbf{O}_{p*}(\mathfrak{G}) = \mathbf{O}_{p',\mathbf{E}}(\mathfrak{G})\mathbf{O}_{p'}(\mathbf{C}_{\mathfrak{G}}(\mathfrak{P})).$$

Proof. Let $\overline{\mathfrak{G}} = \mathfrak{G}/\mathbf{O}_{p'}(\mathfrak{G})$. By 14.4b), $\mathbf{O}_{p*}(\overline{\mathfrak{G}})$ centralizes $\mathbf{O}_p(\overline{\mathfrak{G}})$. By 14.5, $\mathbf{O}_{p*}(\overline{\mathfrak{G}}) = \mathbf{O}_{p*}(\mathfrak{G})/\mathbf{O}_{p'}(\mathfrak{G})$. Also $\overline{\mathfrak{K}} \le \mathbf{O}_{p*}(\overline{\mathfrak{G}})$, by 14.13, where $\mathfrak{K} = \mathbf{O}_{p*}(\mathbf{C}_{\mathfrak{G}}(\mathfrak{P}))$. Thus $\overline{\mathfrak{K}}$ centralizes $\mathbf{O}_p(\overline{\mathfrak{G}}) = \mathbf{F}(\overline{\mathfrak{G}})$.

By 14.17b), $\mathfrak{K}$ normalises any component $\overline{\mathfrak{H}}$ of $\mathbf{E}(\overline{\mathfrak{G}})$. Now $\mathfrak{P} \cap \mathbf{O}_{p',\mathbf{E}}(\mathfrak{G}) \in S_p(\mathbf{O}_{p',\mathbf{E}}(\mathfrak{G}))$, so $\overline{\mathfrak{P}} = ((\mathfrak{P} \cap \mathbf{O}_{p',\mathbf{E}}(\mathfrak{G}))\mathbf{O}_{p'}(\mathfrak{G})/\mathbf{O}_{p'}(\mathfrak{G})) \cap \overline{\mathfrak{H}}$ is a Sylow p-subgroup of $\overline{\mathfrak{H}}$. Of course $\mathfrak{K}$ centralizes $\overline{\mathfrak{P}}$, so $\mathfrak{K}$ centralizes a Sylow p-subgroup of $\overline{\mathfrak{H}}/\mathbf{Z}(\overline{\mathfrak{H}})$. It follows from the hypothesis that $\mathfrak{K}/\mathfrak{K}_0$ is p-soluble, where $\mathfrak{K}_0 = \mathbf{C}_{\mathfrak{K}}(\overline{\mathfrak{H}}/\mathbf{Z}(\overline{\mathfrak{H}}))$. Now $[\mathfrak{K}_0, \mathfrak{H}, \mathfrak{H}] \le \mathbf{O}_{p'}(\mathfrak{G})$, where $\mathfrak{H}/\mathbf{O}_{p'}(\mathfrak{G}) = \overline{\mathfrak{H}}$, so by the three subgroup theorem, $[\mathfrak{H}', \mathfrak{K}_0] \le \mathbf{O}_{p'}(\mathfrak{G})$. Hence $\mathfrak{K}_0$ centralizes $\overline{\mathfrak{H}}' = \overline{\mathfrak{H}}$.

Let $\mathfrak{L}$ be the smallest normal subgroup of $\mathfrak{K}$ for which $\mathfrak{K}/\mathfrak{L}$ is p-soluble. Then $\mathfrak{L} \le \mathfrak{K}_0$, so $\mathfrak{L}$ centralizes $\overline{\mathfrak{H}}$. Hence $\mathfrak{L}$ centralizes $\mathbf{E}(\overline{\mathfrak{G}})$ and $\mathbf{E}(\overline{\mathfrak{G}})\mathbf{F}(\overline{\mathfrak{G}}) = \mathbf{F}^*(\overline{\mathfrak{G}})$. By 13.12, $\mathfrak{L}\mathbf{O}_{p'}(\mathfrak{G})/\mathbf{O}_{p'}(\mathfrak{G})$ is contained in the centre of $\mathbf{F}^*(\overline{\mathfrak{G}})$ and is thus Abelian. Hence $\mathfrak{L}$ is p-soluble and $\mathfrak{K}$ is p-soluble. By 14.3d), $\mathfrak{K}$ is a p'-group.

Thus $\mathfrak{K} = \mathbf{O}_{p'}(\mathbf{C}_{\mathfrak{G}}(\mathfrak{P}))$ and by 14.17,

$$\mathbf{O}_{p*}(\mathfrak{G}) = \mathbf{O}_{p',\mathbf{E}}(\mathfrak{G})\mathbf{O}_{p'}(\mathbf{C}_{\mathfrak{G}}(\mathfrak{P})). \qquad \textbf{q.e.d.}$$

14.19 Remarks. a) The hypothesis about the composition factor $\mathfrak{X}$ in 14.18 holds for all known simple groups and all primes; in fact it is a weak form of the Schreier hypothesis. Moreover, it has been proved to be true for every simple group in the case $p = 2$ by GLAUBERMAN [1], who showed that if $\mathfrak{G}$ is a group, $\mathbf{O}_{2'}(\mathfrak{G}) = 1$, $\mathfrak{S} \in S_2(\mathfrak{G})$, $\mathfrak{A} = \mathbf{Aut}\,\mathfrak{G}$ and $\mathfrak{C} = \mathbf{C}_{\mathfrak{A}}(\mathfrak{S})$, then $\mathfrak{C}$ is 2-nilpotent and the Sylow 2-subgroups of $\mathfrak{C}$ are Abelian. It follows easily from this and the solubility of groups of odd order that a group is perfect and quasinilpotent if and only if it is a p^*-group for every prime p.

b) Let $\mathfrak{L} = \mathbf{O}_{2',\mathbf{E}}(\mathfrak{G})$ and put $\mathbf{D}(\mathfrak{G}) = \bigcap_{n \ge 1} \mathfrak{L}^{(n)}$. Thus $\mathfrak{L} = \mathbf{D}(\mathfrak{G})\mathbf{O}_{2'}(\mathfrak{G})$, since $\mathbf{E}(\mathfrak{G}/\mathbf{O}_{2'}(\mathfrak{G}))$ is perfect. And if $\mathfrak{K} \trianglelefteq \mathfrak{G}$ and $\mathfrak{L} = \mathfrak{K}\mathbf{O}_{2'}(\mathfrak{G})$, then $\mathbf{D}(\mathfrak{G}) \le \mathfrak{K}$ on account of the solubility of groups of odd order. $\mathbf{D}(\mathfrak{G})$ has been called the *2-layer* of $\mathfrak{G}$, and it is easy to deduce from 14.13 and 14.18 that

$$\mathbf{D}(\mathbf{C}_{\mathfrak{G}}(\mathfrak{T})) \le \mathbf{D}(\mathfrak{G})$$

for any 2-subgroup $\mathfrak{T}$ of $\mathfrak{G}$.

c) Put $\mathfrak{G}_0 = \mathbf{C}_{\mathfrak{G}}(\mathbf{E}(\mathfrak{G}))$, $\mathfrak{C} = \mathbf{O}_{2',\mathbf{E}}(\mathfrak{G}_0)$ and

$$\mathbf{B}(\mathfrak{G}) = \bigcap_{n \ge 1} \mathfrak{C}^{(n)}.$$

Then $\mathbf{D}(\mathfrak{G}) = \mathbf{E}(\mathfrak{G})\mathbf{B}(\mathfrak{G})$. It is conjectured by Thompson that

$$\mathbf{B}(\mathbf{C}_{\mathfrak{G}}(t)) \le \mathbf{B}(\mathfrak{G})$$

for any involution t of $\mathfrak{G}$. This implies that for a simple group $\mathfrak{G}$,

$$\mathbf{O}_{2',\mathbf{E}}(\mathbf{C}_{\mathfrak{G}}(t)) = \mathbf{O}_{2'}(\mathbf{C}_{\mathfrak{G}}(t))\mathbf{E}(\mathbf{C}_{\mathfrak{G}}(t)).$$

§ 15. Applications of the Generalized Fitting Subgroup

One of the important ideas in the study of simple groups is the proof that certain subgroups are contained in only one maximal subgroup; (see, for example, 2.14). We shall use the results of § 13 and § 14 to obtain information about subgroups contained in at least two maximal subgroups.

15.1 Lemma. *Suppose that* $\mathfrak{U} \le \mathbf{F}^*(\mathfrak{G})$. *Then* $\mathfrak{U}$ *is a subnormal subgroup of* $\mathbf{F}^*(\mathfrak{G})$ *and* $\mathfrak{U} \ge \mathbf{C}_{\mathbf{F}^*(\mathfrak{G})}(\mathfrak{U})$ *if and only if* $\mathfrak{U} = \mathbf{E}(\mathfrak{G})(\mathfrak{U} \cap \mathbf{F}(\mathfrak{G}))$ *and* $\mathbf{C}_{\mathbf{F}(\mathfrak{G})}(\mathfrak{U} \cap \mathbf{F}(\mathfrak{G})) \le \mathfrak{U}$.

Proof. If $\mathfrak{U} = \mathbf{E}(\mathfrak{G})(\mathfrak{U} \cap \mathbf{F}(\mathfrak{G}))$, then $\mathfrak{U}/\mathbf{E}(\mathfrak{G})$ is a subgroup of the nilpotent group $\mathbf{F}^*(\mathfrak{G})/\mathbf{E}(\mathfrak{G})$ and is therefore subnormal; also by 13.15c), $\mathbf{C}_{\mathbf{F}^*(\mathfrak{G})}(\mathbf{E}(\mathfrak{G})) \le \mathbf{F}(\mathfrak{G})$, so

$$\mathbf{C}_{\mathbf{F}^*(\mathfrak{G})}(\mathfrak{U}) = \mathbf{C}_{\mathbf{F}(\mathfrak{G})}(\mathfrak{U}) = \mathbf{C}_{\mathbf{F}(\mathfrak{G})}(\mathfrak{U} \cap \mathbf{F}(\mathfrak{G})).$$

Hence if $\mathfrak{U} = \mathbf{E}(\mathfrak{G})(\mathfrak{U} \cap \mathbf{F}(\mathfrak{G}))$ and $\mathbf{C}_{\mathbf{F}(\mathfrak{G})}(\mathfrak{U} \cap \mathbf{F}(\mathfrak{G})) \le \mathfrak{U}$, then $\mathfrak{U}$ is subnormal in $\mathbf{F}^*(\mathfrak{G})$ and $\mathbf{C}_{\mathbf{F}^*(\mathfrak{G})}(\mathfrak{U}) \le \mathfrak{U}$.

Conversely, suppose that $\mathfrak{U}$ is subnormal in $\mathbf{F}^*(\mathfrak{G})$ and $\mathbf{C}_{\mathbf{F}^*(\mathfrak{G})}(\mathfrak{U}) \le \mathfrak{U}$. We prove that $\mathbf{E}(\mathfrak{G}) \le \mathfrak{U}$ by induction on $|\mathbf{F}^*(\mathfrak{G}) : \mathfrak{U}|$. Assuming that $\mathfrak{U} < \mathbf{F}^*(\mathfrak{G})$, there exists a subnormal subgroup $\mathfrak{Y}$ of $\mathbf{F}^*(\mathfrak{G})$ such that $\mathfrak{U} \lhd \mathfrak{Y}$. By the inductive hypothesis, $\mathbf{E}(\mathfrak{G}) \le \mathfrak{Y}$. By 13.19,

$$\mathbf{E}(\mathfrak{G}) = [\mathbf{E}(\mathfrak{G}), \mathfrak{U}]\mathbf{C}_{\mathbf{E}(\mathfrak{G})}(\mathfrak{U}).$$

By hypothesis, $\mathbf{C}_{\mathbf{E}(\mathfrak{G})}(\mathfrak{U})$ is contained in $\mathfrak{U}$ and is therefore Abelian. Since $\mathbf{E}(\mathfrak{G})$ is perfect, it follows that $\mathbf{E}(\mathfrak{G}) = [\mathbf{E}(\mathfrak{G}), \mathfrak{U}]$. Hence

$$\mathbf{E}(\mathfrak{G}) = [\mathbf{E}(\mathfrak{G}), \mathfrak{U}] \le [\mathfrak{Y}, \mathfrak{U}] \le \mathfrak{U}.$$

Since $\mathbf{F}^*(\mathfrak{G}) = \mathbf{E}(\mathfrak{G})\mathbf{F}(\mathfrak{G})$, it follows at once that

$$\mathfrak{U} = \mathbf{E}(\mathfrak{G})(\mathfrak{U} \cap \mathbf{F}(\mathfrak{G})).$$

But $[\mathbf{E}(\mathfrak{G}), \mathbf{F}(\mathfrak{G})] = 1$, so

$$\mathbf{C}_{\mathbf{F}(\mathfrak{G})}(\mathfrak{U} \cap \mathbf{F}(\mathfrak{G})) \leq \mathbf{C}_{\mathbf{F}^*(\mathfrak{G})}(\mathfrak{U}) \leq \mathfrak{U}.$$ **q.e.d.**

15.2 Theorem. *Let* $\mathfrak{U}$ *be a subnormal subgroup of* $\mathbf{F}^*(\mathfrak{G})$ *such that* $\mathfrak{U} \geq \mathbf{C}_{\mathbf{F}^*(\mathfrak{G})}(\mathfrak{U})$.

a) *A* p'*-element of* $\mathfrak{G}$ *which centralizes* $\mathbf{O}_p(\mathfrak{U})$ *centralizes* $\mathbf{O}_p(\mathfrak{G})$.

b) *If* $\mathfrak{K}$ *is a* $\mathfrak{U}$*-invariant perfect quasinilpotent subgroup of* $\mathfrak{G}$, $\mathfrak{K} \trianglelefteq \mathbf{E}(\mathfrak{G})$.

c) *If* $\mathfrak{P}$ *is a* $\mathfrak{U}$*-invariant* p*-subgroup of* $\mathfrak{G}$, $\mathfrak{P} \leq \mathbf{C}_{\mathfrak{G}}(\mathbf{O}_{p^*}(\mathfrak{G}))$.

Proof. a) Since $\mathfrak{U}$ is subnormal in $\mathfrak{G}$, $\mathbf{F}(\mathfrak{U}) = \mathfrak{U} \cap \mathbf{F}(\mathfrak{G})$ and $\mathbf{O}_q(\mathfrak{U}) = \mathfrak{U} \cap \mathbf{O}_q(\mathfrak{G})$ for any prime q. Hence $\mathbf{C}_{\mathfrak{G}}(\mathbf{O}_q(\mathfrak{G})) \leq \mathbf{C}_{\mathfrak{G}}(\mathbf{O}_q(\mathfrak{U}))$. Thus if $x \in \mathbf{O}_p(\mathfrak{G})$ and $x \in \mathbf{C}_{\mathfrak{G}}(\mathbf{O}_p(\mathfrak{U}))$, then x centralizes $\mathbf{E}(\mathfrak{G})$ and $\mathbf{F}(\mathfrak{U}) = \mathfrak{U} \cap \mathbf{F}(\mathfrak{G})$. Hence by 15.1, $x \in \mathbf{C}_{\mathbf{F}^*(\mathfrak{G})}(\mathfrak{U}) \leq \mathfrak{U}$ and $x \in \mathfrak{U} \cap \mathbf{O}_p(\mathfrak{G}) = \mathbf{O}_p(\mathfrak{U})$. Therefore

$$\mathbf{C}_{\mathbf{O}_p(\mathfrak{G})}(\mathbf{O}_p(\mathfrak{U})) \leq \mathbf{O}_p(\mathfrak{U}).$$

Thus if y is a p'-element of $\mathfrak{G}$ and y centralizes $\mathbf{O}_p(\mathfrak{U})$, then y centralizes $\mathbf{O}_p(\mathfrak{G})$, by 1.2.

b) We have

$$[\mathfrak{U} \cap \mathbf{F}(\mathfrak{G}), \mathfrak{K}] \leq \mathbf{F}(\mathfrak{G}) \cap \mathfrak{K} \leq \mathbf{Z}(\mathfrak{K}),$$

since $\mathbf{F}(\mathfrak{G}) \cap \mathfrak{K}$ is a nilpotent normal subgroup of the perfect quasinilpotent subgroup $\mathfrak{K}$. Thus $[\mathfrak{U} \cap \mathbf{F}(\mathfrak{G}), \mathfrak{K}, \mathfrak{K}] = 1$, and by the three subgroup theorem (III, 1.10), $\mathfrak{U} \cap \mathbf{F}(\mathfrak{G})$ centralizes $\mathfrak{K}' = \mathfrak{K}$. Thus $\mathfrak{K}$ centralizes $\mathbf{O}_p(\mathfrak{U})$ for any prime p. But $\mathfrak{K}$ is generated by p'-elements, so by a), $\mathfrak{K}$ centralizes $\mathbf{O}_p(\mathfrak{G})$. Hence $\mathbf{F}(\mathfrak{G})$ centralizes $\mathfrak{K}$. By 15.1, $\mathfrak{U} \geq \mathbf{E}(\mathfrak{G})$, so it follows that $\mathbf{F}^*(\mathfrak{G}) \leq \mathbf{N}_{\mathfrak{G}}(\mathfrak{K})$. By 13.19,

$$\mathfrak{K} = [\mathfrak{K}, \mathbf{F}^*(\mathfrak{G})]\mathbf{C}_{\mathfrak{K}}(\mathbf{F}^*(\mathfrak{G})).$$

By 13.12, $\mathfrak{K} \leq \mathbf{F}^*(\mathfrak{G})$. By 13.14, $\mathfrak{K} \leq \mathbf{E}(\mathfrak{G})$, so $\mathfrak{K} \trianglelefteq \mathbf{E}(\mathfrak{G})$.

c) Since $\mathbf{E}(\mathfrak{G})$ normalises $\mathfrak{P}$, $\mathfrak{P}$ centralizes $\mathbf{E}(\mathfrak{G})$, by 14.4. And for $q \neq p$,

$$[\mathfrak{P}, \mathbf{O}_q(\mathfrak{U})] \leq \mathbf{O}_q(\mathfrak{G}) \cap \mathfrak{P} = 1.$$

Thus $\mathfrak{P}$ centralizes $\mathbf{E}(\mathfrak{G})\mathbf{O}^p(\mathbf{F}(\mathfrak{G}))$. Let $\mathfrak{N} = \mathbf{F}^*(\mathbf{O}_{p^*}(\mathfrak{G}))$. Thus $\mathfrak{N} \trianglelefteq \mathfrak{G}$ and $\mathfrak{N} \leq \mathbf{F}^*(\mathfrak{G}) = \mathbf{E}(\mathfrak{G})\mathbf{F}(\mathfrak{G})$, so

$$\mathbf{O}_{p^*}(\mathfrak{N}) \leq \mathbf{O}^p(\mathfrak{N}) \leq \mathbf{E}(\mathfrak{G})\mathbf{O}^p(\mathbf{F}(\mathfrak{G})).$$

Therefore $\mathfrak{P}$ centralizes $\mathbf{O}_{p^*}(\mathfrak{N})$. By 13.12, $\mathbf{C}_{\mathbf{O}_{p^*}(\mathfrak{G})}(\mathfrak{N}) \leq \mathfrak{N}$, so by 14.14c), $\mathfrak{P}$ centralizes $\mathbf{O}_{p^*}(\mathfrak{G})$. **q.e.d.**

15.3 Theorem. *Suppose that $\mathfrak{H}$, $\mathfrak{M}$ are subgroups of $\mathfrak{G}$ for which $\mathbf{E}(\mathfrak{H}) \leq \mathfrak{M}$ and $\mathbf{C}_{\mathbf{F}(\mathfrak{H})}(\mathfrak{M} \cap \mathbf{F}(\mathfrak{H})) \leq \mathfrak{M}$.*

a) *For any prime p*, $[\mathbf{O}_p(\mathfrak{M}) \cap \mathfrak{H}, \mathbf{O}_{p^*}(\mathfrak{H})] = 1$.

b) *If* $\mathbf{O}_p(\mathfrak{H}) = 1$, $\mathbf{O}_p(\mathfrak{M}) \cap \mathfrak{H} = 1$.

c) *Any component of* $\mathbf{E}(\mathfrak{M})$ *contained in* $\mathfrak{H}$ *is a component of* $\mathbf{E}(\mathfrak{H})$.

Proof. Let $\mathfrak{U} = \mathfrak{M} \cap \mathbf{F}^*(\mathfrak{H})$. Then $\mathfrak{U} \cap \mathbf{F}(\mathfrak{H}) = \mathfrak{M} \cap \mathbf{F}(\mathfrak{H})$, so $\mathfrak{U} \geq \mathbf{E}(\mathfrak{H})$ and $\mathfrak{U} \geq \mathbf{C}_{\mathbf{F}(\mathfrak{H})}(\mathfrak{U} \cap \mathbf{F}(\mathfrak{H}))$. By 15.1, $\mathfrak{U}$ satisfies the conditions of 15.2.

a) $\mathbf{O}_p(\mathfrak{M}) \cap \mathfrak{H}$ is a $\mathfrak{U}$-invariant p-subgroup of $\mathfrak{H}$, so by 15.2c),

$$[\mathbf{O}_p(\mathfrak{M}) \cap \mathfrak{H}, \mathbf{O}_{p^*}(\mathfrak{H})] = 1.$$

b) If $\mathbf{O}_p(\mathfrak{H}) = 1$, $\mathbf{F}^*(\mathfrak{H})$ is a p^*-group, so $\mathbf{F}^*(\mathfrak{H}) \leq \mathbf{O}_{p^*}(\mathfrak{H})$. Hence by a) and 13.12,

$$\mathbf{O}_p(\mathfrak{M}) \cap \mathfrak{H} \leq \mathbf{C}_{\mathfrak{H}}(\mathbf{F}^*(\mathfrak{H})) \leq \mathbf{F}(\mathfrak{H}).$$

Thus $\mathbf{O}_p(\mathfrak{M}) \cap \mathfrak{H} \leq \mathbf{O}_p(\mathfrak{H}) = 1$.

c) Let $\mathfrak{E}$ be a component of $\mathbf{E}(\mathfrak{M})$ contained in $\mathfrak{H}$; thus $\mathfrak{E} \leq \mathfrak{H} \cap \mathfrak{M}$. Thus if $x \in \mathfrak{H} \cap \mathfrak{M}$, $\mathfrak{E}^x$ is a component of $\mathbf{E}(\mathfrak{M})$ contained in $\mathfrak{H}$. Put

$$\mathfrak{E}^* = \prod_{x \in \mathfrak{H} \cap \mathfrak{M}} \mathfrak{E}^x.$$

Then $\mathfrak{E}^*$ is a perfect quasinilpotent $\mathfrak{U}$-invariant subgroup of $\mathfrak{H}$. By 15.2b), $\mathfrak{E}^* \trianglelefteq \mathbf{E}(\mathfrak{H})$, so $\mathfrak{E}$ is a perfect quasinilpotent subnormal subgroup of $\mathbf{E}(\mathfrak{H})$. By 13.18d), $\mathfrak{E}$ is a component of $\mathbf{E}(\mathfrak{H})$. **q.e.d.**

15.4 Definition. If $\mathfrak{H}$, $\mathfrak{M}$ are subgroups of $\mathfrak{G}$, we write $\mathfrak{H} \to \mathfrak{M}$ when the following conditions are satisfied.

(i) $\mathbf{E}(\mathfrak{H}) \leq \mathfrak{M}$.

(ii) $\mathbf{C}_{\mathbf{F}(\mathfrak{H})}(\mathfrak{M} \cap \mathbf{F}(\mathfrak{H})) \leq \mathfrak{M}$.

(iii) $\mathbf{N}_{\mathfrak{M}}(\mathfrak{X}) \leq \mathfrak{H}$ for every non-identity characteristic subgroup $\mathfrak{X}$ of $\mathfrak{H}$.

15.5 Lemma. *The following conditions are equivalent.*

a) $\mathbf{F}^*(\mathfrak{G})$ *is a p-group.*

b) $\mathbf{O}_p(\mathfrak{G}) \geq \mathbf{C}_{\mathfrak{G}}(\mathbf{O}_p(\mathfrak{G}))$.

c) $\mathbf{O}_{p^*}(\mathfrak{G}) = 1$.

Proof. If a) holds, $\mathbf{F}^*(\mathfrak{G}) = \mathbf{O}_p(\mathfrak{G})$, so b) holds by 13.12. Since by 14.4, $\mathbf{O}_{p^*}(\mathfrak{G}) \leq \mathbf{C}_{\mathfrak{G}}(\mathbf{O}_p(\mathfrak{G}))$, b) implies that $\mathbf{O}_{p^*}(\mathfrak{G})$ is a p-group, and hence $\mathbf{O}_{p^*}(\mathfrak{G}) = \mathbf{O}^p(\mathbf{O}_{p^*}(\mathfrak{G})) = 1$. Finally, c) implies that $\mathbf{O}_{p'}(\mathfrak{G}) = \mathbf{E}(\mathfrak{G}) = 1$, so $\mathbf{F}^*(\mathfrak{G}) = \mathbf{E}(\mathfrak{G})\mathbf{F}(\mathfrak{G})$ is a p-group. **q.e.d.**

15.6 Lemma. *Suppose that $\mathfrak{H}$, $\mathfrak{M}$ are subgroups of $\mathfrak{G}$ and $\mathfrak{H} \to \mathfrak{M}$. Suppose that $\mathbf{O}_p(\mathfrak{H}) \neq 1$.*

a) $\mathbf{E}(\mathfrak{H})(\mathfrak{M} \cap \mathbf{O}_{p'}(\mathbf{F}(\mathfrak{H}))) \leq \mathbf{O}_{p^*}(\mathfrak{H} \cap \mathfrak{M}) \leq \mathbf{O}_{p^*}(\mathfrak{M})$.

b) *If $\mathbf{F}^*(\mathfrak{H})$ is not a p-group, $\mathbf{O}_p(\mathfrak{M}) \leq \mathfrak{H}$.*

c) $[\mathbf{O}_p(\mathfrak{M}), \mathbf{O}_{p^*}(\mathfrak{H})] = 1$.

Proof. Let $\mathfrak{U} = \mathfrak{M} \cap \mathbf{F}^*(\mathfrak{H})$, so $\mathfrak{U} \geq \mathbf{E}(\mathfrak{H})$ and $\mathfrak{U} \geq \mathbf{C}_{\mathbf{F}(\mathfrak{H})}(\mathfrak{U} \cap \mathbf{F}(\mathfrak{H}))$. Thus by 15.1, $\mathfrak{U}$ is subnormal in $\mathfrak{H}$, so

(1) $$\mathbf{O}_q(\mathfrak{U}) = \mathfrak{U} \cap \mathbf{O}_q(\mathfrak{H}) \text{ for every prime } q.$$

We show that

(2) $$\text{if } \mathbf{O}_q(\mathfrak{H}) \neq 1, \text{ then } \mathbf{C}_{\mathfrak{M}}(\mathbf{O}_q(\mathfrak{U})) \leq \mathfrak{H}.$$

Indeed, we have

$$\mathbf{Z}(\mathbf{O}_q(\mathfrak{H})) \leq \mathbf{C}_{\mathbf{O}_q(\mathfrak{H})}(\mathfrak{U} \cap \mathbf{F}(\mathfrak{H})) \leq \mathfrak{U} \cap \mathbf{O}_q(\mathfrak{H}) = \mathbf{O}_q(\mathfrak{U}),$$

so

$$\mathbf{C}_{\mathfrak{M}}(\mathbf{O}_q(\mathfrak{U})) \leq \mathbf{C}_{\mathfrak{M}}(\mathbf{Z}(\mathbf{O}_q(\mathfrak{H}))) \leq \mathbf{N}_{\mathfrak{M}}(\mathbf{Z}(\mathbf{O}_q(\mathfrak{H}))).$$

But $\mathbf{Z}(\mathbf{O}_q(\mathfrak{H})) \neq 1$, so $\mathbf{N}_{\mathfrak{M}}(\mathbf{Z}(\mathbf{O}_q(\mathfrak{H}))) \leq \mathfrak{H}$ by 15.4 (iii), and (2) follows at once.

a) Since $\mathfrak{U} \trianglelefteq\trianglelefteq \mathfrak{H} \cap \mathfrak{M}$, it follows from (1) that

$$\mathfrak{M} \cap \mathbf{O}_{p'}(\mathbf{F}(\mathfrak{H})) = \mathfrak{U} \cap \mathbf{O}_{p'}(\mathbf{F}(\mathfrak{H})) = \mathbf{O}_{p'}(\mathbf{F}(\mathfrak{U})) \leq \mathbf{O}_{p^*}(\mathfrak{H} \cap \mathfrak{M}).$$

Also $\mathbf{E}(\mathfrak{H}) \trianglelefteq\trianglelefteq \mathfrak{H} \cap \mathfrak{M}$, so $\mathbf{E}(\mathfrak{H}) \leq \mathbf{O}_{p^*}(\mathfrak{H} \cap \mathfrak{M})$.

Since $\mathbf{O}_p(\mathfrak{U}) \trianglelefteq\trianglelefteq \mathfrak{H} \cap \mathfrak{M}$, $\mathbf{O}_{p^*}(\mathfrak{H} \cap \mathfrak{M}) \leq \mathbf{C}_{\mathfrak{M}}(\mathbf{O}_p(\mathfrak{U}))$, by 14.4. But by (2), $\mathbf{C}_{\mathfrak{M}}(\mathbf{O}_p(\mathfrak{U})) \leq \mathfrak{H} \cap \mathfrak{M}$, so

$$\mathbf{O}_{p^*}(\mathfrak{H} \cap \mathfrak{M}) \trianglelefteq \mathbf{C}_{\mathfrak{M}}(\mathbf{O}_p(\mathfrak{U}))$$

and $\mathbf{O}_{p^*}(\mathfrak{H} \cap \mathfrak{M}) \leq \mathbf{O}_{p^*}(\mathbf{C}_{\mathfrak{M}}(\mathbf{O}_p(\mathfrak{U})))$. Using 14.13, it follows that

$$\mathbf{O}_{p^*}(\mathfrak{H} \cap \mathfrak{M}) \leq \mathbf{O}_{p^*}(\mathfrak{M}).$$

b) By 14.4, $\mathbf{O}_p(\mathfrak{M}) \leq \mathbf{C}_{\mathfrak{M}}(\mathbf{O}_{p^*}(\mathfrak{M}))$. Using a),

$$\mathbf{O}_p(\mathfrak{M}) \leq \mathbf{C}_{\mathfrak{M}}(\mathbf{E}(\mathfrak{H})(\mathfrak{M} \cap \mathbf{O}_{p'}(\mathbf{F}(\mathfrak{H})))).$$

If $\mathbf{E}(\mathfrak{H}) \neq 1$, $\mathbf{C}_{\mathfrak{M}}(\mathbf{E}(\mathfrak{H})) \leq \mathfrak{H}$ by 15.4(iii), so $\mathbf{O}_p(\mathfrak{M}) \leq \mathfrak{H}$, as asserted. Suppose then that $\mathbf{E}(\mathfrak{H}) = 1$. Thus $\mathbf{F}^*(\mathfrak{H}) = \mathbf{F}(\mathfrak{H})$. Thus by hypothesis, $\mathbf{F}(\mathfrak{H})$ is not a p-group. Let q be a prime such that $q \neq p$ and $\mathbf{O}_q(\mathfrak{H}) \neq 1$. Then $\mathbf{O}_p(\mathfrak{M})$ centralizes

$$\mathfrak{M} \cap \mathbf{O}_q(\mathfrak{H}) = \mathfrak{U} \cap \mathbf{O}_q(\mathfrak{H}) = \mathbf{O}_q(\mathfrak{U}).$$

Thus $\mathbf{O}_p(\mathfrak{M}) \leq \mathbf{C}_{\mathfrak{M}}(\mathbf{O}_q(\mathfrak{U})) \leq \mathfrak{H}$, by (2).

c) If $\mathbf{F}^*(\mathfrak{H})$ is a p-group, then by 15.5, $\mathbf{O}_{p^*}(\mathfrak{H}) = 1$ and the assertion is obvious. Otherwise $\mathbf{O}_p(\mathfrak{M}) \leq \mathfrak{H}$ by b). By 15.3a),

$$[\mathbf{O}_p(\mathfrak{M}), \mathbf{O}_{p^*}(\mathfrak{H})] = 1. \qquad \textbf{q.e.d.}$$

15.7 Theorem. *Suppose that $\mathfrak{H}$, $\mathfrak{M}$ are distinct subgroups of $\mathfrak{G}$, that $\mathfrak{H} \to \mathfrak{M}$ and that $\mathfrak{M} \to \mathfrak{H}$. Then there exists a prime p for which $\mathbf{F}^*(\mathfrak{H})$ and $\mathbf{F}^*(\mathfrak{M})$ are p-groups.*

Proof. Observe first that if $\mathfrak{X}$ is a characteristic subgroup of both $\mathfrak{H}$ and $\mathfrak{M}$, then $\mathfrak{X} = 1$. For otherwise $\mathbf{N}_{\mathfrak{M}}(\mathfrak{X}) \leq \mathfrak{H}$ since $\mathfrak{H} \to \mathfrak{M}$, but also $\mathfrak{M} \leq \mathbf{N}_{\mathfrak{G}}(\mathfrak{X})$, so $\mathfrak{M} \leq \mathfrak{H}$. By symmetry, $\mathfrak{H} \leq \mathfrak{M}$, so $\mathfrak{H} = \mathfrak{M}$, contrary to hypothesis.

Now $\mathbf{E}(\mathfrak{H}) \leq \mathfrak{M}$ and $\mathbf{E}(\mathfrak{M}) \leq \mathfrak{H}$. Hence by 15.3c), $\mathbf{E}(\mathfrak{H}) = \mathbf{E}(\mathfrak{M})$. By the above remark, $\mathbf{E}(\mathfrak{H}) = \mathbf{E}(\mathfrak{M}) = 1$.

Let p be a prime divisor of $|\mathbf{F}(\mathfrak{M})|$. Then $\mathbf{O}_p(\mathfrak{M}) \neq 1$ and

$$1 \neq \mathbf{Z}(\mathbf{O}_p(\mathfrak{M})) \leq \mathbf{O}_p(\mathfrak{M}) \cap \mathbf{C}_{\mathbf{F}(\mathfrak{M})}(\mathfrak{H} \cap \mathbf{F}(\mathfrak{M})) \leq \mathbf{O}_p(\mathfrak{M}) \cap \mathfrak{H}.$$

Hence $\mathbf{C}_{\mathfrak{H}}(\mathbf{O}_p(\mathfrak{M}) \cap \mathfrak{H}) \leq \mathbf{C}_{\mathfrak{H}}(\mathbf{Z}(\mathbf{O}_p(\mathfrak{M}))) \leq \mathfrak{M}$. But by 15.3a), $\mathbf{O}_{p^*}(\mathfrak{H}) \leq \mathbf{C}_{\mathfrak{H}}(\mathbf{O}_p(\mathfrak{M}) \cap \mathfrak{H})$, so $\mathbf{O}_{p^*}(\mathfrak{H}) \leq \mathfrak{M}$. Thus $\mathbf{O}_{p^*}(\mathfrak{H}) \trianglelefteq \mathfrak{H} \cap \mathfrak{M}$, so $\mathbf{O}_{p^*}(\mathfrak{H}) \leq \mathbf{O}_{p^*}(\mathfrak{H} \cap \mathfrak{M})$. But by 15.3b), $\mathbf{O}_p(\mathfrak{H}) \neq 1$, so by 15.6, $\mathbf{O}_{p^*}(\mathfrak{H} \cap \mathfrak{M}) \leq \mathbf{O}_{p^*}(\mathfrak{M})$. Thus $\mathbf{O}_{p^*}(\mathfrak{H}) \leq \mathbf{O}_{p^*}(\mathfrak{M})$. And since $\mathbf{O}_p(\mathfrak{H}) \neq 1$, it follows by symmetry that $\mathbf{O}_{p^*}(\mathfrak{M}) \leq \mathbf{O}_{p^*}(\mathfrak{H})$. Hence $\mathbf{O}_{p^*}(\mathfrak{M}) = \mathbf{O}_{p^*}(\mathfrak{H})$, and by the above remark, $\mathbf{O}_{p^*}(\mathfrak{M}) = \mathbf{O}_{p^*}(\mathfrak{H}) = 1$. By 15.5, $\mathbf{F}^*(\mathfrak{H})$ and $\mathbf{F}^*(\mathfrak{M})$ are p-groups. **q.e.d.**

15.8 Corollary. *Suppose that $\mathfrak{H}$, $\mathfrak{M}$ are distinct maximal subgroups of the simple non-Abelian group $\mathfrak{G}$. If $\mathbf{F}^*(\mathfrak{H}) \leq \mathfrak{M}$ and $\mathbf{F}^*(\mathfrak{M}) \leq \mathfrak{H}$, there exists a prime p such that $\mathbf{F}^*(\mathfrak{H})$, $\mathbf{F}^*(\mathfrak{M})$ are p-groups.*

Proof. We have $\mathbf{E}(\mathfrak{H}) \leq \mathbf{F}^*(\mathfrak{H}) \leq \mathfrak{M}$ and

$$\mathbf{C}_{\mathbf{F}(\mathfrak{H})}(\mathfrak{M} \cap \mathbf{F}(\mathfrak{H})) \leq \mathbf{F}(\mathfrak{H}) \leq \mathfrak{M}.$$

If $\mathfrak{X}$ is a non-identity normal subgroup of $\mathfrak{H}$, $\mathfrak{H} = \mathbf{N}_{\mathfrak{G}}(\mathfrak{X})$ since $\mathfrak{G}$ is simple, so $\mathbf{N}_{\mathfrak{M}}(\mathfrak{X}) \leq \mathfrak{H}$. Thus $\mathfrak{H} \to \mathfrak{M}$. Similarly $\mathfrak{M} \to \mathfrak{H}$. The assertion thus follows from 15.7. **q.e.d.**

15.9 Theorem. *Suppose that $\mathfrak{H}$, $\mathfrak{M}$ are distinct maximal subgroups of the simple non-Abelian group $\mathfrak{G}$. Suppose that there exists a subnormal subgroup $\mathfrak{A}$ of $\mathbf{F}^*(\mathfrak{H})$ such that $\mathbf{C}_{\mathbf{F}^*(\mathfrak{H})}(\mathfrak{A}) \leq \mathfrak{A}$ and $\mathfrak{A} \leq \mathfrak{M}$. Symmetrically, suppose that there exists a subnormal subgroup $\mathfrak{B}$ of $\mathbf{F}^*(\mathfrak{M})$ such that $\mathbf{C}_{\mathbf{F}^*(\mathfrak{M})}(\mathfrak{B}) \leq \mathfrak{B}$ and $\mathfrak{B} \leq \mathfrak{H}$. Then there exists a prime p such that $\mathbf{F}^*(\mathfrak{H})$ and $\mathbf{F}^*(\mathfrak{M})$ are p-groups.*

Proof. By 15.1, $\mathbf{E}(\mathfrak{H}) \leq \mathfrak{A}$ and $\mathbf{C}_{\mathbf{F}(\mathfrak{H})}(\mathfrak{A} \cap \mathbf{F}(\mathfrak{H})) \leq \mathfrak{A}$. Thus $\mathbf{E}(\mathfrak{H}) \leq \mathfrak{M}$ and

$$\mathbf{C}_{\mathbf{F}(\mathfrak{H})}(\mathfrak{M} \cap \mathbf{F}(\mathfrak{H})) \leq \mathbf{C}_{\mathbf{F}(\mathfrak{H})}(\mathfrak{A} \cap \mathbf{F}(\mathfrak{H})) \leq \mathfrak{A} \leq \mathfrak{M}.$$

If $\mathfrak{X}$ is a non-identity normal subgroup of $\mathfrak{H}$, then $\mathbf{N}_{\mathfrak{G}}(\mathfrak{X}) = \mathfrak{H}$ since $\mathfrak{G}$ is simple, and $\mathbf{N}_{\mathfrak{M}}(\mathfrak{X}) \leq \mathfrak{H}$. Hence $\mathfrak{H} \to \mathfrak{M}$. Similarly $\mathfrak{M} \to \mathfrak{H}$. By 15.7, there exists a prime p for which $\mathbf{F}^*(\mathfrak{H})$ and $\mathbf{F}^*(\mathfrak{M})$ are p-groups. **q.e.d.**

15.10 Remark. These theorems are used in BENDER [4] and GOLDSCHMIDT [4]. As an illustration, consider the situation in 2.10, where $\mathfrak{G}$ is a simple group of order $p^a q^b$ and all proper subgroups of $\mathfrak{G}$ are soluble. Let $\mathfrak{H}$ be a maximal subgroup of $\mathfrak{G}$ for which $|\mathbf{F}(\mathfrak{H})|$ is divisible by both p and q. Suppose that $1 \neq \mathfrak{X} \leq \mathbf{F}(\mathfrak{H})$. Let $\mathfrak{M}$ be a maximal subgroup of $\mathfrak{G}$ containing $\mathbf{N}_{\mathfrak{G}}(\mathfrak{X})$. Then $\mathfrak{H} \to \mathfrak{M}$. It follows from 15.6 that $\mathbf{O}_p(\mathfrak{M}) \leq \mathfrak{H}$ and $\mathbf{O}_q(\mathfrak{M}) \leq \mathfrak{H}$; thus $\mathbf{F}^*(\mathfrak{M}) \leq \mathfrak{H}$. Hence $\mathfrak{M} \to \mathfrak{H}$ and by 15.7, $\mathfrak{M} = \mathfrak{H}$.

§ 16. Signalizer Functors and a Transitivity Theorem

In this section we make some remarks about the set of p'-subgroups normalised by an Abelian p-subgroup.

16.1 Definition. Suppose that $\mathfrak{A}$ is an Abelian p-subgroup of $\mathfrak{G}$. An $\mathfrak{A}$-*signalizer functor* on $\mathfrak{G}$ is a mapping $\mathbf{S}$, defined on the set $\{\mathbf{C}_{\mathfrak{G}}(a) | a \in \mathfrak{A}, a \neq 1\}$, such that $\mathbf{S}(\mathbf{C}_{\mathfrak{G}}(a))$ is an $\mathfrak{A}$-invariant p'-subgroup of $\mathbf{C}_{\mathfrak{G}}(a)$ and

$$(1) \qquad \mathbf{C}_{\mathfrak{G}}(a) \cap \mathbf{S}(\mathbf{C}_{\mathfrak{G}}(b)) = \mathbf{S}(\mathbf{C}_{\mathfrak{G}}(a)) \cap \mathbf{C}_{\mathfrak{G}}(b)$$

for all non-identity elements a, b of $\mathfrak{A}$. If $\mathbf{S}(\mathbf{C}_{\mathfrak{G}}(a))$ is soluble for every $a \in \mathfrak{A} - \{1\}$, we say that $\mathbf{S}$ is *soluble*.

It follows from (1) that $\mathbf{S}(\mathbf{C}_{\mathfrak{G}}(a))$ depends only on $\mathbf{C}_{\mathfrak{G}}(a)$ rather than on a.

16.2 Examples. a) If $\mathfrak{A}$ is an Abelian p-subgroup of $\mathfrak{G}$, $\mathfrak{N}$ is an $\mathfrak{A}$-invariant p'-subgroup and $\mathbf{S}(\mathbf{C}_{\mathfrak{G}}(a)) = \mathfrak{N} \cap \mathbf{C}_{\mathfrak{G}}(a)$ for all $a \in \mathfrak{A} - \{1\}$, then $\mathbf{S}$ is an $\mathfrak{A}$-signalizer functor on $\mathfrak{G}$. If $\mathfrak{N}$ is soluble, $\mathbf{S}$ is soluble.

b) If $\mathfrak{A}$ is an Abelian p-subgroup of $\mathfrak{G}$ and $\mathbf{C}_{\mathfrak{G}}(a)$ is p-soluble for every $a \in \mathfrak{A} - \{1\}$ then $\mathbf{O}_{p'}$ is an $\mathfrak{A}$-signalizer functor. For if a, b are non-identity elements of $\mathfrak{A}$ and $\mathfrak{H} = \mathbf{C}_{\mathfrak{G}}(a)$, then $b \in \mathfrak{A} \leq \mathfrak{H}$ and

$$\begin{aligned} \mathbf{C}_{\mathfrak{G}}(a) \cap \mathbf{O}_{p'}(\mathbf{C}_{\mathfrak{G}}(b)) \leq \mathbf{O}_{p'}(\mathbf{C}_{\mathfrak{H}}(b)) &\leq \mathbf{O}_{p'}(\mathfrak{H}) \cap \mathbf{C}_{\mathfrak{G}}(b) \\ &= \mathbf{O}_{p'}(\mathbf{C}_{\mathfrak{G}}(a)) \cap \mathbf{C}_{\mathfrak{G}}(b), \end{aligned}$$

by 1.6. By symmetry it follows that

$$\mathbf{C}_{\mathfrak{G}}(a) \cap \mathbf{O}_{p'}(\mathbf{C}_{\mathfrak{G}}(b)) = \mathbf{O}_{p'}(\mathbf{C}_{\mathfrak{G}}(a)) \cap \mathbf{C}_{\mathfrak{G}}(b).$$

16.3 Definitions. Let $\mathbf{S}$ be a soluble $\mathfrak{A}$-signalizer functor on $\mathfrak{G}$.

a) $\mathbf{S}$ is called *complete* if there exists a soluble $\mathfrak{A}$-invariant p'-subgroup $\mathfrak{N}$ such that $\mathbf{S}(\mathbf{C}_{\mathfrak{G}}(a)) = \mathfrak{N} \cap \mathbf{C}_{\mathfrak{G}}(a)$ for all $a \in \mathfrak{A} - \{1\}$.

b) Denote by $\mathsf{M}_{\mathbf{S}}(\mathfrak{A})$ the set of all soluble $\mathfrak{A}$-invariant p'-subgroups $\mathfrak{M}$ of $\mathfrak{G}$ for which $\mathfrak{M} \cap \mathbf{C}_{\mathfrak{G}}(a) \leq \mathbf{S}(\mathbf{C}_{\mathfrak{G}}(a))$ for all $a \in \mathfrak{A} - \{1\}$. Thus any $\mathfrak{A}$-invariant subgroup of an element of $\mathsf{M}_{\mathbf{S}}(\mathfrak{A})$ also lies in $\mathsf{M}_{\mathbf{S}}(\mathfrak{A})$.

By 16.1, $\mathbf{S}(\mathbf{C}_{\mathfrak{G}}(b)) \in \mathsf{M}_{\mathbf{S}}(\mathfrak{A})$ for all $b \in \mathfrak{A} - \{1\}$. In fact, $\mathbf{S}(\mathbf{C}_{\mathfrak{G}}(b))$ is the unique maximal element of $\{\mathfrak{X} | \mathfrak{X} \in \mathsf{M}_{\mathbf{S}}(\mathfrak{A}), \mathfrak{X} \leq \mathbf{C}_{\mathfrak{G}}(b)\}$, so

$$\mathbf{S}(\mathbf{C}_{\mathfrak{G}}(b)) = \langle \mathfrak{X} | \mathfrak{X} \in \mathsf{M}_{\mathbf{S}}(\mathfrak{A}), \mathfrak{X} \leq \mathbf{C}_{\mathfrak{G}}(b) \rangle.$$

We may therefore define

$$\mathbf{S}(\mathfrak{H}) = \langle \mathfrak{Y} \mid \mathfrak{Y} \in \mathsf{N}_{\mathbf{S}}(\mathfrak{A}),\ \mathfrak{Y} \leq \mathfrak{H} \rangle$$

for any subgroup $\mathfrak{H}$ of $\mathfrak{G}$. Thus if $\mathfrak{H} \leq \mathfrak{K}$, $\mathbf{S}(\mathfrak{H}) \leq \mathbf{S}(\mathfrak{K})$. In particular,

$$\mathbf{S}(\mathfrak{G}) = \langle \mathfrak{Z} \mid \mathfrak{Z} \in \mathsf{N}_{\mathbf{S}}(\mathfrak{A}) \rangle.$$

16.4 Lemma. *Let* $\mathbf{S}$ *be a soluble* $\mathfrak{A}$*-signalizer functor on* $\mathfrak{G}$.

a) *If* $\mathfrak{H}$ *is an* $\mathfrak{A}$*-invariant subgroup of* $\mathfrak{G}$, $\mathbf{S}(\mathbf{C}_{\mathfrak{G}}(a)) \cap \mathfrak{H} \leq \mathbf{S}(\mathfrak{H})$ *for all* $a \in \mathfrak{A} - \{1\}$.

b) *If* $\mathfrak{H}, \mathfrak{K}$ *are* $\mathfrak{A}$*-invariant and* $\mathbf{S}(\mathfrak{H}), \mathbf{S}(\mathfrak{K})$ *are in* $\mathsf{N}_{\mathbf{S}}(\mathfrak{A})$, *then*

$$\mathfrak{H} \cap \mathbf{S}(\mathfrak{K}) = \mathbf{S}(\mathfrak{H}) \cap \mathfrak{K} = \mathbf{S}(\mathfrak{H}) \cap \mathbf{S}(\mathfrak{K}).$$

In particular, for all $a \in \mathfrak{A} - \{1\}$,

$$\mathfrak{H} \cap \mathbf{S}(\mathbf{C}_{\mathfrak{G}}(a)) = \mathbf{S}(\mathfrak{H}) \cap \mathbf{C}_{\mathfrak{G}}(a).$$

c) *If* $1 \neq \mathfrak{B} \leq \mathfrak{A}$, $\mathbf{S}(\mathbf{C}_{\mathfrak{G}}(\mathfrak{B})) = \bigcap_{b \in \mathfrak{B}-\{1\}} \mathbf{S}(\mathbf{C}_{\mathfrak{G}}(b))$.

d) *If* $\mathfrak{X} \in \mathsf{N}_{\mathbf{S}}(\mathfrak{A})$, $\mathfrak{Y} \in \mathsf{N}_{\mathbf{S}}(\mathfrak{A})$ *and* $\mathfrak{X} \leq \mathbf{N}_{\mathfrak{G}}(\mathfrak{Y})$, *then* $\mathfrak{X}\mathfrak{Y} \in \mathsf{N}_{\mathbf{S}}(\mathfrak{A})$.

Proof. a) Since $\mathbf{S}(\mathbf{C}_{\mathfrak{G}}(a)) \in \mathsf{N}_{\mathbf{S}}(\mathfrak{A})$, $\mathbf{S}(\mathbf{C}_{\mathfrak{G}}(a)) \cap \mathfrak{H} \in \mathsf{N}_{\mathbf{S}}(\mathfrak{A})$ and so $\mathbf{S}(\mathbf{C}_{\mathfrak{G}}(a)) \cap \mathfrak{H} \leq \mathbf{S}(\mathfrak{H})$.

b) Since $\mathbf{S}(\mathfrak{H}) \in \mathsf{N}_{\mathbf{S}}(\mathfrak{A})$, $\mathbf{S}(\mathfrak{H}) \cap \mathfrak{K} \leq \mathbf{S}(\mathfrak{K})$. Similarly, $\mathfrak{H} \cap \mathbf{S}(\mathfrak{K}) \leq \mathbf{S}(\mathfrak{H})$, so $\mathfrak{H} \cap \mathbf{S}(\mathfrak{K}) = \mathbf{S}(\mathfrak{H}) \cap \mathfrak{K}$. Since $\mathbf{S}(\mathbf{C}_{\mathfrak{G}}(a)) \in \mathsf{N}_{\mathbf{S}}(\mathfrak{A})$ for all $a \in \mathfrak{A} - \{1\}$, the other assertion follows.

c) Let $\mathfrak{D} = \bigcap_{b \in \mathfrak{B}-\{1\}} \mathbf{S}(\mathbf{C}_{\mathfrak{G}}(b))$. Then $\mathfrak{D} \in \mathsf{N}_{\mathbf{S}}(\mathfrak{A})$, so $\mathfrak{D} \leq \mathbf{S}(\mathbf{C}_{\mathfrak{G}}(\mathfrak{B}))$. And $\mathbf{C}_{\mathfrak{G}}(\mathfrak{B}) \leq \mathbf{C}_{\mathfrak{G}}(b)$ for all $b \in \mathfrak{B} - \{1\}$, so $\mathbf{S}(\mathbf{C}_{\mathfrak{G}}(\mathfrak{B})) \leq \mathbf{S}(\mathbf{C}_{\mathfrak{G}}(b))$ and $\mathbf{S}(\mathbf{C}_{\mathfrak{G}}(\mathfrak{B})) \leq \mathfrak{D}$.

d) Clearly, $\mathfrak{X}\mathfrak{Y}$ is a soluble $\mathfrak{A}$-invariant p'-subgroup of $\mathfrak{G}$. If $x \in \mathfrak{X}$, $y \in \mathfrak{Y}$ and $xy \in \mathbf{C}_{\mathfrak{G}}(a)$, then

$$[a, x] = a^{-1}a^{x} = a^{-1}a^{xyy^{-1}} = a^{-1}a^{y^{-1}} = [a, y^{-1}] \in \mathfrak{X} \cap \mathfrak{Y} \trianglelefteq \mathfrak{X}.$$

Hence $x(\mathfrak{X} \cap \mathfrak{Y}) \in \mathbf{C}_{\mathfrak{X}/(\mathfrak{X} \cap \mathfrak{Y})}(a) = \mathbf{C}_{\mathfrak{X}}(a)(\mathfrak{X} \cap \mathfrak{Y})/(\mathfrak{X} \cap \mathfrak{Y})$, by IX, 6.11. Thus $x = x_1 z$, where $x_1 \in \mathbf{C}_{\mathfrak{X}}(a)$ and $z \in \mathfrak{X} \cap \mathfrak{Y}$. Then $zy = x_1^{-1}xy \in \mathbf{C}_{\mathfrak{G}}(a) \cap \mathfrak{Y} = \mathbf{C}_{\mathfrak{Y}}(a)$, so $xy = x_1 zy \in \mathbf{C}_{\mathfrak{X}}(a)\mathbf{C}_{\mathfrak{Y}}(a)$. But since $\mathfrak{X}$ and $\mathfrak{Y}$ lie in $\mathsf{N}_{\mathbf{S}}(\mathfrak{A})$, $\mathbf{C}_{\mathfrak{X}}(a)$ and $\mathbf{C}_{\mathfrak{Y}}(a)$ are in $\mathbf{S}(\mathbf{C}_{\mathfrak{G}}(a))$. Then

$$\mathbf{C}_{\mathfrak{X}\mathfrak{Y}}(a) \leq \mathbf{S}(\mathbf{C}_{\mathfrak{G}}(a)),$$

and $\mathfrak{X}\mathfrak{Y} \in \mathsf{N}_{\mathbf{S}}(\mathfrak{A})$. **q.e.d.**

16.5 Lemma. *Let* $\mathbf{S}$ *be a soluble* $\mathfrak{A}$*-signalizer functor on* $\mathfrak{G}$ *and suppose that* $\mathfrak{A}$ *is non-cyclic.*

a) *If* $\mathfrak{A}_0$ *is a non-cyclic subgroup of* $\mathfrak{A}$ *and* $\mathfrak{H}$ *is an* $\mathfrak{A}$*-invariant subgroup of* $\mathfrak{G}$,

$$\mathbf{S}(\mathfrak{H}) = \langle \mathfrak{H} \cap \mathbf{S}(\mathbf{C}_{\mathfrak{G}}(\mathfrak{B})) | \mathfrak{B} < \mathfrak{A}_0, \mathfrak{A}_0/\mathfrak{B} \textit{ is cyclic}\rangle$$
$$= \langle \mathfrak{H} \cap \mathbf{S}(\mathbf{C}_{\mathfrak{G}}(a)) | a \in \mathfrak{A}_0 - \{1\}\rangle.$$

b) *The following assertions are equivalent.*

(i) $\mathbf{S}$ *is complete.*

(ii) $\mathbf{S}(\mathfrak{G}) \in \mathcal{V}_{\mathbf{S}}(\mathfrak{A})$.

(iii) $\mathcal{V}_{\mathbf{S}}(\mathfrak{A})$ *has a unique maximal element.*

Proof. a) Let

$$\mathfrak{S} = \langle \mathfrak{H} \cap \mathbf{S}(\mathbf{C}_{\mathfrak{G}}(\mathfrak{B})) | \mathfrak{B} < \mathfrak{A}_0, \mathfrak{U}_0/\mathfrak{B} \text{ is cyclic}\rangle.$$

Since $\mathbf{S}(\mathbf{C}_{\mathfrak{G}}(\mathfrak{B})) = \bigcap_{b \in \mathfrak{B}-\{1\}} \mathbf{S}(\mathbf{C}_{\mathfrak{G}}(b))$ and $\mathbf{S}(\mathbf{C}_{\mathfrak{G}}(a)) \in \mathcal{V}_{\mathbf{S}}(\mathfrak{A})$,

$$\mathfrak{S} \leq \langle \mathfrak{H} \cap \mathbf{S}(\mathbf{C}_{\mathfrak{G}}(a)) | a \in \mathfrak{A}_0 - \{1\}\rangle \leq \mathbf{S}(\mathfrak{H}).$$

Suppose that $\mathfrak{Z} \in \mathcal{V}_{\mathbf{S}}(\mathfrak{A})$ and $\mathfrak{Z} \leq \mathfrak{H}$. By 1.9,

$$\mathfrak{Z} = \langle \mathbf{C}_{\mathfrak{Z}}(\mathfrak{B}) | \mathfrak{B} < \mathfrak{A}_0, \mathfrak{A}_0/\mathfrak{B} \text{ is cyclic}\rangle.$$

But by 16.4,

$$\mathbf{C}_{\mathfrak{Z}}(\mathfrak{B}) = \mathfrak{Z} \cap \mathbf{C}_{\mathfrak{G}}(\mathfrak{B}) = \mathbf{S}(\mathfrak{Z}) \cap \mathbf{C}_{\mathfrak{G}}(\mathfrak{B})$$
$$= \mathfrak{Z} \cap \mathbf{S}(\mathbf{C}_{\mathfrak{G}}(\mathfrak{B})) \leq \mathfrak{H} \cap \mathbf{S}(\mathbf{C}_{\mathfrak{G}}(\mathfrak{B})),$$

so $\mathfrak{Z} \leq \mathfrak{S}$. Hence $\mathbf{S}(\mathfrak{H}) \leq \mathfrak{S}$.

b) (i) $\Rightarrow$ (ii): If $\mathbf{S}$ is complete, $\mathbf{S}(\mathbf{C}_{\mathfrak{G}}(a)) = \mathfrak{N} \cap \mathbf{C}_{\mathfrak{G}}(a)$ for some soluble $\mathfrak{A}$-invariant p'-subgroup $\mathfrak{N}$ of $\mathfrak{G}$. Thus $\mathbf{S}(\mathfrak{G}) \leq \mathfrak{N}$, by a). But also $\mathfrak{N} \in \mathcal{V}_{\mathbf{S}}(\mathfrak{A})$, so $\mathfrak{N} \leq \mathbf{S}(\mathfrak{G})$. Thus $\mathbf{S}(\mathfrak{G}) = \mathfrak{N} \in \mathcal{V}_{\mathbf{S}}(\mathfrak{A})$.

(ii) $\Rightarrow$ (iii): If $\mathbf{S}(\mathfrak{G}) \in \mathcal{V}_{\mathbf{S}}(\mathfrak{A})$, $\mathbf{S}(\mathfrak{G})$ is the unique maximal element of $\mathcal{V}_{\mathbf{S}}(\mathfrak{A})$.

(iii) $\Rightarrow$ (i): Suppose that $\mathfrak{M}$ is the unique maximal element of $\mathcal{V}_{\mathbf{S}}(\mathfrak{A})$. Since $\mathbf{S}(\mathbf{C}_{\mathfrak{G}}(a)) \in \mathcal{V}_{\mathbf{S}}(\mathfrak{A})$, $\mathbf{S}(\mathbf{C}_{\mathfrak{G}}(a)) \leq \mathfrak{M} \cap \mathbf{C}_{\mathfrak{G}}(a)$. Since $\mathfrak{M} \in \mathcal{V}_{\mathbf{S}}(\mathfrak{A})$, $\mathfrak{M} \cap \mathbf{C}_{\mathfrak{G}}(a) = \mathbf{S}(\mathbf{C}_{\mathfrak{G}}(a))$. Thus $\mathbf{S}$ is complete. **q.e.d.**

16.6 Lemma. *Suppose that* $\mathbf{S}$ *is an* $\mathfrak{A}$*-signalizer functor on* $\mathfrak{G}$.

a) *If* $\mathfrak{A} \leq \mathfrak{H} \leq \mathfrak{G}$ *and, for every* $a \in \mathfrak{A} - \{1\}$, *we put* $\mathbf{T}(\mathbf{C}_{\mathfrak{H}}(a)) = \mathfrak{H} \cap \mathbf{S}(\mathbf{C}_{\mathfrak{G}}(a))$, *then* $\mathbf{T}$ *is an* $\mathfrak{A}$*-signalizer functor on* $\mathfrak{H}$. *If* $\mathbf{S}$ *is soluble, then* $\mathbf{T}$ *is soluble,* $\mathsf{M}_{\mathbf{T}}(\mathfrak{A}) = \{\mathfrak{X} | \mathfrak{X} \in \mathsf{M}_{\mathbf{S}}(\mathfrak{A}), \mathfrak{X} \leq \mathfrak{H}\}$ *and*

$$\mathbf{T}(\mathfrak{H}) = \mathbf{S}(\mathfrak{H}).$$

b) *If* $\mathfrak{N}$ *is a normal* p'*-subgroup of* $\mathfrak{G}$ *and, for every* $a \in \mathfrak{A} - \{1\}$, *we put*

$$\overline{\mathbf{S}}(\mathbf{C}_{\mathfrak{G}/\mathfrak{N}}(a\mathfrak{N})) = \mathbf{S}(\mathbf{C}_{\mathfrak{G}}(a))\mathfrak{N}/\mathfrak{N},$$

then $\overline{\mathbf{S}}$ *is an* $\overline{\mathfrak{A}}$*-signalizer functor on* $\mathfrak{G}/\mathfrak{N}$, *where* $\overline{\mathfrak{A}} = \mathfrak{A}\mathfrak{N}/\mathfrak{N}$. *If* $\mathbf{S}$ *is soluble,* $\overline{\mathbf{S}}$ *is soluble.*

Proof. a) It is clear that $\mathbf{T}(\mathbf{C}_{\mathfrak{H}}(a))$ is an $\mathfrak{A}$-invariant p'-subgroup of $\mathbf{C}_{\mathfrak{H}}(a)$, and if a, b are non-identity elements of $\mathfrak{A}$,

$$\begin{aligned}\mathbf{C}_{\mathfrak{H}}(a) \cap \mathbf{T}(\mathbf{C}_{\mathfrak{H}}(b)) &= \mathfrak{H} \cap \mathbf{C}_{\mathfrak{G}}(a) \cap \mathbf{S}(\mathbf{C}_{\mathfrak{G}}(b))\\ &= \mathfrak{H} \cap \mathbf{S}(\mathbf{C}_{\mathfrak{G}}(a)) \cap \mathbf{C}_{\mathfrak{G}}(b)\\ &= \mathbf{T}(\mathbf{C}_{\mathfrak{H}}(a)) \cap \mathbf{C}_{\mathfrak{H}}(b).\end{aligned}$$

If $\mathbf{S}$ is soluble, so is $\mathbf{T}$. Let $\mathfrak{Z}$ be a soluble $\mathfrak{A}$-invariant p'-subgroup of $\mathfrak{G}$. Then $\mathfrak{Z} \in \mathsf{M}_{\mathbf{T}}(\mathfrak{A})$ if and only if $\mathfrak{Z} \leq \mathfrak{H}$ and $\mathbf{C}_{\mathfrak{Z}}(a) \leq \mathbf{T}(\mathbf{C}_{\mathfrak{H}}(a))$ for all $a \in \mathfrak{A} - \{1\}$. Thus $\mathfrak{Z} \in \mathsf{M}_{\mathbf{T}}(\mathfrak{A})$ if and only if $\mathfrak{Z} \leq \mathfrak{H}$ and $\mathbf{C}_{\mathfrak{Z}}(a) \leq \mathbf{S}(\mathbf{C}_{\mathfrak{G}}(a))$. Hence $\mathsf{M}_{\mathbf{T}}(\mathfrak{A}) = \{\mathfrak{Z} | \mathfrak{Z} \in \mathsf{M}_{\mathbf{S}}(\mathfrak{A}), \mathfrak{Z} \leq \mathfrak{H}\}$ and

$$\mathbf{S}(\mathfrak{H}) = \langle \mathfrak{Y} | \mathfrak{Y} \in \mathsf{M}_{\mathbf{T}}(\mathfrak{A}) \rangle = \mathbf{T}(\mathfrak{H}).$$

b) $\overline{\mathbf{S}}$ is well-defined since $\mathfrak{A} \cap \mathfrak{N} = 1$. It is clear that $\overline{\mathbf{S}}(\mathbf{C}_{\mathfrak{G}/\mathfrak{N}}(a\mathfrak{N}))$ is an $\overline{\mathfrak{A}}$-invariant p'-subgroup of $\mathbf{C}_{\mathfrak{G}/\mathfrak{N}}(a\mathfrak{N})$ for every $a \in \mathfrak{A} - \{1\}$.

Suppose that a, b are non-identity elements of $\mathfrak{A}$, and put

$$\mathfrak{K}/\mathfrak{N} = \mathbf{C}_{\mathfrak{G}/\mathfrak{N}}(a\mathfrak{N}) \cap \overline{\mathbf{S}}(\mathbf{C}_{\mathfrak{G}/\mathfrak{N}}(b\mathfrak{N})).$$

Thus $\mathfrak{K} \leq \mathbf{S}(\mathbf{C}_{\mathfrak{G}}(b))\mathfrak{N}$, so $\mathfrak{K} = \mathfrak{L}\mathfrak{N}$, where $\mathfrak{L} = \mathfrak{K} \cap \mathbf{S}(\mathbf{C}_{\mathfrak{G}}(b))$. Since $\mathbf{C}_{\mathfrak{G}/\mathfrak{N}}(a\mathfrak{N}) \geq \mathfrak{A}\mathfrak{N}/\mathfrak{N}$, $\mathfrak{K}$ is $\mathfrak{A}$-invariant; thus $\mathfrak{L}$ is $\mathfrak{A}$-invariant and $[\mathfrak{L}, \mathfrak{A}] \leq \mathfrak{L}$. But also, $\mathfrak{L}\mathfrak{N}/\mathfrak{N} \leq \mathbf{C}_{\mathfrak{G}/\mathfrak{N}}(a\mathfrak{N})$, so $[\mathfrak{L}, a] \leq \mathfrak{L} \cap \mathfrak{N}$. Now $\mathfrak{N}$ is a p'-group, so $\mathfrak{K}$ and $\mathfrak{L}$ are p'-groups. By IX, 6.11,

$$\mathbf{C}_{\mathfrak{L}}(a)(\mathfrak{L} \cap \mathfrak{N})/(\mathfrak{L} \cap \mathfrak{N}) = \mathbf{C}_{\mathfrak{L}/(\mathfrak{L} \cap \mathfrak{N})}(a) = \mathfrak{L}/(\mathfrak{L} \cap \mathfrak{N}),$$

and

$$\mathfrak{L} = \mathbf{C}_{\mathfrak{L}}(a)(\mathfrak{L} \cap \mathfrak{N}) \leq (\mathbf{C}_{\mathfrak{G}}(a) \cap \mathbf{S}(\mathbf{C}_{\mathfrak{G}}(b)))\mathfrak{N}$$
$$= (\mathbf{S}(\mathbf{C}_{\mathfrak{G}}(a)) \cap \mathbf{C}_{\mathfrak{G}}(b))\mathfrak{N},$$

so since $\mathfrak{K} = \mathfrak{L}\mathfrak{N}$,

$$\mathfrak{K}/\mathfrak{N} \leq \overline{\mathbf{S}}(\mathbf{C}_{\mathfrak{G}/\mathfrak{N}}(a\mathfrak{N})) \cap \mathbf{C}_{\mathfrak{G}/\mathfrak{N}}(b\mathfrak{N}).$$

Thus $\overline{\mathbf{S}}$ is an $\overline{\mathfrak{A}}$-signalizer functor on $\mathfrak{G}/\mathfrak{N}$.

It is clear that if $\mathbf{S}$ is soluble, $\overline{\mathbf{S}}$ is soluble. **q.e.d.**

16.7 Theorem. *Suppose that $\mathfrak{G}$ is soluble and that* $\mathbf{S}$ *is a soluble $\mathfrak{A}$-signalizer functor on $\mathfrak{G}$. If $d(\mathfrak{A}) \geq 3$,* $\mathbf{S}$ *is complete.*

Proof. Suppose that this is false and let $\mathfrak{G}$ be a counterexample of minimal order.

(1) $\mathbf{S}(\mathfrak{G}) \trianglelefteq \mathfrak{G}$ and $\mathfrak{G} = \mathfrak{A}\mathbf{S}(\mathfrak{G})$.

We have $\mathfrak{A} \leq \mathbf{N}_{\mathfrak{G}}(\mathbf{S}(\mathfrak{G}))$, so let $\mathfrak{H} = \mathfrak{A}\mathbf{S}(\mathfrak{G})$ and define the $\mathfrak{A}$-signalizer functor $\mathbf{T}$ on $\mathfrak{H}$ as in 16.6a). Then $\mathscr{U}_{\mathbf{S}}(\mathfrak{A}) = \mathscr{U}_{\mathbf{T}}(\mathfrak{A})$. Since $\mathbf{S}$ is not complete, neither is $\mathbf{T}$. By minimality of $|\mathfrak{G}|$, $\mathfrak{H} = \mathfrak{G}$. Thus $\mathbf{S}(\mathfrak{G}) \trianglelefteq \mathfrak{G}$.

Let $\mathfrak{K}$ be a maximal normal subgroup of $\mathfrak{G}$ for which $\mathfrak{K} \in \mathscr{U}_{\mathbf{S}}(\mathfrak{A})$. Thus $\mathfrak{K} \leq \mathbf{S}(\mathfrak{G})$, but by 16.5b), $\mathbf{S}(\mathfrak{G}) \notin \mathscr{U}_{\mathbf{S}}(\mathfrak{A})$, so $\mathfrak{K} < \mathbf{S}(\mathfrak{G})$. We choose a minimal normal subgroup $\mathfrak{M}$ of $\mathfrak{G}$ for which $\mathfrak{K} < \mathbf{S}(\mathfrak{M})$.

(2) $\mathfrak{G} = \mathfrak{A}\mathfrak{M}$.

Suppose that this is false and put $\mathfrak{H} = \mathfrak{A}\mathfrak{M}$. If $\mathbf{T}$ is defined as in 16.6a), then $\mathbf{T}$ is complete, by minimality of $\mathfrak{G}$. Thus $\mathbf{T}(\mathfrak{H}) \in \mathscr{U}_{\mathbf{T}}(\mathfrak{A})$ and $\mathbf{S}(\mathfrak{H}) \in \mathscr{U}_{\mathbf{S}}(\mathfrak{A})$. But $\mathbf{S}(\mathfrak{H}) = \mathbf{S}(\mathfrak{M}) > \mathfrak{K}$, so by maximality of $\mathfrak{K}$, $\mathbf{S}(\mathfrak{M})$ is not normal in $\mathfrak{G}$. Since $\mathbf{S}(\mathfrak{M})$ is $\mathfrak{A}$-invariant, it follows from (1) that $\mathbf{S}(\mathfrak{M})$ is not normal in $\mathbf{S}(\mathfrak{G})$. By 16.5a), there is a subgroup $\mathfrak{B}$ of $\mathfrak{A}$ such that $\mathfrak{A}/\mathfrak{B}$ is cyclic and $\mathbf{S}(\mathbf{C}_{\mathfrak{G}}(\mathfrak{B})) \not\leq \mathbf{N}_{\mathfrak{G}}(\mathbf{S}(\mathfrak{M}))$. Since $d(\mathfrak{A}) \geq 3$, $d(\mathfrak{B}) \geq 2$. Also, since $\mathbf{S}(\mathfrak{M}) = \mathbf{S}(\mathfrak{H}) \in \mathscr{U}_{\mathbf{S}}(\mathfrak{A})$,

$$\mathbf{C}_{\mathfrak{G}}(b) \cap \mathbf{S}(\mathfrak{M}) = \mathbf{S}(\mathbf{C}_{\mathfrak{G}}(b)) \cap \mathfrak{M} \quad (b \in \mathfrak{B} - \{1\})$$

by 16.4b); hence by 1.9,

$$\mathbf{S}(\mathfrak{M}) = \langle \mathbf{C}_{\mathfrak{G}}(b) \cap \mathbf{S}(\mathfrak{M}) | b \in \mathfrak{B} - \{1\} \rangle$$
$$= \langle \mathbf{S}(\mathbf{C}_{\mathfrak{G}}(b)) \cap \mathfrak{M} | b \in \mathfrak{B} - \{1\} \rangle.$$

However $\mathbf{S}(\mathbf{C}_{\mathfrak{G}}(\mathfrak{B}))$ normalises all $\mathbf{S}(\mathbf{C}_{\mathfrak{G}}(b)) \cap \mathfrak{M}$, which gives a contradiction. Hence (2) holds.

Now let $\mathfrak{L}$ be a maximal normal subgroup of $\mathfrak{G}$ for which $\mathfrak{K} \leq \mathfrak{L} < \mathfrak{M}$.

(3) $\mathbf{S}(\mathfrak{L}) = \mathfrak{K}$.

This follows at once from the minimality of $\mathfrak{M}$.

(4) If $1 \neq \mathfrak{X} \leq \mathfrak{A}$ and $\mathbf{S}(\mathbf{C}_{\mathfrak{G}}(\mathfrak{X})) \nleq \mathfrak{K}$,

$$\mathfrak{M} = \mathfrak{L}\mathbf{S}(\mathbf{C}_{\mathfrak{G}}(\mathfrak{X})).$$

Since $\mathfrak{M}/\mathfrak{L}$ is a chief factor of a soluble group, $\mathfrak{M}/\mathfrak{L}$ is Abelian; hence $\mathfrak{L}\mathbf{S}(\mathbf{C}_{\mathfrak{G}}(\mathfrak{X})) \trianglelefteq \mathfrak{M}$. It follows from (2) that $\mathfrak{L}\mathbf{S}(\mathbf{C}_{\mathfrak{G}}(\mathfrak{X})) \trianglelefteq \mathfrak{G}$, so $\mathfrak{L}\mathbf{S}(\mathbf{C}_{\mathfrak{G}}(\mathfrak{X}))$ is either $\mathfrak{L}$ or $\mathfrak{M}$. In the first case, $\mathbf{S}(\mathbf{C}_{\mathfrak{G}}(\mathfrak{X})) \leq \mathfrak{L}$, but since $\mathbf{S}(\mathbf{C}_{\mathfrak{G}}(\mathfrak{X})) \in \mathsf{U}_{\mathbf{S}}(\mathfrak{A})$, this implies that $\mathbf{S}(\mathbf{C}_{\mathfrak{G}}(\mathfrak{X})) \leq \mathbf{S}(\mathfrak{L}) = \mathfrak{K}$, by (3). This contradicts the hypothesis of (4), so $\mathfrak{L}\mathbf{S}(\mathbf{C}_{\mathfrak{G}}(\mathfrak{X})) = \mathfrak{M}$.

Since $\mathfrak{K} < \mathbf{S}(\mathfrak{G})$, it follows from 16.5a) that there is a subgroup $\mathfrak{B}$ of $\mathfrak{A}$ such that $\mathfrak{A}/\mathfrak{B}$ is cyclic and $\mathfrak{U} = \mathbf{S}(\mathbf{C}_{\mathfrak{G}}(\mathfrak{B})) \nleq \mathfrak{K}$. By (4), $\mathfrak{M} = \mathfrak{L}\mathfrak{U}$. But if $b \in \mathfrak{B} - \{1\}$, $\mathfrak{U} = \mathbf{S}(\mathbf{C}_{\mathfrak{G}}(\mathfrak{B})) \leq \mathbf{S}(\mathbf{C}_{\mathfrak{G}}(b))$ by 16.4c), so

$$\mathbf{S}(\mathbf{C}_{\mathfrak{G}}(b)) = (\mathfrak{L} \cap \mathbf{S}(\mathbf{C}_{\mathfrak{G}}(b)))\mathfrak{U}.$$

Since $\mathbf{S}(\mathfrak{L}) = \mathfrak{K} \in \mathsf{U}_{\mathbf{S}}(\mathfrak{A})$, 16.4b) gives

$$\mathfrak{L} \cap \mathbf{S}(\mathbf{C}_{\mathfrak{G}}(b)) = \mathbf{S}(\mathfrak{L}) \cap \mathbf{C}_{\mathfrak{G}}(b) = \mathbf{C}_{\mathfrak{K}}(b),$$

so $\mathbf{S}(\mathbf{C}_{\mathfrak{G}}(b)) = \mathbf{C}_{\mathfrak{K}}(b)\mathfrak{U}$ for all $b \in \mathfrak{B} - \{1\}$. But since $d(\mathfrak{A}) \geq 3, d(\mathfrak{B}) \geq 2$; thus 16.5a) gives

$$\mathbf{S}(\mathfrak{G}) = \langle \mathbf{S}(\mathbf{C}_{\mathfrak{G}}(b)) | b \in \mathfrak{B} - \{1\} \rangle \leq \mathfrak{K}\mathfrak{U}.$$

Since $\mathfrak{U} \leq \mathbf{S}(\mathfrak{G})$, it follows that $\mathbf{S}(\mathfrak{G}) = \mathfrak{K}\mathfrak{U}$. But $\mathfrak{K}\mathfrak{U} \in \mathsf{U}_{\mathbf{S}}(\mathfrak{A})$, by 16.4d), contrary to $\mathbf{S}(\mathfrak{G}) \notin \mathsf{U}_{\mathbf{S}}(\mathfrak{A})$. **q.e.d.**

16.8 Example (GLAUBERMAN [9]). Theorem 16.7 does not remain true if the condition on $d(\mathfrak{A})$ is weakened so as to allow $d(\mathfrak{A})$ to be 2.

For let p, q be primes such that $q \equiv 1\ (p)$, and let s be a primitive p-th root of unity modulo q. Let $\mathfrak{Q}$ be a non-Abelian group of order q^3 and exponent q. Suppose that $\mathfrak{Q} = \langle a, b \rangle$, $c = [b, a]$. Then $\mathfrak{Q}$ has automorphisms α, β given by

$$a\alpha = a^s, \quad b\alpha = b, \quad c\alpha = c^s,$$

$$a\beta = a, \quad b\beta = b^s, \quad c\beta = c^s,$$

and $\mathfrak{A} = \langle \alpha, \beta \rangle$ is elementary Abelian of order p^2. Let $\mathfrak{G}$ be the semi-direct product $\mathfrak{A}\mathfrak{Q}$. It is easy to check that

$$\mathbf{C}_{\mathfrak{G}}(\alpha^i\beta^j) = \begin{cases} \mathfrak{A}\langle b \rangle & \text{if } i \not\equiv 0\,(p), j \equiv 0\,(p), \\ \mathfrak{A}\langle a \rangle & \text{if } i \equiv 0\,(p), j \not\equiv 0\,(p), \\ \mathfrak{A}\langle c \rangle & \text{if } i \not\equiv 0\,(p), i + j \equiv 0\,(p), \\ \mathfrak{A} & \text{if } i \not\equiv 0\,(p), j \not\equiv 0\,(p), i + j \not\equiv 0\,(p). \end{cases}$$

A soluble $\mathfrak{A}$-signalizer functor $\mathbf{S}$ on $\mathfrak{G}$ is defined as follows:

$$\mathbf{S}(\mathbf{C}_{\mathfrak{G}}(\alpha^i\beta^j)) = \begin{cases} \langle b \rangle & \text{if } i \not\equiv 0\,(p), j \equiv 0\,(p), \\ \langle a \rangle & \text{if } i \equiv 0\,(p), j \not\equiv 0\,(p), \\ 1 & \text{if } i \not\equiv 0\,(p), i + j \equiv 0\,(p), \\ 1 & \text{if } i \not\equiv 0\,(p), j \not\equiv 0\,(p), i + j \not\equiv 0\,(p), \end{cases}$$

since $\mathbf{C}_{\mathfrak{G}}(\xi) \cap \mathbf{S}(\mathbf{C}_{\mathfrak{G}}(\eta)) = 1$ unless $\langle \xi \rangle = \langle \eta \rangle$. But $\mathbf{S}$ is not complete, since $\langle a, b \rangle = \mathfrak{Q}$ and $\mathbf{C}_{\mathfrak{Q}}(\alpha\beta^{-1}) \neq \mathbf{S}(\mathbf{C}_{\mathfrak{G}}(\alpha\beta^{-1}))$.

16.9 Theorem *(Transitivity theorem). Suppose that $\mathfrak{A}$ is an Abelian p-subgroup of $\mathfrak{G}$ for which $d(\mathfrak{A}) \geq 3$ and that $\mathbf{S}$ is an $\mathfrak{A}$-signalizer functor on $\mathfrak{G}$. Let q be a prime distinct from p and let $\mathcal{N}$ be the set of $\mathfrak{A}$-invariant q-subgroups $\mathfrak{X}$ of $\mathfrak{G}$ such that $\mathbf{C}_{\mathfrak{X}}(a) \leq \mathbf{S}(\mathbf{C}_{\mathfrak{G}}(a))$ for all $a \in \mathfrak{A} - \{1\}$. Then $\mathcal{N}$ is invariant under $\mathfrak{C} = \bigcap_{a \in \mathfrak{A}-\{1\}} \mathbf{S}(\mathbf{C}_{\mathfrak{G}}(a))$, and the maximal elements of $\mathcal{N}$ are permuted transitively by $\mathfrak{C}$.*

Proof. If $\mathfrak{Q} \in \mathcal{N}$ and $u \in \mathfrak{C}$, $\mathfrak{Q}^u$ is $\mathfrak{A}$-invariant and $\mathbf{C}_{\mathfrak{Q}^u}(a) = (\mathbf{C}_{\mathfrak{Q}}(a))^u \leq \mathbf{S}(\mathbf{C}_{\mathfrak{G}}(a))$. Thus $\mathfrak{Q}^u \in \mathcal{N}$. Hence $\mathcal{N}$ is invariant under $\mathfrak{C}$.

Suppose that the theorem is false. Then the representation ρ of $\mathfrak{C}$ on $\mathcal{N}$ has more than one orbit on the set of maximal elements of $\mathcal{N}$. From all pairs of maximal elements of $\mathcal{N}$ in distinct orbits, choose $\mathfrak{Q}_1$, $\mathfrak{Q}_2$ such that $|\mathfrak{Q}_1 \cap \mathfrak{Q}_2|$ is maximal. If $\mathfrak{D} = \mathfrak{Q}_1 \cap \mathfrak{Q}_2$, then $\mathfrak{D} < \mathfrak{Q}_1$ and $\mathfrak{D} < \mathfrak{Q}_2$, since $\mathfrak{Q}_1$, $\mathfrak{Q}_2$ are maximal elements of $\mathcal{N}$. Let $\mathfrak{R}_i = \mathbf{N}_{\mathfrak{Q}_i}(\mathfrak{D})$. Thus $\mathfrak{R}_i$ is $\mathfrak{A}$-invariant, and by I, 8.8, $\mathfrak{R}_i > \mathfrak{D}$ $(i = 1, 2)$. Let $\mathfrak{S}_i$ be a minimal $\mathfrak{A}$-invariant subgroup of $\mathfrak{R}_i$ for which $\mathfrak{S}_i > \mathfrak{D}$, and let $\mathfrak{B}_i = \mathbf{C}_{\mathfrak{A}}(\mathfrak{S}_i/\mathfrak{D})$ $(i = 1, 2)$. Then there is a faithful irreducible representation of $\mathfrak{A}/\mathfrak{B}_i$ on $\mathfrak{S}_i/\mathfrak{D}$, and $\mathfrak{A}/\mathfrak{B}_i$ is cyclic (II, 3.10). Thus $d(\mathfrak{A}/(\mathfrak{B}_1 \cap \mathfrak{B}_2)) \leq 2$. Since $d(\mathfrak{A}) \geq 3$, $\mathfrak{B}_1 \cap \mathfrak{B}_2 \neq 1$. Suppose that $1 \neq b \in \mathfrak{B}_1 \cap \mathfrak{B}_2$. Then $\mathfrak{S}_i/\mathfrak{D} = \mathbf{C}_{\mathfrak{S}_i/\mathfrak{D}}(b)$ and, by IX, 6.11, $\mathfrak{S}_i = \mathbf{C}_{\mathfrak{S}_i}(b)\mathfrak{D}$. Thus $\mathbf{C}_{\mathfrak{S}_i}(b) \nleq \mathfrak{D}$ and $\mathbf{C}_{\mathfrak{R}_i}(b) \nleq \mathfrak{D}$ $(i = 1, 2)$.

Let $\mathfrak{H} = \mathbf{N}_{\mathfrak{G}}(\mathfrak{D}) \cap \mathbf{S}(\mathbf{C}_{\mathfrak{G}}(b))$. Since $\mathfrak{D}$ is $\mathfrak{A}$-invariant, $\mathfrak{H}$ is $\mathfrak{A}$-invariant, and since $\mathbf{S}(\mathbf{C}_{\mathfrak{G}}(b))$ is a p'-group, $\mathfrak{H}$ is a p'-group. Since $\mathfrak{Q}_i \in \mathcal{N}$, $\mathbf{C}_{\mathfrak{Q}_i}(b) \leq$

$\mathbf{S}(\mathbf{C}_{\mathfrak{G}}(b))$; hence $\mathbf{C}_{\mathfrak{R}_i}(b)$ is an $\mathfrak{A}$-invariant q-subgroup of $\mathfrak{H}$. By IX, 1.11, $\mathbf{C}_{\mathfrak{R}_i}(b) \leq \mathfrak{T}_i$ for some $\mathfrak{A}$-invariant Sylow q-subgroup $\mathfrak{T}_i$ of $\mathfrak{H}$, and there exists $x \in \mathbf{C}_{\mathfrak{H}}(\mathfrak{A})$ such that $\mathfrak{T}_1 = \mathfrak{T}_2^x$. Thus, if

$$\mathfrak{T} = \langle \mathbf{C}_{\mathfrak{R}_1}(b), \mathbf{C}_{\mathfrak{R}_2}(b)^x \rangle,$$

$\mathfrak{T} \leq \mathfrak{T}_1$. Thus $\mathfrak{T}$ is an $\mathfrak{A}$-invariant q-subgroup of $\mathfrak{H}$. We have $\mathfrak{H} \leq \mathbf{N}_{\mathfrak{G}}(\mathfrak{D})$; put $\mathfrak{Q} = \mathfrak{T}\mathfrak{D}$ and $\mathfrak{G}_0 = \mathfrak{A}\mathfrak{Q}$. For all $a \in \mathfrak{A} - \{1\}$, define $\mathbf{S}_0(\mathbf{C}_{\mathfrak{G}_0}(a)) = \mathfrak{G}_0 \cap \mathbf{S}(\mathbf{C}_{\mathfrak{G}}(a))$. By 16.6, $\mathbf{S}_0$ is a soluble $\mathfrak{A}$-signalizer functor on $\mathfrak{G}_0$. Since

$$\mathbf{C}_{\mathfrak{T}}(a) \leq \mathfrak{G}_0 \cap \mathbf{C}_{\mathfrak{G}}(a) \cap \mathbf{S}(\mathbf{C}_{\mathfrak{G}}(b)) \leq \mathfrak{G}_0 \cap \mathbf{S}(\mathbf{C}_{\mathfrak{G}}(a)) = \mathbf{S}_0(\mathbf{C}_{\mathfrak{G}_0}(a)),$$

$\mathfrak{T} \in \mathscr{M}_{\mathbf{S}_0}(\mathfrak{A})$. Similarly,

$$\mathbf{C}_{\mathfrak{D}}(a) \leq \mathfrak{G}_0 \cap \mathbf{C}_{\mathfrak{Q}_1}(a) \leq \mathfrak{G}_0 \cap \mathbf{S}(\mathbf{C}_{\mathfrak{G}}(a)) = \mathbf{S}_0(\mathbf{C}_{\mathfrak{G}_0}(a)),$$

so $\mathfrak{D} \in \mathscr{M}_{\mathbf{S}_0}(\mathfrak{A})$. By 16.4d), $\mathfrak{Q} = \mathfrak{T}\mathfrak{D} \in \mathscr{M}_{\mathbf{S}_0}(\mathfrak{A})$. Also $\mathfrak{Q}$ is a q-group, so $\mathfrak{Q} \in \mathscr{N}$.

Let $\mathfrak{Q}_0$ be a maximal element of $\mathscr{N}$ such that $\mathfrak{Q}_0 \geq \mathfrak{Q}$. Then $\mathfrak{Q}_0 \cap \mathfrak{Q}_1 \geq \mathbf{C}_{\mathfrak{R}_1}(b)\mathfrak{D} > \mathfrak{D}$, so $\mathfrak{Q}_0$ and $\mathfrak{Q}_1$ lie in the same orbit of ρ on account of the maximality of $|\mathfrak{Q}_1 \cap \mathfrak{Q}_2|$. But also $x \in \mathfrak{C}$, for if $a \in \mathfrak{A} - \{1\}$,

$$x \in \mathbf{C}_{\mathfrak{H}}(\mathfrak{A}) \leq \mathbf{C}_{\mathfrak{G}}(a) \cap \mathbf{S}(\mathbf{C}_{\mathfrak{G}}(b)) \leq \mathbf{S}(\mathbf{C}_{\mathfrak{G}}(a)).$$

Thus $\mathfrak{Q}_2^x = \mathfrak{Q}_2\rho(x) \in \mathscr{N}$, and $\mathfrak{Q}_0 \cap \mathfrak{Q}_2^x \geq \mathbf{C}_{\mathfrak{R}_2}(b)^x\mathfrak{D} > \mathfrak{D}$. Therefore $\mathfrak{Q}_0$ and $\mathfrak{Q}_2$ lie in the same orbit of ρ. But then, so do $\mathfrak{Q}_1$ and $\mathfrak{Q}_2$. This is a contradiction. **q.e.d.**

16.10 Corollary. *In the notation of 16.9 (with $d(\mathfrak{A}) \geq 3$), suppose $\mathfrak{Q} \in \mathscr{N}$ and that $\mathfrak{B}$ is a non-cyclic subgroup of $\mathfrak{A}$. Then $\mathfrak{Q}$ is a maximal element of $\mathscr{N}$ if and only if $\mathbf{C}_{\mathfrak{Q}}(b) \in S_q(\mathbf{S}(\mathbf{C}_{\mathfrak{G}}(b)))$ for all $b \in \mathfrak{B} - \{1\}$.*

Proof. Suppose that $\mathfrak{Q}$ is a maximal element of $\mathscr{N}$ and $b \in \mathfrak{B} - \{1\}$. By IX, 1.11, there exists an $\mathfrak{A}$-invariant Sylow q-subgroup $\mathfrak{Q}_0$ of $\mathbf{S}(\mathbf{C}_{\mathfrak{G}}(b))$, and $\mathfrak{Q}_0 \in \mathscr{N}$. By 16.9, there exists $x \in \mathfrak{C}$ such that $\mathfrak{Q}_0^x \leq \mathfrak{Q}$. Since $\mathfrak{C} \leq \mathbf{S}(\mathbf{C}_{\mathfrak{G}}(b))$, $\mathfrak{Q}_0^x \in S_q(\mathbf{S}(\mathbf{C}_{\mathfrak{G}}(b)))$ and $\mathfrak{Q}_0^x \leq \mathbf{C}_{\mathfrak{Q}}(b)$. But since $\mathfrak{Q} \in \mathscr{N}$, $\mathbf{C}_{\mathfrak{Q}}(b) \leq \mathbf{S}(\mathbf{C}_{\mathfrak{G}}(b))$, so $\mathbf{C}_{\mathfrak{Q}}(b) \in S_q(\mathbf{S}(\mathbf{C}_{\mathfrak{G}}(b)))$.

Conversely, suppose that $\mathbf{C}_{\mathfrak{Q}}(b) \in S_q(\mathbf{S}(\mathbf{C}_{\mathfrak{G}}(b)))$ for all $b \in \mathfrak{B} - \{1\}$. Let $\mathfrak{Q}^*$ be a maximal element of $\mathscr{N}$ containing $\mathfrak{Q}$. By 1.9, $\mathfrak{Q}^* = \langle \mathbf{C}_{\mathfrak{Q}^*}(b) \mid b \in \mathfrak{B} - \{1\} \rangle$, since $\mathfrak{B}$ is not cyclic. Since $\mathfrak{Q}^* \in \mathscr{N}$,

$\mathbf{C}_{\mathfrak{Q}^*}(b) \leq \mathbf{S}(\mathbf{C}_{\mathfrak{G}}(b))$. But $\mathbf{C}_{\mathfrak{Q}}(b) \leq \mathbf{C}_{\mathfrak{Q}^*}(b)$ and $\mathbf{C}_{\mathfrak{Q}}(b) \in S_q(\mathbf{S}(\mathbf{C}_{\mathfrak{G}}(b)))$. Thus $\mathbf{C}_{\mathfrak{Q}}(b) = \mathbf{C}_{\mathfrak{Q}^*}(b)$ for all $b \in \mathfrak{B} - \{1\}$, and $\mathfrak{Q}^* \leq \mathfrak{Q}$. Hence $\mathfrak{Q} = \mathfrak{Q}^*$ is a maximal element of $\mathcal{N}$. **q.e.d.**

In order to apply this with $\mathbf{S} = \mathbf{O}_{p'}$ (see 16.2), we shall need the following lemma.

16.11 Lemma. *Suppose that $\mathfrak{S} \in S_p(\mathfrak{G})$, $\mathfrak{A}$ is a maximal normal Abelian subgroup of $\mathfrak{S}$ and $\mathfrak{A}$ is not cyclic. Suppose that $\mathbf{C}_{\mathfrak{G}}(a)$ is p-soluble for every $a \in \mathfrak{A} - \{1\}$. If $\mathfrak{Q}$ is an $\mathfrak{A}$-invariant q-subgroup, where q is a prime different from p, then $\mathfrak{Q} \leq \mathbf{O}_{p'}(\mathfrak{L})$ for any p-soluble subgroup $\mathfrak{L}$ of $\mathfrak{G}$ containing $\mathfrak{A}\mathfrak{Q}$.*

Proof. Suppose that this is false, and suppose that $\mathfrak{Q}$ is a minimal $\mathfrak{A}$-invariant q-subgroup of $\mathfrak{L}$ for which $\mathfrak{Q} \nleq \mathbf{O}_{p'}(\mathfrak{L})$.

(1) If $x \in \mathbf{Z}(\mathfrak{S}) - \{1\}$, $\mathbf{C}_{\mathfrak{Q}}(x) \leq \mathbf{O}_{p'}(\mathfrak{L})$.

Since $x \in \mathbf{Z}(\mathfrak{S}) \leq \mathbf{C}_{\mathfrak{S}}(\mathfrak{A}) \leq \mathfrak{A}$, $\mathbf{C}_{\mathfrak{Q}}(x)$ is an $\mathfrak{A}$-invariant p'-subgroup of the p-soluble group $\mathbf{C}_{\mathfrak{G}}(x)$. By IX, 1.4 $\mathbf{C}_{\mathfrak{G}}(x)$ is p-constrained, so by 1.11, $\mathbf{C}_{\mathfrak{Q}}(x) \leq \mathbf{O}_{p'}(\mathbf{C}_{\mathfrak{G}}(x))$. Hence

$$\mathbf{C}_{\mathfrak{Q}}(x) \leq \mathbf{O}_{p'}(\mathbf{C}_{\mathfrak{G}}(x)) \cap \mathfrak{L} \leq \mathbf{O}_{p'}(\mathbf{C}_{\mathfrak{L}}(x)).$$

Since $x \in \mathfrak{A} \leq \mathfrak{L}$, $\mathbf{O}_{p'}(\mathbf{C}_{\mathfrak{L}}(x)) \leq \mathbf{O}_{p'}(\mathfrak{L})$, by 1.6. Therefore $\mathbf{C}_{\mathfrak{Q}}(x) \leq \mathbf{O}_{p'}(\mathfrak{L})$.

(2) $\mathbf{Z}(\mathfrak{S})$ is cyclic.

Since $\mathfrak{Q} \nleq \mathbf{O}_{p'}(\mathfrak{L})$, it follows from (1) that $\mathfrak{Q} \neq \langle \mathbf{C}_{\mathfrak{Q}}(x) | x \in \mathbf{Z}(\mathfrak{S}) - \{1\}\rangle$. Since $\mathbf{Z}(\mathfrak{S}) \leq \mathfrak{A}$, $\mathfrak{Q}$ is $\mathbf{Z}(\mathfrak{S})$-invariant. Thus $\mathbf{Z}(\mathfrak{S})$ is cyclic, by 1.9.

Let z be an element of order p in $\mathbf{Z}(\mathfrak{S})$.

(3) $\mathfrak{Q} = [\mathfrak{Q}, \langle z \rangle]$, and the representation of $\mathfrak{A}$ on $\mathfrak{Q}/\Phi(\mathfrak{Q})$ is irreducible.

By 1.8, $[\mathfrak{Q}, \langle z \rangle]\mathbf{C}_{\mathfrak{Q}}(z) = \mathfrak{Q} \nleq \mathbf{O}_{p'}(\mathfrak{L})$. But $\mathbf{C}_{\mathfrak{Q}}(z) \leq \mathbf{O}_{p'}(\mathfrak{L})$ by (1), so $[\mathfrak{Q}, \langle z \rangle] \nleq \mathbf{O}_{p'}(\mathfrak{L})$. But since $z \in \mathfrak{A}$, $[\mathfrak{Q}, \langle z \rangle]$ is $\mathfrak{A}$-invariant. Hence by minimality of $\mathfrak{Q}$, $[\mathfrak{Q}, \langle z \rangle] = \mathfrak{Q}$. If the representation of $\mathfrak{A}$ on $\mathfrak{Q}/\Phi(\mathfrak{Q})$ is reducible, there exist proper $\mathfrak{A}$-invariant subgroups $\mathfrak{Q}_1$, $\mathfrak{Q}_2$ of $\mathfrak{Q}$ such that $\mathfrak{Q} = \mathfrak{Q}_1\mathfrak{Q}_2$, by the Maschke-Schur theorem. Then $\mathfrak{Q}_i \leq \mathbf{O}_{p'}(\mathfrak{L})$, by minimality of $\mathfrak{Q}$, which gives the contradiction $\mathfrak{Q} = \mathfrak{Q}_1\mathfrak{Q}_2 \leq \mathbf{O}_{p'}(\mathfrak{L})$.

Let $\mathfrak{B} = \mathbf{C}_{\mathfrak{A}}(\mathfrak{Q})$.

(4) There exists $u \in \mathfrak{B}$ such that $\langle u, z \rangle$ is a normal subgroup of $\mathfrak{S}$ and is elementary Abelian of order p^2.

$\mathfrak{A}$ is a group of operators on $\mathfrak{Q}$, so by III, 3.18, $\mathbf{C}_{\mathfrak{A}}(\mathfrak{Q}/\Phi(\mathfrak{Q})) = \mathfrak{B}$. Hence by (3), there is a faithful irreducible representation of $\mathfrak{A}/\mathfrak{B}$ on

$\mathfrak{Q}/\Phi(\mathfrak{Q})$. By II, 3.10, $\mathfrak{A}/\mathfrak{B}$ is cyclic. By (3), $z \notin \mathfrak{B}$, so $\Omega_1(\mathfrak{A}/\mathfrak{B}) = \langle z\mathfrak{B}\rangle$. Thus $\Omega_1(\mathfrak{A}) = \langle z\rangle\Omega_1(\mathfrak{B})$. Since $\mathfrak{A}$ is not cyclic, $\langle z\rangle < \Omega_1(\mathfrak{A})$. Since $\langle z\rangle$ and $\Omega_1(\mathfrak{A})$ are normal in $\mathfrak{S}$, there exists $\mathfrak{C} \trianglelefteq \mathfrak{S}$ such that $\langle z\rangle < \mathfrak{C} \leq \Omega_1(\mathfrak{A})$ and $|\mathfrak{C}| = p^2$. Thus $\mathfrak{C} = \langle z\rangle(\mathfrak{C} \cap \Omega_1(\mathfrak{B}))$. If $\mathfrak{C} \cap \Omega_1(\mathfrak{B}) = \langle u\rangle$, u has the properties stated in (4).

Let $\mathfrak{M} = \mathbf{C}_{\mathfrak{G}}(u)$, $\mathfrak{N} = \mathbf{O}_{p',p}(\mathfrak{M})$. By hypothesis, $\mathfrak{M}$ is p-soluble.

(5) $\mathfrak{A}\mathfrak{Q} \leq \mathfrak{M}$, but $\mathfrak{Q} \nleq \mathfrak{N}$.

Since $u \in \mathfrak{B} = \mathbf{C}_{\mathfrak{A}}(\mathfrak{Q})$, $\mathfrak{A}\mathfrak{Q} \leq \mathfrak{M}$. By 1.6, $\mathbf{O}_{p'}(\mathbf{C}_{\mathfrak{L}}(\mathfrak{B})) \leq \mathbf{O}_{p'}(\mathfrak{L})$. Thus $\mathfrak{Q} \nleq \mathbf{O}_{p'}(\mathbf{C}_{\mathfrak{L}}(\mathfrak{B}))$. Since $u \in \mathfrak{B}$, $\mathbf{C}_{\mathfrak{L}}(\mathfrak{B}) \leq \mathbf{C}_{\mathfrak{G}}(u) = \mathfrak{M}$, so

$$\mathbf{O}_{p'}(\mathfrak{M}) \cap \mathbf{C}_{\mathfrak{L}}(\mathfrak{B}) \leq \mathbf{O}_{p'}(\mathbf{C}_{\mathfrak{L}}(\mathfrak{B})).$$

Since $\mathfrak{Q} \leq \mathbf{C}_{\mathfrak{L}}(\mathfrak{B})$, it follows that $\mathfrak{Q} \nleq \mathbf{O}_{p'}(\mathfrak{M})$. Since $\mathfrak{Q}$ is a p'-group, $\mathfrak{Q} \nleq \mathbf{O}_{p',p}(\mathfrak{M}) = \mathfrak{N}$.

(6) Since $\mathfrak{Q} \nleq \mathfrak{N}$, $[\mathfrak{Q}, \langle z\rangle] \nleq \mathfrak{N}$, by (3). Since $z \in \mathfrak{M}$ and $\mathfrak{N} \trianglelefteq \mathfrak{M}$, it follows that $z \notin \mathfrak{N}$. By IX, 1.4, z does not centralize $\mathfrak{N}/\mathbf{O}_{p'}(\mathfrak{M})$. But $z \in \mathbf{Z}(\mathfrak{S})$, so $\mathfrak{N} \neq (\mathfrak{N} \cap \mathfrak{S})\mathbf{O}_{p'}(\mathfrak{M})$. Thus $\mathfrak{N} \cap \mathfrak{S} \notin S_p(\mathfrak{N})$ and $\mathfrak{M} \cap \mathfrak{S} \notin S_p(\mathfrak{M})$. By Sylow's theorem, there exists $\mathfrak{P} \in S_p(\mathfrak{M})$ such that $\mathfrak{P} > \mathfrak{M} \cap \mathfrak{S} = \mathbf{C}_{\mathfrak{S}}(u)$. But by (4), $\langle u, z\rangle$ is a normal subgroup of $\mathfrak{S}$ of order p^2, and $\mathbf{C}_{\mathfrak{S}}(u) = \mathbf{C}_{\mathfrak{S}}(\langle u, z\rangle)$. Thus $|\mathfrak{S} : \mathbf{C}_{\mathfrak{S}}(u)| \leq p$. It follows that $|\mathfrak{P} : \mathbf{C}_{\mathfrak{S}}(u)| = p$. Thus $\mathbf{C}_{\mathfrak{S}}(u) \trianglelefteq \mathfrak{P}$ and $\mathbf{Z}(\mathbf{C}_{\mathfrak{S}}(u)) \trianglelefteq \mathfrak{P}$. But $z \in \mathbf{Z}(\mathbf{C}_{\mathfrak{S}}(u))$, and since $\mathfrak{A} \leq \mathbf{C}_{\mathfrak{S}}(u)$, $\mathbf{Z}(\mathbf{C}_{\mathfrak{S}}(u)) \leq \mathbf{C}_{\mathfrak{S}}(\mathfrak{A}) = \mathfrak{A}$. Thus

$$[\mathfrak{P}, \langle z\rangle] \leq \mathbf{Z}(\mathbf{C}_{\mathfrak{S}}(u)) \leq \mathfrak{A}.$$

Let $\mathfrak{P}_1 = \mathfrak{P} \cap \mathfrak{N}$. Then $\mathfrak{P}_1 \in S_p(\mathfrak{N})$ and $\mathfrak{N} = \mathfrak{P}_1\mathbf{O}_{p'}(\mathfrak{M})$. Thus

$$[\mathfrak{N}, \langle z\rangle] \leq [\mathfrak{P}_1, \langle z\rangle]\mathbf{O}_{p'}(\mathfrak{M}) \leq \mathfrak{A}\mathbf{O}_{p'}(\mathfrak{M})$$

and

$$[\mathfrak{N}, \langle z\rangle, \mathfrak{Q}] \leq [\mathfrak{A}, \mathfrak{Q}]\mathbf{O}_{p'}(\mathfrak{M}) \cap \mathfrak{N} \leq (\mathfrak{Q} \cap \mathfrak{N})\mathbf{O}_{p'}(\mathfrak{M}) = \mathbf{O}_{p'}(\mathfrak{M}),$$

since $\mathfrak{N}/\mathbf{O}_{p'}(\mathfrak{M})$ is a p-group. Thus if $\mathfrak{V} = \mathfrak{N}/\mathbf{O}_{p'}(\mathfrak{M})$, $\mathfrak{M}$ is a group of operators on $\mathfrak{V}$ and $\mathfrak{Q}$ centralizes $[\mathfrak{V}, \langle z\rangle]$. Thus if $\mathfrak{U} = \mathbf{C}_{\mathfrak{V}}(\mathfrak{Q})$, $[\mathfrak{V}, \langle z\rangle] \leq \mathfrak{U}$. Now $\mathfrak{V}$ and $\mathfrak{U}$ are $\mathfrak{A}\mathfrak{Q}$-invariant, so $\mathbf{N}_{\mathfrak{V}}(\mathfrak{U})$ is $\mathfrak{A}\mathfrak{Q}$-invariant. Thus z and z^y centralize $\mathbf{N}_{\mathfrak{V}}(\mathfrak{U})/\mathfrak{U}$ for any $y \in \mathfrak{Q}$. Thus $[\mathfrak{Q}, \langle z\rangle]$ centralizes $\mathbf{N}_{\mathfrak{V}}(\mathfrak{U})/\mathfrak{U}$. It follows from (3) that $\mathfrak{Q}$ centralizes $\mathbf{N}_{\mathfrak{V}}(\mathfrak{U})/\mathfrak{U}$ and $\mathfrak{U}$. By I, 4.4, $\mathfrak{Q}$ centralizes $\mathbf{N}_{\mathfrak{V}}(\mathfrak{U})$; that is, $\mathbf{N}_{\mathfrak{V}}(\mathfrak{U}) \leq \mathbf{C}_{\mathfrak{V}}(\mathfrak{Q}) = \mathfrak{U}$. By I, 8.8, $\mathfrak{U} = \mathfrak{V}$, so $\mathfrak{Q}$ centralizes $\mathfrak{V} = \mathfrak{N}/\mathbf{O}_{p'}(\mathfrak{M})$. By IX, 1.4, $\mathfrak{Q} \leq \mathfrak{N}$, contrary to (5).

q.e.d.

16.12 Theorem. *Suppose that* $\mathfrak{G}$ *is a group,* $\mathfrak{S} \in S_p(\mathfrak{G})$, $\mathfrak{A}$ *is a maximal normal Abelian subgroup of* $\mathfrak{S}$ *and* $d(\mathfrak{A}) \geq 3$. *Suppose that* $\mathbf{C}_{\mathfrak{G}}(a)$ *is p-soluble for every* $a \in \mathfrak{A} - \{1\}$. *If q is a prime different from p and* $\mathfrak{Q}_1$, $\mathfrak{Q}_2$ *are maximal elements of the set of q-subgroups of* $\mathfrak{G}$ *normalised by* $\mathfrak{A}$, *there exists* $x \in \mathbf{O}_{p'}(\mathbf{C}_{\mathfrak{G}}(\mathfrak{A}))$ *such that* $\mathfrak{Q}_1^x = \mathfrak{Q}_2$.

Proof. By 16.2, $\mathbf{O}_{p'}$ is an $\mathfrak{A}$-signalizer functor on $\mathfrak{G}$. Let $\mathcal{N}$ be the set of $\mathfrak{A}$-invariant q-subgroups of $\mathfrak{G}$. If $\mathfrak{X} \in \mathcal{N}$ and $a \in \mathfrak{A} - \{1\}$, we may apply 16.11 to the p-soluble group $\mathbf{C}_{\mathfrak{G}}(a)$ and the $\mathfrak{A}$-invariant q-subgroup $\mathbf{C}_{\mathfrak{X}}(a)$ of $\mathbf{C}_{\mathfrak{G}}(a)$; thus $\mathbf{C}_{\mathfrak{X}}(a) \leq \mathbf{O}_{p'}(\mathbf{C}_{\mathfrak{G}}(a))$. Thus by 16.9, the set of maximal elements of $\mathcal{N}$ is permuted transitively by $\mathfrak{C}$, where $\mathfrak{C} = \bigcap_{a \in \mathfrak{A} - \{1\}} \mathbf{O}_{p'}(\mathbf{C}_{\mathfrak{G}}(a))$. Since $\mathfrak{C} \leq \mathbf{O}_{p'}(\mathbf{C}_{\mathfrak{G}}(\mathfrak{A}))$, the assertion is clear.

q.e.d.

16.13 Theorem. *Suppose that* $\mathfrak{G}$ *is a group,* $\mathfrak{S} \in S_p(\mathfrak{G})$, $\mathfrak{A}$ *is a maximal normal Abelian subgroup of* $\mathfrak{S}$ *and* $d(\mathfrak{A}) \geq 3$. *Suppose that* $\mathbf{C}_{\mathfrak{G}}(a)$ *is p-soluble for every* $a \in \mathfrak{A} - \{1\}$. *If q is a prime different from p, then in the set of maximal* $\mathfrak{A}$*-invariant q-subgroups of* $\mathfrak{G}$, *there is an element* $\mathfrak{X}$ *such that* $\mathfrak{S} \leq \mathbf{N}_{\mathfrak{G}}(\mathfrak{X})$.

Proof. Let $\mathcal{N}$ be the set of maximal $\mathfrak{A}$-invariant q-subgroups of $\mathfrak{G}$. If $\mathfrak{X} \in \mathcal{N}$, then $\mathfrak{X}^u \in \mathcal{N}$ for all $u \in \mathbf{N}_{\mathfrak{G}}(\mathfrak{A})$, so $\mathbf{N}_{\mathfrak{G}}(\mathfrak{A})$ permutes $\mathcal{N}$. By 16.12, $\mathbf{O}_{p'}(\mathbf{C}_{\mathfrak{G}}(\mathfrak{A}))$ permutes $\mathcal{N}$ transitively, so $|\mathcal{N}|$ is prime to p and the stabiliser in $\mathbf{N}_{\mathfrak{G}}(\mathfrak{A})$ of an element of $\mathcal{N}$ contains a Sylow p-subgroup of $\mathbf{N}_{\mathfrak{G}}(\mathfrak{A})$ and of $\mathfrak{G}$. Hence some element $\mathfrak{X}$ of $\mathcal{N}$ is left fixed by $\mathfrak{S}$ and $\mathfrak{S} \leq \mathbf{N}_{\mathfrak{G}}(\mathfrak{X})$. **q.e.d.**

16.14 Remark. If $\mathbf{S}$ is a soluble $\mathfrak{A}$-signalizer functor on $\mathfrak{G}$ and $d(\mathfrak{A}) \geq 3$, $\mathbf{S}$ is complete. This was proved in the case when $d(\mathfrak{A}) \geq 4$ by GOLDSCHMIDT [3], and for $d(\mathfrak{A}) \geq 3$ by GLAUBERMAN [9].

It is a consequence of this that if $\mathfrak{A}$ is an Abelian p-group of operators on a p'-group $\mathfrak{G}$, $d(\mathfrak{A}) \geq 3$ and $\mathbf{C}_{\mathfrak{G}}(a)$ is soluble for every $a \in \mathfrak{A} - \{1\}$, then $\mathfrak{G}$ is soluble, for in the semidirect product $\mathfrak{H} = \mathfrak{A}\mathfrak{G}$, we can take $\mathbf{S}(\mathbf{C}_{\mathfrak{H}}(a)) = \mathbf{C}_{\mathfrak{G}}(a)$.

For the role played by signalizer functors in the theory of simple groups, see GORENSTEIN's article "Centralizers of involutions in finite simple groups" in POWELL and HIGMAN [1].

Notes on Chapter X

We are heavily indebted to G. Glauberman and H. Bender for assistance with this chapter. Both read earlier versions and suggested many improvements.

§ **1:** Lemma 1.1 appeared in FEIT-THOMPSON [1]; the proof given here is due to R. M. Bryant. Lemmas 1.8, 1.9 have their origin in FEIT-THOMPSON [1]. For Theorem 1.12, see THOMPSON [2] and BENDER [1].
§ **2:** The version given here is taken from unpublished notes of R. P. Martineau and incorporates a number of suggestions by H. Bender.
§ **3:** The original version of the replacement theorem is in THOMPSON [7]. Theorem 3.3 follows GLAUBERMAN [3]. The argument in 3.10 is from GOLDSCHMIDT [2].
§ **4:** The first account of the material in this section is in ALPERIN [2]. Our presentation is based on GLAUBERMAN's account in his article "Global and local properties of finite groups" in POWELL and HIGMAN [1]. Theorem 4.4 is taken from DOLAN [1].
§ **5**–§ **9:** These sections mainly follow GLAUBERMAN's article in POWELL and HIGMAN [1]. The treatment of Theorems 6.16–6.18 follows a written communication from Glauberman.
§ **10:** This is based on two papers of GLAUBERMAN [7].
§ **11:** The proof of Theorem 11.18 in the case when $|\mathfrak{A}| = r^2$ is from MARTINEAU [1]. Martineau proved it in the general case in [2] and subsequently obtained the version presented here [3].
§ **12:** This follows HOLT [1]. Theorem 12.8 goes back to work of Evens and Swan.
§ **13:** The original version of this section was based on lectures by Bender in 1971 and on BENDER [4].
§ **14:** The results in this section are due to Bender: see also GORENSTEIN and WALTER [1, 2]. A unified treatment of the characteristic subgroups in § 13, § 14 is given in BENDER [9].
§ **15:** Our sources for this are the same as § 13.
§ **16:** This is based on part of a paper of GOLDSCHMIDT [3]. The proof of Theorem 16.7 was given by Bender.

Chapter XI

Zassenhaus Groups

For $q > 3$, the groups $PSL(2, q)$, regarded as permutation groups in their natural representation of degree $q + 1$ on the projective line over $GF(q)$, have the following properties.

(1) They are doubly transitive.

(2) Any non-identity element has at most two fixed points.

(3) They have no regular normal subgroup.

A permutation group $\mathfrak{G}$ on a set Ω with the properties (1)–(3) is called a *Zassenhaus group.*

Already in 1936, Zassenhaus sought to prove that any Zassenhaus group is a semilinear projective group. He did not succeed, but he was able to obtain important results. Above all he described all sharply triply transitive groups. In 1960 Suzuki discovered another infinite series of Zassenhaus groups; these are denoted by $Sz(2^{2n+1})$ $(n = 1, 2, \ldots)$, and they are called the *Suzuki groups.* This completed the construction of all Zassenhaus groups, but the proof of this fact required three essential steps. In 1960, Feit obtained important results by a subtle character theoretical investigation; among other things he showed that the degree of a Zassenhaus group is always of the form $q + 1$, where q is a prime-power. For Feit's investigation, the fact, proved shortly before by Thompson, that the Frobenius kernel of a Frobenius group is nilpotent (V, 8.7), was of basic importance. Then a distinction between the cases when q is odd or even appeared. For odd q, N. Ito showed in 1962 that $\mathfrak{G}$ has a normal subgroup isomorphic to $PSL(2, q)$ of index 1 or 2. Ito's proof used some modular representation theory, but a simpler proof was given by Glauberman in 1969. For even q, the situation is more difficult, since there are now two series of groups, namely, $PSL(2, 2^n)$ and $Sz(2^{2n+1})$. In 1962, Suzuki proved that these are all the Zassenhaus groups. In doing this he used among other things the classification of 2-groups having a cyclic $2'$-automorphism group which permutes the involutions transitively (VIII, 7.9).

Apart from these results from Chapter VIII, only simple facts about permutation groups and the basic essentials of character theory will be used. Sections 4 and 8 contain results in character theory which are of independent interest and have been of great use in numerous investigations.

§ 1. Elementary Theory of Zassenhaus Groups

In this chapter we shall consider doubly transitive permutation groups $\mathfrak{G}$ having the property that any non-identity element of $\mathfrak{G}$ has at most two fixed points. In order to exclude the case of Frobenius groups (dealt with in V, § 8), we shall assume that $\mathfrak{G}$ has non-identity elements with two fixed points. At first, we denote the set of permuted symbols by $\Omega = \{\infty, 0, 1, \ldots, n-1\}$. The degree of $\mathfrak{G}$ is thus $n+1$. Later however we shall be able to give the set of permuted symbols a richer structure, so that the elements of $\mathfrak{G}$ can be described by suitable operations.

First we deal with a simple special case.

1.1 Theorem (FEIT [1]). *Suppose that $\mathfrak{G}$ is a doubly transitive permutation group of degree $n+1$, in which each non-identity element has at most two fixed points. Suppose that $\mathfrak{G}$ is not a Frobenius group but that $\mathfrak{G}$ has a regular normal subgroup. Then $n+1 = 2^p$ for a prime p, and $\mathfrak{G}$ is the group of all semilinear mappings $x \to a(x\alpha) + b$ ($a, b \in GF(2^p)$, $a \neq 0$, α is an automorphism of $GF(2^p)$). In particular, $|\mathfrak{G}| = 2^p(2^p-1)p$ and $\mathfrak{G}$ is soluble.*

Proof. Let a, b be distinct elements of Ω. Obviously, the stabiliser $\mathfrak{G}_a$ of a is a Frobenius group on $\Omega - \{a\}$. If $\mathfrak{F}$ is the Frobenius kernel of $\mathfrak{G}_a$, then $\mathfrak{G}_a = \mathfrak{F}\mathfrak{G}_{a,b}$ and $|\mathfrak{F}| = n$ (see V, 8.2). Let $\mathfrak{N}$ be a regular normal subgroup of $\mathfrak{G}$. By II, 2.3, $\mathfrak{N}$ is elementary Abelian and $n+1 = |\mathfrak{N}| = q^t$ for some prime q and some t.

Suppose that $g = fx \in \mathfrak{F}\mathfrak{N} \cap \mathfrak{G}_{a,b}$, where $f \in \mathfrak{F}$ and $x \in \mathfrak{N}$. Since $a = ag = (af)x = ax$, if follows that $x = 1$ on account of the regularity of $\mathfrak{N}$. Thus $f \in \mathfrak{G}_{a,b}$ and $bf = b$. Since $\mathfrak{F}$ is regular on $\Omega - \{a\}$, we have $f = 1$. Thus $\mathfrak{F}\mathfrak{N} \cap \mathfrak{G}_{a,b} = 1$ and $\mathfrak{F}\mathfrak{N}$ is a doubly transitive Frobenius group with Frobenius kernel $\mathfrak{N}$. By V, 8.18, if $|\mathfrak{F}|$ is even, $\mathfrak{F}$ has just one involution. This is impossible, since $\mathfrak{F}$ is the Frobenius kernel of $\mathfrak{G}_a$ and $\mathfrak{G}_{a,b} \neq 1$ induces fixed point free automorphisms on $\mathfrak{F}$. Thus $|\mathfrak{F}| = n = q^t - 1$ is odd and $q = 2$.

Since $\mathfrak{F}\mathfrak{N}$ is a Frobenius group, the Sylow subgroups of $\mathfrak{F}$ are cyclic, by V, 8.7. And since $\mathfrak{F}$ is the Frobenius kernel of $\mathfrak{G}_a$, $\mathfrak{F}$ is nilpotent, by V, 8.7. Hence $\mathfrak{F}$ is cyclic. Since $\mathfrak{F}\mathfrak{N}$ is doubly transitive, $\mathfrak{N}$ is a minimal normal subgroup of $\mathfrak{F}\mathfrak{N}$. By II, 3.12, $\mathfrak{G}$ is a subgroup of the group of semilinear mappings on $GF(2^t)$. The group $\mathfrak{G}_{0,1}$, (where $0, 1 \in GF(2^t)$), consists of mappings g of the form $cg = c^{2^i}$ ($c \in GF(2^t)$), and by hypothesis, 0 and 1 are the only fixed points of g. Suppose that g is an element of $\mathfrak{G}_{0,1}$ of prime order p. Then p is a divisor of t. As $GF(2^{t/p})$ is the field of fixed points of g, $GF(2^{t/p}) = GF(2)$, so $t = p$. Since $\mathfrak{G}_{0,1} \neq 1$ and $|\mathfrak{G}_{0,1}|$ divides p, $\mathfrak{G}$ is the group of all semilinear mappings on $GF(2^p)$.

q.e.d.

It is convenient to exclude the groups considered in 1.1 from the definition of Zassenhaus groups.

1.2 Definition. A *Zassenhaus group* is a permutation group $\mathfrak{G}$ in which the following hold.

(1) $\mathfrak{G}$ is doubly transitive.

(2) Each non-identity element of $\mathfrak{G}$ has at most two fixed points.

(3) $\mathfrak{G}$ has no regular normal subgroup.

It follows from (3) and V, 8.2 that $\mathfrak{G}$ is not a Frobenius group. Thus there exist elements of $\mathfrak{G}$ having exactly two fixed points.

1.3 Examples. In the following examples, let Ω be the projective line $\mathsf{P}(1, p^f) = GF(p^f) \cup \{\infty\}$ over $GF(p^f)$.

a) The group $PGL(2, p^f)$ of all projective linear mappings on Ω is sharply triply transitive (see II, 1.18c)). For $p^f > 3$, $PGL(2, p^f)$ contains the simple non-Abelian normal subgroup $PSL(2, p^f)$ of index 1 or 2 (see II, 6.13). Thus for $p^f > 3$, $PGL(2, p^f)$ has no regular normal subgroup and is hence a Zassenhaus group. Since $PGL(2, 2) \cong \mathfrak{S}_3$ and $PGL(2, 3) \cong \mathfrak{S}_4$, $PGL(2, p^f)$ has a regular normal subgroup and is not a Zassenhaus group for $p^f = 2$ or $p^f = 3$.

b) For $p > 2$ and $p^f \neq 3$, $PSL(2, p^f)$ is a simple Zassenhaus group of order

$$\frac{(p^f + 1)p^f(p^f - 1)}{2}.$$

c) Suppose that $p > 2$ and $f = 2m$ is even. Then $\mathsf{K} = GF(p^f)$ has exactly one automorphism α of order 2 and $x\alpha = x^{p^m}$. Let $\mathfrak{H} = PGL(2, p^f)\langle\alpha\rangle$, (where $\infty\alpha = \infty$). Thus $\mathfrak{H}$ is a subgroup of $P\Gamma L(2, p^f)$ and $\mathfrak{H}/PSL(2, p^f)$ is elementary Abelian of order 4. Thus $\mathfrak{H}$ has three subgroups of index 2 which contain $PSL(2, p^f)$. $PGL(2, p^f)$ and $\langle PSL(2, p^f), \alpha\rangle$ are two of them; we denote the third by $M(p^f)$. (The group $M(3^2)$ is the stabiliser of an element in the Mathieu group $\mathfrak{M}_{11}$; see Chapter XII, § 1).

If $h \in \mathfrak{H}$, we have

$$xh = \frac{a(x\beta) + b}{c(x\beta) + d}$$

where $\delta = ad - bc \neq 0$ and $\beta \in \{1, \alpha\}$. The mapping $h \to (\varepsilon, \beta)$, where $\varepsilon = 1$ if $\delta \in \mathsf{K}^2$ and $\varepsilon = -1$ if $\delta \notin \mathsf{K}^2$, is a homomorphism of $\mathfrak{H}$ onto the elementary Abelian group of order 4 with kernel $PSL(2, p^f)$. $M(p^f)$ is the inverse image of $\langle(-1, \alpha)\rangle$, so $M(p^f)$ consists of all mappings of the above form h, with $\beta = \alpha$ if $\delta \notin \mathsf{K}^2$ and $\beta = 1$ if $\delta \in \mathsf{K}^2$. Since $PSL(2, p^f) < M(p^f)$, $M(p^f)$ is certainly doubly transitive on Ω. $M(p^f)_{0,\infty}$ consists of mappings g of the form $xg = a(x\beta)$, where $a \neq 0$, $\beta = \alpha$ if $a \notin \mathsf{K}^2$ and $\beta = 1$ if $a \in \mathsf{K}^2$. Suppose that $x \neq 0$, $x \neq \infty$ and $x = a(x\beta)$. If $\beta = \alpha$, it follows since $p \neq 2$ that

$$a = x(x\alpha)^{-1} = x^{1-p^m} \in \mathsf{K}^2,$$

a contradiction. Thus $\beta = 1$, $a = 1$ and $g = 1$. Thus no non-identity element of $M(p^f)$ has more than two fixed points. Since $PSL(2, p^f)$ is a simple normal subgroup of index 2 in $M(p^f)$, $M(p^f)$ has no regular normal subgroup. Thus $M(p^f)$ is a sharply triply transitive Zassenhaus group.

The Sylow p-subgroups of $PGL(2, p^f)$ and $M(p^f)$ have non-isomorphic normalisers, so $PGL(2, p^f)$ is not isomorphic to $M(p^f)$.

Any Zassenhaus group is either one of the groups mentioned in 1.3 or one of the simple groups of Suzuki, which will be dealt with in § 3. The proof of this deep theorem is the main aim of this chapter.

Throughout the remainder of this section, the following notation will be used. $\mathfrak{G}$ is a Zassenhaus group of degree $n + 1$ on the set $\Omega = \{\infty, 0, 1, \ldots, n - 1\}$. The subgroup $\mathfrak{G}_\infty$ is a Frobenius group of order nd on the set $\{0, 1, \ldots, n - 1\}$. Thus $d = |\mathfrak{G}_{0,\infty}|$, so $d \neq 1$. Also $|\mathfrak{G}| = (n + 1)nd$ and d divides $n - 1$. Let $\mathfrak{F}$ be the Frobenius kernel of $\mathfrak{G}_\infty$, so $|\mathfrak{F}| = n$. We put $\mathfrak{H} = \mathfrak{G}_{0,\infty}$, so $\mathfrak{G}_\infty = \mathfrak{F}\mathfrak{H}$.

In § 1 and § 2, we consider mostly Zassenhaus groups of degree $n + 1$ satisfying $|\mathfrak{G}_{0,\infty}| \geq \frac{1}{2}(n - 1)$ and show that these are groups of projective semilinear mappings of the form described in 1.3.

In the following lemmas we investigate the structure of $\mathfrak{G}_\infty$.

1.4 Lemma. *Suppose that* $\mathfrak{G}$ *is a Zassenhaus group and* $d = |\mathfrak{G}_{0,\infty}|$.

a) *If d is even,* $\mathfrak{F}$ *is Abelian.*

b) *If* $d \geq \frac{1}{2}(n-1)$, $\mathfrak{F}$ *is elementary Abelian of prime-power order* p^f *and the degree of* $\mathfrak{G}$ *is* $p^f + 1$. *In particular, the degree of any sharply triply transitive permutation group is always of the form* $p^f + 1$*; this is the case* $d = n - 1$.

Proof. a) By V, 8.18, $\mathfrak{F}$ is Abelian.

b) Let p be a prime divisor of $n = |\mathfrak{F}|$ and let $\mathfrak{P}$ be the Sylow p-subgroup of the nilpotent group $\mathfrak{F}$. If $1 \neq g \in \Omega_1(\mathbf{Z}(\mathfrak{P}))$, then $g^h \in \Omega_1(\mathbf{Z}(\mathfrak{P}))$ for all $h \in \mathfrak{H}$. Since each non-identity element of $\mathfrak{H}$ transforms the Frobenius kernel $\mathfrak{F}$ of $\mathfrak{F}\mathfrak{H}$ fixed point freely,

$$|\Omega_1(\mathbf{Z}(\mathfrak{P}))| \geq |\{g^h | h \in \mathfrak{H}\}| + 1 = d + 1 \geq \tfrac{1}{2}(n+1).$$

Since $|\Omega_1(\mathbf{Z}(\mathfrak{P}))|$ divides $|\mathfrak{F}| = n$, it follows that $\mathfrak{F} = \Omega_1(\mathbf{Z}(\mathfrak{P}))$ and $\mathfrak{F}$ is an elementary Abelian p-group. **q.e.d.**

1.5 Lemma. *Suppose that* $\mathfrak{G}$ *is a Zassenhaus group with odd d, and suppose that* $\mathfrak{G}$ *is not q-nilpotent for any prime divisor q of d. Then* $\mathfrak{H}$ *is cyclic and* $\mathbf{C}_{\mathfrak{G}}(\mathfrak{H}) = \mathfrak{H}$. *Also, whenever* $s \in \mathfrak{G}$ *and s interchanges* 0 *and* ∞, *then* $h^s = h^{-1}$ *for all* $h \in \mathfrak{H}$ *and* $\mathbf{N}_{\mathfrak{G}}(\mathfrak{H}) = \mathfrak{H}\langle s \rangle$.

Proof. Put $\mathfrak{T} = \mathbf{N}_{\mathfrak{G}}(\mathfrak{H})$. Since $\mathfrak{T}$ at most interchanges the only fixed points 0 and ∞ of $\mathfrak{H}$, $|\mathfrak{T}/\mathfrak{H}| \leq 2$. On the other hand, there exists an element s of $\mathfrak{G}$ which interchanges 0 and ∞, on account of the double transitivity of $\mathfrak{G}$. Thus $s \in \mathfrak{T} - \mathfrak{H}$, $\mathfrak{T} = \mathbf{N}_{\mathfrak{G}}(\mathfrak{H}) = \mathfrak{H}\langle s \rangle$ and $|\mathfrak{T}| = 2d$.

Since d is odd, all Sylow subgroups of $\mathfrak{H}$ are cyclic, by V, 8.7. Thus all Sylow subgroups of $\mathfrak{T}$ are cyclic. Hence by IV, 2.11, $\mathfrak{T}$ is metacyclic and $\mathfrak{T}'$, $\mathfrak{T}/\mathfrak{T}'$ are cyclic groups of coprime orders. Suppose that q is an odd prime divisor of $|\mathfrak{T}/\mathfrak{T}'|$. Let $\mathfrak{Q}$ be a Sylow q-subgroup of $\mathfrak{H}$ and hence also of $\mathfrak{T}$. Since $\mathfrak{G}$ is a Zassenhaus group, 0 and ∞ are the only fixed points of $\mathfrak{Q}$, so 0, ∞ are at most interchanged by $\mathbf{N}_{\mathfrak{G}}(\mathfrak{Q})$. Thus $\mathbf{N}_{\mathfrak{G}}(\mathfrak{Q}) \leq \mathbf{N}_{\mathfrak{G}}(\mathfrak{H}) = \mathfrak{T}$. Since q divides $|\mathfrak{T}/\mathfrak{T}'|$ and $(|\mathfrak{T}'|, |\mathfrak{T}/\mathfrak{T}'|) = 1$, $\mathfrak{Q} \cap \mathfrak{T}' = 1$. It follows that

$$[\mathbf{N}_{\mathfrak{G}}(\mathfrak{Q}), \mathfrak{Q}] \leq \mathfrak{Q} \cap \mathbf{N}_{\mathfrak{G}}(\mathfrak{Q})' \leq \mathfrak{Q} \cap \mathfrak{T}' = 1,$$

and so $\mathfrak{Q} \leq \mathbf{Z}(\mathbf{N}_{\mathfrak{G}}(\mathfrak{Q}))$. Since q is odd and q divides $n - 1$, q does not divide $(n+1)n$. Thus $\mathfrak{Q}$ is a Sylow q-subgroup of $\mathfrak{G}$. By Burnside's transfer theorem (IV, 2.6), $\mathfrak{G}$ is q-nilpotent, contrary to hypothesis. Thus $|\mathfrak{T}/\mathfrak{T}'| = 2$. Hence $\mathfrak{T}' = \mathfrak{H}$ and $\mathfrak{H}$ is cyclic.

Suppose that $\mathfrak{H} = \langle h \rangle$. Since $s^2 \in \mathfrak{H}$, $h^s = h^k$ where $k^2 \equiv 1$ (d). On the other hand, $\mathfrak{H} = \mathbf{N}_{\mathfrak{G}}(\mathfrak{H})' = \langle h^{k-1} \rangle$ and so $(k-1, d) = 1$. This forces $k \equiv -1$ (d), so $h^s = h^{-1}$. Hence $\mathbf{C}_{\mathfrak{G}}(\mathfrak{H}) = \mathfrak{H}$. **q.e.d.**

1.6 Lemma. *Suppose that $\mathfrak{G}$ is a Zassenhaus group of degree $n + 1$.*

a) *If $1 < \mathfrak{N} \trianglelefteq \mathfrak{G}$, then $\mathfrak{N}$ is a Zassenhaus group of degree $n + 1$ and $\mathfrak{F} < \mathfrak{N}$.*

b) *If n is even, $\mathfrak{G}$ is simple.*

Proof. a) Suppose that $1 < \mathfrak{N} \trianglelefteq \mathfrak{G}$. On account of the double transitivity of $\mathfrak{G}$, $\mathfrak{N}$ is transitive. Since $\mathfrak{N}$ is not regular, $1 < \mathfrak{N}_\infty \trianglelefteq \mathfrak{G}_\infty$. By V, 8.16, either $\mathfrak{N}_\infty \leq \mathfrak{F}$ or $\mathfrak{F} < \mathfrak{N}_\infty$, since $\mathfrak{G}_\infty$ is a Frobenius group on $\Omega - \{\infty\}$. In the first case, $\mathfrak{N}_{a,\infty} \leq \mathfrak{F}_a = 1$ whenever $\infty \neq a \in \Omega$. Then $\mathfrak{N}$ is a Frobenius group on Ω. By V, 8.3, the Frobenius kernel $\mathfrak{K}$ of $\mathfrak{N}$ is a characteristic subgroup of $\mathfrak{N}$. Thus $\mathfrak{K}$ is a regular normal subgroup of $\mathfrak{G}$, contrary to hypothesis.

Hence $\mathfrak{F} < \mathfrak{N}_\infty$. Thus $\mathfrak{N}_\infty$ is transitive on $\Omega - \{\infty\}$ and is not regular. Hence $\mathfrak{N}_\infty$ is a Frobenius group on $\Omega - \{\infty\}$.

To show that $\mathfrak{N}$ is a Zassenhaus group, it only remains to prove that $\mathfrak{N}$ has no regular normal subgroup. Suppose that $\mathfrak{R}$ is a regular normal subgroup of $\mathfrak{N}$. Since $\mathfrak{N}$ is doubly transitive, it follows from II, 2.3 that $\mathfrak{R}$ is elementary Abelian. Hence if $\mathfrak{S}$ is the maximal soluble normal subgroup of $\mathfrak{N}$, $\mathfrak{S} \neq 1$, for $\mathfrak{R} \leq \mathfrak{S}$. If $\mathfrak{S}^{(k)} = 1 \neq \mathfrak{S}^{(k-1)}$, $\mathfrak{S}^{(k-1)}$ is a characteristic subgroup of $\mathfrak{N}$ and hence a non-identity normal Abelian subgroup of $\mathfrak{G}$. Since $\mathfrak{G}$ is doubly transitive, $\mathfrak{S}^{(k-1)}$ is transitive. By I, 5.13, $\mathfrak{S}^{(k-1)}$ is regular. But this is not possible, since $\mathfrak{G}$ is a Zassenhaus group.

b) Again suppose that $1 < \mathfrak{N} \trianglelefteq \mathfrak{G}$, and now suppose that n is even. By a), $|\mathfrak{G}| = (n + 1)nd$ and $|\mathfrak{N}| = (n + 1)nt$, where t divides d. For any $g \in \mathfrak{G}$, let $\pi(g)$ be the number of fixed points of g.

If $\pi(g) = 1$, then $g^h \in \mathfrak{G}_\infty$ for some h and so g^h lies in the Frobenius kernel $\mathfrak{F}$ of $\mathfrak{G}_\infty$. By a), $\mathfrak{F} \leq \mathfrak{N}$, so $g \in \mathfrak{N}$.

If $\pi(g) = 0$, let m be the smallest odd integer which is the length of a cycle of g; since $n + 1$ is odd, such integers exist. Then g^m has at least $m \geq 3$ fixed points, so $g^m = 1$. Thus the length of every cycle of g is odd and, indeed, is m. Hence m divides $n + 1$. Since m is odd, it follows that $(m, n - 1) = 1$. Hence $\left(m, \frac{d}{t}\right) = 1$ since d divides $n - 1$. Since $\frac{d}{t} = |\mathfrak{G}/\mathfrak{N}|$ and $(g\mathfrak{N})^m = 1$, it follows that $g \in \mathfrak{N}$.

Thus any element of $\mathfrak{G} - \mathfrak{N}$ has exactly two fixed points. Using V, 20.2a), it follows that

$$|\mathfrak{G}| - |\mathfrak{N}| = \sum_{g \in \mathfrak{G}} \pi(g) - \sum_{h \in \mathfrak{N}} \pi(h) = \sum_{g \in \mathfrak{G} - \mathfrak{N}} \pi(g) = 2(|\mathfrak{G}| - |\mathfrak{N}|).$$

Hence $\mathfrak{N} = \mathfrak{G}$. **q.e.d.**

1.7 Lemma. *Let $\mathfrak{G}$ be a Zassenhaus group of degree $n + 1$.*

a) *Suppose that $s \in \mathfrak{G}$ and s interchanges 0 and ∞. Then $\mathfrak{G} = \mathfrak{G}_\infty \cup \mathfrak{G}_\infty s\mathfrak{F}$ and $\mathfrak{G}_\infty \cap \mathfrak{G}_\infty s\mathfrak{F} = \emptyset$. Each element g of $\mathfrak{G} - \mathfrak{G}_\infty$ is of the form $g = g'sf$ for uniquely determined elements $f \in \mathfrak{F}$ and $g' \in \mathfrak{G}_\infty$.*

b) *If n is even, there exists an involution s in $\mathfrak{G}$ which interchanges 0 and ∞. Any such involution has exactly one fixed point.*

c) *If n is odd, then again there exists an involution s which interchanges 0 and ∞. Any such involution has either 0 or 2 fixed points. If d is even, there exists such an involution with 2 fixed points.*

Proof. a) Since $\mathfrak{G}$ is doubly transitive,

$$\mathfrak{G} = \mathfrak{G}_\infty \cup \mathfrak{G}_\infty x \mathfrak{G}_\infty$$

for any $x \in \mathfrak{G} - \mathfrak{G}_\infty$, by II, 1.12. As usual we put $\mathfrak{H} = \mathfrak{G}_{0,\infty}$. Since s interchanges 0 and ∞, $s \in \mathbf{N}_{\mathfrak{G}}(\mathfrak{H})$. Also $\mathfrak{G}_\infty = \mathfrak{H}\mathfrak{F}$, so $\mathfrak{G}_\infty s \mathfrak{G}_\infty = \mathfrak{G}_\infty s\mathfrak{F}$. Since ∞ is carried into $0f \neq \infty$ by elements of the form $g'sf$ ($g' \in \mathfrak{G}_\infty$, $f \in \mathfrak{F}$), $\mathfrak{G}_\infty \cap \mathfrak{G}_\infty s\mathfrak{F} = \emptyset$.

Suppose that $g'_1 s f_1 = g'_2 s f_2$, where $g'_i \in \mathfrak{G}_\infty$ and $f_i \in \mathfrak{F}$ $(i = 1, 2)$. Then

$$f_1 f_2^{-1} = s^{-1} g_1'^{-1} g'_2 s \in \mathfrak{F} \cap s^{-1}\mathfrak{G}_\infty s = \mathfrak{F} \cap \mathfrak{G}_0 = 1.$$

Thus $f_1 = f_2$ and $g'_1 = g'_2$.

b) Suppose that $s \in \mathfrak{G}$ and s interchanges 0 and ∞. By replacing s by s^k for a suitable odd k if necessary, we may assume that the order of s is a power of 2. The lengths of the cycles of s are thus powers of 2. Since $n + 1 \not\equiv 0$ (2) and s has at most two fixed points, exactly one cycle of length 1 occurs. Thus s^2 has at least three fixed points and $s^2 = 1$.

c) Suppose that d is odd. As in b), we may suppose that the order of s is a power of 2. Then $s^2 \in \mathfrak{G}_{0,\infty} = \mathfrak{H}$ and $|\mathfrak{H}| = d \equiv 1$ (2), so $s^2 = 1$.

If d is even, then $\mathfrak{H}$ contains an involution t. We have $0t = 0$, $\infty t = \infty$. If $at = b \neq a$, there exists $g \in \mathfrak{G}$ such that $ag = \infty$, $bg = 0$. Then $s = t^g$ interchanges 0 and ∞. Also $s^2 = 1$ and s has exactly the two fixed points $0g$, ∞g.

If s is an involution in $\mathfrak{G}$ and s interchanges 0 and ∞, then the number of fixed points of s is even, since $n + 1 \equiv 0\ (2)$. As s has at most two fixed points, the number of them is either 0 or 2. **q.e.d.**

In the following lemmas it is our aim to define addition and multiplication on the set $\Omega - \{\infty\}$ in such a way that $\Omega - \{\infty\}$ is a field K and the Zassenhaus group $\mathfrak{G}$ is a group of projective semilinear mappings on K. We begin by introducing addition on $\Omega - \{\infty\}$ by carrying over to $\Omega - \{\infty\}$ the group structure of $\mathfrak{F}$, which is regular on $\Omega - \{\infty\}$.

1.8 Lemma. *If $\mathfrak{F}$ is Abelian, there exists an addition on $\mathsf{K} = \Omega - \{\infty\}$ such that $0f_1 + 0f_2 = 0f_1f_2$ for all f_1, f_2 in $\mathfrak{F}$. K is an Abelian group with respect to this addition. Also the following hold.*

a) *The elements of $\mathfrak{F}$ are translations of the form $kf = k + 0f$ for $k \in \{\infty, \mathsf{K}\} = \Omega$; here $\infty + a$ is interpreted to be ∞.*

b) *If d is even, $\mathfrak{H}$ has a unique involution t and $kt = -k$ for all $k \in \mathsf{K}$. Also $\infty t = \infty$.*

c) *If d is even and s is an involution in $\mathfrak{G}$ which interchanges 0 and ∞ (see 1.7), then $(-k)s = -(ks)$ for $0 \neq k \in \mathsf{K}$. Further, $(-k)h = -(kh)$ for all $h \in \mathfrak{H}$ and $k \in \mathsf{K}$.*

d) *If $d \geq \frac{1}{2}(n - 1)$ and $\mathfrak{H}$ is Abelian, multiplication may be defined on K in such a way that K is a field and every element of $\mathfrak{H}$ is of the form $x \to ax$ $(a \in \mathsf{K})$.*

Proof. Since $\mathfrak{F}$ is regular on $\mathsf{K} = \Omega - \{\infty\}$, each element of K is of the form $0f$ for a unique $f \in \mathfrak{F}$. We may therefore define addition on K by putting $0f_1 + 0f_2 = 0f_1f_2$. The mapping $f \to 0f$ is thus an isomorphism of $\mathfrak{F}$ onto K and so K is an Abelian group with respect to addition; the zero element of K is $01_{\mathfrak{F}} = 0$ and $-(0f) = 0f^{-1}$.

a) If $k = 0f'$, $kf = 0f'f = 0f' + 0f = k + 0f$. Since $\infty f = \infty$, $\infty f = \infty + 0f$ if $\infty + a$ is defined to be ∞ for all $a \in \mathsf{K}$.

b) By V, 8.18, $\mathfrak{H}$ contains a unique involution t and $f^t = f^{-1}$ for all $f \in \mathfrak{F}$. Thus if $k = 0f'$,

$$kt = 0f't = 0tf'^{-1} = 0f'^{-1} = -k.$$

c) As t is the only involution in $\mathfrak{H}$, it follows from $s \in \mathbf{N}_{\mathfrak{G}}(\mathfrak{H})$ that $t^s = t$. If $0 \neq k \in \mathsf{K}$, then since $ks \in \mathsf{K}$,

$$(-k)s = kts = kst = -(ks),$$

by b). Further, since $t \in \mathbf{Z}(\mathfrak{H})$, it follows that for all $h \in \mathfrak{H}$ and $k \in \mathsf{K}$,

$$(-k)h = kth = kht = -(kh).$$

d) First we observe that $\mathfrak{F}$ is a faithful, irreducible $\mathfrak{H}$-module. For otherwise there exists a normal subgroup $\mathfrak{N}$ of $\mathfrak{F}\mathfrak{H}$ such that $1 < \mathfrak{N} < \mathfrak{F}$. But since $\mathfrak{H}$ transforms $\mathfrak{F}$ fixed point freely,

$$|\mathfrak{N}| \geq 1 + |\mathfrak{H}| \geq \tfrac{1}{2}(n + 1) > \tfrac{1}{2}n,$$

which is impossible.

Since $\mathfrak{H}$ is Abelian, it follows from II, 3.10 that there exists an isomorphism ψ of the additive group of $GF(p^f)$ onto $\mathfrak{F}$ such that given $h \in \mathfrak{H}$, there exists $a(h) \in GF(p^f)$ for which

$$x^h = (a(h)(x\psi^{-1}))\psi$$

for all $x \in \mathfrak{F}$. Thus $x \to 0(x\psi)$ is an additive isomorphism of $GF(p^f)$ onto K. We define multiplication in K in such a way that K becomes a field and $x \to 0(x\psi)$ is a field isomorphism of $GF(p^f)$ onto K. Then

$$0xh = 0x^h = 0((a(h)(x\psi^{-1}))\psi) = (0(a(h)\psi))(0x),$$

and for all $k \in \mathsf{K}$, $kh = b(h)k$ for some $b(h) \in \mathsf{K}$. **q.e.d.**

1.9 Lemma. *Suppose that d is even, (so that $\mathfrak{F}$ is Abelian, by 1.4a)). Suppose that s is an involution which interchanges 0 and ∞. Given $k \in \mathsf{K} - \{0\}$, there exists $h_k \in \mathfrak{H}$ such that the functional equation*

$$(xs + k)s = (xh_k - k)s + ks \quad (x \in \Omega)$$

holds.

Proof. Suppose that $k = 0f$. Then $f \neq 1$ and $xf = x + k$ $(x \in \Omega)$. By 1.7, $sfs = h_k f_1 s f_2$ for suitable $h_k \in \mathfrak{H}$ and $f_i \in \mathfrak{F}$ $(i = 1, 2)$. Then

$$ks = 0fs = \infty sfs = \infty h_k f_1 s f_2 = 0f_2,$$

so $xf_2 = x + ks$. And

$$0 = 0sfs = 0h_k f_1 s f_2 = (0f_1 s)f_2 = 0f_1 s + ks.$$

Thus, using 1.8c),

$$0f_1 s = -(ks) = (-k)s$$

and $0f_1 = -k$. Therefore $xf_1 = x - k$ and

$$(xs + k)s = xsfs = xh_k f_1 s f_2 = (xh_k - k)s + ks. \qquad \textbf{q.e.d.}$$

1.10 Lemma. *Suppose that d is even and that s, h_k have the same meaning as in 1.9.*

a) $h_k = h_{k'}$ *if and only if* $k' = \pm k$.

b) $d \geq \frac{1}{2}(n - 1)$.

c) *If* $d = \frac{1}{2}(n - 1)$, *then for each* $h \in \mathfrak{H}$, *there exists* $k \in \mathsf{K} - \{0\}$ *such that* $ksh = -k$.

Proof. a) Using 1.9 and 1.8c),

$$\begin{aligned}(xh_{-k} + k)s &= (xs - k)s - ((-k)s)\\ &= -((-x)s + k)s + ks\\ &= (xh_k + k)s,\end{aligned}$$

so $h_{-k} = h_k$.

To prove the reverse statement, we show first that $ksh_k = -k$. We have

$$(xs + k)s = (xh_k - k)s + ks.$$

Since $s^2 = 1$, substitution of $x = (-k)s$ gives

$$\infty = 0s = (-k + k)s = ((-k)sh_k - k)s + ks.$$

Since $k \neq 0$, $ks \neq \infty$. It follows that $((-k)sh_k - k)s = \infty$ and so $k = (-k)sh_k = -(ksh_k)$ by 1.8c).

Suppose that k, k' are non-zero elements of K for which $h_k = h_{k'} = h$, say. From above, $ksh = -k$ and $k'sh = -k'$. Now

$$(1) \qquad (xs + k)s = (xh - k)s + ks,$$

$$(2) \qquad (xs + k')s = (xh - k')s + k's$$

for all $x \in \Omega$. Substituting $x = k's$ in (1) and $x = ks$ in (2) gives

$$\begin{aligned}(k' + k)s &= (k'sh - k)s + ks = (-k' - k)s + ks,\\ (k + k')s &= (-k - k')s + k's.\end{aligned}$$

Hence either $ks = k's$ or $(-k - k')s = \infty$. Thus either $k = k'$ or $k' = -k$.

b) By a), the $n - 1$ non-zero elements of K yield $\frac{1}{2}(n - 1)$ distinct elements h_k in $\mathfrak{H}$. Thus $d = |\mathfrak{H}| \geq \frac{1}{2}(n - 1)$.

c) If $d = \frac{1}{2}(n - 1)$, every element h of $\mathfrak{H}$ is of the form h_k for some $k \in \mathsf{K} - \{0\}$. But as was shown in the proof of a), $ksh = ksh_k = -k$.

q.e.d.

We now prove the first classification theorem for Zassenhaus groups.

1.11 Theorem (ZASSENHAUS [2]). *Let $\mathfrak{G}$ be a Zassenhaus group of degree $n + 1$ and of order $(n + 1)nd$. Suppose that d is even and $d \leq \frac{1}{2}(n - 1)$. Then $n = p^f$ for some odd prime p, and $\mathfrak{G}$ is the group $PSL(2, p^f)$ of projective linear mappings of determinant a square in $GF(p^f)^\times$, regarded as a permutation group on $\mathsf{P}(1, p^f)$.*

Proof. Since d is even, $d = \frac{1}{2}(n - 1)$ by 1.10b). Thus by 1.4b), $\mathfrak{F}$ is an elementary Abelian p-group of order p^f and $n = p^f$. Since $p^f = n = 2d + 1$, p is odd.

By 1.7c), there exists an involution s which interchanges 0 and ∞ and has two fixed points. By 1.10c), for each $h \in \mathfrak{H}$, there exists $k \in \mathsf{K}$ such that $k \neq 0$ and $ksh = -k$. As d is even, it follows from 1.8c) that

$$k(shsh) = (-k)sh = -(ksh) = k.$$

Since 0 and ∞ are also fixed points of $(sh)^2$, $(sh)^2 = 1$. Thus $s^{-1}hs = shs = h^{-1}$ for all $h \in \mathfrak{H}$, and $\mathfrak{H}$ is Abelian. By 1.8d), $\mathsf{K} = \Omega - \{\infty\}$ has the structure of a field and every element of $\mathfrak{H}$ is of the form $x \to ax$ ($a \in \mathsf{K}$). Since $|\mathfrak{H}| = \frac{1}{2}(n - 1) = |GF(p^f)^{\times 2}|$, $\mathfrak{H}$ consists of all mappings of the form $x \to a^2x$ ($a \in GF(p^f)^\times$). By 1.8a), $\mathfrak{F}$ consists of all translations on K, so $\mathfrak{G}_\infty$ consists of all mappings of the form $x \to a^2x + b$ on K, with $0 \neq a \in \mathsf{K}$ and $b \in \mathsf{K}$.

We show that if $a \in \mathsf{K}^\times$,

$$(1) \qquad (a^{-2}x)s = a^2(xs)$$

for all $x \in \mathsf{K}$. Indeed, there exists $h \in \mathfrak{H}$ such that $xh = a^2x$ for all $x \in \mathsf{K}$. Since $sh = h^{-1}s$,

$$(a^{-2}x)s = xh^{-1}s = xsh = a^2(xs).$$

It follows from (1) that $(a^{-2})s = a^2(1s)$. Thus s carries the set $\mathsf{K}^{\times 2}$ either onto itself or onto the set of non-squares in $\mathsf{K}^\times$. But in the latter case, s has no fixed point, contrary to the choice of s. Hence $1s = c^2$ for some $c \in \mathsf{K}$. Since $d = \frac{1}{2}(p^f - 1)$ is even, $p^f \equiv 1(4)$ and $-1 = l^2$ for some $l \in \mathsf{K}$. Now there exists $h_0 \in \mathfrak{H}$ such that $xh_0 = c^2 l^2 x$, and

$$(x^2)h_0 s = (c^2 l^2 x^2)s = c^{-2} l^{-2} x^{-2}(1s) = -x^{-2}.$$

Thus if $s_0 = h_0 s$, s_0 is an involution which interchanges 0 and ∞, and

$$(x^2)s_0 = -x^{-2} \quad (x \in \mathsf{K}).$$

In particular, $1s_0 = -1$. Also there exists $k \in \mathsf{K}$ such that

$$ys_0 = ky^{-1}$$

for all $y \in \mathsf{K} - \mathsf{K}^2$. For if y_1, y_2 are elements of $\mathsf{K} - \mathsf{K}^2$, $y_1^{-1} y_2 = x^2$ for some $x \in \mathsf{K}$ and by (1),

$$\begin{aligned}(y_2 s_0) y_2 &= ((y_1 x^2) h_0 s) y_2 = ((c^2 l^2 x^2 y_1) s) y_2 \\ &= (x^{-2}(c^2 l^2 y_1) s) y_2 = (y_1 s_0) y_1.\end{aligned}$$

We show that $k = -1$. To do this we use the fact that since $p > 2$, K^2 is not a subfield of K. Thus there exist a, b in K such that $a^2 + b^2 \notin \mathsf{K}^2$. But

$$a^2 + b^2 = a^2 - l^2 b^2 = (lb)^2(e - 1),$$

where $e \in \mathsf{K}^{\times 2}$ and $e - 1 \notin \mathsf{K}^2$. Now there exists $f \in \mathfrak{F}$ such that $xf = x + 1$ for all $x \in \mathsf{K}$. Then $s_0 f = (0, \infty, 1) \cdots$, so $(s_0 f)^3 = 1$. Since

$$es_0 f = (-e^{-1})f = \frac{e-1}{e}, \quad \left(\frac{e-1}{e}\right)s_0 f = \left(\frac{ke}{e-1}\right)f = \frac{ke + e - 1}{e - 1},$$

it follows that

$$\left(\frac{ke + e - 1}{e - 1}\right)s_0 f = e$$

and

$$\frac{ke + e - 1}{e - 1} = ef^{-1}s_0 = \frac{k}{e - 1}.$$

Thus $k = -1$.

It follows that $xs_0 = -x^{-1}$ for all $x \in \mathsf{K}$. Thus $\mathfrak{G} = \langle \mathfrak{G}_\infty, s_0 \rangle$ is contained in $PSL(2, p^f)$. Equality of orders forces $\mathfrak{G} = PSL(2, p^f)$.

q.e.d.

1.12 Theorem. *Suppose that $\mathfrak{G}$ is a simple Zassenhaus group of degree $n + 1$ and of order $(n + 1)nd$, where $d = \frac{1}{2}(n - 1)$ is odd. Then $n = p^f$ for p odd, and $\mathfrak{G}$ is the group $PSL(2, p^f)$ of projective linear mappings of determinant a square in $GF(p^f)^\times$, regarded as a permutation group on* $\mathsf{P}(1, p^f)$.

Proof. On account of the simplicity of $\mathfrak{G}$, it follows from Lemma 1.5 that $\mathfrak{H}$ is cyclic, say $\mathfrak{H} = \langle h \rangle$. Also $\mathbf{N}_\mathfrak{G}(\mathfrak{H})' = \mathfrak{H}$, and if s is an involution in $\mathfrak{G}$ which interchanges 0 and ∞, $\mathbf{N}_\mathfrak{G}(\mathfrak{H}) = \mathfrak{H}\langle s \rangle$ and $h^s = h^{-1}$.

By 1.8d), multiplication may be defined on K in such a way that $\mathsf{K} = GF(p^f)$ and $xh = a^2x$, where a is a generator of $\mathsf{K}^\times$. It follows from $(h^j)^s = h^{-j}$ that

$$(a^{2j}x)s = xh^js = xsh^{-j} = a^{-2j}(xs).$$

In particular, $1(h^js) = (a^{-2j})(1s)$. Now -1 is not a square in K, since $p^f = 2d + 1 \equiv 3(4)$. Thus we can find j such that $s_0 = h^js$ carries 1 into either 1 or -1. But if $1s_0 = 1$, s_0 has at least two fixed points, since $n + 1$ is even and $s_0^2 = 1$. This is impossible, since $|\mathfrak{G}_{0,\infty}|$ is odd. Thus $1s_0 = -1$ and

$$(a^{2i})s_0 = a^{2(i+j)}s = a^{-2i}(a^{2j}s) = a^{-2i}(1h^js) = a^{-2i}(1s_0) = -a^{-2i}.$$

Since $(-1)s_0 = 1s_0^2 = 1$,

$$\begin{aligned}(-a^{2i})s_0 &= (-a^{2(i+j)})s = a^{-2i}((-a^{2j})s) = a^{-2i}((-1)s_0)\\ &= a^{-2i} = -(-a^{2i})^{-1}.\end{aligned}$$

Thus $xs_0 = -x^{-1}$ for all $x \in \mathsf{K}^\times$. Hence $\mathfrak{G} = \langle \mathfrak{G}_\infty, s_0 \rangle$ is contained in $PSL(2, p^f)$. By comparison of orders, $\mathfrak{G} = PSL(2, p^f)$. **q.e.d.**

§ 2. Sharply Triply Transitive Permutation Groups

Suppose that $\mathfrak{G}$ is a sharply triply transitive group of degree $n + 1$. Obviously any non-identity element of $\mathfrak{G}$ has at most two fixed points.

If in addition $\mathfrak{G}$ is a Frobenius group, then $|\mathfrak{G}| = (n+1)n(n-1)$ divides $(n+1)n$, so $n = 2$ and $\mathfrak{G} = \mathfrak{S}_3 = PGL(2, 2)$. If $\mathfrak{G}$ has a regular normal subgroup but is not a Frobenius group, then by 1.1, $n + 1 = 2^p$ and

$$2^p(2^p - 1)p = |\mathfrak{G}| = 2^p(2^p - 1)(2^p - 2);$$

thus $n + 1 = 4$ and $\mathfrak{G} = \mathfrak{S}_4 = PGL(2, 3)$. All the remaining sharply triply transitive permutation groups of degree $n + 1$ are Zassenhaus groups of order $(n+1)n(n-1)$. By 1.4b), n is a prime-power p^f and the Frobenius kernel $\mathfrak{F}$ of $\mathfrak{G}_\infty$ is an elementary Abelian p-group.

We begin with the simplest case $p = 2$.

2.1 Theorem (ZASSENHAUS [2]). *Suppose that $\mathfrak{G}$ is sharply triply transitive of degree $n + 1$, where $n = 2^f$. Then $\mathfrak{G} = PGL(2, 2^f)$.*

Proof. For $n = 2$, $\mathfrak{G} = \mathfrak{S}_3 = PGL(2, 2)$. Suppose, then, that $n > 2$. By 1.1, $\mathfrak{G}$ has no regular normal subgroup and is thus a Zassenhaus group.

By 1.6b), $\mathfrak{G}$ is simple, so by 1.5, $\mathfrak{H} = \mathfrak{G}_{0,\infty}$ is cyclic. By 1.8, $\mathsf{K} = \Omega - \{\infty\}$ can be given the structure of a field in such a way that $\mathfrak{G}_\infty$ consists of all mappings of the form

$$x \to ax + b, \infty \to \infty \quad (0 \neq a \in \mathsf{K}, b \in \mathsf{K})$$

on the set Ω. Since $\mathfrak{G}$ is triply transitive, there exists $s \in \mathfrak{G}$ of the form $s = (0, \infty)(1) \cdots$. We show that $(ax)s = (as)(xs)$. Indeed, given $a \in \mathsf{K}^\times$, there exists $h \in \mathfrak{H}$ such that $xh = ax$ for all $x \in \mathsf{K}$. But since $s \in \mathbf{N}_{\mathfrak{G}}(\mathfrak{H})$, $h^s \in \mathfrak{H}$ and $xh^s = a'x$ for some $a' \in \mathsf{K}^\times$. Hence $(ax)s = xhs = xsh^s = a'(xs)$. Since $1s = 1$, we see by taking $x = 1$ that $a' = as$, so $(ax)s = (as)(xs)$, as required.

Thus s induces an automorphism on the cyclic group $\mathsf{K}^\times$. Since $|\mathsf{K}^\times|$ is odd and $s^2 = 1$, the number of fixed points of s on $\mathsf{K}^\times$ and thus on Ω is odd. Since $s \neq 1$, this number is less than 3 and is therefore 1. Hence the automorphism of $\mathsf{K}^\times$ induced by s is fixed point free of order 2. Therefore $xs = x^{-1}$ for all $x \in \mathsf{K}^\times$, by V, 8.18. Thus $\mathfrak{G} = \langle \mathfrak{G}_\infty, s \rangle = PGL(2, 2^f)$. **q.e.d.**

The case when p is odd is substantially more difficult, largely because the groups $M(p^f)$ (see 1.3c)) come into play. As far as 2.6, $\mathfrak{G}$ will denote a sharply triply transitive Zassenhaus group of degree $p^f + 1$ with p odd. In particular, $d = p^f - 1$ is even, so $\mathfrak{H} = \mathfrak{G}_{0,\infty}$ contains exactly one involution t. Further, s will always denote the unique element of

$\mathfrak{G}$ of the form $(0, \infty)(1) \cdots$. Thus $s^2 = 1$. First we clarify the structure of $\mathfrak{H}$ in a series of lemmas.

2.2 Lemma (TITS [1]). *$\mathfrak{H}$ has at most one subgroup of order 3.*

Proof. Suppose that $h = (0)(\infty)(x, y, z) \cdots$ and $h' = (0)(\infty)(x, y', z') \cdots$ are distinct elements of $\mathfrak{H}$ of order 3. Since $\mathfrak{G}$ is triply transitive, there exists $g \in \mathfrak{G}$ of the form $g = (x)(y, z) \cdots$. Then $h^g = (x, z, y) \cdots$ and $h^g h = (x)(y)(z) \cdots = 1$, so $h^g = h^{-1}$. Let g' be the element of $\mathfrak{G}$ for which $xg' = x$, $yg' = y'$, $zg' = z'$. Then $h^{g'} = (x, y', z') \cdots$, so $h^{g'} = h'$. Thus $g' \neq 1$. Also g, g' both carry the fixed point set $\{0, \infty\}$ of h and h' into itself. Since neither g nor g' can have more than two fixed points, it follows that $g = (0, \infty)(x) \cdots$ and $g' = (0, \infty)(x) \cdots$; hence $g' = g$. Hence $h' = h^{g'} = h^g = h^{-1}$ and $\langle h \rangle = \langle h' \rangle$. **q.e.d.**

2.3 Lemma. *The centralizer of s in $\mathfrak{H}$ is $\langle t \rangle$.*

Proof. Since $1s = 1$, $s^2 = 1$ and $p^f + 1$ is even, s has just one fixed point $a \neq 1$. Then $\mathbf{C}_{\mathfrak{H}}(s)$ leaves $\{1, a\}$ fixed, so $\mathbf{C}_{\mathfrak{H}}(s)$ is faithfully represented on $\{1, a\}$. Thus $|\mathbf{C}_{\mathfrak{H}}(s)| \leq 2$ and $\mathbf{C}_{\mathfrak{H}}(s) = \langle t \rangle$. **q.e.d.**

2.4 Lemma. *If $\mathfrak{H}$ has a normal subgroup $\mathfrak{N}$ of even index such that $\mathfrak{N}^s = \mathfrak{N}$, then $\mathfrak{N}$ is cyclic and $u^s = u^{-1}$ for all $u \in \mathfrak{N}$.*

Proof. Since $|\mathfrak{H}/\mathfrak{N}|$ is even, s leaves fixed some coset $\mathfrak{N}x \neq \mathfrak{N}$. Let $\mathsf{X} = \mathfrak{N} \cup \mathfrak{N}x$ and given $h \in \mathsf{X}$, put $h\theta = h^s h^{-1}$. Then $h\theta \in \mathfrak{N}$ for all $h \in \mathsf{X}$. If h_1, h_2 are elements of X for which $h_1\theta = h_2\theta$, then $(h_2^{-1}h_1)^s = h_2^{-1}h_1$. By 2.3, $h_2^{-1}h_1$ is either 1 or t. Thus θ carries the $2|\mathfrak{N}|$ elements of X into at least $|\mathfrak{N}|$ distinct elements of $\mathfrak{N}$. Hence

$$\mathfrak{N} = \{h^s h^{-1} \mid h \in \mathsf{X}\}.$$

Since

$$(h^s h^{-1})^s = h^{s^2} h^{-s} = h h^{-s} = (h^s h^{-1})^{-1},$$

$\mathfrak{N}$ is Abelian. But by V, 8.7, the Sylow subgroups of $\mathfrak{H}$ are either cyclic or generalized quaternion, so $\mathfrak{N}$ is cyclic. **q.e.d.**

We know (by 1.4b)) that $\mathfrak{F}$ is Abelian, so by 1.8, $\mathsf{K} = \Omega - \{\infty\}$ can be made into an additive Abelian group. To define a field structure on K, the multiplicative structure of K will be derived from $\mathfrak{H}$. The following lemma gives us the essential information about $\mathfrak{H}$.

2.5 Lemma. $\mathfrak{H}$ *possesses a normal subgroup* $\mathfrak{N}$ *of index 2 such that* $\mathfrak{N}^s = \mathfrak{N}$.

Proof. We have the following information about $\mathfrak{H}$.

(1) The Sylow subgroups of $\mathfrak{H}$ are cyclic or generalized quaternion (by V, 8.7).

(2) $\mathfrak{H}$ has at most one subgroup of order 3 (by 2.2).

We assume that the assertion is false.

a) $\mathbf{O}^2(\mathfrak{H}) = \mathfrak{H}$.

For otherwise, s permutes the maximal subgroups of $\mathfrak{H}/\mathbf{O}^2(\mathfrak{H})$ and the number of them is odd (III, 8.8). Then s leaves a maximal subgroup fixed, contrary to hypothesis.

b) The Sylow 2-subgroups of $\mathfrak{H}$ are generalized quaternion groups and $|\mathfrak{H}|$ is divisible by 3.

If the Sylow 2-subgroup of $\mathfrak{H}$ is cyclic, then $\mathfrak{H}$ is 2-nilpotent, contrary to a). If the order of $\mathfrak{H}$ is not divisible by 3, then $\mathfrak{H}$ is 2-nilpotent by IV, 5.11, since the Sylow 2-subgroups of $\mathfrak{H}$ are certainly metacyclic.

c) $\mathfrak{H}$ is 3-nilpotent.

By b) and 2.2, $\mathfrak{H}$ has precisely one subgroup $\mathfrak{T}$ of order 3. By (1), the Sylow 3-subgroup $\mathfrak{P}$ of $\mathfrak{H}$ is cyclic. Also $\mathfrak{T} \trianglelefteq \mathfrak{H}$ and $|\mathfrak{H}/\mathbf{C}_{\mathfrak{H}}(\mathfrak{T})| \leq 2$, so by a), $\mathbf{C}_{\mathfrak{H}}(\mathfrak{T}) = \mathfrak{H}$. Hence $\mathfrak{T} = \Omega_1(\mathfrak{P})$ is centralized by $\mathbf{N}_{\mathfrak{H}}(\mathfrak{P})$, so by IV, 5.12a), $\mathbf{N}_{\mathfrak{H}}(\mathfrak{P}) = \mathbf{C}_{\mathfrak{H}}(\mathfrak{P})$. By Burnside's transfer theorem (IV, 2.6), $\mathfrak{H}$ is 3-nilpotent.

d) $\mathfrak{H}$ has a cyclic characteristic subgroup $\mathfrak{Z}$ of index 12.

By c), $\mathfrak{H}$ has a normal 3-complement $\mathfrak{L}$. Then $|\mathfrak{L}|$ is not divisible by 3 and the Sylow 2-subgroups of $\mathfrak{L}$ are generalized quaternion groups. By IV, 5.11, $\mathfrak{L}$ is 2-nilpotent. It therefore follows from b) that $\mathfrak{L}$ has a characteristic subgroup $\mathfrak{M}$ such that $\mathfrak{L}/\mathfrak{M}$ is elementary Abelian of order 4. Let $\mathfrak{K}/\mathfrak{M} = \mathbf{C}_{\mathfrak{H}/\mathfrak{M}}(\mathfrak{L}/\mathfrak{M})$. Then $\mathfrak{K}/\mathfrak{M}$ is nilpotent and by a), $|\mathfrak{H}/\mathfrak{K}| = 3$. If $\mathfrak{Z}/\mathfrak{M}$ is the Sylow 3-subgroup of $\mathfrak{K}/\mathfrak{M}$, then $\mathfrak{Z}$ is a characteristic subgroup of $\mathfrak{H}$ of index 12. Thus $\mathfrak{Z}^s = \mathfrak{Z}$, and $\mathfrak{Z}$ is cyclic by 2.4.

e) $\mathfrak{F}$ is a reducible $\mathfrak{Z}$-module.

Otherwise, by II, 3.12, $\mathfrak{H}$ is a group of semilinear mappings $x \rightarrow a(x\alpha)$ over a finite field. Thus $\mathfrak{H}$ has a normal subgroup $\mathfrak{J}$ such that $\mathfrak{J}$ and $\mathfrak{H}/\mathfrak{J}$ are both cyclic. Thus $\mathfrak{H}$ is supersoluble. But then, by VI, 9.1, $\mathfrak{H}$ is 2-nilpotent, contrary to a).

f) Let $\mathfrak{F}_1$ be an irreducible $\mathfrak{Z}$-submodule of $\mathfrak{F}$ and let $|\mathfrak{F}_1| = p^r$. Since $\mathfrak{H}$ operates transitively on $\mathfrak{F} - \{1\}$, $\mathfrak{F}$ is an irreducible $\mathfrak{H}$-module. Thus by Clifford's theorem (V, 17.3), $f = ru$ for some u. Since $\mathfrak{Z}$ transforms $\mathfrak{F}_1$ fixed point freely, we see by counting the $\mathfrak{Z}$-orbits in $\mathfrak{F}_1$ that $p^r = 1 + v|\mathfrak{Z}|$ for some integer v. Further, $|\mathfrak{H}| = p^f - 1 = 12|\mathfrak{Z}|$. It follows that

$$1 + p^r + \cdots + p^{r(u-1)} = \frac{p^f - 1}{p^r - 1} = \frac{12}{v}.$$

Since $u > 1$, it follows that $p^r < 12$. By b), 3 divides $p^f - 1$, so $p \neq 3$. The remaining possiblities for p^r are 5, 7, 11, and $u = 2$. In each case, 3^2 does not divide $p^f - 1 = |\mathfrak{H}|$. Hence by 2.2, $\mathfrak{H}$ has just one Sylow 3-subgroup $\mathfrak{P}$. By IV, 5.11, $\mathfrak{H}/\mathfrak{P}$ is 2-nilpotent. Thus $\mathfrak{H}$ itself is 2-nilpotent, contrary to a). **q.e.d.**

The description of $\mathfrak{H}$ obtained in 2.5 is sufficient for the proof of the following theorem.

2.6 Theorem (ZASSENHAUS [2]). *If $\mathfrak{G}$ is sharply triply transitive of degree $p^f + 1$, where $p > 2$, then either $\mathfrak{G} = PGL(2, p^f)$, or $f = 2m$ is even and $\mathfrak{G} = M(p^{2m})$ (see 1.3c)).*

Proof. If $\mathfrak{G}$ is sharply triply transitive but is not a Zassenhaus group, then $\mathfrak{G} = \mathfrak{S}_4 = PGL(2, 3)$ according to the remarks at the beginning of this section. Suppose then that $\mathfrak{G}$ is a sharply triply transitive Zassenhaus group of degree $p^f + 1$.

By 2.5, $\mathfrak{H}$ has a cyclic normal subgroup $\mathfrak{N}$ of index 2. If $f \in \mathfrak{F} - \{1\}$, then since $\mathfrak{N}$ transforms $\mathfrak{F}$ fixed point freely,

$$1 + |\{f^h | h \in \mathfrak{N}\}| = 1 + \frac{p^f - 1}{2} > p^{f-1};$$

hence $\mathfrak{F}$ is an irreducible $\mathfrak{N}$-module. By 1.8, there exists an addition on $\mathsf{K} = \Omega - \{\infty\}$ such that $\mathfrak{F}$ is the set of translations $x \to x + k$ of K; also $kt = -k$ and $(-k)s = -(ks)$ for all $k \in \mathsf{K}^\times$. The mapping η of $\mathfrak{F}$ onto K given by $e\eta = 0e$ ($e \in \mathfrak{F}$) is an isomorphism and indeed an $\mathfrak{H}$-isomorphism, since if $h \in \mathfrak{H}$,

$$e^h\eta = 0e^h = 0h^{-1}eh = 0eh = (e\eta)h.$$

Thus K is an irreducible $\mathfrak{N}$-module. By II, 3.11, K can be given the structure of the field $GF(p^f)$ in such a way that $\mathfrak{H}$ consists of semilinear mappings $x \to a(x\alpha)$, where α is an automorphism of K. Further $\mathfrak{N}$ consists of linear mappings, and since $|\mathfrak{H}/\mathfrak{N}| = 2$, $\alpha^2 = 1$. If the only automorphism α which occurs is 1, then $\mathfrak{G}_\infty$ consists of all mappings of the form $x \to ax + b$ $(a \neq 0)$ and $\mathfrak{G}_\infty \leq PGL(2, p^f)$. Otherwise, $f = 2m$, K has a unique automorphism τ of order 2 and there exists $h \in \mathfrak{H}$ such that $xh = c(x\tau)$ for all $x \in \mathsf{K}$, where $c \in \mathsf{K}^\times$. Now if c is a square, $x \to cx$ lies in $\mathfrak{N}$, so $x \to x\tau$ lies in $\mathfrak{H}$. But τ leaves each element of $GF(p)$ fixed and $p \geq 3$, contrary to the fact that $\mathfrak{G}$ is a Zassenhaus group. Thus $c \notin \mathsf{K}^2$. If follows that $\mathfrak{G}_\infty \leq M(p^{2m})$ in this case.

The mapping $x \to x^{-1}$ obviously lies in $PGL(2, p^f)$; but it also lies in $M(p^f)$, since -1 is a square in $GF(p^f)$ for f even. It will therefore be sufficient to show that $xs = x^{-1}$, since $\mathfrak{G} = \langle \mathfrak{G}_\infty, s \rangle$ and $|\mathfrak{G}| = |PGL(2, p^f)| = |M(p^f)|$. To do this, we use 1.9. Thus given $a \in \mathsf{K}^\times$, there exists $h_a \in \mathfrak{H}$ such that for all $x \in \Omega$,

$$(1) \qquad (xs + a)s = (xh_a - a)s + as.$$

Given $a \in \mathsf{K}^\times$, there exists $h \in \mathfrak{N}$ such that $xh = a^2x$. But by 2.4, $h^s = h^{-1}$, so for all $x \in \mathsf{K}$,

$$(2) \qquad (a^2x)s = xhs = xsh^{-1} = a^{-2}(xs).$$

In particular, since $1s = 1$, we have

$$(3) \qquad (a^2)s = a^{-2},$$

so by (2),

$$(4) \qquad (a^2x)s = ((a^2)s)(xs).$$

(3) also shows that s carries $\mathsf{K}^{\times 2}$ onto itself. Hence s carries $\mathsf{K}^\times - \mathsf{K}^{\times 2}$ onto itself, so for any $x \in \mathsf{K}^\times$, $(xs)x^{-1} \in \mathsf{K}^{\times 2}$.

Suppose that there exists $a \in \mathsf{K}^\times$ such that h_a is of the form $xh_a = a'(x\tau)$, where $a' \notin \mathsf{K}^2$ and τ is the automorphism of K of order 2; thus $f = 2m$. If $y = (a\tau)(a'\tau)^{-1}$, then $y\tau = aa'^{-1}$ and by (1),

$$(ys + a)s = (a - a)s + as = \infty.$$

Hence $ys = -a$. Since $(ys)y^{-1} \in \mathsf{K}^{\times 2}$, $-a(a\tau)^{-1}(a'\tau) \in \mathsf{K}^{\times 2}$. Thus $-a^{1-p^m}a'^{p^m} \in \mathsf{K}^{\times 2}$. But -1 is a square in $GF(p^{2m})$, so it follows that $a'^{p^m} \in \mathsf{K}^{\times 2}$ and $a' = a'^{1-p^m}a'^{p^m} \in \mathsf{K}^{\times 2}$, a contradiction. Thus $xh_a = a'x$. If $z = aa'^{-1}$, then (1) gives $zs = -a$. By 1.8c),

$$z = zs^2 = (-a)s = -(as).$$

Thus $a' = az^{-1} = -a(as)^{-1}$. If $a \in \mathsf{K}^{\times 2}$, then it follows from (3) that $a' = -a^2$, and by (1),

$$(5) \qquad (xs + a)s = (-a^2x - a)s + as \quad (a \in \mathsf{K}^{\times 2}, x \in \mathsf{K}).$$

Now there exists $b \in \mathsf{K}^{\times 2}$ such that $b + 1 \notin \mathsf{K}^2$; otherwise K^2 would be a subfield of K. By (3), $bs = b^{-1}$. By (5) (with $x = 1$), $(b + 1)s = (-b^2 - b)s + bs$. Using (4) and 1.8c),

$$(b + 1)s = -((b^2 + b)s) + bs = -(bs)((b + 1)s) + bs$$
$$= -b^{-1}((b + 1)s) + b^{-1}.$$

Hence $(b + 1)s = \dfrac{b^{-1}}{1 + b^{-1}} = (b + 1)^{-1}$. Thus if $c \notin \mathsf{K}^2$, $c = (b + 1)a$ for some $a \in \mathsf{K}^{\times 2}$ and by (2),

$$cs = a^{-1}((b + 1)s) = a^{-1}(b + 1)^{-1} = c^{-1}.$$

Thus $xs = x^{-1}$ for all $x \in \mathsf{K}$, as required. **q.e.d.**

We conclude this section by giving another characterization of $PSL(2, 2^n)$.

2.7 Theorem (BRAUER, SUZUKI and WALL [1]). *Let $\mathfrak{G}$ be a group in which the centralizer of any involution is an elementary Abelian 2-group. Then one of the following holds.*

(1) *$|\mathfrak{G}|$ is twice an odd number.*
(2) *The Sylow 2-subgroup of $\mathfrak{G}$ is normal.*
(3) *$\mathfrak{G} = PSL(2, q)$, where q is a power of 2.*

Proof (GOLDSCHMIDT [5]). Let $\mathfrak{T}$ be a Sylow 2-subgroup of $\mathfrak{G}$ and let $q = |\mathfrak{T}|$. We suppose that (1) and (2) are false. Thus $q > 2$ and $\mathfrak{T}$ is not normal.

a) $\mathfrak{T}$ is elementary Abelian and $\mathbf{C}_{\mathfrak{G}}(t) = \mathfrak{T}$ for all $t \in \mathfrak{T} - \{1\}$.

Let x be an involution in $\mathbf{Z}(\mathfrak{T})$. By hypothesis, $\mathbf{C}_{\mathfrak{G}}(x)$ is elementary Abelian, so $\mathfrak{T}$ is elementary Abelian. Thus if $t \in \mathfrak{T} - \{1\}$, $\mathfrak{T} \leq \mathbf{C}_{\mathfrak{G}}(t)$ and $\mathbf{C}_{\mathfrak{G}}(t)$ is elementary Abelian. Since $\mathfrak{T} \in S_2(\mathfrak{G})$, $\mathfrak{T} = \mathbf{C}_{\mathfrak{G}}(t)$.

b) If $\mathfrak{T}_1, \mathfrak{T}_2$ are distinct Sylow 2-subgroups of $\mathfrak{G}$, $\mathfrak{T}_1 \cap \mathfrak{T}_2 = 1$.

Suppose that $t \in \mathfrak{T}_1 \cap \mathfrak{T}_2$ and $t \neq 1$. By a), $\mathfrak{T}_1 = \mathbf{C}_{\mathfrak{G}}(t) = \mathfrak{T}_2$.

c) All involutions of $\mathfrak{G}$ are conjugate.

Let s, t be distinct involutions. By Sylow's theorem, t is conjugate to an element t' of $\mathfrak{T}$. Also, since $\mathfrak{T}$ is not normal, it follows from b) that s is conjugate to an element s' outside $\mathfrak{T}$. Let k be the order of $s't'$; thus $\langle s', t' \rangle$ is a dihedral group of order $2k$ (I, 19.5). If k is even, $\langle s't' \rangle$ contains an involution u. Thus $u \in \mathbf{C}_{\mathfrak{G}}(t')$, so by a), $u \in \mathfrak{T}$. But again, $s' \in \mathbf{C}_{\mathfrak{G}}(u)$, so by a), $s' \in \mathfrak{T}$. This is a contradiction, so k is odd. Thus all involutions in the dihedral group $\langle s', t' \rangle$ are conjugate; in particular, s' and t' are conjugate. Hence s and t are conjugate.

Let $\mathfrak{M} = \mathbf{N}_{\mathfrak{G}}(\mathfrak{T})$. Since $\mathfrak{T}$ is not normal, $\mathfrak{M} < \mathfrak{G}$.

d) $\mathfrak{M}$ is *strongly embedded* in $\mathfrak{G}$; that is, $|\mathfrak{M} \cap \mathfrak{M}^g|$ is odd for all $g \in \mathfrak{G} - \mathfrak{M}$.

Since $S_2(\mathfrak{M}) = \{\mathfrak{T}\}$ and $S_2(\mathfrak{M}^g) = \{\mathfrak{T}^g\}$, $\mathfrak{T} \cap \mathfrak{T}^g$ is the Sylow 2-subgroup of $\mathfrak{M} \cap \mathfrak{M}^g$. Since $g \notin \mathfrak{M}$, $\mathfrak{T}^g \neq \mathfrak{T}$, so by b), $\mathfrak{T} \cap \mathfrak{T}^g = 1$. Thus $|\mathfrak{M} \cap \mathfrak{M}^g|$ is odd.

e) Any two involutions in $\mathfrak{T}$ are conjugate in $\mathfrak{M}$.

If s, t are involutions in $\mathfrak{T}$, $s = t^x$ for some $x \in \mathfrak{G}$, by c). So $s = t^x \in \mathfrak{T} \cap \mathfrak{T}^x$ and $\mathfrak{T} \cap \mathfrak{T}^x \neq 1$. By b), $\mathfrak{T}^x = \mathfrak{T}$, so $x \in \mathfrak{M}$.

f) $|\mathfrak{M}| = q(q-1)$.

If t is an involution in $\mathfrak{M}$, t has $q - 1$ conjugates in $\mathfrak{M}$, by e). Hence $|\mathfrak{M} : \mathbf{C}_{\mathfrak{M}}(t)| = q - 1$. By a), $\mathbf{C}_{\mathfrak{M}}(t) = \mathfrak{T}$, so $|\mathfrak{M} : \mathfrak{T}| = q - 1$ and $|\mathfrak{M}| = q(q-1)$.

For any involution s in $\mathfrak{G} - \mathfrak{M}$, put

$$\mathfrak{H}(s) = \{h \mid h \in \mathfrak{M},\ h^s = h^{-1}\}.$$

(Note that we do not yet know that $\mathfrak{H}(s)$ is a subgroup of $\mathfrak{G}$). Clearly $\mathfrak{H}(s) \subseteq \mathfrak{M} \cap \mathfrak{M}^s$.

g) If s is an involution in $\mathfrak{G} - \mathfrak{M}$, the set of involutions in $\mathfrak{M}s$ is $\{hs \mid h \in \mathfrak{H}(s)\}$.

For if $h \in \mathfrak{M}$, $(hs)^2 = 1$ if and only if $h^s = h^{-1}$.

h) The number of involutions in any right coset of $\mathfrak{M}$ in $\mathfrak{G}$ is $q - 1$.

We show first that the number of involutions in a right coset X of $\mathfrak{M}$ is at most $q - 1$. To do this, we may clearly suppose that $X \neq \mathfrak{M}$ and that X contains an involution s. By g), the number of involutions in $X = \mathfrak{M}s$ is $|\mathfrak{H}(s)|$. But $|\mathfrak{H}(s)| \leq |\mathfrak{M} \cap \mathfrak{M}^s|$. By d) and f), $|\mathfrak{M} \cap \mathfrak{M}^s|$ divides $q - 1$, so $|\mathfrak{H}(s)| \leq q - 1$ and the number of involutions in X is at most $q - 1$.

If there exists a right coset of $\mathfrak{M}$ containing fewer than $q - 1$ involutions, the total number of involutions in $\mathfrak{G}$ is less than $(q-1)|\mathfrak{G} : \mathfrak{M}|$. In fact, however, this is precisely the number of involutions in $\mathfrak{G}$, for if t is an involution in $\mathfrak{T}$, the number of involutions in $\mathfrak{G}$ is $|\mathfrak{G} : \mathbf{C}_{\mathfrak{G}}(t)|$ by c). Using a), this number is

$$|\mathfrak{G} : \mathfrak{T}| = |\mathfrak{M} : \mathfrak{T}||\mathfrak{G} : \mathfrak{M}| = (q-1)|\mathfrak{G} : \mathfrak{M}|,$$

by f). Thus every right coset of $\mathfrak{M}$ contains $q - 1$ involutions.

i) For any involution s in $\mathfrak{G} - \mathfrak{M}$, $\mathfrak{H}(s)$ is an Abelian subgroup of $\mathfrak{M}$ of order $q - 1$. Also $\mathfrak{T}\mathfrak{H}(s) = \mathfrak{M}$ and $\mathfrak{H}(s) \cap \mathfrak{T} = 1$.

By g) and h), $|\mathfrak{H}(s)| = q - 1$. By d) and f), $|\mathfrak{M} \cap \mathfrak{M}^s| \leq q - 1$. Since $\mathfrak{H}(s) \subseteq \mathfrak{M} \cap \mathfrak{M}^s$, it follows that $\mathfrak{H}(s) = \mathfrak{M} \cap \mathfrak{M}^s$. Thus $\mathfrak{H}(s)$ is a subgroup of $\mathfrak{M}$ of order $q - 1$. Thus if x, y are elements of $\mathfrak{H}(s)$, we have $xy \in \mathfrak{H}(s)$ and

$$y^{-1}x^{-1} = (xy)^{-1} = (xy)^s = x^s y^s = x^{-1}y^{-1}.$$

Hence $\mathfrak{H}(s)$ is Abelian. Since the orders of $\mathfrak{H}(s)$, $\mathfrak{T}$ are coprime, $\mathfrak{H}(s) \cap \mathfrak{T} = 1$. Hence $|\mathfrak{T}\mathfrak{H}(s)| = |\mathfrak{T}||\mathfrak{H}(s)| = |\mathfrak{M}|$ and $\mathfrak{T}\mathfrak{H}(s) = \mathfrak{M}$.

j) If s is any involution in $\mathfrak{G} - \mathfrak{M}$ and $1 \neq \mathfrak{X} \leq \mathfrak{H}(s)$, then $\mathbf{N}_{\mathfrak{M}}(\mathfrak{X}) = \mathfrak{H}(s)$.

Since $\mathfrak{H}(s)$ is Abelian, $\mathbf{N}_{\mathfrak{M}}(\mathfrak{X}) \geq \mathfrak{H}(s)$. But $\mathfrak{M} = \mathfrak{T}\mathfrak{H}(s)$, so $\mathbf{N}_{\mathfrak{M}}(\mathfrak{X}) = \mathfrak{Y}\mathfrak{H}(s)$, where $\mathfrak{Y} = \mathfrak{T} \cap \mathbf{N}_{\mathfrak{M}}(\mathfrak{X})$. Since

$$[\mathfrak{Y}, \mathfrak{X}] \leq [\mathfrak{T}, \mathfrak{X}] \cap [\mathbf{N}_{\mathfrak{M}}(\mathfrak{X}), \mathfrak{X}] \leq \mathfrak{T} \cap \mathfrak{X} = 1,$$

$\mathfrak{Y} \leq \mathbf{C}_{\mathfrak{T}}(\mathfrak{X})$. By a), $\mathbf{C}_{\mathfrak{T}}(\mathfrak{X}) = 1$, so $\mathfrak{Y} = 1$ and $\mathbf{N}_{\mathfrak{M}}(\mathfrak{X}) = \mathfrak{H}(s)$.

k) If s is any involution in $\mathfrak{G} - \mathfrak{M}$ and $\mathfrak{H} = \mathfrak{H}(s)$,

$$\mathbf{N}_{\mathfrak{G}}(\mathfrak{H}) = \mathfrak{H} \cup \mathfrak{H}s.$$

Clearly, $\mathfrak{H} \cup \mathfrak{H}s \subseteq \mathbf{N}_{\mathfrak{G}}(\mathfrak{H})$.

Suppose that $g \in \mathbf{N}_{\mathfrak{G}}(\mathfrak{H}) - \mathfrak{H}$. By h), $\mathfrak{M}g$ contains an involution t. If $g = xt$, then $x \in \mathfrak{M}$, and since $\mathfrak{H}^g = \mathfrak{H}$,

$$\mathfrak{H}^{xt} = \mathfrak{H}^g = \mathfrak{H} \leq \mathfrak{M} \cap \mathfrak{M}^t = \mathfrak{H}(t).$$

By i), $|\mathfrak{H}(t)| = |\mathfrak{H}|$, so $\mathfrak{H}(t) = \mathfrak{H}$. Thus $t \in \mathbf{N}_{\mathfrak{G}}(\mathfrak{H})$ and $x = gt^{-1} \in \mathbf{N}_{\mathfrak{M}}(\mathfrak{H})$. By j), $\mathbf{N}_{\mathfrak{M}}(\mathfrak{H}) = \mathfrak{H}$, so $x \in \mathfrak{H}$. Since $\mathfrak{H}$ is Abelian, it follows that for any $h \in \mathfrak{H}$,

$$h^g = h^{xt} = h^t = h^{-1}.$$

Since $q > 2$ and $|\mathfrak{H}| = q - 1$, we have $\mathfrak{H} \neq 1$ and $g \notin \mathbf{C}_{\mathfrak{G}}(\mathfrak{H})$. Thus $\mathbf{C}_{\mathfrak{G}}(\mathfrak{H}) = \mathfrak{H}$. But also $h^{gs^{-1}} = (h^{-1})^s = h$ for all $h \in \mathfrak{H}$, so $gs^{-1} \in \mathbf{C}_{\mathfrak{G}}(\mathfrak{H}) = \mathfrak{H}$ and $g \in \mathfrak{H}s$. Thus $\mathbf{N}_{\mathfrak{G}}(\mathfrak{H}) = \mathfrak{H} \cup \mathfrak{H}s$.

l) $|\mathfrak{G}| = (q + 1)q(q - 1)$.

Let I be the set of involutions in $\mathfrak{G} - \mathfrak{M}$ and let

$$\mathscr{H} = \{\mathfrak{H}(s) | s \in \mathsf{I}\}.$$

If $\mathfrak{H} \in \mathscr{H}$, then it follows from k) that $\mathbf{N}_{\mathfrak{G}}(\mathfrak{H})$ consists of $\mathfrak{H}$ and $q - 1$ involutions t for which $h^t = h^{-1}$ $(h \in \mathfrak{H})$. Hence $\mathfrak{H} = \mathfrak{H}(t)$ for precisely these $q - 1$ involutions in $\mathbf{N}_{\mathfrak{G}}(\mathfrak{H})$. Therefore

$$|\mathsf{I}| = (q - 1)|\mathscr{H}|.$$

Next, let $\mathfrak{H}_1$, $\mathfrak{H}_2$ be two distinct elements of $\mathscr{H}$. Since $\mathfrak{H}_i$ is Abelian, $\mathfrak{H}_i \leq \mathbf{N}_{\mathfrak{M}}(\mathfrak{H}_1 \cap \mathfrak{H}_2)$ for $i = 1, 2$. Thus by j), $\mathfrak{H}_1 \cap \mathfrak{H}_2 \neq 1$ implies that $\mathfrak{H}_1 = \mathfrak{H}_2$. Hence $\mathfrak{H}_1 \cap \mathfrak{H}_2 = 1$. Since any element $\mathfrak{H}$ of $\mathscr{H}$ contains $q - 2$ non-identity elements and $\mathfrak{M}$ contains $q(q - 2)$ elements outside $\mathfrak{T}$, it follows that $|\mathscr{H}| \leq q$. But if $\mathfrak{H} = \mathfrak{H}(s)$ $(s \in \mathsf{I})$, then $\mathfrak{H}^x = \mathfrak{H}(s^x)$ for all $x \in \mathfrak{M}$. Thus if $\mathfrak{H} \in \mathscr{H}$, all the conjugates of $\mathfrak{H}$ in $\mathfrak{M}$ lie in $\mathscr{H}$. By j), $\mathbf{N}_{\mathfrak{M}}(\mathfrak{H}) = \mathfrak{H}$, so $\mathfrak{H}$ has q conjugates in $\mathfrak{M}$. Thus $|\mathscr{H}| = q$ and

$$|\mathsf{I}| = (q - 1)q.$$

The total number of involutions in $\mathfrak{G}$ is thus $(q - 1)(q + 1)$. But if t is an involution in $\mathfrak{G}$, then by c), the number of involutions in $\mathfrak{G}$ is $|\mathfrak{G} : \mathbf{C}_{\mathfrak{G}}(t)| = |\mathfrak{G} : \mathfrak{T}|$. Hence $|\mathfrak{G} : \mathfrak{T}| = (q - 1)(q + 1)$ and

$$|\mathfrak{G}| = (q + 1)q(q - 1).$$

We now choose a fixed involution s outside $\mathfrak{M}$ and put $\mathfrak{H} = \mathfrak{H}(s)$.

m) The $q + 1$ right cosets of $\mathfrak{M}$ in $\mathfrak{G}$ are $\mathfrak{M}$ and all $\mathfrak{M}st$, as t runs through $\mathfrak{T}$.

By f), and l), $|\mathfrak{G} : \mathfrak{M}| = q + 1$. Thus it is sufficient to show that the $\mathfrak{M}st$ $(t \in \mathfrak{T})$ are distinct and different from $\mathfrak{M}$. But if $\mathfrak{M}st_1 = \mathfrak{M}st_2$ $(t_i \in \mathfrak{T})$, then $t_1 t_2^{-1} \in \mathfrak{M}^s \cap \mathfrak{T} = \mathfrak{T}^s \cap \mathfrak{T} = 1$, by b). And $\mathfrak{M}st \neq \mathfrak{M}$, since otherwise $st \in \mathfrak{M}$ and $s \in \mathfrak{M}$.

n) $\mathfrak{G} = PSL(2, q)$.

Let ρ be the permutation representation of $\mathfrak{G}$ on the set of right cosets $\mathfrak{M}x$ of $\mathfrak{M}$ given by

$$\mathfrak{M}x\rho(g) = \mathfrak{M}xg.$$

By m), ρ is doubly transitive, and the stabiliser of $\mathfrak{M}$ and $\mathfrak{M}s$ is $\mathfrak{M} \cap \mathfrak{M}^s = \mathfrak{H}$. But $\mathfrak{H}$ is transitively represented by ρ on the cosets $\mathfrak{M}st$, where $1 \neq t \in \mathfrak{T}$, since if t_1, t_2 are non-identity elements of $\mathfrak{T}$, then by e) and i), $t_1^h = t_2$ for some $h \in \mathfrak{H}$ and

$$\mathfrak{M}st_1\rho(h) = \mathfrak{M}st_1h = \mathfrak{M}sht_1^h = \mathfrak{M}h^{-1}st_2 = \mathfrak{M}st_2.$$

Thus ρ is triply transitive of degree $q + 1$. By l), $|\mathfrak{G}| = (q + 1)q(q - 1)$, so ρ is faithful and sharply triply transitive. By 2.1, $\mathfrak{G} = PSL(2, q)$.

q.e.d.

2.8 Corollary. *If $\mathfrak{G}$ is a non-Abelian simple group and the centralizer of any involution in $\mathfrak{G}$ is elementary Abelian, then $\mathfrak{G} = PSL(2, 2^f)$ for some f.*

2.9 Remark. BRAUER, SUZUKI and WALL [1] also gave a characterization of $PSL(2, q)$ for q odd. In a simplified version by BENDER [7, 8], the following is proved.

Let t be an involution in a group $\mathfrak{G}$ with the following properties.

(1) $\mathfrak{H} = \mathbf{C}_{\mathfrak{G}}(t)$ has an Abelian normal subgroup $\mathfrak{K}$ of index 2 such that $\mathfrak{H} = \mathfrak{K}\langle s \rangle$, where $s^2 = 1$, $\mathbf{C}_{\mathfrak{K}}(s) = \langle t \rangle$ and $x^s = x^{-1}$ for all $x \in \mathfrak{K}$.

(2) $\mathfrak{K} \cap \mathfrak{K}^g = 1$ for all $g \notin \mathfrak{H}$.

(3) All involutions in $\mathfrak{G}$ are conjugate.

Then $\mathfrak{G} \cong PSL(2, q)$ for some odd prime-power q.

The original result of Brauer, Suzuki and Wall, obtained by combining 2.7 and 2.9, is the following.

Let $\mathfrak{G}$ be a group of even order having no subgroup of index 2. Suppose that whenever $\mathfrak{A}, \mathfrak{B}$ are cyclic subgroups of the same even order and $\mathfrak{A} \cap \mathfrak{B} \neq 1$, then $\mathfrak{A} = \mathfrak{B}$. Then either $\mathfrak{G} \cong PSL(2, q)$ for some $q \geq 3$ or the Sylow 2-subgroup of $\mathfrak{G}$ is normal and elementary Abelian.

§ 3. The Suzuki Groups

The groups described in 1.3 were the only known Zassenhaus groups until 1959. In attempting to characterize the groups $PSL(2, p^f)$, Suzuki then found another infinite series of simple Zassenhaus groups, which will be described in this section. We begin with the groups $A(n, \theta)$ introduced in VIII, 6.7; here however we shall need them in a representation of degree 4.

3.1 Lemma. *Let $\mathsf{K} = GF(q)$, where $q = 2^{2m+1}$ and $m > 0$. Then K has exactly one automorphism π such that $x\pi^2 = x^2$ for $x \in \mathsf{K}$, namely $x\pi = x^{2^{m+1}}$. For a, b in K and λ in $\mathsf{K}^\times$, let*

$$S(a, b) = \begin{pmatrix} 1 & 0 & 0 & 0 \\ a & 1 & 0 & 0 \\ b & a\pi & 1 & 0 \\ a^2(a\pi) + ab + b\pi & a(a\pi) + b & a & 1 \end{pmatrix}$$

and

$$M(\lambda) = \begin{pmatrix} \lambda^{1+2^m} & 0 & 0 & 0 \\ 0 & \lambda^{2^m} & 0 & 0 \\ 0 & 0 & \lambda^{-2^m} & 0 \\ 0 & 0 & 0 & \lambda^{-1-2^m} \end{pmatrix}.$$

a) *The set*

$$\mathfrak{F} = \{S(a, b) | a, b \in \mathsf{K}\}$$

is a 2-group of exponent 4, class 2 and order q^2, isomorphic to $A(2m + 1, \pi)$. We have

$$S(a, b)S(a', b') = S(a + a', b + b' + (a\pi)a')$$

and

$$\mathfrak{F}' = \mathbf{Z}(\mathfrak{F}) = \{S(0, b) | b \in \mathsf{K}\}.$$

b) *The set*

$$\mathfrak{H} = \{M(\lambda) | \lambda \in \mathsf{K}^\times\}$$

is a group isomorphic to $\mathsf{K}^\times$

c) *We have*

$$S(a, b)^{M(\lambda)} = S(\lambda a, \lambda(\lambda\pi)b).$$

Thus $\mathfrak{H}$ induces a group of fixed point free automorphisms on $\mathfrak{F}$, and $\mathfrak{F}\mathfrak{H}$ is a Frobenius group with Frobenius kernel $\mathfrak{F}$.

Proof. It is easily verified that

(1) $$S(a, b)S(a', b') = S(a + a', b + b' + (a\pi)a'),$$

$$M(\lambda)M(\lambda') = M(\lambda\lambda')$$

and

(2) $$S(a, b)^{M(\lambda)} = S(\lambda a, \lambda(\lambda\pi)b).$$

Thus $\mathfrak{F}$ and $\mathfrak{H}$ are subgroups of $GL(4, q)$ and $\mathfrak{H} \leq \mathbf{N}(\mathfrak{F})$. Clearly $|\mathfrak{F}| = q^2$. If $M(\lambda)$ is the identity matrix, $\lambda = 1$, so $\mathfrak{H} \cong \mathsf{K}^\times$. Put

$$\mathfrak{F}_1 = \{S(0, b) | b \in \mathsf{K}\}.$$

Then $\mathfrak{F}_1$ is an $\mathfrak{H}$-invariant subgroup of $\mathfrak{F}$. Since the mapping $S(a, b) \to a$ is obviously an epimorphism of $\mathfrak{F}$ onto the additive group of K with kernel $\mathfrak{F}_1$, $\mathfrak{F}' \leq \mathfrak{F}_1$. Also if $a \neq a\pi$,

$$S(a, 0)S(1, 0) = S(a + 1, a\pi) \neq S(a + 1, a) = S(1, 0)S(a, 0),$$

so $\mathfrak{F}$ is non-Abelian and $\mathfrak{F}' \neq 1$.

If $\lambda(\lambda\pi) = 1$, then $\lambda\pi = \lambda^{-1}$ and $\lambda^2 = \lambda\pi^2 = \lambda$, so $\lambda = 1$. It therefore follows from (2) that $\mathfrak{H}$ permutes transitively the sets of non-identity elements of $\mathfrak{F}/\mathfrak{F}_1$ and of $\mathfrak{F}_1$. Hence $\mathfrak{F}' = \mathfrak{F}_1$. Also, $\mathfrak{F}_1 \leq \mathbf{Z}(\mathfrak{F}) < \mathfrak{F}$, so $\mathfrak{F}_1 = \mathbf{Z}(\mathfrak{F})$. Since $\lambda(\lambda\pi) \neq 1$ for $\lambda \neq 1$, each non-identity element of $\mathfrak{H}$ transforms $\mathfrak{F}$ fixed point freely. Since $\mathfrak{F}/\mathfrak{F}_1$ and $\mathfrak{F}_1$ both have exponent 2, the exponent of $\mathfrak{F}$ is 4. Comparison of (1) with VIII, 6.7 shows that $\mathfrak{F} \cong A(2m + 1, \pi)$. **q.e.d.**

We now define the Suzuki group $Sz(q)$ over $\mathsf{K} = GF(q)$.

3.2 Definition. Let $\mathsf{K} = GF(q)$, where $q = 2^{2m+1}$ and $m > 0$. Put

$$T = \begin{pmatrix} 0 & 0 & 0 & 1 \\ 0 & 0 & 1 & 0 \\ 0 & 1 & 0 & 0 \\ 1 & 0 & 0 & 0 \end{pmatrix}.$$

The *Suzuki group* $Sz(q)$ is defined to be the following subgroup of $GL(4, q)$.

$$Sz(q) = \langle S(a, b), M(\lambda), T \,|\, a \in \mathsf{K}, b \in \mathsf{K}, \lambda \in \mathsf{K}^\times \rangle.$$

We obtain information about the structure of $Sz(q)$ from the following permutation representation.

3.3 Theorem. *In the projective space* $\mathsf{P}(3, q)$, *let*

$$p_\infty = [1, 0, 0, 0], \quad p(x, y) = [xy + (x\pi)x^2 + y\pi, y, x, 1]$$

(in homogeneous coordinates). Let

$$\mathcal{O} = \{p_\infty, p(x, y) \,|\, x \in \mathsf{K}, y \in \mathsf{K}\}.$$

Then $Sz(q)$ induces the group of all collineations of $\mathsf{P}(3, q)$ *which carry $\mathcal{O}$ onto itself. Further, $Sz(q)$ is faithfully represented on $\mathcal{O}$ and induces a Zassenhaus group of order $(q^2 + 1)q^2(q - 1)$ on $\mathcal{O}$. The stabiliser in $Sz(q)$ of p_∞ is the group $\mathfrak{F}\mathfrak{H}$ of 3.1.*

Proof. a) $Sz(q)$ is faithfully represented on $P(3, q)$.

Suppose that $S \in Sz(q)$ and that S induces the identity collineation on $\mathsf{P}(3, q)$. Then $S = kI$ for some $k \in \mathsf{K}$. But $Sz(q)$ consists of matrices of determinant 1, so $k^4 = 1$. Since char $\mathsf{K} = 2$, it follows that $k = 1$ and $S = I$.

b) $Sz(q)$ carries $\mathcal{O}$ onto itself.

In fact,

$$p_\infty S(a, b) = p_\infty, \quad p_\infty M(\lambda) = p_\infty,$$

$$p(x, y)S(a, b) = p(x + a, y + b + x(a\pi) + a(a\pi)),$$

$$p(x, y)M(\lambda) = p(\lambda x, \lambda(\lambda\pi)y).$$

Further, $p_\infty T = p(0, 0)$ and $p(0, 0)T = p_\infty$. Given $x \in \mathsf{K}$, $y \in \mathsf{K}$, we put

$$z = xy + (x\pi)x^2 + y\pi.$$

A simple calculation then shows that

$$(y\pi)y = z(x\pi)x + zy + x(z\pi).$$

Thus it follows from $z = 0$ that first $y = 0$ and then $x = 0$. Hence for $(x, y) \neq (0, 0)$, $z \neq 0$. Thus for $(x, y) \neq (0, 0)$,

$$p(x, y)T = [1, x, y, z] = [z^{-1}, xz^{-1}, yz^{-1}, 1] = p(yz^{-1}, xz^{-1}).$$

c) The plane $x_3 = 0$ is the only plane through p_∞ which intersects $\mathcal{O}$ only in the point p_∞.

Let $\mathscr{E}$ be the plane given by $\sum_{i=0}^{3} a_i x_i = 0$. From $p_\infty \in \mathscr{E}$, it follows that $a_0 = 0$. If $(a_1, a_2) \neq (0, 0)$, there exist x, y such that

$$a_1 y + a_2 x + a_3 = 0,$$

and $p(x, y) \in \mathscr{E} \cap \mathscr{O}$. Hence if $\mathscr{E} \cap \mathscr{O} = \{p_\infty\}$, then $a_1 = a_2 = 0$. Conversely, it is easily seen that the plane $x_3 = 0$ has only the point p_∞ in common with $\mathscr{O}$.

d) Suppose that G is a collineation in $PGL(4, q)$ which leaves $\mathscr{O}$, p_∞ and $p(0, 0)$ fixed. Then G is induced by $M(\lambda)$ for some $\lambda \in \mathsf{K}^\times$.

By c), the plane $x_3 = 0$ is left fixed by G. Since T interchanges p_∞ and $p(0, 0)$, $\tilde{T}^{-1}G\tilde{T}$ leaves $\mathscr{O}$, p_∞ and $p(0, 0)$ fixed, where $\tilde{T}$ is the collineation induced by T. Thus $\tilde{T}^{-1}G\tilde{T}$ also leaves the plane $x_3 = 0$ fixed. It follows that G is induced by a matrix of the form

$$\begin{pmatrix} e & 0 & 0 & 0 \\ 0 & a & b & 0 \\ 0 & d & c & 0 \\ 0 & 0 & 0 & 1 \end{pmatrix}.$$

Since $\mathscr{O}G = \mathscr{O}$, for all x, y in K we have

$$p(x, 0)G = [e(x\pi)x^2, dx, cx, 1] \in \mathscr{O}$$

and

$$p(0, y)G = [e(y\pi), ay, by, 1] \in \mathscr{O}.$$

This requires

$$e(x\pi)x^2 = dcx^2 + (d\pi)(x\pi) + (c\pi)c^2(x\pi)x^2$$

and

$$e(y\pi) = aby^2 + (b\pi)b^2(y\pi)y^2 + (a\pi)(y\pi).$$

This means that each of the polynomials f_1, f_2 given by

$$f_1(t) = dct^2 + (d\pi)t^{2^{m+1}} + ((c\pi)c^2 - e)t^{2+2^{m+1}}$$

and

$$f_2(t) = abt^2 + ((a\pi) - e)t^{2^{m+1}} + (b\pi)b^2t^{2+2^{m+1}}$$

has q zeros in $GF(q)$. Since f_1, f_2 are both of degree less than $q = 2^{2m+1}$, it follows that $f_1 = f_2 = 0$. This gives $b = d = 0$ and $e = a\pi = (c\pi)c^2$.

Thus $a = c(c^2\pi^{-1}) = c(c\pi)$. The matrix of G is therefore

$$\begin{pmatrix} (c\pi)c^2 & 0 & 0 & 0 \\ 0 & (c\pi)c & 0 & 0 \\ 0 & 0 & c & 0 \\ 0 & 0 & 0 & 1 \end{pmatrix} = c^{1+2^m} M(c).$$

Hence G is induced by $M(c)$.

e) If $1 \neq \lambda \in \mathsf{K}^\times$, p_∞ and $p(0, 0)$ are the only points of $\mathcal{O}$ left fixed by $M(\lambda)$.

For if $p(x, y)M(\lambda) = p(x, y)$, then $y\lambda^{1+2^{m+1}} = y$ and $x\lambda = x$. Since $\lambda \neq 1$ and $(1 + 2^{m+1}, 2^{2m+1} - 1) = 1$, it follows that $x = y = 0$.

f) $Sz(q)$ is faithfully represented on $\mathcal{O}$ and $Sz(q)_{p_\infty, p(0,0)} = \mathfrak{H}$.

Suppose that $S \in Sz(q)$ and S induces the identity mapping on $\mathcal{O}$. By d), $S = kM(\lambda)$ for some k, λ in $\mathsf{K}^\times$, and by e), $\lambda = 1$. Thus $S = kI$. By a), $S = I$.

If $S \in Sz(q)_{p_\infty, p(0,0)}$, then $S = kM(\lambda)$ and by a), $k = 1$. Thus $S = M(\lambda) \in \mathfrak{H}$. Hence $Sz(q)_{p_\infty, p(0,0)} = \mathfrak{H}$.

g) $Sz(q)$ is doubly transitive on $\mathcal{O}$.

Since

$$p(0, 1)S(a, b + 1 + a(a\pi)) = p(a, b),$$

$Sz(q)_{p_\infty}$ is certainly transitive on $\mathcal{O} - \{p_\infty\}$. Since $p_\infty T = p(0, 0)$, it then follows that $Sz(q)$ is doubly transitive on $\mathcal{O}$.

h) $|Sz(q)| = (q^2 + 1)q^2(q - 1)$, and $Sz(q)$ is the group of all collineations which leave $\mathcal{O}$ fixed.

By f), $|Sz(q)_{p_\infty, p(0,0)}| = q - 1$. Since $Sz(q)$ is doubly transitive by g) and $|\mathcal{O}| = q^2 + 1$, it follows that $|Sz(q)| = (q^2 + 1)q^2(q - 1)$.

If $\mathfrak{S}$ is the group of all collineations in $PGL(4, q)$ which leave $\mathcal{O}$ fixed, then $\mathfrak{S} \geq Sz(q)$, so $\mathfrak{S}$ is doubly transitive. By d), $\mathfrak{S}_{p_\infty, p(0,0)} = \mathfrak{H}$, so $|\mathfrak{S}| = (q^2 + 1)q^2|\mathfrak{H}| = |Sz(q)|$. Thus $\mathfrak{S} = Sz(q)$.

i) $Sz(q)_{p_\infty} = \mathfrak{F}\mathfrak{H}$.

For $Sz(q)_{p_\infty} \geq \mathfrak{F}\mathfrak{H}$ and $|Sz(q)_{p_\infty}| = q^2(q - 1) = |\mathfrak{F}\mathfrak{H}|$.

j) $Sz(q)$ is a Zassenhaus group on $\mathcal{O}$.

By g), $Sz(q)$ is doubly transitive. By f) and e), no non-identity element of $Sz(q)_{p_\infty, p(0,0)}$ leaves fixed any point of $\mathcal{O}$ other than p_∞ and $p(0, 0)$. If $Sz(q)$ has a regular normal subgroup, then $q^2 + 1 = 2^p$ for some prime p, by 1.1. This is obviously impossible, so $Sz(q)$ is a Zassenhaus group. **q.e.d.**

The geometrical nature of the set $\mathcal{O}$ is described in the following theorem.

3.4 Theorem. *$\mathcal{O}$ is an ovoid; that is, any line intersects $\mathcal{O}$ in at most two points.*

Proof. On account of the double transitivity of $Sz(q)$ on $\mathcal{O}$ we may restrict ourselves to the consideration of the line through p_∞ and $p(0, 0)$. The points of this line other than p_∞ have homogeneous coordinates of the form $[x_0, 0, 0, 1]$. This point lies on $\mathcal{O}$ if and only if $x_0 = 0$. **q.e.d.**

3.5 Remark. Over the field of real numbers the quadrics in all dimensions which have the signature of the sphere are ovoids. Over finite fields the situation is quite different. If n is the dimension of the vector space, the index of any quadratic form g is at least $\frac{1}{2}(n - 2)$; (see II, 10.9 for the case when the field has odd characteristic; see DIEUDONNÉ [1, p. 34] in the case of characteristic 2). If $\{\langle v \rangle \mid g(v) = 0\}$ is an ovoid, the isotropic subspaces of g are of dimension at most 1, and hence $n \leq 4$. It is easy to see that for $n = 3$ and $n = 4$, there actually are quadrics which are ovoids. Tits has shown that for $\mathsf{K} = GF(q)$ and char $\mathsf{K} = 2$, if $n = 4$, there are no ovoids $\mathcal{O}$ with $q^2 + 1$ points for which the group of linear collineations leaving $\mathcal{O}$ fixed is doubly transitive, other than quadrics and the ovoid given in 3.4 (see TITS [2]). This perhaps explains the exceptional appearance of the Suzuki groups.

3.6 Theorem. *$Sz(q)$ is simple and 3 does not divide $|Sz(q)|$.*

Proof. By 3.3, $Sz(q)$ is a Zassenhaus group on $\mathcal{O}$, and $|\mathcal{O}|$ is odd. By 1.6, $Sz(q)$ is simple.

Since $q = 2^{2m+1}$, $q - 1 \equiv 1\ (3)$ and $q^2 + 1 \equiv 2\ (3)$. Thus $|Sz(q)| \equiv 2\ (3)$. **q.e.d.**

3.7 Remarks. a) The Suzuki groups are the only simple non-Abelian groups of order prime to 3. This is a very difficult theorem, first proved by Thompson. Subsequently, GLAUBERMAN [10] gave a simpler proof.

b) Since

$$q^2 + 1 = 2^{4m+2} + 1 \equiv 0\ (5),$$

$|Sz(q)|$ is divisible by 5. Together with a), this implies that the order of a non-Abelian simple group is divisible by 3 or 5. This was first proved by Thompson, being an immediate consequence of his theorem on N-groups.

c) COLLINS [1] has proved that if $\mathfrak{G}$ is simple and the Sylow 2-subgroups of $\mathfrak{G}$ are isomorphic to those of $Sz(q)$, then $\mathfrak{G} \cong Sz(q)$. This follows from an important theorem of GOLDSCHMIDT [4], which also implies that the only simple groups in which the Sylow 2-subgroups are Suzuki 2-groups are $Sz(q)$ and $PSU(3, 2^{2n})$ $(n \geq 2)$.

d) In the standard notation for Chevalley groups, $Sz(q)$ is ${}^2B_2(q)$.

3.8 Remark. The discovery of the Suzuki groups caused something of a sensation at the time, since they seemed to be so very different from the classical groups. But it soon became clear that this is not so. The unitary group $U(n, q^2)$ is the set of elements of $GL(n, q^2)$ left fixed by the automorphism α of $GL(n, q^2)$ given by $A\alpha = (\overline{A}^t)^{-1}$, where $^-$ denotes the unique automorphism of $GF(q^2)$ of order 2. The Suzuki groups can be described similarly as the set of fixed points of an automorphism of the symplectic group $Sp(4, q)$.

Let V be a vector space of dimension 4 over $GF(q)$ with a non-singular symplectic scalar product. Let $\{v_1, v_2, v_3, v_4\}$ be a basis of V such that $(v_1, v_4) = (v_2, v_3) = 1$ and $(v_i, v_j) = 0$ for all (i, j) other than $(1, 4), (4, 1), (2, 3), (3, 2)$. Define $S'(a, b)$, $M'(\lambda)$ and T' as follows:

$$v_1 S'(a, b) = v_1, v_2 S'(a, b) = av_1 + v_2, v_3 S'(a, b) = bv_1 + (a\pi)v_2 + v_3,$$

$$v_4 S'(a, b) = (a^2(a\pi) + ab + b\pi)v_1 + (a(a\pi) + b)v_2 + av_3 + v_4,$$

$$v_1 M'(\lambda) = \lambda^{1+2^m} v_1, \quad v_2 M'(\lambda) = \lambda^{2^m} v_2, \quad v_3 M'(\lambda) = \lambda^{-2^m} v_3,$$

$$v_4 M'(\lambda) = \lambda^{-1-2^m} v_4,$$

$$v_1 T' = v_4, v_2 T' = v_3, v_3 T' = v_2, v_4 T' = v_1.$$

It is easy to verify that $S'(a, b)$, $M'(\lambda)$ and T' lie in $Sp(4, q)$, so we have an embedding of $Sz(q)$ in $Sp(4, q)$.

There is an automorphism α of $Sp(4, q)$ the fixed points of which are precisely the elements of $Sz(q)$ (see ONO [1]). (In the case $q = 2$, then by II, 9.21, $Sp(4, 2) \cong \mathfrak{S}_6$, and α is then the non-inner automorphism of $\mathfrak{S}_6$ (II, 5.5b)). For similar cases of 'twisted types', cf. II, 10.17.

3.9 Theorem. *If p is odd, the Sylow p-subgroups of Sz(q) are cyclic.*

Proof. If r is a common divisor of $q^2 + 1$ and $q - 1$, then r divides

$$q^2 + 1 - (q - 1)(q + 1) = 2.$$

If p is odd and p^a is the highest power of p which divides $|Sz(q)|$, then either p^a divides $q - 1$ or p^a divides $q^2 + 1$. In the first case, the Sylow p-subgroups of $Sz(q)$ are cyclic, as $\mathfrak{H}$ contains a cyclic subgroup of order p^a.

Suppose then that p divides $q^2 + 1$. By II, 7.3, $PGL(4, q)$ has a cyclic subgroup of order $q^2 + 1$. It is easily verified that this is a Hall subgroup of $PGL(4, q)$. Thus $PGL(4, q)$, and so certainly $Sz(q)$, has cyclic Sylow p-subgroups. **q.e.d.**

ASCHBACHER [1] has proved that the following are the only simple groups with the property that for every odd prime p, the Sylow p-subgroups are cyclic: $PSL(2, 2^f)$ $(f > 1)$, $PSL(2, p)$ $(p > 3)$, $Sz(q)$ and the simple group of Janko of order $2^3 \cdot 3 \cdot 5 \cdot 7 \cdot 11 \cdot 19$ (see 13.6).

By II, 8.4, $PSL(2, 2^f)$ has a cyclic subgroup of order $2^f + 1$. We prove now a corresponding assertion for $Sz(q)$.

3.10 Theorem. *We put* $q = 2^{2m+1}$, $r = 2^m$ *and* $\mathfrak{G} = Sz(q)$.

a) *$\mathfrak{G}$ possesses cyclic subgroups $\mathfrak{U}_1$ and $\mathfrak{U}_2$ of orders $q + 2r + 1$ and $q - 2r + 1$. $\mathfrak{U}_1$ and $\mathfrak{U}_2$ are Hall subgroups of $\mathfrak{G}$.*

b) *If $1 \neq u \in \mathfrak{U}_i$, $\mathbf{C}_{\mathfrak{G}}(u) = \mathfrak{U}_i$. Also $|\mathbf{N}_{\mathfrak{G}}(\mathfrak{U}_i) : \mathfrak{U}_i| = 4$ and $\mathbf{N}_{\mathfrak{G}}(\mathfrak{U}_i) = \langle \mathfrak{U}_i, t_i \rangle$, where $u^{t_i} = u^q$ for all $u \in \mathfrak{U}_i$. In particular, $\mathbf{N}_{\mathfrak{G}}(\mathfrak{U}_i)$ is a Frobenius group with Frobenius kernel $\mathfrak{U}_i$.*

c) *The conjugates of $\mathfrak{F}$, $\mathfrak{H}$, $\mathfrak{U}_1$ and $\mathfrak{U}_2$ form a partition of $\mathfrak{G}$.*

Proof. Let σ be the set of prime divisors of $q^2 + 1$. Since $(q^2 + 1, q - 1) = 1$, a σ-subgroup of $\mathfrak{G}$ is the same as a subgroup of order a divisor of $q^2 + 1$.

a) If $\mathfrak{U}$ is a σ-subgroup of $\mathfrak{G}$, then $\mathfrak{U} = \langle u \rangle$ is cyclic and $|\mathfrak{U}|$ divides $q^2 + 1$. Also if $x \in \mathbf{N}_{\mathfrak{G}}(\mathfrak{U})$, then $u^x = u^{q^i}$ for some i.

Let $\mathsf{L} = GF(q^4)$; thus L may be regarded as a vector space of dimension 4 over $\mathsf{K} = GF(q)$. Also $\mathsf{L}^\times$ is a cyclic group $\langle a \rangle$, and if we put $xT = ax$ for all $x \in \mathsf{L}$, T is a K-linear transformation (a *Singer cycle*) of L of order $q^4 - 1$. Let $\mathfrak{T} = \langle T \rangle$ and let $\mathfrak{M} = GL(4, q)$. It is proved in II, 7.3 that $\mathbf{C}_{\mathfrak{M}}(\mathfrak{T}) = \mathfrak{T}$ and $\mathbf{N}_{\mathfrak{M}}(\mathfrak{T})/\mathfrak{T}$ is cyclic of order 4. In fact if we put $xS = x^q$ $(x \in \mathsf{L})$, S is K-linear and $S^{-1}TS = T^q$, so $\mathbf{N}_{\mathfrak{M}}(\mathfrak{T}) = \mathfrak{T}\langle S \rangle$. It is also proved in II, 7.3 that if $T_1 \in \mathfrak{T}$ and the order of T_1 does not divide $q^2 - 1$, then $\mathbf{N}_{\mathfrak{M}}(\langle T_1 \rangle) = \mathbf{N}_{\mathfrak{M}}(\mathfrak{T})$, so if $R \in \mathbf{N}_{\mathfrak{M}}(\langle T_1 \rangle)$, $T_1^R = T_1^{q^i}$ for some i. Note that $T^{q-1} \in SL(4, q)$ and $\mathfrak{S} = \langle T^{q^2-1} \rangle$ is a Hall subgroup of $SL(4, q)$ of order $q^2 + 1$.

Now $\mathfrak{G} = Sz(q)$ can be regarded as a group of K-linear transformations of L of determinant 1, so $\mathfrak{G} \leq SL(4, q)$. By a theorem of Wielandt

(III, 5.8), $\mathfrak{U} \leq \mathfrak{S}^t$ for some $t \in SL(4, q)$. Thus $\mathfrak{U} = \langle u \rangle$ is cyclic. From above, if $x \in \mathbf{N}_{\mathfrak{G}}(\mathfrak{U})$, $u^x = u^{q^i}$ for some i.

b) If $\mathfrak{U}$ is a maximal σ-subgroup of $\mathfrak{G}$, $\mathfrak{U} = \mathbf{C}_{\mathfrak{G}}(u)$ for any non-identity element u of $\mathfrak{U}$. Conversely, if u is a non-identity element of $\mathfrak{G}$ and the order of u divides $q^2 + 1$, $\mathbf{C}_{\mathfrak{G}}(u)$ is a maximal σ-subgroup of $\mathfrak{G}$.

We show first that if u is a non-identity element of $\mathfrak{G}$ and the order of u divides $q^2 + 1$, then $\mathbf{C}_{\mathfrak{G}}(u)$ is a maximal σ-subgroup. To do this, suppose that $1 \neq v \in \mathbf{C}_{\mathfrak{G}}(u)$. If z is a fixed point of v on $\mathcal{O}$, zu^j is also a fixed point of v for any j. But u has no fixed points, since the order of any stabiliser is $q^2(q - 1)$. Thus if v has any fixed points, it has at least three and $v = 1$. Hence each non-identity element of $\mathbf{C}_{\mathfrak{G}}(u)$ operates fixed point freely on $\mathcal{O}$. Hence the orbits of $\mathbf{C}_{\mathfrak{G}}(u)$ on $\mathcal{O}$ all have the same length $|\mathbf{C}_{\mathfrak{G}}(u)|$, whence $|\mathbf{C}_{\mathfrak{G}}(u)|$ divides $q^2 + 1$ and $\mathbf{C}_{\mathfrak{G}}(u)$ is a σ-subgroup. If $\mathbf{C}_{\mathfrak{G}}(u) \leq \mathfrak{U}$ and $\mathfrak{U}$ is a σ-subgroup, then $\mathfrak{U}$ is cyclic by a) and $\mathfrak{U} \leq \mathbf{C}_{\mathfrak{G}}(u)$. Hence $\mathbf{C}_{\mathfrak{G}}(u)$ is a maximal σ-subgroup of $\mathfrak{G}$.

Conversely, suppose that $\mathfrak{U}$ is a maximal σ-subgroup of $\mathfrak{G}$ and that $1 \neq u \in \mathfrak{U}$. By a), $\mathfrak{U}$ is cyclic, so $\mathfrak{U} \leq \mathbf{C}_{\mathfrak{G}}(u)$. Since $\mathbf{C}_{\mathfrak{G}}(u)$ is a σ-subgroup, it follows that $\mathfrak{U} = \mathbf{C}_{\mathfrak{G}}(u)$.

c) The maximal σ-subgroups of $\mathfrak{G}$ are Hall subgroups of $\mathfrak{G}$.

Suppose that $\mathfrak{U}$ is a maximal σ-subgroup of $\mathfrak{G}$, p is a prime divisor of $|\mathfrak{U}|$, g is an element of $\mathfrak{U}$ of order p and $\mathfrak{P}$ is a Sylow p-subgroup of $\mathfrak{G}$ containing g. By a), $\mathfrak{P}$ is cyclic, and by b), $\mathfrak{U} = \mathbf{C}_{\mathfrak{G}}(g) \geq \mathfrak{P}$. Hence $\mathfrak{U}$ is a Hall subgroup of $\mathfrak{G}$.

d) Any two distinct maximal σ-subgroups of $\mathfrak{G}$ intersect in 1.

If $\mathfrak{U}$, $\mathfrak{V}$ are maximal σ-subgroups of $\mathfrak{G}$ and $1 \neq g \in \mathfrak{U} \cap \mathfrak{V}$, then by b), $\mathfrak{U} = \mathbf{C}_{\mathfrak{G}}(g) = \mathfrak{V}$.

e) If g leaves no point of $\mathcal{O}$ fixed, then the order of g divides $q^2 + 1$.

Regarded as a permutation of $\mathcal{O}$, g has a cycle of odd length, since $|\mathcal{O}|$ is odd. Let l be the smallest odd length of a cycle of g. Since g is fixed point free, $l \geq 3$. But g^l has at least l fixed points, so $g^l = 1$. Thus l is the length of any cycle of g and l divides $q^2 + 1$.

Let $\mathfrak{U}_1, \ldots, \mathfrak{U}_c$ be a complete set of representatives of the conjugacy classes of the maximal σ-subgroups of $\mathfrak{G}$.

f) $(|\mathfrak{U}_i|, |\mathfrak{U}_j|) = 1$ for $i \neq j$ and $\prod_{i=1}^{c} |\mathfrak{U}_i| = q^2 + 1$.

If p is a common prime divisor of $|\mathfrak{U}_i|$ and $|\mathfrak{U}_j|$, then by c), $\mathfrak{U}_i$, $\mathfrak{U}_j$ both contain Sylow p-subgroups of $\mathfrak{G}$. By Sylow's theorem, $\mathfrak{U}_i \cap \mathfrak{U}_j^g \neq 1$ for some $g \in \mathfrak{G}$, and by d), $\mathfrak{U}_i = \mathfrak{U}_j^g$.

Since the $|\mathfrak{U}_i|$ are coprime, $\prod_{i=1}^{c} |\mathfrak{U}_i|$ is a divisor of $q^2 + 1$. Let p be a prime divisor of $q^2 + 1$, let $\mathfrak{P}$ be Sylow p-subgroup of $\mathfrak{G}$ and let $\mathfrak{U}$ be a maximal σ-subgroup of $\mathfrak{G}$ containing $\mathfrak{P}$. Then $\mathfrak{U} = \mathfrak{U}_j^g$ for some $g \in \mathfrak{G}$, so $|\mathfrak{P}|$ divides $\prod_{i=1}^{c} |\mathfrak{U}_i|$. Hence $\prod_{i=1}^{c} |\mathfrak{U}_i| = q^2 + 1$.

g) The conjugates of $\mathfrak{F}$, $\mathfrak{H}$, $\mathfrak{U}_1, \ldots, \mathfrak{U}_c$ form a partition of $\mathfrak{G}$.

By d), $\mathfrak{U}_i \cap \mathfrak{U}_j^g = 1$ unless $\mathfrak{U}_i = \mathfrak{U}_j^g$. And

$$\mathfrak{F}^g \cap \mathfrak{H}^{g'} = \mathfrak{F}^g \cap \mathfrak{U}_i^{g'} = \mathfrak{H}^g \cap \mathfrak{U}_i^{g'} = 1,$$

since these are all intersections of subgroups of coprime orders. Since any non-identity element of $\mathfrak{F}$ has only the one fixed point p_∞, $\mathfrak{F} \cap \mathfrak{F}^g = 1$ whenever $\mathfrak{F} \neq \mathfrak{F}^g$. If $\mathfrak{H} \neq \mathfrak{H}^g$, $\mathfrak{H} \cap \mathfrak{H}^g$ has at least three fixed points and is therefore 1.

Suppose that $g \in \mathfrak{G}$. We must show that g lies in a conjugate of one of $\mathfrak{F}$, $\mathfrak{H}$, $\mathfrak{U}_1, \ldots, \mathfrak{U}_c$. If g has two fixed points, it lies in a conjugate of $\mathfrak{H}$, and if g has one fixed point, it lies in a conjugate of $\mathfrak{F}$. If g has no fixed points, then by e), g is a σ-element and lies in a maximal σ-subgroup $\mathfrak{U}_i^{g'}$.

h) $|\mathbf{N}_{\mathfrak{G}}(\mathfrak{U}_i) : \mathfrak{U}_i|$ is either 2 or 4.

By a), $\mathbf{N}_{\mathfrak{G}}(\mathfrak{U}_i)/\mathbf{C}_{\mathfrak{G}}(\mathfrak{U}_i)$ is a cyclic group of order a divisor of 4, since if β is the automorphism of $\mathfrak{U}_i$ given by $u\beta = u^q$, then $u\beta^2 = u^{-1}$ and $\beta^4 = 1$. By b), $\mathbf{C}_{\mathfrak{G}}(\mathfrak{U}_i) = \mathfrak{U}_i$. Thus $|\mathbf{N}_{\mathfrak{G}}(\mathfrak{U}_i) : \mathfrak{U}_i|$ is 1, 2 or 4. Suppose that $\mathbf{N}_{\mathfrak{G}}(\mathfrak{U}_i) = \mathfrak{U}_i$. Let $\mathfrak{P}$ be a non-identity Sylow p-subgroup of $\mathfrak{U}_i$ and hence of $\mathfrak{G}$. If $g \in \mathbf{N}_{\mathfrak{G}}(\mathfrak{P})$, then $\mathfrak{P} \leq \mathfrak{U}_i \cap \mathfrak{U}_i^g$. By d), $\mathfrak{U}_i = \mathfrak{U}_i^g$, so $\mathbf{N}_{\mathfrak{G}}(\mathfrak{P}) \leq \mathbf{N}_{\mathfrak{G}}(\mathfrak{U}_i) = \mathfrak{U}_i \leq \mathbf{C}_{\mathfrak{G}}(\mathfrak{P})$. By Burnside's transfer theorem (IV, 2.6), $\mathfrak{G}$ is p-nilpotent, contrary to 3.6. Thus $|\mathbf{N}_{\mathfrak{G}}(\mathfrak{U}_i) : \mathfrak{U}_i|$ is 2 or 4.

i) $c = 2$, $|\mathbf{N}_{\mathfrak{G}}(\mathfrak{U}_i)/\mathfrak{U}_i| = 4$ $(i = 1, 2)$, $|\mathfrak{U}_1| = q + 2r + 1$ and $|\mathfrak{U}_2| = q - 2r + 1$.

On account of the partition in g) and as $|\mathbf{N}_{\mathfrak{G}}(\mathfrak{H})| = 2|\mathfrak{H}|$ (see 1.5)

$$(*) \qquad \begin{aligned}(q^2+1)q^2(q-1) = |\mathfrak{G}| &= 1 + |\mathfrak{G} : \mathbf{N}_{\mathfrak{G}}(\mathfrak{F})|(|\mathfrak{F}| - 1) \\ &\quad + |\mathfrak{G} : \mathbf{N}_{\mathfrak{G}}(\mathfrak{H})|(|\mathfrak{H}| - 1) \\ &\quad + \sum_{i=1}^{c} |\mathfrak{G} : \mathbf{N}_{\mathfrak{G}}(\mathfrak{U}_i)|(|\mathfrak{U}_i| - 1) \\ &= 1 + (q^2+1)(q^2-1) + \frac{(q^2+1)q^2}{2}(q-2) \\ &\quad + \sum_{i=1}^{c} \frac{|\mathfrak{G}|}{k_i|\mathfrak{U}_i|}(|\mathfrak{U}_i| - 1),\end{aligned}$$

where $k_i = |\mathbf{N}_{\mathfrak{G}}(\mathfrak{U}_i) : \mathfrak{U}_i|$. Suppose that a of the k_i are 2 and that b of them are 4. By h), $a + b = c$. Since 3 does not divide $|\mathfrak{G}|$, $|\mathfrak{U}_i| \geq 5$ and so $\dfrac{|\mathfrak{U}_i| - 1}{|\mathfrak{U}_i|} \geq \dfrac{4}{5}$. It follows that

$$(q^2 + 1)(\tfrac{1}{2}q^3 - q^2 + 1) = 1 + \sum_{i=1}^{c} \frac{|\mathfrak{G}|}{k_i} \frac{|\mathfrak{U}_i| - 1}{|\mathfrak{U}_i|}$$
$$> \frac{4}{5}\left(\frac{a}{2} + \frac{b}{4}\right)(q^2 + 1)q^2(q - 1),$$

and so

$$\frac{4}{5}\left(\frac{a}{2} + \frac{b}{4}\right) < \frac{\tfrac{1}{2}q^3 - q^2 + 1}{q^3 - q^2} < \frac{1}{2}.$$

Hence $2a + b < \frac{5}{2}$.

Suppose that $c = 1$. It then follows from f) that $|\mathfrak{U}_1| = q^2 + 1$ and so

$$(q^2 + 1)q^2(q - 1) = q^4 + \frac{(q^2 + 1)q^2}{2}(q - 2) + \frac{q^2(q - 1)}{k_1}q^2.$$

Modulo q^3 this gives

$$-q^2 \equiv \tfrac{1}{2}q^3 - q^2 \quad (q^3),$$

which is impossible.

Thus $c = a + b \geq 2$. From $2a + b \leq 2$, it follows that $a = 0$ and $b = 2$.

By f), $|\mathfrak{U}_1||\mathfrak{U}_2| = q^2 + 1$. On the other hand it follows from $(*)$ that $|\mathfrak{U}_1| + |\mathfrak{U}_2| = 2(q + 1)$. Hence $|\mathfrak{U}_1|$ and $|\mathfrak{U}_2|$ are the zeros of the polynomial

$$t^2 - 2(q + 1)t + (q^2 + 1) = (t - q - 2r - 1)(t - q + 2r - 1).$$

j) $\mathbf{N}_{\mathfrak{G}}(\mathfrak{U}_i) = \langle \mathfrak{U}_i, t_i \rangle$, where $u^{t_i} = u^q$ for all $u \in \mathfrak{U}_i$ $(i = 1, 2)$.

By i), $|\mathbf{N}_{\mathfrak{G}}(\mathfrak{U}_i)/\mathfrak{U}_i| = 4$, and by a), any automorphism of $\mathfrak{U}_i$ induced by an element of $\mathbf{N}_{\mathfrak{G}}(\mathfrak{U}_i)$ is a power of β, where $u\beta = u^q$. Since $\mathbf{C}_{\mathfrak{G}}(\mathfrak{U}_i) = \mathfrak{U}_i$, it follows that $\mathbf{N}_{\mathfrak{G}}(\mathfrak{U}_i)/\mathfrak{U}_i$ is cyclic of order 4. Hence $\mathbf{N}_{\mathfrak{G}}(\mathfrak{U}_i) = \mathfrak{U}_i\mathfrak{T}_i$ for some cyclic subgroup $\mathfrak{T}_i$ of order 4, and $\mathfrak{T}_i$ has a generator t_i of order 4 for which $u^{t_i} = u^q$ $(u \in \mathfrak{U}_i)$. **q.e.d.**

3.11 Theorem. *If* $1 \neq g \in Sz(q) = \mathfrak{G}$, *then* $\mathbf{C}_{\mathfrak{G}}(g)$ *is nilpotent.*

Proof. By 3.10c), we may suppose that g lies in $\mathfrak{U}_i$, $\mathfrak{F}$ or $\mathfrak{H}$.

a) If $g \in \mathfrak{U}_i$, then by 3.10b), $\mathbf{C}_{\mathfrak{G}}(g) = \mathfrak{U}_i$ is cyclic.

b) If $g \in \mathfrak{F}$ and $x \in \mathbf{C}_{\mathfrak{G}}(g)$, then $x \in \mathfrak{G}_{p_\infty} = \mathfrak{F}\mathfrak{H}$. But $\mathfrak{F}\mathfrak{H}$ is a Frobenius group with Frobenius kernel $\mathfrak{F}$, so by V, 8.5, $x \in \mathfrak{F}$. Hence $\mathbf{C}_{\mathfrak{G}}(g)$ is a 2-group.

c) If $g \in \mathfrak{H}$ and $x \in \mathbf{C}_{\mathfrak{G}}(g)$, then x carries the set $\{p_\infty, p(0, 0)\}$ of fixed points of g onto itself. Hence $\mathbf{C}_{\mathfrak{G}}(g) \leq \mathfrak{H}\langle T\rangle$, where T is defined as in 3.2. But by direct calculation, $M(\lambda)^T = M(\lambda^{-1})$ for all $\lambda \in \mathsf{K}^\times$, so $\mathbf{C}_{\mathfrak{G}}(g) = \mathfrak{H}$ is cyclic. **q.e.d.**

We mention some characterizations and other properties of $Sz(q)$ without proof.

3.12 Remarks. a) $\mathfrak{G}$ is called a *CN-group* if $\mathbf{C}_{\mathfrak{G}}(g)$ is nilpotent whenever $1 \neq g \in \mathfrak{G}$. The only simple non-Abelian CN-groups are $Sz(q)$, $PSL(3, 4)$, $PSL(2, 9)$, $PSL(2, 2^n)$ and $PSL(2, p)$ where p is a Fermat or a Mersenne prime (see SUZUKI [3]).

b) Suppose that $\mathfrak{G}$ is a simple non-Abelian group having a non-trivial partition. Then $\mathfrak{G}$ is one of the groups $PSL(2, q)$ $(q > 3)$ (see II, 8.5) or $Sz(q)$ (SUZUKI [2]).

c) Any automorphism of $Sz(q)$ is a product of an inner automorphism and an automorphism induced by a field automorphism of $GF(q)$. In particular, the outer automorphism group of $Sz(q)$ is cyclic, so the Schreier hypothesis holds for $Sz(q)$ (SUZUKI [3]).

d) For $q > 8$, the Schur multiplier of $Sz(q)$ is 1. The multiplier of $Sz(8)$ is elementary Abelian of order 4 (ALPERIN and GORENSTEIN [1]).

e) Let $\mathfrak{U}_1$, $\mathfrak{U}_2$ be the subgroups of $\mathfrak{G} = Sz(q)$ defined in Theorem 3.10 and put $\mathfrak{B}_i = \mathbf{N}_{\mathfrak{G}}(\mathfrak{U}_i)$ $(i = 1, 2)$. $\mathfrak{G}$ also has the subgroup $\mathfrak{F}\mathfrak{H}$ and $\mathbf{N}_{\mathfrak{G}}(\mathfrak{H})$ is a dihedral subgroup of order $2(q - 1)$. In addition, if q is a power of s, $Sz(q)$ has a subgroup isomorphic to $Sz(s)$. SUZUKI [3] has shown that any subgroup of $Sz(q)$ either is isomorphic to some such $Sz(s)$ or is conjugate to a subgroup of $\mathfrak{F}\mathfrak{H}$, $\mathbf{N}_{\mathfrak{G}}(\mathfrak{H})$, $\mathfrak{B}_1$ or $\mathfrak{B}_2$.

f) The complex characters of $Sz(q)$ will be investigated in § 5.

Exercises

1) Prove that any Zassenhaus group is $\frac{5}{2}$-fold transitive.

2) The mapping z given by

$$[x_0, x_1, x_2, x_3]z = [x_0^2, x_1^2, x_2^2, x_3^2]$$

carries the ovoid $\mathcal{O}$ onto itself and normalises $Sz(q)$.

3) (McDermott [1]). Suppose that

$$Sz(q) < \mathfrak{H} \leq Sz(q)\langle z\rangle.$$

Then $\mathfrak{H}$ is not $\frac{5}{2}$-fold transitive, although $Sz(q)$ is.

§ 4. Exceptional Characters

If ϕ is a character of a proper subgroup $\mathfrak{U}$ of $\mathfrak{G}$, then the induced character $\phi^{\mathfrak{G}}$ of $\mathfrak{G}$ is often of large norm, even if ϕ is irreducible. To obtain irreducible characters of $\mathfrak{G}$, it is sometimes more useful to consider $\phi^{\mathfrak{G}}$ when ϕ is a generalized character of $\mathfrak{U}$ vanishing on a large subset of $\mathfrak{U}$. In V, 22.7, we obtained an isometry from a space of generalized characters of this kind into the space of generalized characters of $\mathfrak{G}$. We shall now study the consequences of this in situations which are relevant to the classification of Zassenhaus groups.

4.1 Theorem (Feit [1]). *Let* D *be a subset of* $\mathfrak{G}$ *such that* $\mathsf{D} \cap \mathsf{D}^g = \varnothing$ *whenever* $\mathsf{D} \neq \mathsf{D}^g$, *and let* $\mathfrak{L} = \mathbf{N}_{\mathfrak{G}}(\mathsf{D})$. *Thus* $\mathfrak{L} \supseteq \mathsf{D}$. *Suppose that* $\{\lambda_{ij} | i = 1, \ldots, n; j = 1, \ldots, m_i\}$ *is a set of distinct irreducible characters of* $\mathfrak{L}$ *which all vanish on* $\mathfrak{L} - (\mathsf{D} \cup \{1\})$. *Suppose further that there exist rational integers* z_i $(i = 1, \ldots, n)$ *such that* $1 = z_1 < z_2 \cdots < z_n$ *and* $\lambda_{ij}(1) = z_i\lambda_{11}(1)$. *Finally suppose that* $m_1 \geq 2$ *and that whenever* $1 < k \leq n$,

$$\sum_{i=1}^{k-1} m_i z_i^2 > 2z_k. \tag{1}$$

Then the following hold.

a) *There exist* $\varepsilon = \pm 1$ *and distinct irreducible characters* Λ_{ij} *of* $\mathfrak{G}$ *such that whenever* $\{a_{ij}\}$ *is a set of rational integers such that* $\sum_{i,j} a_{ij}\lambda_{ij}(1) = 0$, *then*

$$\left(\sum_{i,j} a_{ij}\lambda_{ij}\right)^{\mathfrak{G}} = \varepsilon \sum_{i,j} a_{ij}\Lambda_{ij}.$$

The Λ_{ij} *are called the exceptional characters belonging to the* λ_{ij}.

b) *There exist a rational integer* a *and a character* ν *of* $\mathfrak{G}$, *both inde-*

pendent of i and j, such that $(v, \Lambda_{ij})_{\mathfrak{G}} = 0$ *for all i, j and*

$$\lambda_{ij}^{\mathfrak{G}} = \varepsilon\Lambda_{ij} + az_i \sum_{k,l} z_k \Lambda_{kl} + z_i v.$$

Proof. a) First we deal with the case $n = 1$.

Put $\alpha_{ij} = \lambda_{1i} - \lambda_{1j}$; thus α_{ij} is a generalized character of $\mathfrak{L}$, and since $m_1 \geq 2$, there exists a non-zero α_{ij}. Now $\alpha_{ij}(x) = 0$ when $x \in (\mathfrak{L} - \mathsf{D}) \cup \{1\}$. By V, 22.7, it follows that

(2) $$(\alpha_{ij}^{\mathfrak{G}}, \alpha_{kl}^{\mathfrak{G}})_{\mathfrak{G}} = (\alpha_{ij}, \alpha_{kl})_{\mathfrak{L}} = \delta_{ik} - \delta_{il} - \delta_{jk} + \delta_{jl}.$$

In particular, if $i = k \neq j = l$,

(3) $$(\alpha_{ij}^{\mathfrak{G}}, \alpha_{ij}^{\mathfrak{G}})_{\mathfrak{G}} = 2.$$

Hence $\alpha_{ij}^{\mathfrak{G}} = \pm\Delta_{ij} \pm \Delta'_{ij}$, where Δ_{ij}, Δ'_{ij} are two distinct irreducible characters of $\mathfrak{G}$. Since $\alpha_{ij}(1) = 0$, we also have $\alpha_{ij}^{\mathfrak{G}}(1) = 0$, so we may assume that

(4) $$\alpha_{ij}^{\mathfrak{G}} = \Delta_{ij} - \Delta'_{ij}.$$

If $m_1 = 2$, we put $\varepsilon = 1$ and $\Lambda_{11} = \Delta_{12}$, $\Lambda_{12} = \Delta'_{12}$. From $a_{11}\lambda_{11}(1) + a_{12}\lambda_{12}(1) = 0$, it follows that $a_{11} + a_{12} = 0$, so

$$(a_{11}\lambda_{11} + a_{12}\lambda_{12})^{\mathfrak{G}} = a_{11}\alpha_{12}^{\mathfrak{G}} = a_{11}\Lambda_{11} + a_{12}\Lambda_{12}.$$

The theorem is thus established in the case $n = 1$, $m_1 = 2$.

Now suppose that $n = 1$ and $m_1 > 2$. By (4), $\alpha_{1j}^{\mathfrak{G}} = \Delta_{1j} - \Delta'_{1j}$, and $\Delta_{1j} \neq \Delta'_{1j}$. By (2), if $1 \neq j \neq k \neq 1$, then

(5) $$\begin{aligned} 1 &= (\alpha_{1j}^{\mathfrak{G}}, \alpha_{1k}^{\mathfrak{G}})_{\mathfrak{G}} \\ &= (\Delta_{1j} - \Delta'_{1j}, \Delta_{1k} - \Delta'_{1k})_{\mathfrak{G}} \\ &= (\Delta_{1j}, \Delta_{1k})_{\mathfrak{G}} + (\Delta'_{1j}, \Delta'_{1k})_{\mathfrak{G}} - (\Delta_{1j}, \Delta'_{1k})_{\mathfrak{G}} - (\Delta'_{1j}, \Delta_{1k})_{\mathfrak{G}}. \end{aligned}$$

Thus $\Delta_{1j} = \Delta_{1k}$ or $\Delta'_{1j} = \Delta'_{1k}$. If both hold, then $(\alpha_{1j}^{\mathfrak{G}}, \alpha_{1k}^{\mathfrak{G}})_{\mathfrak{G}} = 2$, which is a contradiction.

If $m_1 = 3$, we obtain

$$\alpha_{12}^{\mathfrak{G}} = \Delta_{12} - \Delta'_{12}, \quad \alpha_{13}^{\mathfrak{G}} = \Delta_{13} - \Delta'_{13}$$

with $\Delta_{12} = \Delta_{13}$ or $\Delta'_{12} = \Delta'_{13}$. In the first case we put $\varepsilon = 1$, $\Lambda_{11} = \Delta_{12}$,

$\Lambda_{12} = \Delta'_{12}$ and $\Lambda_{13} = \Delta'_{13}$. In the second case, put $\varepsilon = -1$, $\Lambda_{11} = \Delta'_{12}$, $\Lambda_{12} = \Delta_{12}$ and $\Lambda_{13} = \Delta_{13}$. In either case,

$$(\lambda_{11} - \lambda_{1i})^{\mathfrak{G}} = \varepsilon(\Lambda_{11} - \Lambda_{1i}) \quad (i = 2, 3).$$

From this, a) follows easily.

Now suppose that $m_1 > 3$. We know already that for $1 \neq j \neq k \neq 1$, either $\Delta_{1j} = \Delta_{1k}$ or $\Delta'_{1j} = \Delta'_{1k}$ but not both. We show that either $\Delta_{1j} = \Delta_{12}$ for all $j > 1$ or $\Delta'_{1j} = \Delta'_{12}$ for all $j > 1$. Otherwise there exist j, k such that $\Delta_{1j} \neq \Delta_{12}$ and $\Delta'_{1k} \neq \Delta'_{12}$. Then $\Delta'_{1j} = \Delta'_{12} \neq \Delta'_{1k}$ and $\Delta_{1k} = \Delta_{12} \neq \Delta_{1j}$, so by (2) and (4),

$$\begin{aligned} 1 = (\alpha_{1j}^{\mathfrak{G}}, \alpha_{1k}^{\mathfrak{G}})_{\mathfrak{G}} &= (\Delta_{1j} - \Delta'_{1j}, \Delta_{1k} - \Delta'_{1k})_{\mathfrak{G}} \\ &= -(\Delta_{1j}, \Delta'_{1k})_{\mathfrak{G}} - (\Delta'_{1j}, \Delta_{1k})_{\mathfrak{G}} \leq 0. \end{aligned}$$

Hence there exist a sign $\varepsilon = \pm 1$ and an irreducible character Λ_{11} of $\mathfrak{G}$ such that for all $j > 1$, $(\alpha_{1j}^{\mathfrak{G}}, \Lambda_{11})_{\mathfrak{G}} = \varepsilon$. For $j > 1$, we define the irreducible character Λ_{1j} of $\mathfrak{G}$ by putting

$$\alpha_{1j}^{\mathfrak{G}} = \varepsilon(\Lambda_{11} - \Lambda_{1j}). \tag{6}$$

Since $(\alpha_{1j}^{\mathfrak{G}}, \alpha_{1j}^{\mathfrak{G}})_{\mathfrak{G}} = 2$, $\Lambda_{11} \neq \Lambda_{1j}$, and since $(\alpha_{1j}^{\mathfrak{G}}, \alpha_{1k}^{\mathfrak{G}})_{\mathfrak{G}} = 1$ for $j \neq k$, $\Lambda_{1j} \neq \Lambda_{1k}$.

If $\sum_{j=1}^{m_1} a_{1j}\lambda_{1j}(1) = 0$, then $\sum_{j=1}^{m_1} a_{1j} = 0$ on account of the hypothesis $\lambda_{1j}(1) = \lambda_{11}(1)$. It follows that

$$\begin{aligned} \left(\sum_{j=1}^{m_1} a_{1j}\lambda_{1j}\right)^{\mathfrak{G}} &= \sum_{j=1}^{m_1} a_{1j}(\lambda_{1j} - \lambda_{11})^{\mathfrak{G}} = -\sum_{j=1}^{m_1} a_{1j}\alpha_{1j}^{\mathfrak{G}} \\ &= \varepsilon \sum_{j=1}^{m_1} a_{1j}(\Lambda_{1j} - \Lambda_{11}) = \varepsilon \sum_{j=1}^{m_1} a_{1j}\Lambda_{1j}. \end{aligned}$$

Thus a) is proved in the case $n = 1$.

We prove the general assertion by induction on n. Suppose that $n > 1$ and that irreducible characters Λ_{ij} $(i = 1, \ldots, n-1; j = 1, \ldots, m_i)$ of $\mathfrak{G}$ and a sign $\varepsilon = \pm 1$ with the required properties have already been found. Put

$$\beta_j = z_n\lambda_{11} - \lambda_{nj} \quad (j = 1, \ldots, m_n). \tag{7}$$

Thus β_j is a generalized character of $\mathfrak{L}$ and $\beta_j(1) = 0$. Further we define rational integers b_j by putting

(8) $$(\Lambda_{11}, \beta_j^{\mathfrak{G}})_{\mathfrak{G}} = \varepsilon(z_n - b_j).$$

It then follows that for $1 \leq i \leq n - 1$ and $(i, k) \neq (1, 1)$,

$$\begin{aligned}(\Lambda_{ik}, \beta_j^{\mathfrak{G}})_{\mathfrak{G}} &= (z_i\Lambda_{11}, \beta_j^{\mathfrak{G}})_{\mathfrak{G}} - (z_i\Lambda_{11} - \Lambda_{ik}, \beta_j^{\mathfrak{G}})_{\mathfrak{G}}\\ &= \varepsilon z_i(z_n - b_j) - (z_i\Lambda_{11} - \Lambda_{ik}, \beta_j^{\mathfrak{G}})_{\mathfrak{G}}.\end{aligned}$$

By the inductive hypothesis,

$$z_i\Lambda_{11} - \Lambda_{ik} = \varepsilon(z_i\lambda_{11} - \lambda_{ik})^{\mathfrak{G}}.$$

Using V, 22.7, it therefore follows that

(9) $$\begin{aligned}(\Lambda_{ik}, \beta_j^{\mathfrak{G}})_{\mathfrak{G}} &= \varepsilon z_i(z_n - b_j) - \varepsilon(z_i\lambda_{11} - \lambda_{ik}, z_n\lambda_{11} - \lambda_{nj})_{\mathfrak{L}}\\ &= \varepsilon z_i(z_n - b_j) - \varepsilon z_i z_n = -\varepsilon z_i b_j.\end{aligned}$$

Also from V, 22.7,

(10) $$(\beta_j^{\mathfrak{G}}, \beta_j^{\mathfrak{G}})_{\mathfrak{G}} = (\beta_j, \beta_j)_{\mathfrak{L}} = z_n^2 + 1.$$

We therefore obtain

(11) $$\beta_j^{\mathfrak{G}} = \varepsilon(z_n - b_j)\Lambda_{11} - \varepsilon b_j \sum_{i=2}^{m_1} \Lambda_{1i} - \varepsilon b_j \sum_{k=2}^{n-1} \sum_{l=1}^{m_k} z_k\Lambda_{kl} + v_j,$$

where v_j is a generalized character in which each Λ_{ik} $(1 \leq i \leq n - 1)$ does not occur.

We prove that $b_j = 0$. First suppose that $v_j = 0$. Then (11) gives

$$\beta_j^{\mathfrak{G}}(1) = \varepsilon z_n\Lambda_{11}(1) - \varepsilon b_j \sum_{k=1}^{n-1} \sum_{l=1}^{m_k} z_k\Lambda_{kl}(1).$$

But since $\beta_j(1) = 0$, $\beta_j^{\mathfrak{G}}(1) = 0$. Also, since $z_i\lambda_{11}(1) - \lambda_{ij}(1) = 0$,

$$(z_i\Lambda_{11} - \Lambda_{ij})(1) = \varepsilon(z_i\lambda_{11} - \lambda_{ij})^{\mathfrak{G}}(1) = 0$$

for $i < n$. Hence

$$z_n = b_j \sum_{k=1}^{n-1} m_k z_k^2.$$

Thus $b_j > 0$. But then $z_n \geq \sum_{k=1}^{n-1} m_k z_k^2 > 2z_n$, by (1). This is a contradiction, so $v_j \neq 0$. It therefore follows from (10) and (11) that $z_n^2 + 1 = (\beta_j^{\mathfrak{G}}, \beta_j^{\mathfrak{G}})_{\mathfrak{G}} \geq (z_n - b_j)^2 + b_j^2(m_1 - 1) + b_j^2 \sum_{k=2}^{n-1} m_k z_k^2 + 1$. This gives

$$-2z_n b_j + b_j^2 \sum_{k=1}^{n-1} m_k z_k^2 \leq 0,$$

so by (1), $b_j = 0$.

(11) now reduces to

$$\beta_j^{\mathfrak{G}} = \varepsilon z_n \Lambda_{11} + v_j. \tag{12}$$

By (10), $(v_j, v_j)_{\mathfrak{G}} = 1$. Since $\beta_j^{\mathfrak{G}}(1) = 0$, it follows that $v_j = -\varepsilon\Lambda_{nj}$ for an irreducible character Λ_{nj}. Since no Λ_{ik} with $1 \leq i \leq n - 1$ occurs in v_j, $\Lambda_{ik} \neq \Lambda_{nj}$ for $i < n$. And by V, 22.7, for $j \neq k$,

$$(\beta_j^{\mathfrak{G}}, \beta_k^{\mathfrak{G}})_{\mathfrak{G}} = (\beta_j, \beta_k)_{\mathfrak{L}} = (z_n\lambda_{11} - \lambda_{nj}, z_n\lambda_{11} - \lambda_{nk})_{\mathfrak{L}} = z_n^2.$$

Since by (10), $(\beta_j^{\mathfrak{G}}, \beta_j^{\mathfrak{G}})_{\mathfrak{G}} = z_n^2 + 1$, $\beta_j^{\mathfrak{G}} \neq \beta_k^{\mathfrak{G}}$ for $j \neq k$. Hence $\Lambda_{nj} \neq \Lambda_{nk}$ for $j \neq k$. Thus the Λ_{ij} $(i = 1, \ldots, n; j = 1, \ldots, m_i)$ are distinct irreducible characters of $\mathfrak{G}$, and by (12),

$$\beta_j^{\mathfrak{G}} = \varepsilon(z_n\Lambda_{11} - \Lambda_{nj}).$$

Suppose now that $\sum_{i,j} a_{ij}\lambda_{ij}(1) = 0$. Then $\sum_{i,j} a_{ij}z_i = 0$. We put $v_{ij} = z_i\lambda_{11} - \lambda_{ij}$, (so $v_{nj} = \beta_j$). For $1 \leq i \leq n - 1$, $v_{ij}(1) = 0$, so by the inductive hypothesis,

$$v_{ij}^{\mathfrak{G}} = \varepsilon(z_i\Lambda_{11} - \Lambda_{ij}).$$

Also,

$$v_{nj}^{\mathfrak{G}} = \beta_j^{\mathfrak{G}} = \varepsilon(z_n\Lambda_{11} - \Lambda_{nj}).$$

Thus it follows that

$$\left(\sum_{i,j} a_{ij}\lambda_{ij}\right)^{\mathfrak{G}} = -\sum_{i,j} a_{ij}(z_i\lambda_{11} - \lambda_{ij})^{\mathfrak{G}} = -\varepsilon\sum_{i,j} a_{ij}(z_i\Lambda_{11} - \Lambda_{ij})$$
$$= \varepsilon\sum_{i,j} a_{ij}\Lambda_{ij}.$$

b) We define non-negative integers b_{kl} by putting

(13) $$b_{kl} = (\lambda_{11}^{\mathfrak{G}}, \Lambda_{kl})_{\mathfrak{G}}.$$

Then

(14) $$\lambda_{11}^{\mathfrak{G}} = \sum_{k,l} b_{kl}\Lambda_{kl} + v,$$

where $(v, \Lambda_{kl})_{\mathfrak{G}} = 0$ for all k, l. If χ is an irreducible character of $\mathfrak{G}$ distinct from all the Λ_{ij}, then by a),

$$(\lambda_{ij}^{\mathfrak{G}} - z_i\lambda_{11}^{\mathfrak{G}}, \chi)_{\mathfrak{G}} = \varepsilon(\Lambda_{ij} - z_i\Lambda_{11}, \chi)_{\mathfrak{G}} = 0.$$

Further,

$$\begin{aligned}(\lambda_{ij}^{\mathfrak{G}} - z_i\lambda_{11}^{\mathfrak{G}}, \Lambda_{kl})_{\mathfrak{G}} &= \varepsilon(\Lambda_{ij} - z_i\Lambda_{11}, \Lambda_{kl})_{\mathfrak{G}}\\ &= \begin{cases} 0 & \text{for } (i,j) \neq (k,l) \neq (1,1),\\ \varepsilon & \text{for } (i,j) = (k,l) \neq (1,1),\\ -\varepsilon z_i & \text{for } (i,j) \neq (k,l) = (1,1).\end{cases}\end{aligned}$$

Thus also by a)

$$\lambda_{ij}^{\mathfrak{G}} - z_i\lambda_{11}^{\mathfrak{G}} = \varepsilon\Lambda_{ij} - \varepsilon z_i\Lambda_{11}$$

(even if $(i, j) = (1, 1)$). Adding to (14),

$$\lambda_{ij}^{\mathfrak{G}} = \varepsilon\Lambda_{ij} + z_i\sum_{k,l} b_{kl}\Lambda_{kl} + (a - b_{11})z_i\Lambda_{11} + z_i v,$$

where $a = b_{11} - \varepsilon$.

By a) and V, 22.7, we have for $(k, l) \neq (1, 1)$,

$$\begin{aligned}z_k b_{11} - b_{kl} &= (\lambda_{11}^{\mathfrak{G}}, z_k\Lambda_{11} - \Lambda_{kl})_{\mathfrak{G}} = \varepsilon(\lambda_{11}^{\mathfrak{G}}, z_k\lambda_{11}^{\mathfrak{G}} - \lambda_{kl}^{\mathfrak{G}})_{\mathfrak{G}}\\ &= \varepsilon(\lambda_{11}, z_k\lambda_{11} - \lambda_{kl})_{\mathfrak{L}} = \varepsilon z_k.\end{aligned}$$

This gives $b_{kl} = z_k a$, and so we obtain the assertion

$$\lambda_{ij}^{\mathfrak{G}} = \varepsilon\Lambda_{ij} + az_i\sum_{k,l} z_k\Lambda_{kl} + z_i v.$$ **q.e.d.**

The following is the particular case of 4.1 which is of greatest importance in our study of Zassenhaus groups.

4.2 Hypothesis. a) $\mathfrak{F}$ is a proper subgroup of $\mathfrak{G}$.

b) $\mathfrak{L} = \mathbf{N}_{\mathfrak{G}}(\mathfrak{F})$ is a Frobenius group with Frobenius kernel $\mathfrak{F}$ and Frobenius complement $\mathfrak{H}$.

c) $\mathfrak{F} \cap \mathfrak{F}^g = 1$ whenever $g \notin \mathfrak{L}$.

d) ϕ_{00} is the unit character of $\mathfrak{F}$ and ρ is the regular character of $\mathfrak{F}$.

e) ϕ_{ij} $(i = 1, \ldots, n;\ j = 1, \ldots, m_i)$ are representatives of the $\mathfrak{H}$-orbits of the set of non-identity irreducible characters of $\mathfrak{F}$. The notation is so chosen that $\phi_{ij}(1) = z_i$ depends only on i and

$$z_1 < z_2 < \cdots < z_n.$$

Since $\mathfrak{F}$ is nilpotent, $\mathfrak{F} > \mathfrak{F}'$ and $z_1 = 1$.

f) $\lambda_{ij} = \phi_{ij}^{\mathfrak{L}}$. Thus $\lambda_{ij}(1) = |\mathfrak{L} : \mathfrak{F}|\phi_{ij}(1) = |\mathfrak{H}|z_i$, and λ_{ij} is zero on $\mathfrak{L} - \mathfrak{F}$. By V, 16.13, the λ_{ij} are precisely the irreducible characters of $\mathfrak{L}$ for which $\mathfrak{F}$ is not contained in the kernel of λ_{ij}.

g) $m_1 \geq 2$ and $\sum_{i=1}^{k-1} m_i z_i^2 > 2z_k$ whenever $1 < k \leq n$.

h) Λ_{ij} $(i = 1, \ldots, n; j = 1, \ldots, m_i)$ are irreducible characters of $\mathfrak{G}$ such that

$$\lambda_{ij}^{\mathfrak{G}} = \varepsilon\Lambda_{ij} + az_i \sum_{k,l} z_k \Lambda_{kl} + z_i \nu,$$

where $\varepsilon = \pm 1$, a is an integer and $(\nu, \Lambda_{kl})_{\mathfrak{G}} = 0$ for all k, l. These exist by 4.1, applied with $\mathsf{D} = \mathfrak{F} - \{1\}$. The Λ_{ij} are called the *exceptional characters* of $\mathfrak{G}$ belonging to $\mathfrak{F}$.

4.3 Lemma. *Under Hypothesis 4.2 the following hold.*

a) *The exceptional characters Λ_{ij} are not constant on $\mathfrak{F} - \{1\}$.*

b) *If χ is an irreducible character of $\mathfrak{G}$ which is not constant on $\mathfrak{F} - \{1\}$, then χ is one of the exceptional characters Λ_{ij}.*

Proof. a) The class functions on $\mathfrak{F}$ which are constant on $\mathfrak{F} - \{1\}$ form a vector space V of dimension 2 spanned by ϕ_{00} and ρ. Thus if $(\Lambda_{ij})_{\mathfrak{F}}$ is constant on $\mathfrak{F} - \{1\}$, $(\Lambda_{ij})_{\mathfrak{F}} = c\phi_{00} + d\rho$ for some c, d. It follows from 4.2 and Frobenius reciprocity (V, 16.5) that

$$\varepsilon + az_i^2 = (\lambda_{ij}^{\mathfrak{G}}, \Lambda_{ij})_{\mathfrak{G}} = (\phi_{ij}^{\mathfrak{G}}, \Lambda_{ij})_{\mathfrak{G}} = (\phi_{ij}, c\phi_{00} + d\rho)_{\mathfrak{F}} = dz_i,$$

since $(\phi_{ij}, \rho)_{\mathfrak{F}} = \phi_{ij}(1) = z_i$. Hence z_i divides $\varepsilon = \pm 1$ and $z_i = 1$. Therefore $i = 1$. Now there is a character $\lambda_{1k} \neq \lambda_{1j}$, since $m_1 \geq 2$. It then follows that

$$\begin{aligned}\varepsilon + az_1^2 = d = dz_1 &= (\phi_{1k}, c\phi_{00} + d\rho)_{\mathfrak{F}} \\ &= (\phi_{1k}, (\Lambda_{1j})_{\mathfrak{F}})_{\mathfrak{F}} = (\lambda_{1k}^{\mathfrak{G}}, \Lambda_{1j})_{\mathfrak{G}} = az_1^2.\end{aligned}$$

But this implies that $\varepsilon = 0$. Thus Λ_{ij} is not constant on $\mathfrak{F} - \{1\}$.

b) Suppose that χ is an irreducible character of $\mathfrak{G}$ and $\chi \neq \Lambda_{ij}$ for all i, j. By V, 16.13,

$$\chi_{\mathfrak{F}\mathfrak{H}} = \sum_{i,j} c_{ij}\phi_{ij}^{\mathfrak{F}\mathfrak{H}} + \psi,$$

where ψ is a character constant on $\mathfrak{F}$; further

$$(\phi_{ij}^{\mathfrak{F}\mathfrak{H}})_{\mathfrak{F}} = \sum_{h \in \mathfrak{H}} \phi_{ij}^h.$$

Hence $(\chi_{\mathfrak{F}}, \phi_{ij})_{\mathfrak{F}} = c_{ij}$ for $(i, j) \neq (0, 0)$. By Frobenius reciprocity, $(\chi, \lambda_{ij}^{\mathfrak{G}})_{\mathfrak{G}} = c_{ij}$. But by 4.2,

$$(\chi, \lambda_{ij}^{\mathfrak{G}})_{\mathfrak{G}} = z_i(\chi, \nu)_{\mathfrak{G}} = z_i b,$$

where $b = (\chi, \nu)_{\mathfrak{G}}$, since $\chi \neq \Lambda_{kl}$ for any k, l. Thus $c_{ij} = z_i b$ and

$$\chi_{\mathfrak{F}} = \sum_{i,j}\sum_{h \in \mathfrak{H}} z_i b\phi_{ij}^h + d\phi_{00} = b\rho + (d - b)\phi_{00}.$$

Hence χ is constant on $\mathfrak{F} - \{1\}$. **q.e.d.**

4.4 Lemma. *Under Hypothesis 4.2, the following hold.*

a) $z_i\Lambda_{jk} - z_j\Lambda_{il}$ *is zero on* $\mathfrak{G} - \bigcup_{g \in \mathfrak{G}} (\mathfrak{F} - \{1\})^g$ *for all* i, j, k, l.

b) *The restrictions of the* Λ_{ij} *to* $\mathfrak{F} - \{1\}$ *are linearly independent.*

Proof. a) By 4.2,

$$z_i\Lambda_{jk} - z_j\Lambda_{il} = \varepsilon(z_i\lambda_{jk} - z_j\lambda_{il})^{\mathfrak{G}} = \varepsilon(z_i\phi_{jk} - z_j\phi_{il})^{\mathfrak{G}}.$$

This shows that $z_i\Lambda_{jk} - z_j\Lambda_{il}$ is zero outside $\bigcup_{g \in \mathfrak{G}} \mathfrak{F}^g$. Since

$$z_i\phi_{jk}(1) - z_j\phi_{il}(1) = z_i z_j - z_j z_i = 0,$$

$z_i\Lambda_{jk}(1) - z_j\Lambda_{il}(1) = 0$.

b) Suppose that $\sum_{i,j} a_{ij}\Lambda_{ij}(x) = 0$ for all $x \in \mathfrak{F} - \{1\}$. Using a), it follows that

$$\left(\sum_{i,j} a_{ij}\Lambda_{ij}\right)(z_k\overline{\Lambda}_{11} - \overline{\Lambda}_{kl}) = 0$$

(on all $\mathfrak{G}$). The orthogonality relations then give $a_{kl} = z_k a_{11}$. Thus to within a scalar multiple, the given relation, if non trivial, takes the form

$$\sum_{i,j} z_i \Lambda_{ij}(x) = 0 \quad (x \in \mathfrak{F} - \{1\}).$$

By 4.2,

$$\lambda_{ij}^{\mathfrak{G}} = \varepsilon\Lambda_{ij} + z_i\Delta,$$

where $\Delta = a\sum_{k,l} z_k \Lambda_{kl} + v$ is independent of i and j. Thus for all $x \in \mathfrak{F} - \{1\}$,

$$0 = \sum_{i,j} z_i(\lambda_{ij}^{\mathfrak{G}} - z_i\Delta)(x).$$

But $\lambda_{ij} = \phi_{ij}^{\mathfrak{L}}$. By V, 22.7 and 16.13, if $x \in \mathfrak{F} - \{1\}$,

$$\lambda_{ij}^{\mathfrak{G}}(x) = \lambda_{ij}(x) = \sum_{h \in \mathfrak{H}} \phi_{ij}^{h}(x),$$

so

$$\sum_{i,j} z_i \lambda_{ij}^{\mathfrak{G}}(x) = \sum_{i,j}\sum_{h \in \mathfrak{H}} z_i \phi_{ij}^{h}(x) = \rho(x) - \phi_{00}(x) = -1.$$

Hence $\sum_{i,j} z_i^2 \Delta(x) = -1$ for all $x \in \mathfrak{F} - \{1\}$ and $(\sum_i m_i z_i^2)\Delta(x) = -1$. But then $\Delta(x) = -(\sum_i m_i z_i^2)^{-1}$. This is impossible, since $\Delta(x)$ is an algebraic integer and $m_1 \geq 2$. **q.e.d.**

4.5 Lemma. *Suppose that $\mathfrak{F}_1$, $\mathfrak{F}_2$ are subgroups of $\mathfrak{G}$ for which Hypothesis 4.2 is satisfied. If $\mathfrak{F}_1 \cap \mathfrak{F}_2^g = 1$ for all $g \in \mathfrak{G}$, then no exceptional character belonging to $\mathfrak{F}_1$ is an exceptional character belonging to $\mathfrak{F}_2$.*

Proof. Suppose that Λ is an exceptional character belonging to both $\mathfrak{F}_1$ and $\mathfrak{F}_2$. By 4.3a), Λ is not constant on $\mathfrak{F}_2 - \{1\}$. Let Λ' be an exceptional character belonging to $\mathfrak{F}_1$ distinct from Λ. Then by 4.4a), $\Lambda(1)\Lambda' - \Lambda'(1)\Lambda$ is zero on the set $\mathfrak{G} - \bigcup_{g \in \mathfrak{G}}(\mathfrak{F}_1 - \{1\})^g$, which contains $\mathfrak{F}_2 - \{1\}$. Thus Λ' is not constant on $\mathfrak{F}_2 - \{1\}$; hence Λ' is an exceptional character belonging to $\mathfrak{F}_2$, by 4.3b). But then the relation $\Lambda(1)\Lambda'(x) - \Lambda'(1)\Lambda(x) = 0$ ($x \in \mathfrak{F}_2 - \{1\}$) contradicts 4.4b). **q.e.d.**

We specialize further.

4.6 Theorem. *Let $\mathfrak{A}$ be a proper Abelian subgroup of $\mathfrak{G}$ for which $\mathfrak{A} \cap \mathfrak{A}^g = 1$ for $\mathfrak{A} \neq \mathfrak{A}^g$. Further, suppose that $\mathfrak{L} = \mathbf{N}_{\mathfrak{G}}(\mathfrak{A})$ is a Frobenius group with Frobenius kernel $\mathfrak{A}$. Suppose that $s = \dfrac{|\mathfrak{A}| - 1}{|\mathfrak{L} : \mathfrak{A}|} \geq 2$. Let λ_0 be the character of $\mathfrak{L}$ induced from the unit character of $\mathfrak{A}$, and let $\lambda_1, \dots, \lambda_s$ be the irreducible characters of $\mathfrak{L}$ for which $\mathfrak{A}$ is not contained in the kernel of λ_i. Then there exist distinct irreducible characters $\Lambda_1, \dots, \Lambda_s$ of $\mathfrak{G}$, a sign $\varepsilon = \pm 1$ and a rational integer b with the following properties.*

a) $\lambda_0^{\mathfrak{G}} - \lambda_i^{\mathfrak{G}} = \varepsilon \Lambda_i + b \sum_{j=1}^{s} \Lambda_j + \sum_\chi c(\chi)\chi$,
where the second sum runs over all irreducible characters of $\mathfrak{G}$ other than the Λ_i. Further, $\chi(x) = c(\chi)$ for all $x \in \mathfrak{A} - \{1\}$, and $\lambda_0^{\mathfrak{G}} - \lambda_i^{\mathfrak{G}}$ is zero outside $\bigcup_{g \in \mathfrak{G}} (\mathfrak{A} - \{1\})^g$. Of course, $c(\chi)$ is a rational integer.

b) *$\Lambda_1, \dots, \Lambda_s$ are precisely the irreducible characters of $\mathfrak{G}$ which are not constant on $\mathfrak{A} - \{1\}$.*

c) $(\varepsilon + b)^2 + (s - 1)b^2 + \sum_\chi |c(\chi)|^2 = |\mathfrak{L} : \mathfrak{A}| + 1$.

Proof. Since $\mathfrak{A}$ is Abelian, all the irreducible characters of $\mathfrak{A}$ are of degree 1. By hypothesis,

$$m_1 = \frac{|\mathfrak{A}| - 1}{|\mathfrak{L} : \mathfrak{A}|} \geq 2,$$

so Hypothesis 4.2 is satisfied. Hence there exist $\varepsilon = \pm 1$ and distinct irreducible characters $\Lambda_1, \dots, \Lambda_s$ of $\mathfrak{G}$ such that

(1) $$\lambda_i^{\mathfrak{G}} = \varepsilon \Lambda_i + a \sum_{j=1}^{s} \Lambda_j + v \quad (i = 1, \dots, s),$$

where a is an integer and $(v, \Lambda_j)_{\mathfrak{G}} = 0$ for all j. By 4.3, $\Lambda_1, \dots, \Lambda_s$ satisfy b). By V, 22.7,

$$(\lambda_0^{\mathfrak{G}}, \Lambda_i - \Lambda_j)_{\mathfrak{G}} = \varepsilon(\lambda_0^{\mathfrak{G}}, \lambda_i^{\mathfrak{G}} - \lambda_j^{\mathfrak{G}})_{\mathfrak{G}} = \varepsilon(\lambda_0, \lambda_i - \lambda_j)_{\mathfrak{L}} = 0.$$

Thus $(\lambda_0^{\mathfrak{G}}, \Lambda_i)_{\mathfrak{G}}$ is independent of i. We put $a' = (\lambda_0^{\mathfrak{G}}, \Lambda_i)_{\mathfrak{G}}$. Now

$$(\lambda_0^{\mathfrak{G}}, 1_{\mathfrak{G}})_{\mathfrak{G}} = (\lambda_0, 1_{\mathfrak{L}})_{\mathfrak{L}} = (1_{\mathfrak{A}}, 1_{\mathfrak{A}})_{\mathfrak{A}} = 1.$$

Hence

$$\lambda_0^{\mathfrak{G}} = 1_{\mathfrak{G}} + a' \sum_{j=1}^{s} \Lambda_j + v', \tag{2}$$

where $(v', \Lambda_j)_{\mathfrak{G}} = (v', 1_{\mathfrak{G}})_{\mathfrak{G}} = 0$. Subtracting (1) from (2),

$$\lambda_0^{\mathfrak{G}} - \lambda_i^{\mathfrak{G}} = -\varepsilon \Lambda_i + b \sum_{j=1}^{s} \Lambda_j + \sum_{\chi} c(\chi)\chi$$

for appropriate integers b and $c(\chi)$. Since $\lambda_i = \phi_i^{\mathfrak{L}}$ for an irreducible character ϕ_i of $\mathfrak{A}$,

$$c(\chi) = (\lambda_0^{\mathfrak{G}} - \lambda_i^{\mathfrak{G}}, \chi)_{\mathfrak{G}} = (\lambda_0 - \lambda_i, \chi_{\mathfrak{L}})_{\mathfrak{L}} = (1_{\mathfrak{A}} - \phi_i, \chi_{\mathfrak{A}})_{\mathfrak{A}}.$$

But by 4.3b), χ has a constant value c on $\mathfrak{A} - \{1\}$. Since $\phi_i(1) = 1$, it follows that

$$c(\chi) = c(1_{\mathfrak{A}} - \phi_i, 1_{\mathfrak{A}})_{\mathfrak{A}} = c.$$

And since $\lambda_0^{\mathfrak{G}} - \lambda_i^{\mathfrak{G}} = (1_{\mathfrak{A}} - \phi_i)^{\mathfrak{G}}$, it follows at once that $\lambda_0^{\mathfrak{G}} - \lambda_i^{\mathfrak{G}}$ vanishes outside $\bigcup_{g \in \mathfrak{G}} (\mathfrak{A} - \{1\})^g$.

By a), we have on the one hand

$$(\lambda_0^{\mathfrak{G}} - \lambda_i^{\mathfrak{G}}, \lambda_0^{\mathfrak{G}} - \lambda_i^{\mathfrak{G}})_{\mathfrak{G}} = (b + \varepsilon)^2 + (s - 1)b^2 + \sum_{\chi} |c(\chi)|^2.$$

On the other hand, by V, 22.7,

$$\begin{aligned}(\lambda_0^{\mathfrak{G}} - \lambda_i^{\mathfrak{G}}, \lambda_0^{\mathfrak{G}} - \lambda_i^{\mathfrak{G}})_{\mathfrak{G}} &= (\lambda_0 - \lambda_i, \lambda_0 - \lambda_i)_{\mathfrak{L}} \\ &= (\lambda_0, \lambda_0)_{\mathfrak{L}} + (\lambda_i, \lambda_i)_{\mathfrak{L}} \\ &= (\lambda_0, \lambda_0)_{\mathfrak{L}} + 1.\end{aligned}$$

But since $\lambda_0(y)$ is 0 for $y \notin \mathfrak{A}$ and $|\mathfrak{L} : \mathfrak{A}|$ for $y \in \mathfrak{A}$, $(\lambda_0, \lambda_0)_{\mathfrak{L}} = \dfrac{|\mathfrak{A}||\mathfrak{L} : \mathfrak{A}|^2}{|\mathfrak{L}|}$ $= |\mathfrak{L} : \mathfrak{A}|$. The assertion follows at once. **q.e.d.**

§ 5. Characters of Zassenhaus Groups

First we prove some theorems about characters of Zassenhaus groups which will have important applications for the classification of Zassenhaus groups in § 6 and § 9. In the second half of this section we use these theorems to determine the characters of the groups $PSL(2, q)$ and the Suzuki groups $Sz(2^n)$.

As our investigations are directed in the first place towards the determination of the Zassenhaus groups, we remind the reader of the results already obtained.

Let $\mathfrak{G}$ be a Zassenhaus group of degree $n + 1$ and let $|\mathfrak{G}| = (n + 1)nd$. Then d divides $n - 1$.

(1) If $d = n - 1$, then $\mathfrak{G}$ is sharply triply transitive and is therefore known from § 2.

(2) If $d < n - 1$ and d is even, then by 1.11, $\mathfrak{G} = PSL(2, p^f)$, where p is odd.

It therefore remains to deal with the Zassenhaus groups in which d is odd and $d < n - 1$.

5.1 Hypothesis. a) $\mathfrak{G}$ is a Zassenhaus group of degree $n + 1$ on the set $\Omega = \{\infty, 0, \ldots, n - 1\}$.

b) $\mathfrak{F}$ is the Frobenius kernel of $\mathfrak{G}_\infty$.

c) $\mathfrak{H} = \mathfrak{G}_{0,\infty}$ and $|\mathfrak{H}| = d$. Thus $\mathfrak{H}$ is a Frobenius complement of $\mathfrak{G}_\infty$ and d divides $n - 1$.

d) $\mathfrak{H}$ is cyclic.

e) There exists an involution s which interchanges 0 and ∞ such that $h^s = h^{-1}$ for all $h \in \mathfrak{H}$. (Note that by 1.5, d) and e) are satisfied if d is odd and $\mathfrak{G}$ is simple.)

f) π is the permutation character of $\mathfrak{G}$. Thus if $1 \neq g \in \mathfrak{G}$, $0 \leq \pi(g) \leq 2$.

g) For $i = 0, 1, 2$,

$$\mathfrak{I}_i = \{g \mid g \in \mathfrak{G}, \pi(g) = i\}.$$

Thus $\mathfrak{G} = \{1\} \cup \mathfrak{I}_0 \cup \mathfrak{I}_1 \cup \mathfrak{I}_2$ and

$$\mathfrak{I}_1 = \bigcup_{g \in \mathfrak{G}} (\mathfrak{F} - \{1\})^g, \quad \mathfrak{I}_2 = \bigcup_{g \in \mathfrak{G}} (\mathfrak{H} - \{1\})^g.$$

On the other hand we have no good description of $\mathfrak{I}_0$.

In the next lemmas we describe certain characters of $\mathfrak{G}$.

5.2 Lemma. *Under Hypothesis 5.1, let $\alpha = \pi - 1$. Then α is an irreducible character of $\mathfrak{G}$ and*

$$\alpha(g) = \begin{cases} n & \text{for } g = 1, \\ -1 & \text{for } g \in \mathfrak{J}_0, \\ 0 & \text{for } g \in \mathfrak{J}_1, \\ 1 & \text{for } g \in \mathfrak{J}_2. \end{cases}$$

Proof. Since $\mathfrak{G}$ is doubly transitive, α is an irreducible character of $\mathfrak{G}$, by V, 20.2d). **q.e.d.**

5.3 Lemma. *Under Hypothesis 5.1, let $\psi_1, \ldots, \psi_{d-1}$ be the distinct non-identity irreducible characters of $\mathfrak{G}_\infty = \mathfrak{F}\mathfrak{H}$ for which $\mathfrak{F}$ is in the kernel. Further let ψ_0 be the unit character of $\mathfrak{G}_\infty$. (Thus $\psi_0, \ldots, \psi_{d-1}$ are essentially the irreducible characters of $\mathfrak{H}$.)*

a) *For all $i = 0, 1, \ldots, d-1$,*

$$\psi_i^{\mathfrak{G}}(g) = \begin{cases} n + 1 & \text{for } g = 1, \\ 0 & \text{for } g \in \mathfrak{J}_0, \\ 1 & \text{for } g \in \mathfrak{J}_1, \\ \psi_i(g) + \overline{\psi_i}(g) & \text{for } g \in \mathfrak{J}_2. \end{cases}$$

Further $\psi_i^{\mathfrak{G}} = \overline{\psi_i^{\mathfrak{G}}}$, and $\psi_i^{\mathfrak{G}}(g)$ is real for all $g \in \mathfrak{G}$.

b) *For $0 \leq i, j \leq d-1$,*

$$\begin{aligned} (\psi_i^{\mathfrak{G}}, \psi_j^{\mathfrak{G}})_{\mathfrak{G}} &= \tfrac{1}{2}(\psi_i + \overline{\psi_i}, \psi_j + \overline{\psi_j})_{\mathfrak{H}} \\ &= \begin{cases} 0 & \text{for } \psi_i \neq \psi_j \text{ and } \psi_i \neq \overline{\psi_i}, \\ 1 & \text{for } \psi_i = \psi_j \text{ and } \psi_i \neq \overline{\psi_i}, \\ 2 & \text{for } \psi_i = \psi_j = \overline{\psi_i}. \end{cases} \end{aligned}$$

c) *If $|\mathfrak{H}|$ is odd, the ψ_i $(i > 0)$ may be so numbered that $\overline{\psi_i} = \psi_{i+\frac{1}{2}(d-1)}$, and then the $\beta_i = \psi_i^{\mathfrak{G}}$ $(i = 1, \ldots, \frac{1}{2}(d-1))$ are distinct irreducible characters of $\mathfrak{G}$.*

d) $(\psi_0^{\mathfrak{G}}, \psi_i^{\mathfrak{G}})_{\mathfrak{G}} = 0)$ $(i = 1, \ldots, d-1)$.

e) *If $|\mathfrak{H}|$ is odd, then for $1 \neq h \in \mathfrak{H}$,*

$$\sum_{i=1}^{\frac{1}{2}(d-1)} |\beta_i(h)|^2 = d - 2.$$

Proof. a) Clearly $\psi_i^{\mathfrak{G}}(1) = n + 1$. We have

$$\psi_i^{\mathfrak{G}}(g) = \frac{1}{|\mathfrak{G}_\infty|} \sum_{g^x \in \mathfrak{G}_\infty} \psi_i(g^x).$$

Since $\mathfrak{J}_0 \cap \mathfrak{G}_\infty = \varnothing$, $\psi_i^{\mathfrak{G}}(g) = 0$ for $g \in \mathfrak{J}_0$.

To obtain the value on $\mathfrak{J}_1$ we may suppose that $1 \neq g \in \mathfrak{F}$. Then

$$\psi_i^{\mathfrak{G}}(g) = \frac{1}{|\mathfrak{G}_\infty|} \sum_{g^x \in \mathfrak{F}} \psi_i(g^x) = \frac{1}{|\mathfrak{G}_\infty|} \sum_{x \in \mathfrak{G}_\infty} 1 = 1.$$

For $\mathfrak{J}_2$, suppose that $1 \neq g \in \mathfrak{H}$. If $g^x \in \mathfrak{G}_\infty$, then either $\infty x = \infty$ and $x \in \mathfrak{G}_\infty$, or $0x = \infty$ and $x \in s\mathfrak{G}_\infty$, where s is an involution which interchanges 0 and ∞. For $x \in \mathfrak{G}_\infty$, $\psi_i(g^x) = \psi_i(g)$, and for $x \in s\mathfrak{G}_\infty$,

$$\psi_i(g^x) = \psi_i(g^s) = \psi_i(g^{-1}) = \overline{\psi_i}(g),$$

since $g^s = g^{-1}$. Altogether, then,

$$\psi_i^{\mathfrak{G}}(g) = \frac{1}{|\mathfrak{G}_\infty|} \sum_{x \in \mathfrak{G}_\infty \cup s\mathfrak{G}_\infty} \psi_i(g^x) = \psi_i(g) + \overline{\psi_i}(g).$$

From this formula we see at once that $\psi_i^{\mathfrak{G}} = \overline{\psi_i^{\mathfrak{G}}}$.

b) Let $m = |\mathfrak{G}|$. For $0 \leq i, j \leq d - 1$,

$$(\psi_i^{\mathfrak{G}}, \psi_j^{\mathfrak{G}})_{\mathfrak{G}} = \frac{1}{m}\{(n+1)^2 + |\mathfrak{J}_1| + |\mathfrak{G} : \mathbf{N}_{\mathfrak{G}}(\mathfrak{H})| \sum_{1 \neq g \in \mathfrak{H}} (\psi_i(g) + \overline{\psi_i}(g))(\psi_j(g) + \overline{\psi_j}(g))\}.$$

Since

$$|\mathfrak{J}_1| = (|\mathfrak{F}| - 1)|\mathfrak{G} : \mathbf{N}_{\mathfrak{G}}(\mathfrak{F})| = (n-1)\frac{m}{nd}$$

and $|\mathfrak{G} : \mathbf{N}_{\mathfrak{G}}(\mathfrak{H})| = \dfrac{m}{2d}$,

$$\begin{aligned}(\psi_i^{\mathfrak{G}}, \psi_j^{\mathfrak{G}})_{\mathfrak{G}} &= \frac{1}{m}\left\{(n+1)^2 + (n-1)\frac{m}{nd} + \frac{m}{2d} \sum_{h \in \mathfrak{H}} (\psi_i(h) + \overline{\psi_i}(h))(\psi_j(h) + \overline{\psi_j}(h)) - \frac{4m}{2d}\right\} \\ &= \tfrac{1}{2}(\psi_i + \overline{\psi_i}, \psi_j + \overline{\psi_j})_{\mathfrak{H}}.\end{aligned}$$

c) If $|\mathfrak{H}|$ is odd, then all ψ_i other than ψ_0 are non-real, by V, 13.8. The assertion thus follows from b).

d) This follows from b).

e) As $|\mathfrak{H}|$ is odd, $\psi_1^2, \ldots, \psi_{d-1}^2$ are the distinct non-identity irreducible characters of $\mathfrak{H}$. On account of the orthogonality relations it follows that for $1 \neq h \in \mathfrak{H}$,

$$\sum_{i=1}^{\frac{1}{2}(d-1)} |\beta_i(h)|^2 = \sum_{i=1}^{\frac{1}{2}(d-1)} (\psi_i(h) + \overline{\psi_i}(h))(\overline{\psi_i}(h) + \psi_i(h))$$
$$= \sum_{i=1}^{d-1} \psi_i(h)\overline{\psi_i}(h) + \sum_{i=1}^{d-1} \psi_i(h)^2$$
$$= \sum_{i=0}^{d-1} \psi_i(h)\overline{\psi_i}(h) - 1 + \sum_{i=0}^{d-1} \psi_i(h)^2 - 1 = d - 2. \quad \textbf{q.e.d.}$$

5.4 Lemma. *Suppose that Hypothesis 5.1 holds and that $|\mathfrak{H}|$ is odd. Let χ be an irreducible character of $\mathfrak{G}$ distinct from $1, \alpha, \beta_i$ $(i = 1, \ldots, \frac{1}{2}(d-1))$. Then d divides $\chi(1)$ and $\chi(h) = 0$ whenever $1 \neq h \in \mathfrak{H}$.*

Proof. Suppose that $1 \neq h \in \mathfrak{H}$. By the orthogonality relations,

$$\sum_{\tau} \tau(h)\overline{\tau}(h) = |\mathbf{C}_{\mathfrak{G}}(h)|,$$

where the summation is over all irreducible characters τ of $\mathfrak{G}$. As $\mathbf{C}_{\mathfrak{G}}(h)$ at most interchanges 0 and ∞, $\mathbf{C}_{\mathfrak{G}}(h) \leq \mathfrak{H}\langle s\rangle$. Since $|\mathfrak{H}|$ is odd, $h^s = h^{-1} \neq h$. Hence $\mathbf{C}_{\mathfrak{G}}(h) = \mathfrak{H}$. Thus we obtain with the help of 5.3e),

$$d = 1 + \alpha(h)^2 + \sum_{i=1}^{\frac{1}{2}(d-1)} |\beta_i(h)|^2 + \sum_{\chi \neq 1, \alpha, \beta_i} |\chi(h)|^2$$
$$= 1 + 1 + d - 2 + \sum_{\chi \neq 1, \alpha, \beta_i} |\chi(h)|^2.$$

This forces $\chi(h) = 0$ for $h \in \mathfrak{H} - \{1\}$. From

$$\chi(1) = \sum_{h \in \mathfrak{H}} \chi(h) = |\mathfrak{H}|\, (\chi_{\mathfrak{H}}, 1_{\mathfrak{H}})_{\mathfrak{H}} \equiv 0 \quad (|\mathfrak{H}|),$$

it follows that $|\mathfrak{H}| = d$ is a divisor of $\chi(1)$. **q.e.d.**

We apply these considerations to the groups $PSL(2, q)$ and $Sz(q)$.

5.5 Theorem. *For $f > 1$, $PSL(2, 2^f)$ has precisely the following irreducible characters.*

(1) *The unit character.*

(2) *The character* $\alpha = \pi - 1$; *we have* $\alpha(1) = 2^f$.

(3) $2^{f-1} - 1$ *characters* β_i *of degree* $2^f + 1$; *for* $f > 2$, *these are the exceptional characters belonging to* $\mathfrak{H}$.

(4) 2^{f-1} *characters* γ_i *of degree* $2^f - 1$; *these are the exceptional characters belonging to a regular cyclic subgroup (Singer cycle)* $\mathfrak{S}$ *of* $PSL(2, 2^f)$ *(see II, 8.4).*

Proof. Let $\mathfrak{G} = PSL(2, 2^f)$ $(f > 1)$. Then $\mathfrak{G}$ satisfies Hypothesis 5.1, and $d = 2^f - 1$ is odd. Hence Theorems 5.2 and 5.3 yield the characters α and β_i $(i = 1, \ldots, \frac{1}{2}(d - 1))$ with the given degrees.

Suppose that $f > 2$. It follows from Theorem 5.3a) that the β_i are not constant on $\mathfrak{H} - \{1\}$. By II, 8.3 $\mathbf{N}_{\mathfrak{G}}(\mathfrak{H})$ is a dihedral group of order $2d$. Also

$$\frac{|\mathfrak{H}| - 1}{|\mathbf{N}_{\mathfrak{G}}(\mathfrak{H}) : \mathfrak{H}|} = \tfrac{1}{2}(d - 1) = 2^{f-1} - 1 \geq 2,$$

so the conditions of Theorem 4.6 are satisfied. Hence there exist $\frac{1}{2}(d - 1)$ exceptional characters belonging to $\mathfrak{H}$. By 4.6b), these are the β_i.

By II, 8.4, $PSL(2, 2^f)$ has a cyclic subgroup $\mathfrak{S}$ of order $2^f + 1$, and $\mathbf{N}_{\mathfrak{G}}(\mathfrak{S})$ is a dihedral group of order $2|\mathfrak{S}|$. Since $\dfrac{|\mathfrak{S}| - 1}{2} = 2^{f-1} \geq 2$, we can apply 4.6 for $f > 1$. We thereby obtain 2^{f-1} irreducible exceptional characters γ_i belonging to $\mathfrak{S}$. By 4.4a), these all have the same degree l, and by 5.4, $2^f - 1$ divides l. If $l = k(2^f - 1)$,

$$\begin{aligned}
|PSL(2, 2^f)| &\geq 1 + \alpha(1)^2 + \sum_{i=1}^{\frac{1}{2}(d-1)} \beta_i(1)^2 + \sum_{i=1}^{2^{f-1}} \gamma_i(1)^2 \\
&= 1 + 2^{2f} + (2^{f-1} - 1)(2^f + 1)^2 + k^2(2^f - 1)^2 2^{f-1} \\
&= 1 + 2^{2f} + (2^{f-1} - 1)(2^f + 1)^2 + (2^f - 1)^2 2^{f-1} \\
&\quad + (k^2 - 1)(2^f - 1)^2 2^{f-1} \\
&= (2^f + 1)2^f(2^f - 1) + (k^2 - 1)(2^f - 1)^2 2^{f-1}.
\end{aligned}$$

It follows in the first place that $k = 1$, and in the second place that we have found all the irreducible characters of $PSL(2, 2^f)$. The values of α and of the β_i are found in 5.2 and 5.3. Our procedure yields no information about the values of the γ_i, except $\gamma_i(1) = 2^f - 1$ and $\gamma_i(h) = 0$ for $1 \neq h \in \mathfrak{H}$ (5.4). **q.e.d.**

5.6 Theorem. *Suppose that $p^f \equiv 1\ (4)$ and that $p^f > 5$. Then $PSL(2, p^f)$ has precisely the following irreducible characters.*

(1) *The unit character.*

(2) *The character $\alpha = \pi - 1$; we have $\alpha(1) = p^f$.*

(3) *$\frac{1}{4}(p^f - 5)$ characters β_i of degree $p^f + 1$.*

(4) *$\frac{1}{4}(p^f - 1)$ exceptional characters γ_i of degree $p^f - 1$ belonging to the Singer cycle $\mathfrak{S}$.*

(5) *Two characters δ_1 and δ_2 of degree $\frac{1}{2}(p^f + 1)$; these are the exceptional characters belonging to the Sylow p-subgroup $\mathfrak{F}$ of $PSL(2, p^f)$.*

Proof. Let $\mathfrak{G} = PSL(2, p^f)$ and $\mathfrak{H} = \mathfrak{G}_{0,\infty}$. Then by II, 8.3, $\mathbf{N}_{\mathfrak{G}}(\mathfrak{H}) = \mathfrak{H}\langle s\rangle$, where $s^2 = 1$, s interchanges 0 and ∞ and $h^s = h^{-1}$ for all $h \in \mathfrak{H}$. Hence $\mathfrak{G}$ satisfies Hypothesis 5.1. Theorem 5.2 yields the characters 1, α. The cyclic group $\mathfrak{H}$ of even order $\frac{1}{2}(p^f - 1)$ has exactly two real characters and has therefore $\frac{1}{2}(d - 2) = \frac{1}{4}(p^f - 5)$ pairs of complex conjugate characters. According to 5.3b), these yield $\frac{1}{4}(p^f - 5)$ irreducible characters β_i of degree $p^f + 1$.

Let $\mathfrak{S}$ be a Singer cycle of $PSL(2, p^f)$; thus $|\mathfrak{S}| = \frac{1}{2}(p^f + 1)$. By II, 8.4, $\mathbf{N}_{\mathfrak{G}}(\mathfrak{S})$ is a dihedral group of order $p^f + 1$. Since $\frac{1}{2}(p^f + 1)$ is odd, $\mathbf{N}_{\mathfrak{G}}(\mathfrak{S})$ is a Frobenius group with Frobenius kernel $\mathfrak{S}$. Further, $\mathfrak{S} \cap \mathfrak{S}^g = 1$ for $\mathfrak{S} \neq \mathfrak{S}^g$ (see II, 8.4). Since $\frac{1}{2}(|\mathfrak{S}| - 1) = \frac{1}{4}(p^f - 1) > 1$, 4.6 is applicable and yields $l = \frac{1}{4}(p^f - 1)$ exceptional characters γ_i of $PSL(2, p^f)$ belonging to $\mathfrak{S}$. Since $\alpha = \pi - 1$ takes the constant value -1 on $\mathfrak{S} - \{1\}$, it follows from 4.6 that

$$(1)\quad 0 = (\lambda_0 - \lambda_i)^{\mathfrak{G}}(1) = \varepsilon\gamma_i(1) + b\sum_{j=1}^{l}\gamma_j(1) + 1 - \alpha(1) + \sum_{\chi\neq 1,\alpha} c(\chi)\chi(1)$$

and

$$(2)\quad (\varepsilon + b)^2 + b^2\frac{p^f - 5}{4} + 2 + \sum_{\chi\neq 1,\alpha} |c(\chi)|^2 = |\mathbf{N}_{\mathfrak{G}}(\mathfrak{S}) : \mathfrak{S}| + 1 = 3.$$

Since $c(\chi)$ is a rational integer it follows from (2) that for $p^f > 9$, $b = c(\chi) = 0$, and (1) then yields

$$0 = \varepsilon\gamma_i(1) + 1 - \alpha(1).$$

This shows that $\varepsilon = 1$ and $\gamma_i(1) = p^f - 1$.

For $p^f = 9$, (2) reads

$$(\varepsilon + b)^2 + b^2 + \sum_{\chi\neq 1,\alpha} |c(\chi)|^2 = 1.$$

If $b = 0$, we then obtain $\gamma_i(1) = 8$, as before. If $b \neq 0$, it follows that $b = \pm 1$, $\varepsilon = -b$ and $c(\chi) = 0$. Then from (1),

$$0 = \varepsilon\gamma_i(1) - \varepsilon(\gamma_1(1) + \gamma_2(1)) + 1 - 9,$$

so again $\gamma_i(1) = 8$.

The assertions in (4) are therefore proved.

Since $\dfrac{n-1}{d} = 2$, we can also apply 4.6 to $\mathfrak{F}$ and obtain two exceptional characters δ_1, δ_2 belonging to $\mathfrak{F}$. Certainly, $\delta_1(1) = \delta_2(1)$. By 4.5, δ_1 and δ_2 are distinct from the γ_i. The δ_i are distinct from 1, α, β_i, since by 5.3, all these are constant on $\mathfrak{F} - \{1\}$, whereas by 4.6b) the δ_i are not.

We assert that all irreducible characters of $PSL(2, p^f)$ have now been found. To this end we show that $\frac{1}{2}(p^f + 5)$ is the class number of $PSL(2, p^f)$. In fact, on account of the partition

$$\begin{aligned} &PSL(2, p^f) \\ &= \{1\} \cup \Big(\bigcup_{g \in \mathfrak{G}} (\mathfrak{F} - \{1\})^g\Big) \cup \Big(\bigcup_{g \in \mathfrak{G}} (\mathfrak{H} - \{1\})^g\Big) \cup \Big(\bigcup_{g \in \mathfrak{G}} (\mathfrak{S} - \{1\})^g\Big) \end{aligned}$$

(II, 8.5), we obtain the following conjugacy classes.

$$\{1\},$$

$$\frac{n-1}{d} = 2 \text{ classes in } \bigcup_{g \in \mathfrak{G}} (\mathfrak{F} - \{1\})^g,$$

$$\frac{|\mathfrak{S}| - 1}{2} = \frac{n-1}{4} \text{ classes in } \bigcup_{g \in \mathfrak{G}} (\mathfrak{S} - \{1\})^g,$$

$$1 + \frac{|\mathfrak{H}| - 2}{2} = \frac{n-1}{4} \text{ classes in } \bigcup_{g \in \mathfrak{G}} (\mathfrak{H} - \{1\})^g.$$

(Note that $\mathfrak{H}$ now has an involution central in $\mathbf{N}_{\mathfrak{G}}(\mathfrak{H})$.)

The degrees of δ_1 and δ_2 are now easily determined from

$$|PSL(2, p^f)| = 1 + \alpha(1)^2 + \sum_i \beta_i(1)^2 + \sum_i \gamma_i(1)^2 + \delta_1(1)^2 + \delta_2(1)^2.$$

They are $\frac{1}{2}(p^f + 1)$. **q.e.d.**

Similarly one proves the following.

5.7 Theorem. *Suppose that $p^f \equiv 3$ (4). Then $PSL(2, p^f)$ has precisely the following irreducible characters.*

(1) *The unit character.*

(2) *The character $\alpha = \pi - 1$ of degree p^f.*

(3) $\frac{1}{4}(p^f - 3)$ *characters β_i of degree $p^f + 1$.*

(4) $\frac{1}{4}(p^f - 3)$ *characters γ_i of degree $p^f - 1$. (For $p^f \geq 11$, these are exceptional characters belonging to characters of the Singer cycle in the sense of 4.6. For $p^f = 7$, a different argument is needed.)*

(5) *Two characters δ_1, δ_2 of degree $\frac{1}{2}(p^f - 1)$. For $p^f > 3$, these are exceptional characters belonging to $\mathfrak{F}$.*

Turning now to the Suzuki groups, we prove the following.

5.8 Theorem. *Let $\mathfrak{G}$ be a group with the following properties.*

(i) *$\mathfrak{G}$ is a Zassenhaus group of degree $q^2 + 1$ and order $(q^2 + 1)q^2(q - 1)$, where $q = 2^{2m+1}$ and $m \geq 2$. Write $q = 2r^2$.*

(ii) *$\mathfrak{G}$ has Abelian Hall subgroups $\mathfrak{A}_1$, $\mathfrak{A}_2$ of orders $q + 2r + 1$, $q - 2r + 1$ respectively.*

(iii) *For $i = 1, 2$, $\mathfrak{A}_i \cap \mathfrak{A}_i^g = 1$ whenever $g \in \mathfrak{G}$ and $\mathfrak{A}_i^g \neq \mathfrak{A}_i$.*

(iv) *For $i = 1, 2$, $\mathbf{N}_{\mathfrak{G}}(\mathfrak{A}_i)$ is a Frobenius group of order $4|\mathfrak{A}_i|$ with Frobenius kernel $\mathfrak{A}_i$.*

Then $\mathfrak{G}$ has precisely the following irreducible characters.

(1) *The unit character.*

(2) *An irreducible character α of degree q^2.*

(3) $\frac{1}{2}(q - 2)$ *characters β_i of degree $q^2 + 1$.*

(4) $\frac{1}{4}(q + 2r)$ *exceptional characters $\gamma_j^{(1)}$ belonging to $\mathfrak{A}_1$; these vanish on $\mathfrak{A}_2 - \{1\}$.*

(5) $\frac{1}{4}(q - 2r)$ *exceptional characters $\gamma_j^{(2)}$ belonging to $\mathfrak{A}_2$; these vanish on $\mathfrak{A}_1 - \{1\}$.*

(6) *Two further characters δ_1, δ_2.*

In particular, the class number of $\mathfrak{G}$ is $q + 3$.

Proof. By 1.6b), $\mathfrak{G}$ is simple. Also $|\mathfrak{G}_{0,\infty}| = q - 1$ is odd, so by 1.5, $\mathfrak{H}$ is cyclic. By 1.7b), $\mathfrak{G}$ has an involution s which interchanges 0 and ∞. By 1.5, $h^s = h^{-1}$ for all $h \in \mathfrak{H}$. Hence $\mathfrak{G}$ satisfies Hypothesis 5.1.

Note that since $q \pm 2r + 1$ and $q^2(q - 1)$ are coprime, $\mathfrak{A}_i - \{1\} \subseteq \mathfrak{J}_0$, and since $q + 2r + 1$ and $q - 2r + 1$ are coprime, $\mathfrak{A}_1 \cap \mathfrak{A}_2^g = 1$ for all $g \in \mathfrak{G}$.

a) Since $|\mathfrak{H}| = q - 1$ is odd, the characters given in (1)–(3) are obtained from 5.2 and 5.3c).

Put $m_i = \dfrac{|\mathfrak{A}_i| - 1}{4}$. Since $q \geq 2^5$,

$$m_1 = \frac{q + 2r}{4} \geq 10, \quad m_2 = \frac{q - 2r}{4} \geq 6.$$

Thus we may apply 4.6 to $\mathfrak{A}_1$ and $\mathfrak{A}_2$. This yields exceptional characters $\gamma_j^{(1)} \left(j = 1, \dots, \dfrac{q + 2r}{4}\right)$ and $\gamma_k^{(2)} \left(k = 1, \dots, \dfrac{q - 2r}{4}\right)$. By 4.5, the $\gamma_j^{(1)}$ and the $\gamma_k^{(2)}$ are distinct, since $\mathfrak{A}_1 \cap \mathfrak{A}_2^g = 1$ for all $g \in \mathfrak{G}$. By 5.2 and 5.3a), α and the β_i are constant on $\mathfrak{A}_1 - \{1\}$ and on $\mathfrak{A}_2 - \{1\}$, so by 4.6b), α and the β_i are distinct from the $\gamma_j^{(1)}$ and the $\gamma_k^{(2)}$.

b) We prove the following for $i = 1, 2$. Let $\lambda_1^{(i)}, \dots, \lambda_{m_i}^{(i)}$ be the irreducible characters of $\mathbf{N}_{\mathfrak{G}}(\mathfrak{A}_i)$ which are induced from non-identity irreducible characters of $\mathfrak{A}_i$, and let $\lambda_0^{(i)}$ be the character of $\mathbf{N}_{\mathfrak{G}}(\mathfrak{A}_i)$ induced from the unit character of $\mathfrak{A}_i$. Then

$$(\lambda_0^{(i)} - \lambda_j^{(i)})^{\mathfrak{G}} = \varepsilon_i \gamma_j^{(i)} + 1 - \alpha + \varepsilon_{1i}\delta_1^{(i)} + \varepsilon_{2i}\delta_2^{(i)},$$

where $\varepsilon_i = \pm 1$, $\varepsilon_{ji} = \pm 1$; also $\delta_1^{(i)}$ and $\delta_2^{(i)}$ are irreducible and distinct from $1, \alpha, \beta_l, \gamma_j^{(i)}$. If χ is any irreducible character of $\mathfrak{G}$ distinct from 1, $\alpha, \gamma_j^{(i)}, \delta_1^{(i)}$ and $\delta_2^{(i)}$, then χ is zero on $\mathfrak{A}_i - \{1\}$.

Since α is -1 on $\mathfrak{A}_i - \{1\}$, it follows from 4.6a) that

$$(\lambda_0^{(i)} - \lambda_j^{(i)})^{\mathfrak{G}} = \varepsilon_i \gamma_j^{(i)} + b_i \sum_{k=1}^{m_i} \gamma_k^{(i)} + 1 - \alpha + \sum_{\chi} c_i(\chi)\chi,$$

where χ runs through the irreducible characters of $\mathfrak{G}$ other than $1, \alpha, \gamma_j^{(i)}$, and $c_i(\chi)$ is the constant value of χ on $\mathfrak{A}_i - \{1\}$. By 4.6c),

$$(\varepsilon_i + b_i)^2 + (m_i - 1)b_i^2 + 2 + \sum_{\chi} |c_i(\chi)|^2 = |\mathbf{N}_{\mathfrak{G}}(\mathfrak{A}_i)/\mathfrak{A}_i| + 1 = 5.$$

Since $m_1 \geq 10$ and $m_2 \geq 6$, it follows that $b_i = 0$ and $\sum_{\chi} |c_i(\chi)|^2 = 2$. Since the $c_i(\chi)$ are rational integers, there are exactly two characters $\delta_1^{(i)}$ and $\delta_2^{(i)}$, distinct from $1, \alpha, \gamma_k^{(i)}$, for which $c_i(\delta_j^{(i)}) \neq 0$. By 5.3a), β_j is zero on $\mathfrak{A}_i - \{1\}$, so $\delta_1^{(i)}, \delta_2^{(i)}$ are distinct from β_j.

c) $\delta_1^{(1)}$ and $\delta_2^{(1)}$ are not exceptional characters belonging to $\mathfrak{A}_2$ and are therefore distinct from $\gamma_k^{(2)}$; similarly $\delta_1^{(2)}$ and $\delta_2^{(2)}$ are not exceptional characters belonging to $\mathfrak{A}_1$. Hence by b), $\gamma_k^{(2)}$ is zero on $\mathfrak{A}_1 - \{1\}$ and $\gamma_j^{(1)}$ is zero on $\mathfrak{A}_2 - \{1\}$.

Suppose, say, that $\delta_1^{(1)}$ is an exceptional character belonging to $\mathfrak{A}_2$. By 4.4a), $\delta_1^{(1)} - \gamma_j^{(2)}$ is zero on $\mathfrak{G} - \bigcup_g (\mathfrak{A}_2 - \{1\})^g$. Since $\mathfrak{A}_1 \cap \mathfrak{A}_2^g = 1$, it follows that $\delta_1^{(1)} - \gamma_j^{(2)}$ vanishes on $\mathfrak{A}_1$. This shows that $\gamma_j^{(2)}$ and $\delta_1^{(1)}$ are equal to $c_1(\delta_1^{(1)}) \neq 0$ on $\mathfrak{A}_1 - \{1\}$. Since $\gamma_j^{(2)} \neq \gamma_k^{(1)}$, it follows using

b) that $\gamma_j^{(2)} \in \{\alpha, \delta_1^{(1)}, \delta_2^{(1)}\}$ for all $j = 1, \ldots, m_2$. Since $m_2 \geq 6$, this is impossible.

d) With appropriate numbering, $\delta_j^{(1)} = \delta_j^{(2)}$ for $j = 1, 2$.

We put $\alpha_j^{(i)} = (\lambda_0^{(i)} - \lambda_j^{(i)})^{\mathfrak{G}}$. By 4.6, $\alpha_j^{(i)}$ vanishes outside $\bigcup_g (\mathfrak{A}_i - \{1\})^g$. Since $\mathfrak{A}_1^g \cap \mathfrak{A}_2 = 1$ for all $g \in \mathfrak{G}$, it follows that $\alpha_j^{(1)}\bar{\alpha}_k^{(2)} = 0$ (on all $\mathfrak{G}$). Therefore, by b),

$$0 = (\alpha_j^{(1)}\bar{\alpha}_k^{(2)}, 1_{\mathfrak{G}})_{\mathfrak{G}} = (\alpha_j^{(1)}, \alpha_k^{(2)})_{\mathfrak{G}}$$

$$= (\varepsilon_1\gamma_j^{(1)} + 1 - \alpha + \varepsilon_{11}\delta_1^{(1)} + \varepsilon_{21}\delta_2^{(1)}, \varepsilon_2\gamma_k^{(2)} + 1 - \alpha + \varepsilon_{12}\delta_1^{(2)} + \varepsilon_{22}\delta_2^{(2)})_{\mathfrak{G}}.$$

Since $\delta_j^{(1)} \neq \gamma_k^{(2)}$, $\delta_j^{(2)} \neq \gamma_k^{(1)}$ (by c)) and $\delta_j^{(i)} \neq \alpha$, this gives

$$0 = 2 + (\varepsilon_{11}\delta_1^{(1)} + \varepsilon_{21}\delta_2^{(1)}, \varepsilon_{12}\delta_1^{(2)} + \varepsilon_{22}\delta_2^{(2)})_{\mathfrak{G}}.$$

Since $\varepsilon_{ji} = \pm 1$, the assertion then follows easily.

Henceforth we put $\delta_j = \delta_j^{(1)}$.

e) The characters 1, α, β_l, $\gamma_j^{(i)}$ ($i = 1, 2$; $j = 1, \ldots, m_i$), δ_1 and δ_2 are all the irreducible characters of $\mathfrak{G}$.

Suppose that χ is an irreducible character of $\mathfrak{G}$ distinct from all the given ones. By 5.4, $q - 1$ is a divisor of $\chi(1)$. By b), χ is zero on $\mathfrak{A}_i - \{1\}$, so

$$\chi(1) = \sum_{a \in \mathfrak{A}_i} \chi(a) = |\mathfrak{A}_i| (\chi_{\mathfrak{A}_i}, 1_{\mathfrak{A}_i})_{\mathfrak{A}_i} \equiv 0 \ (|\mathfrak{A}_i|).$$

Since $q - 1$, $|\mathfrak{A}_1|$ and $|\mathfrak{A}_2|$ are coprime in pairs, this forces $(q - 1)|\mathfrak{A}_1||\mathfrak{A}_2| = (q - 1)(q^2 + 1)$ to be a divisor of $\chi(1)$. Thus,

$$(q^2 + 1)q^2(q - 1) = |\mathfrak{G}| \geq \chi(1)^2 \geq (q^2 + 1)^2(q - 1)^2,$$

a contradiction. **q.e.d.**

To find the degrees of the characters of the Suzuki groups, we prove the following lemma.

5.9 Lemma. *Let $\mathfrak{F}$ be a Sylow 2-subgroup of $Sz(q)$, where $q = 2^{2m+1}$. Then there are exactly four $\mathfrak{H}$-orbits of irreducible characters of $\mathfrak{F}$, their lengths being* 1, $q - 1$, $q - 1$, $q - 1$. *The degrees of the characters in these orbits are* 1, 1, 2^m, 2^m *respectively.*

Proof. Since $\mathfrak{F}\mathfrak{H}$ is a Frobenius group with Frobenius kernel $\mathfrak{F}$, $\mathfrak{H}$ permutes the classes of $\mathfrak{F}$ fixed point freely, by V, 8.9c). The involutions in $\mathfrak{F}$ are precisely the elements of $\mathfrak{F}' - \{1\}$, and these form exactly one class under $\mathfrak{H}$. If (in the notation of § 3), $z = S(a, b) \in \mathfrak{F} - \mathfrak{F}'$, then $a \neq 0$. A trivial calculation gives

$$\mathbf{C}_{\mathfrak{F}}(z) = \left\{ S(x, y) \,\middle|\, \left(\frac{x}{a}\right)\pi = \frac{x}{a},\ y \text{ arbitrary} \right\},$$

where π is the automorphism of $GF(q)$ for which $x\pi^2 = x^2$. Since π generates the automorphism group of $GF(q)$, it follows that $|\mathbf{C}_{\mathfrak{F}}(z)| = 2q$. Thus $\mathfrak{F} - \mathfrak{F}'$ is the union of conjugacy classes each having $\frac{1}{2}q$ elements. The number of these classes is thus $\dfrac{q^2 - q}{\frac{1}{2}q} = 2(q - 1)$. Hence there are exactly two conjugacy classes of elements of order 4 in $\mathfrak{F}\mathfrak{H}$, by 3.1c).

By V, 13.5a), $\mathfrak{H}$ also permutes the irreducible characters of $\mathfrak{F}$ fixed point freely, and by V, 13.5b), there are exactly 4 $\mathfrak{H}$-orbits of irreducible characters of $\mathfrak{F}$, their lengths being $1, q - 1, q - 1, q - 1$. Their degrees have the form $1, 1, 2^{f_1}, 2^{f_2}$, where $0 \leq f_1 \leq f_2$. It follows that

$$q^2 = |\mathfrak{F}| = 1 + (q - 1) + (q - 1)(2^{2f_1} + 2^{2f_2}).$$

This shows that $q = 2^{2f_1} + 2^{2f_2}$. Since $q = 2^{2m+1}$, this forces $f_1 = f_2 = m$. **q.e.d.**

5.10 Theorem. *Let $\mathfrak{G} = Sz(q)$, where $q = 2^{2m+1} > 8$. Then the class number of $\mathfrak{G}$ is $q + 3$ and $\mathfrak{G}$ has precisely the following irreducible characters.*

(1) *The unit character.*

(2) *An irreducible character α of degree q^2.*

(3) *$\frac{1}{2}(q - 2)$ characters β_i of degree $q^2 + 1$.*

(4) *$\frac{1}{4}(q + 2r)$ exceptional characters $\gamma_j^{(1)}$ belonging to the cyclic group $\mathfrak{U}_1$ in 3.10 of order $q + 2r + 1$; here $r = 2^m$. The degree of $\gamma_j^{(1)}$ is $(q - 1)(q - 2r + 1)$.*

(5) *$\frac{1}{4}(q - 2r)$ exceptional characters $\gamma_j^{(2)}$ belonging to the cyclic group $\mathfrak{U}_2$ in 3.10 of order $q - 2r + 1$. The degree of $\gamma_j^{(2)}$ is $(q - 1)(q + 2r + 1)$.*

(6) *Two exceptional characters δ_1 and δ_2 belonging to the Sylow 2-subgroup $\mathfrak{F}$ of degree $(q - 1)2^m$.*

Proof. By 3.10, $\mathfrak{G}$ satisfies the conditions of 5.8 with $\mathfrak{U}_1, \mathfrak{U}_2$ in place of $\mathfrak{A}_1, \mathfrak{A}_2$ respectively. Hence the class number of $\mathfrak{G}$ is $q + 3$ and 5.8 yields the characters $1, \alpha, \beta_i, \gamma_j^{(1)}, \gamma_k^{(2)}, \delta_1, \delta_2$ of $\mathfrak{G}$. Thus it only remains to prove that

$$\gamma_j^{(1)}(1) = (q - 1)(q - 2r + 1), \quad \gamma_k^{(2)}(1) = (q - 1)(q + 2r + 1),$$

$$\delta_1(1) = \delta_2(1) = (q - 1)2^m$$

and that δ_1, δ_2 are exceptional characters belonging to $\mathfrak{F}$.

a) We have

$$\gamma_j^{(1)}(1) \equiv 0 \quad ((q - 1)\,|\mathfrak{U}_2|)$$

and

$$\gamma_j^{(2)}(1) \equiv 0 \quad ((q - 1)|\mathfrak{U}_1|).$$

By 5.4, $q - 1$ is a divisor of $\gamma_j^{(i)}(1)$. And by 5.8, $\gamma_j^{(2)}$ vanishes on $\mathfrak{U}_1 - \{1\}$, so

$$\gamma_j^{(2)}(1) = \sum_{u \in \mathfrak{U}_1} \gamma_j^{(2)}(u) = |\mathfrak{U}_1|((\gamma_j^{(2)})_{\mathfrak{U}_1}, 1_{\mathfrak{U}_1})_{\mathfrak{U}_1} \equiv 0 \quad (|\mathfrak{U}_1|).$$

Since $(q - 1, |\mathfrak{U}_1|) = 1$, it follows that $\gamma_j^{(2)}(1) \equiv 0 \;((q - 1)|\mathfrak{U}_1|)$. A similar argument shows that $\gamma_j^{(1)}(1) \equiv 0 \;((q - 1)|\mathfrak{U}_2|)$.

Now by 5.9, there are three $\mathfrak{H}$-orbits of non-identity irreducible characters of $\mathfrak{F}$, each of which is of length $q - 1$. Let $\lambda_0, \lambda_1, \lambda_2$ be characters of $\mathfrak{F}\mathfrak{H}$ induced from characters of $\mathfrak{F}$, one from each of these orbits, with $\lambda_0(1) = q - 1$, $\lambda_1(1) = \lambda_2(1) = 2^m(q - 1)$. Then λ_1, λ_2 vanish on $\mathfrak{F}\mathfrak{H} - \mathfrak{F}$, so by 4.1 (with $\mathsf{D} = \mathfrak{F} - \{1\}$), λ_1 and λ_2 give rise to exceptional characters Λ_1 and Λ_2, and we may assume that

$$(\lambda_1 - \lambda_2)^{\mathfrak{G}} = \Lambda_1 - \Lambda_2.$$

b) $\{\Lambda_1, \Lambda_2\} = \{\delta_1, \delta_2\}$. Thus δ_1, δ_2 are exceptional characters belonging to $\mathfrak{F}$.

It is sufficient to show that the Λ_i are distinct from 1, α, β_i, $\gamma_j^{(i)}$. Certainly $\Lambda_i \neq 1$, since $\Lambda_1(1) = \Lambda_2(1)$ and $\mathfrak{G}$ is simple. As $\Lambda_1 - \Lambda_2 = (\lambda_1 - \lambda_2)^{\mathfrak{G}}$ vanishes outside $\bigcup_g (\mathfrak{F} - \{1\})^g$ and α vanishes on $\bigcup_g (\mathfrak{F} - \{1\})^g$, we have $(\Lambda_1 - \Lambda_2, \alpha)_{\mathfrak{G}} = 0$. Thus $\Lambda_i \neq \alpha$ since $\Lambda_1 \neq \Lambda_2$.

Suppose that $\Lambda_1 = \psi$, where ψ is one of the characters β_i, $\gamma_j^{(i)}$. Let ψ' be another of the β_i or $\gamma_j^{(i)}$ as the case may be. Then

$$\psi - \psi' = (\tau - \tau')^{\mathfrak{G}}$$

for certain characters τ, τ' of $\mathfrak{V}$, where $\mathfrak{V} = \mathfrak{H}$ if ψ is a β_i and $\mathfrak{V} = \mathfrak{U}_i$ if ψ is a $\gamma_j^{(i)}$. This is clear from 4.6 in the latter case. In the former, $\psi = \phi^{\mathfrak{G}}$ and $\psi' = \phi'^{\mathfrak{G}}$, where ϕ, ϕ' are characters of $\mathfrak{F}\mathfrak{H}$ having $\mathfrak{F}$ in their kernels. By direct calculation,

$$(\phi_{\mathfrak{H}} - \phi'_{\mathfrak{H}})^{\mathfrak{F}\mathfrak{H}} = \phi - \phi',$$

so the assertion is clear. It follows from it that $\psi - \psi'$ is zero outside $\bigcup_g (\mathfrak{V} - \{1\})^g$. But $\Lambda_1 - \Lambda_2 = (\lambda_1 - \lambda_2)^{\mathfrak{G}}$ vanishes outside $\bigcup_g (\mathfrak{F} - \{1\})^g$, so $(\Lambda_1 - \Lambda_2)(\overline{\psi} - \overline{\psi'}) = 0$ (on all $Sz(q)$). But this gives the contradiction

$$0 = (\Lambda_1 - \Lambda_2, \psi - \psi')_{\mathfrak{G}} = (\Lambda_1, \psi)_{\mathfrak{G}} + (\Lambda_2, \psi')_{\mathfrak{G}} \geq 1.$$

c) $\delta_i(1) \geq (q - 1)2^m = (q - 1)r$.
By b), we may suppose that

$$(\lambda_1 - \lambda_2)^{\mathfrak{G}} = \delta_1 - \delta_2.$$

It follows that

$$\pm 1 = (\delta_i, (\lambda_1 - \lambda_2)^{\mathfrak{G}})_{\mathfrak{G}} = ((\delta_i)_{\mathfrak{G}_\infty}, \lambda_1 - \lambda_2)_{\mathfrak{G}_\infty}.$$

Thus either λ_1 or λ_2 occurs in $(\delta_i)_{\mathfrak{G}_\infty}$, and it follows that

$$\delta_i(1) \geq \lambda_i(1) = (q - 1)2^m.$$

d) $\delta_i(1) = (q - 1)2^m$, $\gamma_j^{(1)}(1) = (q - 1)(q - 2r + 1)$ and $\gamma_j^{(2)}(1) = (q - 1)(q + 2r + 1)$.
On account of a) and c),

$$\begin{aligned}
&(q^2 + 1)q^2(q - 1) = |Sz(q)| = \textstyle\sum \chi(1)^2 \\
&= 1 + \alpha(1)^2 + \textstyle\sum \beta_i(1)^2 + \sum \gamma_j^{(1)}(1)^2 + \sum \gamma_j^{(2)}(1)^2 + \delta_1(1)^2 + \delta_2(1)^2 \\
&\geq 1 + q^4 + \frac{q-2}{2}(q^2 + 1)^2 + \frac{q + 2r}{4}(q - 1)^2(q - 2r + 1)^2 \\
&\quad + \frac{q - 2r}{4}(q - 1)^2(q + 2r + 1)^2 + 2r^2(q - 1)^2 \\
&= (q^2 + 1)q^2(q - 1).
\end{aligned}$$

Hence equality holds everywhere and the assertion follows. **q.e.d.**

5.11 Remark. For the smallest Suzuki group $Sz(8)$, $|\mathfrak{U}_2| = 5$, so $\frac{|\mathfrak{U}_2| - 1}{4} = 1$. Thus the method of exceptional characters is inapplicable for $\mathfrak{U}_2$. Since

$$|Sz(8)| = 13 \cdot 7 \cdot 5 \cdot 2^6,$$

the character table of $Sz(8)$ can be quickly found by means of deep theorems of BRAUER [3], as $Sz(8)$ has self-centralizing subgroups of orders 13, 7 and 5. However, we shall not go into this.

§ 6. Feit's Theorem

6.1 Theorem (FEIT [1]). *Let $\mathfrak{G}$ be a Zassenhaus group of degree $n + 1$ and order $(n + 1)nd$.*

a) *n is a prime-power p^f.*

b) *If the Frobenius kernel $\mathfrak{F}$ of $\mathfrak{G}_\infty$ is non-Abelian, then $|\mathfrak{F}/\mathfrak{F}'| < 4d^2$.*

c) *If $\mathfrak{F}$ is Abelian, $d \geq \frac{1}{2}(n - 1)$.*

The following proof of 6.1 is that of Feit with simplifications by Bender. We need the following lemmas.

6.2 Lemma. *Let $\mathfrak{P}$ be a p-group of order p^n and let ρ be an irreducible character of $\mathfrak{P}$ for which $\rho(1) > 1$. Then*

$$\sum_\sigma \sigma(1)^2 \equiv 0 \quad (\rho(1)^2),$$

where the summation is over all irreducible characters σ of $\mathfrak{P}$ for which $\sigma(1) < \rho(1)$.

Proof. Let χ_i $(i = 1, \dots, h)$ be all the irreducible characters of $\mathfrak{P}$ and put $\chi_i(1) = p^{s_i}$, where $0 = s_1 \leq s_2 \leq \dots \leq s_h$. Since $\sum_{i=1}^h p^{2s_i} = |\mathfrak{P}| = p^n$, $2s \leq n$, where $\rho(1) = p^s$. Also

$$p^n = \sum_{s_i < s} p^{2s_i} + \sum_{s_i \geq s} p^{2s_i}.$$

This shows that

$$\sum_{s_i < s} p^{2s_i} \equiv 0 \quad (p^{2s}).$$ **q.e.d.**

6.3 Lemma. *Suppose that $\mathfrak{F}\mathfrak{H}$ is a Frobenius group with Frobenius kernel $\mathfrak{F}$, where $\mathfrak{F} \cap \mathfrak{H} = 1$ and $|\mathfrak{H}| = d$. Suppose that d is odd and that $\mathfrak{F}$ has at least one of the following properties.*

(1) *$\mathfrak{F}$ is not of prime-power order.*

(2) $\mathfrak{F}$ *is non-Abelian and* $|\mathfrak{F}/\mathfrak{F}'| \geq 4d^2$.
If χ *is an irreducible character of* $\mathfrak{F}$ *of degree greater than 1, then*

$$\sum_{\psi} \psi(1)^2 > 2d\chi(1),$$

where the summation is over all irreducible characters ψ *of* $\mathfrak{F}$ *for which* $\psi \neq 1_{\mathfrak{F}}$ *and* $\psi(1) < \chi(1)$.

Proof. Being a Frobenius kernel, $\mathfrak{F}$ is nilpotent (V, 8.7). If $\mathfrak{F}$ is Abelian, $\mathfrak{F}$ has no irreducible character of degree greater than 1. Suppose then that $\mathfrak{F}$ is non-Abelian. We put $\chi(1) = z$ and

$$A(\chi) = \sum_{\psi} \psi(1)^2,$$

where the summation is over all irreducible non-unit characters of $\mathfrak{F}$ of degree less than z.

a) Suppose first that $A(\chi) \geq 4d^2$ and that p is a divisor of z. Let $\mathfrak{F}_p$ be the Sylow p-subgroup of the nilpotent group $\mathfrak{F}$. Since z divides $|\mathfrak{F}|$, $\mathfrak{F}_p \neq 1$. Also $\mathfrak{F} = \mathfrak{F}_p \times \mathfrak{F}_{p'}$, where $\mathfrak{F}_{p'}$ is the p-complement of $\mathfrak{F}$. By V, 10.3, $\chi = \chi_p\chi_{p'}$, where χ_p, $\chi_{p'}$ are respectively irreducible characters of $\mathfrak{F}_p$ and $\mathfrak{F}_{p'}$. Clearly, $\chi_p(1)$ is the highest power of p which divides $\chi(1) = z$. Thus $\chi_p(1) > 1$. By 6.2,

$$\sum_{\rho} \rho(1)^2 \geq \chi_p(1)^2,$$

where the sum runs over all irreducible characters ρ of $\mathfrak{F}_p$ of degree less than $\chi_p(1)$. By V, 10.3, the $\rho\chi_{p'}$ are distinct irreducible characters of $\mathfrak{F}$ and

$$\rho(1)\chi_{p'}(1) < \chi_p(1)\chi_{p'}(1) = \chi(1).$$

This gives

$$(1) \qquad A(\chi) + 1 \geq \sum_{\rho} \rho(1)^2\chi_{p'}(1)^2 \geq \chi_p(1)^2\chi_{p'}(1)^2 = z^2.$$

On account of our assumption that $A(\chi) \geq 4d^2$, this gives

$$(A(\chi) + 1)^2 > A(\chi)(A(\chi) + 1) \geq 4d^2z^2,$$

so $A(\chi) + 1 > 2dz$ and $A(\chi) \geq 2dz$.

Suppose that $A(\chi) = 2dz$. Then $2dz \geq 4d^2$ and $z \geq 2d$. Since z

divides $|\mathfrak{F}|$ and $(|\mathfrak{F}|, d) = 1$, $z \geq 2d + 1$. Thus by (1),

$$2dz + 1 = A(\chi) + 1 \geq z^2 \geq z(2d + 1),$$

which is impossible since $z > 1$. Hence $A(\chi) > 2dz$.

b) Suppose now that $A(\chi) < 4d^2$. Since $\mathfrak{F}$ has exactly $|\mathfrak{F}/\mathfrak{F}'|$ characters of degree 1,

$$|\mathfrak{F}/\mathfrak{F}'| \leq A(\chi) + 1 < 4d^2 + 1.$$

Thus $|\mathfrak{F}/\mathfrak{F}'| \leq 4d^2$. Since $(|\mathfrak{F}|, d) = 1$, $|\mathfrak{F}/\mathfrak{F}'| < 4d^2$. Thus by hypothesis, $\mathfrak{F}$ is not of prime-power order. Let $\mathfrak{F}_1, \dots, \mathfrak{F}_r$ be the non-identity Sylow subgroups of $\mathfrak{F}$, so $r > 1$ and $\mathfrak{F} = \mathfrak{F}_1 \times \cdots \times \mathfrak{F}_r$. Also

$$|\mathfrak{F}/\mathfrak{F}'| = \prod_{i=1}^{r} |\mathfrak{F}_i/\mathfrak{F}_i'|.$$

By V, 8.10, $\mathfrak{H}$ transforms $\mathfrak{F}_i/\mathfrak{F}_i'$ fixed point freely, so $|\mathfrak{F}_i/\mathfrak{F}_i'| \equiv 1 (d)$. If $|\mathfrak{F}_i|$ is odd, it follows that $|\mathfrak{F}_i/\mathfrak{F}_i'| \equiv 1 \ (2d)$, since d is odd. Thus if $|\mathfrak{F}|$ has at least two odd prime divisors, we obtain

$$|\mathfrak{F}/\mathfrak{F}'| \geq (1 + 2d)^2 > 4d^2 + 1,$$

a contradiction. Thus we may suppose that $\mathfrak{F}_1$ is a q-group for some odd prime q and that $\mathfrak{F}_2$ is a 2-group. Then

$$|\mathfrak{F}/\mathfrak{F}'| \geq (1 + 2d)(1 + d) > 2d^2.$$

It follows that $A(\chi) \geq |\mathfrak{F}/\mathfrak{F}'| - 1 \geq 2d^2$.

Suppose as in a) that $\chi = \chi_1\chi_2$. We distinguish several cases.

First suppose that $\chi_2(1) = 1$. By 6.2, it follows since $z = \chi_1(1) \geq q \geq 3$ that

$$\sum_{\tau} \tau(1)^2 \geq \chi_1(1)^2 - 1 > 2\chi_1(1) = 2z,$$

where the sum runs over all non-unit irreducible characters τ of $\mathfrak{F}_1$ for which $\tau(1) < \chi_1(1)$. Since $|\mathfrak{F}_2/\mathfrak{F}_2'| \geq 1 + d$, $\mathfrak{F}_2$ has at least $d + 1$ characters λ of degree 1. This gives

$$A(\chi) \geq \sum_{\tau, \lambda} \tau(1)^2 \lambda(1)^2 \geq 2(d + 1)z > 2dz,$$

which is the assertion.

Next suppose that $\chi_1(1) = 1$. As above we obtain from 6.2

$$\sum_\rho \rho(1)^2 \geq \chi_2(1)^2 - 1 = z^2 - 1 \geq z,$$

where the sum runs over all non-unit irreducible characters ρ of $\mathfrak{F}_2$ of degree less than $\chi_2(1)$. Since $|\mathfrak{F}_1/\mathfrak{F}_1'| \geq 1 + 2d$, it then follows that

$$\sum_{\rho,\lambda} \rho(1)^2\lambda(1)^2 \geq z(1 + 2d) > 2dz,$$

where λ runs through the characters of $\mathfrak{F}_1$ of degree 1. Again we have the assertion.

Suppose finally that $\chi_1(1) > 1$ and $\chi_2(1) > 1$. From 6.2,

$$\sum_\tau \tau(1)^2 \geq \chi_1(1)^2 \quad \text{and} \quad \sum_\rho \rho(1)^2 \geq \chi_2(1)^2,$$

where τ runs over all irreducible characters of $\mathfrak{F}_1$ of degree less than $\chi_1(1)$ and ρ runs over all irreducible characters of $\mathfrak{F}_2$ of degree less than $\chi_2(1)$. It follows that

$$\begin{aligned} A(\chi) + 1 &\geq \sum_\rho \rho(1)^2\chi_1(1)^2 + \sum_\tau \tau(1)^2\chi_2(1)^2 + \sum_{\rho,\tau} \rho(1)^2\tau(1)^2 \\ &\geq 3\chi_1(1)^2\chi_2(1)^2 = 3z^2. \end{aligned}$$

Since $A(\chi) \geq 2d^2$, we now obtain

$$(A(\chi) + 1)^2 > A(\chi)(A(\chi) + 1) \geq 6d^2z^2,$$

and hence since $dz \geq 6$,

$$A(\chi) + 1 > \sqrt{6}(dz) > 2dz + 1. \qquad \textbf{q.e.d.}$$

We now begin the proof of Theorem 6.1. Let $\mathfrak{G}$ be a counterexample of minimal order. Thus $\mathfrak{G}$ is a Zassenhaus group of degree $n + 1$ and order $(n + 1)nd$, and $\mathfrak{G}$ satisfies one of the following conditions.

(1) $\mathfrak{F}$ is not of prime-power order.

(2) $\mathfrak{F}$ is non-Abelian and $|\mathfrak{F}/\mathfrak{F}'| \geq 4d^2$.

(3) $\mathfrak{F}$ is Abelian and $d < \frac{1}{2}(n - 1)$.

6.4 Lemma. a) $\mathfrak{G}$ *is simple.*

b) $1 < d < \frac{1}{2}(n - 1)$.

c) $\mathfrak{H}$ *is cyclic of odd order* d.

d) $\mathbf{N}_{\mathfrak{G}}(\mathfrak{H}) = \langle \mathfrak{H}, s \rangle$ *is dihedral of order* $2d$.

Proof. a) Suppose that $1 \neq \mathfrak{N} \lhd \mathfrak{G}$. By 1.6a), $\mathfrak{N}$ is a Zassenhaus group of degree $n + 1$ and $\mathfrak{N} > \mathfrak{F}$. Let $|\mathfrak{N}| = (n + 1)nd'$. Since $\mathfrak{N} < \mathfrak{G}$, $\mathfrak{N}$ is not a counterexample to 6.1. Therefore n is a prime-power, either $\mathfrak{F}$ is Abelian or $|\mathfrak{F}/\mathfrak{F}'| < 4d'^2$, and if $\mathfrak{F}$ is Abelian, $d' \geq \frac{1}{2}(n - 1)$. But $d > d'$, so $\mathfrak{G}$ satisfies none of the conditions (1), (2) or (3). This is a contradiction, so $\mathfrak{G}$ is simple.

b) If $d \geq \frac{1}{2}(n - 1)$, then by 1.4b), none of (1), (2) or (3) holds.

c) If $d = |\mathfrak{H}|$ is even, then by b) and 1.11, $\mathfrak{G} = PSL(2, p^f)$. But then $d \geq \frac{1}{2}(n - 1)$, contrary to b). Thus d is odd. Since $\mathfrak{G}$ is simple, it follows from 1.5 that $\mathfrak{H}$ is cyclic.

d) By 1.7, there exists an involution s which interchanges 0 and ∞. By 1.5, $h^s = h^{-1}$ for all $h \in \mathfrak{H}$, so $\langle \mathfrak{H}, s \rangle$ is dihedral. Clearly $\mathbf{N}_{\mathfrak{G}}(\mathfrak{H}) = \langle \mathfrak{H}, s \rangle$. **q.e.d.**

6.5 Lemma. $\mathfrak{G}$ *satisfies Hypothesis 4.2.*

Proof. Let ϕ_{ij} $(i = 1, \ldots, k; j = 1, \ldots, m_i)$ be representatives of the $\mathfrak{H}$-orbits of the non-unit irreducible characters of $\mathfrak{F}$, the notation being so chosen that $\phi_{ij}(1) = z_i$ depends only on i and

$$1 = z_1 < z_2 < \cdots < z_k.$$

Let $\lambda_{ij} = \phi_{ij}^{\mathfrak{G}_\infty}$ $(i = 1, \ldots, k; j = 1, \ldots, m_i)$.

We show that if $1 < l \leq k$,

$$\sum_{i=1}^{l-1} m_i z_i^2 > 2z_l.$$

Since $k > 1$, $\mathfrak{F}$ is non-Abelian. Since $\mathfrak{H}$ operates fixed point freely on the non-unit characters of $\mathfrak{F}$ (see V, 16.13), the distinct non-unit characters of $\mathfrak{F}$ of degree less than z_l are the ϕ_{ij}^h with $i < l$ and $h \in \mathfrak{H}$. Hence by 6.3,

$$\sum_{i=1}^{l-1} dm_i z_i^2 > 2dz_l \quad (1 < l \leq k).$$

It only remains to prove that $m_1 \geq 2$. We have

$$m_1 d = |\mathfrak{F}/\mathfrak{F}'| - 1.$$

If $\mathfrak{F}$ is not of prime-power order, then since $|\mathfrak{F}_p/\mathfrak{F}_p'| \equiv 1$ (d) for any Sylow subgroup $\mathfrak{F}_p$ of $\mathfrak{F}$,

$$m_1 d \geq (d + 1)^2 - 1 \geq 2d$$

and $m_1 \geq 2$. If $|\mathfrak{F}/\mathfrak{F}'| \geq 4d^2$, $m_1 d \geq 4d^2 - 1$ and $m_1 \geq 4d - d^{-1} > 4d - 1 \geq 2d$. Finally, if $\mathfrak{F}$ is Abelian,

$$m_1 d = |\mathfrak{F}| - 1 = n - 1 > 2d,$$

by 6.4b). Thus $m_1 \geq 2$ in all cases. **q.e.d.**

In the remaining steps, we use a suggestion of Bender.

6.6 Lemma. $\mathfrak{G}$ *has at least two conjugacy classes of elements having no fixed points.*

Proof. As in 5.1, let $\mathfrak{J}_i$ be the set of elements of $\mathfrak{G}$ which have precisely i fixed points ($i = 0, 1, 2$). Thus

$$\mathfrak{G} = \{1\} \cup \mathfrak{J}_0 \cup \mathfrak{J}_1 \cup \mathfrak{J}_2.$$

Since the number of elements having the one fixed point ∞ is $|\mathfrak{F}| - 1 = n - 1$, $|\mathfrak{J}_1| = (n + 1)(n - 1) = n^2 - 1$. Similarly,

$$|\mathfrak{J}_2| = \frac{(n + 1)n}{2}(|\mathfrak{H}| - 1) = \frac{(n + 1)n(d - 1)}{2}.$$

Thus

$$\begin{aligned} |\mathfrak{J}_0| &= (n + 1)nd - 1 - (n^2 - 1) - \tfrac{1}{2}(n + 1)n(d - 1) \\ &= \tfrac{1}{2}(n + 1)n(d + 1) - n^2. \end{aligned}$$

Suppose that $g \in \mathfrak{J}_0$. We show that $\mathbf{C}_{\mathfrak{G}}(g)$ permutes each of its orbits regularly. Otherwise, there exists $x \in \mathbf{C}_{\mathfrak{G}}(g)$ such that $x \neq 1$ and $x \in \mathfrak{G}_a$ for some a. Then a, ag, ag^2 are fixed points of x. Since $x \neq 1$ and $a \neq ag$, $ag^2 = a$. Hence g interchanges a and ag, and

$$g \in \mathbf{N}_{\mathfrak{G}}(\mathfrak{G}_{a,ag}) - \mathfrak{G}_{a,ag}.$$

By 6.4, $\mathbf{N}_{\mathfrak{G}}(\mathfrak{G}_{a,ag})$ is dihedral. Since $x \in \mathfrak{G}_{a,ag}$, it follows that $x^g = x^{-1}$. But $x^g = x$, so $x^2 = 1$. Since d is odd, this gives $x = 1$, a contradiction.

It follows that $|\mathbf{C}_{\mathfrak{G}}(g)|$ divides $n + 1$; write $n + 1 = |\mathbf{C}_{\mathfrak{G}}(g)|u$. If the assertion is false, $|\mathfrak{J}_0| = |\mathfrak{G} : \mathbf{C}_{\mathfrak{G}}(g)|$, so

$$\tfrac{1}{2}(n + 1)n(d + 1) - n^2 = \frac{(n + 1)ndu}{n + 1} = ndu.$$

As u divides $n + 1$ and $\frac{1}{2}(d + 1)$ is an integer, it follows that u also divides n^2 and $u = 1$. Thus

$$\tfrac{1}{2}(n + 1)(d + 1) = n + d$$

and $n = 1$, a contradiction. **q.e.d.**

6.7 Proof of Theorem 6.1. By 6.5, $\mathfrak{G}$ satisfies Hypothesis 4.2, so $\mathfrak{G}$ has the exceptional irreducible characters Λ_{ij} $(i = 1, \ldots, k; j = 1, \ldots, m_i)$, and by 4.2h),

$$\lambda_{ij}^{\mathfrak{G}} - z_i\lambda_{11}^{\mathfrak{G}} = \varepsilon(\Lambda_{ij} - z_i\Lambda_{11}),$$

since $z_1 = 1$. The number of Λ_{ij} is the same as the number of ϕ_{ij}, which is $\dfrac{h(\mathfrak{F}) - 1}{d}$. Further, $\mathfrak{G}$ satisfies Hypothesis 5.1, by 6.4. Thus by 5.2 and 5.3, $\mathfrak{G}$ has irreducible characters 1, α, β_i $(i = 1, \ldots, \frac{1}{2}(d - 1))$. The number of irreducible characters constructed so far is at most

$$a = \frac{h(\mathfrak{F}) - 1}{d} + 2 + \tfrac{1}{2}(d - 1).$$

But $\mathfrak{G}$ has more than a conjugacy classes. In fact $\dfrac{h(\mathfrak{F}) - 1}{d}$ is the number of classes in $\mathfrak{J}_1$ and $\frac{1}{2}(d - 1)$ is the number of classes in $\mathfrak{J}_2$. By 6.6, the number of classes in $\mathfrak{J}_0$ is at least 2, and we have also the class containing only the unit element. Thus $\mathfrak{G}$ has an irreducible character χ distinct from 1, α, β_i and the Λ_{ij}. We compute $\psi = \chi_{\mathfrak{G}_\infty}$. The irreducible characters of $\mathfrak{G}_\infty$ are $\psi_0, \ldots, \psi_{d-1}$ and the λ_{ij}.

Suppose that $(\psi, \lambda_{11})_{\mathfrak{G}_\infty} = v$. Then

$$\begin{aligned}(\psi, \lambda_{ij})_{\mathfrak{G}_\infty} &= z_iv + (\psi, \lambda_{ij} - z_i\lambda_{11})_{\mathfrak{G}_\infty}\\ &= z_iv + (\chi, \lambda_{ij}^{\mathfrak{G}} - z_i\lambda_{11}^{\mathfrak{G}})_{\mathfrak{G}}\\ &= z_iv + \varepsilon(\chi, \Lambda_{ij} - z_i\Lambda_{11})_{\mathfrak{G}}\\ &= z_iv,\end{aligned}$$

since χ is distinct from the Λ_{ij}. Again

$$(\psi, \psi_0)_{\mathfrak{G}_\infty} = (\chi, \psi_0^{\mathfrak{G}})_{\mathfrak{G}} = (\chi, \pi)_{\mathfrak{G}} = (\chi, 1 + \alpha)_{\mathfrak{G}} = 0.$$

For $i = 1, \ldots, \frac{1}{2}(d - 1)$,

$$(\psi, \psi_i)_{\mathfrak{G}_\infty} = (\chi, \psi_i^{\mathfrak{G}})_{\mathfrak{G}} = (\chi, \beta_i)_{\mathfrak{G}} = 0,$$

and

$$(\psi, \psi_{i+\frac{1}{2}(d-1)})_{\mathfrak{G}_\infty} = (\psi, \overline{\psi_i})_{\mathfrak{G}_\infty} = (\chi, \overline{\beta}_i)_{\mathfrak{G}} = (\chi, \beta_i)_{\mathfrak{G}} = 0.$$

Hence

$$\psi = \sum_{i,j} z_i v \lambda_{ij}.$$

Therefore,

$$\chi(1) = \psi(1) = v \sum_{i,j} z_i \lambda_{ij}(1) = \frac{v}{d} \sum_{i,j} \lambda_{ij}(1)^2,$$

since $\lambda_{ij}(1) = dz_i$. Since the λ_{ij} are precisely the irreducible characters of $\mathfrak{G}_\infty$ for which $\mathfrak{F}$ is not contained in the kernel,

$$\chi(1) = \frac{v}{d}(|\mathfrak{G}_\infty| - |\mathfrak{G}_\infty : \mathfrak{F}|) = v(n - 1).$$

As $\chi(1)$ divides $|\mathfrak{G}|$, $n - 1$ divides $(n + 1)nd$. It follows that $n - 1$ divides $2d$ and $\frac{1}{2}(n - 1) \leq d$. This contradicts 6.4b). **q.e.d.**

The following is an important consequence of 6.1.

6.8 Theorem. *Suppose that $\mathfrak{G}$ is a Zassenhaus group and that the Frobenius kernel of $\mathfrak{G}_\infty$ is Abelian. Then either $\mathfrak{G}$ is sharply triply transitive, or $\mathfrak{G} = PSL(2, p^f)$ for p odd.*

Proof. By 6.1, $d \geq \frac{1}{2}(n - 1)$. If $d = n - 1$, $\mathfrak{G}$ is sharply triply transitive. If $d = \frac{1}{2}(n - 1)$ and d is even, $\mathfrak{G} = PSL(2, p^f)$, by 1.11. If $d = \frac{1}{2}(n - 1)$ and d is odd, then $\mathfrak{G}$ is simple. For if $1 < \mathfrak{N} \trianglelefteq \mathfrak{G}$, then by 1.6a), $\mathfrak{N}$ is a Zassenhaus group of degree $n + 1$. By 6.1, $|\mathfrak{N}| = (n + 1)nd'$, where $d' \geq \frac{1}{2}(n - 1)$. Thus $d' = d$ and $\mathfrak{G}$ is simple. The assertion now follows from 1.12. **q.e.d.**

The only Zassenhaus groups in which $\mathfrak{F}$ is non-Abelian are in fact the Suzuki groups (§ 3), as we shall see in § 11.

The following is a generalization of a theorem of FROBENIUS [2]. The original proof contains the first application of the functional equation which appeared in 1.9–1.11 and 2.6.

6.9 Theorem. *Let p be a prime and let $\mathfrak{G}$ be a transitive permutation group of degree $p + 1$ and order $(p + 1)pd$, where $d < p$. Then one of the following holds.*

(1) $d = p - 1$ and $\mathfrak{G} = PGL(2, p)$.

(2) $d = \frac{1}{2}(p - 1)$ and $\mathfrak{G} = PSL(2, p)$.

(3) $p + 1 = 2^q$ for some prime q. Either $|\mathfrak{G}| = (p + 1)p$ or $|\mathfrak{G}| = (p + 1)pq$, and $\mathfrak{G}$ consists of semilinear mappings on $GF(2^q)$ of the form $x \to a(x\alpha) + b$.

Proof. Denote one of the symbols permuted by ∞. Since $|\mathfrak{G}_\infty| = pd$, $\mathfrak{G}_\infty$ is transitive on $\Omega - \{\infty\}$, so $\mathfrak{G}$ is doubly transitive. Since $|\mathfrak{G}_\infty| = pd$ and $d < p$, $\mathfrak{G}_\infty$ has a normal Sylow subgroup. Thus by II, 3.6, $\mathfrak{G}_\infty$ is a soluble Frobenius group. This shows that all non-identity elements of $\mathfrak{G}$ have at most two fixed points. If $\mathfrak{G}$ has no regular normal subgroup, then $\mathfrak{G}$ is a Zassenhaus group and $|\mathfrak{F}| = p$. By 6.8, $\mathfrak{G} = PSL(2, p)$ or $\mathfrak{G} = PGL(2, p)$.

Suppose that $\mathfrak{G}$ has a regular normal subgroup. If $d > 1$, then by 1.1, $p + 1 = 2^q$, $d = q$ is a prime and $\mathfrak{G}$ consists of all semilinear mappings on $GF(2^q)$. If $d = 1$, $\mathfrak{G}$ is soluble and doubly primitive. By II, 3.13, it follows that $\mathfrak{G}$ is the group of all mappings on $GF(2^q)$ of the form $x \to ax + b$. **q.e.d.**

§ 7. Non-Regular Normal Subgroups of Multiply Transitive Permutation Groups

In II, 2.3 we studied multiply transitive permutation groups having a regular normal subgroup; we restate the result, which will be used in this section.

Let $\mathfrak{G}$ be a permutation group of degree n and let $\mathfrak{N}$ be a regular normal subgroup of $\mathfrak{G}$.

a) If $\mathfrak{G}$ is doubly transitive, $\mathfrak{N}$ is elementary Abelian and n is a prime-power.

b) If $\mathfrak{G}$ is doubly primitive or $\frac{5}{2}$-transitive, then $n = 2^m$ or $n = 3$.

c) If $\mathfrak{G}$ is triply primitive or $\frac{7}{2}$-transitive, then $n = 3$ or $n = 4$.

d) $\mathfrak{G}$ is never 5-fold transitive.

With the help of our knowledge of Frobenius and Zassenhaus groups we can now investigate non-regular normal subgroups of multiply transitive permutation groups. First we remind the reader of a result proved in Chapter VIII.

7.1 Theorem (WIELANDT [4]). *If $\mathfrak{G}$ is $\frac{3}{2}$-transitive, then either $\mathfrak{G}$ is primitive or $\mathfrak{G}$ is a Frobenius group.*

Proof. This is VIII, 3.1. **q.e.d.**

7.2 Theorem. *Suppose that $\mathfrak{G}$ is a $\frac{5}{2}$-transitive soluble permutation group of degree greater than 3. Then $\mathfrak{G}$ is the group $\Gamma(2^p)$ of all semilinear mappings on $GF(2^p)$, where p is a prime.*

Proof. Suppose that Ω is the set of symbols permuted and let $0 \in \Omega$.

If $\mathfrak{G}_0$ is primitive, then $\mathfrak{G}$ is doubly primitive. Then by II, 3.13a), $\mathfrak{G}$ is the group of all linear or all semilinear mappings over $GF(2)$, $GF(3)$ or $GF(2^p)$, where $2^p - 1$ (and hence p) is a prime. The groups of linear mappings are however sharply doubly transitive and are therefore not $\frac{5}{2}$-transitive.

Suppose then that $\mathfrak{G}_0$ is imprimitive. By 7.1, $\mathfrak{G}_0$ is a Frobenius group on $\Omega - \{0\}$. Hence no non-identity element of $\mathfrak{G}$ leaves more than two symbols fixed. If $\mathfrak{G}$ is itself a Frobenius group, $\mathfrak{G}$ is not $\frac{5}{2}$-transitive, since $\mathfrak{G}_{0,1} = 1$. Let $\mathfrak{N}$ be a minimal normal subgroup of $\mathfrak{G}$. Since $\mathfrak{G}$ is soluble, it follows from II, 3.2 that $\mathfrak{N}$ is regular. Hence by 1.1 $n = 2^p$ for some prime p and $\mathfrak{G}$ is the group of all semilinear mappings on $GF(2^p)$. **q.e.d.**

7.3 Theorem (HUPPERT and WIELANDT [1]). *Let k be a natural number and let $\mathfrak{G}$ be a k-fold transitive permutation group of degree n. Suppose that $\mathfrak{G} \neq \mathfrak{S}_n$. If $1 < \mathfrak{N} \trianglelefteq \mathfrak{G}$ and $\mathfrak{N}$ is not regular, then $\mathfrak{N}$ is $(k - \frac{1}{2})$-transitive.*

Proof. Let Ω be the set of symbols permuted, with $1 \in \Omega$. The theorem will be proved by induction on k.

a) For $k = 1$, if $\mathfrak{N}$ is intransitive, the orbits of $\mathfrak{N}$ form an imprimitivity system for $\mathfrak{G}$, by II, 1.5. Thus all the orbits have the same length and $\mathfrak{N}$ is $\frac{1}{2}$-transitive.

b) Suppose that $k = 2$. Since $\mathfrak{N}$ is not regular, $1 < \mathfrak{N}_1 \trianglelefteq \mathfrak{G}_1$. Thus by a), $\mathfrak{N}_1$ is $\frac{1}{2}$-transitive on $\Omega - \{1\}$. Since $\mathfrak{G}$ is primitive, $\mathfrak{N}$ is transitive on Ω, so $\mathfrak{N}$ is $\frac{3}{2}$-transitive on Ω.

c) For $k \geq 3$, $\mathfrak{G}_1$ is $(k-1)$-fold transitive on $\Omega - \{1\}$, and $1 < \mathfrak{N}_1 \leqq \mathfrak{G}_1 \neq \mathfrak{S}_{n-1}$. If $\mathfrak{N}_1$ is not regular on $\Omega - \{1\}$, then by the inductive hypothesis, $\mathfrak{N}_1$ is $(k - \frac{3}{2})$-transitive on $\Omega - \{1\}$. Hence $\mathfrak{N}$ is $(k - \frac{1}{2})$-transitive on Ω.

Suppose now that $\mathfrak{N}_1$ is regular on $\Omega - \{1\}$. Then $\mathfrak{N}$ is a Frobenius group on Ω. By V, 8.3 the Frobenius kernel $\mathfrak{F}$ of $\mathfrak{N}$ is a characteristic subgroup of $\mathfrak{N}$ and hence a regular normal subgroup of $\mathfrak{G}$. Since $\mathfrak{G}$ is triply transitive and $\mathfrak{G} \neq \mathfrak{S}_n$, $n > 4$. Hence by II, 2.3, $k = 3$, $n = 2^m$ and $\mathfrak{F}$ is elementary Abelian. Since the regular normal subgroup $\mathfrak{N}_1$ of $\mathfrak{G}_1$ is transitive on $\Omega - \{1\}$, $\mathfrak{G}_1$ is represented irreducibly on $\mathfrak{F}$, by II, 2.2. As $\mathfrak{G}_1$ is doubly transitive, $\mathfrak{N}_1$ is Abelian, by II, 2.3a). Thus by II, 3.12, $\mathfrak{G}$ is contained in the group of semilinear mappings on $GF(2^m)$. Since $\mathfrak{G}$ is triply transitive, this is impossible by II, 1.18d), since $n > 4$. **q.e.d.**

7.4 Definition. Suppose that $\mathfrak{G}$ is a permutation group on Ω.

a) If $\Lambda \subseteq \Omega$, we put

$$\mathfrak{G}_{(\Lambda)} = \{g \mid g \in \mathfrak{G}, \Lambda g = \Lambda\}.$$

b) Suppose that $k < |\Omega|$. $\mathfrak{G}$ is called *generously k-fold transitive* if, whenever $\Lambda \subseteq \Omega$ and $|\Lambda| = k + 1$, $\mathfrak{G}_{(\Lambda)}$ induces the full symmetric group on Λ. In particular, $\mathfrak{G}$ is *generously transitive* if, given distinct elements a, b of Ω, there exists $g \in \mathfrak{G}$ such that g interchanges a and b.

7.5 Lemma. a) *If $\mathfrak{G}$ is $(k+1)$-fold transitive, $\mathfrak{G}$ is generously k-fold transitive.*

b) *If $\mathfrak{G}$ is generously k-fold transitive, $\mathfrak{G}$ is k-fold transitive.*

c) *If $\mathfrak{G}$ is generously k-fold transitive, $\Delta \subset \Omega$ and $|\Delta| = k$, then $\mathfrak{G}_\Delta$ and $\mathfrak{G}_{(\Delta)}$ have the same orbits on $\Omega - \Delta$.*

Proof. a) This is trivial.

b) This is proved by induction on k. For $k = 1$, it is clear from the definition. For $k > 1$, choose $1 \in \Omega$. Suppose that $\Delta \subseteq \Omega - \{1\}$, and $|\Delta| = k$. We put $\Lambda = \Delta \cup \{1\}$. By hypothesis, $\mathfrak{G}_{(\Lambda)}$ induces the full symmetric group on Λ. Therefore $(\mathfrak{G}_1)_{(\Delta)}$ induces the full symmetric group on Δ. Hence $\mathfrak{G}_1$ is generously $(k-1)$-fold transitive on $\Omega - \{1\}$. Hence by the inductive hypothesis, $\mathfrak{G}_1$ is $(k-1)$-fold transitive and $\mathfrak{G}$ is k-fold transitive.

c) Obviously, $\mathfrak{G}_\Delta \leqq \mathfrak{G}_{(\Delta)}$. Suppose that a, b are elements of $\Omega - \Delta$ and that $ag = b$, where $g \in \mathfrak{G}_{(\Delta)}$. Since $\mathfrak{G}_{(\Delta \cup \{a\})}$ induces the full symmetric group on $\Delta \cup \{a\}$, there exists $h \in \mathfrak{G}$ such that $ah = a$ and

$ih = ig^{-1}(\in \Delta)$ for all $i \in \Delta$. Then $ahg = ag = b$ and $ihg = i$ for all $i \in \Delta$. Thus a and b lie in the same orbit of $\mathfrak{G}_\Delta$. **q.e.d.**

7.6 Lemma. *Suppose that $\mathfrak{G}$ is generously k-fold transitive on Ω. Suppose that p is a prime and $p \leq k$. If $\Gamma \subset \Omega$ and $|\Gamma| = k$, $\mathfrak{G}_\Gamma$ has at most $p - 1$ orbits on $\Omega - \Gamma$ of length prime to p.*

Proof. Write $\Gamma = \{1, \ldots, p, p + 1, \ldots, k\}$. Suppose that $\Delta_1, \ldots, \Delta_t$ are the orbits of $\mathfrak{G}_\Gamma$ on $\Omega - \Gamma$ of length prime to p, and suppose that $t \geq p$. Let

$$\begin{aligned}\mathfrak{H} &= (\mathfrak{G}_{(\{1,\ldots,p\})})_{p+1,\ldots,k} \\ &= \{g \mid g \in \mathfrak{G}, \{1, \ldots, p\}g = \{1, \ldots, p\}, \\ &\quad \text{and } ig = i \text{ for } p + 1 \leq i \leq k\}.\end{aligned}$$

Thus $\mathfrak{H} \leq \mathfrak{G}_{(\Gamma)}$. Let $\mathfrak{P}$ be a Sylow p-subgroup of $\mathfrak{H}$. By 7.5c), the Δ_i are orbits of $\mathfrak{G}_{(\Gamma)}$, so Δ_i is left fixed by $\mathfrak{P}$. Hence $\mathfrak{P}$ has a fixed point in each Δ_i, since $|\Delta_i|$ is prime to p. The total number of fixed points of $\mathfrak{P}$ is therefore at least $k - p + t \geq k$. But by 7.5b), $\mathfrak{G}$ is k-fold transitive. Thus $\mathfrak{P}^g \leq \mathfrak{G}_\Gamma$ for some $g \in \mathfrak{G}$. Thus p does not divide $|\mathfrak{H} : \mathfrak{G}_\Gamma|$. But since $\mathfrak{G}$ is generously k-fold transitive, $\mathfrak{H}$ induces the full symmetric group on $\{1, \ldots, p\}$, so p divides $|\mathfrak{H} : \mathfrak{G}_\Gamma|$, a contradiction. **q.e.d.**

7.7 Theorem (A. WAGNER [1]). *Suppose that $\mathfrak{G}$ is a triply transitive group of degree n, where $n \geq 5$ and n is odd. If $1 < \mathfrak{N} \trianglelefteq \mathfrak{G}$, $\mathfrak{N}$ is triply transitive.*

Proof. First suppose that $\mathfrak{N}$ is regular. Then by II, 2.3b), n is 2^m or 3, contrary to hypothesis. Thus $\mathfrak{N}$ is not regular.

By 7.3, either $\mathfrak{G} = \mathfrak{S}_n$ or $\mathfrak{N}$ is $\frac{5}{2}$-transitive. If $\mathfrak{G} = \mathfrak{S}_n$, $\mathfrak{N} \geq \mathfrak{A}_n$ and $\mathfrak{N}$ is triply transitive since $n \geq 5$. In any case, $|\mathfrak{N}|$ is even.

Let t be an involution in $\mathfrak{N}$. Since n is odd, we may suppose that

$$t = (1)(2, 3) \cdots.$$

Put $\Delta = \{1, 2, 3\}$. Since $\mathfrak{G}$ is triply transitive, $\mathfrak{G}^\Delta_{(\Delta)} = \mathfrak{S}^\Delta$. Thus

$$(2, 3) \in \mathfrak{N}^\Delta_{(\Delta)} \trianglelefteq \mathfrak{G}^\Delta_{(\Delta)} = \mathfrak{S}^\Delta,$$

so $\mathfrak{N}^\Delta_{(\Delta)} = \mathfrak{S}^\Delta$. Again since $\mathfrak{G}$ is triply transitive, it follows that $\mathfrak{N}^\Lambda_{(\Lambda)} = \mathfrak{S}^\Lambda$ for every subset Λ of Ω with three elements. Thus $\mathfrak{N}$ is generously 2-fold transitive. We apply 7.6 with $p = k = 2$. Thus if $|\Gamma| = 2$, $\mathfrak{N}_\Gamma$ has at most one orbit of odd length on $\Omega - \Gamma$. Since $\mathfrak{N}_\Gamma$ is normal in $\mathfrak{G}_\Gamma$

and $\mathfrak{G}_\Gamma$ is transitive on $\Omega - \Gamma$, all orbits of $\mathfrak{N}_\Gamma$ on $\Omega - \Gamma$ are of the same length m. Thus m divides $n - 2$ and is odd. Hence $\mathfrak{N}_\Gamma$ is transitive on $\Omega - \Gamma$ and $\mathfrak{N}$ is triply transitive. **q.e.d.**

7.8 Remarks. a) It follows immediately from 7.7 that $PGL(2, 2^f)$ is simple for $f \geq 2$.

Theorem 7.7 is false if the condition that n be odd is omitted, as is seen by taking $\mathfrak{G} = PGL(2, p^f)$ and $\mathfrak{N} = PSL(2, p^f)$, where p is odd.

b) Using deeper methods, ITO [4] has proved the following theorem.

Suppose that $\mathfrak{G}$ is 4-fold transitive of degree $n > 5$, where n is not divisible by 3. If $1 < \mathfrak{N} \trianglelefteq \mathfrak{G}$, then $\mathfrak{N}$ is 4-fold transitive.

7.9 Theorem (HUPPERT and WIELANDT [1]). *Let k be a natural number and let $\mathfrak{G}$ be a $(k + \frac{1}{2})$-transitive permutation group of degree n. Suppose that $\mathfrak{G} \neq \mathfrak{S}_n$. If $1 < \mathfrak{N} \trianglelefteq \mathfrak{G}$ and $\mathfrak{N}$ is not regular, then $\mathfrak{N}$ is k-fold transitive, except possibly in the following cases.*

(1) $k = 1$; $\mathfrak{G}$ is then a Frobenius group.

(2) $k = 2$; $\mathfrak{G}$ is then the group of all semilinear mappings on $GF(2^p)$, where p is a prime but $2^p - 1$ is not.

Proof. Let Ω be the set of symbols permuted and suppose that $1 \in \Omega$. We proceed by induction on k.

a) Suppose that $k = 1$. If $\mathfrak{G}$ is primitive, $\mathfrak{N}$ is transitive. If $\mathfrak{G}$ is imprimitive, then $\mathfrak{G}$ is a Frobenius group by 7.1, since $\mathfrak{G}$ is $\frac{3}{2}$-transitive.

b) Suppose that $k > 1$. Then $1 < \mathfrak{N}_1 \trianglelefteq \mathfrak{G}_1$ and $\mathfrak{G}_1$ is $(k - \frac{1}{2})$-transitive on $\Omega - \{1\}$. If $\mathfrak{N}_1$ is regular on $\Omega - \{1\}$, then $\mathfrak{N}$ is a Frobenius group and the Frobenius kernel of $\mathfrak{N}$ is a regular normal subgroup of $\mathfrak{G}$, so by II, 2.3, $k = 2$, since $\mathfrak{G} \neq \mathfrak{S}_n$. If $\mathfrak{N}_1$ is not regular on $\Omega - \{1\}$, then by the inductive hypothesis, $\mathfrak{N}_1$ is $(k - 1)$-fold transitive on $\Omega - \{1\}$ and $\mathfrak{N}$ is k-fold transitive on Ω, except possibly when $k = 2$, or $k = 3$ and $\mathfrak{G}_1$ is the group $\Gamma(2^p)$ of all semilinear mappings on $GF(2^p)$, where p is a prime. But in this last case, $n = 2^p + 1$ is odd and $n \geq 5$, so $\mathfrak{N}$ is triply transitive by 7.7.

c) It remains to deal with the case when $k = 2$ and $\mathfrak{N}_1$ is intransitive on $\Omega - \{1\}$. Then $\mathfrak{G}_1$ is imprimitive on $\Omega - \{1\}$. By 7.1, $\mathfrak{G}_1$ is a Frobenius group on $\Omega - \{1\}$. If $\mathfrak{F}$ is the Frobenius kernel of $\mathfrak{G}_1$, $|\mathfrak{F}| = n - 1$ and $\mathfrak{F}$ is regular on $\Omega - \{1\}$. Since $\mathfrak{N}_1$ is intransitive on $\Omega - \{1\}$, $\mathfrak{F} \not\leq \mathfrak{N}_1$. But $\mathfrak{N}_1 \trianglelefteq \mathfrak{G}_1$, so by V, 8.16, $\mathfrak{N}_1 < \mathfrak{F}$. Hence $\mathfrak{N}_{1,i} \leq \mathfrak{F}_i = 1$ for all $i \neq 1$, so $\mathfrak{N}$ is a Frobenius group on Ω. If $\mathfrak{K}$ is the Frobenius kernel of $\mathfrak{N}$, $\mathfrak{K}$ is a regular normal subgroup of $\mathfrak{G}$. By II, 2.3b), $n = 3$ or $n = 2^f$, since $\mathfrak{G}$ is $\frac{5}{2}$-transitive. The first case is excluded, since $\mathfrak{G} \neq \mathfrak{S}_3$.

$\mathfrak{K}\mathfrak{F}$ is a sharply doubly transitive permutation group on Ω. Since $|\mathfrak{F}| = 2^f - 1$ is odd, the Sylow subgroups of $\mathfrak{F}$ are all cyclic, by V, 8.7b). By V, 8.7a), $\mathfrak{F}$ is nilpotent, so $\mathfrak{F}$ is cyclic. By II, 2.2, $\mathfrak{F}$ transforms $\mathfrak{K} - \{1\}$ transitively, so $\mathfrak{K}$ is elementary Abelian. Also, ($\mathfrak{F}$ on $\mathfrak{K}$) is irreducible, so by II, 3.12, $\mathfrak{G}$ is contained in the group of all semilinear mappings on $GF(2^f)$. Thus $\mathfrak{G}$ is soluble. From 7.2 it follows that $\mathfrak{G}$ is the group of all semilinear mappings on $GF(2^p)$, where p is a prime. Since $1 < \mathfrak{N}_1 < \mathfrak{F}$ and $|\mathfrak{F}| = 2^p - 1$, $2^p - 1$ is not a prime. **q.e.d.**

Theorems 7.3 and 7.9 show that for non-regular normal subgroups, the degree of transitivity is as a rule at most $\frac{1}{2}$ less than that of the whole group. In addition to $(k + \frac{1}{2})$-transitivity, we have another refinement of the transitivity scale, namely, multiple primitivity (II, 1.14a)). Theorems analogous to 7.3 and 7.9 hold for this. The first is considerably deeper.

7.10 Theorem (ITO [2]). *Suppose that k is a natural number and that $\mathfrak{G}$ is a $(k + 1)$-fold transitive permutation group of degree n. Suppose that $\mathfrak{G} \neq \mathfrak{S}_n$. If $1 < \mathfrak{N} \trianglelefteq \mathfrak{G}$ and $\mathfrak{N}$ is not regular, then either $\mathfrak{N}$ is k-fold primitive, or $k = 1$ and $\mathfrak{N}$ is a Frobenius group.*

Proof. We use induction on k.

a) If $k = 1$, $\mathfrak{G}$ is doubly transitive, so by 7.3, $\mathfrak{N}$ is $\frac{3}{2}$-transitive. If $\mathfrak{N}$ is imprimitive, then $\mathfrak{N}$ is a Frobenius group by 7.1.

b) Suppose that $k = 2$. We have $1 < \mathfrak{N}_1 \trianglelefteq \mathfrak{G}_1$.

If $\mathfrak{N}_1$ is regular on $\Omega - \{1\}$, $\mathfrak{N}$ is a Frobenius group. The Frobenius kernel $\mathfrak{K}$ of $\mathfrak{N}$ is thus a regular normal subgroup of $\mathfrak{G}$. It then follows from II, 2.3 that $n = 3$ or $n = 2^f$. Since $\mathfrak{G} \neq \mathfrak{S}_3$, the case $n = 3$ cannot occur. Since $\mathfrak{N}_1$ transforms $\mathfrak{K} - \{1\}$ transitively, $\mathfrak{K}$ is an elementary Abelian group and an irreducible $\mathfrak{N}_1$-module. On account of the double transitivity of $\mathfrak{G}_1$, $\mathfrak{N}_1$ is Abelian, by II, 2.3. Thus by II, 3.12, $\mathfrak{G}$ is a group of semilinear mappings on $GF(2^f)$. But $\mathfrak{G}$ is triply transitive, so by II, 1.18d), $2^f = 4$ and $\mathfrak{G} = \mathfrak{S}_4$, contrary to hypothesis.

Hence $\mathfrak{N}_1$ is not regular. If $\mathfrak{N}_1$ is primitive on $\Omega - \{1\}$, then $\mathfrak{N}$ is doubly primitive and there is nothing more to prove. If $\mathfrak{N}_1$ is imprimitive on $\Omega - \{1\}$, then by a), $\mathfrak{N}_1$ is a Frobenius group on $\Omega - \{1\}$.

If $\mathfrak{N}$ has a regular normal subgroup $\mathfrak{M}$, then by 1.1, $\mathfrak{N}$ is the group of all semilinear mappings on a field $GF(2^p)$, since $\mathfrak{N}$ is not a Frobenius group. It follows that $\mathfrak{M}$ is characteristic in $\mathfrak{N}$ and normal in $\mathfrak{G}$. Also $\mathfrak{G}_1$ has a normal Abelian subgroup which is irreducibly represented on $\mathfrak{M}$. Thus by II, 3.12, $\mathfrak{G}$ consists of semilinear mappings on $GF(2^p)$ and $\mathfrak{N} = \mathfrak{G}$. Hence in this case, $\mathfrak{N}$ is 2-fold primitive.

If, on the other hand, $\mathfrak{N}$ has no regular normal subgroup, $\mathfrak{N}$ is a

Zassenhaus group of degree n. Let $\mathfrak{F}$ be the Frobenius kernel of the Frobenius group $\mathfrak{N}_1$. Since $\mathfrak{F} \trianglelefteq \mathfrak{G}_1$ and $\mathfrak{G}_1$ is doubly transitive, $\mathfrak{F}$ is elementary Abelian, by II, 2.3a). Hence by Theorem 6.1c), $|\mathfrak{N}_{1,i}| \geq \frac{1}{2}(n - 2)$ whenever $1 \neq i \in \Omega$; (note that the degree of $\mathfrak{N}$ is n, not $n + 1$).

Suppose that Δ is a domain of imprimitivity of $\mathfrak{N}_1$ on $\Omega - \{1\}$, and let 2, 3 be elements of Δ. Then

$$|\Delta| \geq |\{2, 3x | x \in \mathfrak{N}_{1,2}\}| = 1 + |\mathfrak{N}_{1,2}| \geq \tfrac{1}{2}n.$$

On the other hand, $|\Delta|$ divides $n - 1$, so $|\Delta| = n - 1$. Hence $\mathfrak{N}_1$ is not imprimitive, a contradiction.

c) Suppose that $k \geq 3$. If $\mathfrak{N}_1$ is not regular, $\mathfrak{N}_1$ is $(k - 1)$-fold primitive by the inductive hypothesis, and $\mathfrak{N}$ is k-fold primitive. But if $\mathfrak{N}_1$ is regular on $\Omega - \{1\}$, $\mathfrak{N}$ is a Frobenius group, the Frobenius kernel of $\mathfrak{N}$ is a regular normal subgroup of $\mathfrak{G}$ and by II, 2.3, $n = 4$ and $\mathfrak{G} = \mathfrak{S}_4$, a contradiction. **q.e.d.**

Whereas the proof of 7.10 uses the deep Theorem 6.1, the proof of the following theorem follows directly from the earlier results.

7.11 Theorem (WIELANDT [4]). *Let k be a natural number and let $\mathfrak{G}$ be a k-fold primitive permutation group of degree n. Suppose that $\mathfrak{G} \neq \mathfrak{S}_n$. If $1 < \mathfrak{N} \trianglelefteq \mathfrak{G}$ and $\mathfrak{N}$ is not regular, then $\mathfrak{N}$ is k-fold transitive.*

Proof. We prove the theorem by induction on k. For $k = 1$, it follows from II, 1.5. For $k = 2$, it follows from $1 < \mathfrak{N}_1 \trianglelefteq \mathfrak{G}_1$ and the primitivity of $\mathfrak{G}_1$ on $\Omega - \{1\}$ that $\mathfrak{N}_1$ is transitive on $\Omega - \{1\}$. Hence $\mathfrak{N}$ is doubly transitive on Ω.

For $k \geq 3$, if $\mathfrak{N}_1$ is not regular on $\Omega - \{1\}$, it follows from the inductive hypothesis that $\mathfrak{N}_1$ is $(k - 1)$-fold transitive on $\Omega - \{1\}$ and $\mathfrak{N}$ is k-fold transitive on Ω. If $\mathfrak{N}_1$ is regular on $\Omega - \{1\}$, $\mathfrak{N}$ is a Frobenius group. The Frobenius kernel of $\mathfrak{N}$ is a regular normal subgroup of $\mathfrak{G}$. Since $\mathfrak{G} \neq \mathfrak{S}_n$, it follows from II, 2.3 that $k \leq 2$, a contradiction.

q.e.d.

The following theorem makes an assertion about non-regular minimal normal subgroups of doubly transitive permutation groups.

7.12 Theorem (BURNSIDE). *Suppose that $\mathfrak{G}$ is doubly transitive and $\mathfrak{N}$ is a minimal normal subgroup of $\mathfrak{G}$. If $\mathfrak{N}$ is not regular, $\mathfrak{N}$ is simple and primitive, and $\mathbf{C}_{\mathfrak{G}}(\mathfrak{N}) = 1$.*

Proof. Since $\mathfrak{N}$ is a minimal normal subgroup of $\mathfrak{G}$, $\mathfrak{N}$ is characteristically simple. Hence $\mathfrak{N}$ is not a Frobenius group, so by 7.10, $\mathfrak{N}$ is primitive. Since $\mathfrak{N}$ is not regular, $\mathfrak{N}$ is non-Abelian, by II, 3.2. By I, 9.12, $\mathfrak{N}$ is the direct product of simple, non-Abelian isomorphic groups $\mathfrak{K}$, $\mathfrak{L}, \ldots$. Since $\mathfrak{N}$ is primitive, $\mathfrak{K}$ is transitive. Thus $|\mathfrak{K}|$ is divisible by the degree n.

We show that if $1 \neq x \in \mathbf{C}_{\mathfrak{G}}(\mathfrak{K})$, x has no fixed point on Ω. For if $ix = i$, then for any $j \in \Omega$, there exists $y \in \mathfrak{K}$ such that $iy = j$, since $\mathfrak{K}$ is transitive. Then $jx = iyx = ixy = iy = j$, so $x = 1$, a contradiction. Thus $|\mathbf{C}_{\mathfrak{G}}(\mathfrak{K})| \leq n$. But if $\mathfrak{N}$ is not simple, $|\mathfrak{L}| = |\mathfrak{K}|$ is divisible by n and $\mathfrak{L} \leq \mathbf{C}_{\mathfrak{G}}(\mathfrak{K})$. Thus $\mathfrak{L} = \mathbf{C}_{\mathfrak{G}}(\mathfrak{K})$ and $|\mathfrak{L}| = |\mathfrak{K}| = n$. Further, $\mathfrak{N} = \mathfrak{K} \times \mathfrak{L}$, so $|\mathfrak{N}| = n^2$ and $|\mathfrak{N}_1| = n$. But since $\mathfrak{G}_1$ is transitive on $\Omega - \{1\}$, all the orbits of $\mathfrak{N}_1$ on $\Omega - \{1\}$ are of equal length m. Hence m divides $n - 1$ and m divides n. Thus $m = 1$ and $\mathfrak{N}_1 = 1$, a contradiction. Therefore $\mathfrak{N}$ is simple, and if $1 \neq x \in \mathbf{C}_{\mathfrak{G}}(\mathfrak{N})$, x has no fixed point on Ω. But if $\mathbf{C}_{\mathfrak{G}}(\mathfrak{N}) \neq 1$, then $\mathbf{C}_{\mathfrak{G}}(\mathfrak{N})$ is transitive, since $\mathbf{C}_{\mathfrak{G}}(\mathfrak{N}) \trianglelefteq \mathfrak{G}$ and $\mathfrak{G}$ is primitive. Thus $\mathbf{C}_{\mathfrak{G}}(\mathfrak{N})$ is regular. By II, 2.3, $\mathbf{C}_{\mathfrak{G}}(\mathfrak{N})$ is Abelian. By II, 3.1, $\mathbf{C}_{\mathfrak{G}}(\mathfrak{N}) = \mathbf{C}_{\mathfrak{G}}(\mathbf{C}_{\mathfrak{G}}(\mathfrak{N})) \geq \mathfrak{N}$, a contradiction. **q.e.d.**

§ 8. Real Characters

All representations considered in this section are representations over the field of either real or complex numbers.

8.1 Theorem. *Let* V *be a finite-dimensional real (complex) vector space and suppose that* V *is a representation module for the finite group* $\mathfrak{G}$. *Then there exists a positive definite symmetric (Hermitean) bilinear form* (,) *on* V *such that* $(v_1 g, v_2 g) = (v_1, v_2)$ *for all* v_1, v_2 *in* V *and all* $g \in \mathfrak{G}$.

Proof. Let [,] be any positive definite symmetric (Hermitean) bilinear form on V, and put

$$(v_1, v_2) = \sum_{g \in \mathfrak{G}} [v_1 g, v_2 g].$$

Then (,) is symmetric (Hermitean), and $(v_1 g, v_2 g) = (v_1, v_2)$ for all v_1, v_2 in V and all $g \in \mathfrak{G}$. For $v \neq 0$, $(v, v) \geq [v, v] > 0$, so (,) is positive definite. **q.e.d.**

8.2 Remark. The original proof of Maschke's theorem in the case of the real or complex field was as follows.

Let V be a $\mathfrak{G}$-module and let W be a $\mathfrak{G}$-submodule of V. By 8.1, there exists a positive definite symmetric (Hermitean) bilinear form on V which is $\mathfrak{G}$-invariant. Then (,) is certainly non-singular. Hence if $W^\perp$ is the subspace of V orthogonal to W with respect to (,) $V = W \oplus W^\perp$ and $W^\perp$ is $\mathfrak{G}$-invariant.

8.3 Theorem (FROBENIUS and SCHUR [1]). *Let* V *be an irreducible* $\mathbb{C}\mathfrak{G}$*-module and let* χ *be the character of* V. *Then precisely one of the following occurs.*

(1) *Not all* $\chi(g)$ $(g \in \mathfrak{G})$ *are real, and*

$$\sum_{g \in \mathfrak{G}} \chi(g^2) = 0.$$

There is no non-zero $\mathfrak{G}$*-invariant bilinear form on* V. *(We then say that* χ *is of the first kind.)*

(2) *For all* $g \in \mathfrak{G}$, $\chi(g)$ *is real, and*

$$\sum_{g \in \mathfrak{G}} \chi(g^2) = |\mathfrak{G}|.$$

Also χ *is the character of an* $\mathbb{R}\mathfrak{G}$*-module. Further, there exists a non-zero* $\mathfrak{G}$*-invariant bilinear form* f *on* V, *unique to within a scalar multiple, and* f *is symmetric and non-singular. (*χ *is then said to be of the second kind.)*

(3) *For all* $g \in \mathfrak{G}$, $\chi(g)$ *is real and*

$$\sum_{g \in \mathfrak{G}} \chi(g^2) = -|\mathfrak{G}|.$$

There is no $\mathbb{R}\mathfrak{G}$*-module with character* χ. *There exists a non-zero* $\mathfrak{G}$*-invariant bilinear form* f *on* V, *unique to within a scalar multiple, and* f *is skew-symmetric and non-singular. (*χ *is then said to be of the third kind.)*

The proof of Theorem 8.3 will be given in the form of three lemmas.

8.4 Lemma. *Let* V *be an irreducible* $\mathbb{C}\mathfrak{G}$*-module and let* χ *be the character of* V. *Then precisely one of the following occurs.*

(1) *Not all* $\chi(g)$ $(g \in \mathfrak{G})$ *are real, and*

$$\sum_{g \in \mathfrak{G}} \chi(g^2) = 0.$$

There is no non-zero $\mathfrak{G}$-invariant bilinear form on V.

(2) *For all $g \in \mathfrak{G}$, $\chi(g)$ is real, and*

$$\sum_{g \in \mathfrak{G}} \chi(g^2) = |\mathfrak{G}|.$$

There exists a non-zero $\mathfrak{G}$-invariant bilinear form f on V, *unique to within a scalar multiple, and f is symmetric and non-singular.*

(3) *For all $g \in \mathfrak{G}$, $\chi(g)$ is real, and*

$$\sum_{g \in \mathfrak{G}} \chi(g^2) = -|\mathfrak{G}|.$$

There exists a non-zero $\mathfrak{G}$-invariant bilinear form f on V, *unique to within a scalar multiple, and f is skew-symmetric and non-singular.*

Proof. a) Let $\mathbf{B}(\mathsf{V})$ be the vector space of bilinear forms on V. By VII, 8.9a), $\mathbf{B}(\mathsf{V})$ is a $\mathbb{C}\mathfrak{G}$-module in which

$$(bg)(v_1, v_2) = b(v_1 g^{-1}, v_2 g^{-1})$$

for all $g \in \mathfrak{G}$ and v_1, v_2 in V. If V* denotes the dual module $\mathrm{Hom}_{\mathbb{C}}(\mathsf{V}, \mathbb{C})$, then by VII, 8.9b),

$$\mathbf{B}(\mathsf{V}) \cong \mathsf{V}^* \otimes_{\mathbb{C}} \mathsf{V}^*.$$

If β is the character of the $\mathbb{C}\mathfrak{G}$-module $\mathbf{B}(\mathsf{V})$, it follows that

$$\beta(g) = \chi(g^{-1})^2.$$

Let

$$\mathbf{B}_0(\mathsf{V}) = \{b \mid b \in \mathbf{B}(\mathsf{V}),\ bg = b \quad \text{for all } g \in \mathfrak{G}\}.$$

Then the dimension of $\mathbf{B}_0(\mathsf{V})$ is

$$\begin{aligned}(\beta, 1)_{\mathfrak{G}} &= \frac{1}{|\mathfrak{G}|} \sum_{g \in \mathfrak{G}} \chi(g^{-1})^2 \\ &= \frac{1}{|\mathfrak{G}|} \sum_{g \in \mathfrak{G}} \overline{\chi(g)}\chi(g^{-1}) \\ &= (\bar{\chi}, \chi)_{\mathfrak{G}},\end{aligned}$$

where $\overline{\chi}$ is the complex conjugate character of χ, so

$$\overline{\chi}(g) = \chi(g^{-1}) = \overline{\chi(g)}.$$

It follows from the orthogonality relations that if $\overline{\chi} \neq \chi$, then $(\beta, 1)_{\mathfrak{G}} = 0$, whereas if $\overline{\chi} = \chi$, then $\dim_{\mathbb{C}} \mathbf{B}_0(\mathsf{V}) = 1$. Thus in the second case, there exists a non-zero $b \in \mathbf{B}_0(\mathsf{V})$, unique to within a scalar multiple. Thus $\operatorname{rad} \mathsf{V} \neq \mathsf{V}$, where

$$\operatorname{rad} \mathsf{V} = \{v \mid b(v, w) = 0 \quad \text{for all } w \in \mathsf{V}\}.$$

But $\operatorname{rad} \mathsf{V}$ is a $\mathfrak{G}$-submodule of V, so $\operatorname{rad} \mathsf{V} = 0$ on account of the irreducibility of V. Hence b is non-singular.

b) Let

$$\mathbf{S}(\mathsf{V}) = \{b \mid b \in \mathbf{B}(\mathsf{V}), b \text{ is symmetric}\},$$

and let

$$\mathbf{A}(\mathsf{V}) = \{b \mid b \in \mathbf{B}(\mathsf{V}), b \text{ is skew-symmetric}\}.$$

Then

$$\mathbf{B}(\mathsf{V}) = \mathbf{S}(\mathsf{V}) \oplus \mathbf{A}(\mathsf{V})$$

is a direct decomposition of $\mathbf{B}(\mathsf{V})$ into $\mathbb{C}\mathfrak{G}$-submodules. Let σ be the character of $\mathbf{S}(\mathsf{V})$ and let α be the character of $\mathbf{A}(\mathsf{V})$. Thus $\beta = \sigma + \alpha$. We calculate $\sigma(g)$ and $\alpha(g)$. Instead of working in $\mathbf{B}(\mathsf{V})$, we use the isomorphic $\mathbb{C}\mathfrak{G}$-module $\mathsf{V}^* \otimes_{\mathbb{C}} \mathsf{V}^*$, in which $\mathbf{S}(\mathsf{V})$ corresponds to the space S of symmetric tensors and $\mathbf{A}(\mathsf{V})$ corresponds to the space A of skew-symmetric tensors. Suppose that $g \in \mathfrak{G}$. Then there is a basis $\{f_1, \ldots, f_n\}$ of V^* such that

$$f_i g = a_i f_i \quad (i = 1, \ldots, n)$$

for some $a_i \in \mathbb{C}$. Then

$$\{f_i \otimes f_j + f_j \otimes f_i \mid i \leq j\}$$

is a $\mathbb{C}$-basis of S and

$$(f_i \otimes f_j + f_j \otimes f_i)g = a_i a_j (f_i \otimes f_j + f_j \otimes f_i).$$

As $\bar{\chi}$ is the character of the dual module V* (see VII, 8.2), we obtain

$$\sigma(g) = \sum_{i \le j} a_i a_j = \tfrac{1}{2}\left\{\left(\sum_{i=1}^{n} a_i\right)^2 + \sum_{i=1}^{n} a_i^2\right\}$$
$$= \tfrac{1}{2}\{\chi(g^{-1})^2 + \chi(g^{-2})\}.$$

Further

$$\alpha(g) = \chi(g^{-1})^2 - \sigma(g) = \tfrac{1}{2}\{\chi(g^{-1})^2 - \chi(g^{-2})\}.$$

c) Suppose that $\chi \neq \bar{\chi}$. Then by a),

$$0 = (\beta, 1)_{\mathfrak{G}} = (\sigma + \alpha, 1)_{\mathfrak{G}},$$

so $(\alpha, 1)_{\mathfrak{G}} = (\sigma, 1)_{\mathfrak{G}} = 0$, and hence

$$0 = (\sigma, 1)_{\mathfrak{G}} - (\alpha, 1)_{\mathfrak{G}} = \frac{1}{|\mathfrak{G}|} \sum_{g \in \mathfrak{G}} \chi(g^{-2}).$$

Thus case (1) occurs.

d) Now suppose that $\chi = \bar{\chi}$. By a),

$$1 = (\beta, 1)_{\mathfrak{G}} = (\sigma + \alpha, 1)_{\mathfrak{G}}.$$

Hence either

(2′) $1 = (\sigma, 1)_{\mathfrak{G}}$ and $0 = (\alpha, 1)_{\mathfrak{G}}$, whence

$$1 = (\sigma - \alpha, 1)_{\mathfrak{G}} = \frac{1}{|\mathfrak{G}|} \sum_{g \in \mathfrak{G}} \chi(g^{-2}),$$

or

(3′) $0 = (\sigma, 1)_{\mathfrak{G}}$ and $1 = (\alpha, 1)_{\mathfrak{G}}$, whence

$$-1 = (\sigma - \alpha, 1)_{\mathfrak{G}} = \frac{1}{|\mathfrak{G}|} \sum_{g \in \mathfrak{G}} \chi(g^{-2}).$$

In case (2′), $\dim_{\mathbb{C}}(\mathbf{S}(\mathsf{V}) \cap \mathbf{B}_0(\mathsf{V})) = 1$, and in case (3′), $\dim_{\mathbb{C}}(\mathbf{A}(\mathsf{V}) \cap \mathbf{B}_0(\mathsf{V})) = 1$. **q.e.d.**

8.5 Lemma. *Let* V *be an irreducible* $\mathbb{C}\mathfrak{G}$*-module and let* χ *be the character of* V. *Suppose that* χ *is the character of an* $\mathbb{R}\mathfrak{G}$*-module* W. *(Then* $\mathsf{W} \otimes_{\mathbb{R}} \mathbb{C}$

has character χ, *so* $\mathsf{W} \otimes_{\mathbb{R}} \mathbb{C} \cong \mathsf{V}$.) *Then*

$$\sum_{g \in \mathfrak{G}} \chi(g^2) = |\mathfrak{G}|.$$

Proof. By 8.4, it is sufficient to prove that there is a non-zero $\mathfrak{G}$-invariant symmetric $\mathbb{C}$-bilinear form b on V. By 8.1, there is a $\mathfrak{G}$-invariant symmetric positive definite $\mathbb{R}$-bilinear form b_0 on W. Since $\mathsf{V} \cong \mathsf{W} \otimes_{\mathbb{R}} \mathbb{C}$, this yields a $\mathfrak{G}$-invariant symmetric $\mathbb{C}$-bilinear form $b \neq 0$ on V. **q.e.d.**

The following is the last step in the proof of 8.3.

8.6 Lemma. *Suppose that* V *is an irreducible* $\mathbb{C}\mathfrak{G}$*-module and that* χ *is the character of* V. *Suppose that*

$$\sum_{g \in \mathfrak{G}} \chi(g^2) = |\mathfrak{G}|.$$

Then there is an $\mathbb{R}\mathfrak{G}$*-module* W *such that* $\mathsf{V} \cong \mathsf{W} \otimes_{\mathbb{R}} \mathbb{C}$.

Proof (LORENZ [1]). We seek $\rho \in \mathrm{Hom}_{\mathbb{R}\mathfrak{G}}(\mathsf{V}, \mathsf{V})$ such that $\rho^2 = 1$ and

$$(cv)\rho = \bar{c}(v\rho)$$

for all $c \in \mathbb{C}$, $v \in \mathsf{V}$. If we have such a ρ, put

$$\mathsf{W} = \{v \mid v \in \mathsf{V}, v\rho = v\}.$$

Obviously W is an $\mathbb{R}\mathfrak{G}$-module. We show that

$$\mathsf{V} = \mathsf{W} \oplus i\mathsf{W} \cong \mathsf{W} \otimes_{\mathbb{R}} \mathbb{C}.$$

Indeed, if w, w' are elements of W for which $w = iw'$, then

$$w = w\rho = (iw')\rho = -i(w'\rho) = -iw' = -w,$$

so $w = 0$. Thus $\mathsf{W} \cap i\mathsf{W} = 0$. Suppose that $v \in \mathsf{V}$. Since $\rho^2 = 1$, $v = w + iw'$, where

$$w = \tfrac{1}{2}(v + v\rho) \in \mathsf{W}, \quad w' = \frac{1}{2i}(v - v\rho) \in \mathsf{W}.$$

Hence $\mathsf{V} = \mathsf{W} \oplus i\mathsf{W}$.

Thus we only have to establish the existence of ρ.

Since $\sum_{g \in \mathfrak{G}} \chi(g^2) = |\mathfrak{G}|$, there exists a $\mathfrak{G}$-invariant $\mathbb{C}$-bilinear form b on V and b is symmetric and non-singular, by 8.4. By 8.1, there exists a $\mathfrak{G}$-invariant positive definite Hermitean form h on V. Given $v' \in \mathsf{V}$, there is a linear form f on V defined by

$$vf = b(v, v').$$

As h is non-singular, it follows that there exists a unique $v'' \in \mathsf{V}$ such that

$$vf = h(v, v''),$$

as is well-known. Put $v'' = v'\tau$. Thus τ is a mapping of V into itself and

$$b(v, v') = h(v, v'\tau). \tag{1}$$

It is easily verified that $(v_1 + v_2)\tau = v_1\tau + v_2\tau$. For $a \in \mathbb{C}$,

$$\begin{aligned} h(v, (av')\tau) &= b(v, av') = a(b(v, v')) \\ &= a(h(v, v'\tau)) = h(v, \bar{a}(v'\tau)). \end{aligned}$$

As h is non-singular, it follows that

$$(av')\tau = \bar{a}(v'\tau). \tag{2}$$

Hence $\tau \in \mathrm{Hom}_{\mathbb{R}}(\mathsf{V}, \mathsf{V})$.

Since b and h are $\mathfrak{G}$-invariant,

$$h(vg, (v'\tau)g) = h(v, v'\tau) = b(v, v') = b(vg, v'g) = h(vg, (v'g)\tau)$$

for all $g \in \mathfrak{G}$. It follows that $(v'\tau)g = (v'g)\tau$, so

$$\tau \in \mathrm{Hom}_{\mathbb{R}\mathfrak{G}}(\mathsf{V}, \mathsf{V}). \tag{3}$$

We define b_1 by putting

$$b_1(v, v') = b(v, v'\tau^2).$$

As τ^2 is $\mathbb{C}$-linear, b_1 is $\mathbb{C}$-bilinear and $\mathfrak{G}$-invariant. Hence by 8.4, there exists $c \in \mathbb{C}$ such that $b_1 = cb$. From

$$b(v, v'\tau^2) = b_1(v, v') = b(v, cv'),$$

it follows that

(4) $$\tau^2 = c.$$

By (1), $\tau \neq 0$, so there exists $u \in \mathsf{V}$ such that $u\tau \neq 0$. Since b is symmetric, we obtain

$$h(u\tau, u\tau) = b(u\tau, u) = b(u, u\tau) = h(u, u\tau^2) = h(u, cu) = \bar{c}h(u, u).$$

Since h is positive definite, it follows that c is real and positive. Thus there exists $d \in \mathbb{R}$ such that $d^2 = c$. We put $\rho = d^{-1}\tau$. Then ρ satisfies the conditions (2) and (3). Further, since $d \in \mathbb{R}$, it follows that for all $v \in \mathsf{V}$,

$$v\rho^2 = d^{-1}(d^{-1}(v\tau))\tau = d^{-2}(v\tau^2) = d^{-2}cv = v.$$

Thus $\rho^2 = 1$. **q.e.d.**

Thus 8.3 is completely proved. The following consequence will play a central role in the investigation of Zassenhaus groups of even degree in § 9.

8.7 Theorem (FROBENIUS and SCHUR [1]). *Let $\chi_1, \ldots, \chi_h$ be the complex irreducible characters of $\mathfrak{G}$. We put*

$$\varepsilon(\chi_i) = \frac{1}{|\mathfrak{G}|} \sum_{g \in \mathfrak{G}} \chi_i(g^2).$$

Thus by 8.3,

$$\varepsilon(\chi_i) = \begin{cases} 0 & \text{if } \chi_i \text{ is of the first kind,} \\ 1 & \text{if } \chi_i \text{ is of the second kind,} \\ -1 & \text{if } \chi_i \text{ is of the third kind.} \end{cases}$$

For $g' \in \mathfrak{G}$, let $t(g')$ denote the number of elements g of $\mathfrak{G}$ such that $g^2 = g'$. Then

$$t(g') = \sum_{i=1}^{h} \varepsilon(\chi_i)\chi_i(g'^{-1}).$$

In particular,

$$t(1) = \sum_{i=1}^{h} \varepsilon(\chi_i)\chi_i(1)$$

is the number of $g \in \mathfrak{G}$ such that $g^2 = 1$.

Proof. On account of the orthogonality relations,

$$\sum_{i=1}^{h} \varepsilon(\chi_i)\chi_i(g'^{-1}) = \frac{1}{|\mathfrak{G}|}\sum_{i=1}^{h}\sum_{g\in\mathfrak{G}} \chi_i(g^2)\chi_i(g'^{-1}) = \sum_{g\in\mathfrak{G}} \frac{1}{|\mathfrak{G}|}\sum_{i=1}^{h} \chi_i(g^2)\chi_i(g'^{-1})$$

$$= \frac{|\{g \mid g^2 \text{ is conjugate to } g'\}|}{|\{x^{-1}g'x \mid x \in \mathfrak{G}\}|} = t(g').$$

q.e.d.

We shall now use Theorem 8.3 to discuss representation theory over the field of real numbers $\mathbb{R}$. First we collect some simple facts.

8.8 Lemma. *Let* V *be an irreducible* $\mathbb{C}\mathfrak{G}$*-module and let* χ *be the character of* V.

a) *Let* V_0 *be the* $\mathbb{R}\mathfrak{G}$*-module obtained from* V *by restricting the domain of operators to* $\mathbb{R}\mathfrak{G}$; *thus* $\dim_{\mathbb{R}}\mathsf{V}_0 = 2\dim_{\mathbb{C}}\mathsf{V}$. *Then*

$$\mathsf{V}_0 \otimes_{\mathbb{R}} \mathbb{C} \cong \mathsf{V} \oplus \mathsf{V}^*.$$

In particular, $\chi + \bar{\chi}$ *is the character of the* $\mathbb{R}\mathfrak{G}$*-module* V_0.

b) *If there is no* $\mathbb{R}\mathfrak{G}$*-module with character* χ, *then* V_0 *is an irreducible* $\mathbb{R}\mathfrak{G}$*-module.*

c) *For any* $\mathbb{R}\mathfrak{G}$*-module* W,

$$\dim_{\mathbb{R}}\mathrm{Hom}_{\mathbb{R}\mathfrak{G}}(\mathsf{W}, \mathsf{W}) = \dim_{\mathbb{C}}\mathrm{Hom}_{\mathbb{C}\mathfrak{G}}(\mathsf{W} \otimes_{\mathbb{R}} \mathbb{C}, \mathsf{W} \otimes_{\mathbb{R}} \mathbb{C}).$$

Proof. a) This is VII, 1.16.

b) If V_0 is reducible, $\mathsf{V}_0 = \mathsf{W}_1 \oplus \mathsf{W}_2$, so

$$\mathsf{V}_0 \otimes_{\mathbb{R}} \mathbb{C} = \mathsf{W}_1 \otimes_{\mathbb{R}} \mathbb{C} \oplus \mathsf{W}_2 \otimes_{\mathbb{R}} \mathbb{C}.$$

By a), the character of $\mathsf{V}_0 \otimes_{\mathbb{R}} \mathbb{C}$ is $\chi + \bar{\chi}$, so either W_1 or W_2 has character χ.

c) This is VII, 1.12. **q.e.d.**

8.9 Theorem. *Let* W *be an irreducible* $\mathbb{R}\mathfrak{G}$*-module. Then one of the following occurs.*

(1) $\mathsf{W} \otimes_{\mathbb{R}} \mathbb{C}$ *is an irreducible* $\mathbb{C}\mathfrak{G}$*-module, and* $\mathrm{Hom}_{\mathbb{R}\mathfrak{G}}(\mathsf{W}, \mathsf{W}) = \mathbb{R}$.

(2) $\mathsf{W} \otimes_{\mathbb{R}} \mathbb{C} \cong \mathsf{V} \oplus \mathsf{V}^*$ *for some irreducible* $\mathbb{C}\mathfrak{G}$*-module* V *such that* $\mathsf{V} \not\cong \mathsf{V}^*$, *and* $\mathrm{Hom}_{\mathbb{R}\mathfrak{G}}(\mathsf{W}, \mathsf{W}) \cong \mathbb{C}$.

(3) $\mathsf{W} \otimes_{\mathbb{R}} \mathbb{C} \cong \mathsf{V} \oplus \mathsf{V}$ *for some irreducible* $\mathbb{C}\mathfrak{G}$*-module* V, *and* $\mathrm{Hom}_{\mathbb{R}\mathfrak{G}}(\mathsf{W}, \mathsf{W})$ *is the division ring of quaternions.*

Proof. (1) Suppose that $\mathsf{W} \otimes_{\mathbb{R}} \mathbb{C}$ is irreducible. Then by 8.8c),

$$\dim_{\mathbb{R}}\mathrm{Hom}_{\mathbb{R}\mathfrak{G}}(\mathsf{W}, \mathsf{W}) = \dim_{\mathbb{C}}\mathrm{Hom}_{\mathbb{C}\mathfrak{G}}(\mathsf{W} \otimes_{\mathbb{R}} \mathbb{C}, \mathsf{W} \otimes_{\mathbb{R}} \mathbb{C}) = 1,$$

so $\mathrm{Hom}_{\mathbb{R}\mathfrak{G}}(\mathsf{W}, \mathsf{W}) = \mathbb{R}$.

Now suppose that $\mathsf{W} \otimes_{\mathbb{R}} \mathbb{C}$ is reducible. Let χ be the character of an irreducible direct summand of $\mathsf{W} \otimes_{\mathbb{R}} \mathbb{C}$. By 8.8a), $\chi + \bar{\chi}$ is the character of an $\mathbb{R}\mathfrak{G}$-module V_0. Now if ψ is the character of W,

$$(\chi + \bar{\chi}, \psi)_{\mathfrak{G}} \geq (\chi, \psi)_{\mathfrak{G}} \geq 1.$$

Hence by V, 5.10a), ψ is the character of an irreducible direct summand of V_0 isomorphic to W. But $\mathsf{W} \otimes_{\mathbb{R}} \mathbb{C}$ is reducible and $\mathsf{V}_0 \otimes_{\mathbb{R}} \mathbb{C}$ is the direct sum of two irreducible $\mathbb{C}\mathfrak{G}$-modules. Hence $\psi = \chi + \bar{\chi}$, and $\mathsf{W} \otimes_{\mathbb{R}} \mathbb{C} \cong \mathsf{V} \oplus \mathsf{V}^*$ for some irreducible $\mathbb{C}\mathfrak{G}$-module V with character χ.

(2) If $\chi \neq \bar{\chi}$, then $\mathsf{V} \not\cong \mathsf{V}^*$, so

$$\begin{aligned}\dim_{\mathbb{R}}\mathrm{Hom}_{\mathbb{R}\mathfrak{G}}(\mathsf{W}, \mathsf{W}) &= \dim_{\mathbb{C}}\mathrm{Hom}_{\mathbb{C}\mathfrak{G}}(\mathsf{V} \oplus \mathsf{V}^*, \mathsf{V} \oplus \mathsf{V}^*)\\ &= \dim_{\mathbb{C}}(\mathrm{Hom}_{\mathbb{C}\mathfrak{G}}(\mathsf{V}, \mathsf{V}) \oplus \mathrm{Hom}_{\mathbb{C}\mathfrak{G}}(\mathsf{V}^*, \mathsf{V}^*))\\ &= 2.\end{aligned}$$

Since $\mathrm{Hom}_{\mathbb{R}\mathfrak{G}}(\mathsf{W}, \mathsf{W})$ is a division ring, it follows that $\mathrm{Hom}_{\mathbb{R}\mathfrak{G}}(\mathsf{W}, \mathsf{W}) \cong \mathbb{C}$.

(3) If $\chi = \bar{\chi}$, then $\mathsf{V}^* \cong \mathsf{V}$ and

$$\begin{aligned}\dim_{\mathbb{R}}\mathrm{Hom}_{\mathbb{R}\mathfrak{G}}(\mathsf{W}, \mathsf{W}) &= \dim_{\mathbb{C}}\mathrm{Hom}_{\mathbb{C}\mathfrak{G}}(\mathsf{V} \oplus \mathsf{V}, \mathsf{V} \oplus \mathsf{V})\\ &= 4\dim_{\mathbb{C}}\mathrm{Hom}_{\mathbb{C}\mathfrak{G}}(\mathsf{V}, \mathsf{V}) = 4\dim_{\mathbb{C}} \mathbb{C} = 4.\end{aligned}$$

As is well known, it follows from this that the division ring $\mathrm{Hom}_{\mathbb{R}\mathfrak{G}}(\mathsf{W}, \mathsf{W})$ is isomorphic to the ring of quaternions; (see, for example, VAN DER WAERDEN [3]). **q.e.d.**

8.10 Examples. a) Let $\mathfrak{G} = \langle a, b \rangle$ be a dihedral group of order $2n > 4$, where

$$a^n = b^2 = 1, \quad a^b = a^{-1}.$$

By V, 21.5, $\mathfrak{G}$ has the following irreducible representations ρ_j of degree 2 over $\mathbb{C}$:

$$\rho_j(a) = \begin{pmatrix} \omega^j & 0 \\ 0 & \omega^{-j} \end{pmatrix}, \quad \rho_j(b) = \begin{pmatrix} 0 & 1 \\ 1 & 0 \end{pmatrix},$$

where ω is a primitive n-th root of unity, and $1 \leq j \leq \frac{1}{2}(n-1)$ for odd n, but $1 \leq j \leq \frac{1}{2}(n-2)$ for even n. Let χ_j be the character of ρ_j. Since $(a^i b)^2 = 1$,

$$\begin{aligned} \sum_{g \in \mathfrak{G}} \chi_j(g^2) &= \sum_{i=0}^{n-1} (\omega^{2ij} + \omega^{-2ij}) + 2n \\ &= 2\frac{\omega^{2jn} - 1}{\omega^{2j} - 1} + 2n = 2n = |\mathfrak{G}|. \end{aligned}$$

Thus each χ_j is of the second kind and is therefore the character of an $\mathbb{R}\mathfrak{G}$-module. (The real form of ρ_j can easily be given in the form of orthogonal mappings of $\mathbb{R}^2$ which map a regular polygon with n sides into itself.) The remaining irreducible characters of $\mathfrak{G}$ are all of degree 1 and can be realized in $\mathbb{R}$, since $\mathfrak{G}/\mathfrak{G}'$ is elementary Abelian. For n odd, there are 2 characters of degree 1, for n even, 4.

By 8.7, the number of elements g of $\mathfrak{G}$ for which $g^2 = 1$ is

$$\begin{aligned} t(1) &= \sum_{i=1}^{h} \varepsilon(\chi_i)\chi_i(1) = \sum_{i=1}^{h} \chi_i(1) \\ &= \begin{cases} 2 + \frac{1}{2}(n-1)\cdot 2 = n + 1 & \text{for odd } n, \\ 4 + \frac{1}{2}(n-2)\cdot 2 = n + 2 & \text{for even } n. \end{cases} \end{aligned}$$

Of course, this can be easily confirmed by trivial calculations.

b) Let $\mathfrak{G}$ be an extraspecial 2-group of order 2^{2m+1}. By V, 16.14, $\mathfrak{G}$ has 2^{2m} irreducible characters of degree 1, and these are all of the second kind, since $\mathfrak{G}/\mathfrak{G}'$ is elementary Abelian. Also, $\mathfrak{G}$ has one other irreducible character χ, which is of degree 2^m. Hence by 8.7, the number of solutions of $g^2 = 1$ in $\mathfrak{G}$ is

$$t(1) = 2^{2m} + \varepsilon(\chi)2^m.$$

By III, 13.8, either

(1) $$\mathfrak{G} = \mathfrak{D}_1 \curlyvee \cdots \curlyvee \mathfrak{D}_m$$

or

(2) $$\mathfrak{G} = \mathfrak{D}_1 \curlyvee \cdots \curlyvee \mathfrak{D}_{m-1} \curlyvee \mathfrak{Q},$$

where the $\mathfrak{D}_i$ are dihedral groups of order 8 and $\mathfrak{Q}$ is a quaternion group of order 8.

In case (1), $\mathfrak{G}$ has a maximal Abelian normal subgroup $\mathfrak{A}$ of order 2^{m+1}, and $\mathfrak{A}$ is elementary Abelian. By V, 16.14, $\chi = \mu^{\mathfrak{G}}$ for some irreducible character μ of $\mathfrak{A}$. Since μ can be realized in $\mathbb{R}$, so can χ. Thus χ is of the second kind and $t(1) = 2^{2m} + 2^m$ in case (1). This is easily confirmed, using VIII, 5.2.

In case (2), write $\mathfrak{H} = \mathfrak{D}_1 \curlyvee \cdots \curlyvee \mathfrak{D}_{m-1}$; then the number of solutions of $h^2 = 1$ in $\mathfrak{H}$ is $2^{2(m-1)} + 2^{m-1}$ and $\mathfrak{G} = \mathfrak{H} \curlyvee \mathfrak{Q}$. By VIII, 5.2, the number of solutions of $g^2 = 1$ in $\mathfrak{G}$ is $2^{2m} - 2^m$. Hence $\varepsilon(\chi) = -1$ and χ is of the third kind.

Exercises

4) Investigate the characters of the generalized quaternion groups and the semidihedral groups as in 8.10a).

5) Describe the structure of $\mathbb{R}\mathfrak{G}$ when $\mathfrak{G}$ is a dihedral group, a semidihedral group or a generalized quaternion group.

6) Let V be an irreducible $\mathbb{C}\mathfrak{G}$-module of the second or the third kind. Let $\mathfrak{Z} = \langle z \rangle$ be a cyclic group of order 2 and let $\mathfrak{H} = \mathfrak{G} \wr \mathfrak{Z}$ be the wreath product of $\mathfrak{G}$ and $\mathfrak{Z}$.

a) $V \otimes_{\mathbb{C}} V$ becomes an irreducible $\mathbb{C}\mathfrak{H}$-module of the second kind if we put

$$(v_1 \otimes v_2)(g_1, g_2) = v_1 g_1 \otimes v_2 g_2,$$

$$(v_1 \otimes v_2)(g_1, g_2)z = v_2 g_2 \otimes v_1 g_1.$$

b) Let $\mathfrak{G}$ be $SL(2,5)$ and let V be an irreducible $\mathbb{C}\mathfrak{G}$-module of dimension 2 over $\mathbb{C}$. The procedure in a) yields a group of order $5^2 \cdot 3^2 \cdot 2^7$ of orthogonal mappings of a 4-dimensional real Euclidean space, which is generated by reflections. (This is the reflection group $\mathfrak{H}_4$; see HUPPERT [5]).

§ 9. Zassenhaus Groups of Even Degree

In 6.1, we have seen that the degree of a Zassenhaus group is always of the form $p^f + 1$ for some prime p. In this section we shall complete the case when p is odd. We prove the following.

9.1 Theorem (ITO [3]). *Let $\mathfrak{G}$ be a Zassenhaus group of degree $p^f + 1$, where p is an odd prime. Then the Frobenius kernel $\mathfrak{F}$ of $\mathfrak{G}_\infty$ is Abelian.*

Using 6.8, it then follows that $\mathfrak{G}$ is either sharply triply transitive or $\mathfrak{G} = PSL(2, p^f)$.

To prove 9.1, let $\mathfrak{G}$ be a counterexample of minimal order. We begin by collecting together some simple facts.

9.2 Lemma. a) $|\mathfrak{H}| = d$ *is odd.*

b) $\mathfrak{G}$ *is simple.*

c) $\mathfrak{H}$ *is cyclic. Also, if* $1 \neq h \in \mathfrak{H}$ *and s interchanges* 0 *and* ∞, $h^s = h^{-1}$.

d) *No element of* $\mathfrak{F} - \{1\}$ *is a product of two involutions in* $\mathfrak{G}$.

e) $\mathfrak{G}$ *has just one class of involutions, and this has nd elements.*

Proof. a) If d is even, then $\mathfrak{F}' = 1$ by 1.4a).

b) If $\mathfrak{N}$ is a minimal normal subgroup of $\mathfrak{G}$, then $\mathfrak{N}$ is also a Zassenhaus group, by 1.6a), and $\mathfrak{F} < \mathfrak{N}$. Since $\mathfrak{G}$ is a minimal counterexample to the theorem, $\mathfrak{N} = \mathfrak{G}$.

c) By 1.5, this follows from a) and b).

d) Suppose that t_1, t_2 are involutions in $\mathfrak{G}$ and $t_1 t_2 \in \mathfrak{F} - \{1\}$. Since

$$(t_1 t_2)^{t_1} = t_2 t_1 = (t_1 t_2)^{-1},$$

t_1 permutes the fixed points of $t_1 t_2$. But ∞ is the only fixed point of

$t_1 t_2$, so $t_1 \in \mathfrak{G}_\infty$. This is impossible, since $n = p^f$ is odd and, by a), d is odd.

e) If t_1, t_2 are involutions and $t_1\mathfrak{F} = t_2\mathfrak{F}$, then by d), $t_1 = t_2$. Hence each coset $x\mathfrak{F}$ contains at most one involution. If $x\mathfrak{F} \subseteq \mathfrak{H}\mathfrak{F}$, then $x\mathfrak{F}$ contains no involution, since $|\mathfrak{H}\mathfrak{F}| = dn$ is odd. Hence the number of involutions in $\mathfrak{G}$ is at most

$$|\mathfrak{G}:\mathfrak{F}| - |\mathfrak{H}\mathfrak{F}:\mathfrak{F}| = (n+1)d - d = nd.$$

Let t be an involution in $\mathfrak{G}$. Since $|\mathfrak{G}_\infty|$ is odd, t has no fixed points. We show that if $1 \neq x \in \mathbf{C}_\mathfrak{G}(t)$, then x also has no fixed points. Indeed, if $ax = a$, then $(at)x = axt = at$ and $at \neq a$. Hence $x \in \mathfrak{G}_{a,at}$. It follows from c) that $x^t = x^{-1}$. Thus $x = x^{-1}$. But this is impossible, since d is odd and $x \neq 1$. Hence $\mathbf{C}_\mathfrak{G}(t)$ permutes each of its orbits regularly and $|\mathbf{C}_\mathfrak{G}(t)|$ divides $n+1$. This shows that the number of conjugates of t is at least $|\mathfrak{G}|/(n+1) = nd$. Hence $\mathfrak{G}$ has exactly nd involutions, and they are all conjugate in $\mathfrak{G}$. **q.e.d.**

It follows that $\mathfrak{G}$ satisfies Hypothesis 5.1.

9.3 Lemma. *Let $\psi_0 = 1, \psi_1, \ldots, \psi_{d-1}$ be the irreducible characters of $\mathfrak{G}_\infty$ for which $\mathfrak{F}$ is contained in the kernel. Since $|\mathfrak{G}_\infty|$ is odd, $\psi_1, \ldots, \psi_{d-1}$ are not real (see V, 13.8), and we may therefore suppose that $\overline{\psi_i} = \psi_{i+\frac{1}{2}(d-1)}$ for $i = 1, \ldots, \frac{1}{2}(d-1)$. Further, let $\phi_0 = 1, \phi_1, \ldots, \phi_m$ be the irreducible characters of $\mathfrak{F}$. Since $\mathfrak{F} > \mathfrak{F}'$, we may also suppose that $\phi_1(1) = 1$. Suppose that the irreducible characters $\alpha = \pi - 1$ and $\beta_i = \psi_i^\mathfrak{G}$ ($i = 1, \ldots, \frac{1}{2}(d-1)$) of $\mathfrak{G}$ are defined as in 5.2 and 5.3. Suppose that $j > 0$ and $\phi_j(1) = 1$.*

a) $(\phi_j^\mathfrak{G}, \phi_j^\mathfrak{G})_\mathfrak{G} = d + 1$.
b) $(\phi_j^\mathfrak{G}, \beta_i)_\mathfrak{G} = 1$ *for* $i = 1, \ldots, \frac{1}{2}(d-1)$.
c) $\psi_0^\mathfrak{G} = 1_\mathfrak{G} + \alpha$.
d) $(\phi_j^\mathfrak{G}, \alpha)_\mathfrak{G} = 1$.
e) $(\phi_j^\mathfrak{G}, 1_\mathfrak{G})_\mathfrak{G} = 0$.
f) $\phi_0^\mathfrak{G} = 1_\mathfrak{G} + \alpha + 2\sum_{i=1}^{\frac{1}{2}(d-1)} \beta_i$.

Proof. a) Let $\lambda_j = \phi_j^{\mathfrak{G}_\infty}$. Thus by V, 16.13, λ_j is an irreducible character of $\mathfrak{G}_\infty$. Also $\phi_j^\mathfrak{G}$ is zero outside $\bigcup_{g \in \mathfrak{G}} \mathfrak{F}^g$. If T is a transversal of $\mathfrak{G}_\infty$ in $\mathfrak{G}$ for which $\mathsf{T} \cap \mathfrak{G}_\infty = 1$, then for $1 \neq x \in \mathfrak{F}$,

$$\phi_j^\mathfrak{G}(x) = \lambda_j^\mathfrak{G}(x) = \sum_{t \in \mathsf{T},\, txt^{-1} \in \mathfrak{F}} \lambda_j(x^{t^{-1}}) = \lambda_j(x),$$

since if $1 \neq t \in \mathsf{T}$, $x^{t^{-1}} \notin \mathfrak{F}$. This gives

$$\begin{aligned}(\phi_j^{\mathfrak{G}}, \phi_j^{\mathfrak{G}})_{\mathfrak{G}} &= \frac{1}{|\mathfrak{G}|}\Big(d^2(n+1)^2 + \frac{|\mathfrak{G}|}{nd} \sum_{1 \neq x \in \mathfrak{F}} |\lambda_j(x)|^2\Big) \\ &= \frac{1}{|\mathfrak{G}|}\Big(d^2(n+1)^2 - \frac{|\mathfrak{G}|}{nd} d^2 + |\mathfrak{G}|(\lambda_j, \lambda_j)_{\mathfrak{G}_\infty}\Big) \\ &= \frac{d(n+1)}{n} - \frac{d}{n} + 1 = d + 1.\end{aligned}$$

b) We have

$$(\phi_j^{\mathfrak{G}}, \beta_i)_{\mathfrak{G}} = (\phi_j, (\beta_i)_{\mathfrak{F}})_{\mathfrak{F}} = \frac{1}{|\mathfrak{F}|}\Big(\phi_j(1)\beta_i(1) + \sum_{1 \neq f \in \mathfrak{F}} \phi_j(f)\beta_i(f^{-1})\Big).$$

It follows from 5.3a) that

$$(\phi_j^{\mathfrak{G}}, \beta_i)_{\mathfrak{G}} = \frac{1}{n}\Big(n + 1 + \sum_{1 \neq f \in \mathfrak{F}} \phi_j(f)\Big) = 1 + (\phi_j, \phi_0)_{\mathfrak{F}} = 1.$$

c) Since $\psi_0^{\mathfrak{G}} = \pi$, the assertion follows from 5.2.

d) Since $\alpha(f) = 0$ for $f \in \mathfrak{F} - \{1\}$ (see 5.2),

$$(\phi_j^{\mathfrak{G}}, \alpha)_{\mathfrak{G}} = (\phi_j, \alpha_{\mathfrak{F}})_{\mathfrak{F}} = \frac{\phi_j(1)\alpha(1)}{|\mathfrak{F}|} = 1.$$

e) We have

$$(\phi_j^{\mathfrak{G}}, 1_{\mathfrak{G}})_{\mathfrak{G}} = (\phi_j, \phi_0)_{\mathfrak{F}} = 0.$$

f) $\phi_0^{\mathfrak{G}_\infty}$ is the character of the regular representation of $\mathfrak{G}_\infty/\mathfrak{F}$, so

$$\phi_0^{\mathfrak{G}} = (\phi_0^{\mathfrak{G}_\infty})^{\mathfrak{G}} = (\psi_0 + \psi_1 + \cdots + \psi_{d-1})^{\mathfrak{G}} = 1_{\mathfrak{G}} + \alpha + 2 \sum_{i=1}^{\frac{1}{2}(d-1)} \beta_i.$$

q.e.d.

Ito's proof of Theorem 9.1 made use of exceptional characters, but it also depended upon an assertion which has only been proved by modular character theory. We give here a simpler proof, due to Glauberman, which uses the considerations of § 8.

9.4 Proof of 9.1 (GLAUBERMAN [4]).
The proof consists of several steps.

a) There exists an irreducible character χ_1 of $\mathfrak{G}$, distinct from $1_{\mathfrak{G}}$, α and $\beta_1, \ldots, \beta_{\frac{1}{2}(d-1)}$, with the following properties.

(i) $(\phi_1^{\mathfrak{G}}, \chi_1)_{\mathfrak{G}} = 1$ and $\varepsilon(\chi_1) = 0$, that is, χ_1 is not real-valued.

(ii) If $\chi \neq \chi_1$ is any irreducible character of $\mathfrak{G}$ for which $(\phi_1^{\mathfrak{G}}, \chi)_{\mathfrak{G}} > 0$, then $(\phi_1^{\mathfrak{G}}, \chi)_{\mathfrak{G}} = 1$ and $\varepsilon(\chi) = 1$; in particular, χ is real-valued.

Let A be the set of elements of $\mathfrak{G}$ of odd order, and let B be the set of elements of $\mathfrak{G}$ of even order. If $b \in \mathsf{B}$, then $b \notin \bigcup_{g \in \mathfrak{G}} \mathfrak{F}^g$ since $\mathfrak{F}$ is of odd order, so $\phi_1^{\mathfrak{G}}(b) = 0$. If $g \in \mathfrak{G}$ and $g^2 \in \mathfrak{F} - \{1\}$, then it follows easily that $g \in \mathbf{C}_{\mathfrak{G}}(g^2) \leq \mathfrak{F}$, for the only fixed point of g^2 is ∞ and $\mathfrak{H}$ operates fixed point freely on $\mathfrak{F}$. Then g is of odd order, since $p > 2$. Hence

$$\sum_{g \in \mathfrak{G}} \phi_1^{\mathfrak{G}}(g^2) = \sum_{g^2 = 1 \neq g} \phi_1^{\mathfrak{G}}(g^2) + \sum_{g \in \mathsf{A}} \phi_1^{\mathfrak{G}}(g^2).$$

As $a \to a^2$ is a permutation of A, it follows with the help of 9.2e) that

(1)
$$\begin{aligned}\sum_{g \in \mathfrak{G}} \phi_1^{\mathfrak{G}}(g^2) &= nd\,\phi_1^{\mathfrak{G}}(1) + \sum_{a \in \mathsf{A}} \phi_1^{\mathfrak{G}}(a) \\ &= nd^2(n+1) + \sum_{g \in \mathfrak{G}} \phi_1^{\mathfrak{G}}(g) \\ &= |\mathfrak{G}|d + |\mathfrak{G}|(\phi_1^{\mathfrak{G}}, 1_{\mathfrak{G}})_{\mathfrak{G}} \\ &= |\mathfrak{G}|d \quad \text{(by 9.3e)).}\end{aligned}$$

Let $\phi_1^{\mathfrak{G}} = \sum_\chi a(\chi)\chi$ be the decomposition of $\phi_1^{\mathfrak{G}}$ into irreducible characters χ of $\mathfrak{G}$. Then it follows from (1) that

(2)
$$d = \frac{1}{|\mathfrak{G}|} \sum_{g \in \mathfrak{G}} \phi_1^{\mathfrak{G}}(g^2) = \frac{1}{|\mathfrak{G}|} \sum_\chi a(\chi) \sum_{g \in \mathfrak{G}} \chi(g^2) = \sum_\chi a(\chi)\varepsilon(\chi),$$

where $\varepsilon(\chi)$ has the same meaning as in 8.7. Further, we find, using 9.3, that

(3)
$$d + 1 = (\phi_1^{\mathfrak{G}}, \phi_1^{\mathfrak{G}})_{\mathfrak{G}} = \sum_\chi a(\chi)^2.$$

From (2) and (3),

$$1 = \sum_\chi a(\chi)(a(\chi) - \varepsilon(\chi)).$$

Since $|\varepsilon(\chi)| \leq 1$ (see 8.7), $a(\chi)(a(\chi) - \varepsilon(\chi))$ is a non-negative rational integer. Thus there is exactly one χ_1 for which $a(\chi_1)(a(\chi_1) - \varepsilon(\chi_1)) = 1$, and for $\chi \neq \chi_1$, $a(\chi)(a(\chi) - \varepsilon(\chi)) = 0$. It follows that $a(\chi_1) = 1$ and $\varepsilon(\chi_1) = 0$. By 8.7), χ_1 is not real-valued. If $\chi \neq \chi_1$ and $a(\chi) > 0$, then necessarily $a(\chi) = \varepsilon(\chi)$, which forces $a(\chi) = \varepsilon(\chi) = 1$, so χ is real-valued by 8.7). Since χ_1 is not real-valued, but on the other hand $1_{\mathfrak{G}}$, α and the β_i are real-valued (see 5.3 for the β_i), χ_1 is certainly different from $1_{\mathfrak{G}}$, α and the β_i.

b) Let Λ be the set of irreducible characters χ of $\mathfrak{G}$ such that χ is distinct from $\chi_1, \alpha, \beta_1, \ldots, \beta_{\frac{1}{2}(d-1)}$ and

$$(\phi_1^{\mathfrak{G}}, \chi)_{\mathfrak{G}} > 0.$$

Then

$$\phi_1^{\mathfrak{G}} = \chi_1 + \alpha + \sum_{i=1}^{\frac{1}{2}(d-1)} \beta_i + \sum_{\chi \in \Lambda} \chi \tag{4}$$

and $1_{\mathfrak{G}} \notin \Lambda$.

For by 9.3b), d), e) and 9.4a),

$$(\phi_1^{\mathfrak{G}}, \beta_i)_{\mathfrak{G}} = (\phi_1^{\mathfrak{G}}, \alpha)_{\mathfrak{G}} = (\phi_1^{\mathfrak{G}}, \chi_1)_{\mathfrak{G}} = 1 \text{ and } (\phi_1^{\mathfrak{G}}, 1_{\mathfrak{G}})_{\mathfrak{G}} = 0.$$

If $\chi \in \Lambda$, then by a), $(\phi_1^{\mathfrak{G}}, \chi)_{\mathfrak{G}} = 1$.

c) Let $\chi_2 = \overline{\chi}_1$ and $\phi_2 = \overline{\phi_1}$. Then $\chi_1 \neq \chi_2$, $\chi_2 \notin \Lambda$ and

$$\phi_2^{\mathfrak{G}} = \chi_2 + \alpha + \sum_{i=1}^{\frac{1}{2}(d-1)} \beta_i + \sum_{\chi \in \Lambda} \chi. \tag{5}$$

By a), $\chi = \overline{\chi}$ for $\chi \in \Lambda$ so by (4),

$$\begin{aligned} \phi_2^{\mathfrak{G}} = \overline{\phi_1}^{\mathfrak{G}} = \overline{\phi_1^{\mathfrak{G}}} &= \overline{\chi}_1 + \alpha + \sum_{i=1}^{\frac{1}{2}(d-1)} \beta_i + \sum_{\chi \in \Lambda} \chi \\ &= \chi_2 + \alpha + \sum_{i=1}^{\frac{1}{2}(d-1)} \beta_i + \sum_{\chi \in \Lambda} \chi. \end{aligned}$$

Since χ_1 is not real-valued, $\chi_1 \neq \chi_2$. But all characters in Λ are real-valued, so $\chi_2 \notin \Lambda$.

d) Let $\lambda_1, \ldots, \lambda_m$ be the irreducible characters of the Frobenius group $\mathfrak{G}_\infty$ for which $\mathfrak{F}$ is not contained in the kernel. By V, 16.13,

$\lambda_i = \phi_i^{\mathfrak{G}_\infty}$ for irreducible characters $\phi_i \neq \phi_0$ of $\mathfrak{F}$. (From each class of $\mathfrak{H}$-conjugate irreducible characters of $\mathfrak{F}$ one ϕ_i is chosen.) We choose the numbering so that $\phi_i(1) = 1$ for $i = 1, \ldots, t$ and $\phi_i(1) > 1$ for $i = t + 1, \ldots, m$. Since $|\mathfrak{G}_\infty|$ is odd, none of the characters λ_i is real, by V, 13.8. Thus we may suppose that $\lambda_{2i} = \overline{\lambda}_{2i-1}$. Since

$$\overline{\phi_{2i}^{\mathfrak{G}_\infty}} = \overline{\phi_{2i}}^{\mathfrak{G}_\infty} = \overline{\lambda}_{2i} = \lambda_{2i-1},$$

we may also suppose that $\overline{\phi}_{2i} = \phi_{2i-1}$.

e) For $3 \leq i \leq m$,

$$(\chi_1, \phi_i^{\mathfrak{G}})_{\mathfrak{G}} = (\chi_2, \phi_i^{\mathfrak{G}})_{\mathfrak{G}} \tag{6}$$

and for $2 \leq i \leq \dfrac{m}{2}$

$$(\chi_1, \phi_{2i-1}^{\mathfrak{G}})_{\mathfrak{G}} = (\chi_1, \phi_{2i}^{\mathfrak{G}})_{\mathfrak{G}}. \tag{7}$$

Clearly, $(\phi_1 - \phi_2)^{\mathfrak{G}_\infty}(x) = 0$ for $x = 1$ and for $x \in \mathfrak{G}_\infty - \mathfrak{F}$. Also, if $f \in \mathfrak{F} - \{1\}$,

$$\phi_i^{\mathfrak{G}}(f) = \frac{1}{|\mathfrak{F}|} \sum_{\substack{f^y \in \mathfrak{F} \\ y \in \mathfrak{G}}} \phi_i(f^y) = \frac{1}{|\mathfrak{F}|} \sum_{y \in \mathfrak{G}_\infty} \phi_i(f^y) = \phi_i^{\mathfrak{G}_\infty}(f).$$

It follows now that

$$\begin{aligned}(\chi_1 - \chi_2, \phi_i^{\mathfrak{G}})_{\mathfrak{G}} &= (\phi_1^{\mathfrak{G}} - \phi_2^{\mathfrak{G}}, \phi_i^{\mathfrak{G}})_{\mathfrak{G}} \quad \text{(see (4) and (5))} \\ &= (\phi_1^{\mathfrak{G}_\infty} - \phi_2^{\mathfrak{G}_\infty}, (\phi_i^{\mathfrak{G}})_{\mathfrak{G}_\infty})_{\mathfrak{G}_\infty} \\ &= (\phi_1^{\mathfrak{G}_\infty} - \phi_2^{\mathfrak{G}_\infty}, \phi_i^{\mathfrak{G}_\infty})_{\mathfrak{G}_\infty} = 0\end{aligned}$$

for $i \geq 3$, for then, $\phi_1^{\mathfrak{G}_\infty}, \phi_2^{\mathfrak{G}_\infty}, \phi_i^{\mathfrak{G}_\infty}$ are distinct irreducible characters of $\mathfrak{G}_\infty$. This shows that

$$(\chi_1, \phi_i^{\mathfrak{G}})_{\mathfrak{G}} = (\chi_2, \phi_i^{\mathfrak{G}})_{\mathfrak{G}}$$

for $3 \leq i \leq m$. Finally it follows that

$$(\chi_2, \phi_i^{\mathfrak{G}})_{\mathfrak{G}} = \overline{(\chi_2, \phi_i^{\mathfrak{G}})_{\mathfrak{G}}} = (\overline{\chi_2}, \overline{\phi_i^{\mathfrak{G}}})_{\mathfrak{G}} = (\chi_1, \overline{\phi_i}^{\mathfrak{G}})_{\mathfrak{G}}.$$

By d), $\overline{\phi}_{2i-1} = \phi_{2i}$, so this gives

$$(\chi_1, \phi_{2i-1}^{\mathfrak{G}})_{\mathfrak{G}} = (\chi_1, \phi_{2i}^{\mathfrak{G}})_{\mathfrak{G}}$$

for $2 \leq i \leq \dfrac{m}{2}$.

f) $(\chi_1, \phi_i^{\mathfrak{G}})_{\mathfrak{G}} = 0$ and $((\chi_1)_{\mathfrak{G}_\infty}, \lambda_i)_{\mathfrak{G}_\infty} = 0$ for $3 \leq i \leq t$.

For $3 \leq i \leq t$, $\phi_i(1) = 1$. Thus, as in a) and b),

$$\phi_i^{\mathfrak{G}} = \chi_i + \sum_j \tau_j,$$

where χ_i is an irreducible non-real character of $\mathfrak{G}$, but the τ_j are irreducible, real-valued characters of $\mathfrak{G}$. By e),

$$(\chi_1, \phi_i^{\mathfrak{G}})_{\mathfrak{G}} = (\chi_2, \phi_i^{\mathfrak{G}})_{\mathfrak{G}}.$$

Since χ_1, χ_2 are distinct and not real-valued, $\chi_1 \neq \chi_i \neq \chi_2$. Thus $((\chi_1)_{\mathfrak{G}_\infty}, \lambda_i)_{\mathfrak{G}_\infty} = (\chi_1, \phi_i^{\mathfrak{G}})_{\mathfrak{G}} = 0$.

g) $(\chi_1)_{\mathfrak{G}_\infty} = \lambda_1 + \sum_{\frac{1}{2}t < i \leq \frac{1}{2}m} a_{2i}(\lambda_{2i-1} + \lambda_{2i})$.

Since $\chi_1 \neq \beta_i$ $(i = 1, \ldots, \frac{1}{2}(d-1))$,

$$((\chi_1)_{\mathfrak{G}_\infty}, \psi_i)_{\mathfrak{G}_\infty} = (\chi_1, \psi_i^{\mathfrak{G}})_{\mathfrak{G}} = (\chi_1, \beta_i)_{\mathfrak{G}} = 0 \quad \text{and}$$

$$((\chi_1)_{\mathfrak{G}_\infty}, \overline{\psi_i})_{\mathfrak{G}_\infty} = (\chi_1, \overline{\psi_i}^{\mathfrak{G}})_{\mathfrak{G}} = (\chi_1, \beta_i)_{\mathfrak{G}} = 0.$$

Since $1_{\mathfrak{G}} \neq \chi_1 \neq \alpha$, it follows further that

$$((\chi_1)_{\mathfrak{G}_\infty}, \psi_0)_{\mathfrak{G}_\infty} = (\chi_1, \psi_0^{\mathfrak{G}})_{\mathfrak{G}} = (\chi_1, 1_{\mathfrak{G}} + \alpha)_{\mathfrak{G}} = 0.$$

By (4),

$$((\chi_1)_{\mathfrak{G}_\infty}, \lambda_1)_{\mathfrak{G}_\infty} = (\chi_1, \lambda_1^{\mathfrak{G}})_{\mathfrak{G}} = (\chi_1, \phi_1^{\mathfrak{G}})_{\mathfrak{G}} = 1,$$

and by (5),

$$((\chi_1)_{\mathfrak{G}_\infty}, \lambda_2)_{\mathfrak{G}_\infty} = (\chi_1, \lambda_2^{\mathfrak{G}})_{\mathfrak{G}} = (\chi_1, \phi_2^{\mathfrak{G}})_{\mathfrak{G}} = 0.$$

Since $\psi_0, \psi_1, \ldots, \psi_{d-1}, \lambda_1, \ldots, \lambda_m$ are all the irreducible characters of $\mathfrak{G}_\infty$, and $\lambda_i = \phi_i^{\mathfrak{G}_\infty}$, it follows using f) that

$$(\chi_1)_{\mathfrak{G}_\infty} = \lambda_1 + \sum_{i=t+1}^{m} a_i \lambda_i$$

for suitable a_i. But by (7),

$$((\chi_1)_{\mathfrak{G}_\infty}, \lambda_{2i-1})_{\mathfrak{G}_\infty} = (\chi_1, \phi_{2i-1}^{\mathfrak{G}})_{\mathfrak{G}} = (\chi_1, \phi_{2i}^{\mathfrak{G}})_{\mathfrak{G}} = ((\chi_1)_{\mathfrak{G}_\infty}, \lambda_{2i})_{\mathfrak{G}_\infty},$$

so $a_{2i-1} = a_{2i}$.

h) If s is an involution in $\mathfrak{G}$, then $\chi_1(s)^2 \leq \chi_1(1)$.

Put $\mu = \phi_1 - \phi_0$, so μ is a generalized character of $\mathfrak{F}$. Then by (4) and 9.3f),

$$\mu^{\mathfrak{G}} = \phi_1^{\mathfrak{G}} - \phi_0^{\mathfrak{G}} = \chi_1 - 1_{\mathfrak{G}} - \sum_{i=1}^{\frac{1}{2}(d-1)} \beta_i + \sum_{\chi \in \Lambda} \chi. \tag{8}$$

By 9.2d), no element of $\mathfrak{F} - \{1\}$ is a product of two involutions. If k denotes the sum of the involutions of $\mathfrak{G}$ (formed in the group-ring $\mathbb{C}\mathfrak{G}$), then in the formula

$$k^2 = \sum_i b_i k_i$$

(see V, 22.5), no class-sum k_i containing elements of $\mathfrak{F} - \{1\}$ occurs. Hence by V, 22.5, for $f \in \mathfrak{F} - \{1\}$,

$$\sum_\chi \frac{\chi(k)^2 \chi(f^{-1})}{\chi(1)} = 0,$$

where the summation is over all irreducible characters χ of $\mathfrak{G}$. Since, by 9.2e), all nd involutions of $\mathfrak{G}$ are conjugate to s, $\chi(k) = nd\chi(s)$. This gives

$$\sum_\chi \frac{\chi(s)^2 \chi(f^{-1})}{\chi(1)} = 0 \quad (f \in \mathfrak{F} - \{1\}).$$

Since $\mu(1) = 0$, it follows that

$$0 = \sum_\chi \frac{\chi(s)^2}{\chi(1)} \frac{1}{|\mathfrak{F}|} \sum_{f \in \mathfrak{F}} \mu(f)\chi(f^{-1}) = \sum_\chi \frac{\chi(s)^2}{\chi(1)} (\mu, \chi_{\mathfrak{F}})_{\mathfrak{F}} = \sum_\chi \frac{\chi(s)^2}{\chi(1)} (\mu^{\mathfrak{G}}, \chi)_{\mathfrak{G}}.$$

Using (8), this gives

$$0 = \frac{\chi_1(s)^2}{\chi_1(1)} - 1 - \sum_{i=1}^{\frac{1}{2}(d-1)} \frac{\beta_i(s)^2}{\beta_i(1)} + \sum_{\chi \in \Lambda} \frac{\chi(s)^2}{\chi(1)}.$$

Since $s \notin \bigcup_{g \in \mathfrak{G}} \mathfrak{G}_\infty^g$, $\beta_i(s) = 0$ by 5.3. Thus

$$1 = \frac{\chi_1(s)^2}{\chi_1(1)} + \sum_{\chi \in \Lambda} \frac{\chi(s)^2}{\chi(1)}.$$

But $\chi(s)$ is real, since it is a sum of square roots of 1. Hence $\chi_1(s)^2 \leq \chi_1(1)$.

i) $\chi_1(1) \leq d^2$.

For $t < i \leq m$, $z_i = \phi_i(1) > 1$. Naturally, z_i is a power of p. Since $\lambda_1(1) = d\phi_1(1) = d$ and $\lambda_i(1) = d\phi_i(1) = dz_i$, we see from g) that

$$\chi_1(1) \equiv 0\,(d),$$

$$\chi_1(1) \equiv d \not\equiv 0\,(p),$$

$$\chi_1(1) \equiv d \not\equiv 0\,(2).$$

Let s be an involution in $\mathfrak{G}$. Suppose that in the representation with character χ_1, s is represented by a matrix having a eigenvalues equal to 1 and b eigenvalues equal to -1. Then

$$\chi_1(s) = a - b \equiv a + b = \chi_1(1) \not\equiv 0\,(2).$$

In particular, $\chi_1(s) \neq 0$. Since s has exactly nd conjugates,

$$nd\frac{\chi_1(s)}{\chi_1(1)}$$

is an algebraic integer, by V, 5.11. Putting $\chi_1(1) = dx$, it follows that x divides $n\chi_1(s)$. On the other hand, $\chi_1(1)$ divides $|\mathfrak{G}|$, so dx divides $(n + 1)nd$. Since n is a power of p and $\chi_1(1) \not\equiv 0\,(p)$, it follows that x divides $n + 1$. This shows that x is a divisor of

$$(n + 1)\chi_1(s) - n\chi_1(s) = \chi_1(s).$$

By h), $\chi_1(s)^2 \leq \chi_1(1)$. Since $\chi_1(s) \neq 0$, this forces

$$x^2 \leq \chi_1(s)^2 \leq \chi_1(1) = xd,$$

so $x \leq d$ and $\chi_1(1) \leq d^2$.

j) $\chi_1(1)^2 \geq 1 + \frac{1}{2}(d - 1)(n + 1)$.

We consider the character $\chi_1\bar{\chi}_1 = \chi_1\chi_2$ of $\mathfrak{G}$. We have

$$(\chi_1\bar{\chi}_1, 1_{\mathfrak{G}})_{\mathfrak{G}} = (\chi_1, \chi_1)_{\mathfrak{G}} = 1.$$

We show that $(\chi_1\bar{\chi}_1, \beta_i)_{\mathfrak{G}} > 0$ for $i = 1, \ldots, \frac{1}{2}(d - 1)$. The assertion then follows, for

$$\chi_1(1)^2 = \chi_1(1)\bar{\chi}_1(1) \geq 1 + \tfrac{1}{2}(d - 1)\beta_1(1) = 1 + \tfrac{1}{2}(d - 1)(n + 1).$$

By (4) and (5), $\chi_1 - \chi_2 = (\phi_1 - \phi_2)^{\mathfrak{G}}$, so certainly $(\chi_1 - \chi_2)(x) = 0$ for $x \notin \bigcup_{g \in \mathfrak{G}} (\mathfrak{F} - \{1\})^g$. But by 5.3a), $(\beta_i - 1)(x) = 0$ for $x \in \bigcup_{g \in \mathfrak{G}} (\mathfrak{F} - \{1\})^g$, so $(\chi_1 - \chi_2)(\beta_i - 1) = 0$. Thus

$$\chi_1\beta_i + \chi_2 = \chi_2\beta_i + \chi_1.$$

Since $\chi_1 \neq \chi_2$, we have

$$0 \neq (\chi_1, \chi_1\beta_i)_{\mathfrak{G}} = (\chi_1\bar{\chi}_1, \beta_i)_{\mathfrak{G}}.$$

k) If $n = p^f$, f is the smallest positive integer for which $p^f - 1$ is divisible by d.

Certainly, d divides $p^f - 1$. Let e be the smallest positive integer such that $p^e - 1$ is divisible by d. Then $f = ke$ for some integer k. Since p and d are both odd, $2d$ divides $p^e - 1$. It follows from i) and j) that

$$d^4 \geq \chi_1(1)^2 \geq 1 + \tfrac{1}{2}(d - 1)(n + 1),$$

and hence, since $d > 2$,

$$d^4 > \tfrac{1}{4}d(n + 1) > \tfrac{1}{4}dn.$$

It follows that

$$4d^3 > n = p^f = p^{ke} \geq (1 + 2d)^k.$$

This forces $k \leq 2$. And if $k = 2$, then $n = p^f = p^{2e} \equiv 1$ (4), so $n + 1 \equiv 2$ (4). Since $|\mathfrak{G}_\infty| = nd$ is odd, the involutions in $\mathfrak{G}$ have no fixed points and are therefore products of $\frac{1}{2}(n + 1)$ transpositions. Hence the involutions are odd permutations, since $\frac{1}{2}(n + 1) \equiv 1$ (2). By 9.2b), $\mathfrak{G}$ is simple, so this is a contradiction. Thus $k = 1$ and $e = f$.

l) Since $\mathfrak{H}$ operates fixed point freely on $\mathfrak{F}/\mathfrak{F}'$, d divides $|\mathfrak{F}/\mathfrak{F}'| - 1$. By k), $|\mathfrak{F}/\mathfrak{F}'| = |\mathfrak{F}|$, so $\mathfrak{F}' = 1$. **q.e.d.**

§ 10. Zassenhaus Groups of Odd Degree and a Characterization of $PGL(2, 2^f)$

In this section we begin the determination of Zassenhaus groups of odd degree $n + 1$. Among other things we shall be concerned with elements having the following property.

10.1 Definition. An element g of a group is called *strongly real* if it is the product of two involutions j_1 and j_2. Then $g^{j_1} = j_2 j_1 = g^{-1}$, so any strongly real element is real (see V, 13.6).

For the remainder of this section, we shall suppose that $\mathfrak{G}$ is a Zassenhaus group of order $(n + 1)nd$, where n is even. By 6.1, $n = 2^f$ for some f. Every involution in $\mathfrak{G}$ has exactly one fixed point, and we denote by s a fixed involution of the form $(0, \infty)(1) \ldots$.

10.2 Lemma. *Let $\mathfrak{G}$ be a Zassenhaus group of degree $2^f + 1$.*

a) *$\mathfrak{G}$ is simple.*

b) *$\mathfrak{H}$ is cyclic of odd order, and $h^s = h^{-1}$ for all $h \in \mathfrak{H}$. We have $\mathbf{N}_{\mathfrak{G}}(\mathfrak{H}) = \langle \mathfrak{H}, s \rangle$ and $\mathbf{C}_{\mathfrak{G}}(\mathfrak{H}) = \mathfrak{H}$.*

c) *If $1 \neq x \in \mathfrak{F}$, then $\mathbf{C}_{\mathfrak{G}}(x) \leq \mathfrak{F}$. Thus the centralizer of a 2-element is always a 2-group, and the centralizer of an element of odd order is always of odd order.*

d) *$\mathfrak{G}$ has exactly one conjugacy class of involutions, and this has $(n + 1)d$ elements.*

e) *$\mathfrak{F}$ has exactly d involutions. These are permuted regularly by $\mathfrak{H}$ and all lie in $\mathbf{Z}(\mathfrak{F})$. In particular, $d = 2^l - 1$ for some l.*

Proof. a) Since $n = 2^f$ is even, $\mathfrak{G}$ is simple by 1.6b).

b) Since d divides $n - 1$, d is odd. Thus the assertion follows from a) and 1.5.

c) For $1 \neq x \in \mathfrak{F}$, ∞ is the only fixed point of x. Hence $\mathbf{C}_{\mathfrak{G}}(x) \leq \mathfrak{G}_\infty$. But $\mathfrak{F}$ is the Frobenius kernel of $\mathfrak{G}_\infty$, so $\mathbf{C}_{\mathfrak{G}}(x) \leq \mathfrak{F}$.

d) If t, t' are involutions with $t \in \mathfrak{F}$, $t' \in \mathfrak{F}^g \neq \mathfrak{F}$, then by c),

$$\mathbf{C}_{\mathfrak{G}}(t) \cap \mathbf{C}_{\mathfrak{G}}(t') \leq \mathfrak{F} \cap \mathfrak{F}^g = 1.$$

By I, 19.5, $\mathfrak{D} = \langle t, t' \rangle$ is a dihedral group, and since

$$\mathbf{Z}(\mathfrak{D}) \leq \mathbf{C}_{\mathfrak{G}}(t) \cap \mathbf{C}_{\mathfrak{G}}(t') = 1,$$

the order of tt' is odd. Thus $\langle t \rangle$ and $\langle t' \rangle$ are conjugate, being Sylow 2-subgroups of $\mathfrak{D}$.

If t is an involution in $\mathbf{Z}(\mathfrak{F})$, $\mathbf{C}_{\mathfrak{G}}(t) = \mathfrak{F}$. The number of involutions in $\mathfrak{G}$ is thus $|\mathfrak{G} : \mathbf{C}_{\mathfrak{G}}(t)| = (n + 1)d$.

e) Since $\mathfrak{G}$ has $n + 1$ Sylow 2-subgroups $\mathfrak{F}^g$ and any two of these intersect in $\{1\}$, it follows from d) that $\mathfrak{F}$ contains d involutions. These are all conjugate under $\mathfrak{H}$, since if $1 \neq t \in \mathbf{Z}(\mathfrak{F})$, $\mathbf{C}_{\mathfrak{G}}(t) = \mathfrak{F}$. Thus all involutions in $\mathfrak{F}$ lie in $\mathbf{Z}(\mathfrak{F})$. Hence

$$d + 1 = |\Omega_1(\mathbf{Z}(\mathfrak{F}))|$$

and $d + 1 = 2^l$ for some l. **q.e.d.**

In the following two lemmas, we investigate the strongly real elements of $\mathfrak{G}$ and their centralizers.

10.3 Lemma. a) *Suppose that g is a strongly real element of $\mathfrak{G}$ and $g^2 \neq 1$. Then g is of odd order and is conjugate to an element js, where j is an involution in $\mathfrak{F}$.*

b) *Suppose that j_1, j_2 are involutions in $\mathfrak{F}$. Then $j_1 s, j_2 s$ are conjugate in $\mathfrak{G}$ if and only if $j_1 = j_2$.*

Proof. a) Suppose that $1 \neq g = j_1 j_2$, where j_1, j_2 are involutions. Since $n + 1 \equiv 1\ (2)$, each involution has just one fixed point.

If j_1, j_2 have the same fixed point, which may be supposed to be ∞, then j_1, j_2 both lie in $\mathfrak{G}_\infty$ and hence in $\mathfrak{F}$. Thus by 10.2e), j_1, j_2 lie in $\mathbf{Z}(\mathfrak{F})$, so $g = j_1 j_2$ is an involution.

Hence j_1, j_2 have distinct fixed points. On account of the double transitivity of $\mathfrak{G}$, we may suppose that the fixed point of j_1 is ∞ and that of j_2 is 1. As $\mathfrak{G}_{1,\infty}$ permutes the involutions of $\mathfrak{G}_1$ transitively and j_2, s both lie in $\mathfrak{G}_1$, there exists $h \in \mathfrak{G}_{1,\infty}$ such that $j_2^h = s$. Then $j_1^h \in \mathfrak{G}_\infty^h = \mathfrak{G}_\infty$, so $j = j_1^h \in \mathfrak{F}$ and $g^h = (j_1 j_2)^h = j_1^h s = js$.

Suppose that js has even order $2m$. Then $\langle j, s\rangle$ is a dihedral group of order $4m$. Hence $j' = (js)^m$ is an involution in the centre of $\langle j, s\rangle$. It follows that

$$j' \in \mathbf{C}_{\mathfrak{G}}(j) \cap \mathbf{C}_{\mathfrak{G}}(s) \leq \mathfrak{G}_\infty \cap \mathfrak{G}_1 = \mathfrak{G}_{1,\infty}.$$

This is a contradiction, since $|\mathfrak{G}_{1,\infty}| = d$ is odd.

b) Suppose that j_1, j_2 are involutions in $\mathfrak{F}$ and $(j_1 s)^g = j_2 s$. As $1 \neq s j_1 s \in \mathfrak{G}_0$, $(s j_1)^2 \neq 1$. Thus by a), $j_1 s$ has odd order. Since

$$(j_1 s)^s = s j_1 = (j_1 s)^{-1},$$

$s \in \mathbf{C}^*_{\mathfrak{G}}(j_1 s) = \{x | x \in \mathfrak{G}, (j_1 s)^x = (j_1 s)^{\pm 1}\}$. Using 10.2c), it follows that

$\langle s \rangle$ is a Sylow 2-subgroup of $\mathbf{C}^*_{\mathfrak{G}}(j_1 s)$ and similarly of $\mathbf{C}^*_{\mathfrak{G}}(j_2 s)$. Since $(j_1 s)^g = j_2 s$, $\langle s^g \rangle$ is a Sylow 2-subgroup of $\mathbf{C}^*_{\mathfrak{G}}(j_2 s)$. From $\mathbf{C}^*_{\mathfrak{G}}(j_2 s) = \langle s \rangle \mathbf{C}_{\mathfrak{G}}(j_2 s)$, it follows that there exists $x \in \mathbf{C}_{\mathfrak{G}}(j_2 s)$ such that $s^{gx} = s$. Then

$$j_2 s = (j_2 s)^x = (j_1 s)^{gx} = j_1^{gx} s^{gx} = j_1^{gx} s.$$

Hence $j_1^{gx} = j_2$. As ∞ is the only fixed point of j_1 and j_2, $gx \in \mathfrak{G}_\infty \cap \mathbf{C}_{\mathfrak{G}}(s)$. By 10.2c), $\mathbf{C}_{\mathfrak{G}}(s) = \mathfrak{F}^h$, where $\infty h = 1$. Thus

$$gx \in \mathfrak{G}_\infty \cap \mathfrak{F}^h = 1,$$

and hence $j_1 = j_2$. **q.e.d.**

10.4 Lemma. *Suppose that g is a strongly real element and $g^2 \neq 1$.*

a) $\mathfrak{A} = \mathbf{C}_{\mathfrak{G}}(g)$ *is Abelian of odd order. If* $1 \neq a \in \mathfrak{A}$, *then a is strongly real and* $\mathbf{C}_{\mathfrak{G}}(a) = \mathfrak{A}$.

b) $\mathfrak{A}$ *is a Hall subgroup of* $\mathfrak{G}$ *and* $\mathbf{N}_{\mathfrak{G}}(\mathfrak{A})/\mathfrak{A}$ *is a 2-group with just one involution.*

c) *For each* $g \in \mathfrak{G} - \mathbf{N}_{\mathfrak{G}}(\mathfrak{A})$, $\mathfrak{A} \cap \mathfrak{A}^g = 1$.

Proof. a) Suppose that $g = jj'$ for involutions j, j'. Then $g^j = g^{-1}$ and so $j \in \mathbf{N}_{\mathfrak{G}}(\mathbf{C}_{\mathfrak{G}}(g)) = \mathbf{N}_{\mathfrak{G}}(\mathfrak{A})$. We show that j induces a fixed point free automorphism on $\mathfrak{A}$. By 10.2c), $\mathbf{C}_{\mathfrak{G}}(j)$ is a 2-group. By 10.3a), $g = jj'$ has odd order. Hence by 10.2c), $|\mathbf{C}_{\mathfrak{G}}(g)|$ is odd, so no non-identity element of $\mathfrak{A} = \mathbf{C}_{\mathfrak{G}}(g)$ commutes with j. Hence j induces a fixed point free automorphism of order 2 on $\mathfrak{A}$, and $\mathfrak{A}$ is Abelian.

Suppose that $1 \neq a \in \mathfrak{A}$. Then $a^j = a^{-1}$. Hence since $j \notin \mathfrak{A}$, ja is an involution and $a = j(ja)$ is strongly real. Since $a^2 \neq 1$, the above argument now shows that also $\mathbf{C}_{\mathfrak{G}}(a)$ is an Abelian group. From $g \in \mathfrak{A} \leq \mathbf{C}_{\mathfrak{G}}(a)$, it then follows that $\mathbf{C}_{\mathfrak{G}}(a) \leq \mathbf{C}_{\mathfrak{G}}(g) = \mathfrak{A}$. Thus $\mathfrak{A} = \mathbf{C}_{\mathfrak{G}}(a)$ whenever $1 \neq a \in \mathfrak{A}$.

b) Suppose that p is a common prime divisor of $|\mathfrak{A}|$ and $|\mathfrak{G} : \mathfrak{A}|$. Then there is a Sylow p-subgroup $\mathfrak{P} \neq 1$ of $\mathfrak{A}$ and a Sylow p-subgroup $\mathfrak{Q}$ of $\mathfrak{G}$ such that $\mathfrak{P} < \mathfrak{Q}$. Then $\mathfrak{P} < \mathbf{N}_{\mathfrak{Q}}(\mathfrak{P})$. Suppose that $1 \neq a \in \mathbf{Z}(\mathbf{N}_{\mathfrak{Q}}(\mathfrak{P})) \cap \mathfrak{P}$ and $b \in \mathbf{N}_{\mathfrak{Q}}(\mathfrak{P}) - \mathfrak{P}$. From a), we then obtain $b \in \mathbf{C}_{\mathfrak{G}}(a) = \mathfrak{A}$. Hence $b \in \mathfrak{A} \cap \mathbf{N}_{\mathfrak{Q}}(\mathfrak{P}) = \mathfrak{P}$, a contradiction. Thus $\mathfrak{A}$ is a Hall subgroup of $\mathfrak{G}$.

Since $(|\mathfrak{A}|, |\mathbf{N}_{\mathfrak{G}}(\mathfrak{A})/\mathfrak{A}|) = 1$, there is a complement $\mathfrak{B}$ of $\mathfrak{A}$ in $\mathbf{N}_{\mathfrak{G}}(\mathfrak{A})$. As $\mathfrak{A} = \mathbf{C}_{\mathfrak{G}}(a)$ for all non-identity elements a of $\mathfrak{A}$, $\mathfrak{B}$ induces a group of fixed point free automorphisms on $\mathfrak{A}$. Since $j \in \mathbf{N}_{\mathfrak{G}}(\mathfrak{A})$ (where j is the involution used in a)) and since $|\mathfrak{A}|$ is odd, $|\mathfrak{B}|$ is even. Thus by

V, 8.18a), $\mathfrak{B}$ has just one involution and this is in $\mathbf{Z}(\mathfrak{B})$. By 10.2c), it then follows that $\mathfrak{B}$ is a 2-group.

c) Suppose that $1 \neq a \in \mathfrak{A} \cap \mathfrak{A}^g$. By a), we then obtain $\mathfrak{A} = \mathbf{C}_{\mathfrak{G}}(a) = \mathfrak{A}^g$. **q.e.d.**

10.5 Lemma. *Suppose that $g \in \mathfrak{G}_\infty$ and $x \in \mathfrak{F}$. Then gsx is an involution if and only if $xg \in \mathfrak{H}$.*

Proof. Suppose first that gsx is an involution. Then $gsxgsx = 1$, and

$$xg = sg^{-1}x^{-1}s \in \mathfrak{G}_\infty^s \cap \mathfrak{G}_\infty = \mathfrak{G}_{0,\infty} = \mathfrak{H}.$$

Conversely, suppose that $xg \in \mathfrak{H}$. Since $|\mathfrak{H}|$ is odd, there exists $h \in \mathfrak{H}$ such that $xg = h^{-2}$. Since $h^s = h^{-1}$, it follows that

$$gsx = x^{-1}h^{-2}sx = x^{-1}h^{-1}shx = (hx)^{-1}s(hx)$$

and hence $(gsx)^2 = 1$. **q.e.d.**

10.6 Lemma. a) *$\mathfrak{F}$ contains an involution j and an element $g \neq 1$ such that*

$$sjs = gsg^{-1}.$$

(This is the so-called structure equation for $\mathfrak{G}$.)

b) *With s fixed, j and g are uniquely determined by the equation $sjs = gsg^{-1}$.*

c) *$gjsg^{-1} = (js)^2$, so the order of js is odd.*

d) *j is a power of g.*

Proof. a) Since $\mathfrak{F}$ is regular on $\Omega - \{\infty\}$, there exists $g \in \mathfrak{F}$ such that $1g = 0$. Thus $g \neq 1$ and, since $s = (0, \infty)\ (1)\ \ldots$, $1gs = \infty$. Hence $j = s^{gs}$ is an involution with fixed point ∞. Thus $j \in \mathfrak{F}$ and $sjs = g^{-1}sg$.

b) Suppose that j' is an involution in $\mathfrak{F}$, $g' \in \mathfrak{F}$ and $sj's = g'sg'^{-1}$. By 10.2e), there exists $h \in \mathfrak{H}$ such that $j' = j^h$. It follows that

$$g'sg'^{-1} = sj's = sh^{-1}jhs = hsjsh^{-1} = hgsg^{-1}h^{-1} = (hgh^{-1})h^2s(hg^{-1}h^{-1}).$$

On account of the uniqueness of the decomposition in 1.7a), it follows that $hgh^{-1} = g'$ and $h^2 = 1$. Since $|\mathfrak{H}|$ is odd, we obtain $h = 1$, $g = g'$ and $j = j'$.

c) Since $j \in \mathbf{Z}(\mathfrak{F})$ and $g \in \mathfrak{F}$, it follows from $sjs = gsg^{-1}$ that

$$(js)^2 = jgsg^{-1} = gjsg^{-1}.$$

Thus js has odd order.

d) Put $\mathfrak{U} = \langle js \rangle$. By 10.2c), $|\mathbf{C}_{\mathfrak{G}}(\mathfrak{U})|$ is odd. Thus $\mathbf{C}_{\mathfrak{G}}(\mathfrak{U}) \cap \mathfrak{F} = 1$. Hence $\mathbf{N}_{\mathfrak{G}}(\mathfrak{U}) \cap \mathfrak{F}$ is isomorphic to a subgroup of the automorphism group of $\mathfrak{U}$. Since $\mathfrak{U}$ is cyclic, it follows that $\mathbf{N}_{\mathfrak{G}}(\mathfrak{U}) \cap \mathfrak{F}$ is Abelian. Suppose that $u \in \mathfrak{U}$ and

$$1 \neq x \in \mathbf{N}_{\mathfrak{G}}(\mathfrak{U}) \cap \mathfrak{F} \cap \mathbf{C}_{\mathfrak{G}}(u).$$

Then $u \in \mathbf{C}_{\mathfrak{G}}(x) \leq \mathfrak{F}$ and so $u \in \mathbf{C}_{\mathfrak{G}}(\mathfrak{U}) \cap \mathfrak{F} = 1$. Hence $\mathbf{N}_{\mathfrak{G}}(\mathfrak{U}) \cap \mathfrak{F}$ induces a group of fixed point free automorphisms on $\mathfrak{U}$. Thus the Abelian 2-group $\mathbf{N}_{\mathfrak{G}}(\mathfrak{U}) \cap \mathfrak{F}$ has exactly one involution and is therefore cyclic. Obviously j is the involution in $\mathbf{N}_{\mathfrak{G}}(\mathfrak{U}) \cap \mathfrak{F}$, and
$1 \neq g \in \mathbf{N}_{\mathfrak{G}}(\mathfrak{U}) \cap \mathfrak{F}$, so j is a power of g. **q.e.d.**

10.7 Examples. a) Let $\mathfrak{G} = PGL(2, 2^f)$. Then we may assume that $zs = z^{-1}$ for all $z \in \mathsf{P}(1, 2^f)$. If we put $zj = z + 1$, we see easily that $sjs = jsj$. Thus this is the structure equation of $PGL(2, 2^f)$ with $g = j$.

b) Let $\mathfrak{G} = Sz(2^m)$. In the notation of § 3 (pp. 182, 184), put

$$j = S(0,1) = \begin{pmatrix} 1 & 0 & 0 & 0 \\ 0 & 1 & 0 & 0 \\ 1 & 0 & 1 & 0 \\ 1 & 1 & 0 & 1 \end{pmatrix},$$

$$g = S(1,0) = \begin{pmatrix} 1 & 0 & 0 & 0 \\ 1 & 1 & 0 & 0 \\ 0 & 1 & 1 & 0 \\ 1 & 1 & 1 & 1 \end{pmatrix},$$

and

$$T = \begin{pmatrix} 0 & 0 & 0 & 1 \\ 0 & 0 & 1 & 0 \\ 0 & 1 & 0 & 0 \\ 1 & 0 & 0 & 0 \end{pmatrix}.$$

It is easily verified that $TjT = gTg^{-1}$. Thus this is the structure equation of $Sz(2^m)$. We have $g^2 = j$, in accordance with 10.6d).

The assertion of 10.7a) characterizes the groups $PGL(2, 2^f)$.

10.8 Theorem (SUZUKI [3]). *Let $\mathfrak{G}$ be a Zassenhaus group of degree $2^f + 1$. Then $\mathfrak{G} \cong PGL(2, 2^f)$ if and only if $j = g$ in the structure equation of $\mathfrak{G}$.*

Proof. For $\mathfrak{G} = PGL(2, 2^f)$, $j = g$, by 10.7a). Conversely, suppose that $j = g$, so

$$(1) \qquad sjs = jsj.$$

We prove that $\mathfrak{G} \cong PGL(2, 2^f)$ in several steps. Put $\mathfrak{J} = \Omega_1(\mathbf{Z}(\mathfrak{F}))$. Then by 10.2e), $|\mathfrak{J}| = 2^l = d + 1$.

a) The set $\tilde{\mathfrak{G}} = \mathfrak{J}\mathfrak{H} \cup \mathfrak{J}\mathfrak{H}s\mathfrak{J}$ is a subgroup of $\mathfrak{G}$.

As $\mathfrak{J}$ is a characteristic subgroup of $\mathfrak{F}$ and $\mathfrak{H} \leq \mathbf{N}_{\mathfrak{G}}(\mathfrak{F})$, $\mathfrak{J}\mathfrak{H}$ is certainly a subgroup of $\mathfrak{G}$. Obviously

$$(\mathfrak{J}\mathfrak{H})(\mathfrak{J}\mathfrak{H}s\mathfrak{J}) = \mathfrak{J}\mathfrak{H}s\mathfrak{J}$$

and

$$(\mathfrak{J}\mathfrak{H}s\mathfrak{J})(\mathfrak{J}\mathfrak{H}) = \mathfrak{J}\mathfrak{H}s\mathfrak{H}\mathfrak{J} = \mathfrak{J}\mathfrak{H}\mathfrak{H}^s s\mathfrak{J} = \mathfrak{J}\mathfrak{H}s\mathfrak{J}.$$

Thus it only remains to show that

$$(\mathfrak{J}\mathfrak{H}s\mathfrak{J})(\mathfrak{J}\mathfrak{H}s\mathfrak{J}) \subseteq \tilde{\mathfrak{G}}.$$

Since $\mathfrak{H}^s = \mathfrak{H}$ and $(\mathfrak{J}\mathfrak{H})\tilde{\mathfrak{G}} = \tilde{\mathfrak{G}} = \tilde{\mathfrak{G}}(\mathfrak{J}\mathfrak{H})$, it suffices to show that $s\mathfrak{J}s \subseteq \tilde{\mathfrak{G}}$. Suppose that $y \in \mathfrak{J}$. By 10.2e), there exists $h \in \mathfrak{H}$ such that $y = j^h$. Since $h^s = h^{-1}$ (10.2b)), it follows from (1) that

$$sys = sh^{-1}jhs = hsjsh^{-1} = hjsjh^{-1} = hjh^{-1}h^2shjh^{-1} \in \mathfrak{J}\mathfrak{H}s\mathfrak{J} \subseteq \tilde{\mathfrak{G}}.$$

b) $\tilde{\mathfrak{G}} = PGL(2, 2^l)$, (where $d = 2^l - 1$).

By 1.7,

$$|\tilde{\mathfrak{G}}| = |\mathfrak{J}||\mathfrak{H}| + |\mathfrak{J}|^2|\mathfrak{H}| = (2^l + 1)2^l(2^l - 1).$$

Putting $\tilde{\mathfrak{G}}_\infty = \tilde{\mathfrak{G}} \cap \mathfrak{G}_\infty$, we have $|\tilde{\mathfrak{G}}_\infty| = 2^l(2^l - 1)$, $|\tilde{\mathfrak{G}} : \tilde{\mathfrak{G}}_\infty| = 2^l + 1$ and $\tilde{\mathfrak{G}} = \tilde{\mathfrak{G}}_\infty \cup \tilde{\mathfrak{G}}_\infty s \tilde{\mathfrak{G}}_\infty$. Thus $\tilde{\mathfrak{G}}$ is represented as a doubly transitive permutation group on the cosets $\tilde{\mathfrak{G}}_\infty$, $\tilde{\mathfrak{G}}_\infty s$, $\tilde{\mathfrak{G}}_\infty sj_1, \dots, \tilde{\mathfrak{G}}_\infty sj_d$, where $j_1, \dots, j_d$ are the involutions in $\mathfrak{F}$. We show that $\tilde{\mathfrak{G}}$ is represented on the cosets of $\tilde{\mathfrak{G}}_\infty$ as a sharply triply transitive group; to do this it is sufficient to show that the stabiliser of $\tilde{\mathfrak{G}}_\infty$, $\tilde{\mathfrak{G}}_\infty s$ and $\tilde{\mathfrak{G}}_\infty s j_k$ is 1 ($k = 1, \ldots, d$). Suppose then that $g \in \tilde{\mathfrak{G}}_\infty$ and

$$\tilde{\mathfrak{G}}_\infty s g = \tilde{\mathfrak{G}}_\infty s, \tilde{\mathfrak{G}}_\infty s j_k g = \tilde{\mathfrak{G}}_\infty s j_k.$$

It then follows that $g \in \tilde{\mathfrak{G}}_\infty \cap \tilde{\mathfrak{G}}_\infty^s = \tilde{\mathfrak{G}}_\infty \cap \tilde{\mathfrak{G}} \cap \mathfrak{G}_0 = \mathfrak{H}$ and hence

$$\tilde{\mathfrak{G}}_\infty s j_k = \tilde{\mathfrak{G}}_\infty s j_k g = \tilde{\mathfrak{G}}_\infty s g(g^{-1} j_k g) = \tilde{\mathfrak{G}}_\infty g^{-1} s(g^{-1} j_k g) = \tilde{\mathfrak{G}}_\infty s j_k^g.$$

This forces $j_k = j_k^g$. As $\mathfrak{H}$ permutes the involutions of $\mathfrak{F}$ regularly, it follows that $g = 1$. Hence $\tilde{\mathfrak{G}}$ is a sharply triply transitive group of degree $2^l + 1$. By 2.1, $\tilde{\mathfrak{G}} \cong PGL(2, 2^l)$.

c) $\mathfrak{G}$ has exactly $d + 1 = 2^l$ conjugacy classes of strongly real non-identity elements, and each one intersects $\tilde{\mathfrak{G}}$.

If g is strongly real, then by 10.3a), either $g^2 = 1$ or g is conjugate to an element is, where $1 \neq i \in \mathfrak{F}$ and $i^2 = 1$. By 10.3b), g is conjugate to only one such element. This implies the assertion.

From now on, let $\mathfrak{S} = \langle a \rangle$ denote a Singer cycle of $\tilde{\mathfrak{G}} = PGL(2, 2^l)$. Thus by II, 8.4), $|\mathfrak{S}| = 2^l + 1$ and $\mathbf{N}_{\tilde{\mathfrak{G}}}(\mathfrak{S})$ is a dihedral group of order $2|\mathfrak{S}|$. In particular, each element of $\mathfrak{S}$ is strongly real. Let $\mathfrak{A} = \mathbf{C}_{\mathfrak{G}}(\mathfrak{S}) = \mathbf{C}_{\mathfrak{G}}(a)$. By 10.4a), $\mathfrak{A}$ is an Abelian subgroup of odd order, and if $1 \neq b \in \mathfrak{A}$, b is strongly real and $\mathbf{C}_{\mathfrak{G}}(b) = \mathfrak{A}$.

d) $\mathfrak{G}$ has precisely the following classes of strongly real non-identity elements.

(1) The class of involutions.
(2) $2^{l-1} - 1$ classes in $\bigcup_{g \in \mathfrak{G}} \mathfrak{H}^g$.
(3) 2^{l-1} classes in $\bigcup_{g \in \mathfrak{G}} \mathfrak{S}^g$.

By c), each strongly real non-identity element of $\mathfrak{G}$ is conjugate to an element of $\tilde{\mathfrak{G}}$. By II, 8.5a), $\tilde{\mathfrak{G}}$ has the partition

$$\tilde{\mathfrak{G}} = \{1\} \cup \left(\bigcup_{g \in \tilde{\mathfrak{G}}} \mathfrak{J}^g\right) \cup \left(\bigcup_{g \in \tilde{\mathfrak{G}}} \mathfrak{H}^g\right) \cup \left(\bigcup_{g \in \tilde{\mathfrak{G}}} \mathfrak{S}^g\right),$$

so every strongly real element g of $\mathfrak{G}$ for which $g^2 \neq 1$ is conjugate to an element of either $\mathfrak{H}$ or $\mathfrak{S}$. Since $h^s = h^{-1}$ for all $h \in \mathfrak{H}$, the elements of $\mathfrak{H}$ are all strongly real. Since $\mathbf{N}_{\mathfrak{G}}(\mathfrak{H}) = \mathfrak{H}\langle s \rangle$ and $\mathfrak{H} \cap \mathfrak{H}^g = 1$ for $g \notin \mathbf{N}_{\mathfrak{G}}(\mathfrak{H})$, the elements of $\mathfrak{H}$ split into $\frac{1}{2}(d - 1) = 2^{l-1} - 1$ classes. By c), $\mathfrak{G}$ has 2^l classes of strongly real non-identity elements altogether, so this leaves 2^{l-1} classes which intersect $\mathfrak{S} - \{1\}$.

e) Suppose that $b \in \mathfrak{S}$ and $g \in \mathfrak{G}$. If $b^g \in \mathfrak{S}$, then $b^g = b^{\pm 1}$.

As a and a^{-1} are conjugate in the dihedral group $\mathbf{N}_{\mathfrak{G}}(\mathfrak{S})$, $\{a, a^{-1}\}$, $\{a^2, a^{-2}\}, \ldots, \{a^{2^{l-1}}, a^{-2^{l-1}}\}$ are pairs of conjugate elements in $\mathfrak{S} - \{1\}$. By d), $\mathfrak{S} - \{1\}$ contains representatives of 2^{l-1} classes, so there can be no other pairs of non-identity elements of $\mathfrak{S}$ conjugate in $\mathfrak{G}$. Thus it follows from $b^g \in \mathfrak{S}$ and $b \in \mathfrak{S}$ that $b^g = b^{\pm 1}$.

f) If $g \in \mathfrak{A} = \mathbf{C}_{\mathfrak{G}}(\mathfrak{S})$, then g is conjugate to an element of $\mathfrak{S}$.

By 10.4a), g is strongly real, and by 10.2c), $g^2 \neq 1$. Hence by d), g is conjugate to an element of $\mathfrak{H}$ or $\mathfrak{S}$. By 10.4a), $\mathbf{C}_{\mathfrak{G}}(g) = \mathfrak{A} \geq \mathfrak{S}$, so $|\mathfrak{S}| = 2^l + 1$ is a divisor of $|\mathbf{C}_{\mathfrak{G}}(g)|$. But if g is conjugate to an element of $\mathfrak{H}$, $|\mathbf{C}_{\mathfrak{G}}(g)| = |\mathfrak{H}| = 2^l - 1$. This is impossible, so g is conjugate to an element of $\mathfrak{S}$.

g) If $g \in \mathfrak{G} - \mathbf{N}_{\mathfrak{G}}(\mathfrak{S})$, $\mathfrak{S} \cap \mathfrak{S}^g = 1$.

Suppose that $\mathfrak{T} = \mathfrak{S} \cap \mathfrak{S}^g \neq 1$. Since $\mathfrak{S}$ is cyclic, $\mathfrak{T} = \langle t \rangle$ for some $t \in \mathfrak{S} \cap \mathfrak{S}^g$, and t, $t^{g^{-1}}$ generate subgroups of $\mathfrak{S}$ of the same order, so $\langle t \rangle = \langle t^{g^{-1}} \rangle$. Hence $t^g \in \mathfrak{S}$. By e), $t^g = t^{\pm 1}$. Now by 10.4a), $\mathbf{C}_{\mathfrak{G}}(t) = \mathbf{C}_{\mathfrak{G}}(\mathfrak{S}) \leq \mathbf{N}_{\mathfrak{G}}(\mathfrak{S})$. Since $g \notin \mathbf{N}_{\mathfrak{G}}(\mathfrak{S})$, it follows that $g \notin \mathbf{C}_{\mathfrak{G}}(t)$, so $t^g = t^{-1}$. By 10.4, $\mathbf{N}_{\mathfrak{G}}(\mathfrak{A})$ contains an involution i which induces a fixed point free automorphism on $\mathfrak{A}$. Then $b^i = b^{-1}$ for all $b \in \mathfrak{A}$. Hence $i \in \mathbf{N}_{\mathfrak{G}}(\mathfrak{S})$ and $t^i = t^{-1}$. Thus $t^{ig} = t$ and $ig \in \mathbf{C}_{\mathfrak{G}}(t) \leq \mathbf{N}_{\mathfrak{G}}(\mathfrak{S})$. Thus we again obtain the contradiction $g \in \mathbf{N}_{\mathfrak{G}}(\mathfrak{S})$.

h) $\mathbf{C}_{\mathfrak{G}}(\mathfrak{S}) = \mathfrak{S}$ and $|\mathbf{N}_{\mathfrak{G}}(\mathfrak{S})| = 2|\mathfrak{S}|$.

Suppose that $\mathfrak{S} < \mathbf{C}_{\mathfrak{G}}(\mathfrak{S}) = \mathfrak{A}$ and $b \in \mathfrak{A} - \mathfrak{S}$. By 10.4a), $\mathbf{C}_{\mathfrak{G}}(b) = \mathfrak{A}$. By f), $b^{g^{-1}} \in \mathfrak{S}$ for some g. Since $b \notin \mathfrak{S}$, $g \notin \mathbf{N}_{\mathfrak{G}}(\mathfrak{S})$, so by g), $\mathfrak{S} \cap \mathfrak{S}^g = 1$. Also $b \in \mathfrak{S}^g$, so $\mathfrak{S}^g \leq \mathbf{C}_{\mathfrak{G}}(b) = \mathfrak{A}$.

We show now that $|\mathfrak{S}|$ is a prime. For suppose not, and let p be a prime divisor of $|\mathfrak{S}|$. Let c be an element of $\mathfrak{S}^g$ of order p. Since $\mathfrak{S}^g < \mathfrak{A}$, $1 \neq a^p = (ac)^p$. Thus $\langle ac \rangle \cap \langle a \rangle \neq 1$. But by f), $ac \in \mathfrak{S}^h$ for some h, so $\mathfrak{S} \cap \mathfrak{S}^h \neq 1$. By g), $h \in \mathbf{N}_{\mathfrak{G}}(\mathfrak{S})$, so $ac \in \mathfrak{S}$ and $c \in \mathfrak{S}$. Then $\mathfrak{S} \cap \mathfrak{S}^g \neq 1$ and $\mathfrak{S}^g = \mathfrak{S}$, a contradiction. Hence $|\mathfrak{S}|$ is a prime p.

By f), every element of $\mathfrak{A}$ is conjugate to an element of $\mathfrak{S}$. Hence $\mathfrak{A}$ is an elementary Abelian p-group and, using d), $\mathfrak{A}$ contains elements of precisely 2^{l-1} conjugacy classes of $\mathfrak{G}$ other than $\{1\}$. But by 10.4c), $\mathfrak{A} \cap \mathfrak{A}^g = 1$ if $g \notin \mathbf{N}_{\mathfrak{G}}(\mathfrak{A})$. Hence if $k = |\mathbf{N}_{\mathfrak{G}}(\mathfrak{A}) : \mathfrak{A}|$, each non-identity element of $\mathfrak{A}$ has k conjugates which lie in $\mathfrak{A}$. Then $|\mathfrak{A}| - 1 = k \cdot 2^{l-1}$. By 10.4b), k is a power of 2. If $|\mathfrak{A}| = p^m$, $m > 1$ since $\mathfrak{A} > \mathfrak{S}$. From $p^m - 1 = k \cdot 2^{l-1}$, it follows that $p^m - 1$ and $\dfrac{p^m - 1}{p - 1}$ are powers of 2.

Since

$$\frac{p^m - 1}{p - 1} = 1 + p + \cdots + p^{m-1},$$

m is even. If $m = 2m'$, $p^m - 1 = (p^{m'} - 1)(p^{m'} + 1)$, so $p^{m'} - 1, p^{m'} + 1$ are both powers of 2. Then $p^{m'} = 3$ and $3 = |\mathfrak{S}| = 2^l + 1$. Hence $d = 2^l - 1 = 1$, a contradiction. Hence $\mathbf{C}_{\mathfrak{G}}(\mathfrak{S}) = \mathfrak{S}$ and by e), $|\mathbf{N}_{\mathfrak{G}}(\mathfrak{S})| = 2|\mathfrak{S}|$.

i) $\mathfrak{G} = \tilde{\mathfrak{G}} \cong PGL(2, 2^f)$.

We count the number of elements of $\mathfrak{G}$ which we now have. Since $\mathfrak{G}_\infty = \mathbf{N}_{\mathfrak{G}}(\mathfrak{F})$, $\mathfrak{F}$ has just $n + 1$ conjugates in $\mathfrak{G}$, and these together contain $n^2 - 1$ non-identity elements. The group $\mathfrak{H} = \mathfrak{G}_{0,\infty}$ has exactly $\frac{1}{2}n(n + 1)$ conjugates in $\mathfrak{G}$, which together contain $\frac{1}{2}(n + 1)n(d - 1)$ non-identity elements. Finally, by h), $\mathfrak{S}$ has exactly

$$\frac{|\mathfrak{G}|}{2(2^l + 1)} = \frac{(n + 1)n(2^l - 1)}{2(2^l + 1)}$$

conjugates, and by g), any two of them intersect in the identity subgroup. Altogether, it follows that

$$\begin{aligned}|\mathfrak{G}| &\geq n^2 + \frac{(n + 1)n}{2}(2^l - 2) + \frac{(n + 1)n(2^l - 1)2^l}{2(2^l + 1)} \\ &= n^2 + |\mathfrak{G}|\frac{2^{2l} - 2^l - 1}{2^{2l} - 1} = n^2 + |\mathfrak{G}| - |\mathfrak{G}|\frac{2^l}{2^{2l} - 1}.\end{aligned}$$

This shows that $n + 1 \geq n\frac{2^l + 1}{2^l}$ and hence $2^l \geq n$. On the other hand, $2^l - 1 = d$ is a divisor of $n - 1$. It follows that $n = 2^l$, so $|\mathfrak{G}| = |\tilde{\mathfrak{G}}|$. Hence $\mathfrak{G} = \tilde{\mathfrak{G}}$. **q.e.d.**

10.9 Theorem (SUZUKI [3]). *Suppose that $\mathfrak{G}$ is a Zassenhaus group of degree $2^f + 1$ and order $(2^f + 1)2^f d$. If 3 divides d, then $\mathfrak{G} = PGL(2, 2^f)$.*

Proof. By 10.2b), the subgroup $\mathfrak{H}$ is cyclic of order d. Let h be an element of $\mathfrak{H}$ of order 3. By 10.2b), $(sh)^2 = 1$, so $h = s(sh)$ is strongly real. Thus by 10.3a), h is conjugate to an element js, where j is an involution in $\mathfrak{F}$. Hence $(js)^3 = 1$ and so $sjs = jsj$. By 10.8, $\mathfrak{G} = PGL(2, 2^f)$. **q.e.d.**

§ 11. The Characterization of the Suzuki Groups

To determine all the Zassenhaus groups, we still have to deal with the following cases. $\mathfrak{G}$ is a Zassenhaus group of degree $2^f + 1$ and order

$(2^f + 1)2^f(2^l - 1)$ for some l. Put $q = 2^l$. The Frobenius kernel $\mathfrak{F}$ of $\mathfrak{G}_\infty$ is non-Abelian and 3 does not divide $q - 1$ (see 6.8, 9.1 and 10.9). It follows that l is odd. We suppose now that $\mathfrak{G}$ satisfies these conditions and will show that $\mathfrak{G}$ is then a Suzuki group. The first step is to show that the Sylow 2-subgroup $\mathfrak{F}$ of $\mathfrak{G}$ is the group $A(l, \theta)$ of VIII, 6.7.

11.1 Lemma. *$\mathfrak{F}$ is of exponent 4 and either $|\mathfrak{F}| = q^3$ or $\mathfrak{F} \cong A(l, \theta)$ for some non-identity automorphism θ of $GF(2^l)$ of odd order.*

Proof. By 10.2e), $\mathfrak{F}$ has more than one involution and the set of involutions in $\mathfrak{F}$ is permuted transitively by $\mathfrak{H}$. Hence $\mathfrak{F}$ is a Suzuki 2-group (see VIII, 7.1). The assertions thus follow at once from VIII, 7.9. **q.e.d.**

11.2 Lemma. *We have $|\mathfrak{F}| = q^2$ and $|\mathfrak{G}| = (q^2 + 1)q^2(q - 1)$. The Sylow 2-subgroup $\mathfrak{F}$ of $\mathfrak{G}$ is isomorphic to $A(l, \theta)$ for some non-identity automorphism θ of $GF(2^l)$ of odd order, and $|\mathfrak{G}|$ is not divisible by* 3.

Proof. Suppose that $\mathfrak{F}$ is not isomorphic to $A(l, \theta)$. Then by 11.1, $\mathfrak{F}$ is of exponent 4 and $|\mathfrak{F}| = q^3$. Thus $|\mathfrak{G}| = (q^3 + 1)q^3(q - 1)$.

By 10.6, $\mathfrak{F}$ contains an element $g \neq 1$ such that if j is the involution in $\langle g \rangle$, $sjs = gsg^{-1}$ *(structure equation)*. Since $g \in \mathfrak{F}$, $g^4 = 1$. If $g^2 = 1$, then $j = g$ and by 10.8, $\mathfrak{G} = PGL(2, 2^f)$, contrary to our assumption that $\mathfrak{F}$ is non-Abelian. Thus the order of g is 4 and $j = g^2$. Hence $jsj = g^2sg^{-2} = g(gsg^{-1})g^{-1} = gsjsg^{-1}$ and

$$jsj = (gsg^{-1})(gjg^{-1})(gsg^{-1}) = sjsjsjs.$$

Thus $(sj)^5 = 1$, but $sj \neq 1$, since $s \notin \mathfrak{F}$. Thus $|\mathfrak{G}| = (q^3 + 1)q^3(q - 1)$ is divisible by 5.

Since l is odd, 5 does not divide $2^l - 1$. Hence 5 divides $q^3 + 1 = 2^{3l} + 1$. This forces $3l \equiv 2\,(4)$, which is not the case since l is odd. Thus $\mathfrak{F}$ is isomorphic to $A(l, \theta)$ for some non-identity θ of odd order. Hence $|\mathfrak{F}| = q^2$ and $|\mathfrak{G}| = (q^2 + 1)q^2(q - 1)$. Since l is odd, $q^2 + 1 = 4^l + 1 \equiv 2\,(3)$, so 3 does not divide $|\mathfrak{G}|$. **q.e.d.**

11.3 Lemma. *If x, y are elements of $\mathfrak{F}$ such that $x^y = x^{-1}$, then $x^2 = 1$.*

Proof. By 11.2, $\mathfrak{F} \cong A(l, \theta)$. Suppose that x corresponds to $u(a, b)$ and y to $u(c, d)$ in the notation of VIII, 6.7. Then

$$\begin{aligned} u(a + c, b + d + c(a\theta)) &= xy = yx^{-1} \\ &= u(a + c, b + d + a(a\theta) + a(c\theta)), \end{aligned}$$

so $a(a\theta) + a(c\theta) + c(a\theta) = 0$. Hence $c(c\theta) = (a + c)((a + c)\theta)$. Since θ is of odd order, the mapping $t \to t(t\theta)$ is injective, by VIII, 6.9. Hence $a + c = c$ and $a = 0$, whence $x^2 = 1$. **q.e.d.**

In the next lemmas, we shall derive information about the real and strongly real elements of $\mathfrak{G}$, which will finally lead us to a partition of $\mathfrak{G}$.

11.4 Lemma. *Suppose that y is a strongly real element of $\mathfrak{G}$ and $y^2 \neq 1$. Let $\mathfrak{A} = \mathbf{C}_{\mathfrak{G}}(y)$. Then $\mathfrak{A}$ is Abelian of odd order and $\mathbf{N}_{\mathfrak{G}}(\mathfrak{A})/\mathfrak{A}$ is cyclic of order 2 or 4.*

Proof. By 10.4, $\mathfrak{A}$ is Abelian of odd order and $\mathbf{N}_{\mathfrak{G}}(\mathfrak{A})/\mathfrak{A}$ is a 2-group with just one involution. If $\mathfrak{B}$ is a complement of $\mathfrak{A}$ in $\mathbf{N}_{\mathfrak{G}}(\mathfrak{A})$, then $\mathfrak{B}$ is isomorphic to a subgroup of $\mathfrak{F}$. Thus by 11.2, the exponent of $\mathfrak{B}$ is at most 4. By 11.3, $\mathfrak{B}$ cannot be the quaternion group of order 8. But $\mathfrak{B}$ has only one involution, so $\mathfrak{B}$ is cyclic of order 2 or 4. **q.e.d.**

11.5 Lemma. *Any real element y of $\mathfrak{G}$ is strongly real.*

Proof. This is trivial if $y^2 = 1$, since $\mathfrak{F}$ has more than one involution. Suppose then that $y^2 \neq 1$. As y is real, there exists $x \in \mathfrak{G}$ such that $y^x = y^{-1}$. Obviously x is of even order, and we may assume that the order of x is a power of 2. Suppose that $x^2 \neq 1$. Then by 10.2c), $\mathbf{C}_{\mathfrak{G}}(x^2)$ is a 2-group containing y and x. But by 11.3, this implies that $y^2 = 1$. Hence $x^2 = 1$ and $y = (yx)x$ is strongly real. **q.e.d.**

11.6 Lemma. a) *If $y \in \mathfrak{G}$ and y is of odd order, y is strongly real.*

b) *Write $q = 2r^2$. $\mathfrak{G}$ has Abelian subgroups $\mathfrak{A}_1$ and $\mathfrak{A}_2$, where $|\mathfrak{A}_1| = q + 2r + 1$ and $|\mathfrak{A}_2| = q - 2r + 1$. Also $|\mathbf{N}_{\mathfrak{G}}(\mathfrak{A}_i)/\mathfrak{A}_i| = 4$ and $\mathbf{N}_{\mathfrak{G}}(\mathfrak{A}_i)$ is a Frobenius group with Frobenius kernel $\mathfrak{A}_i$.*

c) *The decomposition*

$$\mathfrak{G} = \left(\bigcup_{y \in \mathfrak{G}} \mathfrak{F}^y\right) \cup \left(\bigcup_{y \in \mathfrak{G}} \mathfrak{H}^y\right) \cup \left(\bigcup_{y \in \mathfrak{G}} \mathfrak{A}_1^y\right) \cup \left(\bigcup_{y \in \mathfrak{G}} \mathfrak{A}_2^y\right)$$

is a partition of $\mathfrak{G}$.

Proof. The proof is a counting argument based on Lemma 10.4.

Let $\mathscr{X}$ be the set of all subgroups $\mathfrak{X}$ of $\mathfrak{G}$ for which (1) $\mathfrak{X} = \mathbf{C}_{\mathfrak{G}}(x)$ for some element x of odd order, and (2) $\mathfrak{X}$ contains a non-identity strongly real element y. Thus by 10.2b), $\mathfrak{H} \in \mathscr{X}$. By 10.2c), such a subgroup $\mathfrak{X}$ is of odd order, and by 10.4a), x is strongly real since $x \in \mathbf{C}_{\mathfrak{G}}(y)$; thus if $1 \neq z \in \mathfrak{X}$, z is strongly real and $\mathfrak{X} = \mathbf{C}_{\mathfrak{G}}(z)$. This implies that $\mathbf{N}_{\mathfrak{G}}(\mathfrak{X})$ is a Frobenius group with Abelian Frobenius kernel $\mathfrak{X}$ and that $\mathfrak{X} \cap \mathfrak{Y} = 1$ for $\mathfrak{X} \in \mathscr{X}$, $\mathfrak{Y} \in \mathscr{X}$, $\mathfrak{X} \neq \mathfrak{Y}$. Also $\mathsf{A} = \bigcup_{\mathfrak{X} \in \mathscr{X}} \mathfrak{X}$ is the set of strongly real elements of $\mathfrak{G}$ of odd order. Let B be the set of elements of odd order which are not strongly real. By 10.2c), every element of $\mathfrak{G}$ is either a 2-element or an element of odd order, so

$$\mathfrak{G} = \left(\bigcup_{y \in \mathfrak{G}} \mathfrak{F}^y\right) \cup \mathsf{A} \cup \mathsf{B}.$$

Now $\mathscr{X}$ is a union of conjugacy classes of subgroups; let $\mathfrak{H} = \mathfrak{A}_0$, $\mathfrak{A}_1, \ldots, \mathfrak{A}_c$ be a complete set of representatives of these classes. Thus

$$\mathsf{A} = \left(\bigcup_{y \in \mathfrak{G}} \mathfrak{H}^y\right) \cup \left(\bigcup_{y \in \mathfrak{G}} \mathfrak{A}_1^y\right) \cup \cdots \cup \left(\bigcup_{y \in \mathfrak{G}} \mathfrak{A}_c^y\right).$$

Since $\mathfrak{A}_i$ is the centralizer of any of its non-identity elements, $\mathfrak{A}_i^y \cap \mathfrak{A}_j^{y'} = 1$ unless $\mathfrak{A}_i^y = \mathfrak{A}_j^{y'}$; also $\mathfrak{H}^y \cap \mathfrak{A}_i^{y'} = 1$ and $\mathfrak{H}^y \cap \mathfrak{H}^{y'} = 1$ unless $\mathfrak{H}^y = \mathfrak{H}^{y'}$. Further $\mathfrak{F}^y \cap \mathfrak{F}^{y'} = 1$ unless $\mathfrak{F}^y = \mathfrak{F}^{y'}$, and $\mathfrak{F}^y \cap \mathfrak{H}^{y'} = \mathfrak{F}^y \cap \mathfrak{A}_i^{y'} = 1$. Thus to prove the lemma, it is sufficient to prove that

(i) B is the empty set,

(ii) $c = 2$, $|\mathfrak{A}_1| = q + 2r + 1$ and $|\mathfrak{A}_2| = q - 2r + 1$,

(iii) $|\mathbf{N}_{\mathfrak{G}}(\mathfrak{A}_i) : \mathfrak{A}_i| = 4 \quad (i = 1, 2)$.

Put $a_i = |\mathfrak{A}_i|$, $k_i = |\mathbf{N}_{\mathfrak{G}}(\mathfrak{A}_i) : \mathfrak{A}_i|$. Thus $a_0 = |\mathfrak{H}| = d = q - 1$ and $k_0 = 2$, so we see from the above decomposition of A that

$$|\mathsf{A}| = 1 + \frac{(q-2)|\mathfrak{G}|}{2(q-1)} + \sum_{j=1}^{c} \frac{(a_j - 1)|\mathfrak{G}|}{a_j k_j}.$$

Hence

$$|\mathfrak{G}| = 1 + \frac{(q^2-1)|\mathfrak{G}|}{q^2(q-1)} + \frac{(q-2)|\mathfrak{G}|}{2(q-1)} + \sum_{j=1}^{c} \frac{(a_j - 1)|\mathfrak{G}|}{a_j k_j} + |\mathsf{B}|.$$

It is easily checked that this reduces to

(1) $$\sum_{j=1}^{c} \frac{(a_j - 1)}{a_j k_j} = \frac{q(q-1)}{2(q^2+1)} - \frac{|\mathsf{B}|}{|\mathfrak{G}|}.$$

By 10.3, the number of conjugacy classes of strongly real elements

y such that $y^2 \neq 1$ is $q - 1$. Since $\mathfrak{A}_j \cap \mathfrak{A}_j^y = 1$ when $y \notin \mathbf{N}_{\mathfrak{G}}(\mathfrak{A}_j)$, the number of classes of non-identity elements which intersect $\mathfrak{A}_j$ is $\dfrac{a_j - 1}{k_j}$, for $\mathbf{N}_{\mathfrak{G}}(\mathfrak{A}_j)$ is a Frobenius group ($j = 0, 1, \ldots, c$). Since $\mathfrak{A}_i^y \cap \mathfrak{A}_j^{y'} = 1$ for $i \neq j$, it follows that

$$q - 1 = \sum_{j=0}^{c} \frac{a_j - 1}{k_j} = \sum_{j=1}^{c} \frac{a_j - 1}{k_j} + \frac{q - 2}{2},$$

or

$$\sum_{j=1}^{c} \frac{a_j - 1}{k_j} = \frac{q}{2}. \tag{2}$$

By 11.4, k_i is 2 or 4. Suppose that u of the k_i are 2 and that v of them are 4, so $u + v = c$. Since 3 does not divide $|\mathfrak{G}|$, $a_j \geq 5$ and $\dfrac{a_j - 1}{a_j} \geq \dfrac{4}{5}$. Hence by (1),

$$\frac{1}{2} > \frac{q(q - 1)}{2(q^2 + 1)} \geq \frac{4}{5}\left(\frac{u}{2} + \frac{v}{4}\right) = \frac{2}{5}u + \frac{1}{5}v.$$

Thus $u \leq 1$. If $u = 1$, then $v = 0$ and $c = 1$, whence (2) gives $a_1 - 1 = q$. Thus 3 divides neither $a_1 = q + 1$ nor $q - 1$, which is impossible. Hence $u = 0$ and $v \leq 2$. By (2), $v > 0$. If $v = 1$, (2) gives $a_1 = 2q + 1$, so $2q + 1$ divides $|\mathfrak{G}| = (q^2 + 1)q^2(q - 1)$. Since $\mathfrak{G}$ is a $3'$-group, it follows that $2q + 1$ divides $q^2 + 1$. If $q^2 + 1 = b(2q + 1)$, then $b > 1$ since $q > 2$; also $b \equiv 1\,(q)$, so $b \geq q + 1$ and $q^2 + 1 \geq (q + 1)(2q + 1)$, a contradiction. Thus $v = 2$, $c = 2$ and $k_1 = k_2 = 4$. (2) now gives

$$a_1 + a_2 = 2(q + 1). \tag{3}$$

We prove next that if $y \in \mathfrak{G}$ and $(|\mathbf{C}_{\mathfrak{G}}(y)|, |\mathfrak{A}_i|) \neq 1$ for some $i \geq 0$, then $\mathbf{C}_{\mathfrak{G}}(y) = \mathfrak{A}_i^{y'}$ for some y'. Indeed, suppose that p is a prime divisor of $(|\mathbf{C}_{\mathfrak{G}}(y)|, |\mathfrak{A}_i|)$ and that x is an element of $\mathbf{C}_{\mathfrak{G}}(y)$ of order p. Thus p is odd and by 10.2c), y is of odd order. By 10.4b), $\mathfrak{A}_i$ is a Hall subgroup of $\mathfrak{G}$, so $\mathfrak{A}_i$ contains a Sylow p-subgroup $\mathfrak{P}$ of $\mathfrak{G}$. By Sylow's theorem, $x \in \mathfrak{P}^{y'}$ for some y'. Then $x \in \mathfrak{A}_i^{y'}$ and x is strongly real. Hence $\mathbf{C}_{\mathfrak{G}}(y) \in \mathscr{X}$ and $\mathbf{C}_{\mathfrak{G}}(y) = \mathfrak{A}_i^{y'}$, since $\mathbf{C}_{\mathfrak{G}}(y) \cap \mathfrak{A}_i^{y'} \neq 1$.

In particular, $(q - 1, |\mathfrak{A}_i|) = 1$ ($i = 1, 2$) since $q - 1 = |\mathfrak{A}_0|$, and since $(|\mathfrak{A}_1|, |\mathfrak{A}_2|) = 1$, $a_1 a_2$ divides $|\mathfrak{G}| = (q^2 + 1)q^2(q - 1)$. Thus $a_1 a_2$

divides $q^2 + 1$ and $(q - 1)q^2 a_1 a_2$ divides $|\mathfrak{G}|$. Further, if $y \in \mathsf{B}$, $|\mathbf{C}_{\mathfrak{G}}(y)|$ and $(q - 1)a_1 a_2$ are coprime. Since also $|\mathbf{C}_{\mathfrak{G}}(y)|$ is odd, $(q - 1)q^2 a_1 a_2$ divides $|\mathfrak{G} : \mathbf{C}_{\mathfrak{G}}(y)|$, which is the number of conjugates of y. But B is a union of conjugacy classes, so $(q - 1)q^2 a_1 a_2$ divides $|\mathsf{B}|$. Put $|\mathsf{B}| = (q - 1)q^2 a_1 a_2 e$. Then (1) gives

$$\frac{a_1 - 1}{4a_1} + \frac{a_2 - 1}{4a_2} = \frac{q(q - 1)}{2(q^2 + 1)} - \frac{a_1 a_2 e}{q^2 + 1};$$

hence by (3),

$$(4) \qquad \frac{2(q + 1)}{a_1 a_2} = \frac{1}{a_1} + \frac{1}{a_2} = 2 - \frac{2q(q - 1)}{q^2 + 1} + \frac{4a_1 a_2 e}{q^2 + 1} = \frac{2(1 + q)}{q^2 + 1} + \frac{4a_1 a_2 e}{q^2 + 1}.$$

Since $q + 1$ and $q^2 + 1$ are coprime and $a_1 a_2$ divides $q^2 + 1$, we see that $q + 1$ divides e. But by (4),

$$2(q + 1) > \frac{4a_1^2 a_2^2 e}{q^2 + 1}.$$

Thus if $e \neq 0$, $e \geq q + 1$ and $q^2 + 1 > 2a_1^2 a_2^2$. But, supposing that $a_1 \geq a_2$, (3) gives $a_1 \geq q + 1$, so $q^2 + 1 > 2(q + 1)^2$, which is impossible. Thus $e = 0$ and B is empty. By (4), $q^2 + 1 = a_1 a_2$, and by (3), $2(q + 1) = a_1 + a_2$. Hence $a_1 = q + 1 + \sqrt{2q}$ and $a_2 = q + 1 - \sqrt{2q}$. All the assertions are therefore proved. **q.e.d.**

The information in 11.6 is sufficient for the determination of the class number of $\mathfrak{G}$ for $q > 8$.

11.7 Lemma. *For $q > 8$, $\mathfrak{G}$ has exactly $q + 3$ conjugacy classes, exactly two of which consist of elements of order 4.*

Proof. By 11.6, the following are the conjugacy classes of $\mathfrak{G}$.

(1) $\{1\}$.

(2) The class of involutions.

(3) $\frac{1}{2}(|\mathfrak{H}| - 1) = \frac{1}{2}q - 1$ classes in $\bigcup (\mathfrak{H} - \{1\})^y$.

(4) $\frac{1}{4}(|\mathfrak{A}_1| - 1) = \frac{1}{4}(q + 2r)$ classes in $\bigcup (\mathfrak{A}_1 - \{1\})^y$.

(5) $\frac{1}{4}(|\mathfrak{A}_2| - 1) = \frac{1}{4}(q - 2r)$ classes in $\bigcup (\mathfrak{A}_2 - \{1\})^y$.

(6) The classes of elements of order 4. Suppose that the number of these is v.

Thus $h(\mathfrak{G}) = q + v + 1$. By 11.6 and 5.8, $h(\mathfrak{G}) = q + 3$, so $v = 2$.

q.e.d.

The class number of $Sz(8)$ is 11, but we shall not need this.

The automorphism θ of $GF(q)$ which occurs in the description of the Sylow 2-subgroup $\mathfrak{F} \cong A(l, \theta)$ of $\mathfrak{G}$ in 11.2 is still unknown. In the end we shall show that $x\theta^2 = x^2$. To do so, we first describe the notation we shall use for the elements of $\mathfrak{F}$ and $\mathfrak{H}$.

11.8 Notation. By 11.2, $\mathfrak{F} \cong A(l, \theta)$. Let $\mathfrak{A}$ be the group of automorphisms of $\mathfrak{F}$ which induce the identity automorphism on $\mathbf{Z}(\mathfrak{F})$ and $\mathfrak{F}/\mathbf{Z}(\mathfrak{F})$. If ξ is the automorphism of $\mathfrak{F}$ induced by a generator of $\mathfrak{H}$, then ξ permutes the set of involutions of $\mathfrak{F}$ transitively. Hence by VIII, 6.8, there exists $\lambda \in GF(q)$ such that $\xi\xi_\lambda^{-1} \in \mathfrak{A}$; ($\xi_\lambda$ is defined in VIII, 6.7). Then $\langle \xi_\lambda \rangle$, $\langle \xi \rangle$ are subgroups of $\langle \xi, \mathfrak{A} \rangle$ of order $q - 1$, and $\mathfrak{A}$ is a 2-group. By I, 17.5, $\langle \xi_\lambda \rangle = \psi^{-1}\langle \xi \rangle \psi$ for some $\psi \in \mathfrak{A}$. Thus $\xi_\lambda^c = \psi^{-1}\xi\psi$ for some c. But then

$$\xi_\lambda^{c-1} = (\psi^{-1}\xi\psi\xi^{-1})(\xi\xi_\lambda^{-1}) \in \mathfrak{A} \cap \langle \xi_\lambda \rangle = 1,$$

so $\xi_\lambda = \psi^{-1}\xi\psi$.

By VIII, 6.7, then, we can identify the elements of $\mathfrak{F}$ with elements $u(a, b)$, where $a \in GF(q)$, $b \in GF(q)$ and

$$u(a, b)u(a', b') = u(a + a', b + b' + (a\theta)a'),$$

$$u(a, b)\psi^{-1}\xi\psi = u(\lambda a, \lambda(\lambda\theta)b).$$

Now by 10.6, $\mathfrak{G}$ contains a non-identity element g such that if j is the involution in $\langle g \rangle$, $\mathfrak{G}$ has the structure equation $sjs = gsg^{-1}$. Since ξ permutes the set of involutions of $\mathfrak{F}$ transitively, $j = u(0, 1)\psi^{-1}\xi^e$ for some e. We put

$$(a, b) = u(a, b)\psi^{-1}\xi^e$$

for all a, b in $GF(q)$. Thus $j = (0, 1)$, $\mathfrak{F}$ is the set of all (a, b) with a, b in $GF(q)$, and

$$(a, b)(a', b') = (a + a', b + b' + (a\theta)a'),$$

$$(a, b)^{-1} = (a, b + (a\theta)a).$$

Also

$$(a, b)\xi = u(a, b)\psi^{-1}\xi^{1+e} = u(\lambda a, \lambda(\lambda\theta)b)\psi^{-1}\xi^{e} = (\lambda a, \lambda(\lambda\theta)b).$$

If $h \in \mathfrak{H}$, the automorphism of $\mathfrak{F}$ induced by h is ξ^i for some i; we shall write $h = (\lambda^i)$. Hence $(\mu)(\nu) = (\mu\nu)$ and

$$(a, b)^{(\mu)} = (\mu a, \mu(\mu\theta)b).$$

11.9 Lemma. *The only fixed points of θ are 0 and 1.*

Proof. For $q = 8$, this is obvious, since $\theta \neq 1$. Suppose that $q > 8$. By 11.7, $\mathfrak{G}$ has just two classes of elements of order 4; hence this is also true of $\mathfrak{G}_\infty$. Let x be an element of $\mathfrak{F}$ of order 4. Since $\mathbf{C}_{\mathfrak{G}_\infty}(x) \geq \langle x \rangle \mathbf{Z}(\mathfrak{F})$, $|\mathbf{C}_{\mathfrak{G}_\infty}(x)| \geq 2q$. It follows that the number of conjugates of x in $\mathfrak{G}_\infty$ is

$$|\mathfrak{G}_\infty : \mathbf{C}_{\mathfrak{G}_\infty}(x)| \leq \frac{q^2(q-1)}{2q} = \tfrac{1}{2}q(q-1).$$

But $\mathfrak{G}_\infty$ has only $q^2 - q$ elements of order 4 and these lie in two classes, so each class has $\frac{1}{2}q(q-1)$ elements. In particular, if $y = (1, 0)$, $|\mathbf{C}_{\mathfrak{G}_\infty}(y)| = 2q$. But (a, b) lies in $\mathbf{C}_{\mathfrak{F}}(y)$ if and only if $a = a\theta$, since

$$(a, b)y = (a + 1, b + a\theta), \quad y(a, b) = (a + 1, b + a).$$

Thus $a\theta = a$ has just two solutions, which must be 0 and 1. **q.e.d.**

We turn now to the investigation of the functional equation which appeared in 1.9.

11.10 Lemma. a) *Suppose that $1 \neq x \in \mathfrak{F}$ and put*

$$sxs = \beta(x)\gamma(x)s\alpha(x),$$

where $\alpha(x)$, $\beta(x)$ are in $\mathfrak{F}$ and $\gamma(x) \in \mathfrak{H}$. Then α, β are mappings of $\mathfrak{F} - \{1\}$ into $\mathfrak{F} - \{1\}$ and γ is a mapping of $\mathfrak{F} - \{1\}$ into $\mathfrak{H}$.

b) *The following hold for all $x \in \mathfrak{F} - \{1\}$ and $h \in \mathfrak{H}$.*

(1) $\alpha(\alpha(x)) = x$.

(2) $\beta(x)^{-1} = \alpha(x^{-1})$ *and* $\gamma(x^{-1}) = \gamma(x)$.

(3) $\alpha(x^h) = \alpha(x)^{h^{-1}}$, $\beta(x^h) = \beta(x)^{h^{-1}}$.

(4) $\gamma(x^h) = h^2\gamma(x)$.

(5) $\alpha(\beta(x)) = \gamma(x)^{-1}\alpha(x)^{-1}\gamma(x)$.

(6) *For* x_1, x_2 *in* $\mathfrak{F} - \{1\}$ *with* $x_1 x_2 \neq 1$,

$$\alpha(x_1 x_2) = \gamma(x_2)\alpha(\alpha(x_1)\beta(x_2))\gamma(x_2)^{-1}\alpha(x_2).$$

Proof. a) If $1 \neq x \in \mathfrak{F}$, $sxs \notin \mathfrak{G}_\infty$. By 1.7, $sxs = y'sx'$ for uniquely determined $y' \in \mathfrak{G}_\infty$ and $x' \in \mathfrak{F}$. We put $x' = \alpha(x)$ and $y' = \beta(x)\gamma(x)$, with $\alpha(x)$, $\beta(x)$ in $\mathfrak{F}$ and $\gamma(x) \in \mathfrak{H}$.

If $\alpha(x) = 1$, then $sxs = \beta(x)\gamma(x)s$ and hence $sx \in \mathfrak{G}_\infty$, a contradiction. Thus $\alpha(x) \neq 1$. If $\beta(x) = 1$, then by 10.2b),

$$sxs = \gamma(x)s\alpha(x) = s\gamma(x)^{-1}\alpha(x),$$

and this implies the contradiction $xs \in \mathfrak{G}_\infty$.

b) (1) From $sxs = \beta(x)\gamma(x)s\alpha(x)$, it follows that

$$s\alpha(x)s = \gamma(x)^{-1}\beta(x)^{-1}sx.$$

This shows that $\alpha(\alpha(x)) = x$.

(2) By 10.2b), $h^s = h^{-1}$ for $h \in \mathfrak{H}$. Thus we obtain

$$sx^{-1}s = (sxs)^{-1} = \alpha(x)^{-1}s\gamma(x)^{-1}\beta(x)^{-1} = \alpha(x)^{-1}\gamma(x)s\beta(x)^{-1}.$$

It follows that $\alpha(x^{-1}) = \beta(x)^{-1}$ and $\gamma(x^{-1}) = \gamma(x)$.

(3), (4) We have

$$\begin{aligned} sx^h s &= sh^{-1}xhs = hsxsh^{-1} \\ &= (h\beta(x)h^{-1})(h^2\gamma(x))h^{-1}s\alpha(x)h^{-1} \quad (\text{since } \mathfrak{H}' = 1) \\ &= (h\beta(x)h^{-1})(h^2\gamma(x))s(h\alpha(x)h^{-1}). \end{aligned}$$

This yields $\alpha(x^h) = \alpha(x)^{h^{-1}}$, $\beta(x^h) = \beta(x)^{h^{-1}}$ and $\gamma(x^h) = h^2\gamma(x)$.

(5) From $sxs = \beta(x)\gamma(x)s\alpha(x)$, we obtain

$$s\beta(x)s = xs\alpha(x)^{-1}\gamma(x) = x\gamma(x)^{-1}s\gamma(x)^{-1}\alpha(x)^{-1}\gamma(x).$$

It follows that

$$\alpha(\beta(x)) = \gamma(x)^{-1}\alpha(x)^{-1}\gamma(x).$$

(6) Put $x = \alpha(x_1)\beta(x_2)$. Then

$$\begin{aligned} sx_1x_2s = sx_1ssx_2s &= \beta(x_1)\gamma(x_1)s\alpha(x_1)\beta(x_2)\gamma(x_2)s\alpha(x_2) \\ &= \beta(x_1)\gamma(x_1)sxs\gamma(x_2)^{-1}\alpha(x_2). \end{aligned}$$

Now $x \neq 1$, since otherwise $sx_1x_2s \in \mathfrak{G}_\infty$, which is not the case since $x_1x_2 \neq 1$. Hence

$$\begin{aligned} sx_1x_2s &= \beta(x_1)\gamma(x_1)\beta(x)\gamma(x)s\alpha(x)\gamma(x_2)^{-1}\alpha(x_2) \\ &= (\beta(x_1)\gamma(x_1)\beta(x)\gamma(x)\gamma(x_2))s(\gamma(x_2)\alpha(x)\gamma(x_2)^{-1}\alpha(x_2)). \end{aligned}$$

Since

$$\beta(x_1)\gamma(x_1)\beta(x)\gamma(x)\gamma(x_2) \in \mathfrak{G}_\infty$$

and

$$\gamma(x_2)\alpha(x)\gamma(x_2)^{-1}\alpha(x_2) \in \mathfrak{F},$$

it then follows that

$$\alpha(x_1x_2) = \gamma(x_2)\alpha(x)\gamma(x_2)^{-1}\alpha(x_2).$$ **q.e.d.**

11.11 Lemma. *Suppose that $sjs = gsg^{-1}$ is the structure equation of $\mathfrak{G}$.*

a) *If h_1, h_2 are distinct elements of $\mathfrak{H}$, then $g^{-1}h_1^{-1}gh_1 \neq g^{-1}h_2^{-1}gh_2$.*

b) *If $x \in \mathfrak{F} - \{1\}$, x belongs to the same $\mathfrak{H}$-orbit as precisely one element of the set*

$$\{j, g, g^{-1}\} \cup \{g^{-1}h^{-1}gh \mid 1 \neq h \in \mathfrak{H}\}.$$

c) *For the function α occurring in the functional equation in 11.10, if $h \in \mathfrak{H} - \{1\}$,*

$$\alpha(g^{-1}h^{-1}gh) = kg^{-1}k^{-1}hgh^{-1},$$

where $k = (\nu)$ is determined from $h = (\mu)$ by the equation

$$\nu(\nu\theta) = \mu^2(\mu\theta)^2 + \mu(\mu\theta).$$

Proof. From $g^{-1}h_1^{-1}gh_1 = g^{-1}h_2^{-1}gh_2$, it follows that $(h_1h_2^{-1})^{-1}g(h_1h_2^{-1}) = g$. Since $\mathfrak{F}\mathfrak{H}$ is a Frobenius group with Frobenius kernel $\mathfrak{F}$, this implies that $h_1 = h_2$.

By 11.10a), for any $x \in \mathfrak{F} - \{1\}$,

$$sxs = \beta(x)\gamma(x)s\alpha(x),$$

where $\alpha(x)$, $\beta(x)$ are in $\mathfrak{F}$ and $\gamma(x) \in \mathfrak{H}$. By 11.10b) (4),

$$\gamma(x^h) = h^2\gamma(x)$$

for any $h \in \mathfrak{H}$. Since $\gamma(x) \in \mathfrak{H}$ and $|\mathfrak{H}|$ is odd, it follows that for each $x \in \mathfrak{F} - \{1\}$, there is exactly one $h \in \mathfrak{H}$ such that $\gamma(x^h) = 1$. We put $i(x) = x^h$. Then x, y lie in the same $\mathfrak{H}$-orbit if and only if $i(x) = i(y)$.

Let

$$\mathsf{M} = \{j, g, g^{-1}\} \cup \{g^{-1}h^{-1}gh | 1 \neq h \in \mathfrak{H}\}.$$

Thus it is to be proved that each $\mathfrak{H}$-orbit of $\mathfrak{F} - \{1\}$ contains an element of M. Now there are $q + 1$ such orbits and there are $q + 1$ elements in M. Thus it is sufficient to prove that if u, v are distinct elements of M, then $i(u) \neq i(v)$.

We have

(1) $$sjs = gsg^{-1}.$$

Since $g \in \mathfrak{F}$, it follows that $\gamma(j) = 1$, so

(2) $$i(j) = j.$$

From

(3) $$sgs = jsg$$

and

(4) $$sg^{-1}s = g^{-1}sj,$$

we obtain $\gamma(g) = \gamma(g^{-1}) = 1$, so

(5) $$i(g) = g, i(g^{-1}) = g^{-1}.$$

It remains to determine $i(g^{-1}h^{-1}gh)$ for $1 \neq h \in \mathfrak{H}$. From (3), (4) and $h^s = h^{-1}$, it follows that

(6) $$\begin{aligned} s(g^{-1}h^{-1}gh)s &= (sg^{-1}s)(sh^{-1})g(hs) = (g^{-1}sj)(hs)g(sh^{-1}) \\ &= (g^{-1}sj)h(jsg)h^{-1} = g^{-1}h^{-2}s(h^{-1}(h^{-1}jhj)h)shgh^{-1}. \end{aligned}$$

Since $h \neq 1$ and $j \in \mathbf{Z}(\mathfrak{F})$, $h^{-1}jhj$ is an element of $\mathbf{Z}(\mathfrak{F}) - \{1\}$ and is therefore an involution in $\mathbf{Z}(\mathfrak{F})$. Hence $h^{-1}(h^{-1}jhj)h$ is an involution in

$\mathbf{Z}(\mathfrak{F})$. Since the representation of $\mathfrak{H}$ on $\mathbf{Z}(\mathfrak{F}) - \{1\}$ is regular, there exists a unique $k \in \mathfrak{H}$ such that

(7) $$h^{-1}(h^{-1}jhj)h = j^k.$$

Since $k^s = k^{-1}$, substitution of (7) in (6) gives

$$\begin{aligned} s(g^{-1}h^{-1}gh)s &= g^{-1}h^{-2}s(k^{-1}jk)shgh^{-1} = g^{-1}h^{-2}k(sjs)k^{-1}hgh^{-1} \\ &= g^{-1}h^{-2}k(gsg^{-1})k^{-1}hgh^{-1} = g^{-1}h^{-2}k(gkskg^{-1})k^{-1}hgh^{-1} \\ &= (g^{-1}h^{-2}kgk^{-1}h^2)(h^{-2}k^2)s(kg^{-1}k^{-1}hgh^{-1}). \end{aligned}$$

Since $g^{-1}(h^{-2}k)g(k^{-1}h^2) \in \mathfrak{F}$, $h^{-2}k^2 \in \mathfrak{H}$ and $(kg^{-1}k^{-1})(hgh^{-1}) \in \mathfrak{F}$, it follows from 11.10a) that

(8) $$\alpha(g^{-1}h^{-1}gh) = kg^{-1}k^{-1}hgh^{-1},$$

(9) $$\beta(g^{-1}h^{-1}gh) = g^{-1}h^{-2}kgk^{-1}h^2,$$

(10) $$\gamma(g^{-1}h^{-1}gh) = h^{-2}k^2.$$

Hence by 11.10b) (4),

(11) $$\gamma(h^{-1}k(g^{-1}h^{-1}gh)k^{-1}h) = (k^{-1}h)^2\gamma(g^{-1}h^{-1}gh) = 1.$$

It follows that

$$i(g^{-1}h^{-1}gh) = h^{-1}k(g^{-1}h^{-1}gh)k^{-1}h.$$

Further, by 11.10b) (3),

(12) $$\begin{aligned} \beta(i(g^{-1}h^{-1}gh)) &= hk^{-1}\beta(g^{-1}h^{-1}gh)kh^{-1} \\ &= (hk^{-1}g^{-1}kh^{-1})(h^{-1}gh). \end{aligned}$$

We have therefore now found $i(u)$ for all $u \in \mathsf{M}$. To show that they are distinct, we use the notation of 11.8. Let $h = (\mu)$, $k = (\nu)$. Since $j = (0, 1)$, $g \in \mathfrak{F}$ and $g^2 = j$, $g = (1, \rho)$ for some ρ (VIII, 6.9). Then

$$g^{-1}h^{-1}gh = (1, 1 + \rho)(\mu, \mu(\mu\theta)\rho) = (1 + \mu, 1 + \rho + \mu(\mu\theta)\rho + \mu)$$

and

$$(13)\qquad i(g^{-1}h^{-1}gh) = h^{-1}k(1 + \mu, *)k^{-1}h = (v^{-1}\mu(1 + \mu), *).$$

Also, by (12),

$$(14)\qquad \beta(i(g^{-1}h^{-1}gh)) = (\mu^{-1}v, *)(\mu, *) = (\mu^{-1}v + \mu, *).$$

Equation (7), which determines k, gives

$$(0, v(v\theta)) = (0, 1)^k = k^{-1}jk = j^{h^2}j^h = (0, \mu^2(\mu\theta)^2 + \mu(\mu\theta)),$$

so

$$(15)\qquad v(v\theta) = \mu^2(\mu\theta)^2 + \mu(\mu\theta).$$

Since the mapping $a \to a(a\theta)$ is injective, v is determined by (15). Thus c) is proved.

By (2), (5) and (13), then, the values of $i(u)$ $(u \in \mathsf{M})$ are as follows.

$$i(j) = j = (0, 1),$$

$$i(g) = g = (1, \rho),$$

$$i(g^{-1}) = g^{-1} = (1, 1 + \rho),$$

$$i(g^{-1}h^{-1}gh) = (v^{-1}(\mu + \mu^2), *).$$

Since $h \neq 1$, $\mu \neq 1$ and so $v^{-1}(\mu + \mu^2) \neq 0$. And using (15), it follows from $v^{-1}(\mu + \mu^2) = 1$ that

$$\mu^2(\mu\theta)^2 + \mu(\mu\theta) = (\mu + \mu^2)((\mu\theta) + (\mu\theta)^2).$$

Hence

$$\mu^2(\mu\theta) + \mu(\mu\theta)^2 = 0,$$

so $\mu\theta = \mu$. It follows from 11.9 that $\mu = 1$, a contradiction. Thus we have shown that $i(g^{-1}h^{-1}gh)$ is distinct from $i(j)$, $i(g)$ and $i(g^{-1})$. It remains to show that if h_1, h_2 are non-identity elements of $\mathfrak{H}$ and $i(g^{-1}h_1^{-1}gh_1) = i(g^{-1}h_2^{-1}gh_2)$, then $h_1 = h_2$. Put

$$\alpha_i = v_i^{-1}(\mu_i + \mu_i^2), \quad \beta_i = \mu_i^{-1}v_i + \mu_i \quad (i = 1, 2).$$

Thus by (13) and (14), $\alpha_1 = \alpha_2$ and $\beta_1 = \beta_2$. Now, eliminating v_i,

$$\alpha_i\beta_i = \mu_i^{-1}(\mu_i + \mu_i^2) + \alpha_i\mu_i = 1 + \mu_i(1 + \alpha_i).$$

Since $\alpha_i \neq 1$, it follows that $\mu_1 = \mu_2$, so $h_1 = h_2$. **q.e.d.**

11.12 Lemma. *Replacing θ by θ^{-1} if necessary, we may suppose that* $g = (1, 0)$.

Proof. Since $g^2 = j = (0, 1)$, we have $g = (1, \rho)$ for some ρ. Suppose that $\rho \neq 0$. If $1 \neq h \in \mathfrak{H}$, then by 11.11, $g^{-1}h^{-1}gh$ and g are not in the same $\mathfrak{H}$-orbit of elements of $\mathfrak{F}$. Now if $h = (\mu)$,

$$g^{-1}h^{-1}gh = (1 + \mu, 1 + \rho + \mu(\mu\theta)\rho + \mu)$$

and

$$g^{(1+\mu)} = (1 + \mu, (1 + \mu)(1 + \mu\theta)\rho),$$

so

$$\rho(\mu\theta) + \rho\mu \neq 1 + \mu$$

for all elements μ of $GF(q)$ other than 0, 1. Since $\rho \neq 0$,

$$(1) \qquad 1 + \rho^{-1} \neq ((1 + \mu) + (\mu + \mu\theta))(1 + \mu)^{-1} = (1 + \mu\theta)(1 + \mu)^{-1}.$$

But by 11.9, θ has no fixed points other than 0 and 1, so $(1 + \mu\theta)(1 + \mu)^{-1}$ is distinct from 0 and 1. Hence as μ runs through $GF(q) - \{0, 1\}$, $(1 + \mu\theta)(1 + \mu)^{-1}$ runs through the same set. From (1), it therefore follows that $1 + \rho^{-1} \in \{0, 1\}$ and hence $\rho = 1$.

Put $(a, b)' = (a\theta^{-1}, b + a(a\theta^{-1}))$ and $(\lambda)' = (\lambda\theta^{-1})$. Thus $(1, 0)' = (1, 1) = g$. Also

$$\begin{aligned}&(a, b)'(a_1, b_1)'\\ &= (a\theta^{-1} + a_1\theta^{-1}, b + b_1 + a(a\theta^{-1}) + a(a_1\theta^{-1}) + a_1(a_1\theta^{-1}))\\ &= (a + a_1, b + b_1 + (a\theta^{-1})a_1)'\end{aligned}$$

and

$$(a, b)'^{(\lambda)'} = (a\theta^{-1}, b + a(a\theta^{-1}))^{(\lambda\theta^{-1})}$$
$$= ((a\lambda)\theta^{-1}, (\lambda\theta^{-1})\lambda(b + a(a\theta^{-1}))) = (a\lambda, \lambda(\lambda\theta^{-1})b)'.$$

Thus if we replace $(a, b)'$ by (a, b) and $(\lambda)'$ by (λ), the notation of 11.8 holds for the elements of $\mathfrak{F}$ and $\mathfrak{H}$ with θ replaced by θ^{-1}, and $g = (1, 0)$.

q.e.d.

11.13 Lemma. *The function α occurring in the functional equation 11.10 has the following properties.*

a) $\alpha((0, b(b\theta))) = (b^{-1}, b^{-1}(b\theta)^{-1})$ for $b \neq 0$.
$\alpha((a, 0)) = (a^{-1}, 0)$ for $a \neq 0$.
$\alpha((a, a(a\theta))) = (0, a^{-1}(a\theta)^{-1})$ for $a \neq 0$.

b) *Suppose that $a \neq 0$, $b \neq 0$ and $b \neq a(a\theta)$. Then*
$\alpha((a, b)) = (\varepsilon a^{-1}, \varepsilon(v\theta)^{-1}b^{-1})$,

where

$$\lambda\theta = 1 + a(a\theta)b^{-1}, \quad v(v\theta) = \lambda^2(\lambda\theta)^2 + \lambda(\lambda\theta)$$

and

$$\varepsilon = (1 + \lambda)(\lambda^{-1} + v^{-1}).$$

Proof. a) By 11.12, $j = (0, 1)$, $g = (1, 0)$ and $g^{-1} = (1, 1)$. From the structure equation $sjs = gsg^{-1}$, we have

(1) $$\alpha((0, 1)) = (1, 1),$$

(2) $$\alpha((1, 0)) = (1, 0),$$

(3) $$\alpha((1, 1)) = (0, 1).$$

Thus, using 11.10b) (3), we obtain

$$\begin{aligned}\alpha((0, b(b\theta))) &= \alpha((b)^{-1}(0, 1)(b))\\ &= (b)\alpha((0, 1))(b)^{-1} = (b)(1, 1)(b)^{-1}\\ &= (b^{-1}, b^{-1}(b\theta)^{-1})\end{aligned}$$

and

$$\begin{aligned}\alpha((a, 0)) &= \alpha((a)^{-1}(1, 0)(a)) = (a)\alpha((1, 0))(a)^{-1}\\ &= (a)(1, 0)(a)^{-1} = (a^{-1}, 0).\end{aligned}$$

Finally,

$$\begin{aligned}\alpha((a, a(a\theta))) &= \alpha((a)^{-1}(1, 1)(a)) = (a)\alpha((1, 1))(a)^{-1} \\ &= (a)(0, 1)(a)^{-1} = (0, a^{-1}(a\theta)^{-1}).\end{aligned}$$

Thus a) is proved.

b) If $a \neq 0$, $b \neq 0$ and $b \neq a(a\theta)$, then (a, b) does not lie in any of the $\mathfrak{H}$-orbits containing j, g or g^{-1}. Hence by 11.11, there exist elements h, h' in $\mathfrak{H}$, with $h \neq 1$, such that

(4) $$h'^{-1}(a, b)h' = g^{-1}h^{-1}gh.$$

Put $h = (\lambda)$, $h' = (\mu)$. By 11.11c),

$$\alpha(g^{-1}h^{-1}gh) = (\nu)g^{-1}(\nu)^{-1}(\lambda)g(\lambda)^{-1},$$

where ν is the element of $GF(q)$ uniquely determined by

$$\nu(\nu\theta) = \lambda^2(\lambda\theta)^2 + \lambda(\lambda\theta).$$

Thus

$$\begin{aligned}\alpha(g^{-1}h^{-1}gh) &= (\nu^{-1}, \nu^{-1}(\nu\theta)^{-1})(\lambda^{-1}, 0) \\ &= (\nu^{-1} + \lambda^{-1}, \nu^{-1}(\nu\theta)^{-1} + (\nu\theta)^{-1}\lambda^{-1}),\end{aligned}$$

and by 11.10b) (3),

$$\begin{aligned}\alpha((a, b)) &= h'^{-1}\alpha(g^{-1}h^{-1}gh)h' \\ &= (\mu(\nu^{-1} + \lambda^{-1}), \mu(\mu\theta)(\nu\theta)^{-1}(\nu^{-1} + \lambda^{-1})).\end{aligned}$$

Now by (4),

$$\begin{aligned}(a, b) &= (\mu)(1, 1)(\lambda^{-1})(1, 0)(\lambda)(\mu^{-1}) \\ &= (\mu)(1, 1)(\lambda, 0)(\mu^{-1}) \\ &= (\mu)(1 + \lambda, 1 + \lambda)(\mu^{-1}) \\ &= (\mu^{-1}(1 + \lambda), \mu^{-1}(\mu\theta)^{-1}(1 + \lambda)).\end{aligned}$$

Hence

$$a = \mu^{-1}(1 + \lambda), \quad b = \mu^{-1}(\mu\theta)^{-1}(1 + \lambda).$$

Eliminating μ,

$$b(1+\lambda\theta) = \mu^{-1}(1+\lambda)((\mu^{-1}(1+\lambda))\theta) = a(a\theta),$$

which gives the stated formula for $\lambda\theta$. And

$$\alpha((a, b)) = ((\lambda^{-1}+v^{-1})(1+\lambda)a^{-1}, (\lambda^{-1}+v^{-1})(v\theta)^{-1}(1+\lambda)b^{-1}).$$

q.e.d.

11.14 Lemma. *For all $\lambda \in GF(q)$, $\lambda\theta^2 = \lambda^2$.*

Proof. For $q = 8$, let a be an element of $GF(8)$ such that $a^3 + a + 1 = 0$, and for $q > 8$, let a be any element which generates the field $GF(q)$ over $GF(2)$. We have

$$(a\theta^{-1}, 1 + a(a\theta^{-1})) = (a\theta^{-1}, a(a\theta^{-1}))(0, 1) = (a\theta^{-1}, a(a\theta^{-1}))j.$$

Hence by 11.10b) (6),

$$\alpha((a\theta^{-1}, 1 + a(a\theta^{-1}))) = \gamma(j)\alpha(x)\gamma(j)^{-1}\alpha(j), \tag{1}$$

where

$$x = \alpha((a\theta^{-1}, a(a\theta^{-1})))\beta(j).$$

Hence by 11.13a),

$$x = (0, (a\theta^{-1})^{-1}a^{-1})\beta(j).$$

Since $sjs = gsg^{-1}$, $\alpha(j) = g^{-1}$, $\beta(j) = g$ and $\gamma(j) = 1$. Thus

$$x = (0, (a\theta^{-1})^{-1}a^{-1})(1, 0) = (1, (a\theta^{-1})^{-1}a^{-1}).$$

Since $(a\theta^{-1})^{-1}a^{-1} = ((a(a\theta))\theta^{-1})^{-1} \neq 1$, it follows from 11.13b) that

$$\alpha(x) = (\eta, \eta(\mu\theta)^{-1}a(a\theta^{-1})),$$

where

$$\eta = (1+\kappa)(\kappa^{-1}+\mu^{-1}), \tag{2}$$

$$\kappa\theta = 1 + a(a\theta^{-1}), \tag{3}$$

(4) $$\mu(\mu\theta) = \kappa^2(\kappa\theta)^2 + \kappa(\kappa\theta).$$

Hence by (1),

$$\begin{aligned}\alpha((a\theta^{-1}, 1 + a(a\theta^{-1}))) &= (\eta, \eta(\mu\theta)^{-1}a(a\theta^{-1}))(1, 1)\\ &= (\eta + 1, 1 + \eta(\mu\theta)^{-1}a(a\theta^{-1}) + \eta\theta).\end{aligned}$$

Thus by 11.13b),

(5) $$\eta + 1 = \varepsilon(a\theta^{-1})^{-1},$$

(6) $$1 + \eta(\mu\theta)^{-1}a(a\theta^{-1}) + \eta\theta = \varepsilon(\nu\theta)^{-1}(1 + a(a\theta^{-1}))^{-1},$$

where

(7) $$\varepsilon = (1 + \xi)(\xi^{-1} + \nu^{-1}),$$

(8) $$\xi\theta = 1 + (a\theta^{-1})a(1 + a(a\theta^{-1}))^{-1} = (1 + a(a\theta^{-1}))^{-1},$$

(9) $$\nu(\nu\theta) = \xi^2(\xi\theta)^2 + \xi(\xi\theta).$$

(8) and (3) give

$$\xi\theta = (1 + a(a\theta^{-1}))^{-1} = \kappa^{-1}\theta,$$

so $\xi = \kappa^{-1}$. Hence by (4) and (9),

$$\begin{aligned}\mu(\mu\theta) &= \xi^{-2}(\xi\theta)^{-2} + \xi^{-1}(\xi\theta)^{-1}\\ &= \xi^{-3}(\xi\theta)^{-3}(\xi(\xi\theta) + \xi^2(\xi\theta)^2) = \xi^{-3}(\xi\theta)^{-3}\nu(\nu\theta).\end{aligned}$$

Thus $(\mu\xi^3\nu^{-1})(\mu\xi^3\nu^{-1})\theta = 1$. Since the mapping $x \to x(x\theta)$ is injective, it follows that $\nu = \mu\xi^3$. Thus by (6) and (8),

$$\mu\theta + \eta a(a\theta^{-1}) + (\eta\theta)(\mu\theta) = \varepsilon(\xi\theta)^{-2},$$

or

$$\eta + \eta(\xi\theta)^{-1} + (1 + \eta\theta)(\mu\theta) = \varepsilon(\xi\theta)^{-2}.$$

Using (2) and (7) to eliminate η and ε,

$$(\xi + \mu^{-1} + 1 + \xi^{-1}\mu^{-1})(1 + (\xi\theta)^{-1}) + (\xi\theta)(\mu\theta) + 1 + (\xi\theta)^{-1}$$
$$= (\xi^{-1} + \mu^{-1}\xi^{-3} + 1 + \mu^{-1}\xi^{-2})(\xi\theta)^{-2}.$$

Multiplying by μ and using (4),

$$(\xi\mu + 1 + \xi^{-1})(1 + (\xi\theta)^{-1}) + (\xi\theta)^{-1}\xi^{-2} + \xi^{-1}$$
$$= (\xi^{-1}\mu + \xi^{-3} + \mu + \xi^{-2})(\xi\theta)^{-2}.$$

Collecting the terms involving μ on the left hand side,

$$\mu(\xi + \xi(\xi\theta)^{-1} + (\xi^{-1} + 1)(\xi\theta)^{-2})$$
$$= 1 + (\xi\theta)^{-1} + \xi^{-1}(\xi\theta)^{-1} + \xi^{-2}(\xi\theta)^{-1} + (\xi^{-3} + \xi^{-2})(\xi\theta)^{-2},$$

or

$$\mu(\xi + (\xi\theta)^{-1})(1 + (\xi\theta)^{-1} + \xi^{-1}(\xi\theta)^{-1})$$
$$= (1 + \xi^{-2}(\xi\theta)^{-1})(1 + (\xi\theta)^{-1} + \xi^{-1}(\xi\theta)^{-1}).$$

If $1 + (\xi\theta)^{-1} + \xi^{-1}(\xi\theta)^{-1} \neq 0$, it follows that

$$\mu(\xi + (\xi\theta)^{-1}) = 1 + \xi^{-2}(\xi\theta)^{-1}.$$

We apply θ, multiply and use (4). This gives

$$(\xi^{-2}(\xi\theta)^{-2} + \xi^{-1}(\xi\theta)^{-1})(\xi + (\xi\theta)^{-1})(\xi\theta + (\xi\theta^2)^{-1})$$
$$= (1 + \xi^{-2}(\xi\theta)^{-1})(1 + (\xi\theta)^{-2}(\xi\theta^2)^{-1}).$$

Multiplying by $\xi^2(\xi\theta)^3$, we get

$$(1 + \xi(\xi\theta))^2(\xi\theta + (\xi\theta^2)^{-1}) = (1 + \xi^2(\xi\theta))((\xi\theta)^2 + (\xi\theta^2)^{-1}),$$

which reduces to

$$(\xi\theta^2 + \xi^2)(1 + \xi\theta) = 0.$$

By (8), $\xi\theta \neq 1$. Hence $\xi\theta^2 = \xi^2$. Now again by (8),

$$a(a\theta^{-1}) = 1 + (\xi\theta)^{-1}.$$

Applying θ,

$$(a\theta)a = 1 + \xi^{-2} = (1 + \xi^{-1})^2,$$

and applying θ again,

$$(a\theta^2)(a\theta) = (1 + \xi^{-1}\theta)^2 = a^2(a\theta^{-1})^2.$$

So if $b = (a\theta^2)a^{-2}$,

$$b(b\theta^{-1}) = (a\theta^2)a^{-2}(a\theta)(a\theta^{-1})^{-2} = 1.$$

Since $x \to x(x\theta)$ is injective, $b = 1$ and $a\theta^2 = a^2$.

Now suppose that $1 + (\xi\theta)^{-1} + \xi^{-1}(\xi\theta)^{-1} = 0$. Thus $\xi\theta = 1 + \xi^{-1}$. It follows using (8) that

$$(1 + \xi^{-1})(1 + a(a\theta^{-1})) = 1.$$

Thus

$$\xi = (1 + \xi)(1 + a(a\theta^{-1})) = (\xi\theta)^{-1} + \xi + \xi a(a\theta^{-1}),$$

and $\xi^{-1}(\xi\theta)^{-1} = a(a\theta^{-1})$. Since $x \to x(x\theta)$ is injective, it follows that $a\theta^{-1} = \xi^{-1}$. Thus $a = (\xi\theta)^{-1} = (1 + \xi^{-1})^{-1} = \xi(1 + \xi)^{-1}$. Since $\xi\theta = 1 + \xi^{-1}$, it follows that

$$a\theta = (1 + \xi^{-1})\xi = 1 + \xi.$$

Hence $a\theta^2 = \xi^{-1} = a\theta^{-1}$ and $a\theta^3 = a$. Since a generates $GF(q)$, it follows that $\theta^3 = 1$. Since 0, 1 are the only fixed points of θ, $q = 8$. Thus by choice of a, $a^3 + a + 1 = 0$, and

$$a\theta^2 = \xi^{-1} = (1 + a)a^{-1} = a^2.$$

In either case, then, $a\theta^2 = a^2$. Since a generates $GF(q)$, it follows that $\lambda\theta^2 = \lambda^2$ for all $\lambda \in GF(q)$. **q.e.d.**

11.15 Theorem (SUZUKI [3]). *Let $\mathfrak{G}$ be a Zassenhaus group of degree $2^f + 1$ and order $(2^f + 1)2^f d$. Suppose that the Frobenius kernel $\mathfrak{F}$ of $\mathfrak{G}_\infty$ is non-Abelian. Then f is even, so $2^f = q^2$ for some q, and $\mathfrak{G}$ is $Sz(q)$.*

Proof. By 10.2, $d = 2^l - 1$ for some l, and by 10.9, 3 does not divide d. Thus $\mathfrak{G}$ satisfies the conditions of this section. By 11.2, $f = 2l$ and $\mathfrak{F} \cong A(l, \theta)$. Hence $2^f = q^2$ for $q = 2^l$. By 11.14, $\lambda\theta^2 = \lambda^2$ for all

$\lambda \in GF(q)$. Hence by 11.8 and 3.1, there exists an isomorphism ξ of $\mathfrak{G}_\infty$ onto $(Sz(q))_{p_\infty}$ for which

$$(a, b)\xi = S(a, b), \quad (\lambda)\xi = M(\lambda).$$

In particular, $g\xi = S(1, 0)$ and $j\xi = S(0, 1)$.

We extend ξ to a mapping of $\mathfrak{G}$ into $Sz(q)$ in the following way. If $u \in \mathfrak{G} - \mathfrak{G}_\infty$, then by 1.7, $u = ysz$ for uniquely defined elements y of $\mathfrak{G}_\infty$ and z of $\mathfrak{F}$. We put $u\xi = (y\xi)T(z\xi)$ (see 3.2). Thus ξ is an injective mapping of $\mathfrak{G}$ onto $Sz(q)$. We show that ξ is an isomorphism. Thus suppose that u_1, u_2 are elements of $\mathfrak{G}$.

If $u_1 \in \mathfrak{G}_\infty$ and $u_2 \in \mathfrak{G}_\infty$, then clearly $(u_1u_2)\xi = (u_1\xi)(u_2\xi)$. If $u_1 \in \mathfrak{G}_\infty$ and $u_2 = y_2sz_2$ with $y_2 \in \mathfrak{G}_\infty$, $z_2 \in \mathfrak{F}$, then $u_1u_2 = u_1y_2sz_2$ and

$$(u_1u_2)\xi = (u_1y_2)\xi T(z_2\xi) = (u_1\xi)(y_2\xi)T(z_2\xi) = (u_1\xi)(u_2\xi).$$

If $u_1 = y_1sz_1$ with $y_1 \in \mathfrak{G}_\infty$ and $z_1 \in \mathfrak{F}$, and if $u_2 \in \mathfrak{G}_\infty$, then $z_1u_2 = hx$ for some $h \in \mathfrak{H}$, $x \in \mathfrak{F}$. Thus

$$u_1u_2 = y_1sz_1u_2 = y_1shx = y_1h^{-1}sx,$$

so

$$(u_1u_2)\xi = (y_1h^{-1})\xi T(x\xi) = (y_1\xi)(h\xi)^{-1}T(x\xi).$$

Since $M(\lambda)^{-1}T = TM(\lambda)$, it follows that

$$(u_1u_2)\xi = (y_1\xi)T(h\xi)(x\xi) = (y_1\xi)T(z_1\xi)(u_2\xi) = (u_1\xi)(u_2\xi).$$

To deal with the remaining case, we use the equation

$$sxs = \beta(x)\gamma(x)s\alpha(x),$$

which holds for all $x \in \mathfrak{F} - \{1\}$. In $Sz(q)$, there holds a similar equation

$$TXT = B(X)\Gamma(X)TA(X),$$

of course. Now in 11.13, we have calculated the value of $\alpha(x)$ in terms of the labelling of elements of $\mathfrak{F}$ and $\mathfrak{H}$ in the notation of 11.8 with $g = (1, 0)$. By means of ξ, the same labelling holds in $Sz(q)$, so $A(x\xi) = \alpha(x)\xi$ for all $x \in \mathfrak{F} - \{1\}$. By 11.10b) (2),

$$B(x\xi) = A((x\xi)^{-1})^{-1} = (\alpha(x^{-1})^{-1})\xi = \beta(x)\xi.$$

And by 11.10b) (5),

$$\Gamma(x\xi)^{-1}A(x\xi)\Gamma(x\xi) = A(B(x\xi))^{-1} = A(\beta(x)\xi)^{-1} = (\alpha(\beta(x))\xi)^{-1}$$
$$= (\gamma(x)^{-1}\alpha(x)\gamma(x))\xi = (\gamma(x)\xi)^{-1}A(x\xi)(\gamma(x)\xi).$$

Thus $\Gamma(x\xi)(\gamma(x)\xi)^{-1}$ lies in the centralizer of $A(x\xi)$. Since $A(x\xi) \neq 1$, it follows that $\Gamma(x\xi)(\gamma(x)\xi)^{-1}$ is a 2-element. But it also lies in $(Sz(q))_{p_\infty, p_{(0,0)}}$, which is of odd order. Hence $\Gamma(x\xi) = \gamma(x)\xi$ for all $x \in \mathfrak{F} - \{1\}$. Thus

$$(sxs)\xi = (\beta(x)\gamma(x))\xi T(\alpha(x)\xi) = B(x\xi)\Gamma(x\xi)TA(x\xi) = T(x\xi)T.$$

Now suppose that $u_1 = y_1sz_1$, $u_2 = y_2sz_2$ with y_1, y_2 in $\mathfrak{G}_\infty$ and z_1, z_2 in $\mathfrak{F}$. Write $z_1y_2 = hx$, with $h \in \mathfrak{H}$, $x \in \mathfrak{F}$. Then

$$u_1u_2 = y_1shxsz_2 = y_1h^{-1}sxsz_2.$$

If $x = 1$, this gives

$$(u_1u_2)\xi = (y_1h^{-1}z_2)\xi = (y_1\xi)(h\xi)^{-1}T(x\xi)T(z_2\xi)$$
$$= (y_1\xi)T((hx)\xi)T(z_2\xi) = (y_1\xi)T(z_1\xi)(y_2\xi)T(z_2\xi)$$
$$= (u_1\xi)(u_2\xi).$$

If $x \neq 1$, then

$$u_1u_2 = y_1h^{-1}\beta(x)\gamma(x)s\alpha(x)z_2,$$

so

$$(u_1u_2)\xi = (y_1\xi)(h\xi)^{-1}(\beta(x)\xi)(\gamma(x)\xi)T(\alpha(x)\xi)(z_2\xi)$$
$$= (y_1\xi)(h\xi)^{-1}B(x\xi)\Gamma(x\xi)TA(x\xi)(z_2\xi)$$
$$= (y_1\xi)(h\xi)^{-1}T(x\xi)T(z_2\xi)$$
$$= (y_1\xi)T(h\xi)(x\xi)T(z_2\xi)$$
$$= (y_1\xi)T(z_1\xi)(y_2\xi)T(z_2\xi)$$
$$= (u_1\xi)(u_2\xi).$$

Hence ξ is an isomorphism of $\mathfrak{G}$ onto $Sz(q)$. **q.e.d.**

This completes the proof of the main theorem of this chapter.

11.16 Theorem. *Suppose that $\mathfrak{G}$ is a Zassenhaus group of degree $n + 1$ and order $(n + 1)nd$. Then $n = p^f$ is a prime-power, and the following are the only possibilities.*

a) $\mathfrak{G} = PGL(2, p^f)$.

b) *p is odd and $\mathfrak{G} = PSL(2, p^f)$.*

c) *p is odd, f is even and $\mathfrak{G}$ is one of the sharply triply transitive groups $M(p^f)$ given in 1.3.*

d) *$p = 2$ and $\mathfrak{G}$ is a Suzuki group $Sz(q)$.*

Proof. If $\mathfrak{G}$ is sharply triply transitive, then $\mathfrak{G}$ is $PGL(2, p^f)$ or $M(p^f)$ (p odd), by 2.1 and 2.6. If $\mathfrak{F}$ is Abelian, the only other possibility is $PSL(2, p^f)$ (p odd), by 6.8. In the remaining case, $n = p^f$ by 6.1 and $p = 2$ by 9.1. By 11.15, $\mathfrak{G} = Sz(q)$. **q.e.d.**

§ 12. Order Formulae

In this section we illustrate further the counting techniques introduced earlier in this chapter.

12.1 Lemma*. *Suppose that $\mathfrak{G}$ is a group of operators on a group $\mathfrak{H}$ and that $(|\mathfrak{G}|, |\mathfrak{H}|) = 1$. If $\mathfrak{S}$ is a $\mathfrak{G}$-invariant Sylow p-subgroup of $\mathfrak{H}$, then $\mathfrak{S} \cap \mathbf{C}_{\mathfrak{H}}(\mathfrak{X})$ is a Sylow p-subgroup of $\mathbf{C}_{\mathfrak{H}}(\mathfrak{X})$ for any subgroup $\mathfrak{X}$ of $\mathfrak{G}$.*

Proof. Suppose that $\mathfrak{P} \in S_p(\mathbf{C}_{\mathfrak{H}}(\mathfrak{X}))$. By IX, 1.11, $\mathfrak{P}$ is contained in an $\mathfrak{X}$-invariant Sylow p-subgroup $\mathfrak{S}_1$ of $\mathfrak{H}$ and $\mathfrak{S}_1^x = \mathfrak{S}$ for some $x \in \mathbf{C}_{\mathfrak{H}}(\mathfrak{X})$. Thus $\mathfrak{P} = \mathfrak{S}_1 \cap \mathbf{C}_{\mathfrak{H}}(\mathfrak{X})$ and $\mathfrak{P}^x = \mathfrak{S} \cap \mathbf{C}_{\mathfrak{H}}(\mathfrak{X}) \in S_p(\mathbf{C}_{\mathfrak{H}}(\mathfrak{X}))$. **q.e.d.**

12.2 Lemma*. *Suppose that $\mathfrak{G}$ is a group of operators on a group $\mathfrak{H}$ and that $(|\mathfrak{G}|, |\mathfrak{H}|) = 1$. If $\mathfrak{K}$ is a $\mathfrak{G}$-invariant normal subgroup of $\mathfrak{H}$,*

$$|\mathbf{C}_{\mathfrak{H}}(\mathfrak{G})| = |\mathbf{C}_{\mathfrak{K}}(\mathfrak{G})|\,|\mathbf{C}_{\mathfrak{H}/\mathfrak{K}}(\mathfrak{G})|.$$

Proof. By I, 18.6, $\mathbf{C}_{\mathfrak{H}/\mathfrak{K}}(\mathfrak{G}) = \mathbf{C}_{\mathfrak{H}}(\mathfrak{G})\mathfrak{K}/\mathfrak{K} \cong \mathbf{C}_{\mathfrak{H}}(\mathfrak{G})/\mathbf{C}_{\mathfrak{K}}(\mathfrak{G})$. **q.e.d.**

12.3 Lemma. *Let V be a $\mathsf{K}\mathfrak{G}$-module, where $\mathsf{K} = GF(p)$ and p does not divide $|\mathfrak{G}|$. For any positive integer m, there exists an $\mathsf{L}\mathfrak{G}$-module W, where $\mathsf{L} = \mathbb{Z}/p^m\mathbb{Z}$, such that $\mathsf{W}/p\mathsf{W}$, V are isomorphic $\mathsf{K}\mathfrak{G}$-modules and the additive group W is homocyclic of exponent p^m.*

*The proofs of Lemmas 12.1 and 12.2 both utilize the solubility of groups of odd order.

Proof. Since V is necessarily a completely reducible $K\mathfrak{G}$-module, there is no loss of generality in supposing that V is irreducible.

Now the group ring $L\mathfrak{G}$ may be regarded as a homocyclic additive p-group with $\mathfrak{G}$ as a group of operators. By VIII, 5.9, $L\mathfrak{G} = W_1 \oplus \cdots \oplus W_r$, where each W_i is a homocyclic $\mathfrak{G}$-invariant group and $V_i = W_i/pW_i$ is an irreducible $K\mathfrak{G}$-module. Thus

$$K\mathfrak{G} \cong V_1 \oplus \cdots \oplus V_r.$$

Thus by the Jordan-Hölder theorem, V is $K\mathfrak{G}$-isomorphic to some V_i and W_i has the required properties. **q.e.d.**

12.4 Theorem (Wielandt's order formula [2]). *Let $\mathfrak{G}$ be a group of operators on a group $\mathfrak{H}$ for which $(|\mathfrak{G}|, |\mathfrak{H}|) = 1$. Suppose that $\mathfrak{G}_1, \ldots, \mathfrak{G}_m$ are subgroups of $\mathfrak{G}$ for which there exists a relation*

$$\sum_{i=1}^{m} n_i \underline{\mathfrak{G}_i} = 0 \quad (n_i \in \mathbb{Z}),$$

where $\underline{\mathfrak{G}_i} = \sum_{x \in \mathfrak{G}_i} x$. Then if $|\mathfrak{G}_i| = g_i$,

$$\prod_{i=1}^{m} |\mathbf{C}_{\mathfrak{H}}(\mathfrak{G}_i)|^{n_i g_i} = 1.$$

Proof. This is proved by induction on $|\mathfrak{H}|$. If $\mathfrak{H}$ is not a p-group, write $|\mathfrak{H}| = p_1^{a_1} \cdots p_l^{a_l}$, where $p_1, \ldots, p_l$ are distinct primes and $a_1, \ldots, a_l$ are positive integers. By 12.1, there exists a $\mathfrak{G}$-invariant Sylow p_j-subgroup $\mathfrak{S}_j$ of $\mathfrak{H}$ such that

$$\mathfrak{S}_j \cap \mathbf{C}_{\mathfrak{H}}(\mathfrak{G}_i) \in S_{p_j}(\mathbf{C}_{\mathfrak{H}}(\mathfrak{G}_i)) \quad (j = 1, \ldots, l).$$

By the inductive hypothesis applied to $\mathfrak{S}_j$,

$$\prod_{i=1}^{m} |\mathfrak{S}_j \cap \mathbf{C}_{\mathfrak{H}}(\mathfrak{G}_i)|^{n_i g_i} = 1.$$

Since

$$|\mathbf{C}_{\mathfrak{H}}(\mathfrak{G}_i)| = \prod_{j=1}^{l} |\mathfrak{S}_j \cap \mathbf{C}_{\mathfrak{H}}(\mathfrak{G}_i)|,$$

the assertion follows at once.

Suppose then that $\mathfrak{H}$ is a p-group and put $\mathfrak{K} = \Phi(\mathfrak{H})$. By 12.2,

$$|\mathbf{C}_{\mathfrak{H}}(\mathfrak{G}_i)| = |\mathbf{C}_{\mathfrak{K}}(\mathfrak{G}_i)|\,|\mathbf{C}_{\mathfrak{H}/\mathfrak{K}}(\mathfrak{G}_i)|,$$

so

$$\prod_{i=1}^{m} |\mathbf{C}_{\mathfrak{H}}(\mathfrak{G}_i)|^{n_i g_i} = \prod_{i=1}^{m} |\mathbf{C}_{\mathfrak{K}}(\mathfrak{G}_i)|^{n_i g_i} |\mathbf{C}_{\mathfrak{H}/\mathfrak{K}}(\mathfrak{G}_i)|^{n_i g_i}.$$

Hence if $\mathfrak{K} \neq 1$, the result again follows from the inductive hypothesis.

Now suppose that $\mathfrak{K} = 1$ and $\mathfrak{H}$ is elementary Abelian. Then $\mathfrak{H}$ is a $\mathsf{K}\mathfrak{G}$-module, where $\mathsf{K} = GF(p)$. Let k be any positive integer. By 12.3, there exists an $\mathsf{L}\mathfrak{G}$-module W, where $\mathsf{L} = \mathbb{Z}/p^k\mathbb{Z}$, such that $\overline{\mathsf{W}} = \mathsf{W}/p\mathsf{W}$ and $\mathfrak{H}$ are isomorphic $\mathsf{K}\mathfrak{G}$-modules and the additive group W is homocyclic of exponent p^k.

Let $|\mathbf{C}_{\mathfrak{H}}(\mathfrak{G}_i)| = p^{f_i}$ $(i = 1, \dots, m)$. Then $|\mathbf{C}_{\overline{\mathsf{W}}}(\mathfrak{G}_i)| = p^{f_i}$. By I, 18.6, $\mathbf{C}_{\overline{\mathsf{W}}}(\mathfrak{G}_i) = (\mathbf{C}_{\mathsf{W}}(\mathfrak{G}_i) + p\mathsf{W})/p\mathsf{W}$. But by III, 13.4,

$$\mathsf{W} = [\mathsf{W}, \mathfrak{G}_i] \oplus \mathbf{C}_{\mathsf{W}}(\mathfrak{G}_i) \quad (i = 1, \dots, m).$$

Hence

$$\mathbf{C}_{\overline{\mathsf{W}}}(\mathfrak{G}_i) = (\mathbf{C}_{\mathsf{W}}(\mathfrak{G}_i) + p\mathsf{W})/p\mathsf{W} \cong \mathbf{C}_{\mathsf{W}}(\mathfrak{G}_i)/p\mathbf{C}_{\mathsf{W}}(\mathfrak{G}_i),$$

and $\mathbf{C}_{\mathsf{W}}(\mathfrak{G}_i)$ is homocyclic of exponent p^k with f_i invariants.

We define the L-linear mapping γ_i of W into W by putting

$$a\gamma_i = \sum_{x \in \mathfrak{G}_i} ax = a\underline{\mathfrak{G}_i}.$$

We calculate $\operatorname{tr}\gamma_i$. If $b \in [\mathsf{W}, \mathfrak{G}_i]$, then $b\gamma_i = 0$, for

$$\begin{aligned}(a(1 - x))\gamma_i &= a\gamma_i - (ax)\gamma_i \\ &= a\underline{\mathfrak{G}_i} - ax\underline{\mathfrak{G}_i} = a\underline{\mathfrak{G}_i} - a\underline{\mathfrak{G}_i} = 0 \quad (a \in \mathsf{W}, x \in \mathfrak{G}_i).\end{aligned}$$

But if $c \in \mathbf{C}_{\mathsf{W}}(\mathfrak{G}_i)$, $c\gamma_i = g_i c$. Thus

$$\operatorname{tr}\gamma_i = g_i f_i.$$

But by hypothesis, $\sum_i n_i \gamma_i = 0$, so

$$\sum_i n_i g_i f_i \equiv 0 \ (p^k).$$

Since this holds for all k,

$$\sum_i n_i g_i f_i = 0,$$

and

$$\prod_{i=1}^{m} |\mathbf{C}_{\mathfrak{H}}(\mathfrak{G}_i)|^{n_i g_i} = p^{\sum n_i g_i f_i} = 1. \qquad \textbf{q.e.d.}$$

12.5 Corollary. *Suppose that $\mathfrak{G}_i \leq \mathfrak{G}$ $(i = 1, \ldots, n)$ and that $\mathfrak{G} = \bigcup_{i=1}^{n} \mathfrak{G}_i$. For $i_1 < \cdots < i_k$, put*

$$\mathfrak{G}_{i_1 \cdots i_k} = \mathfrak{G}_{i_1} \cap \cdots \cap \mathfrak{G}_{i_k}, \quad g_{i_1 \cdots i_k} = |\mathfrak{G}_{i_1 \cdots i_k}|.$$

Suppose that $\mathfrak{G}$ is a group of operators on a group $\mathfrak{H}$ and that $(|\mathfrak{G}|, |\mathfrak{H}|) = 1$. If $f_{i_1 \cdots i_k} = |\mathbf{C}_{\mathfrak{H}}(\mathfrak{G}_{i_1 \cdots i_k})|$, then

$$\prod_{\substack{i_1 < \cdots < i_k \\ k \geq 0}} f_{i_1 \cdots i_k}^{(-1)^k g_{i_1 \cdots i_k}} = 1.$$

Proof. If the element x of $\mathfrak{G}$ lies in precisely r of the $\mathfrak{G}_i$, then x lies in precisely $\binom{r}{k}$ of the $\mathfrak{G}_{i_1 \cdots i_k}$ $(0 \leq k \leq r)$. Thus

$$\underline{\mathfrak{G}} - \sum \underline{\mathfrak{G}_i} + \sum \underline{\mathfrak{G}_{ij}} - \cdots (-1)^n \underline{\mathfrak{G}_{12\cdots n}} = 0,$$

since the coefficient of x on the left hand side is

$$1 - r + \binom{r}{2} - \cdots + (-1)^r = 0.$$

The assertion thus follows at once from the order formula. **q.e.d.**

12.6 Examples. The formula given in the corollary becomes simplest if $\mathfrak{G}$ has a partition. In this case, $\mathfrak{G}$ has a set of subgroups $\mathfrak{G}_1, \ldots, \mathfrak{G}_n$ satisfying

$$\mathfrak{G} = \mathfrak{G}_1 \cup \cdots \cup \mathfrak{G}_n, \quad \mathfrak{G}_i \cap \mathfrak{G}_j = 1 \quad (i \neq j).$$

Then for $k \geq 2$, $g_{i_1 \cdots i_k} = 1$ and $f_{i_1 \cdots i_k} = |\mathfrak{H}|$. The formula thus becomes

$$|\mathbf{C}_{\mathfrak{H}}(\mathfrak{G})|^{|\mathfrak{G}|} f_1^{-g_1} \cdots f_n^{-g_n} |\mathfrak{H}|^{\binom{n}{2} - \binom{n}{3} + \cdots + (-1)^n} = 1,$$

or

$$|\mathbf{C}_{\mathfrak{H}}(\mathfrak{G})|^{|\mathfrak{G}|}|\mathfrak{H}|^{n-1} = f_1^{g_1} \cdots f_n^{g_n}.$$

It follows, for example, that if $\mathfrak{H} \neq 1$, then some $f_i \neq 1$, so $\mathbf{C}_{\mathfrak{H}}(\mathfrak{G}_i) \neq 1$ for some i. (V, 8.12 follows at once from this.)

If $\mathfrak{G}$ is elementary Abelian of order 4, the formula becomes

$$|\mathbf{C}_{\mathfrak{H}}(\mathfrak{G})|^2|\mathfrak{H}| = f_1 f_2 f_3 = \prod_{1 \neq x \in \mathfrak{G}} |\mathbf{C}_{\mathfrak{H}}(x)|.$$

It is this case, which is due to R. Brauer, that has been most frequently applied.

We give another order formula, which has several applications in characterizations of simple groups.

12.7 Theorem (THOMPSON's order formula). *Suppose that $\mathfrak{G}$ has more than one class of involutions and let* R *be a complete set of representatives of the classes of involutions in $\mathfrak{G}$. Let a, b be non-conjugate involutions in $\mathfrak{G}$, and for each $t \in \mathsf{R}$, let $n(t)$ denote the number of ordered pairs of involutions (x, y) for which x is conjugate to a, y is conjugate to b and $t \in \langle xy \rangle$. Then*

$$|\mathfrak{G}| = |\mathbf{C}_{\mathfrak{G}}(a)||\mathbf{C}_{\mathfrak{G}}(b)| \sum_{t \in \mathsf{R}} \frac{n(t)}{|\mathbf{C}_{\mathfrak{G}}(t)|}.$$

Proof. Let $g = |\mathfrak{G}|$, $c_1 = |\mathbf{C}_{\mathfrak{G}}(a)|$, $c_2 = |\mathbf{C}_{\mathfrak{G}}(b)|$. Let J be the set of ordered pairs (x, y) for which x, y are conjugate to a, b respectively. Thus $|\mathsf{J}| = \dfrac{g^2}{c_1 c_2}$.

Let I be the set of involutions in $\mathfrak{G}$. For $u \in \mathsf{I}$, let J_u be the set of all $(x, y) \in \mathsf{J}$ for which $u \in \langle xy \rangle$. Thus if $z \in \mathfrak{G}$,

$$\mathsf{J}_{u^z} = \{(x^z, y^z) | (x, y) \in \mathsf{J}_u\},$$

so $|\mathsf{J}_u| = n(t)$, where t is the element of R conjugate to u.

Suppose that $(x, y) \in \mathsf{J}$. Then x, y are not conjugate in $\mathfrak{G}$. But $\mathfrak{D} = \langle x, y \rangle$ is a dihedral group and $|\mathfrak{D} : \langle xy \rangle| = 2$. Therefore xy is of even order $2m$, and if $u = (xy)^m$, $(x, y) \in \mathsf{J}_u$. Hence

$$\mathsf{J} = \bigcup_{u \in \mathsf{I}} \mathsf{J}_u.$$

This union is clearly disjoint, so

$$\frac{g^2}{c_1 c_2} = |\mathsf{J}| = \sum_{u \in \mathsf{I}} |\mathsf{J}_u|$$

$$= \sum_{t \in \mathsf{R}} |\mathfrak{G} : \mathbf{C}_{\mathfrak{G}}(t)| n(t) = g \sum_{t \in \mathsf{R}} \frac{n(t)}{|\mathbf{C}_{\mathfrak{G}}(t)|}.$$

Hence

$$g = c_1 c_2 \sum_{t \in \mathsf{R}} \frac{n(t)}{|\mathbf{C}_{\mathfrak{G}}(t)|}.$$ **q.e.d.**

§ 13. Survey of Ree Groups

In view of the determination of the Zassenhaus groups, it is desirable to widen the definition. This leads for example to the following problem.

13.1 Problem. Find all permutation groups $\mathfrak{G}$ on $\Omega = \{1, \dots, n\}$ with the following properties.

(1) $\mathfrak{G}$ is doubly transitive.

(2) The stabiliser $\mathfrak{G}_1$ of 1 in $\mathfrak{G}$ has a regular normal subgroup.

(3) $\mathfrak{G}$ has no regular normal subgroup.

Obviously, every Zassenhaus group has these properties. The following are further examples of groups with them.

a) $P\Gamma L(2, p^f)$, represented as a permutation group on the projective line on $GF(p^f)$.

b) The unitary groups $PSU(3, q^2)$ and $PU(3, q^2)$, represented as permutation groups of degree $q^3 + 1$ on

$$\Omega = \{\langle v \rangle | 0 \neq v \in \mathsf{V}(3, q^2), (v, v) = 0\}$$

(see II, 10.12).

However, there are other essentially different groups with the properties in 13.1. In 3.8 we have seen how the Suzuki group $Sz(q)$ arises from the symplectic group $PSp(4, q)$, where $q = 2^{2n+1}$. In a similar way, Ree obtained simple groups from the Chevalley groups of type G_2 over $GF(q)$, where $q = 3^{2n+1}$ (REE [1], see II, 10. 17b)). These are denoted by ${}^2G_2(3^{2n+1})$. We describe the properties of these groups.

13.2 Theorem. *Suppose that* $q = 3^{2n+1}$, *where* $n \geq 1$. *Then there exist groups* $R(q)$ *of order* $(q^3 + 1)q^3(q - 1)$ *with the following properties.*

a) $R(q)$ *is doubly transitive of degree* $q^3 + 1$.

b) $R(q)$ *is simple.*

c) *The stabiliser* $R(q)_1$ *has a regular normal subgroup* $\mathfrak{P}$ *isomorphic to the group of all triples* (x, y, z) *with* x, y, z *in* $GF(q)$ *and the following multiplication rule.*

$$(x_1, y_1, z_1)(x_2, y_2, z_2)$$
$$= (x_1 + x_2, y_1 + y_2 + x_1(x_2\sigma), z_1 + z_2 - x_1y_2 + y_1x_2 - x_1(x_1\sigma)x_2).$$

Here σ *is the automorphism of* $GF(q)$ *for which* $x\sigma^2 = x^3$ *for all* $x \in GF(q)$. *We have*

$$\mathfrak{P}' = \Phi(\mathfrak{P}) = \{(0, y, z) \mid y, z \in GF(q)\}$$

and

$$\mathbf{Z}(\mathfrak{P}) = \gamma_3(\mathfrak{P}) = \{(0, 0, z) \mid z \in GF(q)\}.$$

d) *The stabiliser* $\mathfrak{H} = R(q)_{1,2}$ *of two symbols is cyclic of order* $q - 1$. $\mathfrak{H}$ *is isomorphic to the multiplicative group of all* (λ) *with* $\lambda \in GF(q)^\times$ *and*

$$(\lambda_1)(\lambda_2) = (\lambda_1\lambda_2), \quad (x, y, z)^{(\lambda)} = (\lambda x, \lambda(\lambda\sigma)y, \lambda^2(\lambda\sigma)z).$$

The stabiliser of three symbols is of order 2.

e) *The Sylow 2-subgroup* $\mathfrak{S}$ *of* $R(q)$ *is elementary Abelian of order 8. Also* $\mathbf{C}_{\mathfrak{G}}(\mathfrak{S}) = \mathfrak{S}$ *and* $|\mathbf{N}_{\mathfrak{G}}(\mathfrak{S})| = 8 \cdot 7 \cdot 3$. *Thus all involutions in* $R(q)$ *are conjugate.*

f) $R(q)$ *has Abelian Hall subgroups* $\mathfrak{H}_i$ $(i = 1, 2)$ *of orders* $q + 1 \pm 3m$, *where* $m = 3^n$. $\mathfrak{H}_i$ *is the centralizer of each of its non-identity elements and* $|\mathbf{N}_{\mathfrak{G}}(\mathfrak{H}_i)/\mathfrak{H}_i| = 6$. *(Note that* $(q + 1 + 3m)(q + 1 - 3m)(q + 1) = q^3 + 1$; *cf. 3.10.)*

g) *For* $p > 3$, *the Sylow* p*-subgroups of* $R(q)$ *are cyclic.*

h) *The centralizer in* $R(q)$ *of an involution* j *is* $\langle j \rangle \times PSL(2, q)$.

i) $R(q)$ *has a faithful representation of degree 7 in* $GF(q)$.

A geometrical description of $R(q)$ independent of the Lie theory has been given by Tits (TITS [2]). It is analogous to the construction of $Sz(q)$ as the group of collineations which leave an ovoid fixed (3.3).

Attempts to characterize the Ree groups led to the following definition.

13.3 Definition. A group $\mathfrak{G}$ is said to be of *Ree type* if $\mathfrak{G}$ has the following properties.

(1) The Sylow 2-subgroups of $\mathfrak{G}$ are Abelian.

(2) $\mathfrak{G}$ has no normal subgroup of index 2.

(3) $\mathfrak{G}$ possesses an involution j for which $\mathbf{C}_{\mathfrak{G}}(j) \cong \langle j \rangle \times PSL(2, q)$, where $q > 5$.

The question whether any group of Ree type is one of the Ree groups described in 13.2 remained open for many years. After partial results by H. N. WARD [1] and JANKO and THOMPSON [1], Bombieri recently proved the following final result (BOMBIERI [1]).

13.4 Theorem. *Every group $\mathfrak{G}$ of Ree type is one of the simple Ree groups described in 13.2. In particular, if j is an involution in $\mathfrak{G}$, then $\mathbf{C}_{\mathfrak{G}}(j) \cong \langle j \rangle \times PSL(2, 3^{2n+1})$ for some natural number n.*

In the definition of the groups of Ree type, the condition $q > 5$ is imposed. For $q \leq 5$, we have the following results.

13.5 Theorem (JANKO [2]). *Suppose that $\mathfrak{G}$ is a group with the following properties.*

(1) *The Sylow 2-subgroups of $\mathfrak{G}$ are Abelian.*

(2) *$\mathfrak{G}$ has no normal subgroup of index 2.*

(3) *$\mathfrak{G}$ possesses an involution j for which $\mathbf{C}_{\mathfrak{G}}(j) \cong \langle j \rangle \times PSL(2, 3)$. Then $\mathfrak{G} \cong P\Gamma L(2, 2^3)$. (This group, which is not simple, may be thought of as $R(3)$.)*

13.6 Theorem (JANKO [1]). *Suppose that $\mathfrak{G}$ is a group with the following properties.*

(1) *The Sylow 2-subgroups of $\mathfrak{G}$ are Abelian.*

(2) *$\mathfrak{G}$ has no normal subgroup of index 2.*

(3) *$\mathfrak{G}$ possesses an involution j for which $\mathbf{C}_{\mathfrak{G}}(j) \cong \langle j \rangle \times PSL(2, 5)$. Then $\mathfrak{G}$ is simple and is uniquely determined to within isomorphism; in particular,*

$$|\mathfrak{G}| = 2^3 \cdot 3 \cdot 5 \cdot 7 \cdot 11 \cdot 19.$$

The Schur multiplier of $\mathfrak{G}$ is 1 and every automorphism of $\mathfrak{G}$ is inner.

The construction of this group $\mathfrak{G}$ given by Janko rested on detailed matrix calculations, but D. Livingstone established the existence of $\mathfrak{G}$ by a permutation representation (LIVINGSTONE [1]). This goes as follows.

Let $\mathfrak{H} = PSL(2, 11)$ and let Λ_1 denote the projective line on $GF(11)$ on which $\mathfrak{H}$ operates. By II, 8.27, $\mathfrak{H}$ has a subgroup isomorphic to $\mathfrak{A}_5$. Hence $\mathfrak{H}$ also has a transitive representation on a second set Λ_2 with 11 elements. Now let

$$\Lambda_3 = \Lambda_1 \times \Lambda_2$$

and

$$\Lambda_4 = \{(a, b) \mid a \in \Lambda_2, b \in \Lambda_2, a \neq b\}.$$

Thus $\mathfrak{H}$ operates in a natural way on Λ_3 and Λ_4. Finally, put

$$\Omega = \{\infty\} \cup \Lambda_1 \cup \Lambda_2 \cup \Lambda_3 \cup \Lambda_4.$$

Thus $|\Omega| = 266$ and $\mathfrak{H}$ operates on Ω with five orbits. There is a permutation g of Ω such that $\mathfrak{G} = \langle \mathfrak{H}, g \rangle$ is a transitive permutation group on Ω and is isomorphic to the simple group of Janko described above. This group is denoted by $\mathfrak{J}_1$.

The Ree groups and the Janko group $\mathfrak{J}_1$ also occur in the following important theorem (cf. IX, 4.16).

13.7 Theorem (WALTER [2]). *Suppose that the Sylow 2-subgroups of the group $\mathfrak{G}$ are Abelian. Then $\mathfrak{G}$ has normal subgroups $\mathfrak{M}$, $\mathfrak{N}$ with $\mathfrak{M} \leq \mathfrak{N}$, having the following properties.*

(1) *$\mathfrak{M}$ and $\mathfrak{G}/\mathfrak{N}$ are of odd order.*

(2) *$\mathfrak{N}/\mathfrak{M} = \mathfrak{G}_0 \times \mathfrak{G}_1 \times \cdots \times \mathfrak{G}_k$, where $\mathfrak{G}_0$ is an Abelian 2-group and $\mathfrak{G}_1, \ldots, \mathfrak{G}_k$ are simple non-Abelian groups.*

Each of $\mathfrak{G}_1, \ldots, \mathfrak{G}_k$ is isomorphic to one of the following.

a) *$PSL(2, 2^f)$.*

b) *$PSL(2, q)$, where $q \equiv 3 \ (8)$ or $q \equiv 5 \ (8)$.*

c) *$\mathfrak{J}_1$.*

d) *A Ree group.*

It follows from 13.7 that if the Sylow 2-subgroup $\mathfrak{T}$ of a simple non-Abelian group is Abelian, then $\mathfrak{T}$ is elementary Abelian. Another proof of this fact was given by Goldschmidt.

Substantial simplifications of part of the proof of 13.7 were given by BENDER [4].

Finally, we return to the problem of 13.1.

13.8 Theorem (HERING, KANTOR and SEITZ [1]). *Suppose that $\mathfrak{G}$ is a doubly transitive permutation group having no regular normal subgroup. Suppose that the stabiliser of a point has a regular normal subgroup. Then there exists a group $\mathfrak{H}$ such that $\mathfrak{H} \trianglelefteq \mathfrak{G} \leq$ **Aut** $\mathfrak{H}$, and $\mathfrak{H}$ is isomorphic to one of the following groups.*

a) *$PSL(2, q)$.*
b) *$Sz(q)$, where $q = 2^{2n+1}$.*
c) *$PSU(3, q^2)$.*
d) *A Ree group.*

In each case the degree is of the form $1 + p^m$ for some prime p.

Of the numerous contributions of Suzuki, we mention only the following.

13.9 Theorem (SUZUKI [6]). *Suppose that $\mathfrak{G}$ is a simple non-Abelian group and that the intersection of any two different Sylow 2-subgroups of $\mathfrak{G}$ is $\{1\}$. Then $\mathfrak{G}$ is one of the groups $PSL(2, q)$, $PSU(3, q)$ or $Sz(q)$, where q is a power of 2.*

Notes on Chapter XI

§ **1:** Theorem 1.1 is from FEIT [1]; the remaining results of this section appear in the basic paper of ZASSENHAUS [1].
§ **2:** Theorems 2.1 and 2.6 come from ZASSENHAUS [2]. Another proof was given by TITS [1], and Lemma 2.2 is taken from his paper. The proof of 2.6 follows HUPPERT [2].
§ **3:** The construction of the Suzuki groups and most of their properties are found in SUZUKI [3]. The description of $Sz(q)$ as the group of collineations leaving an ovoid fixed (3.3) is taken from TITS [3]. We are indebted to McDermott for pointing out that $Sz(q)$ is $\frac{5}{2}$-fold transitive in a written communication. The description of $Sz(q)$ as a twisted type is in ONO [1].
§ **4:** The treatment of exceptional characters follows FEIT [2]. These methods have been refined in many subsequent papers (see, for example, SIBLEY [1]).
§ **5:** This section also follows FEIT [2]. The character tables of the $PSL(2, q)$ can of course be determined in many other ways; the case when q is a prime was known to FROBENIUS [1] already. The character table of $Sz(q)$ was given by SUZUKI [3]; part of it is given in 5.10.
§ **6:** Here we again follow FEIT [1]. Theorem 6.15 is a result of FROBENIUS [2], given here with a different proof.
§ **7:** Theorems 7.1 and 7.11 are in WIELANDT [4]. The rest of the section follows mainly HUPPERT and WIELANDT [1], with simplifications in 7.9 which come from Wagner's Theorem 7.7. 7.10 was proved at greater length in ITÔ [2]; the proof here is much simplified by using Feit's Theorem 1.12. Theorem 7.12 is in BURNSIDE [1], page 202.
§ **8:** The main results in this section are due to FROBENIUS and SCHUR [1]; the proofs follow partially a lecture of F. LORENZ [1].
§ **9:** The main theorem of this section (9.1) was first proved in ITÔ [3], using modular representation theory. The proof reproduced here is taken from GLAUBERMAN [4].
§ **10:** The proofs follow SUZUKI [3].
§ **11:** The proofs here also follow SUZUKI [3].

Chapter XII

Multiply Transitive Permutation Groups

Since the beginnings of finite group theory, the multiply transitive permutation groups have exercised a certain fascination. This is mainly due to the fact that apart from the symmetric and alternating groups not many of them were known. Only very recently final results about multiply transitive permutation groups have been proved, using the classification of all finite simple groups (see 7.5).

In § 1, the five simple multiply transitive Mathieu groups are constructed and some of their most important properties are derived. After some preparations on transitive extensions in § 2, the sharply k-fold transitive permutation groups for $k \geq 4$ are determined in § 3; besides the symmetric and alternating groups, there are only the Mathieu groups $\mathfrak{M}_{11}$ and $\mathfrak{M}_{12}$, which are respectively sharply 4-fold transitive of degree 11 and sharply 5-fold transitive of degree 12. (In § 3, some simple results from the beginning of Chapter XI are used.) In § 4, it is proved that any 7-fold transitive permutation group is symmetric or alternating under the assumption that the Schreier hypothesis on the automorphism groups of simple groups holds. In § 5, a characterization of $\mathfrak{M}_{11}$ and $PSL(3, 3)$ in terms of the centralizer of an involution is given. The methods used there are largely character theoretical and are typical of a number of investigations which have led to the characterization and also to the discovery of previously unknown finite simple groups. In § 6, we consider permutation groups of degree n which permute transitively the k-element subsets of the set permuted for some given k ($\leq n$). This is an obvious weakening of the concept of k-fold transitivity, but the result, which is due to Livingstone and Wagner, is very surprising; it states that for $5 \leq k \leq \frac{1}{2}n$, the group is k-fold transitive. In § 7, the doubly transitive soluble permutation groups are found; apart from a finite number of exceptions, these groups consist of semilinear mappings on finite fields. In § 8, it is proved that $SL(2, 5)$ is the only perfect group which operates fixed point freely on another group. This theorem essentially completes earlier results about Frobenius groups, but the proof is unfortunately very wearisome. However, with the aid of this result, the sharply doubly transitive permutation groups can then be easily determined in § 9; this also contains the determination of the finite near-fields. The last section is on permutation groups of prime degree and is only loosely related to the earlier parts of the chapter. It contains new, very elementary proofs for theorems of Burnside, P. M. Neumann and Wielandt, thereby improving the results of V, § 21.

This chapter presents only a small selection of the many investigations on multiply transitive permutation groups. It may now be that the determination of all multiply transitive permutation groups can take place after the discovery of all the simple groups. (It is to a large extent proved that the simple groups only rarely admit representations by multiply transitive permutation groups.)

§ 1. The Mathieu Groups

1.1 Definition. Let $\mathfrak{G}$ be a not necessarily transitive permutation group on a set Ω. Further, let $\Omega' = \Omega \cup \{v\}$, where $v \notin \Omega$. A transitive permutation group $\mathfrak{K}$ on Ω' is called a *transitive extension* of $\mathfrak{G}$ if the restriction to Ω of the stabiliser $\mathfrak{K}_v$ is equal to $\mathfrak{G}$.

There is a construction procedure for the determination of the transitive extensions of a given permutation group $\mathfrak{G}$, but this entails the solution of complicated functional equations which are hardly understood even in very special cases. If, however, $\mathfrak{G}$ is at least doubly transitive on Ω, then there is a construction which is useful in practice. We shall describe this procedure and several of its applications.

1.2 Theorem (WITT [2]). *Suppose that $\mathfrak{G}$ is doubly transitive on the set $\Omega = \{p_1, \dots, p_n, q_1, q_2\}$. Suppose that $s_1 \in \mathfrak{G}$ and s_1 interchanges q_1, q_2. Let $\mathfrak{H} = \mathfrak{G}_{q_1, q_2}$. Suppose further that $t \geq 2$ and $\Omega' = \{p_1, \dots, p_n, q_1, \dots, q_t\}$. We make $\mathfrak{G}$ operate on Ω' by putting $q_i g = q_i$ for all $g \in \mathfrak{G}$, $i > 2$. Finally, let $s_2, \dots, s_{t-1}$ be permutations of Ω' with the following properties.*

(1) $s_i = (q_1) \cdots (q_{i-1})(q_i, q_{i+1})(q_{i+2}) \cdots (q_t) \cdots$, *and s_i permutes $p_1, \dots, p_n$ among themselves (possibly non-trivially).*

(2) $s_i^2 \in \mathfrak{H}$ *for* $i = 1, \dots, t - 1$.

(3) $(s_i s_{i+1})^3 \in \mathfrak{H}$ *for* $i = 1, \dots, t - 2$.

(4) $(s_i s_j)^2 \in \mathfrak{H}$ *for* $i - j \neq \pm 1$ *and* $i, j = 1, \dots, t - 1$.

(5) s_i *normalises* $\mathfrak{G}_{q_2}$ *for* $i \geq 2$.

(The conditions (2)–(4) correspond to the defining relations of the symmetric group $\mathfrak{S}_t$ given in I, 19.7.)

We define sets $\mathfrak{G}(i)$ of permutations of Ω' by induction on i, putting $\mathfrak{G}(1) = \mathfrak{G}_{q_2}$ and $\mathfrak{G}(j) = \mathfrak{G}(j-1) \cup \mathfrak{G}(j-1)s_{j-1}\mathfrak{G}(j-1)$. Then $\mathfrak{G}(t)$ is a t-fold transitive permutation group on Ω'. Also $\mathfrak{G} = \mathfrak{G}(2)$ and $\mathfrak{G}(t)_{q_1,\ldots,q_t} = \mathfrak{H}$.

Proof. We use induction on t. For $t = 2$, the assertion is correct on account of the double transitivity of $\mathfrak{G}$. For $t > 2$, $\mathfrak{G}(t-2)$ is a group and

$$\mathfrak{G}(t-2) = \langle \mathfrak{G}_{q_2}, s_1, \ldots, s_{t-3} \rangle.$$

Since $\mathfrak{H} \leq \mathfrak{G}(t-2)$, it follows from the conditions (2) and (4) that for $i \leq t-3$,

$$s_{t-1}s_is_{t-1}^{-1} = (s_{t-1}s_i)^2 s_i^{-1}s_{t-1}^{-2} \in \mathfrak{G}(t-2).$$

By (5), $s_{t-1} \in \mathbf{N}(\mathfrak{G}_{q_2})$; (the normalisers in this proof are taken in the full symmetric group on Ω'). Hence $s_{t-1} \in \mathbf{N}(\mathfrak{G}(t-2))$.

We have to show that the set

$$\mathfrak{G}(t) = \mathfrak{G}(t-1) \cup \mathfrak{G}(t-1)s_{t-1}\mathfrak{G}(t-1)$$

is a group. For this, it is obviously sufficient to verify that

$$s_{t-1}\mathfrak{G}(t-1)s_{t-1} \subseteq \mathfrak{G}(t).$$

Now it follows from (2) and $s_{t-1} \in \mathbf{N}(\mathfrak{G}(t-2))$ that for $g \in \mathfrak{G}(t-2)$,

$$s_{t-1}gs_{t-1} = s_{t-1}gs_{t-1}^{-1}s_{t-1}^2 \in \langle \mathfrak{G}(t-2), \mathfrak{H} \rangle = \mathfrak{G}(t-2) \subseteq \mathfrak{G}(t).$$

Next suppose that $g = xs_{t-2}y$, where x, y are in $\mathfrak{G}(t-2)$. Then

$$\begin{aligned}
&s_{t-1}xs_{t-2}ys_{t-1}\\
&= s_{t-1}xs_{t-1}^{-1}s_{t-1}s_{t-2}s_{t-1}s_{t-1}^{-1}ys_{t-1}\\
&\in \mathfrak{G}(t-2)s_{t-1}s_{t-2}s_{t-1}\mathfrak{G}(t-2) \quad (\text{since } s_{t-1} \in \mathbf{N}(\mathfrak{G}(t-2)))\\
&= \mathfrak{G}(t-2)s_{t-2}^{-1}s_{t-1}^{-1}s_{t-2}^{-1}\mathfrak{G}(t-2) \quad (\text{by (3); also } \mathfrak{H} \leq \mathfrak{G}(t-2))\\
&\subseteq \mathfrak{G}(t-1)s_{t-1}\mathfrak{G}(t-1)
\end{aligned}$$

on account of (2) and $\mathfrak{G}(t-1) = \langle \mathfrak{G}(t-2), s_{t-2} \rangle$. Hence

$$s_{t-1}xs_{t-2}ys_{t-1} \in \mathfrak{G}(t),$$

and $\mathfrak{G}(t)$ is a group.

Finally, q_t is carried into an element of $\{p_1, \ldots, p_n, q_1, \ldots, q_{t-1}\}$ by any element of $\mathfrak{G}(t-1)s_{t-1}\mathfrak{G}(t-1)$, so $\mathfrak{G}(t)_{q_t} = \mathfrak{G}(t-1)$. It follows easily that $\mathfrak{G}(t)_{q_1,\ldots,q_t} = \mathfrak{H}$. **q.e.d.**

It is easily seen that any t-fold transitive group can be constructed by using the procedure in 1.2.

In the next two theorems, 1.2 will be used to construct the five Mathieu groups.

1.3 Theorem. *Let $\mathfrak{G}$ be the sharply triply transitive permutation group $M(3^2)$ on $\Omega = GF(3^2) \cup \{\infty\}$, given in XI, 1.3c). Suppose that $j \in GF(3^2)$ and $j^2 = 1 - j$; thus $\{1, j\}$ is a basis of $GF(3^2)$ over $GF(3)$. Let s_1, s_2, s_3 be the following permutations of $GF(3^2) \cup \{\infty, v, w\}$.*

$$vs_1 = v, ws_1 = w, zs_1 = z^{-1} \ (z \in GF(3^2) \cup \{\infty\}),$$

$$\infty s_2 = v, vs_2 = \infty, ws_2 = w, (x + jy)s_2 = x - jy \ (x, y \in GF(3)),$$

$$\infty s_3 = \infty, vs_3 = w, ws_3 = v, zs_3 = z^3 \ (z \in GF(3^2)).$$

Then $\mathfrak{M}_{11} = \langle M(3^2), s_2 \rangle$ is a sharply 4-fold transitive permutation group of degree 11 on $\Omega_1 = GF(3^2) \cup \{\infty, v\}$, and $\mathfrak{M}_{12} = \langle \mathfrak{M}_{11}, s_3 \rangle$ is a sharply 5-fold transitive permutation group of degree 12 on $\Omega_2 = GF(3^2) \cup \{\infty, v, w\}$. The groups $\mathfrak{M}_{11}$ and $\mathfrak{M}_{12}$ are called the Mathieu groups of degree 11 and 12. The restriction of $(\mathfrak{M}_{12})_w$ to Ω_1 is $\mathfrak{M}_{11}$, and the restriction of $(\mathfrak{M}_{11})_v$ to Ω is $M(3^2)$.

Proof. We wish to apply 1.2 with $\{p_1, \ldots, p_8\} = GF(3^2)^\times$, $q_1 = 0$, $q_2 = \infty$, $q_3 = v$ and $q_4 = w$. The condition (1) obviously holds. The determinant of the linear fractional mapping s_1 is -1, which is a square in $GF(3^2)$. Thus $s_1 \in M(3^2)$. Since $s_i^2 = 1$ $(i = 1, 2, 3)$, condition (2) of 1.2 holds. Simple calculation shows that

$$s_1s_2 = (0, v, \infty)(w)(1)(-1)(j, 1 - j, -1 - j)(-j, -1 + j, 1 + j),$$

$$\begin{aligned} s_1s_3 &= s_3s_1 \\ &= (v, w)(0, \infty)(1)(-1)(-1 + j)(1 - j)(j, -j)(1 + j, -1 - j), \end{aligned}$$

$$s_2s_3 = (\infty, w, v)(0)(1)(-1)(j, 1 + j, -1 + j)(-j, -1 - j, 1 - j).$$

Thus $(s_1s_2)^3 = (s_2s_3)^3 = (s_1s_3)^2 = (s_3s_1)^2 = 1$ and conditions (3), (4) hold. It remains to verify that $s_i \in \mathbf{N}(\mathfrak{G}_\infty)$ $(i = 2, 3)$.

We have $\mathfrak{G}_\infty = \mathfrak{F}\mathfrak{Q}$, where $\mathfrak{F}$ is the group of translations c_f given by $zc_f = z + f$ $(f \in GF(3^2))$, and $\mathfrak{Q}$ is the quaternion group of order 8 generated by a, b, where

$$za = j^2z, \quad zb = jz^3.$$

(Clearly, $j \notin GF(3^2)^{\times 2}$, since $\langle j \rangle = GF(3^2)^\times$ is cyclic of order 8.) Simple calculation gives

$$s_2^{-1}as_2 = ab, \quad s_2^{-1}bs_2 = b^{-1}, \quad s_2^{-1}c_fs_2 = c_{f'} \quad \text{for } f' = fs_2.$$

Thus $s_2 \in \mathbf{N}(\mathfrak{G}_\infty)$. Also, $s_3 \in \mathbf{N}(\mathfrak{G}_\infty)$, for $\langle s_3, \mathfrak{G}_\infty \rangle$ is isomorphic to the group of all semilinear mappings $z \to c(z\alpha) + d$ on $GF(3^2)$, so $\mathfrak{G}_\infty$ is of index 2 in $\langle \mathfrak{G}_\infty, s_3 \rangle$.

By 1.2, $\mathfrak{M}_{11} = \mathfrak{G} \cup \mathfrak{G}s_2\mathfrak{G}$ and $\mathfrak{M}_{12} = \mathfrak{M}_{11} \cup \mathfrak{M}_{11}s_3\mathfrak{M}_{11}$; also

$$(\mathfrak{M}_{11})_{0,\infty,v} = (\mathfrak{M}_{12})_{0,\infty,v,w} = \mathfrak{G}_{0,\infty}.$$

Since $\mathfrak{G}_{0,\infty}$ is transitive on $GF(3^2)^\times$, $\mathfrak{M}_{11}$ is 4-fold transitive on Ω_1 and $\mathfrak{M}_{12}$ is 5-fold transitive on Ω_2. Since $|\mathfrak{G}_{0,\infty}| = 8$,

$$|\mathfrak{M}_{11}| = 11 \cdot 10 \cdot 9 \cdot 8, \quad |\mathfrak{M}_{12}| = 12 \cdot 11 \cdot 10 \cdot 9 \cdot 8.$$

Hence $\mathfrak{M}_{11}$, $\mathfrak{M}_{12}$ are sharply 4-fold and 5-fold transitive respectively.

q.e.d.

1.4 Theorem. *Let* P *be the projective plane over* $GF(4)$; *we represent the points of* P *by triples* $[x, y, z] \neq [0, 0, 0]$. *Let* $\mathfrak{G} = PSL(3, 4)$, *regarded as a doubly transitive permutation group of degree* 21 *on the points of* P. *Let* k *be an element of* $GF(4)$, *with* $0 \neq k \neq 1$. *We put* $\Omega' = \mathsf{P} \cup \{u, v, w\}$ *and define the following mappings* s_1, s_2, s_3, s_4 *of* Ω' *into* Ω'.

$$us_1 = u, \quad vs_1 = v, \quad ws_1 = w, \quad [x, y, z]s_1 = [y, x, z].$$

$$[1, 0, 0]s_2 = u, \quad us_2 = [1, 0, 0], \quad vs_2 = v, \quad ws_2 = w,$$

$$[x, y, z]s_2 = [x^2 + yz, y^2, z^2] \text{ for } [x, y, z] \neq [1, 0, 0].$$

$$us_3 = v, \quad vs_3 = u, \quad ws_3 = w, \quad [x, y, z]s_3 = [x^2, y^2, kz^2].$$

$$us_4 = u, \quad vs_4 = w, \quad ws_4 = v, \quad [x, y, z]s_4 = [x^2, y^2, z^2].$$

Then s_1, s_2, s_3, s_4 *are permutations of* Ω' *and we obtain the following groups.*

a) $\mathfrak{M}_{22} = \langle \mathfrak{G}, s_2 \rangle$ *is a triply transitive permutation group of degree* 22 *on* $\mathsf{P} \cup \{u\}$, *and* $|\mathfrak{M}_{22}| = 48 \cdot 20 \cdot 21 \cdot 22$.

b) $\mathfrak{M}_{23} = \langle \mathfrak{M}_{22}, s_3 \rangle$ *is a 4-fold transitive permutation group of degree* 23 *on* $\mathsf{P} \cup \{u, v\}$ *and* $|\mathfrak{M}_{23}| = 48 \cdot 20 \cdot 21 \cdot 22 \cdot 23$.

c) $\mathfrak{M}_{24} = \langle \mathfrak{M}_{23}, s_4 \rangle$ *is a 5-fold transitive permutation group of degree* 24 *on* $\mathsf{P} \cup \{u, v, w\}$ *and* $|\mathfrak{M}_{24}| = 48 \cdot 20 \cdot 21 \cdot 22 \cdot 23 \cdot 24$.

The groups $\mathfrak{M}_{22}, \mathfrak{M}_{23}, \mathfrak{M}_{24}$ *are called the Mathieu groups of degree* 22, 23, 24 *respectively. The restriction of* $(\mathfrak{M}_{24})_w$ *to* $\mathsf{P} \cup \{u, v\}$ *is* $\mathfrak{M}_{23}$, *that of* $(\mathfrak{M}_{23})_v$ *to* $\mathsf{P} \cup \{u\}$ *is* $\mathfrak{M}_{22}$ *and that of* $(\mathfrak{M}_{22})_u$ *to* P *is* $PSL(3, 4)$.

Proof. First we must show that s_1, s_2, s_3, s_4 are permutations of Ω'. Clearly they are mappings of Ω' into Ω'; that is, $[0, 0, 0]$ never appears as an image and proportional triples are mapped into proportional triples. Thus s_1 is a permutation. Since squaring is an automorphism of $GF(4)$, it follows easily that s_2, s_3, s_4 are mappings of Ω' onto itself. Since Ω' is finite, they are permutations of Ω'.

We wish to apply 1.2 with $q_1 = [0, 1, 0]$, $q_2 = [1, 0, 0]$, $q_3 = u$, $q_4 = v$, $q_5 = w$ and $\{p_1, \ldots, p_{19}\} = \mathsf{P} - \{q_1, q_2\}$. Since $\det s_1 = 1$, $s_1 \in \mathfrak{G} = PSL(3, 4)$.

We check the conditions (1)–(5) of 1.2.

(1) This is obviously satisfied.

(2) Since $x^4 = x$ for all $x \in GF(4)$, $k^3 = 1$. Since also char $GF(4) = 2$, it is easy to check that $s_i^2 = 1$ for $i = 1, 2, 3, 4$.

(3) Since $k^2 + k + 1 = 0$, it is easy to check directly that $(s_2 s_3)^3 = (s_3 s_4)^3 = 1$. Also $s_1 s_2$ permutes u, $[1, 0, 0]$, $[0, 1, 0]$ cyclically and leaves v, w fixed. If $[x, y, z]$ is distinct from $[1, 0, 0]$ and $[0, 1, 0]$, then

$$[x, y, z](s_1 s_2)^3 = [y^2 + y^2 z^3 + xz, x^2 + yz + x^2 z^3, z^2].$$

For $z \neq 0$, $z^3 = 1$ and $[x, y, z](s_1 s_2)^3 = [xz, yz, z^2] = [x, y, z]$. If $z = 0$ but x, y are both non-zero, $[x, y, 0](s_1 s_2)^3 = [y^2, x^2, 0] = [x^2 y^2 x, x^2 y^2 y, 0] = [x, y, 0]$. This exhausts all possibilities, so $(s_1 s_2)^3 = 1$.

(4) By direct calculation, $(s_i s_j)^2 = 1$ for $i - j \neq \pm 1$.

(5) It remains to verify that $s_i \in \mathbf{N}(\mathfrak{G}_{q_2})$ for $i = 2, 3, 4$. The group $\mathfrak{G}_{q_2} = (PSL(3, 4))_{q_2}$ consists of collineations r belonging to matrices of the form

$$\begin{pmatrix} f & 0 & 0 \\ g & a & b \\ h & c & d \end{pmatrix}$$

with determinant 1. A simple calculation shows that $s_2 r s_2$ leaves q_2, u, v, w fixed and

$$[x, y, z]s_2 r s_2 = [x', a^2 y + c^2 z, b^2 y + d^2 z],$$

where

$$x' = f^2 x + (g^2 + ab)y + (h^2 + cd)z + (f^2 + ad - bc)y^2 z^2.$$

Since $\det r = 1$, the coefficient of $y^2 z^2$ is

$$f^2 + f^{-1} = f^{-1}(f^3 + 1) = 0.$$

Thus $s_2 r s_2$ is a linear mapping. The corresponding matrix is

$$\begin{pmatrix} f^2 & 0 & 0 \\ g^2 + ab & a^2 & b^2 \\ h^2 + cd & c^2 & d^2 \end{pmatrix}.$$

Since this is of determinant 1, $s_2 r s_2 \in \mathfrak{G}_{q_2}$.

Similarly, $s_3 r s_3$ and $s_4 r s_4$ are linear and the corresponding matrices are respectively

$$\begin{pmatrix} f^2 & 0 & 0 \\ g^2 & a^2 & b^2 k \\ h^2 k^2 & c^2 k^2 & d^2 \end{pmatrix}, \quad \begin{pmatrix} f^2 & 0 & 0 \\ g^2 & a^2 & b^2 \\ h^2 & c^2 & d^2 \end{pmatrix}.$$

These are of determinant 1 (since $k^3 = 1$), so $s_3 r s_3$, $s_4 r s_4$ lie in $\mathfrak{G}_{q_2}$.

It follows from 1.2 that

$$(\mathfrak{M}_{22})_{q_1, q_2, u} = (\mathfrak{M}_{23})_{q_1, q_2, u, v} = (\mathfrak{M}_{24})_{q_1, q_2, u, v, w} = \mathfrak{G}_{q_1, q_2}.$$

By II, 6.2, $|PSL(3, 4)| = 48 \cdot 20 \cdot 21$. The orders of $\mathfrak{M}_{22}$, $\mathfrak{M}_{23}$, $\mathfrak{M}_{24}$ are therefore as stated. **q.e.d.**

The Mathieu groups $\mathfrak{M}_{11}$, $\mathfrak{M}_{12}$, $\mathfrak{M}_{23}$, $\mathfrak{M}_{24}$ are the only known 4-fold transitive permutation groups other than the symmetric and alternating groups. Numerous attempts have been made to find further groups related to these, but these have not been successful. Here we will show that $\mathfrak{M}_{12}$ and $\mathfrak{M}_{24}$ have no transitive extensions.

1.5 Theorem. a) *There exists no transitive group of degree* 13 *and order* $13 \cdot 12 \cdot 11 \cdot 10 \cdot 9 \cdot 8$.

b) *If the order of a triply transitive permutation group of degree* 25 *is divisible by* 11, *then it is divisible by* 11^2.

c) *The Mathieu groups* $\mathfrak{M}_{12}$ *and* $\mathfrak{M}_{24}$ *possess no transitive extension.*

Proof. a) Suppose that $\mathfrak{G}$ is transitive of degree 13 and of order $13 \cdot 12 \cdot 11 \cdot 10 \cdot 9 \cdot 8$. By V, 21.1b), $|\mathfrak{G}| = 13d(1 + 13k)$, where d divides 12 and k is a non-negative integer. Putting $12 = de$,

$$11 \cdot 10 \cdot 9 \cdot 8e = 1 + 13k.$$

Hence $3e \equiv 1\ (13)$ and $e \equiv 9\ (13)$. But then e cannot divide 12.

b) Suppose that $\mathfrak{G}$ is triply transitive of degree 25 and that $|\mathfrak{G}|$ is divisible by 11. If $\mathfrak{G}$ contains a cycle of length 11, then by II, 4.5a), $\mathfrak{G}$ is at least 15-fold transitive. Thus $|\mathfrak{G}|$ is divisible by $25 \cdot 24 \cdot \cdots \cdot 11$ and hence by 11^2. Otherwise, with appropriate choice of the symbols permuted, the Sylow 11-subgroup of $\mathfrak{G}$ is generated by an element of the form

$$z = (1)(2)(3)(4, 5, \ldots, 14)(15, 16, \ldots, 25).$$

Put $\mathfrak{Z} = \langle z \rangle$. By II, 1.13c), $\mathbf{N}_{\mathfrak{G}}(\mathfrak{Z})$ is triply transitive on the set $\{1, 2, 3\}$ of fixed points of $\mathfrak{Z}$. Hence $\mathbf{N}_{\mathfrak{G}}(\mathfrak{Z})$ contains a permutation of the form $a = (1, 2, 3)b$, where b is a permutation on $\Delta = \{4, 5, \ldots, 25\}$ and b lies in the normaliser $\mathfrak{N}$ of the restriction to Δ of $\langle z \rangle$. But, as is easily seen, $|\mathfrak{N}| = 11^2 \cdot 10 \cdot 2$. Hence 3 does not divide the order m of b. Thus

$$a^m = (1, 2, 3)^{\pm 1}$$

and $(1, 2, 3) \in \mathfrak{G}$. But then, by II, 4.5c), $\mathfrak{G}$ contains the alternating group $\mathfrak{A}_{25}$, and $|\mathfrak{G}|$ is divisible by 11^2.

c) It follows at once from a) and b) that $\mathfrak{M}_{12}$ and $\mathfrak{M}_{24}$ have no transitive extension. **q.e.d.**

1.6 Theorem. *The five Mathieu groups are simple.*

Proof. a) Let $\mathfrak{P}$ be a Sylow 11-subgroup of $\mathfrak{M}_{11}$. Then by II, 3.6, $|\mathbf{N}_{\mathfrak{M}_{11}}(\mathfrak{P})| = 11t$, where t is a divisor of 10. If $\mathfrak{M}_{11}$ has k Sylow 11-subgroups, $|\mathfrak{M}_{11}| = 11tk$ and $k \equiv 1\ (11)$. Since $|\mathfrak{M}_{11}| = 11 \cdot 10 \cdot 9 \cdot 8$, $tk = 720$ and we must have $t = 5$, $k = 144$.

Let $\mathfrak{N}$ be a minimal normal subgroup of $\mathfrak{M}_{11}$. Then $\mathfrak{N}$ is transitive (II, 1.5), so $\mathfrak{N}$ contains all 144 Sylow 11-subgroups of $\mathfrak{M}_{11}$. Then $|\mathfrak{N}|$ is divisible by $144 \cdot 11 = 11 \cdot 2 \cdot 9 \cdot 8$ and $|\mathfrak{M}_{11} : \mathfrak{N}|$ divides 5. So if $\mathfrak{N} \neq \mathfrak{M}_{11}$, $|\mathfrak{N}| = 144 \cdot 11$ and $\mathbf{N}_{\mathfrak{N}}(\mathfrak{P}) = \mathfrak{P}$. By a theorem of Burnside (IV, 2.6), $\mathfrak{N}$ is 11-nilpotent. But the normal 11-complement of $\mathfrak{N}$ is normal in $\mathfrak{M}_{11}$, so by minimality of $\mathfrak{N}$, $|\mathfrak{N}| = 11$. This is a contradiction. Hence $\mathfrak{N} = \mathfrak{M}_{11}$ and $\mathfrak{M}_{11}$ is simple.

b) The group $PSL(3, 4)$ is simple, by II, 6.13.

c) Now let $\mathfrak{G}$ be one of the groups $\mathfrak{M}_{12}$, $\mathfrak{M}_{22}$, $\mathfrak{M}_{23}$, $\mathfrak{M}_{24}$; we may suppose that it is already proved that the stabiliser $\mathfrak{G}_1$ of 1 in $\mathfrak{G}$ is simple. Suppose that $1 < \mathfrak{N} \lhd \mathfrak{G}$. Then $\mathfrak{N}$ is transitive, so $|\mathfrak{N} : \mathfrak{N}_1| = |\mathfrak{G} : \mathfrak{G}_1|$ is the degree of $\mathfrak{G}$. Thus $\mathfrak{N}_1 < \mathfrak{G}_1$. But $\mathfrak{N}_1 \trianglelefteqq \mathfrak{G}_1$, so $\mathfrak{N}_1 = 1$. Hence $\mathfrak{N}$ is a regular normal subgroup of $\mathfrak{G}$. Since $\mathfrak{G}$ is triply transitive, it follows from II, 2.3 that the degree is either 2^m or 3. This is not the case, so the Mathieu groups are all simple. **q.e.d.**

1.7 Lemma (FRASCH [1]). *Suppose that p is an odd prime and $GF(p)^\times = \langle \omega \rangle$. Consider the following relations.*

(1) $a^p = b^{\frac{1}{2}(p-1)} = c^2 = 1$.

(2) $a^b = a^{\omega^2}$.

(3) $b^c = b^{-1}$.

(4) $(ac)^3 = 1$.

(5) $(a^\omega cb)^3 = 1$.

For $p \not\equiv 1$ (4), (1)–(4) are defining relations for $PSL(2, p)$.
For $p \equiv 1$ (4), (1)–(5) are defining relations for $PSL(2, p)$.

Proof. Let a', b', c' be the following elements of $PSL(2, p)$.

$$xa' = x + 1, \quad xb' = \omega^2 x, \quad xc' = -\frac{1}{x}.$$

Then $\langle a', b' \rangle$ is the stabiliser of ∞, which is a maximal subgroup of $PSL(2, p)$ and does not contain c'. Hence $PSL(2, p) = \langle a', b', c' \rangle$. It is easily checked that a', b', c' satisfy the relations (1)–(5). Thus if $\mathfrak{G}$ is the group generated by a, b, c with defining relations (1)–(4) for $p \not\equiv 1$ (4) or (1)–(5) for $p \equiv 1$ (4), there is an epimorphism of $\mathfrak{G}$ onto $PSL(2, p)$. It is therefore sufficient to prove that $|\mathfrak{G}| \leq |PSL(2, p)|$.

To do this, let $\mathfrak{H}$ be the subgroup of $\mathfrak{G}$ generated by a, b. By (1) and (2), $|\mathfrak{H}| \leq \frac{1}{2}p(p - 1)$. Let

$$\mathfrak{G}^* = \mathfrak{H} \cup \left(\bigcup_{i=0}^{p-1} \mathfrak{H} c a^i \right).$$

Then $\mathfrak{G}^*$ is the union of at most $p + 1$ cosets of $\mathfrak{H}$; hence

$$|\mathfrak{G}^*| \leq (p + 1) \cdot \tfrac{1}{2}p(p - 1) = |PSL(2, p)|.$$

It is therefore sufficient to show that $\mathfrak{G}^* = \mathfrak{G}$. We do this by showing that $\mathfrak{G}^*\mathfrak{G} = \mathfrak{G}^*$. Since $\mathfrak{G} = \langle a, b, c\rangle$, it is sufficient to show that $\mathfrak{G}^*x = \mathfrak{G}^*$ for $x = a, b, c$. Clearly,

$$\mathfrak{H}a = \mathfrak{H} \subseteq \mathfrak{G}^*, \quad \mathfrak{H}b = \mathfrak{H} \subseteq \mathfrak{G}^*, \quad \mathfrak{H}c = \mathfrak{H}ca^0 \subseteq \mathfrak{G}^*.$$

Also

$$(\mathfrak{H}ca^i)a = \mathfrak{H}ca^{i+1} \subseteq \mathfrak{G}^*.$$

And by (2) and (3),

$$\mathfrak{H}ca^i b = \mathfrak{H}cb(a^b)^i = \mathfrak{H}b^{-1}c(a^b)^i = \mathfrak{H}ca^{i\omega^2} \subseteq \mathfrak{G}^*.$$

It remains to show that $\mathfrak{H}ca^i c \subseteq \mathfrak{G}^*$. In doing this, we shall need to calculate

$$(\mathfrak{H}ca^i c)^{b^j} = \mathfrak{H}c^{b^j}(a^{b^j})^i c^{b^j}.$$

By (3), $c^{b^j} = cb^{2j}$, and by (2), $a^{b^j} = a^{\omega^{2j}}$. Hence

$$\begin{aligned}(\mathfrak{H}ca^i c)^{b^j} &= \mathfrak{H}cb^{2j}a^{i\omega^{2j}}cb^{2j}\\ &= \mathfrak{H}b^{-2j}ca^{i\omega^{2j}}cb^{2j}\\ &= \mathfrak{H}ca^{i\omega^{2j}}cb^{2j}.\end{aligned}$$

By (4) and (1), $cac = a^{-1}ca^{-1}$. Hence

$$\mathfrak{H}cac = \mathfrak{H}ca^{-1}.$$

Transforming by b^j and using the above,

$$\mathfrak{H}ca^{\omega^{2j}}cb^{2j} = b^{-j}\mathfrak{H}ca^{-1}b^j.$$

Thus

$$\begin{aligned}\mathfrak{H}ca^{\omega^{2j}}c &= \mathfrak{H}ca^{-1}b^{-j} = \mathfrak{H}cb^{-j}a^{-\omega^{-2j}}\\ &= \mathfrak{H}b^j ca^{-\omega^{-2j}} = \mathfrak{H}ca^{-\omega^{-2j}} \subseteq \mathfrak{G}^*.\end{aligned}$$

Hence it remains to show that $\mathfrak{H}ca^{\omega^{1+2j}}c \subseteq \mathfrak{G}^*$.

If $p \not\equiv 1$ (4), then $-1 \notin GF(p)^{\times 2}$, so it is sufficient to show that $\mathfrak{H}ca^{-\omega^{2j}}c \subseteq \mathfrak{G}^*$. To do this, we observe that $ca^{-1}c = aca$, whence

$$\mathfrak{H}ca^{-1}c = \mathfrak{H}ca.$$

Transforming by b^j, we obtain

$$\mathfrak{H}ca^{-\omega^{2j}}cb^{2j} = b^{-j}\mathfrak{H}cab^j = \mathfrak{H}cab^j.$$

Hence

$$\mathfrak{H}ca^{-\omega^{2j}}c = \mathfrak{H}cab^{-j} = \mathfrak{H}cb^{-j}a^{\omega^{-2j}} = \mathfrak{H}ca^{\omega^{-2j}} \subseteq \mathfrak{G}^*,$$

as required.

If $p \equiv 1$ (4), we have $(a^{\omega}cb)^3 = 1$, so

$$\begin{aligned}\mathfrak{H} &= \mathfrak{H}a^{\omega}cb \cdot a^{\omega}cb \cdot a^{\omega}cb\\ &= \mathfrak{H}b^{-1}c \cdot a^{\omega}cb \cdot a^{\omega}cb\\ &= \mathfrak{H}ca^{\omega}c \cdot ba^{\omega}cb.\end{aligned}$$

Hence

$$\begin{aligned}\mathfrak{H}ca^{\omega}c &= \mathfrak{H}b^{-1}ca^{-\omega}b^{-1}\\ &= \mathfrak{H}b^{-1}cb^{-1} \cdot ba^{-\omega}b^{-1}\\ &= \mathfrak{H}cb \cdot b^{-1} \cdot (a^{-\omega})^{\omega^{-2}}\\ &= \mathfrak{H}ca^{-\omega^{-1}}.\end{aligned}$$

Transformation by b^j gives

$$\mathfrak{H}ca^{\omega^{2j+1}}c = \mathfrak{H}ca^{-\omega^{-1}}b^{-j} = \mathfrak{H}cb^{-j}a^{-\omega^{-2j-1}} = \mathfrak{H}ca^{-\omega^{-2j-1}} \subseteq \mathfrak{G}^*. \quad \textbf{q.e.d.}$$

By II, 8.13, $PSL(2, 11)$ has a subgroup $\mathfrak{A}$ isomorphic to $\mathfrak{A}_5$, and the representation of $PSL(2, 11)$ on the cosets of $\mathfrak{A}$ is of degree 11. We show now that $PSL(2, 11)$ is isomorphic to a subgroup of $\mathfrak{M}_{11}$.

1.8 Theorem. $\mathfrak{M}_{11}$ *contains a doubly transitive subgroup isomorphic to* $PSL(2, 11)$.

Proof. a) We show first that $\mathfrak{M}_{11}$ contains elements b, c such that $b^5 = c^2 = 1$, $b^c = b^{-1}$, b has two orbits of length 5 and c has exactly one fixed point in each.

By 1.3, $(\mathfrak{M}_{11})_v$ is the group $M(3^2)$, in which $PSL(2, 3^2)$ is a subgroup of index 2. Hence $\mathfrak{M}_{11}$ contains elements b, c given by

$$xb = \frac{jx + j + 1}{(j - 1)x + j} \quad (x \in GF(3^2) \cup \{\infty\}), \quad vb = v,$$

$$xc = -x \quad (x \in GF(3^2)), \quad \infty c = \infty, \quad vc = v,$$

where j is an element of $GF(3^2)$ for which $j^2 = 1 - j$. Then

$$b = (v)(0, -1 + j, -j, j, 1 - j)(\infty, -1 - j, -1, 1, 1 + j),$$

$$c = (v)(0)(\infty)(1, -1)(j, -j)(1 + j, -1 - j)(-1 + j, 1 - j),$$

$$bc = (v)(-j)(-1)(0, 1 - j)(j, -1 + j)(\infty, 1 + j)(1, -1 - j),$$

so b, c have the required properties.

b) As we saw in the proof of 1.6, the order of the normaliser of a Sylow 11-subgroup of $\mathfrak{M}_{11}$ is 55, so b normalises some Sylow 11-subgroup $\langle a \rangle$ of $\mathfrak{M}_{11}$. Since $\langle a \rangle$ is self-centralizing (in $\mathfrak{S}_{11}$), b induces an automorphism of order 5 on $\langle a \rangle$. The subgroup of order 5 of the group of automorphisms of $\langle a \rangle$ is generated by $a \to a^4$. Thus, replacing b by a power of b if necessary, we may suppose that $a^b = a^4$; this possible replacement does not affect the validity of a).

We now relabel the elements of the set permuted by $\mathfrak{M}_{11}$ and denote them by the elements of $GF(11)$ in the following way. First, b has just one fixed point, which we denote by 0. We relabel in such a way that $xa = x + 1$ for all $x \in GF(11)$. Then

$$xb = 0a^x b = 0b(a^b)^x = 0a^{4x} = 4x.$$

Since c leaves fixed one point in each orbit of b of length 5, $1b^ic = 1b^i$ for some i. Replacing c by $b^icb^{-i} = cb^{-2i}$, we may suppose that $1c = 1$. Also $0c = 0$, since 0 is the only fixed point of b and $b^c = b^{-1}$. From this it also follows that $4(xc) = xcb = xb^{-1}c = (3x)c$ for all $x \in GF(11)$. Hence

$$c = (0)(1)(3, 4)(9, 5) \cdots,$$

and $6c = 4(2c) = 5(8c) = 9(10c) = 3(7c)$. Thus there are the following possibilities for c.

(i) $c = (0)(1)(3, 4)(9, 5)(6, 2)(8, 7)(10)$.

(ii) $c = (0)(1)(3, 4)(9, 5)(6, 8)(2)(10, 7)$.

(iii) $c = (0)(1)(3, 4)(9, 5)(6, 10)(2, 8)(7)$.

(iv) $c = (0)(1)(3, 4)(9, 5)(6, 7)(2, 10)(8)$.

(v) $c = (0)(1)(3, 4)(9, 5)(6)(2, 7)(8, 10)$.

In case (i), $a^3c = (0, 4, 8)(1, 3, 2, 9)(5, 7, 10, 6)$, so $(a^3c)^4 = (0, 4, 8)$ has 8 fixed points. This is impossible in a sharply 4-fold transitive group. Similarly, in case (ii), $(a^2c)^6 = (0, 3)(2, 9)$ has 7 fixed points, and in case (v), $(ac)^6 = (0, 7)(1, 10)$. Thus these cases are also impossible. In case (iv), $(ac)^3 = 1$ and in case (iii), $(a^4c)^3 = 1$. Hence by 1.7, $\langle \bar{a}, b, c \rangle$ is an epimorphic image of $PSL(2, 11)$, where $\bar{a} = a$ in case (iv) and $\bar{a} = a^4$ in case (iii). Since $PSL(2, 11)$ is simple, $\langle \bar{a}, b, c \rangle$ is a subgroup of $\mathfrak{M}_{11}$ isomorphic to $PSL(2, 11)$. Since $\langle b, c \rangle$ is transitive on $GF(11) - \{0\}$, this subgroup is doubly transitive. **q.e.d.**

1.9 Theorem. $\mathfrak{M}_{11}$ *has a triply transitive representation of degree* 12 *in which the stabiliser of 3 elements is isomorphic to* $\mathfrak{S}_3$.

Proof. By 1.8, $\mathfrak{M}_{11}$ has a subgroup $\mathfrak{U}$ isomorphic to $PSL(2, 11)$. The representation ρ of $\mathfrak{M}_{11}$ on the cosets of $\mathfrak{U}$ is transitive of degree 12. Let σ be the representation of $\mathfrak{U}$ on the cosets $\mathfrak{U}x \neq \mathfrak{U}$. Then σ is faithful of degree 11. Since $|\mathfrak{U}|$ is divisible by 11, it follows that σ is transitive. But $\mathfrak{U}$ is not soluble, so it follows from a theorem of Burnside (V, 21.3) that σ is doubly transitive. Hence ρ is triply transitive. The stabiliser of $\mathfrak{U}$ in ρ is $\mathfrak{U}$, and the stabiliser of a 2-point subset is a subgroup $\mathfrak{A}$ of $\mathfrak{U}$ of order 60. By II, 8.27, the only subgroup of $PSL(2, 11)$ of order 60 is isomorphic to $\mathfrak{A}_5$. The stabiliser of a 3-point subset in ρ is a subgroup of order 6. But a subgroup of $\mathfrak{A}_5$ of order 6 is isomorphic to $\mathfrak{S}_3$.

q.e.d.

1.10 Remark. The following table gives the conjugacy classes of $\mathfrak{M}_{11}$ and the values of the characters π_{11}, π_{12} of the permutation representations of $\mathfrak{M}_{11}$ of degree 11, 12 respectively.

Representative of class	1	c	e	f, f^{-1}	d	g	b	a, a^{-1}
Order of representative	1	2	4	8	3	6	5	11
Length of class	1	165	990	990	440	1320	1584	720
Cycle type in degree 11	1^{11}	$1^3, 2^4$	$1^3, 4^2$	1, 2, 8	$1^2, 3^3$	2, 3, 6	$1, 5^2$	11
π_{11}	11	3	3	1	2	0	1	0
Cycle type in degree 12	1^{12}	$1^4, 2^4$	$2^2, 4^2$	4, 8	$1^3, 3^3$	1, 2, 3, 6	$1^2, 5^2$	1, 11
π_{12}	12	4	0	0	3	1	2	1

1.11 Theorem. *$\mathfrak{M}_{12}$ has a transitive subgroup which is isomorphic to $PSL(2, 11)$.*

Proof. By the proof of 1.8, we may regard $\mathfrak{M}_{12}$ as a permutation group on $GF(11) \cup \{\infty\}$; also $\mathfrak{M}_{12}$ contains elements a, b given by

$$xa = x + 1, \quad \infty a = \infty, \quad xb = 4x, \quad \infty b = \infty \quad (x \in GF(11)).$$

Thus $a^{11} = b^5 = 1$, $a^b = a^4$ and

$$b = (0)(\infty)(1, 4, 5, 9, 3)(2, 8, 10, 7, 6).$$

Certainly $\mathfrak{P} = \langle b \rangle$ is a Sylow 5-subgroup of $\mathfrak{M}_{12}$. Now the normaliser of $\mathfrak{P}$ in $\mathfrak{S}_{12}$ is of order $25 \cdot 16$, so the normaliser $\mathfrak{K}$ of $\mathfrak{P}$ in $\mathfrak{A}_{12}$ is of order $25 \cdot 8$. So if $\mathfrak{N} = \mathfrak{K} \cap \mathfrak{M}_{12}$ is the normaliser of $\mathfrak{P}$ in $\mathfrak{M}_{12}$, $|\mathfrak{N} : \mathfrak{P}| = 2^i$ for some $i \leq 3$ and $|\mathfrak{M}_{12}|/2^i \cdot 5 \equiv 1 \ (5)$. Thus

$$2^i \equiv 12 \cdot 11 \cdot 2 \cdot 9 \cdot 8 \equiv 8 \ (5),$$

$i = 3$, $|\mathfrak{N}| = 5 \cdot 2^3$ and $|\mathfrak{K} : \mathfrak{N}| = 5$. But if

$$c' = (0, \infty)(1, 2)(3, 8)(4, 6)(5, 7)(9, 10),$$

then $c' \in \mathfrak{K}$, $c'^2 = 1$ and $b^{c'} = b^{-1}$. Thus there exists $u \in \mathfrak{K}$ such that $c = c'^u \in \mathfrak{N}$. Then $b^c = b^{u^{-1}c'u} = b^{-1}$, $c^2 = 1$ and c has no fixed points.

Since $b^c = b^{-1}$, c interchanges the fixed points $0, \infty$ of b. Also,

$$4^i(xc) = xcb^i = xb^{-i}c = (3^ix)c \text{ for all } i.$$

Since c has no fixed points, it follows that for all i and x, $3^ix \neq 4^i(xc)$, or $xc \neq 9^ix$; thus

$$xc \neq x, 9x, 4x, 3x, 5x.$$

In particular, $1c$ is one of 2, 6, 7, 8 or 10. Since $3^ic = 4^i(1c)$, it is easily checked that the only possibilities are the following.

$$c = (0, \infty)(1, 2)(3, 8)(9, 10)(5, 7)(4, 6), \quad (a^3c)^3 = 1.$$

$$c = (0, \infty)(1, 6)(3, 2)(9, 8)(5, 10)(4, 7), \quad (a^4c)^3 = 1.$$

$$c = (0, \infty)(1, 7)(3, 6)(9, 2)(5, 8)(4, 10), \quad (a^2c)^3 = 1.$$

$$c = (0, \infty)(1, 8)(3, 10)(9, 7)(5, 6)(4, 2), \quad (a^5c)^3 = 1.$$

$$c = (0, \infty)(1, 10)(3, 7)(9, 6)(5, 2)(4, 8), \quad (ac)^3 = 1.$$

Thus $\mathfrak{M}_{12}$ has a subgroup $\mathfrak{H} = \langle \bar{a}, b, c \rangle$, where $\bar{a}$ is some a^j and $\bar{a}, b, c$ satisfy the defining relations 1.7 of $PSL(2, 11)$. Clearly $\mathfrak{H}$ is transitive.

q.e.d.

1.12 Remark. $\mathfrak{M}_{12}$ has an outer automorphism α of order 2, and α carries the stabiliser of an element (which is $\mathfrak{M}_{11}$ in its representation of degree 11) into a transitive group of degree 12.

1.13 Lemma. *The number of fixed points of any non-identity element of* $\mathfrak{M}_{24}$ *is at most 8.*

Proof. Since $\mathfrak{M}_{24}$ is 5-fold transitive, it is sufficient to prove the assertion for a non-identity element g of

$$(\mathfrak{M}_{24})_{u,v,w,[1,0,0],[0,1,0]} = (PSL(3, 4))_{[1,0,0],[0,1,0]}.$$

Now if $A \in SL(3, 4)$ and $0 \neq \lambda \in GF(4)$, $\det(\lambda A) = \lambda^3 \det A = 1$, so g corresponds to a matrix of the form

$$\begin{pmatrix} a & 0 & 0 \\ 0 & a^{-1} & 0 \\ b & c & 1 \end{pmatrix},$$

where $a \neq 0$. If $[x, y, z]$ denotes a fixed point of g in $\mathsf{P} = \mathsf{P}(2, 4)$, then

$$ax + bz = dx,$$

$$a^{-1}y + cz = dy,$$

$$z = dz,$$

for some $d \neq 0$. If $a \neq 1$, the solutions of these equations are $[b, ac, a - 1]$, $[1, 0, 0]$ and $[0, 1, 0]$, so g has 6 fixed points. If $a = 1$, then b, c are not both 0 since $g \neq 1$, so the only solutions are $[1, 0, 0]$ and all $[\lambda, 1, 0]$. In this case, g has 8 fixed points. **q.e.d.**

We can use this to give a more conceptual proof of II, 2.5.

1.14 Theorem. $GL(4, 2) \cong \mathfrak{A}_8$.

Proof (WITT [2]). Let $\mathfrak{G} = \mathfrak{M}_{24}$ and let

$$\mathfrak{H} = (\mathfrak{M}_{24})_{u,v,w,[1,0,0],[0,1,0]} = (PSL(3, 4))_{[1,0,0],[0,1,0]}.$$

Thus an element g of $\mathfrak{H}$ corresponds to a unique matrix of the form

$$\begin{pmatrix} a & 0 & 0 \\ 0 & a^{-1} & 0 \\ b & c & 1 \end{pmatrix}.$$

Hence $|\mathfrak{H}| = 4^2 \cdot 3$, and if $\mathfrak{K}$ is the set of elements of $\mathfrak{H}$ corresponding to matrices of the form

$$\begin{pmatrix} 1 & 0 & 0 \\ 0 & 1 & 0 \\ b & c & 1 \end{pmatrix},$$

$\mathfrak{K}$ is a normal Sylow 2-subgroup of $\mathfrak{H}$ and $\mathfrak{K}$ is elementary Abelian of order 16. The set of fixed points of $\mathfrak{K}$ is

$$\Delta = \{u, v, w, [1, 0, 0], [\lambda, 1, 0] | \lambda \in GF(4)\}.$$

Thus $\mathbf{N}_{\mathfrak{G}}(\mathfrak{K})$ leaves the set Δ fixed. By II, 1.13c), $\mathbf{N}_{\mathfrak{G}}(\mathfrak{K})$ induces on Δ a 5-fold transitive group $\mathfrak{L}$ of permutations. But since $|\Delta| = 8$, $|\mathfrak{L}| \geq 8 \cdot 7 \cdot 6 \cdot 5 \cdot 4$ and $|\mathfrak{S}_8 : \mathfrak{L}| \leq 6$. By II, 5.3a), $\mathfrak{L}$ is $\mathfrak{S}_8$ or $\mathfrak{A}_8$.

The kernel of the restriction homomorphism of $\mathbf{N}_{\mathfrak{G}}(\mathfrak{K})$ onto $\mathfrak{L}$ is $\mathbf{N}_{\mathfrak{G}}(\mathfrak{K})_\Delta$. Now $\mathfrak{K} = \mathfrak{K}_\Delta \leq \mathbf{N}_{\mathfrak{G}}(\mathfrak{K})_\Delta \leq \mathfrak{G}_\Delta \leq \mathfrak{H}$ and $|\mathfrak{H} : \mathfrak{K}| = 3$. But since $GF(4)$ has an element $a \notin \{0, 1\}$ and

$$(1, 1, 0)\begin{pmatrix} a & 0 & 0 \\ 0 & a^{-1} & 0 \\ 0 & 0 & 1 \end{pmatrix} = (a, a^{-1}, 0),$$

$\mathfrak{H} \neq \mathbf{N}_{\mathfrak{G}}(\mathfrak{K})_\Delta$. Thus $\mathbf{N}_{\mathfrak{G}}(\mathfrak{K})_\Delta = \mathfrak{K}$ and $\mathbf{N}_{\mathfrak{G}}(\mathfrak{K})/\mathfrak{K} \cong \mathfrak{L}$.

We show that $\mathbf{C}_{\mathfrak{G}}(\mathfrak{K}) = \mathfrak{K}$. To do this, we observe that since

$$(0, 1, 1)\begin{pmatrix} 1 & 0 & 0 \\ 0 & 1 & 0 \\ b & c-1 & 1 \end{pmatrix} = (b, c, 1),$$

the set Σ of points of P of the form $[b, c, 1]$ is an orbit of $\mathfrak{K}$ of length 16. Thus Σ is the only orbit of $\mathfrak{K}$ of length 16. Hence if $g \in \mathbf{C}_{\mathfrak{G}}(\mathfrak{K})$, g leaves Σ invariant and induces a permutation g^Σ on Σ. Now g^Σ centralizes the group $\mathfrak{K}^\Sigma$ of permutations of Σ induced by $\mathfrak{K}$. But $\mathfrak{K}^\Sigma$ is regular and Abelian, so $\mathfrak{K}^\Sigma$ is its own centralizer (for example, by I, 5.13). Thus $g^\Sigma \in \mathfrak{K}^\Sigma$ and there exists $h \in \mathfrak{K}$ such that gh^{-1} leaves every element of Σ fixed. But $|\Sigma| = 16$, so by 1.13, $gh^{-1} = 1$ and $g = h \in \mathfrak{K}$. Thus $\mathbf{C}_{\mathfrak{G}}(\mathfrak{K}) = \mathfrak{K}$.

Hence $\mathbf{N}_{\mathfrak{G}}(\mathfrak{K})/\mathbf{C}_{\mathfrak{G}}(\mathfrak{K}) \cong \mathfrak{L}$. Therefore **Aut** $\mathfrak{K}$ has a subgroup isomorphic to $\mathfrak{S}_8$ or $\mathfrak{A}_8$. Since $|\mathbf{Aut}\,\mathfrak{K}| = |\mathfrak{A}_8|$, $\mathbf{Aut}\,\mathfrak{K} \cong \mathfrak{A}_8$. **q.e.d.**

1.15 Remarks. a) For $i = 11, 23, 24$, $\mathbf{Aut}\,\mathfrak{M}_i \cong \mathfrak{M}_i$, and for $i = 12, 22$, $|\mathbf{Aut}\,\mathfrak{M}_i : \mathbf{Inn}\,\mathfrak{M}_i| = 2$.

b) The Schur multipliers $\mathbf{M}(\mathfrak{M}_i)$ are as follows. For $i = 11, 23, 24$, $|\mathbf{M}(\mathfrak{M}_i)| = 1$. And $|\mathbf{M}(\mathfrak{M}_{12})| = 2$, $|\mathbf{M}(\mathfrak{M}_{22})| = 12$ (see BURGOYNE and FONG [1], MAZET [1]; the correct information about $\mathbf{M}(\mathfrak{M}_{22})$ is only very recent).

c) The complex characters of $\mathfrak{M}_{12}$ and $\mathfrak{M}_{24}$ were determined by FROBENIUS [3]. Frobenius used the fact that some of the irreducible characters of the symmetric group remain irreducible under restriction to multiply transitive groups. (We have encountered special cases in V, 20.6.) The irreducible complex characters of $\mathfrak{M}_{11}$, $\mathfrak{M}_{22}$ and $\mathfrak{M}_{23}$ were given by Burgoyne and Fong (without proof).

The modular representation theory of the Mathieu groups was investigated by G. D. JAMES [1].

d) For defining relations of $\mathfrak{M}_{11}$, $\mathfrak{M}_{12}$ and $\mathfrak{M}_{22}$, see COXETER and MOSER [1], pp. 99–100.

e) $\mathfrak{M}_{24}$ has subgroups isomorphic to $\mathfrak{M}_{12}$ and $PSL(2, 23)$ (see WITT [2] and LÜNEBURG [3], p. 94).

1.16 Remarks. The Mathieu groups were described by Witt as the automorphism groups of certain finite geometries in the following sense. A *t-(v, k, λ) block design* is a set B of v points and k-point subsets of B (called *blocks*), such that any t distinct points are contained in exactly λ blocks. If b is the number of blocks in B and r is the number of blocks containing an arbitrary point, then $vr = bk$ and

$$r(k-1)(k-2)\cdots(k-t+1) = \lambda(v-1)(v-2)\cdots(v-t+1).$$

a) Starting from the affine plane over $GF(3)$, a 4-(11, 5, 1) block design $\mathsf{B}(11)$ may be constructed. $\mathsf{B}(11)$ has 11 points and 66 blocks. The group of all automorphisms of $\mathsf{B}(11)$ is $\mathfrak{M}_{11}$.

b) From $\mathsf{B}(11)$, a 5-(12, 6, 1) block design $\mathsf{B}(12)$ may be constructed with 12 points and 132 blocks. The Mathieu group $\mathfrak{M}_{12}$ is the group of automorphisms of $\mathsf{B}(12)$.

c) Starting from the projective plane over $GF(4)$, three block designs may be constructed successively. They are the following.

$\mathsf{B}(22)$ of type 3-(22, 6, 1), with 22 points and 77 blocks.

$\mathsf{B}(23)$ of type 4-(23, 7, 1), with 23 points and 253 blocks.

$\mathsf{B}(24)$ of type 5-(24, 8, 1), with 24 points and 759 blocks.

$\mathfrak{M}_{23}$ and $\mathfrak{M}_{24}$ are the full automorphism groups of $\mathsf{B}(23)$ and $\mathsf{B}(24)$ respectively, but on the other hand, $\mathfrak{M}_{22}$ is a subgroup of index 2 in the group of automorphisms of $\mathsf{B}(22)$. (See WITT [2] and [3], and LÜNEBURG [3], § 7 and §§ 9–14.)

1.17 Remarks. a) We shall see in 2.6 that doubly transitive permutation groups in which a stabiliser has a regular normal subgroup often have no transitive extensions.

b) D. HUGHES [1] has shown that the group $PSL(n, q)$, regarded as a permutation group on the points of the projective space $\mathsf{P}(n-1, q)$ has transitive extensions only if $(n, q) = (3, 4)$ or if $q = 2$. For $(n, q) = (3, 4)$, the only transitive extension is $\mathfrak{M}_{22}$. For each n, let $\mathfrak{G}_n$ be the group of permutations of $\mathsf{V}(n, 2)$ of the form

$$v \to vA + w \quad (w \in \mathsf{V}(n, 2), \quad A \in SL(n, 2)).$$

Then $\mathfrak{G}_n$ is a transitive extension of $PSL(n, 2) = SL(n, 2)$. Thus $\mathfrak{G}_n$ is triply transitive and the group of translations $v \to v + w$ is a regular normal subgroup of $\mathfrak{G}_n$.

c) Suppose that $\mathfrak{M}_{22}$ is intransitively represented on the set $\mathsf{P} \cup \mathsf{B}$, where P is the set of 22 points and B is the set of 77 blocks of B(22). D. HIGMAN and SIMS [1] have shown that $\mathfrak{M}_{22}$ then has a transitive extension $\mathfrak{G}$ and that $\mathfrak{G}$ is simple. The degree of $\mathfrak{G}$ is $1 + 22 + 77 = 100$, so $|\mathfrak{G}| = 100|\mathfrak{M}_{22}|$. $\mathfrak{G}$ is called the *Higman-Sims* group. G. HIGMAN [3] has shown that $\mathfrak{G}$ has a doubly transitive representation of degree 176; the stabiliser of a symbol in this representation is an extension of $PSU(3, 5^2)$ by a group of order 2.

1.18 Characterizations of the Mathieu groups. a) In § 5 we shall prove the following theorem of W. Wong.

If the centralizer of an involution in the centre of a Sylow 2-subgroup of a simple group $\mathfrak{G}$ is isomorphic to $GL(2, 3)$, then $\mathfrak{G} \cong \mathfrak{M}_{11}$ or $\mathfrak{G} \cong PSL(3, 3)$.

b) Suppose that $\mathfrak{G}$ is a simple group and that $\mathfrak{P}$ is a Sylow 2-subgroup of $\mathfrak{G}$. Suppose:

(1) $\mathbf{Z}(\mathfrak{P})$ is cyclic,

(2) If j is an involution in $\mathbf{Z}(\mathfrak{P})$, then $\mathbf{C}_{\mathfrak{G}}(j)$ is an extension of an elementary Abelian group $\mathfrak{A}$ of order 2^4 by the symmetric group $\mathfrak{S}_4$. Then $\mathbf{C}_{\mathfrak{G}}(j)$ splits over $\mathfrak{A}$ and either $\mathfrak{G} \cong \mathfrak{M}_{22}$ or $\mathfrak{G} \cong \mathfrak{A}_{10}$. (See HELD [1] and JANKO [3].)

c) Again suppose that $\mathfrak{P}$ is a Sylow 2-subgroup of the simple group $\mathfrak{G}$. Suppose:

(1) $\mathbf{Z}(\mathfrak{P})$ is cyclic.

(2) If j is an involution in $\mathbf{Z}(\mathfrak{P})$, then $\mathbf{C}_{\mathfrak{G}}(j)$ is an extension of an elementary Abelian group $\mathfrak{A}$ of order 2^4 by $PSL(2, 7)$.
Then $\mathbf{C}_{\mathfrak{G}}(j)$ splits over $\mathfrak{A}$ and $\mathfrak{G} \cong \mathfrak{M}_{23}$ (JANKO [3]).

d) Let $\mathfrak{A}$ be an elementary Abelian group of order 2^4, let $\mathfrak{H}$ be the split extension of $\mathfrak{A}$ by its automorphism group $GL(4, 2)$, and let $\mathfrak{H}_0$ be the centralizer in $\mathfrak{H}$ of a non-identity element of $\mathfrak{A}$. If $\mathfrak{G}$ is a simple group, j is an involution in the centre of a Sylow 2-subgroup of $\mathfrak{G}$ and $\mathbf{C}_{\mathfrak{G}}(j) \cong \mathfrak{H}_0$, then either $\mathfrak{G} \cong \mathfrak{M}_{24}$, $\mathfrak{G} \cong PSL(5, 2)$ or $\mathfrak{G}$ is the simple group of order $2^{10} \cdot 3^3 \cdot 5^2 \cdot 7^3 \cdot 17$ discovered by Held (HELD [3]).

e) If $\mathfrak{G}$ is a simple group and $|\mathfrak{G}| = |\mathfrak{M}_i|$, then $\mathfrak{G} \cong \mathfrak{M}_i$. This was proved by STANTON [1] for $\mathfrak{M}_{12}$, $\mathfrak{M}_{24}$, by PARROTT [1] for $\mathfrak{M}_{11}$, $\mathfrak{M}_{22}$ and by BRYCE [1] for $\mathfrak{M}_{23}$.

§ 2. Transitive Extensions of Groups of Suzuki Type

In this section we deal with the question of the existence of transitive extensions of certain doubly transitive permutation groups.

2.1 Definition. A permutation group $\mathfrak{G}$ is said to be of *Suzuki type* if the following conditions hold.

a) $\mathfrak{G}$ is doubly transitive on a set $\Omega \cup \{\infty\}$, where $|\Omega| \geq 2$.

b) $\mathfrak{G}$ has no regular normal subgroup.

c) The stabiliser $\mathfrak{G}_\infty$ has a normal subgroup $\mathfrak{F}$ which operates regularly on Ω. Further, $|\mathfrak{F}|$ is a prime-power and $\mathfrak{F}$ is a characteristic subgroup of $\mathfrak{G}_\infty$.

Many doubly transitive permutation groups are of Suzuki type.

2.2 Examples. a) Suppose that $\mathfrak{G}$ is a permutation group on $GF(p^f) \cup \{\infty\}$ such that $PSL(2, p^f) \leq \mathfrak{G} \leq P\Gamma L(2, p^f)$. If $p^f \geq 4$, $\mathfrak{G}$ is a group of Suzuki type.

To see this, first observe that $\mathfrak{G}$ is doubly transitive since $\mathfrak{G} \geq PSL(2, p^f)$. Since $p^f \geq 4$, $PSL(2, p^f)$ is simple. Thus if $\mathfrak{N}$ is a regular normal subgroup of $\mathfrak{G}$, $\mathfrak{N} \cap PSL(2, p^f) = 1$. Thus $|\mathfrak{N}| = p^f + 1$ divides $|\mathfrak{G} : PSL(2, p^f)|$. Hence $p^f + 1$ divides $|P\Gamma L(2, p^f) : PSL(2, p^f)| \leq 2f$. This is impossible, so $\mathfrak{G}$ has no regular normal subgroup.

Let $\mathfrak{F}$ be the group of all permutations of $GF(p^f) \cup \{\infty\}$ of the form $x \to x + a$. If $x\alpha = x + a$ and $x\beta = b(x\theta) + c$, then

$$x\beta^{-1}\alpha\beta = x + b(a\theta).$$

Thus $\mathfrak{F}$ is a regular normal subgroup of $\mathfrak{G}_\infty$. Moreover, if $x\gamma = b^2 x$ and $x\gamma' = c^2 x$, then $x(\gamma^{-1}\alpha\gamma)(\gamma'^{-1}\alpha\gamma') = x + (b^2 + c^2)a$. Since every element of $GF(p^f)$ is a sum of two squares (II, 10.6), it follows that $\mathfrak{F}$ is a minimal normal subgroup of $\mathfrak{G}_\infty$. But also, since $\mathfrak{F}$ is regular, $\mathbf{C}_{\mathfrak{G}_\infty}(\mathfrak{F}) = \mathfrak{F}$, so $\mathfrak{F}$ is the only minimal normal subgroup of $\mathfrak{G}_\infty$. Thus $\mathfrak{F}$ is a characteristic subgroup of $\mathfrak{G}_\infty$.

b) If $\mathfrak{G}$ is a Zassenhaus group of degree $n + 1$, then n is a prime-power, by XI, 6.1. If $\mathfrak{F}$ is the Frobenius kernel of $\mathfrak{G}_\infty$, then $\mathfrak{F}$ is a characteristic subgroup of $\mathfrak{G}_\infty$, by V, 8.3. Thus any Zassenhaus group is of Suzuki type.

c) The Ree groups $R(q)$ (see XI, § 13), where $q = 3^{2n+1}$ and $n \geq 1$, are simple doubly transitive permutation groups of degree $q^3 + 1$, and

$$|R(q)| = (q^3 + 1)q^3(q - 1).$$

$R(q)_\infty$ has a regular normal subgroup of order q^3, which is clearly the normal Sylow 3-subgroup of $R(q)_\infty$. Thus $R(q)$ is of Suzuki type.

d) Let $\mathfrak{G}$ be $PU(3, p^{2f})$ or $PSU(3, p^{2f})$. By II, 10.12, $\mathfrak{G}$ is a doubly transitive permutation group of degree $p^{3f} + 1$. We have

$$|PU(3, p^{2f})| = (p^{3f} + 1)p^{3f}(p^{2f} - 1),$$

$$|PSU(3, p^{2f})| = (p^{3f} + 1)p^{3f}\frac{(p^{2f} - 1)}{d}, \quad d = (3, p^f + 1).$$

Further, $\mathfrak{G}_\infty$ has a regular normal subgroup on p^{3f} points; this is characteristic, since it is a normal Sylow p-subgroup of $\mathfrak{G}_\infty$. If $p^f > 2$, then $PSU(3, p^{2f})$ is simple (II, 10.13), from which it follows easily that $PSU(3, p^{2f})$ and $PU(3, p^{2f})$ have no regular normal subgroup. Thus for $p^f > 2$, $PSU(3, p^{2f})$ and $PU(3, p^{2f})$ are of Suzuki type. On the other hand, $PSU(3, 4)$ is a Frobenius group of degree 9, by II, 10.14, so $PSU(3, 4)$ has a regular normal subgroup. This is also a normal subgroup of $PU(3, 4)$.

For the proof of Theorem 2.6, we shall need the following simple theorem about 2-groups.

2.3 Lemma (SUZUKI [9]). *Let $\mathfrak{G}$ be a 2-group of order greater than 4. If $\mathfrak{G}$ has an involution t such that $|\mathbf{C}_\mathfrak{G}(t)| = 4$, then $\mathfrak{G}$ is dihedral or semidihedral.*

Proof. Suppose that $|\mathfrak{G}| = 2^n$. Then $|\mathfrak{G}'| \leq 2^{n-2}$. Since $t^g = t[t, g]$, the set of conjugates of t is contained in $t\mathfrak{G}'$. Hence

$$2^{n-2} = |\mathfrak{G} : \mathbf{C}_\mathfrak{G}(t)| \leq |t\mathfrak{G}'| = |\mathfrak{G}'|,$$

and $|\mathfrak{G}'| = 2^{n-2}$. By a theorem of Taussky (III, 11.9a)), $\mathfrak{G}$ is dihedral, semidihedral or generalized quaternion. The last case cannot occur, since the only involution in a generalized quaternion group is in the centre. **q.e.d.**

2.4 Lemma. *Let $\mathfrak{G}$ be a permutation group of Suzuki type on $\Omega \cup \{\infty\}$ and let 0 be an element of Ω. Put $\mathfrak{L} = \mathfrak{G}_\infty$ and $\mathfrak{H} = \mathfrak{G}_{0,\infty}$. Let $\mathfrak{F}$ be the regular characteristic subgroup of $\mathfrak{L}$, and let s be an element of $\mathfrak{G}$ which interchanges 0 and ∞.*

a) $\mathfrak{H}^s = \mathfrak{H}$.

b) $|\mathfrak{G}| = (|\Omega| + 1)|\Omega||\mathfrak{H}|$.

c) $\mathfrak{G} = \mathfrak{L} \cup \mathfrak{L}s\mathfrak{F}$. *Each element of $\mathfrak{G} - \mathfrak{L}$ can be written uniquely in the form xsy with $x \in \mathfrak{L}$ and $y \in \mathfrak{F}$.*

d) $\mathbf{C}_\mathfrak{H}(\mathfrak{F}) = 1$.

Proof. a) This is trivial.

b) This follows at once from the double transitivity of $\mathfrak{G}$.

c) Certainly $\mathfrak{G} = \mathfrak{L} \cup \mathfrak{L}s\mathfrak{L}$, since $\mathfrak{G}$ is doubly transitive and $s \notin \mathfrak{L}$. As $\mathfrak{F}$ operates regularly on Ω, $\mathfrak{L} = \mathfrak{H}\mathfrak{F}$. Thus

$$\mathfrak{G} = \mathfrak{L} \cup \mathfrak{L}s\mathfrak{H}\mathfrak{F} = \mathfrak{L} \cup \mathfrak{L}\mathfrak{H}s\mathfrak{F} = \mathfrak{L} \cup \mathfrak{L}s\mathfrak{F}.$$

Since $|\mathfrak{L}| + |\mathfrak{L}||\mathfrak{F}| = |\mathfrak{L}|(1 + |\mathfrak{F}|) = \dfrac{|\mathfrak{G}|}{|\mathfrak{G} : \mathfrak{L}|}(1 + |\Omega|) = |\mathfrak{G}|$, it follows that each element of $\mathfrak{G} - \mathfrak{L}$ is uniquely expressible as xsy with $x \in \mathfrak{L}$ and $y \in \mathfrak{F}$.

d) Suppose that $u \in \mathbf{C}_{\mathfrak{H}}(\mathfrak{F})$. For each $i \in \Omega$ there exists $y \in \mathfrak{F}$ such that $0y = i$. Thus

$$iu = 0yu = 0uy = 0y = i.$$

Hence $u = 1$. **q.e.d.**

2.5 Lemma. *Suppose that $\mathfrak{G}$ is doubly transitive on $\Omega \cup \{\infty\}$ and that 0 is an element of Ω. Let $\tilde{\mathfrak{G}}$ be a transitive extension of $\mathfrak{G}$ on $\Omega \cup \{\infty, v\}$.*

a) *Let s be an involution in $\mathfrak{G}$ which interchanges 0 and ∞. Then there is a conjugate t of s in $\tilde{\mathfrak{G}}$ such that $\tilde{\mathfrak{G}} = \langle \mathfrak{G}, t\rangle$, $t = (0)(v, \infty)\cdots$, $\mathfrak{G}_\infty^t = \mathfrak{G}_\infty$, $\mathfrak{G}_{\infty,0}^t = \mathfrak{G}_{\infty,0}$ and $(ts)^3 \in \mathfrak{G}_{\infty,0}$.*

b) *If $\mathfrak{G}$ is of Suzuki type,*

$$\tilde{\mathfrak{G}} = \mathfrak{G} \cup \mathfrak{G}t \cup \left(\bigcup_{x \in \mathfrak{F}} \mathfrak{G}tsx\right)$$

and this decomposition is disjoint. Also the number of fixed points of t on $\Omega \cup \{\infty, v\}$ is $|\mathbf{C}_{\mathfrak{F}}(t)|$.

Proof. a) Since $s = (v)(0, \infty)\cdots$ and $\tilde{\mathfrak{G}}$ is triply transitive, there is a conjugate t of s of the form $t = (0)(v, \infty)\cdots$. Since $\mathfrak{G}$ is a maximal subgroup of $\tilde{\mathfrak{G}}$, $\tilde{\mathfrak{G}} = \langle \mathfrak{G}, t\rangle$. Further,

$$\mathfrak{G}_\infty^t = \tilde{\mathfrak{G}}_{v,\infty}^t = \tilde{\mathfrak{G}}_{\infty,v} = \mathfrak{G}_\infty,$$

$$\mathfrak{G}_{\infty,0}^t = \tilde{\mathfrak{G}}_{v,\infty,0}^t = \tilde{\mathfrak{G}}_{\infty,v,0} = \mathfrak{G}_{\infty,0}.$$

Also, $(v, 0, \infty)$ is a cycle of ts, so $(ts)^3 \in \tilde{\mathfrak{G}}_{\infty,0,v} = \mathfrak{G}_{\infty,0}$.

b) Put $\mathfrak{L} = \mathfrak{G}_\infty$, $\mathfrak{H} = \mathfrak{G}_{\infty,0}$. By 2.4, $\mathfrak{G} = \mathfrak{L} \cup \mathfrak{L}s\mathfrak{F}$, so

$$\tilde{\mathfrak{G}} = \mathfrak{G} \cup \mathfrak{G}t\mathfrak{G} = \mathfrak{G} \cup \mathfrak{G}t(\mathfrak{L} \cup \mathfrak{L}s\mathfrak{F}) = \mathfrak{G} \cup \mathfrak{G}t\mathfrak{L} \cup \mathfrak{G}t\mathfrak{L}s\mathfrak{F}.$$

By a), $\mathfrak{L}^t = \mathfrak{L}$, so

$$\tilde{\mathfrak{G}} = \mathfrak{G} \cup \mathfrak{G}\mathfrak{L}t \cup \mathfrak{G}\mathfrak{L}ts\mathfrak{F} = \mathfrak{G} \cup \mathfrak{G}t \cup \left(\bigcup_{x \in \mathfrak{F}} \mathfrak{G}tsx\right).$$

There are $|\mathfrak{F}| + 2$ terms in this decomposition and each contains $|\mathfrak{G}| = |\tilde{\mathfrak{G}}|/(|\Omega| + 2)$ elements. Since $|\mathfrak{F}| = |\Omega|$, it follows that the decomposition is disjoint.

Since $\mathfrak{F}$ is a characteristic subgroup of $\mathfrak{L}$ and $\mathfrak{L}^t = \mathfrak{L}$, $\mathfrak{F}^t = \mathfrak{F}$. Thus the mapping $x \to x^t$ $(x \in \mathfrak{F})$ is a permutation of $\mathfrak{F}$ which, by II, 2.2, is similar to the restriction of $t = (0)(v, \infty) \cdots$ to Ω. Hence the number of fixed points of t on Ω or on $\Omega \cup \{\infty, v\}$ is $|\mathbf{C}_{\mathfrak{F}}(t)|$. **q.e.d.**

2.6 Theorem (SUZUKI [9]). *Suppose that $\mathfrak{G}$ is a group of Suzuki type. If $\mathfrak{G}$ has a transitive extension, $|\mathfrak{F}|$ is 4 or 9, and in the case when $|\mathfrak{F}| = 9$, $\mathfrak{G}_{\infty,0}$ has just one involution.*

Proof. Let $\tilde{\mathfrak{G}}$ be a transitive extension of $\mathfrak{G}$ on $\Omega \cup \{\infty, v\}$. For $g \in \tilde{\mathfrak{G}}$, we denote by $\pi(g)$ the number of fixed points of g on $\Omega \cup \{\infty, v\}$. Choose an element 0 of Ω and put $\mathfrak{L} = \mathfrak{G}_\infty = \tilde{\mathfrak{G}}_{\infty,v}$, $\mathfrak{H} = \mathfrak{G}_{\infty,0} = \tilde{\mathfrak{G}}_{\infty,0,v}$. Since $\mathfrak{G}$ has no regular normal subgroup, $\mathfrak{G}$ is not a Frobenius group. Thus $\mathfrak{H} \neq 1$. And $|\mathfrak{F}|$ is a power of a prime p. We distinguish three cases.

Case 1. p is odd and $|\mathfrak{H}|$ is even.

In this case, $\mathfrak{H}$ contains an involution $j = (\infty)(0) \cdots$. As $\mathfrak{F} \trianglelefteq \mathfrak{G}_\infty = \mathfrak{F}\mathfrak{H}$, $\mathfrak{F}$ is normalised by j. Hence the mapping $x \to x^j$ $(x \in \mathfrak{F})$ is a permutation of $\mathfrak{F}$ which, by II, 2.2, is similar to the restriction of j to Ω. Hence

$$\pi(j) = 2 + |\mathbf{C}_{\mathfrak{F}}(j)|.$$

Since $\mathfrak{G}$ is doubly transitive, j has a conjugate s in $\mathfrak{G}$ which interchanges 0 and ∞. Thus by 2.5a), there is a conjugate $t = (0)(v, \infty) \cdots$ of s in $\tilde{\mathfrak{G}}$ such that $\tilde{\mathfrak{G}} = \langle \mathfrak{G}, t \rangle$, $\mathfrak{L}^t = \mathfrak{L}$, $\mathfrak{H}^t = \mathfrak{H}$ and $(ts)^3 \in \mathfrak{H}$. By 2.5b), $\pi(t) = |\mathbf{C}_{\mathfrak{F}}(t)|$, so

$$|\mathbf{C}_{\mathfrak{F}}(t)| = \pi(t) = \pi(j) = 2 + |\mathbf{C}_{\mathfrak{F}}(j)|.$$

Since $\mathfrak{F}$ is a p-group, $\mathbf{C}_{\mathfrak{F}}(t)$ and $\mathbf{C}_{\mathfrak{F}}(j)$ are p-groups. Since p is odd, it follows that $p = 3$, $|\mathbf{C}_{\mathfrak{F}}(j)| = 1$ and $|\mathbf{C}_{\mathfrak{F}}(t)| = 3$. Thus j induces a fixed point free automorphism on $\mathfrak{F}$. Hence $\mathfrak{F}$ is Abelian and $x^j = x^{-1}$ for all $x \in \mathfrak{F}$. By 2.4d), $\mathbf{C}_{\mathfrak{H}}(\mathfrak{F}) = 1$, so $\mathfrak{H}$ is faithfully represented on $\mathfrak{F}$. But we have shown that $x^j = x^{-1}$ for all involutions j in $\mathfrak{H}$, so it follows that $\mathfrak{H}$ has only one involution.

If $|\mathfrak{F}| = 3$, $\mathfrak{G}$ is a doubly transitive group of degree 4, so $\mathfrak{G}$ has a regular normal subgroup. This is not the case, so $|\mathfrak{F}| > 3$. Now $\mathfrak{F}$ is Abelian and $\mathfrak{F}^t = \mathfrak{F}$. Hence $\mathfrak{F} = \mathfrak{F}_1 \times \mathfrak{F}_{-1}$, where $\mathfrak{F}_1 = \mathbf{C}_{\mathfrak{F}}(t)$ and $z^t = z^{-1}$ for all $z \in \mathfrak{F}_{-1}$. Since $|\mathfrak{F}_1| = 3$ and $|\mathfrak{F}| > 3$, $|\mathfrak{F}_{-1}| \geq 3$. Also $z^j = z^{-1}$ for all $z \in \mathfrak{F}_{-1}$, so $jt \in \mathbf{C}_{\mathfrak{G}}(\mathfrak{F}_{-1})$. As j is the only involution in $\mathfrak{H} = \mathfrak{H}^s = \mathfrak{H}^t$, $j = j^s = j^t$. Hence

$$(jt)^2 = jj^t = 1.$$

Since $\mathfrak{F}^{jt} = \mathfrak{F}$ and $jt = (0)(v, \infty) \cdots$, it follows from II, 2.2 that $\pi(jt) = |\mathbf{C}_{\mathfrak{F}}(jt)|$. Thus $\pi(jt) \geq |\mathfrak{F}_{-1}| \geq 3$. But on account of the triple transitivity of $\tilde{\mathfrak{G}}$, there exists $g \in \tilde{\mathfrak{G}}$ such that $(jt)^g \in \tilde{\mathfrak{G}}_{v,0,\infty} = \mathfrak{H}$. Since j is the only involution in $\mathfrak{H}$, it follows that $(jt)^g = j$ and then

$$3 = \pi(j) = \pi(jt) \geq |\mathfrak{F}_{-1}|.$$

This forces $|\mathfrak{F}_{-1}| = 3$ and $|\mathfrak{F}| = 9$.

Case 2. p is odd and $|\mathfrak{H}|$ is odd.

In this case let s be any involution in $\mathfrak{G}$ which interchanges 0 and ∞. Since $|\mathfrak{H}|$ is odd, $|\mathfrak{G}_\infty| = |\mathfrak{F}||\mathfrak{H}|$ is odd, so s has no fixed points in $\Omega \cup \{\infty\}$. Thus $\pi(s) = 1$. By 2.5, s has a conjugate t in $\tilde{\mathfrak{G}}$ such that $\tilde{\mathfrak{G}} = \langle \mathfrak{G}, t \rangle$, $t = (0)(v, \infty) \cdots$, $\mathfrak{L}^t = \mathfrak{L}$, $\mathfrak{H}^t = \mathfrak{H}$, $(ts)^3 \in \mathfrak{H}$ and $\pi(t) = |\mathbf{C}_{\mathfrak{F}}(t)|$. Hence $\mathbf{C}_{\mathfrak{F}}(t) = 1$. Thus t induces a fixed point free auto morphism on $\mathfrak{F}$, $\mathfrak{F}$ is Abelian and

(1) $$x^t = x^{-1} \text{ for all } x \in \mathfrak{F}.$$

Suppose that $h \in \mathfrak{H} = \mathfrak{H}^t$. Then $[h, t] \in \mathfrak{H}$. Since $x^t = x^{-1}$ and $\mathfrak{F}^h = \mathfrak{F}$, $[h, t] \in \mathbf{C}_{\mathfrak{H}}(\mathfrak{F})$. But by 2.4d), $\mathbf{C}_{\mathfrak{H}}(\mathfrak{F}) = 1$, so

(2) $$h^t = h \text{ for all } h \in \mathfrak{H}.$$

Put $h_0 = (ts)^3 \in \mathfrak{H}$. Thus $h_0^t = (st)^3 = h_0^{-1}$, so by (2), $h_0^2 = 1$. Since $|\mathfrak{H}|$ is odd, $h_0 = 1$. Thus

(3) $$(ts)^3 = 1.$$

Now s, ts are elements of $\mathbf{N}_{\tilde{\mathfrak{G}}}(\mathfrak{H})$ and by (2), $s \equiv ts \bmod \mathbf{C}_{\tilde{\mathfrak{G}}}(\mathfrak{H})$. Thus

$$s = s^3 \equiv (ts)^3 = 1 \mod \mathbf{C}_{\tilde{\mathfrak{G}}}(\mathfrak{H}).$$

Hence

$$h^s = h \text{ for all } h \in \mathfrak{H}. \tag{4}$$

Now suppose that $1 \neq x \in \mathfrak{F}$. Since $\mathfrak{L}^s \cap \mathfrak{F} = \mathfrak{G}_0 \cap \mathfrak{F} = 1$, $sxs \notin \mathfrak{L}$. Hence by 2.4c),

$$sxs = x_1 hsx_2, \tag{5}$$

where $h \in \mathfrak{H}$ and x_1, x_2 are in $\mathfrak{F}$. By (1), (2) and (3),

$$(x_1 hsx_2)^t = x_1^{-1}htstx_2^{-1} = x_1^{-1}hstsx_2^{-1}. \tag{6}$$

On the other hand,

$$\begin{aligned}
(sxs)^t &= tstx^{-1}tst && \text{(by (1))}\\
&= stsx^{-1}sts && \text{(by (3))}\\
&= st(sxs)^{-1}ts && \\
&= st(x_2^{-1}sh^{-1}x_1^{-1})ts && \text{(by (5))}\\
&= sx_2tsh^{-1}tx_1s && \text{(by (1))}\\
&= sx_2tsth^{-1}x_1s && \text{(by (2))}\\
&= sx_2stsh^{-1}x_1s && \text{(by (3))}\\
&= (sx_2s)h^{-1}t(sx_1s) && \text{(by (2) and (4))}\\
&= (h^{-1}x_1^{-1}sx)h^{-1}t(xsx_2^{-1}h^{-1}) && \text{(by (5))}\\
&= h^{-1}x_1^{-1}sxh^{-1}x^{-1}tsx_2^{-1}h^{-1} && \text{(by (1))}\\
&= h^{-1}x_1^{-1}sxh^{-1}x^{-1}h^{-1}tshx_2^{-1}h^{-1} && \text{(by (2) and (4)).}
\end{aligned}$$

Using (5) and (6), this gives

$$x_1^{-1}hst \cdot sx_2^{-1} = h^{-1}x_1^{-1}sxh^{-1}x^{-1}h^{-1}t \cdot shx_2^{-1}h^{-1}, \tag{7}$$

so

$$\begin{aligned}
sx_2^{-1}hx_2h^{-1}s &= t(sh^{-1}x_1h^{-1}x_1^{-1}sxh^{-1}x^{-1}h^{-1})t \in \mathfrak{G} \cap \mathfrak{G}^t\\
&= \mathfrak{G}_v \cap \mathfrak{G}_\infty = \mathfrak{L}.
\end{aligned}$$

From this we obtain

$$x_2^{-1}hx_2h^{-1} \in \mathfrak{L} \cap s\mathfrak{L}s = \mathfrak{G}_\infty \cap \mathfrak{G}_0 = \mathfrak{H}.$$

Since $\mathfrak{F} \trianglelefteq \mathfrak{F}\mathfrak{H}$, it follows that

$$x_2^{-1}hx_2h^{-1} \in [\mathfrak{F}, \mathfrak{H}] \cap \mathfrak{H} \leq \mathfrak{F} \cap \mathfrak{H} = 1.$$

This shows that

$$(8) \qquad x_2^h = x_2.$$

Now (7) gives

$$\begin{aligned} x_1^{-1}hs \cdot tsx_2^{-1} &= h^{-1}x_1^{-1}sxh^{-1}x^{-1}h^{-1}tshx_2^{-1}h^{-1} \\ &= h^{-1}x_1^{-1}sxh^{-1}x^{-1}h^{-1} \cdot tsx_2^{-1}. \end{aligned}$$

There remains

$$x_1^{-1}hs = h^{-1}x_1^{-1}s \cdot xh^{-1}x^{-1}h^{-1}.$$

This gives

$$(9) \qquad s(x_1hx_1^{-1}h)s = xh^{-1}x^{-1}h^{-1} \in \mathfrak{L}^s \cap \mathfrak{L} = \mathfrak{H}.$$

Thus we also have

$$xh^{-1}x^{-1}h \in [\mathfrak{F}, \mathfrak{H}] \cap \mathfrak{H} \leq \mathfrak{F} \cap \mathfrak{H} = 1.$$

Thus

$$(10) \qquad x^h = x.$$

By (4), (8) and (10), h commutes with s, x_2 and x, so by (5),

$$(11) \qquad x_1^h = x_1.$$

From (9) we now obtain $sh^2s = h^{-2}$. On the other hand, $sh^2s = h^2$ by (4), so $h^4 = 1$. Since $|\mathfrak{H}|$ is odd, $h = 1$. Hence $sxs = x_1sx_2$ and

$$s\mathfrak{F}s \subseteq \mathfrak{F}s\mathfrak{F}.$$

It follows that the set $\mathfrak{N} = \mathfrak{F} \cup \mathfrak{F}s\mathfrak{F}$ is a subgroup of $\mathfrak{G}$. Since $\mathfrak{F}$ is normalised by $\mathfrak{H}$ and, by (4), $s \in \mathbf{C}_{\mathfrak{G}}(\mathfrak{H})$,

$$\mathfrak{N} \trianglelefteq \langle \mathfrak{N}, \mathfrak{H} \rangle = \langle \mathfrak{F}, \mathfrak{H}, s \rangle = \mathfrak{G}.$$

As $\mathfrak{G}$ is doubly transitive on $\Omega \cup \{\infty\}$, $\mathfrak{N}$ is transitive on $\Omega \cup \{\infty\}$. Since

$$\infty x_1 s x_2 = \infty s x_2 = 0 x_2 \neq \infty,$$

$\mathfrak{N}_\infty = \mathfrak{F}$ and $\mathfrak{N}_{\infty,0} = \mathfrak{F}_0 = 1$. Thus $\mathfrak{N}$ is a Frobenius group on $\Omega \cup \{\infty\}$. The Frobenius kernel $\mathfrak{K}$ of $\mathfrak{N}$ is a characteristic subgroup of $\mathfrak{N}$ and is thus a regular normal subgroup of $\mathfrak{G}$. But this contradicts the hypothesis on $\mathfrak{G}$, so Case 2 does not occur.

Case 3. $p = 2$ and $|\mathfrak{F}| \geq 8$.

Let i be an involution in $\mathfrak{F}$. Since $\mathfrak{F}$ is regular, i has precisely two fixed points, namely, v and ∞. Thus $\pi(i) = 2$. Since $\mathfrak{G}$ is doubly transitive, i has a conjugate s in $\mathfrak{G}$ which interchanges 0 and ∞. Let t be the involution defined as in 2.5. Thus $|\mathbf{C}_{\mathfrak{F}}(t)| = \pi(t) = 2$.

(1) $\mathfrak{F}$ is cyclic, dihedral or generalized quaternion.

Since $\mathfrak{F}^t = \mathfrak{F}$, $\mathfrak{Q} = \mathfrak{F}\langle t \rangle$ is a 2-group. Since $|\mathbf{C}_{\mathfrak{F}}(t)| = 2$,

$$|\mathbf{C}_{\mathfrak{Q}}(t)| = |\mathbf{C}_{\mathfrak{F}}(t)\langle t \rangle| = 4.$$

Thus $\mathfrak{Q}$ is dihedral or semidihedral, by 2.3. Since $\mathfrak{F}$ is a normal subgroup of $\mathfrak{Q}$ of index 2, it follows easily that $\mathfrak{F}$ is cyclic, dihedral or generalized quaternion.

(2) $|\mathfrak{F}| > 8$ and $\mathfrak{H}$ is a non-identity 2-group.

By 2.4d), $\mathbf{C}_{\mathfrak{H}}(\mathfrak{F}) = 1$, so $\mathfrak{H}$ is isomorphic to a subgroup of the group $\mathfrak{A}$ of automorphisms of $\mathfrak{F}$.

First suppose that $|\mathfrak{F}| = 8$. Then $|\mathfrak{A}|$ is a power of 2 or, if $\mathfrak{F}$ is the quaternion group, $|\mathfrak{A}| = 3 \cdot 2^3$. Now $|\mathfrak{G}| = (|\mathfrak{F}| + 1)|\mathfrak{F}||\mathfrak{H}|$. Since $|\mathfrak{H}|$ divides $|\mathfrak{A}|$, it follows that $|\mathfrak{G}|$ is of the form $2^a 3^b$. Thus $\mathfrak{G}$ is soluble and, by II, 3.2, any minimal normal subgroup of $\mathfrak{G}$ is regular. This is contrary to the hypothesis.

Thus $|\mathfrak{F}| > 8$. Then $\mathfrak{A}$ is a 2-group on account of (1). Thus $\mathfrak{H}$ is a 2-group. Since $\mathfrak{G}$ is not a Frobenius group, $\mathfrak{H} \neq 1$.

(3) There is an involution j such that $\mathfrak{H} = \langle j \rangle$ and $\pi(j) = 4$.

Let j be any involution in $\mathfrak{H}$. Then j fixes v and ∞. Since $\mathfrak{F}$ is regular, it follows from II, 2.2 that

$$\pi(j) = 2 + |\mathbf{C}_{\mathfrak{F}}(j)|.$$

But also j has a conjugate $s' = (v)(0, \infty) \cdots$, so by 2.5, j has a conjugate $t' = (0)(v, \infty) \cdots$ for which $\pi(t') = |\mathbf{C}_{\mathfrak{F}}(t')|$. Hence

$$|\mathbf{C}_{\mathfrak{F}}(t')| = \pi(j) = 2 + |\mathbf{C}_{\mathfrak{F}}(j)|.$$

As $|\mathbf{C}_{\mathfrak{F}}(t')|$ and $|\mathbf{C}_{\mathfrak{F}}(j)|$ are powers of 2, it follows that $|\mathbf{C}_{\mathfrak{F}}(j)| = 2$ and $\pi(j) = 4$.

It follows from (1) and $|\mathfrak{F}| > 8$ that $\mathfrak{F}$ has a cyclic characteristic subgroup $\langle z \rangle$ of order 4. Since $z^2 \in \mathbf{C}_{\mathfrak{F}}(j)$ and $|\mathbf{C}_{\mathfrak{F}}(j)| = 2$, $\mathbf{C}_{\mathfrak{F}}(j) = \langle z^2 \rangle$. Thus $z^j = z^{-1}$ for any involution j in $\mathfrak{H}$. Hence $\mathfrak{H}$ is faithfully represented on $\langle z \rangle$. Since the order of the group of automorphisms of $\langle z \rangle$ is 2, $|\mathfrak{H}| = 2$.

(4) $\mathbf{C}_{\mathfrak{G}}(j)/\langle j \rangle \cong \mathfrak{S}_3$.

By (3), $\pi(j) = 4$. Thus the set of fixed points of j consists of v, ∞, 0 and one other point of Ω which we denote by 1. Thus $\{0, 1, \infty\}$ is the full set of fixed points of $\mathfrak{H}$ on $\Omega \cup \{\infty\}$. By II, 1.13b), $\mathbf{N}_{\mathfrak{G}}(\mathfrak{H}) = \mathbf{C}_{\mathfrak{G}}(j)$ is doubly transitive on $\{0, 1, \infty\}$. Thus $\mathbf{C}_{\mathfrak{G}}(j)/\mathbf{C}_{\mathfrak{G}}(j)_{0,1,\infty} \cong \mathfrak{S}_3$. But $\langle j \rangle \leq \mathbf{C}_{\mathfrak{G}}(j)_{0,1,\infty} \leq \mathfrak{G}_{0,\infty} = \mathfrak{H} = \langle j \rangle$. Hence $\mathbf{C}_{\mathfrak{G}}(j)/\langle j \rangle \cong \mathfrak{S}_3$.

(5) $\mathbf{C}_{\mathfrak{G}}(j)$ has exactly 7 involutions.

Since $\mathfrak{H}^s = \mathfrak{H}$ and $\mathfrak{H} = \langle j \rangle$, $\langle j, s \rangle$ is a Sylow 2-subgroup of $\mathbf{C}_{\mathfrak{G}}(j)$. Thus $\mathbf{C}_{\mathfrak{G}}(j)$ has no element of order 4, and since the inverse image of $\mathfrak{A}_3$ in $\mathbf{C}_{\mathfrak{G}}(j)$ is cyclic of order 6, $\mathbf{C}_{\mathfrak{G}}(j)$ is the dihedral group of order 12. Hence $\mathbf{C}_{\mathfrak{G}}(j)$ has 7 involutions.

To conclude the proof, write

$$j - (v)(\infty)(0)(1)(a_1, b_1) \qquad (a_r, b_r).$$

Thus

$$2r + 4 = |\Omega| + 2 = |\mathfrak{F}| + 2,$$

and $r = \frac{1}{2}|\mathfrak{F}| - 1$. Now since $\mathfrak{G}$ is triply transitive, $\mathfrak{G}_{v,a_i,b_i} = \langle j_i \rangle$ for some conjugate j_i of j. Thus $\pi(j_i) = \pi(j) = 4$. Since $\mathfrak{G}^j_{v,a_i,b_i} = \mathfrak{G}_{v,a_i,b_i}$, $j_i \in \mathbf{C}_{\mathfrak{G}}(j)$. Also $j_i \neq j_k$ for $i \neq k$, since otherwise j_i has the five fixed points v, a_i, b_i, a_k, b_k. Hence $\mathbf{C}_{\mathfrak{G}}(j)$ has at least r involutions in addition to j. By (5), $r \leq 6$. Thus $|\mathfrak{F}| \leq 14$, contrary to (2). **q.e.d.**

We combine our results as follows.

2.7 Theorem (SUZUKI [9]). *In their natural representations as permutation groups, the following groups have no transitive extensions.*

a) *Any group* $\mathfrak{G}$ *for which* $PSL(2, p^f) \leq \mathfrak{G} \leq P\Gamma L(2, p^f)$, *where* p^f *is not equal to* 2, 3, 4 *or* 9.

b) *The Suzuki groups* $Sz(q)$ *of degree* $q^2 + 1$.

c) *The Ree groups* $R(q)$ *of degree* $q^3 + 1$.

d) *The unitary groups* $PSU(3, q^2)$ *and* $PU(3, q^2)$ *of degree* $q^3 + 1$, *where* $q > 2$.

Proof. This follows at once from 2.2 and 2.6. **q.e.d.**

2.8 Remarks. a) Since $PSL(2, 2) = PGL(2, 2) \cong \mathfrak{S}_3$, $PSL(2, 2)$ has a transitive extension, namely $\mathfrak{S}_4$.

b) Since $PSL(2, 3) \cong \mathfrak{A}_4$ and $PGL(2, 3) \cong \mathfrak{S}_4$, $PSL(2, 3)$ and $PGL(2, 3)$ have the transitive extensions $\mathfrak{A}_5$, $\mathfrak{S}_5$ respectively.

c) Since $PSL(2, 4) = PGL(2, 4) \cong \mathfrak{A}_5$ and $P\Gamma L(2, 4) \cong \mathfrak{S}_5$, $PSL(2, 4)$ and $P\Gamma L(2, 4)$ have the transitive extensions $\mathfrak{A}_6$, $\mathfrak{S}_6$ respectively.

d) As $P\Gamma L(2, 9)/PSL(2, 9)$ is elementary Abelian of order 4, $P\Gamma L(2, 9)$ has exactly 3 subgroups of index 2. One of these is the sharply triply transitive group $M(9)$ (see XI, 1.3), and $M(9)$ does have a transitive extension, namely the Mathieu group $\mathfrak{M}_{11}$. The group $PGL(2, 9)$ has no transitive extension, as we shall see in 3.3. Further, we shall see in 3.6 that the Zassenhaus group $PSL(2, 9)$ also has no transitive extension. If $\mathfrak{G}$ is $P\Gamma L(2, 9)$ or the extension of $PSL(2, 9)$ by the permutation $x \to x^3$, then $\mathfrak{G}_{0,\infty}$ has two distinct involutions, namely $x \to -x$ and $x \to x^3$. Hence by 2.6, these groups have no transitive extension.

e) The special unitary group $PSU(3, 4)$ is a doubly transitive group of degree 9 and order 72, and $M(9)$ is a transitive extension of $PSU(3, 4)$.

On the other hand, $PU(3, 4)$ has no transitive extension, as may be seen as follows. If $\tilde{\mathfrak{G}}$ is a transitive extension of $\mathfrak{G} = PU(3, 4)$, then $|\tilde{\mathfrak{G}}| = 10 \cdot 9 \cdot 8 \cdot 3$ and $|\tilde{\mathfrak{G}}_{1,2,3}| = 3$ (with suitable labelling of the symbols permuted). Suppose that $1 \neq a \in \tilde{\mathfrak{G}}_{1,2,3}$. If a is a cycle of length 3, then $\tilde{\mathfrak{G}}$, being triply transitive, contains all cycles of length 3 and so $\tilde{\mathfrak{G}} \geq \mathfrak{A}_{10}$. Thus with suitable labelling,

$$a = (1)(2)(3)(4)(5, 6, 7)(8, 9, 10).$$

By II, 1.13b), $\mathbf{N}_{\tilde{\mathfrak{G}}}(\tilde{\mathfrak{G}}_{1,2,3})$ is triply transitive on $\{1, 2, 3, 4\}$. Thus $\mathbf{N}_{\tilde{\mathfrak{G}}}(\langle a \rangle)/\langle a \rangle = \mathfrak{S}_4$. Hence $\mathbf{C}_{\tilde{\mathfrak{G}}}(a)$ contains an involution j in which $(1, 2)$, $(3, 4)$ are cycles. Since $|\tilde{\mathfrak{G}}_{1,2,3}| = 3$, j has at most two fixed points. But if x is a fixed point of j, so are xa, xa^2. Thus j has no fixed point and is therefore an odd permutation. Thus $\mathfrak{N} = \tilde{\mathfrak{G}} \cap \mathfrak{A}_{10}$ is a normal subgroup of $\tilde{\mathfrak{G}}$ of index 2. Hence

$$2 = |\tilde{\mathfrak{G}}/\mathfrak{N}| = |\mathfrak{N}\mathfrak{G}/\mathfrak{N}| = |\mathfrak{G}/(\mathfrak{G} \cap \mathfrak{N})|.$$

But it is easily seen that $\mathfrak{G} = PU(3, 4)$ has no normal subgroup of index 2.

§ 3. Sharply Multiply Transitive Permutation Groups

In XI, § 2, we determined all sharply triply transitive permutation groups. In this section we shall find all sharply k-fold transitive permutation groups for $k \geq 4$. It turns out that the only groups which occur, apart from the alternating and symmetric groups, are the Mathieu groups $\mathfrak{M}_{11}$ and $\mathfrak{M}_{12}$.

First, we prove the following lemmas.

3.1 Lemma. *Let $\mathfrak{G}$ be a group, and for each $g \in \mathfrak{G}$, define a permutation ρ_g of $\mathfrak{G}$ by $x\rho_g = xg$. Then the ρ_g form a group $\mathfrak{R}$. If α is a permutation of $\mathfrak{G}$ for which $\alpha^{-1}\mathfrak{R}\alpha = \mathfrak{R}$ and $1\alpha = 1$, then α is an automorphism of $\mathfrak{G}$.*

Proof. The ρ_g form a group since $\rho_g\rho_h^{-1} = \rho_{gh^{-1}}$. Given $g \in \mathfrak{G}$, $\alpha^{-1}\rho_g\alpha = \rho_{g'}$ for some $g' \in \mathfrak{G}$, since $\alpha^{-1}\mathfrak{R}\alpha = \mathfrak{R}$. Then

$$g\alpha = 1\rho_g\alpha = 1\alpha\rho_{g'} = 1\rho_{g'} = g',$$

so for any $x \in \mathfrak{G}$,

$$(xg)\alpha = x\rho_g\alpha = x\alpha\rho_{g'} = (x\alpha)g' = (x\alpha)(g\alpha).$$

Hence α is an automorphism of $\mathfrak{G}$. **q.e.d.**

3.2 Lemma. *Suppose that $\mathfrak{G}$ is a transitive extension of the group $M(9)$ on $\mathsf{K} \cup \{\infty, v\}$, where $\mathsf{K} = GF(3^2)$. Suppose that $j \in \mathsf{K}$ and $j^2 = 1 - j$; thus $j^4 = -1$ and $\mathsf{K}^\times = \langle j \rangle$. Then $\mathfrak{G}$ is similar to $\langle M(9), c\rangle$, where*

$$c = (\infty, v)(0)(1)(-1)(j, j^5)(j^2, j^7)(j^3, j^6).$$

In particular, $\mathfrak{M}_{11}$ is similar to $\langle M(9), c\rangle$.

Proof. Since $M(9)$ is sharply 3-fold transitive, $\mathfrak{G}$ is sharply 4-fold transitive. Hence there exists $c \in \mathfrak{G}$ such that

$$c = (\infty, v)(0)(1) \cdots,$$

and $c^2 = 1$. Also c normalises $\mathfrak{G}_{v,\infty}$. Now the Frobenius kernel $\mathfrak{F}$ of $\mathfrak{G}_{v,\infty}$ is a characteristic subgroup of $\mathfrak{G}_{v,\infty}$ and consists of all permutations $x \to x + a$ $(a \in \mathsf{K})$. Thus $c^{-1}\mathfrak{F}c = \mathfrak{F}$; hence by 3.1, the restriction of c to K is an additive homomorphism and is therefore linear over $GF(3)$. In particular, $(-y)c = -(yc)$ for all $y \in \mathsf{K}$, and $(-1)c = -1$. Since $\mathfrak{G}$ is sharply 4-fold transitive, c can have no fixed points other than 0, 1 and -1. But $c^2 = 1$, so c has an eigenvector z in K such that $zc = -z$. Thus $z \notin GF(3)$ and every element of K is uniquely expressible in the form $\lambda + \mu z$, with λ, μ in $GF(3)$. We have

$$(\lambda + \mu z)c = \lambda - \mu z.$$

Suppose that $z^4 = 1$. Since $z \notin GF(3)$, $z^2 \neq 1$, so $z^2 = -1$ and $z^3 = -z$. Thus

$$(\lambda + \mu z)c = \lambda + \mu z^3 = (\lambda + \mu z)^3$$

and $xc = x^3$ for all $x \in \mathsf{K}$. But now if d is the permutation of $\mathsf{K} \cup \{\infty, v\}$ given by $xd = -x^{-1}$, $vd = v$, then $d \in \mathfrak{G}_v$ since $\det\begin{pmatrix} 0 & -1 \\ 1 & 0 \end{pmatrix} = 1$. Thus $cd \in \mathfrak{G}$ and $x(cd) = -x^{-3}$ for all $x \in \mathsf{K}$. Then $x(cd)^2 = x$ for all $x \in \mathsf{K}^\times$. Since no non-identity element of $\mathfrak{G}$ has more than three fixed points, $(cd)^2 = 1$. But $0(cd)^2 = v$, a contradiction. Thus $z^4 \neq 1$.

Since $\mathsf{K}^\times = \langle j \rangle$, z is one of j, j^3, j^5, j^7. It follows easily from $(\lambda + \mu z)c = \lambda - \mu z$ that c is one of c_1, c_2, where

$$c_1 = (\infty, v)(0)(1)(-1)(j, j^5)(j^2, j^7)(j^3, j^6),$$

$$c_2 = (\infty, v)(0)(1)(-1)(j, j^2)(j^3, j^7)(j^5, j^6).$$

But $M(9)$ is left invariant and c_1 is transformed into c_2 by the automorphism $x \to x^3$ of K, so $\langle M(9), c_1 \rangle$ and $\langle M(9), c_2 \rangle$ are similar permutation groups. Since $\mathfrak{G} = \langle M(9), c \rangle$, it follows that $\mathfrak{G}$ is similar to $\langle M(9), c_1 \rangle$. **q.e.d.**

3.3 Theorem (JORDAN). *If $\mathfrak{G}$ is sharply 4-fold transitive, $\mathfrak{G}$ is one of the groups $\mathfrak{S}_4$, $\mathfrak{S}_5$, $\mathfrak{A}_6$, $\mathfrak{M}_{11}$.*

Proof. Suppose that $\mathfrak{G}$ is sharply 4-fold transitive on the set $\Omega = \Delta \cup \{v\}$, where $v \notin \Delta$ and $|\Omega| = n + 2$. Then $\mathfrak{G}_v$ is a sharply triply transitive permutation group on Δ.

If $\mathfrak{G}_v$ is a Frobenius group on Δ, then

$$(n+1)n(n-1) = |\mathfrak{G}_v| = (n+1)n,$$

so $n = 2$ and $\mathfrak{G} = \mathfrak{S}_4$.

If $\mathfrak{G}_v$ is not a Frobenius group but has a regular normal subgroup, then by XI, 1.1, $n + 1 = 2^p$ for some prime p and

$$2^p(2^p - 1)(2^p - 2) = |\mathfrak{G}_v| = 2^p(2^p - 1)p.$$

Thus $p = 2$, $n + 2 = 5$ and $\mathfrak{G} = \mathfrak{S}_5$.

If $\mathfrak{G}_v$ has no regular normal subgroup, $\mathfrak{G}_v$ is a Zassenhaus group. By XI, 1.4, $n = p^f$ for some prime p. By 2.1, $\mathfrak{G}_v$ is of Suzuki type. Since $\mathfrak{G}_v$ has the transitive extension $\mathfrak{G}$, it follows from 2.6 that p^f is 4 or 9. If $p^f = 4$, $|\Omega| = 6$, $|\mathfrak{G}| = 6 \cdot 5 \cdot 4 \cdot 3$ and $\mathfrak{G} = \mathfrak{A}_6$.

Suppose henceforth that $p^f = 9$. By XI, 2.6, $\mathfrak{G}_v = PGL(2, 9)$ or $\mathfrak{G}_v = M(9)$.

Suppose first that $\mathfrak{G}_v$ is $PGL(2, 9)$ in its natural permutation representation; thus $\Delta = \mathsf{K} \cup \{\infty\}$, where $\mathsf{K} = GF(9)$. Hence $\mathfrak{G}$ is a permutation group of degree 11 and $\mathfrak{Z} = \mathfrak{G}_{v,0,\infty}$ is the set of the 8 permutations of the form $z \to az$ ($a \in \mathsf{K}^\times$). Hence $\mathfrak{Z}$ is cyclic. We denote by $\mathfrak{Z}^*$ the restriction of $\mathfrak{Z}$ to $\mathsf{K}^\times$, and let $\mathbf{N}(\mathfrak{Z}^*)$, $\mathbf{C}(\mathfrak{Z}^*)$ denote the normaliser and centralizer of $\mathfrak{Z}^*$ in the symmetric group on $\mathsf{K}^\times$. By II, 3.1, $\mathbf{C}(\mathfrak{Z}^*) = \mathfrak{Z}^*$, since $\mathfrak{Z}^*$ is a regular Abelian permutation group on $\mathsf{K}^\times$. Since $\mathfrak{Z}^*$ is cyclic, $\mathbf{N}(\mathfrak{Z}^*)/\mathbf{C}(\mathfrak{Z}^*)$ is Abelian, so $\mathbf{N}(\mathfrak{Z}^*)' \leq \mathfrak{Z}^*$. But now $\mathbf{N}_\mathfrak{G}(\mathfrak{Z})$ is precisely the set of permutations in $\mathfrak{G}$ which leave the set $\Lambda = \{0, v, \infty\}$ fixed, since Λ is the set of all fixed points of $\mathfrak{Z}$. It follows that if $g \in \mathbf{N}_\mathfrak{G}(\mathfrak{Z})'$, then $g = g_1 g_2$, where $g_1 \in \mathfrak{Z}$ and g_2 leaves each element of $\mathsf{K}^\times$ fixed. But then $g_2 = g_1^{-1} g \in \mathfrak{G}$ and g_2 has at least eight fixed points. Since $\mathfrak{G}$ is sharply 4-fold transitive, it follows that $g_2 = 1$. Hence $\mathbf{N}_\mathfrak{G}(\mathfrak{Z})' \leq \mathfrak{Z}$.

On the other hand, since $\mathfrak{G}$ is 4-fold transitive, any permutation π of Λ is induced by some element g of $\mathfrak{G}$; indeed, since g leaves Λ fixed, $g \in \mathbf{N}_\mathfrak{G}(\mathfrak{Z})$. Now $(0, v, \infty)$ lies in $\mathfrak{A}_3 = \mathfrak{S}_3'$, so $(0, v, \infty)$ is a cycle of an element of $\mathbf{N}_\mathfrak{G}(\mathfrak{Z})'$, contrary to the above. Hence $\mathfrak{G}_v$ cannot be $PGL(2, 9)$.

Thus $\mathfrak{G}_v = M(9)$. But by 3.2, there is only one transitive extension of $M(9)$ to within similarity. Since $\mathfrak{M}_{11}$ is a transitive extension of $M(9)$, $\mathfrak{G} \cong \mathfrak{M}_{11}$. **q.e.d.**

3.4 Theorem (JORDAN). *If $\mathfrak{G}$ is sharply 5-fold transitive, $\mathfrak{G}$ is one of the groups $\mathfrak{S}_5$, $\mathfrak{S}_6$, $\mathfrak{A}_7$, $\mathfrak{M}_{12}$.*

Proof. Let w be one of the symbols permuted. By 3.3, $\mathfrak{G}_w$ is one of the sharply 4-fold transitive groups $\mathfrak{S}_4$, $\mathfrak{S}_5$, $\mathfrak{A}_6$ or $\mathfrak{M}_{11}$. In the first three cases, it follows at once that $\mathfrak{G}$ is $\mathfrak{S}_5$, $\mathfrak{S}_6$ or $\mathfrak{A}_7$. Suppose then that $\mathfrak{G}_w$ is $\mathfrak{M}_{11}$. By 3.2, we may suppose that $\mathfrak{G}$ contains the element

$$c = (w)(\infty, v)(0)(1)(-1)(j, j^5)(j^2, j^7)(j^3, j^6),$$

where $j^4 = -1$ and $\mathsf{K}^\times = \langle j \rangle$.

Since $\mathfrak{G}$ is sharply 5-fold transitive, there exists $b \in \mathfrak{G}$ such that

$$b = (v, w)(\infty)(0)(1) \cdots,$$

and $b^2 = 1$. Since $b \in \mathbf{N}_{\mathfrak{G}}(\mathfrak{G}_{v, w, \infty})$, b normalises the Frobenius kernel $\mathfrak{F}$ of $\mathfrak{G}_{v, w, \infty}$. Also $\mathfrak{F}$ consists of the permutations $x \to x + a$ ($a \in \mathsf{K}$), so by 3.1, the restriction of b to K is linear over $GF(3)$. In particular, $(-x)b = -(xb)$ and $(-1)b = -1$. Thus $\infty, 0, 1, -1$ are all the fixed points of b. Also,

$$bc = (w, \infty, v)(0)(1)(-1) \cdots,$$

so $(bc)^3 = 1$. The number of fixed points of bc is therefore a multiple of 3, so $0, 1, -1$ are the only fixed points of bc.

Let $\mathfrak{Q} = \mathfrak{G}_{v, w, 0, \infty}$, so $|\mathfrak{Q}| = 8$. Let r, s be the permutations which leave v, w, ∞ fixed and satisfy

$$xr = j^2 x, \quad xs = j^5 x^3, \quad xrs = j^3 x^3$$

for $x \in \mathsf{K}$. Since these permutations lie in $M(9)$, r and s are in $\mathfrak{Q}$. Thus $\mathfrak{Q} = \langle r, s \rangle$ is a quaternion group of order 8. Note that b normalises $\mathfrak{Q} = \mathfrak{G}_{v, w, 0, \infty}$. But $\mathbf{C}_{\mathfrak{Q}}(b) = \langle r^2 \rangle$, for if $a \in \mathfrak{Q} - \langle r^2 \rangle$, $1a$ is not a fixed point of b, so $1b^{-1}ab \neq 1a$. Thus b normalises some subgroup $\mathfrak{Z}$ of $\mathfrak{Q}$ of order 4, but b does not centralize $\mathfrak{Z}$.

First suppose that $\mathfrak{Z} = \langle s \rangle$. Since $s^c = s^{-1}$, it follows that $s^{bc} = s$. But $1s^{bc} = j^5(bc) \neq j^5 = 1s$, a contradiction.

Next suppose that $\mathfrak{Z} = \langle rs \rangle$. Then $(rs)^b = (rs)^{-1}$. Now the element d given by

$$vd = v, \quad wd = w, \quad xd = -x^{-1} \quad (x \in \mathsf{K} \cup \{\infty\}),$$

lies in $\mathfrak{M}_{11}$ and hence in $\mathfrak{G}$. Since $db = (v, w)(0, \infty)(1, -1) \cdots$ and $bd = (v, w)(0, \infty)(1, -1) \cdots$, $db = bd$. Thus if $x \in \mathsf{K}$, $(-x^{-1})b = -(xb)^{-1}$ and $x^{-1}b = (xb)^{-1}$. However,

$$j^{-1}b = 1(rs)^{-1}b = 1brs = 1rs = j^3 = -j^{-1},$$

so $jb = -j$. Thus

$$(\lambda j + \mu j^{-1})b = \lambda(jb) + \mu(j^{-1}b) = -(\lambda j + \mu j^{-1})$$

for any λ, μ in $GF(3)$. Since j, j^{-1} span K over $GF(3)$, it follows that $xb = -x$ for all $x \in \mathsf{K}$. Since $1b = 1$, this is a contradiction.

Hence $\mathfrak{Z} = \langle r \rangle$ and $r^b = r^{-1}$. Hence

$$(j^2)b = 1rb = 1br^{-1} = 1r^{-1} = j^6.$$

Thus

$$(\lambda + \mu j^2)b = \lambda + \mu j^6 = (\lambda + \mu j^2)^3$$

for all λ, μ in $GF(3)$, and $xb = x^3$ for all $x \in \mathsf{K}$. Thus $\mathfrak{G} = \langle b, \mathfrak{G}_w \rangle$ is uniquely determined. Since $\mathfrak{M}_{12}$ is a transitive extension of $\mathfrak{M}_{11}$, $\mathfrak{G} \cong \mathfrak{M}_{12}$. **q.e.d.**

3.5 Theorem. *If $k \geq 6$ and $\mathfrak{G}$ is sharply k-fold transitive, $\mathfrak{G}$ is one of the groups $\mathfrak{S}_k$, $\mathfrak{S}_{k+1}$, $\mathfrak{A}_{k+2}$.*

Proof. This is proved by induction on k. If $k = 6$, it follows from 3.4, since by 1.5, $\mathfrak{M}_{12}$ has no transitive extension. If $k > 6$, it follows at once from the inductive hypothesis. **q.e.d.**

We can generalize the concepts of Frobenius and Zassenhaus groups and ask for all k-fold transitive permutation groups in which the stabiliser of any $k + 1$ symbols is 1, for $k \geq 3$. We show that this class of groups is very small.

3.6 Theorem (GORENSTEIN and HUGHES [1]). *Suppose that $k \geq 3$ and that $\mathfrak{G}$ is a k-fold transitive permutation group which is not sharply k-fold transitive. If the stabiliser in $\mathfrak{G}$ of any $k + 1$ symbols is 1, then $\mathfrak{G}$ is one of the following groups in its natural permutation representation.*

a) *For $k = 3$, $\mathfrak{G}$ is $\mathfrak{A}_6$, $\mathfrak{M}_{11}$ or $P\Gamma L(2, 2^p)$, where p is a prime.*

b) *For $k = 4$, $\mathfrak{G}$ is $\mathfrak{A}_7$, $\mathfrak{S}_6$ or $\mathfrak{M}_{12}$.*

c) *For $k \geq 5$, $\mathfrak{G}$ is $\mathfrak{A}_{k+3}$ or $\mathfrak{S}_{k+2}$.*

Proof. a) Suppose that $k = 3$. We distinguish two cases.

Case 1. The stabiliser of an element has a regular normal subgroup.

Denote one of the symbols permuted by ∞. By XI, 1.1, $\mathfrak{G}_\infty$ is the group of all semilinear mappings $x \to a(x\alpha) + b$ on $GF(2^p)$, where p is a prime. We prove that $\mathfrak{G} = P\Gamma L(2, 2^p)$. If $p = 2$, $\mathfrak{G}$ is of degree 5, so $\mathfrak{G} = \mathfrak{S}_5 = P\Gamma L(2, 4)$. Henceforth suppose that $p > 2$.

$\mathfrak{G}$ contains a 2-element g of the form $g = (0, \infty)(1) \cdots$. Then $g^2 \in \mathfrak{G}_{0,\infty,1}$. But $\mathfrak{G}_{0,\infty,1}$ is the set of permutations $x \to x\alpha$, so $|\mathfrak{G}_{0,\infty,1}| = p$. Thus $g^2 = 1$. Now g normalises $\mathfrak{G}_{0,\infty}$ and $\mathfrak{G}_{0,\infty}$ consists of the mappings $x \to a(x\alpha)$. The set $\mathfrak{H}$ of mappings $x \to ax$ ($a \in GF(2^p)^\times$) is a normal Hall subgroup of $\mathfrak{G}_{0,\infty}$ and is thus a characteristic subgroup of $\mathfrak{G}_{0,\infty}$. Hence g normalises $\mathfrak{H}$. We show that g operates fixed point freely on $\mathfrak{H}$. For suppose that g commutes with $x \to ax$, where $a \neq 1$. Then $a(xg) = (ax)g$ for all $x \in GF(2^p)$. Since $1g = 1$, it follows that $a^i g = a^i$. Since g has at most 3 fixed points, it follows that $a^2 = 1$ or $a^3 = 1$. But $2^p - 1$ is not divisible by either 2 or 3, since

$$2^p - 1 \equiv (-1)^p - 1 \equiv 1\ (3).$$

Thus we have a contradiction. Hence g operates fixed point freely on $\mathfrak{H}$ and $h^g = h^{-1}$ for all $h \in \mathfrak{H}$. Hence if $a \in GF(2^p)$, $ag = a^{-1}(1g) = a^{-1}$, and $g \in P\Gamma L(2, 2^p)$. Then $\mathfrak{G} = \langle \mathfrak{G}_\infty, g \rangle = P\Gamma L(2, 2^p)$.

Case 2. The stabiliser of an element has no regular normal subgroup.

Suppose that v is one of the elements permuted. Then $\mathfrak{G}_v$ is a Zassenhaus group. By 2.2b), $\mathfrak{G}_v$ is of Suzuki type, and by 2.6, the degree of $\mathfrak{G}_v$ is 5 or 10.

If the degree of $\mathfrak{G}_v$ is 5, $\mathfrak{G}$ is triply transitive of degree 6. Thus by II, 4.7, either $\mathfrak{G}$ is $\mathfrak{S}_6$ or $\mathfrak{A}_6$, or $\mathfrak{G}$ is isomorphic to $\mathfrak{S}_5$ or $\mathfrak{A}_5$. $\mathfrak{S}_6$ is excluded, since it has a non-identity element with 4 fixed points, and $\mathfrak{A}_5$ is impossible, since its order is not divisible by $6 \cdot 5 \cdot 4$ and it has therefore no triply transitive representation of degree 6. Finally, $\mathfrak{S}_5$ is excluded, since the representation of $\mathfrak{S}_5$ of degree 6 is sharply triply transitive. Thus $\mathfrak{G} = \mathfrak{A}_6$.

Henceforth suppose that the degree of $\mathfrak{G}_v$ is 10, and put $\mathfrak{H} = \mathfrak{G}_{v,0,\infty}$. By 2.6, $\mathfrak{H}$ has just one involution. If $1 \neq h \in \mathfrak{H}$ and h leaves i fixed, then i is v, 0 or ∞. Thus every orbit of $\mathfrak{H}$ of length greater than 1 is of length $|\mathfrak{H}|$, and $|\mathfrak{H}|$ is 2, 4 or 8. Now since $\mathfrak{G}$ has more than one subgroup of index 11, $\mathfrak{G}$ is not 11-nilpotent. By IV, 2.6, the order of the normaliser of a Sylow 11-subgroup $\mathfrak{Q}$ of $\mathfrak{G}$ is $11n$, where n is 10, 5 or 2. And $|\mathfrak{G} : \mathbf{N}_{\mathfrak{G}}(\mathfrak{Q})| = \dfrac{11 \cdot 10 \cdot 9 |\mathfrak{H}|}{11n} \equiv 1\ (11)$, so $n \equiv 2|\mathfrak{H}|\ (11)$. Hence $|\mathfrak{H}| = 8$ and $n = 5$. Thus $\mathfrak{H}$ is regular on $\Omega - \{v, \infty, 0\}$ and $\mathfrak{G}$ is sharply 4-fold transitive. By 3.3, $\mathfrak{G} = \mathfrak{M}_{11}$.

b) Suppose that $k = 4$. By a), $\mathfrak{G}_v$ is one of $\mathfrak{A}_6$, $\mathfrak{M}_{11}$ or $P\Gamma L(2, 2^p)$. If $\mathfrak{G}_v$ is $P\Gamma L(2, 2^p)$, then by 2.7, $2^p = 4$ and $\mathfrak{G}_v$ is $\mathfrak{S}_5$. Hence in all cases, $\mathfrak{G}_v$ is sharply 4-fold transitive and by 3.4, $\mathfrak{G}$ is one of $\mathfrak{S}_5$, $\mathfrak{S}_6$, $\mathfrak{A}_7$ or $\mathfrak{M}_{12}$. $\mathfrak{S}_5$ is excluded since it is sharply 4-fold transitive.

c) For $k \geq 5$, the result follows by an easy induction, since by 1.5c), $\mathfrak{M}_{12}$ has no transitive extension. **q.e.d.**

For another generalization of Theorem 3.3 we need the following.

3.7 Theorem (JORDAN). *Let $\mathfrak{G}$ be a primitive permutation group of degree $n = p + k$, where p is a prime and $k \geq 3$. If $\mathfrak{G}$ contains a cycle of length p, then $\mathfrak{G} \geq \mathfrak{A}_n$.*

Proof. Let g be a cycle of length p contained in $\mathfrak{G}$ and let Δ be the set of fixed points of g. Thus $|\Delta| = k$, and if $\Gamma = \Omega - \Delta$, $|\Gamma| = p$. By II, 4.5, $\mathfrak{G}$ is $(k + 1)$-fold transitive. Put $\mathfrak{P} = \langle g \rangle$, $\mathfrak{N} = \mathbf{N}_{\mathfrak{G}}(\mathfrak{P})$. Then $\mathfrak{P} \in S_p(\mathfrak{G}_\Delta)$, so by II, 1.13c), $\mathfrak{N}$ is k-fold transitive on Δ and $\mathfrak{N}$ induces the full symmetric group $\mathfrak{S}_k$ on Δ.

Suppose that $i \in \Gamma$. If $x \in \mathfrak{N}$, then $ix \in \Gamma$, so $ix = ig^l$ for some l. Since $g \in \mathfrak{N}$, $xg^{-l} \in \mathfrak{N}_i$, or $x \in \mathfrak{N}_i g^l$. Then $\mathfrak{N} = \mathfrak{N}_i\langle g \rangle$ and $\mathfrak{N}_i$ induces $\mathfrak{S}_k$ on Δ.

Now $\mathfrak{P}^\Gamma$ is a regular normal subgroup of $\mathfrak{N}^\Gamma$. By II, 2.2, the group of permutations of $\Gamma - \{i\}$ induced by $\mathfrak{N}_i$ is similar to the representation of $\mathfrak{N}_i$ on $\mathfrak{P} - \{1\}$. Since $\mathfrak{P}$ is cyclic, it follows that $\mathfrak{N}_i^\Gamma$ is Abelian. Thus $\mathfrak{N}_i'$ induces the trivial group on Γ and the alternating group $\mathfrak{A}_k$ on Δ. Since $k \geq 3$, it follows that $\mathfrak{G}$ contains a cycle of length 3. By II, 4.5, $\mathfrak{G} \geq \mathfrak{A}_n$. **q.e.d.**

3.8 Theorem (M. HALL [2]). *Suppose that $\mathfrak{G}$ is 4-fold transitive on $\Omega = \{1, 2, 3, 4, \ldots, n\}$. If $|\mathfrak{G}_{1,2,3,4}|$ is odd, then $\mathfrak{G}$ is $\mathfrak{S}_4$, $\mathfrak{S}_5$, $\mathfrak{A}_6$, $\mathfrak{A}_7$ or $\mathfrak{M}_{11}$.*

Proof. If $\mathfrak{G}$ is sharply 4-fold transitive, then by 3.3, $\mathfrak{G}$ is $\mathfrak{S}_4$, $\mathfrak{S}_5$, $\mathfrak{A}_6$ or $\mathfrak{M}_{11}$. Suppose then that $|\mathfrak{G}_{1,2,3,4}| > 1$. Since $|\mathfrak{G}_{1,2,3,4}|$ is odd, $|\Omega| = n \geq 7$. If $n = 7$, $|\mathfrak{G}| = 7!/2$ and $\mathfrak{G}$ is $\mathfrak{A}_7$. Suppose henceforth that $n > 7$.

(1) $\mathfrak{G}$ has exactly one class of involutions.

Since $|\mathfrak{G}_{1,2,3,4}|$ is odd, any involution in $\mathfrak{G}$ has at most three fixed points. Since $n \geq 7$, it follows that every involution i of $\mathfrak{G}$ has the form

$$i = (a, b)(c, d) \cdots .$$

If $i' = (a', b')(c', d') \cdots$ is another involution in $\mathfrak{G}$, then since $\mathfrak{G}$ is 4-fold transitive, there exists $g \in \mathfrak{G}$ such that $i^g = (a', b')(c', d') \cdots$. Then a', b', c', d' are all fixed points of $i^g i'$, so the order of $i^g i'$ is odd. Hence $\mathfrak{D} = \langle i^g, i' \rangle$ is a dihedral group of twice odd order, and i^g, i' are conjugate in $\mathfrak{D}$.

(2) If n is even, all the involutions in $\mathfrak{G}$ have just two fixed points; if n is odd, they have just three fixed points.

By (1), all the involutions in $\mathfrak{G}$ have the same number of fixed points, say m. We have $m \leq 3$, and since

$$|\mathfrak{G}_{1,2}| = (n-2)(n-3)|\mathfrak{G}_{1,2,3,4}|$$

is even, $m \geq 2$. Since $m \equiv n$ (2), $m = 3$ if n is odd and $m = 2$ if n is even.

(3) $\mathfrak{G}$ contains an elementary Abelian subgroup $\mathfrak{V} = \langle i, j \rangle$, where, with suitable choice of notation,

$$i = (1)(2)(3,4)(5,6)\cdots, \quad j = (1,2)(3,4)(5)(6)\cdots.$$

Further, if n is even, $\mathfrak{V}$ operates regularly on each of its orbits in $\Omega - \{1, \ldots, 6\}$; if n is odd, $\mathfrak{V}$ has a fixed point, say 7, and operates regularly on each of its orbits in $\Omega - \{1, \ldots, 7\}$.

Since $\mathfrak{G}$ is 4-fold transitive and $|\mathfrak{G}_{1,2,3,4}|$ is odd, $\mathfrak{G}$ contains involutions

$$i = (1)(2)(3,4)\cdots, \quad j' = (1,2)(3,4)\cdots.$$

Thus $ij' = (1,2)(3)(4)\cdots$. Hence $(ij')^2$ is in $\mathfrak{G}_{1,2,3,4}$ and therefore has odd order m. Thus $(ij')^m$ is an involution in the centre of the dihedral group $\langle i, j' \rangle$. Putting $j = i(ij')^m$ and $\mathfrak{V} = \langle i, j \rangle$, we have

$$i = (1)(2)(3,4)\cdots, \quad j = (1,2)(3,4)\cdots,$$

and $\mathfrak{V}$ is elementary Abelian of order 4. Denote the fixed points of j by 5, 6 when n is even and by 5, 6, 7 when n is odd (see (2)). Since $ij = ji$, i permutes the fixed points of j, so we may suppose the notation so chosen that

$$i = (1)(2)(3,4)(5,6)\cdots, \qquad j = (1,2)(3,4)(5)(6)\cdots \qquad (n \text{ even}),$$

$$i = (1)(2)(3,4)(5,6)(7)\cdots, \quad j = (1,2)(3,4)(5)(6)(7)\cdots \quad (n \text{ odd}).$$

By (2), all the fixed points of i, j, ij lie in $\{1, \ldots, 6\}$ or $\{1, \ldots, 7\}$. Hence $\mathfrak{V}$ operates regularly on all the remaining orbits of $\mathfrak{V}$.

Let l be the number of orbits of $\mathfrak{V}$ of length 4. Thus $n = 6 + 4l$ or $n = 7 + 4l$, and $l > 0$.

(4) $l \geq 2$.

Suppose not; then $l = 1$ and $n = 10$ or $n = 11$. If $n = 10$, $|\mathfrak{G}|$ is divisible by $10 \cdot 9 \cdot 8 \cdot 7$, so $\mathfrak{G}$ possesses a cycle of length 7. By 3.7, $\mathfrak{G} \geq \mathfrak{A}_{10}$, and $|\mathfrak{G}_{1,2,3,4}|$ is even, a contradiction. If $n = 11$, the same argument can be applied if $|\mathfrak{G}_{1,2,3,4}|$ is divisible by 5 or 7. Otherwise, $|\mathfrak{G}_{1,2,3,4}|$ is 3 or 3^2. Let $\mathfrak{P}$ be a Sylow 11-subgroup of $\mathfrak{G}$. Thus $\mathfrak{P}$ is regular and $\mathbf{C}_{\mathfrak{G}}(\mathfrak{P}) = \mathfrak{P}$. Let $|\mathbf{N}_{\mathfrak{G}}(\mathfrak{P})| = 11k$. Then k divides 10. Also

$$|\mathfrak{G} : \mathbf{N}_{\mathfrak{G}}(\mathfrak{P})| = \frac{10}{k} \cdot 9 \cdot 8 |\mathfrak{G}_{1,2,3,4}| \equiv 1 \ (11),$$

so $k \equiv 5|\mathfrak{G}_{1,2,3,4}| \ (11)$. It follows that $|\mathfrak{G}_{1,2,3,4}| = 3^2$ and $k = 1$. Thus $\mathbf{N}_{\mathfrak{G}}(\mathfrak{P}) = \mathfrak{P}$. By Burnside's transfer theorem, $\mathfrak{G}$ has a normal subgroup $\mathfrak{N}$ of index 11. Since $\mathfrak{G}$ is 4-fold transitive, $\mathfrak{N}$ is transitive, so 11 divides $|\mathfrak{N}|$. This is a contradiction.

(5) Let Δ be an orbit of $\mathfrak{V}$ of length 4, let $\mathfrak{G}_{(\Delta)}$ be the set of elements g of $\mathfrak{G}$ for which $\Delta g = \Delta$, and let $\mathfrak{G}_\Delta$ be the set of elements g of $\mathfrak{G}$ which leave each element of Δ fixed. Then $\mathfrak{G}_{(\Delta)} = \mathbf{N}_{\mathfrak{G}_{(\Delta)}}(\mathfrak{V})\mathfrak{G}_\Delta$. In particular, $\mathbf{N}_{\mathfrak{G}_{(\Delta)}}(\mathfrak{V})$ induces the full symmetric group $\mathfrak{S}_4$ on Δ.

Since $\mathfrak{G}$ is 4-fold transitive, $\mathfrak{G}_{(\Delta)}/\mathfrak{G}_\Delta \cong \mathfrak{S}_4$. Since $\mathfrak{V}$ operates regularly on Δ, $\mathfrak{V}\mathfrak{G}_\Delta/\mathfrak{G}_\Delta$ corresponds to the regular normal subgroup of $\mathfrak{S}_4$ and is therefore normal in $\mathfrak{G}_{(\Delta)}/\mathfrak{G}_\Delta$. Since $|\mathfrak{G}_\Delta|$ is odd, $\mathfrak{V}$ is a Sylow 2-subgroup of $\mathfrak{V}\mathfrak{G}_\Delta$. By the Frattini argument, $\mathfrak{G}_{(\Delta)} = \mathbf{N}_{\mathfrak{G}_{(\Delta)}}(\mathfrak{V})\mathfrak{G}_\Delta$.

(6) Corresponding to each orbit Δ of $\mathfrak{V}$ of length 4, there is an involution $g \in \mathfrak{G}_{(\Delta)}$ such that $i^g = i$ and $j^g = ij$. Further, g can be so chosen that g has exactly two fixed points on Δ.

The permutation of Δ induced by i lies in the centre of some Sylow 2-subgroup of the group of all permutations of Δ. If t is a transposition in this Sylow 2-subgroup, then by (5), t is induced by some element g of $\mathbf{N}_{\mathfrak{G}_{(\Delta)}}(\mathfrak{V})$, and we may suppose that the order of g is a power of 2. Since g^2 leaves each element of Δ fixed, $g^2 = 1$. Also j^g and ij induce the same permutation on Δ. Since $\mathfrak{V}$ operates regularly on Δ, it follows that $j^g = ij$ and $i^g = i$.

Let $\mathfrak{C} = \mathfrak{G}_{1,2,3,4} \cap \mathbf{C}_{\mathfrak{G}}(\mathfrak{V})$. Thus $\mathfrak{C}$ permutes the orbits of $\mathfrak{V}$ of length 4.

(7) The orbits of $\mathfrak{V}$ of length 4 are permuted transitively by $\mathfrak{C}$.

Let Δ_1, Δ_2 be orbits of $\mathfrak{V}$ of length 4. By (6), there is an involution g_k $(k = 1, 2)$ in $\mathfrak{G}_{(\Delta_k)}$ such that $i^{g_k} = i$, $j^{g_k} = ij$ and g_k has exactly two fixed points on Δ_k. In the case when n is odd, $7g_k = 7$, since $g_k \in \mathbf{N}_{\mathfrak{G}}(\mathfrak{V})$. Hence, since $i^{g_k} = i$, $\{1, 2\}g_k = \{1, 2\}$. Since g_k already has two fixed points in Δ_k, g_k cannot leave 1 and 2 fixed. Thus $g_k = (1, 2) \cdots$. Since $j^{g_k} = ij$, $\{5, 6\}g_k = \{3, 4\}$. Hence

$$g_k = (1, 2)(3, 5)(4, 6) \cdots \quad \text{or} \quad g_k = (1, 2)(4, 5)(3, 6) \cdots.$$

But now, if $g_k = (1, 2)(3, 5)(4, 6) \cdots$, then $ig_k = (1, 2)(4, 5)(3, 6) \cdots$, so there exists $g_1' \in \{g_1, ig_1\}$ such that g_1' and g_2 have the same effect on $1, \ldots, 6$. Thus

$$g_1'g_2 \in \mathfrak{G}_{1,2,3,4} \cap \mathbf{C}_{\mathfrak{G}}(\mathfrak{V}) = \mathfrak{C}.$$

Hence $\langle g_1', g_2 \rangle$ is a dihedral group of twice odd order, so $g_1'^c = g_2$ for some element $c = (g_1'g_2)^s$. Now g_1' has two fixed points in Δ_1, for, in the case when $g_1' = ig_1$, it follows from $ig_1 = g_1 i$ that i, g_1 induce permutations on Δ_1 of the form $(a_1, a_2)(a_3, a_4)$, (a_1, a_2) respectively. Since 7 is fixed by $\mathfrak{V}$ in the case when n is odd, 7 is also fixed by g_1', g_2 and c. Hence it follows from $g_1'^c = g_2$ that c maps the set of fixed points of g_1' in Δ_1 onto the set of fixed points of g_2 in Δ_2. Hence $\Delta_1 c \cap \Delta_2 \neq \varnothing$. Since $c \in \mathbf{N}_{\mathfrak{G}}(\mathfrak{V})$, it follows that $\Delta_1 c = \Delta_2$.

(8) Each orbit of $\mathfrak{C}$ on $\{7, \ldots, n\}$ or $\{8, \ldots, n\}$ respectively is of length l.

Let Γ be the orbit of $\mathfrak{C}$ containing x, where the orbit Δ of $\mathfrak{V}$ containing x is of length 4. By (7), $|\Gamma| \geq l$. On the other hand, suppose that $c \in \mathfrak{C}$ and $xc \in \Delta$. Then $\Delta c = \Delta$. But $|\mathfrak{C}|$ is odd, so the order of c is odd and c has a fixed point in Δ. But $\mathfrak{V}$ is transitive on Δ and $c \in \mathbf{C}_{\mathfrak{G}}(\mathfrak{V})$, so $yc = y$ for all $y \in \Delta$. Hence $xc = x$ and $\Gamma \cap \Delta = \{x\}$. It follows that $|\Gamma \cap \Delta'| \leq 1$ for all orbits Δ' of $\mathfrak{V}$ of length 4 and so $|\Gamma| \leq l$. Thus $|\Gamma| = l$.

(9) $\mathbf{N}_{\mathfrak{G}}(\mathfrak{V})$ contains an element of the form $(1)(2)(3, 5, 4, 6) \cdots$.

Let Δ be an orbit of $\mathfrak{V}$ of length 4 and put $\mathfrak{N}_0 = \mathbf{N}_{\mathfrak{G}_\Delta}(\mathfrak{V})$, $\mathfrak{N} = \mathbf{N}_{\mathfrak{G}_{(\Delta)}}(\mathfrak{V})$. Thus $\mathfrak{N}_0 \trianglelefteq \mathfrak{N}$ and by (5), $\mathfrak{N}/\mathfrak{N}_0 \cong \mathfrak{S}_4$. In this isomorphism, $\mathfrak{V}\mathfrak{N}_0/\mathfrak{N}_0$ corresponds to the Klein subgroup. Let $\mathfrak{M}/\mathfrak{N}_0$ be the subgroup corresponding to $\mathfrak{A}_4$. Since $|\mathfrak{N}_0|$ is odd, $\mathfrak{M}$ is the only subgroup of $\mathfrak{N}$ of index 2. Note that $\mathfrak{N}$ permutes $1, \ldots, 6$ among themselves. Let $\mathfrak{K}$ be the kernel of the mapping which carries each element of $\mathfrak{N}$ into the permutation of $\{1, \ldots, 6\}$ induced by it. If $\mathfrak{K} \not\leq \mathfrak{N}_0$, $\mathfrak{K}\mathfrak{N}_0 \geq \mathfrak{V}$. But then since $|\mathfrak{N}_0|$ is odd, $\mathfrak{K} \geq \mathfrak{V}$, and this is not the case. So $\mathfrak{K} \leq \mathfrak{N}_0$.

Any transposition of Δ is induced by some involution t in $\mathfrak{N}$. Since t has two fixed points in Δ, it can have none in $\{1, \ldots, 6\}$. Thus t induces an odd permutation of $\{1, \ldots, 6\}$. It follows that all elements of $\mathfrak{N} - \mathfrak{M}$ induce odd permutations on $\{1, \ldots, 6\}$.

Let $\mathfrak{N}_0 g_1, \ldots, \mathfrak{N}_0 g_6$ be the six elements of $\mathfrak{N}/\mathfrak{N}_0$ of order 4; we may suppose that each g_i is of order 4. Then $g_i \notin \mathfrak{M}$, $g_i^2 \in \mathfrak{B}\mathfrak{N}_0 = \mathfrak{B} \times \mathfrak{N}_0$ and $g_i^2 \in \mathfrak{B} - \{1\}$. Thus g_i induces an odd permutation π_i of order 4 on $\{1, \ldots, 6\}$. Also $\pi_1, \ldots, \pi_6$ are distinct, since $\mathfrak{K} \leq \mathfrak{N}_0$ and, by (3), π_i^2 is one of the permutations (3, 4)(5, 6), (1, 2)(3, 4) or (1, 2)(5, 6). Hence $\pi_1, \ldots, \pi_6$ are (3, 5, 4, 6), (3, 6, 4, 5), (1, 3, 2, 4), (1, 4, 2, 3), (1, 5, 2, 6) or (1, 6, 2, 5) (in some order). Thus some g_i is of the stated form.

Henceforth, let Σ denote the set $\{1, \ldots, 6\}$ if n is even or the set $\{1, \ldots, 7\}$ if n is odd.

(10) Let p be a prime divisor of l. If $\mathfrak{P} \in S_p(\mathfrak{C})$, Σ is the set of all fixed points of $\mathfrak{P}$.

If $a \notin \Sigma$, then $|\mathfrak{C} : \mathfrak{C}_a| = l \equiv 0 \ (p)$, by (8), so $\mathfrak{P} \nleq \mathfrak{C}_a$. Thus $\mathfrak{P}$ has no fixed points outside Σ. Since $\mathfrak{C} \leq \mathfrak{G}_{1,2,3,4}$, $\mathfrak{C}$ is of odd order and leaves 1, 2, 3, 4 fixed. Since $\mathfrak{C} \leq \mathbf{N}_{\mathfrak{G}}(\mathfrak{B})$, $\mathfrak{C}$ also leaves $\{1, \ldots, 6\}$ fixed, so $\mathfrak{C}$ leaves 5 and 6 fixed. If n is odd, $\mathfrak{C}$ leaves 7 fixed, since 7 is the only fixed point of $\mathfrak{B}$.

(11) To complete the proof, let p be a prime divisor of l (see (4)). Suppose that $\mathfrak{P} \in S_p(\mathfrak{C})$ and $\mathfrak{P} \leq \mathfrak{Q} \in S_p(\mathfrak{G}_{1,2,3,4})$. If $\mathfrak{Q} \leq \mathfrak{G}_\Sigma$, then Σ is the set of all fixed points of $\mathfrak{Q}$, by (10). By II, 1.13, $\mathbf{N}_{\mathfrak{G}}(\mathfrak{Q})$ is 4-fold transitive on Σ. Thus the order of the group $\mathfrak{A}$ of permutations of Σ induced by $\mathfrak{G}_{(\Sigma)}$ is of the form $6 \cdot 5 \cdot 4 \cdot 3k$ or $7 \cdot 6 \cdot 5 \cdot 4k$, where k is an odd divisor of 2 or 6 respectively. By (9), $\mathfrak{A}$ has a subgroup $\mathfrak{A}_0$ of index 2. But this implies that the only possibility is the second case with $k = 1$. By 3.3, however, this is also impossible. Hence $\mathfrak{Q} \nleq \mathfrak{G}_\Sigma$. Thus there exists $y \in \Sigma$ such that the length of the orbit Γ of $\mathfrak{Q}$ containing y is greater than 1. Clearly $y \in \{5, 6\}$ or $\{5, 6, 7\}$. Since $\mathfrak{P} \leq \mathfrak{Q}$, Γ is a union of orbits of $\mathfrak{P}$. By (10), the number k of fixed points of $\mathfrak{P}$ on Γ is at most 2 or 3, so $|\Gamma| = k + pm$ for some integer m. Also $k > 0$ since y is a fixed point of $\mathfrak{P}$. Since $|\Gamma| > 1$ and $|\Gamma|$ is a power of p, $k \equiv 0 \ (p)$. Since p divides $|\mathfrak{C}|$, p is odd. Since also $k \leq 3$, $k = p = 3$; further n is odd and $\Gamma \supseteq \{5, 6, 7\}$. Hence $l = 3^q$ for some $q > 0$. By (8), the Sylow 3-subgroup $\mathfrak{P}$ of $\mathfrak{C}$ is transitive on each orbit of $\mathfrak{C}$ in $\{8, \ldots, n\}$, and the length of each orbit of $\mathfrak{P}$ in $\{8, \ldots, n\}$ is 3^q. If Γ contains k' such orbits of $\mathfrak{P}$,

$$|\Gamma| = 3^q k' + k = 3^q k' + 3.$$

Hence either $k' = 0$ or $q = 1$.

Suppose first that $k' = 0$. Then $\{5, 6, 7\}$ is an orbit of $\mathfrak{Q}$ and $\mathfrak{Q}$ contains an element of the form

$$h = (1)(2)(3)(4)(5, 6, 7) \cdots.$$

By (9), $\mathfrak{G}$ contains an element g of the form

$$g = (1)(2)(3, 5, 4, 6)(7) \cdots.$$

Then

$$hg = (1)(2)(3, 5)(4, 6, 7) \cdots, \quad (hg)^3 = (1)(2)(3, 5)(4)(6)(7) \cdots.$$

This is impossible.

Thus $k' \neq 0$, $l = 3$ and $n = 7 + 4l = 19$. Also $|\Gamma| = 3k' + 3$ is a power of 3, so $|\Gamma| = 9$. Thus with suitable numbering of the symbols permuted, $\Gamma = \{5, 6, \ldots, 13\}$ is an orbit of $\mathfrak{Q}$, and $\{5\}, \{6\}, \{7\}, \{8, 9, 10\}$ and $\{11, 12, 13\}$ are the orbits of $\mathfrak{P}$ in Γ.

Since all maximal subgroups of $\mathfrak{Q}$ are of index 3, $\mathfrak{Q}$ is imprimitive on Γ. Now $\mathfrak{P}$ leaves fixed the domain of imprimitivity of $\mathfrak{Q}$ containing 5, and since $|\mathfrak{P}|$ is odd, $\mathfrak{P}$ cannot permute the other two; hence $\mathfrak{P}$ leaves fixed all the domains of imprimitivity of $\mathfrak{Q}$ in Γ. Thus $\Gamma_1 = \{5, 6, 7\}$, $\Gamma_2 = \{8, 9, 10\}$ and $\Gamma_3 = \{11, 12, 13\}$ are the domains of imprimitivity of $\mathfrak{Q}$ in Γ. Hence there exists $h_1 \in \mathfrak{Q}$ such that $\Gamma_2 h_1 = \Gamma_1$. Now $\mathfrak{P}$ contains an element of the form

$$h_2 = (1)(2) \cdots (7)(8, 9, 10) \cdots,$$

and either $h_2^{h_1}$ or $h_2^{-h_1}$ is of the form

$$h = (1)(2)(3)(4)(5, 6, 7) \cdots.$$

As in the case $k' = 0$, $\mathfrak{G}$ contains an element g for which

$$(hg)^3 = (1)(2)(3, 5)(4)(6)(7) \cdots,$$

which is impossible. **q.e.d.**

3.9 Remark. The following theorem is due to Bender.

Suppose that $\mathfrak{G}$ is a doubly transitive permutation group with no regular normal subgroup. If $|\mathfrak{G}_{1,2}|$ is odd, $\mathfrak{G}$ has a normal subgroup $\mathfrak{N}$, which is similar to one of the following groups in its natural permutation representation.

a) $PSL(2, q)$, where $q \geq 4$, $q = 2^f$ or $q \equiv 3\ (4)$.

b) A Suzuki group $Sz(q)$, where $q = 2^{2n+1}$.

c) A unitary group $PSU(3, q^2)$, where $q = 2^f$.

The proof of this falls into two parts. If $|\mathfrak{G}_1|$ is even, $\mathfrak{G}_1$ is a strongly embedded subgroup, so the theorem stated without proof in X, 4.14, may be applied. The case when $|\mathfrak{G}_1|$ is odd was treated in BENDER [5].

We shall conclude this section by finding the primitive groups of degree 8 and 9. To do this, we use the following theorem, which we shall not prove here.

3.10 Theorem (WIELANDT [1]). *Suppose that $\mathfrak{G}$ is a primitive permutation group of degree $n = 2p$, where p is a prime. If $n - 1$ is not a square, $\mathfrak{G}$ is doubly transitive.*

3.11 Theorem. *Suppose that $\mathfrak{G}$ is a primitive permutation group of degree 8.*

a) *If $\mathfrak{G}$ has a regular normal subgroup $\mathfrak{N}$, $\mathfrak{N}$ is elementary Abelian, $\mathfrak{G} = \mathfrak{N}\mathfrak{G}_1$ and $\mathfrak{G}_1$ is similar to the subgroup of* **Aut** $\mathfrak{N}$ *induced by it. Further, $|\mathfrak{G}_1|$ is 7, 21 or 168.*

b) *If $\mathfrak{G}$ has no regular normal subgroup, $\mathfrak{G}$ is one of the groups $PSL(2, 7)$, $PGL(2, 7)$, $\mathfrak{A}_8$ or $\mathfrak{S}_8$.*

Proof. a) By II, 3.2, $\mathfrak{N}$ is a minimal normal subgroup of $\mathfrak{G}$, $\mathfrak{N}$ is elementary Abelian and $\mathfrak{G} = \mathfrak{N}\mathfrak{G}_1$. By II, 2.2, $\mathfrak{G}_1$ is similar to the subgroup of **Aut** $\mathfrak{N}$ induced by it. Thus $\mathfrak{G}_1$ is isomorphic to a subgroup of $GL(3, 2)$. Since $\mathfrak{N}$ is minimal, $\mathfrak{G}_1$ operates irreducibly on $\mathfrak{N}$. It follows that $\mathfrak{G}_1$ permutes $\mathfrak{N} - \{1\}$ transitively; otherwise $\mathfrak{G}_1$ has an orbit $\{x_1, \ldots, x_k\}$ of length $k \leq 3$ and then either $\langle x_1, \ldots, x_k\rangle$ is a $\mathfrak{G}_1$-invariant subgroup of order at most 4 or $x_1 \cdots x_k$ is a non-identity fixed point of $\mathfrak{G}_1$. Hence $|\mathfrak{G}_1|$ is divisible by 7. Either $\mathfrak{G}_1 = GL(3, 2)$ or $\mathfrak{G}_1$ is soluble. In the latter case, $\mathfrak{G}_1$ cannot contain all 8 Sylow 7-subgroups of $GL(3, 2)$, so the Sylow 7-subgroup of $\mathfrak{G}_1$ is normal and $|\mathfrak{G}_1|$ is either 7 or 21.

b) First suppose that $|\mathfrak{G}|$ is divisible by 5. Then $\mathfrak{G}$ contains a cycle of length 5. By 3.7, $\mathfrak{G}$ is $\mathfrak{A}_8$ or $\mathfrak{S}_8$.

Suppose henceforth that $|\mathfrak{G}|$ is not divisible by 5. By II, 3.2, $\mathfrak{G}$ is not soluble, since $\mathfrak{G}$ has no regular normal subgroup. Then $|\mathfrak{G}|$ is not of the form 2^a3^b. Hence $|\mathfrak{G}|$ is divisible by 7, and indeed, $|\mathfrak{G}_1|$ is divisible by 7. $\mathfrak{G}_1$ is thus a transitive group of degree 7 on $\Omega - \{1\}$. We use the classification of such groups in V, Exercise 38. Note that $\mathfrak{G}$ is doubly transitive.

If $\mathfrak{G}_1$ is $\mathfrak{A}_7$ or $\mathfrak{S}_7$, then $\mathfrak{G}$ is $\mathfrak{A}_8$ or $\mathfrak{S}_8$ respectively. If $\mathfrak{G}_1$ is $GL(3, 2)$, $\mathfrak{G}$ is simple, for a proper normal subgroup $\mathfrak{N}$ would satisfy $\mathfrak{N} \cap \mathfrak{G}_1 = 1$ and hence be regular. Thus $\mathfrak{G}$ is a simple group of index 15 in $\mathfrak{A}_8$. Since $\mathfrak{A}_8$ is simple, $\mathfrak{G}$ is a maximal subgroup of $\mathfrak{A}_8$ and $\mathfrak{G}^x = \mathfrak{G}$ implies $x \in \mathfrak{G}$ $(x \in \mathfrak{A}_8)$. Let ρ be the permutation representation of $\mathfrak{G}$ on the

14 cosets of $\mathfrak{G}$ in $\mathfrak{A}_8$ other than $\mathfrak{G}$ itself. Then ρ has no orbit of length 1, since $\mathfrak{G}x\mathfrak{G} = \mathfrak{G}x$ implies that $\mathfrak{G} = \mathfrak{G}^x$ and hence $\mathfrak{G}x = \mathfrak{G}$. Thus $\mathfrak{G}$ is faithfully represented on each orbit of ρ. But since 2^6 does not divide $7!$, it follows that the degree of any faithful representation of $\mathfrak{G}$ is at least 8. Hence ρ is transitive and indeed primitive. By 3.10, ρ is doubly transitive. But this implies that $|\mathfrak{G}|$ is divisible by 13, a contradiction.

In the remaining case, $\mathfrak{G}_1$ is soluble. By V, 21.1, $\mathfrak{G}_1 = \mathfrak{P}\mathfrak{G}_{1,2}$, where $\mathfrak{P}$ is a normal subgroup of $\mathfrak{G}_1$ of order 7; also $\mathfrak{G}_{1,2}$ is cyclic and is generated by a power of a cycle of length 6. Thus $|\mathfrak{G}_{1,2}| = d$, where d is a divisor of 6. If d is 1 or 2, $|\mathfrak{G}| = 7 \cdot 2^a$, which is impossible since $\mathfrak{G}$ is not soluble. Thus d is 3 or 6. If $d = 3$, $\mathfrak{G}$ is $PSL(2, 7)$; this follows, for instance, from II, 6.15. If $d = 6$, $\mathfrak{G}$ contains an odd permutation, namely, the cycle of length 6. Hence $\mathfrak{G}$ has a normal subgroup $\tilde{\mathfrak{G}}$ of index 2. As a transitive insoluble group of order 168 and degree 8, $\tilde{\mathfrak{G}}$ is $PSL(2, 7)$. With suitable numbering of the symbols, the permutation $x \rightarrow x + 1$ generates a Sylow 7-subgroup $\mathfrak{P}$ of $\tilde{\mathfrak{G}}$. Since the permutation π given by $x\pi = 3x$ generates the normaliser of $\mathfrak{P}$ in $\mathfrak{S}_8$, π lies in $\mathfrak{G} - \tilde{\mathfrak{G}}$. Also $\pi \in PGL(2, 7)$, so $\mathfrak{G} = PGL(2, 7)$. These last cases can, of course, also be settled by observing that $\mathfrak{G}$ is a Zassenhaus group.

q.e.d.

3.12 Theorem. *Suppose that $\mathfrak{G}$ is a primitive permutation group of degree 9.*

a) *$\mathfrak{G}$ is soluble if and only if $\mathfrak{G}$ has a regular normal subgroup.*

b) *If $\mathfrak{G}$ is insoluble, $\mathfrak{G}$ is one of the groups $PGL(2, 8)$, $P\Gamma L(2, 8)$, $\mathfrak{A}_9$ or $\mathfrak{S}_9$.*

Proof. a) If $\mathfrak{G}$ is soluble, $\mathfrak{G}$ has a regular normal subgroup, by II, 3.2. Conversely, suppose that $\mathfrak{N}$ is a regular normal subgroup of $\mathfrak{G}$. By II, 3.2, $\mathfrak{N}$ is a minimal normal subgroup and $\mathbf{C}_{\mathfrak{G}}(\mathfrak{N}) = \mathfrak{N}$, so $\mathfrak{G}/\mathfrak{N}$ is isomorphic to a subgroup of $\mathbf{Aut}\,\mathfrak{N} = GL(2, 3)$. Since $GL(2, 3)$ is soluble, $\mathfrak{G}$ is soluble.

b) First, suppose that $|\mathfrak{G}|$ is divisible by 5. Then $\mathfrak{G}$ contains a cycle of length 5. By 3.7, $\mathfrak{G}$ is $\mathfrak{A}_9$ or $\mathfrak{S}_9$.

Suppose henceforth that $|\mathfrak{G}|$ is not divisible by 5. Since $\mathfrak{G}$ is insoluble and $9! = 2^7 \cdot 3^4 \cdot 5 \cdot 7$, $|\mathfrak{G}|$ is divisible by 7. Indeed, $|\mathfrak{G}_1|$ is divisible by 7. Thus if $\mathfrak{G}_1$ is intransitive on $\Omega - \{1\}$, the orbits of $\mathfrak{G}_1$ on Ω are of length 7, 1, 1. But if the orbits of length 1 are $\{1\}$ and $\{2\}$, $\mathfrak{G}_1 = \mathfrak{G}_{1,2} = \mathfrak{G}_2$. Now if $g \in \mathfrak{G}$ and $1g = 2$, $\mathfrak{G}_1^g = \mathfrak{G}_2 = \mathfrak{G}_1$, so $g \in \mathbf{N}_{\mathfrak{G}}(\mathfrak{G}_1)$. Hence $\mathbf{N}_{\mathfrak{G}}(\mathfrak{G}_1) > \mathfrak{G}_1$. By II, 1.4, $\mathfrak{G}_1$ is a maximal subgroup of $\mathfrak{G}$, so $\mathbf{N}_{\mathfrak{G}}(\mathfrak{G}_1) = \mathfrak{G}$, $\mathfrak{G}_1 \trianglelefteq \mathfrak{G}$ and $\mathfrak{G}_1 = 1$, a contradiction. Hence $\mathfrak{G}_1$ is transitive on $\Omega - \{1\}$. If $\mathfrak{G}_1$ is imprimitive, $|\mathfrak{G}_1|$ is of the form $2^a 3^b$ and hence so is

$|\mathfrak{G}|$. Since $\mathfrak{G}$ is insoluble, this is not the case, so $\mathfrak{G}_1$ is primitive of degree 8.

If $\mathfrak{G}_1$ has no regular normal subgroup, then by 3.11b), $\mathfrak{G}_1$ is one of the groups $PSL(2, 7)$, $PGL(2, 7)$, $\mathfrak{A}_8$ or $\mathfrak{S}_8$. The first two cases are impossible, by 2.7a). Hence $\mathfrak{G}$ is $\mathfrak{A}_9$ or $\mathfrak{S}_9$.

If $\mathfrak{G}_1$ has a regular normal subgroup $\mathfrak{N}$, then by 3.11a), $\mathfrak{G}_1 = \mathfrak{N}\mathfrak{G}_{1,2}$, $\mathfrak{G}_{1,2}$ is similar to the subgroup of **Aut** $\mathfrak{N}$ induced by it and $|\mathfrak{G}_{1,2}|$ is 7, 21 or 168. Thus $\mathfrak{G}_{1,2}$ is transitive on $\Omega - \{1, 2\}$ and $\mathfrak{G}$ is triply transitive.

If $|\mathfrak{G}_{1,2}| = 7$, $\mathfrak{G}$ is sharply triply transitive. By XI, 2.1, $\mathfrak{G}$ is $PGL(2, 8)$.

If $|\mathfrak{G}_{1,2}| = 21$, $\mathfrak{G}_{1,2}$ is a Frobenius group on $\Omega - \{1, 2\}$. Thus no non-identity element of $\mathfrak{G}$ leaves 4 elements of Ω fixed. By 3.6, $\mathfrak{G}$ is $P\Gamma L(2, 8)$.

If $|\mathfrak{G}_{1,2}| = 168$, $\mathfrak{G}_{1,2}$ is doubly transitive on $\Omega - \{1, 2\}$ and $\mathfrak{G}$ is 4-fold transitive. Also the stabiliser of two non-identity elements x, y of $\mathfrak{N}$ in **Aut** $\mathfrak{N}$ has precisely three fixed points, namely, x, y and xy. Since $|\mathfrak{G}_{1,2}| = |\mathbf{Aut}\,\mathfrak{N}|$, it follows that $\mathfrak{G}_{1,2,3,4}$ has five fixed points in Ω. By II, 1.13, $\mathbf{N}_{\mathfrak{G}}(\mathfrak{G}_{1,2,3,4})$ is 4-fold transitive on these five fixed points. Thus $\mathbf{N}_{\mathfrak{G}}(\mathfrak{G}_{1,2,3,4})$ induces the full symmetric group $\mathfrak{S}_5$ on them and 5 divides $|\mathfrak{G}|$. This is a contradiction. **q.e.d.**

§ 4. On the Existence of 6- and 7-Fold Transitive Permutation Groups

The only known 4-fold transitive groups are $\mathfrak{S}_n$ $(n \geq 4)$, $\mathfrak{A}_n$ $(n \geq 6)$, $\mathfrak{M}_{11}$, $\mathfrak{M}_{12}$, $\mathfrak{M}_{23}$ and $\mathfrak{M}_{24}$. It has not yet been proved that no further 4-fold transitive groups exist. However, Wielandt showed in 1960 that if the Schreier hypothesis is valid, the only 8-fold transitive permutation groups are symmetric and alternating groups (WIELANDT [3]). Later, Suzuki and Nagao improved this to 7-fold transitivity. Here we present Suzuki's proof [9]. To do this, we prove the following generalization of Theorem 3.3.

4.1 Theorem. *Suppose that $\mathfrak{G}$ is a 4-fold transitive permutation group on the set $\Omega = \{1, \ldots, n\}$. If $\mathfrak{G}_{1,2}$ has a regular normal subgroup on $\Omega - \{1, 2\}$, then $\mathfrak{G}$ is one of the groups $\mathfrak{S}_4$, $\mathfrak{S}_5$, $\mathfrak{S}_6$, $\mathfrak{A}_6$ or $\mathfrak{M}_{11}$.*

Proof. $\mathfrak{G}_{1,2}$ is doubly transitive on $\Omega - \{1, 2\}$ and has a regular normal subgroup $\mathfrak{F}$. By II, 2.3a), $\mathfrak{F}$ is elementary Abelian of order p^f for some prime p. By II, 3.2, $\mathbf{C}_{\mathfrak{G}_{1,2}}(\mathfrak{F}) = \mathfrak{F}$ and $\mathfrak{F}$ is the only minimal normal subgroup of $\mathfrak{G}_{1,2}$. Hence $\mathfrak{F}$ is a characteristic subgroup of $\mathfrak{G}_{1,2}$. We distinguish two cases.

Case 1. $\mathfrak{G}_1$ has a regular normal subgroup $\mathfrak{N}$ on $\Omega - \{1\}$.

$\mathfrak{G}_1$ is triply transitive, so by II, 2.3, $|\mathfrak{N}|$ is 2^m or 3. If $|\mathfrak{N}| = 3$, the degree of $\mathfrak{G}$ is 4 and $\mathfrak{G} = \mathfrak{S}_4$. Otherwise, $|\mathfrak{N}| = 2^m = p^f + 1$. By IX, 2.7, $f = 1$. Thus $\mathfrak{G}_{1,2}$ is a doubly transitive permutation group of prime degree p with normal Sylow p-subgroup $\mathfrak{F}$. By a theorem of Galois (II, 3.6), $\mathfrak{G}_{1,2}$ is sharply doubly transitive. Thus $\mathfrak{G}$ is sharply 4-fold transitive. By 3.3, $\mathfrak{G} = \mathfrak{S}_5$, since the degree is $2^m + 1$.

Case 2. $\mathfrak{G}_1$ has no regular normal subgroup on $\Omega - \{1\}$.

In this case, $\mathfrak{G}_1$ is a group of Suzuki type in the sense of 2.1. Since $\mathfrak{G}$ is a transitive extension of $\mathfrak{G}_1$, it follows from 2.6 that either $p^f = 4$ or $p^f = 9$ and $\mathfrak{G}_{1,2,3}$ has just one involution. If $p^f = 4$, the degree is 6 and, since $\mathfrak{G}$ is 4-fold transitive, $\mathfrak{G}$ is $\mathfrak{S}_6$ or $\mathfrak{A}_6$.

Suppose henceforth that $p^f = 9$, so the degree of $\mathfrak{G}$ is 11. Since $\mathbf{C}_{\mathfrak{G}_{1,2}}(\mathfrak{F}) = \mathfrak{F}$, $\mathfrak{G}_{1,2,3}$ is isomorphic to a subgroup of $GL(2, 3)$. Thus $|\mathfrak{G}_{1,2,3}|$ divides 48. In fact, since the Sylow 2-subgroups of $GL(2, 3)$ have more than one involution, $|\mathfrak{G}_{1,2,3}|$ divides 24. Since $\mathfrak{G}_{1,2,3}$ is transitive on $\Omega - \{1, 2, 3\}$, it follows that $|\mathfrak{G}_{1,2,3}|$ is either 8 or 24. If $|\mathfrak{G}_{1,2,3}| = 8$, $\mathfrak{G}$ is sharply 4-fold transitive and by 3.3, $\mathfrak{G} = \mathfrak{M}_{11}$.

If $|\mathfrak{G}_{1,2,3}| = 24$, $|\mathfrak{G}| = 11 \cdot 10 \cdot 9 \cdot 8 \cdot 3$. Let $\mathfrak{Q}$ be a Sylow 11-subgroup of $\mathfrak{G}$. Then $|\mathbf{N}_{\mathfrak{G}}(\mathfrak{Q})| = 11k$, where k is a divisor of 10. By Sylow's theorem,

$$|\mathfrak{G} : \mathbf{N}_{\mathfrak{G}}(\mathfrak{Q})| = \frac{10}{k} \cdot 9 \cdot 8 \cdot 3 \equiv 1\ (11),$$

so $k \equiv 4\ (11)$. Since k divides 10, this is impossible, so this case does not arise. **q.e.d.**

4.2 Theorem. *Suppose that $\mathfrak{G}$ is a 5-fold transitive permutation group on the set $\Omega = \{1, \dots, n\}$. If $\mathfrak{G}_{1,2,3}$ has a regular normal subgroup on $\Omega - \{1, 2, 3\}$, then $\mathfrak{G}$ is one of the groups $\mathfrak{S}_5$, $\mathfrak{S}_6$, $\mathfrak{S}_7$, $\mathfrak{A}_7$ or $\mathfrak{M}_{12}$.*

Proof. $\mathfrak{G}_1$ satisfies the conditions of 4.1, so $\mathfrak{G}_1$ is one of the groups $\mathfrak{S}_4$, $\mathfrak{S}_5$, $\mathfrak{S}_6$, $\mathfrak{A}_6$ or $\mathfrak{M}_{11}$. By 3.4, $\mathfrak{M}_{12}$ is the only transitive extension of $\mathfrak{M}_{11}$, so the assertion follows. **q.e.d.**

4.3 Theorem (WIELANDT [3], SUZUKI [9]). a) *Suppose that $\mathfrak{G}$ is a 7-fold transitive permutation group. If, for any simple non-Abelian subgroup $\mathfrak{X}$ of $\mathfrak{G}$,* **Aut** $\mathfrak{X}/$**Inn** $\mathfrak{X}$ *is soluble, $\mathfrak{G} \geq \mathfrak{A}_n$.*

b) *Suppose that* $\mathfrak{G}$ *is a 6-fold transitive permutation group. If, for any simple non-Abelian subgroup* $\mathfrak{X}$ *of* $\mathfrak{G}$ *there is no epimorphism of any subgroup of* **Aut** $\mathfrak{X}/$**Inn** $\mathfrak{X}$ *onto* $\mathfrak{S}_4$, $\mathfrak{G} \geq \mathfrak{A}_n$.

Proof. Suppose that $\mathfrak{G}$ operates on the set $\Omega = \{1, \ldots, n\}$. Put $k = 5$ in a) and $k = 4$ in b). Thus $\mathfrak{G}$ is $(k + 2)$-fold transitive. Let $\Delta = \{1, \ldots, k\}$, $\Gamma = \Omega - \Delta$ and $\mathfrak{H} = \mathfrak{G}_{1,\ldots,k}$. Also let $\mathfrak{L}$ be the set of all $g \in \mathfrak{G}$ for which $\Delta g = \Delta$. Thus $\mathfrak{H}$ is the kernel of the restriction of $\mathfrak{L}$ to Δ and $\mathfrak{L}/\mathfrak{H} \cong \mathfrak{S}_k$. Suppose that $\mathfrak{G} \ngeq \mathfrak{A}_n$.

By 1.5, $\mathfrak{M}_{12}$ has no transitive extension, so $\mathfrak{G}_{1,\ldots,k-3}$ is not $\mathfrak{M}_{12}$. Neither is it $\mathfrak{S}_5$, $\mathfrak{S}_6$, $\mathfrak{S}_7$ or $\mathfrak{A}_7$. But $\mathfrak{G}_{1,\ldots,k-3}$ is 5-fold transitive on $\{k - 2, \ldots, n\}$, so by 4.2, $\mathfrak{H}^\Gamma$ has no regular normal subgroup on Γ. Of course, $\mathfrak{H} \neq 1$. Let $\mathfrak{N}$ be a minimal normal subgroup of $\mathfrak{L}$ contained in $\mathfrak{H}$. Then $\mathfrak{N}^\Gamma$ is not regular on Γ. But $\mathfrak{L}^\Gamma$ is doubly transitive and $\mathfrak{N}^\Gamma$ is a minimal normal subgroup of $\mathfrak{L}^\Gamma$. Therefore, by XI, 7.12, $\mathfrak{N}^\Gamma$ is simple and $\mathbf{C}_{\mathfrak{L}^\Gamma}(\mathfrak{N}^\Gamma) = 1$. Thus $\mathfrak{N}$ is simple and non-Abelian. By hypothesis, there is no epimorphism of any subgroup of **Aut** $\mathfrak{N}/$**Inn** $\mathfrak{N}$ onto $\mathfrak{S}_k$. In particular, there is no epimorphism of $\mathfrak{L}/\mathfrak{N}\mathbf{C}_{\mathfrak{L}}(\mathfrak{N})$ onto $\mathfrak{L}/\mathfrak{H}$. Hence $\mathbf{C}_{\mathfrak{L}}(\mathfrak{N}) \nleq \mathfrak{H}$. Suppose that $x \in \mathbf{C}_{\mathfrak{L}}(\mathfrak{N})$ and $x \notin \mathfrak{H}$. The permutation induced on Γ by x lies in $\mathbf{C}_{\mathfrak{L}^\Gamma}(\mathfrak{N}^\Gamma)$ and is therefore 1. Thus x permutes $1, \ldots, k$ non-trivially among themselves and leaves $k + 1, \ldots, n$ fixed. If $f \in \mathfrak{S}_n$, there exists $g \in \mathfrak{G}$ such that $if = ig$ for $i = 1, \ldots, k$; thus $x^{gf^{-1}} = x$ and $x^f = x^g \in \mathfrak{G}$. Hence $\mathfrak{G}$ contains every conjugate of x in $\mathfrak{S}_n$ and $\mathfrak{G} \geq \mathfrak{A}_n$. **q.e.d.**

The Schreier hypothesis is that **Aut** $\mathfrak{X}/$**Inn** $\mathfrak{X}$ is soluble for every simple non-Abelian group $\mathfrak{X}$. If this holds, then by 4.3a), the only 7-fold transitive permutation groups are the alternating and symmetric groups.

§ 5. A Characterization of $\mathfrak{M}_{11}$ and $PSL(3, 3)$

Since BRAUER and FOWLER [1] showed that for any given finite group $\mathfrak{H}$, there are only finitely many simple groups $\mathfrak{G}$ having an involution j for which $\mathbf{C}_{\mathfrak{G}}(j) \cong \mathfrak{H}$, numerous simple groups have been characterized by the centralizer of an involution. In this section we give such a characterization of $\mathfrak{M}_{11}$ and $PSL(3, 3)$.

5.1 Lemma. *Let* j *be an involution in the centre of a Sylow 2-subgroup of* $\mathfrak{G}$, *where* $\mathfrak{G}$ *is* $\mathfrak{M}_{11}$ *or* $PSL(3, 3)$. *Then* $\mathbf{C}_{\mathfrak{G}}(j) \cong GL(2, 3)$.

Proof. Any involution in $SL(3, 3)$ is conjugate in $GL(3, 3)$ to

$$t = \begin{pmatrix} -1 & 0 & 0 \\ 0 & -1 & 0 \\ 0 & 0 & 1 \end{pmatrix},$$

and any matrix in $SL(3, 3)$ which commutes with t is of the form

$$\begin{pmatrix} c_{11} & c_{12} & 0 \\ c_{21} & c_{22} & 0 \\ 0 & 0 & c^{-1} \end{pmatrix},$$

where $c = c_{11}c_{22} - c_{12}c_{21}$. Since $PSL(3, 3) \cong SL(3, 3)$, it follows that $\mathbf{C}_{\mathfrak{G}}(j) = GL(2, 3)$ for any involution j in $\mathfrak{G} = PSL(3, 3)$.

To deal with $\mathfrak{M}_{11}$, we first find a Sylow 2-subgroup of the sharply triply transitive group $M(3^2)$. Let $GF(3^2)^\times = \langle \omega \rangle$, and let s, t be the elements of $P\Gamma L(2, 9)$ given by

$$xs = \omega(x\alpha)^{-1}, \quad xt = x^{-1},$$

where α is the unique non-identity automorphism $x \to x^3$ of $GF(3^2)$. Since $-1 \in GF(3^2)^{\times 2}$ and $-\omega \notin GF(3^2)^{\times 2}$, s and t are in $M(3^2)$. Then

$$xs^2 = \omega^{-2}x, \quad xs^3 = (\omega(x\alpha))^{-1},$$

so the order of s is 8, $t^2 = 1$ and $s^t = s^3$. Hence $\langle s, t \rangle$ is a semidihedral group of order 16 and is thus a Sylow 2-subgroup of $M(3^2)$ and of $\mathfrak{G} = \mathfrak{M}_{11}$. As in § 1, let $\mathfrak{M}_{11}$ operate on $\Omega = GF(3^2) \cup \{v, \infty\}$. Now $j = s^4$ is an involution in the centre of $\langle s, t \rangle$, and $xj = -x$. The set of fixed points of j is $\{v, 0, \infty\}$. Thus if $\mathfrak{H} = \mathbf{C}_{\mathfrak{G}}(j)$ and $\mathfrak{Q} = \mathfrak{G}_{v,0,\infty}$, then $\mathfrak{H} \leq \mathbf{N}_{\mathfrak{G}}(\mathfrak{Q})$. Now $|\mathfrak{Q}| = 8$ and $\mathfrak{Q}$ is the quaternion group $\langle s^2, st \rangle$. Also $\mathbf{Z}(\mathfrak{Q}) = \langle j \rangle$, so $\mathfrak{H} = \mathbf{N}_{\mathfrak{G}}(\mathfrak{Q})$. But by II, 1.13, $\mathbf{N}_{\mathfrak{G}}(\mathfrak{Q})$ is triply transitive on $\{v, 0, \infty\}$. Thus $\mathfrak{H}$ induces the full symmetric group on $\{v, 0, \infty\}$ and

$$|\mathfrak{H}/\mathfrak{H}_{v,0,\infty}| = 6.$$

But $\mathfrak{H}_{v,0,\infty} = \mathfrak{Q}$, so $|\mathfrak{H}| = 48$.

We show that $\mathbf{C}_{\mathfrak{G}}(\mathfrak{Q}) = \langle j \rangle$. If not, there exists $a \in \mathbf{C}_{\mathfrak{G}}(\mathfrak{Q})$ such that $a \notin \mathfrak{Q}$. The permutation of $\{v, 0, \infty\}$ induced by a is a cycle of length 2 or 3. If 2, there is an odd integer l such that $a^l \notin \mathfrak{Q}$ and the order of

a^l is a power of 2. Thus $\mathfrak{R} = \langle a^l, \mathfrak{Q} \rangle$ is a Sylow 2-subgroup of $\mathfrak{M}_{11}$, and $\mathfrak{R}$ is semidihedral. This is impossible, since a^l and j both lie in the centre of $\mathfrak{R}$. Hence the permutation of $\{v, 0, \infty\}$ induced by a is a cycle of length 3 and, for some k, a^k is a non-identity element of order a power of 3. But $\mathfrak{Q}$ is regular on $GF(3^2)^\times$, so the group of permutations induced on $GF(3^2)^\times$ by $\mathbf{C}_{\mathfrak{G}}(\mathfrak{Q})$ is isomorphic to a subgroup of $\mathfrak{Q}$. Hence a^k induces the identity permutation on $GF(3^2)^\times$ and is simply a cycle of length 3 on $\{v, 0, \infty\}$. But $\mathfrak{M}_{11}$ obviously contains no cycle of length 3, so $\mathbf{C}_{\mathfrak{G}}(\mathfrak{Q}) = \langle j \rangle$.

Now $|\mathfrak{H}/\mathfrak{Q}| = 6$, so if $\mathfrak{P} \in S_3(\mathfrak{H})$, $\mathfrak{P}\mathfrak{Q} \trianglelefteq \mathfrak{H}$. By the Frattini argument,

$$\mathfrak{H} = \mathbf{N}_{\mathfrak{H}}(\mathfrak{P})\mathfrak{Q}.$$

Since $\mathfrak{P} \not\leq \mathbf{C}_{\mathfrak{G}}(\mathfrak{Q})$, $\mathfrak{P}$ induces an automorphism of order 3 on $\mathfrak{Q}$ and $\mathbf{N}_{\mathfrak{H}}(\mathfrak{P}) \cap \mathfrak{Q} = \langle j \rangle$. Thus $|\mathbf{N}_{\mathfrak{H}}(\mathfrak{P})| = 12$. Now the elements of $\langle s, t \rangle$ of order 4 are s^2, s^6, st, s^3t, s^5t, s^7t, and $\mathfrak{Q}$ is the group generated by these elements. Hence $\mathbf{N}_{\mathfrak{H}}(\mathfrak{P})$ contains no element of order 4 and

$$\mathbf{N}_{\mathfrak{H}}(\mathfrak{P}) = \mathfrak{S} \times \langle j \rangle$$

for some $\mathfrak{S}$. Thus $\mathfrak{H} = \mathfrak{S}\mathfrak{Q}$ and $\mathfrak{S} \cap \mathfrak{Q} = 1$. Since $\mathfrak{S}$ operates faithfully on $\mathfrak{Q}$, $\mathfrak{S} \cong \mathfrak{S}_3$. The isomorphism type of $\mathfrak{H}$ is thus uniquely determined and $\mathfrak{H} \cong GL(2, 3)$. **q.e.d.**

Our aim in this section is the proof of the following theorem.

5.2 Theorem (BRAUER [5], W. WONG [1]). *Suppose that $\mathfrak{G}$ is a simple group and j is an involution in the centre of a Sylow 2-subgroup of $\mathfrak{G}$. If $\mathbf{C}_{\mathfrak{G}}(j) \cong GL(2, 3)$, then either $\mathfrak{G} \cong \mathfrak{M}_{11}$ or $\mathfrak{G} \cong PSL(3, 3)$.*

First we derive some properties of $GL(2, 3)$.

5.3 Lemma. *The elements*

$$a = \begin{pmatrix} 0 & 1 \\ 1 & 1 \end{pmatrix}, \quad b = \begin{pmatrix} -1 & 0 \\ -1 & 1 \end{pmatrix}, \quad c = \begin{pmatrix} 1 & 0 \\ 1 & 1 \end{pmatrix}$$

generate $GL(2, 3)$, and the following relations hold.

$$a^8 = 1, \quad a^b = a^3, \quad b^2 = 1, \quad c^3 = 1,$$

$$(a^2)^c = ab, \quad (ab)^c = aba^2, \quad c^b = c^{-1}.$$

Thus $\langle a, b\rangle$ is a Sylow 2-subgroup of $GL(2, 3)$ and is a semidihedral group of order 16. Also $\mathfrak{Q} = \langle a^2, ab\rangle$ is a quaternion group of order 8 and a normal subgroup of $GL(2, 3)$. Further, $\mathfrak{Q}\langle c\rangle \lhd GL(2, 3)$. Finally, $a^4 = -1$ is the unique central involution in $GL(2, 3)$.

Proof. The relations are easily verified and all the assertions follow at once. **q.e.d.**

5.4 Lemma. *The following is the character table of $GL(2, 3)$.*

Order of element	1	2	4	8	8	2	3	6
Length of class	1	1	6	6	6	12	8	8
Representative of class	1	a^4	a^2	a	a^{-1}	b	c	a^4c
ϕ_0	1	1	1	1	1	1	1	1
ϕ_1	1	1	1	-1	-1	-1	1	1
ϕ_2	2	2	2	0	0	0	-1	-1
ϕ_3	3	3	-1	1	1	-1	0	0
ϕ_4	3	3	-1	-1	-1	1	0	0
ϕ_5	4	-4	0	0	0	0	1	-1
ϕ_6	2	-2	0	$\sqrt{-2}$	$-\sqrt{-2}$	0	-1	1
ϕ_7	2	-2	0	$-\sqrt{-2}$	$\sqrt{-2}$	0	-1	1

Proof. The lengths of the classes containing the given representatives are easily checked by calculating centralizers. Since a, a^{-1} have different traces, they are not conjugate; hence no two of the representatives are conjugate. Thus the classes are as stated.

Since the natural representation of $PGL(2, 3)$ is of degree 4, there is an epimorphism of $GL(2, 3)$ onto $\mathfrak{S}_4$ with kernel $\langle -1\rangle$, given by

$$a \to (0, 1, -1, \infty), \quad b \to (1, \infty), \quad c \to (1, -1, \infty).$$

Hence every character of $\mathfrak{S}_4$ (see V, 20.8) gives rise to a character of $GL(2, 3)$ and ϕ_0 to ϕ_4 are the characters which arise in this way. ϕ_1 is the sign character, ϕ_4 is $\pi - 1$, where π is the character of the natural permutation representation of $\mathfrak{S}_4$, and $\phi_3 = \phi_1\phi_4$. ϕ_2 corresponds to the representation with kernel the Klein subgroup, which is essentially a representation of $\mathfrak{S}_3$.

The normal subgroup $\mathfrak{Q} = \langle a^2, ab\rangle$ has an irreducible representation ρ of degree 2 given by

$$\rho(a^2) = \begin{pmatrix} i & 0 \\ 0 & -i \end{pmatrix}, \quad \rho(ab) = \begin{pmatrix} 0 & 1 \\ -1 & 0 \end{pmatrix}.$$

This admits three extensions to $\langle c \rangle \mathfrak{Q}$, namely, one in which

$$\rho_1(c) = \frac{i-1}{2}\begin{pmatrix} i & 1 \\ -i & 1 \end{pmatrix},$$

and two more given by $\rho_2(c) = \omega\rho_1(c)$, $\rho_3(c) = \omega^2\rho_1(c)$, where ω is a primitive cube root of unity (see V, 17.12). Let $\chi_i = \operatorname{tr} \rho_i$ $(i = 1, 2, 3)$. Then we have the following table.

	1	a^4	a^2	c	ca^2	c^2	c^2a^4
χ_1	2	-2	0	-1	1	-1	1
χ_2	2	-2	0	$-\omega$	ω	$-\omega^2$	ω^2
χ_3	2	-2	0	$-\omega^2$	ω^2	$-\omega$	ω

Thus χ_1 is invariant under $GL(2, 3)$, whereas $\chi_2(x^b) = \chi_3(x)$ for all $x \in \langle c \rangle \mathfrak{Q}$. As $GL(2, 3)/\langle c \rangle \mathfrak{Q}$ is cyclic of order 2, it follows from V, 17.12 that χ_1 has two continuations to $\mathfrak{H} = GL(2, 3)$ which are of the form ϕ_6 and $\phi_7 = \phi_6\phi_1$. And by V, 17.11, $\chi_2^{\mathfrak{H}} = \chi_3^{\mathfrak{H}}$ is an irreducible character ϕ_5 of degree 4. Since the restrictions to $\langle c \rangle \mathfrak{Q}$ of ϕ_5 and ϕ_6 are respectively $\chi_2 + \chi_3$ and χ_1, we obtain the following further part of the character table of $GL(2, 3)$.

	1	a^4	a^2	a	a^{-1}	b	c	a^4c
ϕ_5	4	-4	0	v	$\bar{v}$	y	1	-1
ϕ_6	2	-2	0	u	$\bar{u}$	z	-1	1
ϕ_7	2	-2	0	$-u$	$-\bar{u}$	$-z$	-1	1

It remains to determine the complex numbers u, v, y, z. We have

$$0 = |\mathfrak{H}|(\phi_0, \phi_6)_{\mathfrak{H}} = 6u + 6\bar{u} + 12\bar{z},$$

and

$$0 = |\mathfrak{H}|(\phi_3, \phi_6)_{\mathfrak{H}} = 6u + 6\bar{u} - 12\bar{z},$$

so $u + \bar{u} = z = 0$. But from $(\phi_6, \phi_6)_{\mathfrak{H}} = 1$, it follows that $u\bar{u} = 2$. Hence $u = \sqrt{-2}$. And since $(\phi_5, \phi_5)_{\mathfrak{H}} = 1$,

$$48 = 4^2 + 4^2 + 12v\bar{v} + 12y\bar{y} + 8 + 8,$$

so $v = y = 0$. The character table of $GL(2, 3)$ is thus completely determined. **q.e.d.**

For the rest of this section we suppose that $\mathfrak{G}$ is a simple group, j is an involution in the centre of a Sylow-2-subgroup of $\mathfrak{G}$ and $\mathbf{C}_{\mathfrak{G}}(j) = GL(2, 3)$.

5.5 Lemma. *All involutions of $\mathfrak{G}$ are conjugate and all elements of $\mathfrak{G}$ of order 4 are conjugate.*

Proof. We use the notation of 5.3 for the elements of $\mathbf{C}_{\mathfrak{G}}(j)$. Since $\mathfrak{H} = \mathbf{C}_{\mathfrak{G}}(j)$ contains a Sylow 2-subgroup of $\mathfrak{G}$, $\mathfrak{P} = \langle a, b \rangle$ is a Sylow 2-subgroup of $\mathfrak{G}$. By the focal subgroup theorem (X, 6.2),
$\mathfrak{P} \cap \mathfrak{G}' = \langle xy^{-1} | x \in \mathfrak{P}, y \in \mathfrak{P}, x \text{ and } y \text{ conjugate in } \mathfrak{G} \rangle$.

We calculate the right-hand side under the assumption that the elements of $\mathfrak{G}$ of order 4 are not all conjugate. The set of elements of $\mathfrak{P}$ of order 8 is $\{a, a^3, a^5, a^7\}$. Thus if x, y are of order 8, $xy^{-1} \in \langle a^2 \rangle$. $\mathfrak{P}$ has two conjugacy classes of elements of order 4, namely, $\{a^2, a^6\}$ and $\{ab, a^3b, a^5b, a^7b\}$. Since these classes are not fused in $\mathfrak{G}$, all xy^{-1} with conjugate elements x, y of $\mathfrak{P}$ of order 4 lie in $\langle a^2 \rangle$. Finally, the set of involutions in $\mathfrak{P}$ is $\{a^4, b, a^2b, a^4b, a^6b\}$, so all xy^{-1} with conjugate involutions x, y of $\mathfrak{P}$ lie in $\langle a^2, b \rangle$. Hence

$$\mathfrak{P} \cap \mathfrak{G}' \leq \langle a^2, b \rangle.$$

But since $\mathfrak{G}$ is simple, $\mathfrak{P} \cap \mathfrak{G}' = \mathfrak{P}$, so we have a contradiction. Thus all elements of $\mathfrak{G}$ of order 4 are conjugate in $\mathfrak{G}$.

Again, all xy^{-1} with conjugate elements x, y of $\mathfrak{P}$ of order 4 or 8 lie in $\langle a^2, ab \rangle$. Since $\mathfrak{P} \cap \mathfrak{G}' = \mathfrak{P}$, there exist conjugate involutions x, y in $\mathfrak{P}$ such that $xy^{-1} \notin \langle a^2, ab \rangle$. Hence a^4 is conjugate to one of b, a^2b, a^4b, a^6b. Since b, a^2b, a^4b, a^6b are already conjugate in $\mathfrak{P}$, all involutions in $\mathfrak{G}$ are conjugate. (The conjugacy of the involutions also follows from Thompson's transfer lemma (8.2).) **q.e.d.**

We make way for the application of character theory by finding a TI-set.

5.6 Lemma. *Let $j = a^4$ and*

$$\mathsf{D} = \{g | g \in \mathfrak{G}, g^m = j \quad \textit{for some } m\}.$$

a) D *consists of the conjugates in* $\mathfrak{H} = \mathbf{C}_{\mathfrak{G}}(j)$ *of* j, a^2, a, a^{-1}, a^4c.

b) $\mathsf{D}^h = \mathsf{D}$ *for all* $h \in \mathfrak{H}$ *and* $\mathsf{D}^g \cap \mathsf{D} = \varnothing$ *for all* $g \notin \mathfrak{H}$.

c) *The following is a basis over* $\mathbb{C}$ *of the vector space* V *of all complex-valued class-functions on* $\mathfrak{H}$ *which vanish on* $\mathfrak{H} - \mathsf{D}$.

$$\Phi_1 = \phi_0 + \phi_2 - \phi_4,$$

$$\Phi_2 = \phi_2 - \phi_6,$$

$$\Phi_3 = \phi_6 - \phi_7,$$

$$\Phi_4 = \phi_1 + \phi_4 - \phi_5,$$

$$\Phi_5 = \phi_1 + \phi_2 - \phi_3.$$

d) *The subspace* $\mathsf{V}^\perp$ *of class-functions* ϕ *on* $\mathfrak{H}$ *for which* $(\phi, \psi)_{\mathfrak{H}} = 0$ *for all* $\psi \in \mathsf{V}$ *is of dimension 3 and consists precisely of those class-functions on* $\mathfrak{H}$ *which vanish on* D.

Proof. a) Obviously, $\mathsf{D} \subseteq \mathbf{C}_{\mathfrak{G}}(j)$. The assertion may therefore be easily checked from the list of representatives of the classes in the character table of $\mathfrak{H}$.

b) Suppose that $x \in \mathsf{D} \cap \mathsf{D}^g$ and the order of x is $2l$. Then $x^l = j$ and $x^l = j^g$, so $g \in \mathbf{C}_{\mathfrak{G}}(j)$.

c) $\mathfrak{H} - \mathsf{D}$ consists of the conjugacy classes of $\mathfrak{H}$ containing 1, b and c. From the character table one sees that

$$\Phi_i(1) = \Phi_i(b) = \Phi_i(c) = 0 \quad (i = 1, \dots, 5).$$

Obviously, the Φ_i are linearly independent. The dimension of V is equal to the number of conjugacy classes of $\mathfrak{H}$ in D and is therefore 5. Thus $\Phi_1, \dots, \Phi_5$ is a basis of V.

d) The space of class functions on $\mathfrak{H}$ which vanishes on D is of dimension 3 and is contained in $\mathsf{V}^\perp$. Since $\dim \mathsf{V}^\perp = 8 - \dim \mathsf{V} = 3$, the assertion follows. **q.e.d.**

We obtain characters of $\mathfrak{G}$ by using exceptional characters.

5.7 Lemma. a) *There exist distinct non-identity irreducible characters* $\chi_1, \dots, \chi_7$ *of* $\mathfrak{G}$ *with the following properties.*

(1) $\Phi_1^{\mathfrak{G}} = 1_{\mathfrak{G}} + \varepsilon(\chi_1 - \chi_2)$.

(2) $\Phi_2^{\mathfrak{G}} = \varepsilon(\chi_1 - \chi_3)$.

(3) $\Phi_3^{\mathfrak{G}} = \varepsilon(\chi_3 - \chi_4)$.

(4) $\Phi_4^{\mathfrak{G}} = \varepsilon\chi_2 + \varepsilon_1\chi_5 + \varepsilon_2\chi_6$.
(5) $\Phi_5^{\mathfrak{G}} = \varepsilon\chi_1 + \varepsilon_1\chi_5 + \varepsilon_3\chi_7$.
Here ε *and the* ε_i *are* ± 1.

b) *We have*

$$\varepsilon(\chi_1)_{\mathfrak{H}} - \phi_0 + \phi_3 + \phi_6 + \phi_7 \in \mathsf{V}^\perp$$

and

$$\varepsilon_2(\chi_6)_{\mathfrak{H}} + \phi_5 \in \mathsf{V}^\perp.$$

Proof. a) To prove this we use the following three facts.
(i) $\Phi_i^{\mathfrak{G}}(1) = 0 \quad (i = 1, \ldots, 5)$.
This is clear since $\Phi_i(1) = 0$.
(ii) $(\Phi_1^{\mathfrak{G}}, 1_{\mathfrak{G}})_{\mathfrak{G}} = 1$ and $(\Phi_i^{\mathfrak{G}}, 1_{\mathfrak{G}})_{\mathfrak{G}} = 0$ for $i = 2, \ldots, 5$.
For by Frobenius reciprocity, $(\Phi_i^{\mathfrak{G}}, 1_{\mathfrak{G}})_{\mathfrak{G}} = (\Phi_i, 1_{\mathfrak{H}})_{\mathfrak{H}} = (\Phi_i, \phi_0)_{\mathfrak{H}}$.
(iii) $(\Phi_i^{\mathfrak{G}}, \Phi_j^{\mathfrak{G}})_{\mathfrak{G}} = (\Phi_i, \Phi_j)_{\mathfrak{H}}$
This follows from V, 22.7.
From (iii), we obtain the following table for $(\Phi_i^{\mathfrak{G}}, \Phi_j^{\mathfrak{G}})_{\mathfrak{G}}$.

$i \backslash j$	1	2	3	4	5
1	3	1	0	−1	1
2	1	2	−1	0	1
3	0	−1	2	0	0
4	−1	0	0	3	1
5	1	1	0	1	3

Since $(\Phi_1^{\mathfrak{G}}, \Phi_2^{\mathfrak{G}})_{\mathfrak{G}} = 1$, there is an irreducible character $\chi_1 \neq 1_{\mathfrak{G}}$ of $\mathfrak{G}$ and a sign $\varepsilon = \pm 1$ such that

$$\Phi_1^{\mathfrak{G}} = \varepsilon\chi_1 + a\xi + b\eta + \cdots, \Phi_2^{\mathfrak{G}} = \varepsilon\chi_1 + a'\xi' + b'\eta' + \cdots,$$

where $\{\xi, \eta, \ldots\}$, $\{\xi', \eta', \ldots\}$ are disjoint sets of irreducible characters of $\mathfrak{G}$ neither of which contains χ_1. Since $(\Phi_1^{\mathfrak{G}}, 1_{\mathfrak{G}})_{\mathfrak{G}} = 1$ and $(\Phi_1^{\mathfrak{G}}, \Phi_1^{\mathfrak{G}})_{\mathfrak{G}} = 3$, $\Phi_1^{\mathfrak{G}} = 1_{\mathfrak{G}} + \varepsilon\chi_1 \pm \xi$. Since $\Phi_1^{\mathfrak{G}}(1) = 0$, we have

$$\Phi_1^{\mathfrak{G}} = 1_{\mathfrak{G}} + \varepsilon(\chi_1 - \chi_2) \tag{1}$$

for an irreducible character χ_2 distinct from $1_{\mathfrak{G}}, \chi_1$. And since $(\Phi_2^{\mathfrak{G}}, \Phi_2^{\mathfrak{G}})_{\mathfrak{G}} = 2$, $(\Phi_2^{\mathfrak{G}}, 1_{\mathfrak{G}})_{\mathfrak{G}} = 0$ and $\Phi_2(1) = 0$,

(2) $$\Phi_2^{\mathfrak{G}} = \varepsilon(\chi_1 - \chi_3),$$

where χ_3 is distinct from $1_{\mathfrak{G}}, \chi_1, \chi_2$.

If $(\Phi_3^{\mathfrak{G}}, \chi_1)_{\mathfrak{G}} = c$, it follows from $(\Phi_3^{\mathfrak{G}}, 1_{\mathfrak{G}})_{\mathfrak{G}} = (\Phi_3^{\mathfrak{G}}, \Phi_1^{\mathfrak{G}})_{\mathfrak{G}} = 0$ and (1) that $(\Phi_3^{\mathfrak{G}}, \chi_2)_{\mathfrak{G}} = c$. Suppose that $c \neq 0$. Then it follows from $(\Phi_3^{\mathfrak{G}}, \Phi_3^{\mathfrak{G}})_{\mathfrak{G}} = 2$ that $\Phi_3^{\mathfrak{G}} = c(\chi_1 + \chi_2)$. This is impossible, since $\Phi_3^{\mathfrak{G}}(1) = 0$. Thus $c = 0$. But $(\Phi_3^{\mathfrak{G}}, \Phi_2^{\mathfrak{G}})_{\mathfrak{G}} = -1$, so

(3) $$\Phi_3^{\mathfrak{G}} = \varepsilon(\chi_3 - \chi_4),$$

where χ_4 is distinct from $1_{\mathfrak{G}}, \chi_1, \chi_2, \chi_3$.

Let $(\Phi_5^{\mathfrak{G}}, \chi_3)_{\mathfrak{G}} = d$. Since $(\Phi_3^{\mathfrak{G}}, \Phi_5^{\mathfrak{G}})_{\mathfrak{G}} = 0$, (3) gives $(\Phi_5^{\mathfrak{G}}, \chi_4)_{\mathfrak{G}} = d$. If $d \neq 0$, it follows from $(\Phi_5^{\mathfrak{G}}, \Phi_5^{\mathfrak{G}})_{\mathfrak{G}} = 3$ that $d = \pm 1$ and

$$\Phi_5^{\mathfrak{G}} = d\chi_3 + d\chi_4 + e\xi'',$$

where ξ'' is a non-identity irreducible character of $\mathfrak{G}$ distinct from χ_3, χ_4. But $(\Phi_5^{\mathfrak{G}}, \Phi_1^{\mathfrak{G}})_{\mathfrak{G}} = (\Phi_5^{\mathfrak{G}}, \Phi_2^{\mathfrak{G}})_{\mathfrak{G}} = 1$, so $e\xi'' = -\varepsilon\chi_2$. Since $\Phi_5^{\mathfrak{G}}(1) = 0$, $\Phi_5^{\mathfrak{G}} = -\varepsilon(\chi_2 + \chi_3 + \chi_4)$. But by (2), $\chi_1(1) = \chi_3(1)$, so $\chi_4(1) = 0$, which is impossible. Hence $d = 0$. Since $(\Phi_5^{\mathfrak{G}}, \Phi_2^{\mathfrak{G}})_{\mathfrak{G}} = 1$, (2) gives $\Phi_5^{\mathfrak{G}} = \varepsilon\chi_1 + \cdots$. Now $(\Phi_5^{\mathfrak{G}}, 1_{\mathfrak{G}})_{\mathfrak{G}} = 0$ and $(\Phi_5^{\mathfrak{G}}, \Phi_1^{\mathfrak{G}})_{\mathfrak{G}} = 1$ give $(\Phi_5^{\mathfrak{G}}, \chi_2)_{\mathfrak{G}} = 0$. Finally, by (2), (3) and (1),

$$(\Phi_4^{\mathfrak{G}}, \chi_1)_{\mathfrak{G}} = (\Phi_4^{\mathfrak{G}}, \chi_3)_{\mathfrak{G}} = (\Phi_4^{\mathfrak{G}}, \chi_4)_{\mathfrak{G}} = (\Phi_4^{\mathfrak{G}}, \chi_2)_{\mathfrak{G}} - \varepsilon.$$

Since $\Phi_4^{\mathfrak{G}}(1) = 0$ and $(\Phi_4^{\mathfrak{G}}, \Phi_4^{\mathfrak{G}})_{\mathfrak{G}} = 3$, $\Phi_4^{\mathfrak{G}} = \varepsilon\chi_2 + \cdots$. And $(\Phi_4^{\mathfrak{G}}, \Phi_5^{\mathfrak{G}})_{\mathfrak{G}} = 1$ shows that we may write

(4) $$\Phi_4^{\mathfrak{G}} = \varepsilon\chi_2 + \varepsilon_1\chi_5 + \varepsilon_2\chi_6,$$

(5) $$\Phi_5^{\mathfrak{G}} = \varepsilon\chi_1 + \varepsilon_1\chi_5 + \varepsilon_3\chi_7,$$

where $1_{\mathfrak{G}}, \chi_1, \ldots, \chi_7$ are distinct irreducible characters of $\mathfrak{G}$.

b) We have

$$\begin{aligned}
&(\varepsilon(\chi_1)_{\mathfrak{H}} - \phi_0 + \phi_3 + \phi_6 + \phi_7, \Phi_1)_{\mathfrak{H}} \\
&= \varepsilon(\chi_1, \Phi_1^{\mathfrak{G}})_{\mathfrak{G}} + (-\phi_0 + \phi_3 + \phi_6 + \phi_7, \phi_0 + \phi_2 - \phi_4)_{\mathfrak{H}} \\
&= \varepsilon^2 - 1 = 0.
\end{aligned}$$

Similarly,

$$(\varepsilon(\chi_1)_{\mathfrak{H}} - \phi_0 + \phi_3 + \phi_6 + \phi_7, \Phi_i)_{\mathfrak{H}} = 0$$

for $i = 2, \ldots, 5$, and

$$(\varepsilon_2(\chi_6)_{\mathfrak{H}} + \phi_5, \Phi_i)_{\mathfrak{H}} = 0$$

for $i = 1, \ldots, 5$. **q.e.d.**

Next we find some values of these characters.

5.8 Lemma. (1) $\chi_2(1) = \varepsilon + \chi_1(1)$.
(2) $\chi_1(j) = 2\varepsilon$.
(3) $\chi_2(j) = 3\varepsilon$.
(4) $\chi_5(j) = \varepsilon_1$.
(5) $\chi_6(j) = 4\varepsilon_2$.

Proof. (1) follows from

$$0 = \Phi_1^{\mathfrak{G}}(1) = 1 + \varepsilon(\chi_1(1) - \chi_2(1)).$$

Since $j = a^4 \in \mathsf{D}$, it follows from 5.6d) that $\psi(j) = 0$ for all $\psi \in \mathsf{V}^{\perp}$. Hence by 5.7b),

$$\varepsilon\chi_1(j) = \phi_0(j) - \phi_3(j) - \phi_6(j) - \phi_7(j),$$

$$\varepsilon_2\chi_6(j) = -\phi_5(j).$$

Using the character table of $\mathfrak{H}$ yields (2) and (5). By V, 22.7, $\Phi_1^{\mathfrak{G}}(j) = \Phi_1(j) = 1 + \phi_2(j) - \phi_4(j) = 0$. Hence by 5.7, $\chi_1(j) - \chi_2(j) = -\varepsilon$, and using (2) gives (3). Similarly, $\Phi_4^{\mathfrak{G}}(j) = \Phi_4(j)$ gives

$$\varepsilon\chi_2(j) + \varepsilon_1\chi_5(j) + \varepsilon_2\chi_6(j) = 8.$$

Using (3) and (5) gives (4). **q.e.d.**

5.9 Lemma. *Write* $g = |\mathfrak{G}|$ *and* $f = \chi_1(1)$. *Then*

$$g = \frac{2^9 \cdot 3^2 \cdot f(f + \varepsilon)}{(f - 2\varepsilon)^2}.$$

Proof. We apply the Suzuki formula of V, 22.8 to Φ_1. This gives

$$\frac{1}{|\mathfrak{G}|}\sum_{\chi}\frac{(\chi, \Phi_1^{\mathfrak{G}})_{\mathfrak{G}}}{\chi(1)}\left(\sum_t \chi(t)\right)^2 = \frac{1}{|\mathfrak{H}|}\sum_{\psi}\frac{(\psi, \Phi_1)_{\mathfrak{H}}}{\psi(1)}\left(\sum_{t'}\psi(t')\right)^2,$$

where χ, ψ run through the irreducible characters and t, t' run through the involutions of $\mathfrak{G}$, $\mathfrak{H}$ respectively. Since $\Phi_1 = \phi_0 + \phi_2 - \phi_4$ and $\Phi_1^{\mathfrak{G}} = 1_{\mathfrak{G}} + \varepsilon(\chi_1 - \chi_2)$, this gives

$$\frac{1}{|\mathfrak{G}|}\left(a_0^2 + \frac{\varepsilon}{\chi_1(1)}a_1^2 - \frac{\varepsilon}{\chi_2(1)}a_2^2\right) = \frac{1}{|\mathfrak{H}|}\left(b_0^2 + \frac{1}{\phi_2(1)}b_1^2 - \frac{1}{\phi_4(1)}b_2^2\right),$$

where, since all involutions in $\mathfrak{G}$ are conjugate (5.5) and the involutions in $\mathfrak{H}$ are j and the conjugates of b, the values of $a_0, \ldots, b_2$ are as follows.

$$a_0 = \text{number of involutions in } \mathfrak{G} = |\mathfrak{G} : \mathfrak{H}|,$$

$$a_1 = \sum_t \chi_1(t) = |\mathfrak{G} : \mathfrak{H}|\chi_1(j) = 2\varepsilon|\mathfrak{G} : \mathfrak{H}|,$$

$$a_2 = \sum_t \chi_2(t) = |\mathfrak{G} : \mathfrak{H}|\chi_2(j) = 3\varepsilon|\mathfrak{G} : \mathfrak{H}|,$$

$$b_0 = \text{number of involutions in } \mathfrak{H} = 1 + 12 = 13,$$

$$b_1 = \sum_{t'} \phi_2(t') = 2 + 12 \cdot 0 = 2,$$

$$b_2 = \sum_{t'} \phi_4(t') = 3 + 12 \cdot 1 = 15.$$

Thus, since $\chi_1(1) = f$ and (by 5.8), $\chi_2(1) = \varepsilon + f$,

$$|\mathfrak{G} : \mathfrak{H}|\left(1 + \frac{4\varepsilon}{f} - \frac{9\varepsilon}{\varepsilon + f}\right) = 169 + \frac{4}{2} - \frac{225}{3} - 96$$

and

$$g\frac{f^2 + 4 - 4\varepsilon f}{f(\varepsilon + f)} = 48 \cdot 96 = 2^9 \cdot 3^2,$$

which gives the result. **q.e.d.**

5.10 Lemma. a) $\varepsilon_2\chi_6(1) = \dfrac{8(1 + \varepsilon f)}{\varepsilon f - 8}$.

b) $\varepsilon = 1$ *and either* $f = 10$, $g = 2^4 \cdot 3^2 \cdot 5 \cdot 11$ *or* $f = 26$, $g = 2^4 \cdot 3^3 \cdot 13$.

c) *If χ is an irreducible character of $\mathfrak{G}$ distinct from $1_{\mathfrak{G}}, \chi_1, \ldots, \chi_7$, then $\chi(1)$ is divisible by 16.*

Proof. a) The Suzuki formula applied to Φ_4 gives

$$|\mathfrak{G}:\mathfrak{H}|\left(\frac{\varepsilon}{\chi_2(1)}\chi_2(j)^2 + \frac{\varepsilon_1}{\chi_5(1)}\chi_5(j)^2 + \frac{\varepsilon_2}{\chi_6(1)}\chi_6(j)^2\right)$$
$$= \frac{1}{\phi_1(1)}(1 + 12(-1))^2 + \frac{1}{\phi_4(1)}(3 + 12)^2 - \frac{1}{\phi_5(1)}(-4 + 12\cdot 0)^2$$
$$= 192.$$

Thus

$$\frac{g}{48}\left(\frac{9\varepsilon}{f + \varepsilon} + \frac{\varepsilon_1}{\chi_5(1)} + \frac{16\varepsilon_2}{\chi_6(1)}\right) = 192.$$

Put $e = \varepsilon_2\chi_6(1)$. Then

$$0 = \Phi_4^{\mathfrak{G}}(1) = \varepsilon\chi_2(1) + \varepsilon_1\chi_5(1) + e = \varepsilon f + 1 + \varepsilon_1\chi_5(1) + e.$$

Hence $\varepsilon_1\chi_5(1) = -1 - \varepsilon f - e$ and

$$\frac{g}{48}\left(\frac{9\varepsilon}{f + \varepsilon} - \frac{1}{1 + \varepsilon f + e} + \frac{16}{e}\right) = 192.$$

Using 5.9, this becomes

$$\frac{9\varepsilon}{f + \varepsilon} - \frac{1}{1 + \varepsilon f + e} + \frac{16}{e} = \frac{2^{10}\cdot 3^2}{g} = \frac{2(f - 2\varepsilon)^2}{f(f + \varepsilon)}.$$

This yields a quadratic equation for e:

$$(\varepsilon f - 8)(2\varepsilon f - 1)e^2 + 2(1 + \varepsilon f)(f^2 - 16\varepsilon f + 4)e - 16\varepsilon f(1 + \varepsilon f)^2 = 0.$$

Solving,

$$e = -\frac{2\varepsilon f(1 + \varepsilon f)}{2\varepsilon f - 1} \quad \text{or} \quad e = \frac{8(1 + \varepsilon f)}{\varepsilon f - 8}.$$

Suppose that

$$e = -\frac{2\varepsilon f(1 + \varepsilon f)}{2\varepsilon f - 1}.$$

Then $1 + \varepsilon f = u(2\varepsilon f - 1)$ for some integer u and

$$3 = 2(\varepsilon f + 1) - (2\varepsilon f - 1) = (2u - 1)(2\varepsilon f - 1).$$

Thus $2\varepsilon f - 1$ divides 3 and εf is 2, 1, 0 or -1. Since $\mathfrak{G}$ is simple and $\chi_1 \neq 1_{\mathfrak{G}}$, $f \neq 1$; thus $f = 2$ and $\varepsilon = 1$. This is also impossible by 5.9. Therefore

$$e = \frac{8(1 + \varepsilon f)}{\varepsilon f - 8}.$$

b) By a),

$$e = \frac{8(1 + \varepsilon f)}{\varepsilon f - 8} = \frac{72}{\varepsilon f - 8} + 8.$$

Thus $\varepsilon f - 8$ divides 72. Also

$$g = \frac{2^9 \cdot 3^2 \cdot f(f + \varepsilon)}{(f - 2\varepsilon)^2}.$$

But 2^4 is the highest power of 2 which divides g, so $(f - 2\varepsilon)^2$ is divisible by 2^5 and $f - 2\varepsilon \equiv 0\,(8)$. If $f = 2\varepsilon + 8k$,

$$g - \frac{2^{10} \cdot 3^2(\varepsilon + 4k)(3\varepsilon + 8k)}{2^6 k^2},$$

so k is odd. Indeed, k^2 divides $3^2(3\varepsilon + 8k)$, so 3 is the only possible prime divisor of k and k is either 1 or 3. Thus we have the following possibilities.

(1) $k = \varepsilon = 1, f = 10$.
(2) $k = 1, \varepsilon = -1, f = 6$.
(3) $k = 3, \varepsilon = 1, f = 26$.
(4) $k = 3, \varepsilon = -1, f = 22$.

Cases (2) and (4) cannot occur, since $\varepsilon f - 8$ divides 72. Thus $\varepsilon = 1$ and either $f = 10, g = 2^4 \cdot 3^2 \cdot 5 \cdot 11$ or $f = 26, g = 2^4 \cdot 3^3 \cdot 13$.

c) Since $(\chi, 1_{\mathfrak{G}})_{\mathfrak{G}} = (\chi, \chi_i)_{\mathfrak{G}} = 0$ for $i = 1, \ldots, 7$,

$$0 = (\chi, \Phi_i^{\mathfrak{G}})_{\mathfrak{G}} = (\chi_{\mathfrak{H}}, \Phi_i)_{\mathfrak{H}}.$$

Thus $\chi_{\mathfrak{H}} \in \mathsf{V}^{\perp}$, and by 5.6d), $\chi_{\mathfrak{H}}$ vanishes on D. By 5.5, all involutions in $\mathfrak{G}$ are conjugate, so D contains a representative of every conjugacy class of non-identity 2-elements of $\mathfrak{G}$. Thus χ vanishes on $\mathfrak{P} - \{1\}$, where $\mathfrak{P}$ is a Sylow 2-subgroup of $\mathfrak{G}$. Hence

$$\frac{\chi(1)}{2^4} = (\chi_{\mathfrak{P}}, 1_{\mathfrak{P}})_{\mathfrak{P}}$$

is an integer, and $\chi(1)$ is divisible by 16. **q.e.d.**

The two cases appearing in 5.10 will be dealt with separately in 5.11–5.16 and 5.17–5.27. The first case leads to $\mathfrak{G} \cong \mathfrak{M}_{11}$, the second to $\mathfrak{G} \cong PSL(3, 3)$. We obtain the specific groups by finding faithful permutation representations of degree 11 and 13 respectively, and this is done in the following steps.

Step 1: Determination of the degrees of the irreducible characters of $\mathfrak{G}$ (5.11 and 5.17).
Step 2: Determination of the classes of $\mathfrak{G}$ (5.12 and 5.18–5.21).
Step 3: Finding a subgroup of $\mathfrak{G}$ of index 11 or 13 (5.14–5.15 and 5.22–5.25).
Step 4: Proof that $\mathfrak{G} \cong \mathfrak{M}_{11}$ or $\mathfrak{G} \cong PSL(3, 3)$ (5.16 and 5.27).

In 5.11–5.16, suppose, then, that $f = 10$, $g = 2^4 \cdot 3^2 \cdot 5 \cdot 11$.

5.11 Lemma. *Besides* $1_{\mathfrak{G}}, \chi_1, \ldots, \chi_7$, $\mathfrak{G}$ *has exactly two other irreducible characters* χ_8, χ_9. *If* $f_i = \chi_i(1)$, then $f_1 = 10$, $f_2 = 11$, $f_3 = f_4 = 10$, $f_5 = 55$, $f_6 = 44$, $f_7 = 45$, $f_8 = f_9 = 16$.

Proof. We already know $f_1 = f = 10$. By 5.10,

$$\varepsilon_2 f_6 = \frac{8(1 + \varepsilon f)}{\varepsilon f - 8} = 44.$$

Thus $f_6 = 44$ and $\varepsilon_2 = 1$. By 5.8, $f_2 = 11$. By 5.7,

$$0 = \Phi_2^{\mathfrak{G}}(1) = \varepsilon(f_1 - f_3) = \varepsilon(10 - f_3),$$

$$0 = \Phi_3^{\mathfrak{G}}(1) = \varepsilon(f_3 - f_4),$$

$$0 = \Phi_4^{\mathfrak{G}}(1) = \varepsilon f_2 + \varepsilon_1 f_5 + \varepsilon_2 f_6 = 11 + \varepsilon_1 f_5 + 44,$$

$$0 = \Phi_5^{\mathfrak{G}}(1) = \varepsilon f_1 + \varepsilon_1 f_5 + \varepsilon_3 f_7 = 10\varepsilon + \varepsilon_1 f_5 + \varepsilon_3 f_7.$$

It follows that $f_1 = f_3 = f_4 = 10$, $f_5 = 55$ and $f_7 = 45$.

Let $\chi_8, \ldots, \chi_n$ be the remaining irreducible characters of $\mathfrak{G}$. By 5.10c), $\chi_i(1) = 16m_i$ for some integer m_i ($i \geq 8$). Thus

$$2^4 \cdot 3^2 \cdot 5 \cdot 11 = g = 1 + \sum_{i=1}^{7} f_i^2 + 2^8(m_8^2 + \cdots + m_n^2).$$

This gives

$$2^9 = 2^8(m_8^2 + m_9^2 + \cdots),$$

so $n = 9$, $m_8 = m_9 = 1$, $f_8 = f_9 = 16$. **q.e.d.**

5.12 Lemma. a) *$\mathfrak{G}$ has 10 conjugacy classes, and the following are representatives of them.*

Representative	1	j	a^2	a	a^{-1}	c	a^4c	x	y_1	y_2
Order	1	2	4	8	8	3	6	5	11	11

b) *If $\mathfrak{Q}$ is a Sylow 11-subgroup of $\mathfrak{G}$, then $|\mathbf{N}_{\mathfrak{G}}(\mathfrak{Q})| = 55$ and $\mathbf{C}_{\mathfrak{G}}(\mathfrak{Q}) = \mathfrak{Q}$.*

Proof. By 5.11, the class number of $\mathfrak{G}$ is 10. The elements 1, j, a^2, a, c and a^4c are in different classes, since they have distinct orders. And a is not conjugate to a^{-1}, since if $a^g = a^{-1}$, $(a^4)^g = (a^4)^{-1}$ and $g \in \mathbf{C}_{\mathfrak{G}}(j)$, whereas by 5.4, a and a^{-1} are not conjugate in $\mathfrak{H}$. Thus we have found representatives of 7 classes, and 3 remain to be found. These must contain elements of order 5 and 11.

Now $|\mathbf{C}_{\mathfrak{G}}(\mathfrak{Q})|$ is odd, since otherwise $\mathfrak{Q}$ is contained in the centralizer of an involution and $|\mathfrak{H}|$ is divisible by 11, which is not the case. Also $|\mathbf{N}_{\mathfrak{G}}(\mathfrak{Q}) : \mathbf{C}_{\mathfrak{G}}(\mathfrak{Q})|$ is a divisor of 10, so $|\mathbf{N}_{\mathfrak{G}}(\mathfrak{Q})|$ is not divisible by 4. Thus $|\mathfrak{G} : \mathbf{N}_{\mathfrak{G}}(\mathfrak{Q})|$ is divisible by 8 and is congruent to 1 modulo 11. It follows that $|\mathbf{N}_{\mathfrak{G}}(\mathfrak{Q})| = 55$. By Burnside's theorem, $\mathbf{N}_{\mathfrak{G}}(\mathfrak{Q}) \neq \mathbf{C}_{\mathfrak{G}}(\mathfrak{Q})$, so $\mathbf{C}_{\mathfrak{G}}(\mathfrak{Q}) = \mathfrak{Q}$. Hence $\mathfrak{G}$ has 1440 elements of order 11, each having 720 conjugates. Thus $\mathfrak{G}$ has 2 classes of elements of order 11 and one class of elements of order 5. **q.e.d.**

5.13 Lemma. *If $\mathfrak{R}$ is a Sylow 5-subgroup of $\mathfrak{G}$, $|\mathbf{N}_{\mathfrak{G}}(\mathfrak{R})| = 20$ and $\mathbf{C}_{\mathfrak{G}}(\mathfrak{R}) = \mathfrak{R}$.*

Proof. By 5.12, $\mathfrak{G}$ has no element of order 10, 15 or 55. Thus $\mathbf{C}_{\mathfrak{G}}(\mathfrak{R}) = \mathfrak{R}$. Since all elements of $\mathfrak{G}$ of order 5 are conjugate in $\mathfrak{G}$, $\mathfrak{G}$ has 1584 elements of order 5 and 396 Sylow 5-subgroups. **q.e.d.**

5.14 Lemma. *Let* V *be a* $\mathbb{C}\mathfrak{G}$*-module with character* χ_1*. For any subgroup* $\mathfrak{X}$ *of* $\mathfrak{G}$*, let* $V^{\mathfrak{X}}$ *be the set of fixed points of* V *under* $\mathfrak{X}$.

a) $\dim V^{\mathfrak{Z}} = 4$, *where* $\mathfrak{Z}$ *is any cyclic subgroup of* $\mathfrak{G}$ *of order 4.*

b) $\dim V^{\mathfrak{Q}} = 3$, *where* $\mathfrak{Q}$ *is a quaternion subgroup of* $\mathfrak{G}$ *of order 8.*

c) $\dim V^{\mathbf{N}_{\mathfrak{G}}(\mathfrak{R})} \geq 2$.

Proof. We use the fact that

$$\dim V^{\mathfrak{X}} = ((\chi_1)_{\mathfrak{X}}, 1_{\mathfrak{X}})_{\mathfrak{X}} = \frac{1}{|\mathfrak{X}|} \sum_{x \in \mathfrak{X}} \chi_1(x).$$

Note that by 5.8, $\chi_1(j) = 2$, and by 5.7b),

$$\chi_1(a^2) = \phi_0(a^2) - \phi_3(a^2) - \phi_6(a^2) - \phi_7(a^2) = 2.$$

Hence by 5.5, $\chi_1(x) = 2$ if x is of order 2 or 4.

a) $\dim V^{\mathfrak{Z}} = \dfrac{1}{4} \sum_{x \in \mathfrak{Z}} \chi_1(x) = 4.$

b) $\dim V^{\mathfrak{Q}} = \dfrac{1}{8} \sum_{x \in \mathfrak{Q}} \chi_1(x) = \dfrac{1}{8}(\chi_1(1) + \chi_1(j) + 6\chi_1(a^2)) = 3.$

c) $\mathbf{N}_{\mathfrak{G}}(\mathfrak{R})$ is a Frobenius group of order 20, so if μ is a faithful character of $\mathfrak{R}$ of degree 1, $\mu^* = \mu^{\mathbf{N}_{\mathfrak{G}}(\mathfrak{R})}$ is an irreducible character of $\mathbf{N}_{\mathfrak{G}}(\mathfrak{R})$ (V, 16.13). Let $\mathfrak{Z} = \langle z \rangle$ be a Frobenius complement in $\mathbf{N}_{\mathfrak{G}}(\mathfrak{R})$. Then the remaining irreducible characters of $\mathbf{N}_{\mathfrak{G}}(\mathfrak{R})$ are $\mu_0, \mu_1, \mu_2, \mu_3$, where $\mu_r(z) = i^r$ $(r = 0, 1, 2, 3)$ and $\mu_r(y) = 1$ for all $y \in \mathfrak{R}$. Now

$$(\chi_1)_{\mathbf{N}_{\mathfrak{G}}(\mathfrak{R})} = l\mu^* + l_0\mu_0 + l_1\mu_1 + l_2\mu_2 + l_3\mu_3$$

for non-negative integers $l, l_0, \ldots, l_3$, and $\dim V^{\mathbf{N}_{\mathfrak{G}}(\mathfrak{R})} = l_0$. Since $\mu^*_{\mathfrak{Z}} = (\mu^{\mathbf{N}_{\mathfrak{G}}(\mathfrak{R})})_{\mathfrak{Z}} = (\mu_{\mathfrak{R} \cap \mathfrak{Z}})^{\mathfrak{Z}}$ is the regular character of $\mathfrak{Z}$, we obtain for $\lambda_r = (\mu_r)_{\mathfrak{Z}}$,

$$\mu^*_{\mathfrak{Z}} = \lambda_0 + \lambda_1 + \lambda_2 + \lambda_3.$$

Thus

$$(\chi_1)_{\mathfrak{Z}} = (l + l_0)\lambda_0 + (l + l_1)\lambda_1 + (l + l_2)\lambda_2 + (l + l_3)\lambda_3.$$

By a), $l + l_0 = 4$. Also $4l + l_0 + \cdots + l_3 = \chi_1(1) = 10$, so $l \leq 2$. Hence $l_0 \geq 2$. **q.e.d.**

5.15 Lemma. $\mathfrak{G}$ *has a subgroup* $\mathfrak{T}$ *of index 11.*

Proof. a^2 lies in a quaternion subgroup $\mathfrak{Q}$ of $\mathfrak{G}$ of order 8. Write $\mathfrak{Z} = \langle a^2 \rangle$. Since all elements of order 4 are conjugate, $\mathfrak{Z} \leq \mathbf{N}_{\mathfrak{G}}(\mathfrak{N})$ for some Sylow 5-subgroup $\mathfrak{N}$ of $\mathfrak{G}$, and $\mathfrak{Z} = \mathfrak{Q} \cap \mathbf{N}_{\mathfrak{G}}(\mathfrak{N})$. Let $\mathfrak{T} = \langle \mathbf{N}_{\mathfrak{G}}(\mathfrak{N}), \mathfrak{Q} \rangle$. We show that $|\mathfrak{G} : \mathfrak{T}| = 11$.

Clearly, $\mathsf{V}^{\mathfrak{T}} = \mathsf{V}^{\mathbf{N}_{\mathfrak{G}}(\mathfrak{N})} \cap \mathsf{V}^{\mathfrak{Q}}$. Hence by 5.14,

$$\begin{aligned} 4 &= \dim \mathsf{V}^{\mathfrak{Z}} \geq \dim(\mathsf{V}^{\mathbf{N}_{\mathfrak{G}}(\mathfrak{N})} + \mathsf{V}^{\mathfrak{Q}}) \\ &= \dim \mathsf{V}^{\mathbf{N}_{\mathfrak{G}}(\mathfrak{N})} + \dim \mathsf{V}^{\mathfrak{Q}} - \dim(\mathsf{V}^{\mathbf{N}_{\mathfrak{G}}(\mathfrak{N})} \cap \mathsf{V}^{\mathfrak{Q}}) \\ &\geq 2 + 3 - \dim \mathsf{V}^{\mathfrak{T}}. \end{aligned}$$

Thus $\dim \mathsf{V}^{\mathfrak{T}} \geq 1$. But $\mathsf{V}^{\mathfrak{G}} = 0$, so $\mathfrak{T} \neq \mathfrak{G}$.

By Sylow's theorem,

$$|\mathfrak{T}| = |\mathbf{N}_{\mathfrak{G}}(\mathfrak{N})|(1 + 5n) = 20(1 + 5n)$$

for some n. Since $\mathfrak{Q} \leq \mathfrak{T}$, 8 divides $|\mathfrak{T}|$, so n is odd.

If $n \geq 13$, then

$$|\mathfrak{G} : \mathfrak{T}| \leq \frac{2^4 \cdot 3^2 \cdot 5 \cdot 11}{20 \cdot 66} = 6.$$

Hence $\mathfrak{G}$ has a transitive permutation representation of degree at most 6. But this must be faithful since $\mathfrak{T} < \mathfrak{G}$ and $\mathfrak{G}$ is simple. This is impossible, since $|\mathfrak{G}|$ is divisible by 11. Hence $n \leq 11$.

If $n - 1$, $|\mathfrak{T}| = 120$. Let ρ be the transitive representation of $\mathfrak{T}$ on the 6 conjugates of $\mathfrak{N}$ in $\mathfrak{T}$ and let $\mathfrak{K} = \ker \rho$. Since $\mathfrak{K} \lhd \mathfrak{T}$ and $\mathfrak{K} \leq \mathbf{N}_{\mathfrak{G}}(\mathfrak{N})$, $|\mathfrak{K}|$ is not divisible by 5 and $\mathfrak{K} \leq \mathbf{C}_{\mathfrak{G}}(\mathfrak{N})$. Now $\mathfrak{Q}$ has no faithful permutation representation of degree less than 8, so $j = a^4 \in \mathfrak{K}$. Thus $j \in \mathfrak{K} \leq \mathbf{C}_{\mathfrak{G}}(\mathfrak{N})$. By 5.13, $\mathbf{C}_{\mathfrak{G}}(\mathfrak{N}) = \mathfrak{N}$, so this is impossible. Thus $3 \leq n \leq 11$.

Since $|\mathfrak{T}|$ divides $|\mathfrak{G}|$, $1 + 5n$ divides 396. It follows that $n = 7$ and $|\mathfrak{T}| = 720$. Thus $|\mathfrak{G} : \mathfrak{T}| = 11$. **q.e.d.**

This brings us to the concluding step of the first case.

5.16 Lemma. $\mathfrak{G} \cong \mathfrak{M}_{11}$.

Proof. By 5.15, $\mathfrak{G}$ has a transitive permutation representation ρ of degree 11; let π be the character of ρ. By V, 20.2, $(\pi, 1_{\mathfrak{G}})_{\mathfrak{G}} = 1$, so by 5.11, $\pi = 1_{\mathfrak{G}} + \chi$ for some irreducible character χ, and χ must be one of the characters χ_1, χ_3 or χ_4. Thus by V, 20.2, ρ is doubly transitive.

By V, 22.7, $\Phi_i^{\mathfrak{G}}(j) = \Phi_i(j)$. Since $j \in \mathsf{D}$ and $\varepsilon = 1$, 5.7 and 5.8 give

$$\Phi_2^{\mathfrak{G}}(j) = \chi_1(j) - \chi_3(j) = 2 - \chi_3(j),$$

$$\Phi_3^{\mathfrak{G}}(j) = \chi_3(j) - \chi_4(j).$$

By 5.6 and 5.4,

$$\Phi_2(j) = \phi_2(j) - \phi_6(j) = 4, \quad \Phi_3(j) = \phi_6(j) - \phi_7(j) = 0.$$

Hence $\chi_3(j) = \chi_4(j) = -2$. But $\pi(j) \geq 0$, so $\pi = 1_{\mathfrak{G}} + \chi_1$. Thus $\pi(j) = 1 + \chi_1(j) = 3$. Every involution in $\mathfrak{G}$ is conjugate to j and therefore has exactly 3 fixed points. In particular, each non-identity 2-element in $\mathfrak{G}_{1,2}$ has at most 3 fixed points. Thus any Sylow 2-subgroup $\mathfrak{Q}$ of $\mathfrak{G}_{1,2}$ leaves one further symbol, say 3, fixed, and $\mathfrak{Q}_i = 1$ for $i \notin \{1, 2, 3\}$. But since ρ is doubly transitive,

$$|\mathfrak{G}_{1,2}| = \frac{|\mathfrak{G}|}{11 \cdot 10} = 9 \cdot 8.$$

Thus $|\mathfrak{Q}| = 8$ and $\mathfrak{Q}$ operates regularly on $\{4, 5, \ldots, 11\}$.

If $\mathfrak{G}_{1,2}$ operates intransitively on $\{3, \ldots, 11\}$, its orbits in this set are of lengths 1 and 8, so $\mathfrak{G}_{1,2} = \mathfrak{G}_{1,2,3}$ and $\{1, 2, 3\}$ is the set of all fixed points of $\mathfrak{G}_{1,2}$. By II, 1.13, $\mathbf{N}_{\mathfrak{G}}(\mathfrak{G}_{1,2})$ permutes $\{1, 2, 3\}$ doubly transitively. Hence 3 divides $|\mathbf{N}_{\mathfrak{G}}(\mathfrak{G}_{1,2,3})/\mathfrak{G}_{1,2,3}|$. But $|\mathfrak{G} : \mathfrak{G}_{1,2,3}| = |\mathfrak{G} : \mathfrak{G}_{1,2}| = 11 \cdot 10$, so this is impossible. Thus $\mathfrak{G}_{1,2}$ operates transitively on $\{3, \ldots, 11\}$, and $\mathfrak{G}_{1,2,3} = \mathfrak{Q}$ operates transitively on $\{4, \ldots, 11\}$. Hence ρ is 4-fold transitive and $\mathfrak{G}$ is isomorphic to a sharply 4-fold transitive permutation group. By 3.3, $\mathfrak{G} \cong \mathfrak{M}_{11}$. **q.e.d.**

We now proceed to the second case of 5.10 and assume for the remainder of this section that $\chi_1(1) = 26$, $|\mathfrak{G}| = 2^4 \cdot 3^3 \cdot 13$.

5.17 Lemma. *Besides* $1_{\mathfrak{G}}, \chi_1, \ldots, \chi_7$, $\mathfrak{G}$ *has exactly four other irreducible characters* $\chi_8, \chi_9, \chi_{10}, \chi_{11}$. *If* $f_i = \chi_i(1)$, $f_1 = 26$, $f_2 = 27$, $f_3 = f_4 = 26$, $f_5 = 39$, $f_6 = 12$, $f_7 = 13$, $f_8 = f_9 = f_{10} = f_{11} = 16$. *Also* $\varepsilon_2 = 1$.

Proof. By 5.10a),

$$\varepsilon_2 \chi_6(1) = \frac{8(1 + \varepsilon f)}{\varepsilon f - 8} = 12.$$

Then by 5.8 and 5.7a),

$$f_2 = \varepsilon + f = 27,$$

$$f_3 = f - \varepsilon\Phi_2^{\mathfrak{G}}(1) = 26,$$

$$f_4 = f_3 - \varepsilon\Phi_3^{\mathfrak{G}}(1) = 26,$$

$$\varepsilon_1 f_5 = \Phi_4^{\mathfrak{G}}(1) - \varepsilon f_2 - \varepsilon_2 f_6 = -39,$$

$$\varepsilon_3 f_7 = \Phi_5^{\mathfrak{G}}(1) - \varepsilon f - \varepsilon_1 f_5 = 13.$$

The values of ε_2 and $f_1, \ldots, f_7$ are therefore as stated.

If $\chi_8, \chi_9, \ldots$ are the remaining irreducible characters of $\mathfrak{G}$, $\chi_i(1) = 2^4 m_i$ for some integer m_i, by 5.10c). Then

$$2^8(m_8^2 + \cdots) = g - 1 - \sum_{i=1}^{7} f_i^2 = 1024,$$

so $m_8^2 + \cdots = 4$. If $m_8 = 2$, then $\chi_8(1) = 2^5$. This is impossible, since 2^5 does not divide $|\mathfrak{G}|$. Hence $m_8 = m_9 = m_{10} = m_{11} = 1$. **q.e.d.**

5.18 Lemma. *$\mathfrak{G}$ has 12 conjugacy classes, and the following are representatives of 10 of them.*

Representative	1	j	a^2	a	a^{-1}	a^4c	d_i $(i = 1, 2, 3, 4)$
Order	1	2	4	8	8	6	13
Length of class	1	$3^2 \cdot 13$	$2 \cdot 3^3 \cdot 13$	$2 \cdot 3^3 \cdot 13$	$2 \cdot 3^3 \cdot 13$	$2^3 \cdot 3^2 \cdot 13$	$2^4 \cdot 3^3$

The two remaining classes together contain 728 elements, and one of them contains the element c of order 3.

Proof. By 5.17, the class number of $\mathfrak{G}$ is 12.

Let $\mathfrak{P}$ be a Sylow 13-subgroup of $\mathfrak{G}$. Then

$$|\mathfrak{G} : \mathbf{N}_{\mathfrak{G}}(\mathfrak{P})| = 2^u \cdot 3^v \equiv 1\ (13),$$

where $u \leq 4$, $v \leq 3$. The only possibilities for $2^u \cdot 3^v$ are 1, 27 and 144. But $2^u \cdot 3^v \neq 1$, since $\mathfrak{G}$ is simple. If $2^u \cdot 3^v = 27$, $\mathfrak{P}$ is normalised by a Sylow 2-subgroup $\mathfrak{T}$ of $\mathfrak{G}$. But then since $\mathfrak{P}$ is cyclic, $\mathfrak{T}'$ centralizes $\mathfrak{P}$ and $|\mathbf{C}_{\mathfrak{G}}(j)|$ is therefore divisible by 13. This is not the case, so

$2^u \cdot 3^v = 2^4 \cdot 3^2$. Hence $|\mathbf{N}_{\mathfrak{G}}(\mathfrak{P})| = 3 \cdot 13$. Since $\mathfrak{G}$ is simple, $\mathbf{N}_{\mathfrak{G}}(\mathfrak{P}) \neq \mathbf{C}_{\mathfrak{G}}(\mathfrak{P})$. Hence $\mathbf{C}_{\mathfrak{G}}(\mathfrak{P}) = \mathfrak{P}$. Thus there are four classes of elements of order 13, each of length

$$|\mathfrak{G} : \mathbf{C}_{\mathfrak{G}}(\mathfrak{P})| = \frac{g}{13} = 2^4 \cdot 3^3.$$

The elements $1, j, a^2, a, a^4c$ represent five further classes, since the orders of these elements are distinct. If x is j, a^2, a or a^4c, j is a power of x, so $\mathbf{C}_{\mathfrak{G}}(x) \leq \mathbf{C}_{\mathfrak{G}}(j) = \mathfrak{H}$ and the length of the class containing x is $|\mathfrak{G} : \mathfrak{H}||\mathfrak{H} : \mathbf{C}_{\mathfrak{H}}(a)| = 3^2 \cdot 13|\mathfrak{H} : \mathbf{C}_{\mathfrak{H}}(a)|$. Also a^{-1} is not conjugate to a, since otherwise a, a^{-1} are conjugate in $\mathfrak{H}$; further, the number of conjugates of a^{-1} is $3^2 \cdot 13|\mathfrak{H} : \mathbf{C}_{\mathfrak{H}}(a)|$. It is easy to verify that 728 elements remain, and c must be one of them since the order of c is 3. **q.e.d.**

5.19 Lemma. *$\mathfrak{G}$ has no element of order 9.*

Proof. Suppose that this is false. Then by 5.18, $\mathfrak{G}$ has just one conjugacy class of elements of order 9 and has no elements of order $2 \cdot 9$ or $13 \cdot 9$. Hence if x is an element of order 9, $|\mathbf{C}_{\mathfrak{G}}(x)|$ is not divisible by either 2 or 13. Hence $|\mathbf{C}_{\mathfrak{G}}(x)|$ is either 3^2 or 3^3. In the latter case, x has $16 \cdot 13 = 208$ conjugates, leaving c with $728 - 208 = 520$ conjugates. This is impossible, since 520 does not divide $|\mathfrak{G}|$. Thus $\mathbf{C}_{\mathfrak{G}}(x) = \langle x \rangle$.

Suppose that $\langle x \rangle \leq \mathfrak{P} \in S_3(\mathfrak{G})$. Since $\mathbf{C}_{\mathfrak{P}}(x) = \langle x \rangle$, $\mathfrak{P}$ is non-Abelian of order 27 and exponent 9. Thus $\mathfrak{P}$ has 3 normal cyclic subgroups of order 9 and these are permuted by $\mathbf{N}_{\mathfrak{G}}(\mathfrak{P})/\mathfrak{P}$. Since $|\mathbf{N}_{\mathfrak{G}}(\mathfrak{P})/\mathfrak{P}|$ is not divisible by 3, some subgroup $\langle y \rangle$ of $\mathfrak{P}$ of order 9 is normalised by $\mathbf{N}_{\mathfrak{G}}(\mathfrak{P})$. Thus $\mathbf{N}_{\mathfrak{G}}(\mathfrak{P}) \leq \mathbf{N}_{\mathfrak{G}}(\langle y \rangle)$.

Since y is conjugate to x, $\mathbf{C}_{\mathfrak{G}}(y) = \langle y \rangle$. Thus $|\mathbf{N}_{\mathfrak{G}}(\langle y \rangle)/\langle y \rangle|$ divides $|\mathbf{Aut}\langle y \rangle| = 6$. Thus $|\mathbf{N}_{\mathfrak{G}}(\langle y \rangle)|$ divides $2 \cdot 3^3$ and $|\mathbf{N}_{\mathfrak{G}}(\langle y \rangle) : \mathfrak{P}| \leq 2$. Then $\mathfrak{P} \trianglelefteq \mathbf{N}_{\mathfrak{G}}(\langle y \rangle)$ and $\mathbf{N}_{\mathfrak{G}}(\langle y \rangle) = \mathbf{N}_{\mathfrak{G}}(\mathfrak{P})$. Hence $|\mathbf{N}_{\mathfrak{G}}(\mathfrak{P}) : \mathfrak{P}| \leq 2$. Since $|\mathfrak{G} : \mathbf{N}_{\mathfrak{G}}(\mathfrak{P})| \equiv 1\ (3)$, $|\mathbf{N}_{\mathfrak{G}}(\mathfrak{P}) : \mathfrak{P}| = 1$ and $\mathbf{N}_{\mathfrak{G}}(\langle y \rangle) = \mathbf{N}_{\mathfrak{G}}(\mathfrak{P}) = \mathfrak{P}$. But y and y^{-1} are conjugate, so $y^g = y^{-1}$ for some $g \in \mathfrak{G}$. Thus $y^{g^2} = y$, and g is an element of $\mathbf{N}_{\mathfrak{G}}(\langle y \rangle)$ of even order. This gives a contradiction. **q.e.d.**

5.20 Lemma. *$\mathfrak{G}$ has two conjugacy classes of elements of order 3.*

Proof. It is to be shown that if $x \in \mathfrak{G}$ and x does not lie in any of the 10 classes found in 5.18, then the order of x is 3. If x is of even order, x centralizes an involution, so x is conjugate to an element of $\mathfrak{H}$. Thus by 5.4 and 5.5, x is conjugate to c. Hence x is of odd order. As $\mathbf{C}_{\mathfrak{G}}(y) = \langle y \rangle$

for any element y of order 13, the order of x cannot be divisible by 13. Thus x is a 3-element and by 5.19, the order of x is 3. **q.e.d.**

5.21 Lemma. a) *The number of conjugates of c in $\mathfrak{G}$ is $2^3 \cdot 13$. The other conjugacy class of elements of order 3, represented say by c', contains $2^4 \cdot 3 \cdot 13$ elements.*

b) *The Sylow 3-subgroups of $\mathfrak{G}$ are non-Abelian of order 3^3 and exponent 3.*

c) $|\mathbf{N}_{\mathfrak{G}}(\langle c \rangle)| = 2^2 \cdot 3^3$.

Proof. a) Let z be an element of order 3 in the centre of a Sylow 3-subgroup of $\mathfrak{G}$. Since $\mathbf{C}_{\mathfrak{G}}(x) = \langle x \rangle$ for any element x of $\mathfrak{G}$ of order 13, $|\mathbf{C}_{\mathfrak{G}}(z)|$ is not divisible by 13. Hence $|\mathbf{C}_{\mathfrak{G}}(z)| = 2^u \cdot 3^3$ for some u. If $u \geq 2$, then $\mathbf{C}_{\mathfrak{G}}(z)$ contains an Abelian subgroup $\mathfrak{X}$ of order 12. But then $\mathfrak{X}$ contains an involution and some conjugate of $\mathfrak{X}$ is contained in $\mathfrak{H}$. This is impossible, since $\mathfrak{H}$ has no Abelian subgroup of order 12. Hence $u \leq 1$. If $u = 0$, $|\mathfrak{G} : \mathbf{C}_{\mathfrak{G}}(z)| = 2^4 \cdot 13 = 208$. But then by 5.18, the other class C' of elements of order 3 contains $728 - 208 = 520$ elements, and this is impossible, since 520 does not divide $|\mathfrak{G}|$. Thus $u = 1$, z has $2^3 \cdot 13 = 104$ conjugates and C' contains $624 = 2^4 \cdot 3 \cdot 13$ elements. Hence if $c' \in \mathsf{C}'$, $|\mathbf{C}_{\mathfrak{G}}(c')| = 3^2$. Since $j \in \mathbf{C}_{\mathfrak{G}}(c)$, c lies in the class containing z.

b) Since $|\mathbf{C}_{\mathfrak{G}}(c')| = 3^2$, c' does not lie in the centre of any Sylow 3-subgroup of $\mathfrak{G}$. Hence the Sylow 3-subgroups are non-Abelian. They are of exponent 3 by 5.19.

c) By 5.3, $c^b = c^{-1}$, so $|\mathbf{N}_{\mathfrak{G}}(\langle c \rangle)/\mathbf{C}_{\mathfrak{G}}(c)| = 2$. By a), $|\mathbf{C}_{\mathfrak{G}}(c)| = 2 \cdot 3^3$, so $|\mathbf{N}_{\mathfrak{G}}(\langle c \rangle)| = 2^2 \cdot 3^3$. **q.e.d.**

In order to obtain a subgroup of $\mathfrak{G}$ of index 13 and to investigate the corresponding permutation representation of degree 13, we shall need the values of the character χ_6. (In 5.26, it will turn out that $1 + \chi_6$ is the character of a permutation representation of degree 13.)

5.22 Lemma. *The following is the table of values of χ_6.*

	1	j	a^2	a	a^{-1}	$a^4 c$	$d_i\ (1 \leq i \leq 4)$	c	c'
χ_6	12	4	0	0	0	1	-1	3	0

Proof. a) By 5.17, $\chi_6(1) = f_6 = 12$ and $\varepsilon_2 = 1$. By 5.7b) and 5.6d), $\chi_6 + \phi_5$ vanishes on D. By 5.6, j, a^2, a, a^{-1}, $a^4 c$ lie in D, so the values of χ_6 at them may be read off from 5.4.

b) Let d be any element of order 13 and let ω^{n_i} $(i = 1, \ldots, 12)$ be the eigenvalues of $\rho(d)$, where ρ is a representation of $\mathfrak{G}$ with character

χ_6, ω is a primitive 13th root of unity and $0 \leq n_i \leq 12$ $(i = 1, \ldots, 12)$. Since ρ is faithful, $n_i \neq 0$ for at least one i. Hence if

$$\psi(t) = t^{n_1} + \cdots + t^{n_{12}},$$

ψ is a non-constant polynomial and $\psi(\omega) = \chi_6(d)$. Thus ω is a zero of $\psi(t) - \chi_6(d)$. But χ_6 is rational-valued, since it is the only irreducible character of $\mathfrak{G}$ of degree 12. Since the minimal polynomial of ω over $\mathbb{Q}$ is the cyclotomic polynomial, it follows that

$$\psi(t) - \chi_6(d) = k(1 + t + \cdots + t^{12})$$

for some k. Thus $k = 1$ and $\chi_6(d) = -1$.

c) We determine the values of $\chi_6(c) = \alpha$ and $\chi_6(c') = \beta$ from the relations

$$|\mathfrak{G}| = \sum_{x \in \mathfrak{G}} |\chi_6(x)|^2, \quad 0 = \sum_{x \in \mathfrak{G}} \chi_6(x),$$

using 5.18, 5.21 and the values of χ_6 already found. This gives

$$\alpha^2 + 6\beta^2 = 9, \quad \alpha + 6\beta = 3.$$

Hence

$$\alpha^2 + 6\beta^2 = 9 = (\alpha + 6\beta)^2 = \alpha^2 + 12\alpha\beta + 36\beta^2.$$

If $\beta \neq 0$, this gives

$$12\alpha + 30\beta = 0, \quad 5\alpha + 30\beta = 15,$$

whence $\alpha = -\frac{15}{7}$, which is impossible, as the character value α is an algebraic integer. Hence $\beta = 0$ and $\alpha = 3$. **q.e.d.**

5.23 Lemma. a) $\mathbf{N}_{\mathfrak{G}}(\langle c \rangle)$ *contains a subgroup* $\mathfrak{M} = \langle c, \bar{c}, j, b \rangle$ *of order 36 with a normal Sylow 3-subgroup.*

b) $\mathfrak{M}$ *has 8 elements of order 3, 12 elements of order 6 and 15 involutions.*

c) *All elements of* $\mathfrak{M}$ *of order 3 are conjugate to c in* $\mathfrak{G}$.

d) $\frac{1}{|\mathfrak{M}|} \sum_{y \in \mathfrak{M}} \chi_6(y) = 3.$

Proof. a) Let $\mathfrak{Z} = \langle c \rangle$, $\mathfrak{N} = \mathbf{N}_{\mathfrak{G}}(\mathfrak{Z})$. By 5.21, $|\mathfrak{N}| = 2^2 \cdot 3^3$ and $|\mathfrak{N} : \mathbf{C}_{\mathfrak{G}}(\mathfrak{Z})| = 2$. Thus if $\mathfrak{P} \in S_3(\mathbf{C}_{\mathfrak{G}}(\mathfrak{Z}))$, $\mathfrak{P} \trianglelefteq \mathbf{C}_{\mathfrak{G}}(\mathfrak{Z})$. Since $\mathbf{C}_{\mathfrak{G}}(\mathfrak{Z}) \trianglelefteq \mathfrak{N}$, it follows that $\mathfrak{P} \trianglelefteq \mathfrak{N}$. Clearly, $\mathfrak{S} = \langle b, j \rangle \in S_2(\mathfrak{N})$. Now $\mathfrak{P}/\mathfrak{Z}$ is a completely reducible $\mathfrak{S}$-module. But the degree of an irreducible representation of $\mathfrak{S}$ in $GF(3)$ is 1, so $\mathfrak{P}/\mathfrak{Z} = \langle x_1 \mathfrak{Z} \rangle \times \langle x_2 \mathfrak{Z} \rangle$, where for any $s \in \mathfrak{S}$,

$$x_i^s \in x_i^{\pm 1}\mathfrak{Z} \quad (i = 1, 2).$$

If $x_i^j = x_i c^{v_i}$, then $x_i = x_i^{j^2} = x_i c^{2v_i}$, so $v_i \equiv 0\ (3)$ and $x_i \in \mathbf{C}_{\mathfrak{G}}(j) \cap \mathfrak{P} = \mathfrak{Z}$, a contradiction. Thus $x_i^j \in x_i^{-1}\mathfrak{Z}$ $(i = 1, 2)$. If

$$x_1^b = x_1^{u_1} c^{v_1}, \quad x_2^b = x_2^{u_2} c^{v_2},$$

then $[x_1, x_2]^b = [x_1^{u_1}, x_2^{u_2}]$. By 5.21b), $\mathfrak{Z} = \mathbf{Z}(\mathfrak{P}) = \mathfrak{P}'$, so $[x_1, x_2] = c^w \neq 1$, $[x_1, x_2]^b = [x_1, x_2]^{u_1 u_2}$ and $c^b = c^{u_1 u_2}$. Since $c^b = c^{-1}$, $u_1 u_2 \equiv -1\ (3)$. We may suppose the notation so chosen that $u_1 = 1$. Let $\mathfrak{M} = \langle c, x_1 \rangle \mathfrak{S}$. Then $\langle c, x_1 \rangle$ is a completely reducible $\mathfrak{S}$-module and $\langle c \rangle$ is $\mathfrak{S}$-invariant, so there exist $\bar{c}$ such that $\langle c, x_1 \rangle = \langle c \rangle \times \langle \bar{c} \rangle$ and

$$c^j = c, \quad c^b = c^{-1}, \quad \bar{c}^j = \bar{c}^{-1}, \quad \bar{c}^b = \bar{c}.$$

b) Since $\langle c, \bar{c} \rangle$ is the only Sylow 3-subgroup of $\mathfrak{M}$, $\mathfrak{M}$ has 8 elements of order 3. The involutions in $\mathfrak{M}$ are $j\bar{c}^v$, bc^u, $jbc^u\bar{c}^v$ $(u, v = 0, 1, 2)$. Thus $\mathfrak{M}$ has 15 involutions. The remaining 12 elements of $\mathfrak{M}$ are of order 6.

c, d) Let n be the number of elements of $\mathfrak{M}$ conjugate to c in $\mathfrak{G}$. Then

$$\begin{aligned}\sum_{y \in \mathfrak{M}} \chi_6(y) &= \chi_6(1) + 15\chi_6(j) + 12\chi_6(a^4 c) + n\chi_6(c) + (8 - n)\chi_6(c') \\ &= 3n + 84,\end{aligned}$$

by 5.22. Hence $3n + 84 \equiv 0\ (36)$. Thus $n = 8$ and

$$\frac{1}{|\mathfrak{M}|} \sum_{y \in \mathfrak{M}} \chi_6(y) = 3. \qquad \textbf{q.e.d.}$$

5.24 Lemma. a) $\dfrac{1}{|\mathfrak{H}|} \sum_{h \in \mathfrak{H}} \chi_6(h) = 2.$

b) *Put* $\mathfrak{N} = \langle c, j, b \rangle = \mathfrak{M} \cap \mathbf{C}_{\mathfrak{G}}(j)$. *Then*

$$\frac{1}{|\mathfrak{N}|} \sum_{x \in \mathfrak{N}} \chi_6(x) = 4.$$

Proof. a) By 5.22 and 5.4,

$$\begin{aligned}\sum_{h\in\mathfrak{H}} \chi_6(h) &= \chi_6(1) + \chi_6(j) + 6\chi_6(a^2) + 6\chi_6(a) + 6\chi_6(a^{-1}) \\ &\quad + 12\chi_6(b) + 8\chi_6(c) + 8\chi_6(a^4c) \\ &= 96 = 2|\mathfrak{H}|.\end{aligned}$$

b) $\mathfrak{N}$ has 2 elements of order 3 conjugate to c, the 7 involutions j, bc^u, jbc^u $(u = 0, 1, 2)$ and the 2 elements jc, jc^{-1} of order 6. Hence

$$\sum_{x\in\mathfrak{N}} \chi_6(x) = \chi_6(1) + 7\chi_6(j) + 2\chi_6(c) + 2\chi_6(a^4c) = 48 = 4|\mathfrak{N}|.$$

q.e.d.

5.25 Lemma. *$\mathfrak{G}$ has a subgroup $\mathfrak{B}$ of index 13.*

Proof. Let V be a $\mathbb{C}\mathfrak{G}$-module with character χ_6. For any subgroup $\mathfrak{X}$ of $\mathfrak{G}$, let $\mathsf{V}^{\mathfrak{X}}$ be the set of fixed points of V under $\mathfrak{X}$. Since $\mathfrak{M} \cap \mathbf{C}_{\mathfrak{G}}(j) = \mathfrak{N}$, $\mathsf{V}^{\mathfrak{M}} + \mathsf{V}^{\mathfrak{H}} \subseteq \mathsf{V}^{\mathfrak{N}}$. By 5.23 and 5.24, $\dim \mathsf{V}^{\mathfrak{M}} = 3$, $\dim \mathsf{V}^{\mathfrak{N}} = 4$, and $\dim \mathsf{V}^{\mathfrak{H}} = 2$. Hence

$$\begin{aligned}4 = \dim \mathsf{V}^{\mathfrak{N}} \geq \dim(\mathsf{V}^{\mathfrak{M}} + \mathsf{V}^{\mathfrak{H}}) &= \dim \mathsf{V}^{\mathfrak{M}} + \dim \mathsf{V}^{\mathfrak{H}} - \dim(\mathsf{V}^{\mathfrak{M}} \cap \mathsf{V}^{\mathfrak{H}}) \\ &= 5 - \dim(\mathsf{V}^{\mathfrak{M}} \cap \mathsf{V}^{\mathfrak{H}}).\end{aligned}$$

Thus $\dim(\mathsf{V}^{\mathfrak{M}} \cap \mathsf{V}^{\mathfrak{H}}) \geq 1$, and if $\mathfrak{B} = \langle \mathfrak{M}, \mathfrak{H} \rangle$, $\mathsf{V}^{\mathfrak{B}} \neq 0$. Hence $\mathfrak{B} \neq \mathfrak{G}$.

$|\mathfrak{B}|$ is divisible by $2^4 \cdot 3^2$, so $|\mathfrak{G} : \mathfrak{B}|$ divides $3 \cdot 13$. Since $\mathfrak{G}$ is simple, $|\mathfrak{G} : \mathfrak{B}| > 3$. Thus $|\mathfrak{G} : \mathfrak{B}|$ is 13 or $3 \cdot 13$. If $|\mathfrak{G} : \mathfrak{B}| = 3 \cdot 13$, $|\mathfrak{B} : \mathfrak{H}| = 3$, so $|\mathfrak{H} : \mathfrak{J}| \leq 2$, where $\mathfrak{J} = \bigcap_{u\in\mathfrak{B}} \mathfrak{H}^u$. Hence $\mathfrak{J}$ is isomorphic to $GL(2, 3)$ or $SL(2, 3)$ and $\langle j \rangle = \mathbf{Z}(\mathfrak{J})$. But $\mathfrak{J} \trianglelefteq \mathfrak{B}$, so $j \in \mathbf{Z}(\mathfrak{B})$, a contradiction. Therefore $|\mathfrak{G} : \mathfrak{B}| = 13$. **q.e.d.**

5.26 Lemma. *$\mathfrak{G}$ has a doubly transitive permutation representation of degree* 13. *If* B *is the set of fixed points of c, $|\mathsf{B}| = 4$, and if $\mathfrak{U} = \{g \mid g \in \mathfrak{G}, \mathsf{B}g = \mathsf{B}\}$, then $|\mathfrak{G} : \mathfrak{U}| = 13$.*

Proof. By 5.25, $\mathfrak{G}$ has a transitive permutation representation of degree 13. Let π be the character of this representation. By V, 20.2, $\pi = 1_{\mathfrak{G}} + \chi$ for some character χ satisfying $(\chi, 1_{\mathfrak{G}})_{\mathfrak{G}} = 0$. Now χ is of degree 12, so we see from the degrees of the irreducible characters of $\mathfrak{G}$ (5.17) that $\chi = \chi_6$. By V, 20.2, $\mathfrak{G}$ is doubly transitive. By 5.22, $|\mathsf{B}| = \pi(c) = 4$.

Denote two elements of B by 1, 2. Thus

$$|\mathfrak{G}_{1,2}| = \frac{2^4 \cdot 3^3 \cdot 13}{13 \cdot 12} = 36.$$

Now $\langle c \rangle = \mathbf{Z}(\mathfrak{P})$ for some $\mathfrak{P} \in S_3(\mathfrak{G})$, so $\mathfrak{P}$ permutes the elements of B among themselves. Either $\mathfrak{P}$ permutes B trivially or the lengths of the orbits of $\mathfrak{P}$ on B are 1, 3. Thus $|\mathfrak{P} : \mathfrak{P}_{1,2}| \leq 3$ and $|\mathfrak{P}_{1,2}| \geq 9$. Since $|\mathfrak{G}_{1,2}| = 36$, $|\mathfrak{P}_{1,2}| = 9$ and $\mathfrak{P}_{1,2} \in S_3(\mathfrak{G}_{1,2})$. By II, 1.13, $\mathbf{N}_{\mathfrak{G}}(\mathfrak{P}_{1,2})$ is doubly transitive on the set of fixed points of $\mathfrak{P}_{1,2}$, which is B. It follows that $\mathfrak{U}$ is doubly transitive on B.

Let $\mathfrak{N} = \mathbf{N}_{\mathfrak{G}}(\langle c \rangle)$. Then $\mathfrak{P} \leq \mathfrak{N} \leq \mathfrak{U}$. Since $\mathfrak{P}_{1,2} \neq \mathfrak{P}$, $\mathfrak{P}$ operates non-trivially on B. Thus if $\mathfrak{K}$ is the kernel of the restriction of $\mathfrak{U}$ to B, $\mathfrak{P} \nleq \mathfrak{K}$. But $\mathfrak{P} \trianglelefteq \mathfrak{N}$, for $|\mathfrak{N}| = 2^2 \cdot 3^3$ and $|\mathfrak{N} : \mathbf{C}_{\mathfrak{G}}(c)| = 2$ (5.21). Thus $\mathfrak{N}\mathfrak{K}/\mathfrak{K}$ is isomorphic to a subgroup of $\mathfrak{S}_4$ which has a normal subgroup of order 3. Thus either $|\mathfrak{N}\mathfrak{K}/\mathfrak{K}| = 3$, or $|\mathfrak{N}\mathfrak{K}/\mathfrak{K}| = 6$ and an odd permutation of B is induced by some element of $\mathfrak{N}\mathfrak{K}$. But since $\mathfrak{U}$ is doubly transitive on B, the group of permutations of B induced by $\mathfrak{U}$ is at least the alternating group $\mathfrak{A}_4$. It follows in either case that $|\mathfrak{U} : \mathfrak{N}\mathfrak{K}|$ is divisible by 4. Hence $|\mathfrak{U}|$ is divisible by $4|\mathfrak{N}| = 2^4 \cdot 3^3$. Thus $|\mathfrak{G} : \mathfrak{U}|$ is 1 or 13. But $\mathfrak{U} \neq \mathfrak{G}$, since $\mathfrak{G}$ is transitive. Therefore $|\mathfrak{G} : \mathfrak{U}| = 13$.

q.e.d.

5.27 Lemma. $\mathfrak{G} \cong PSL(3, 3)$.

Proof. We define an incidence structure $\mathfrak{B}$ as follows.

(1) The points are the 13 symbols permuted by $\mathfrak{G}$ in the doubly transitive representation of $\mathfrak{G}$ defined in 5.26.

(2) The lines are the sets $\mathsf{B}g$ as g runs through $\mathfrak{G}$. (Thus the lines are the sets of fixed points of conjugates of c.)

(3) Incidence is defined by inclusion.

Clearly $\mathfrak{G}$ is faithfully represented as a group of automorphisms of $\mathfrak{B}$. We show that $\mathfrak{B}$ is a projective plane.

Since $\mathfrak{G}$ is doubly transitive, the number of lines through two distinct points is constant, say λ. We count the number of triples (B', P_1, P_2), where B' is a line, and P_1, P_2 are distinct points on B'. By 5.26, there are 13 lines; each contains 4 points and therefore 12 ordered pairs of distinct points. Thus the number of triples is $13 \cdot 12$. On the other hand, there are $13 \cdot 12$ ordered pairs of distinct points altogether, and each one lies on λ lines. The number of triples is thus $13 \cdot 12\lambda$, so $\lambda = 1$. Thus each pair of points lies on just one line.

Again, the number of lines through a point is constant, say s. The number of pairs (B', P) with P on B' is thus both $4 \cdot 13$ and $13s$. Hence $s = 4$ and there are just 4 lines through any point.

Finally, we show that any two distinct lines intersect in at least one (and therefore exactly one) point. Suppose that B', B'' are lines with no common point. Let P be a point on B'' and let P_1, P_2, P_3, P_4 be the four points of B'. From above, there exists a unique line B_i containing P and P_i $(i = 1, 2, 3, 4)$, since P does not lie on B' and is therefore distinct from P_i. Thus $\mathsf{B}_i \neq \mathsf{B}''$, since P_i does not lie on B''. But there are only four lines containing P, so $\mathsf{B}_i = \mathsf{B}_j$ for some $i \neq j$. Thus P_i, P_j both lie on B_i and $\mathsf{B}_i = \mathsf{B}'$. This is impossible, since P is not on B'.

Thus $\mathfrak{B}$ is a projective plane with 4 points on each line. Any such plane is Desarguesian (see LÜNEBURG [3], 1.7) and is therefore the projective plane over $GF(3)$. Thus $\mathfrak{G}$ is faithfully represented as a group of collineations of this plane. Since $|\mathfrak{G}| = |PSL(3, 3)|$, $\mathfrak{G} \cong PSL(3, 3)$.

q.e.d.

Theorem 5.2 is therefore completely proved.

§ 6. Multiply Homogeneous Groups

6.1 Definition. Suppose that Ω is a set and k is a positive integer. We denote the set of all subsets of Ω having k elements by $\Omega^{(k)}$.

If $\mathfrak{G}$ is a group of permutations of Ω, $\mathfrak{G}$ induces a group of permutations on $\Omega^{(k)}$. We say that $\mathfrak{G}$ is *k-homogeneous* if $\mathfrak{G}$ permutes $\Omega^{(k)}$ transitively.

6.2 Lemma. a) *If $\mathfrak{G}$ is k-fold transitive, $\mathfrak{G}$ is k-homogeneous.*

b) *If $\mathfrak{G}$ is k-homogeneous of degree n, then $\mathfrak{G}$ is $(n - k)$-homogeneous.*

c) *If $\mathfrak{G}$ is 2-homogeneous and $|\mathfrak{G}|$ is even, then $\mathfrak{G}$ is doubly transitive.*

d) *If $\mathfrak{G} \neq 1$ is 2-homogeneous, then $\mathfrak{G}$ is primitive.*

Proof. a) and b) are trivial.

c) Since $|\mathfrak{G}|$ is even, $\mathfrak{G}$ contains an involution j. Suppose that j interchanges a and b. Given distinct elements c, d of Ω, there exists $g \in \mathfrak{G}$ such that $\{a, b\}g = \{c, d\}$. Then either g or jg carries a into c and b into d. Hence $\mathfrak{G}$ is doubly transitive.

d) First we show that $\mathfrak{G}$ is transitive. As $\mathfrak{G} \neq 1$, there is an orbit $\Gamma = \{a, b, \dots\}$ of $\mathfrak{G}$ of length at least 2. Now for any c, there exists $g \in \mathfrak{G}$ such that

$$\{a, b\}g = \{a, c\}.$$

Then c is either ag or bg, so $c \in \Gamma$.

Now suppose that $\Delta = \{a, b, \ldots\}$ is a domain of imprimitivity of $\mathfrak{G}$. Again, for any c, there exists $g \in \mathfrak{G}$ such that

$$\{a, b\}g = \{a, c\}.$$

Since $a \in \Delta$, it follows that $\Delta g \cap \Delta \neq \varnothing$, so $\Delta g = \Delta$ and $c \in \Delta$. Thus $\mathfrak{G}$ is primitive. **q.e.d.**

6.3 Lemma (WIELANDT [6]). *Suppose that $k \geq 2$ and that $\mathfrak{G}$ is k-homogeneous of degree n on Ω. Suppose further that $k + p^a - 1 \leq n$ for any prime-power divisor p^a of k. Then $\mathfrak{G}$ is $(k - 1)$-homogeneous.*

Proof. Suppose that B is an orbit of $\mathfrak{G}$ on $\Omega^{(k-1)}$. It is to be proved that B contains all subsets of Ω having $k - 1$ elements. Clearly, it is sufficient to show that if T is a subset of Ω with k elements, then $B \supseteq T^{(k-1)}$. In fact, $|B \cap T^{(k-1)}|$ is independent of T, since $\mathfrak{G}$ is k-homogeneous. Thus, if $|B \cap T^{(k-1)}| = l$, we must prove that $l = k$.

Let p^a be any prime-power divisor of k, and put $s = k + p^a - 1$. By hypothesis, $s \leq n$, so there exists a subset Δ of Ω with s elements. Then Δ has $\binom{s}{k}$ subsets Σ for which $|\Sigma| = k$, so there exist $\binom{s}{k} l$ pairs (Λ, Σ), where $\Lambda \in B$, $\Sigma \subseteq \Delta$, $|\Sigma| = k$ and $\Lambda \subseteq \Sigma$. But if $\Lambda \in B$ and $\Lambda \subseteq \Delta$, there exist $s - k + 1$ elements Σ of $\Delta^{(k)}$ for which $\Lambda \subseteq \Sigma$, so

$$\binom{s}{k} l = (s - k + 1)|B \cap \Delta^{(k-1)}| = p^a|B \cap \Delta^{(k-1)}|.$$

But

$$\binom{s}{k} = \binom{s}{s-k} = \binom{k + p^a - 1}{p^a - 1} = \frac{k + p^a - 1}{1}\,\frac{k + p^a - 2}{2} \cdots \frac{k + 1}{p^a - 1}$$

is prime to p. Hence p^a divides l. Since this is true for all prime-power divisors p^a of k, k divides l. Thus $k = l$, as required. **q.e.d.**

6.4 Examples. a) If $\mathfrak{G}$ is the permutation group $PGL(2, 8)$ or $P\Gamma L(2, 8)$ of degree 9 on $\Omega = GF(8) \cup \{\infty\}$, then $\mathfrak{G}$ is k-homogeneous for all $k = 1, \ldots, 9$.

In fact, $\mathfrak{G}$ is k-fold transitive for $k = 1, 2, 3$. Thus by 6.2, $\mathfrak{G}$ is k-homogeneous for $k = 1, 2, 3, 6, 7, 8, 9$. $\mathfrak{G}$ is 5-homogeneous by 6.3, so $\mathfrak{G}$ is also 4-homogeneous by 6.2.

b) $\mathfrak{M}_{11}$ is k-homogeneous for $k = 1, 2, 3, 4$ and therefore also for $k = 7, 8, 9, 10, 11$. But $\mathfrak{M}_{11}$ is not 5-homogeneous, since $|\mathfrak{M}_{11}| = 11 \cdot 10 \cdot 9 \cdot 8$ is not divisible by $\binom{11}{5} = 11 \cdot 7 \cdot 6$. Thus also $\mathfrak{M}_{11}$ is not 6-homogeneous.

c) For p odd, let $\mathfrak{G}$ be the permutation group $PSL(2, p^f)$ of degree $p^f + 1$ on $\Omega = GF(p^f) \cup \{\infty\}$. For $p^f \equiv 3\ (4)$, $\mathfrak{G}$ is 3-homogeneous, but not for $p^f \equiv 1\ (4)$.

Let $\Lambda = \{0, 1, \infty\}$ and let $\mathfrak{U} = \mathfrak{G}_{(\Lambda)}$. Then $|\mathfrak{G} : \mathfrak{U}|$ is the length of the orbit containing Λ. Hence

$$|\mathfrak{G} : \mathfrak{U}| \leq \binom{p^f + 1}{3}$$

and

$$|\mathfrak{U}| \geq \frac{(p^f + 1)p^f(p^f - 1)}{2} \frac{6}{(p^f + 1)p^f(p^f - 1)} = 3,$$

with equality if and only if $\mathfrak{G}$ is 3-homogeneous.

Let $\mathfrak{V} = (PGL(2, p^f))_{(\Lambda)}$. Since $PGL(2, p^f)$ is sharply triply transitive on Ω, $|\mathfrak{V}| = 6$. Since $\mathfrak{V} \cap \mathfrak{G} = \mathfrak{U}$, $\mathfrak{G}$ is 3-homogeneous if and only if $\mathfrak{V} \nleq \mathfrak{G}$. Now if s, t are defined by $xs = x^{-1}$, $xt = -x + 1$, then $s^2 = t^2 = 1$ and $\mathfrak{V} = \langle s, t \rangle$. But s, t are both of determinant -1, so $\mathfrak{G}$ is 3-homogeneous if and only if -1 is not a square in $GF(p^f)$, that is, if and only if $p^f \equiv 3\ (4)$.

d) Suppose that $p^f \equiv 3\ (4)$, and let $\mathfrak{G}$ be the group of permutations of $GF(p^f)$ of the form $x \rightarrow a^2x + b$, where a, b are in $GF(p^f)$ and $a \neq 0$. Then $\mathfrak{G}$ is not doubly transitive, since

$$|\mathfrak{G}| = \tfrac{1}{2}p^f(p^f - 1) \not\equiv 0\ (2).$$

But $\mathfrak{G}$ is 2-homogeneous. For since -1 is not a square in $GF(p^f)$, no element of $\mathfrak{G}$ interchanges 0 and 1. Thus the stabiliser of the set $\{0, 1\}$ is 1 and the length of the orbit containing $\{0, 1\}$ is $|\mathfrak{G}| = \tfrac{1}{2}p^f(p^f - 1) = |GF(p^f)^{(2)}|$.

6.5 Theorem (KANTOR [2]). *Suppose that $\mathfrak{G} \neq \{1\}$ is 2-homogeneous of degree n and that $\mathfrak{G}$ is not doubly transitive.*

a) *$|\mathfrak{G}|$ is odd.*

b) *$n = p^f \equiv 3\ (4)$ for some odd prime p.*

c) $\mathfrak{G}$ *is similar to a primitive group of semilinear mappings* $x \to a(x\alpha) + b$ *on* $GF(p^f)$ *containing all translations* $x \to x + b$.

Proof. By 6.2c), $|\mathfrak{G}|$ is odd. Thus by the Feit-Thompson theorem, $\mathfrak{G}$ is soluble. By 6.2d), $\mathfrak{G}$ is primitive. By II, 3.2, the degree of $\mathfrak{G}$ is a prime-power p^f and $\mathfrak{G} = \mathfrak{N}\mathfrak{G}_0$, where $\mathfrak{N}$ is a regular, elementary Abelian normal subgroup of $\mathfrak{G}$ and $\mathfrak{N} \cap \mathfrak{G}_0 = 1$. By II, 2.2, $\mathfrak{G}$ is similar to a group $\mathfrak{L}$ of permutations of $\mathfrak{N}$ in which $\mathfrak{G}_0$ corresponds to a subgroup of $GL(f, p)$ and $\mathfrak{N}$ to the group $\mathfrak{T}$ of all translations.

Since $\mathfrak{G}$ is 2-homogeneous, $|\mathfrak{G}|$ is divisible by p^f and $\frac{1}{2}p^f(p^f - 1)$. Thus p is odd and $p^f \equiv 3\ (4)$. Let

$$\mathfrak{H} = \langle \mathfrak{L}, -I \rangle = \mathfrak{T}(\mathfrak{L}_0 \times \langle -I \rangle).$$

Then $\mathfrak{H}$ is 2-homogeneous of even order and is therefore doubly transitive. Clearly $\mathfrak{H}$ is soluble. Since $p^f \equiv 3\ (4)$, p^f is not equal to 3^2, 5^2, 7^2, 11^2, 23^2 or 3^4. We shall see in 7.3 that it follows that $\mathfrak{H}$ is similar to a group of semilinear mappings $x \to a(x\alpha) + b$ on $GF(p^f)$. Thus so is $\mathfrak{G}$. **q.e.d.**

6.6 Remark. KANTOR [2] has also shown that $\mathfrak{G} \neq \{1\}$ is 2-homogeneous but not doubly transitive if and only if $\mathfrak{G}$ is transitive and a stabiliser $\mathfrak{G}_a$ has just 2 orbits on $\Omega - \{a\}$ of equal odd lengths.

6.7 Theorem (LIVINGSTONE, WAGNER [1]). *Suppose that* $\mathfrak{G}$ *is k-homogeneous of degree n. If* $2k \leq n$, $\mathfrak{G}$ *is* $(k - 1)$*-homogeneous.*

Proof. If p^a is a prime-power divisor of k, then

$$p^a + k - 1 \leq 2k - 1 < n.$$

Thus $\mathfrak{G}$ is $(k - 1)$-homogeneous, by 6.3. **q.e.d.**

6.8 Lemma. *Suppose that* $\mathfrak{G}$ *is a permutation group on* Ω, *where* $|\Omega| = n$. *Suppose that* $1 \leq k < n$. *If the stabiliser of any* $k - 1$ *elements of* Ω *is transitive on the remaining elements of* Ω, $\mathfrak{G}$ *is k-fold transitive on* Ω.

Proof. This is proved by induction on k. First we show that $\mathfrak{G}$ is transitive. Suppose that a, b are elements of Ω. Since $k - 1 \leq n - 2$, there exist $c_1, \ldots, c_{k-1}$ in Ω such that a and b are not in $\{c_1, \ldots, c_{k-1}\}$. By hypothesis, there exists $g \in \mathfrak{G}_{c_1, \ldots, c_{k-1}}$ such that $ag = b$. Thus $\mathfrak{G}$ is transitive. There is nothing further to be proved for $k = 1$. If $k > 1$, choose $a \in \Omega$. The stabiliser in $\mathfrak{G}_a$ of any $k - 2$ elements of $\Omega - \{a\}$ is transitive on the remaining elements of $\Omega - \{a\}$, so by the inductive hypothesis, $\mathfrak{G}_a$ is $(k - 1)$-fold transitive on $\Omega - \{a\}$. Hence $\mathfrak{G}$ is k-fold transitive on Ω. **q.e.d.**

6.9 Lemma (GLEASON [1]). *Suppose that $\mathfrak{G}$ is a permutation group on Ω, where $|\Omega| = n$. Suppose that $1 \leq k < n$ and that p is a prime. If, for any k distinct elements $a_1, \ldots, a_k$ of Ω, there is a p-subgroup of $\mathfrak{G}$ the fixed points of which are precisely $a_1, \ldots, a_k$, then $\mathfrak{G}$ is k-fold transitive.*

Proof. Let $a_1, \ldots, a_{k-1}$ be distinct elements of Ω and let $\mathfrak{H} = \mathfrak{G}_{a_1, \ldots, a_{k-1}}$, $\Omega' = \Omega - \{a_1, \ldots, a_{k-1}\}$. Let Δ be an orbit of $\mathfrak{H}$ on Ω'. For $a \in \Delta$, there exists a p-subgroup $\mathfrak{P}$ of $\mathfrak{H}$ such that a is the only fixed point of $\mathfrak{P}$ in Ω'. Thus $\Delta - \{a\}$ is the disjoint union of orbits of $\mathfrak{P}$ of length a power of p greater than 1. Thus $|\Delta| \equiv 1$ (p). On the other hand, if $\Delta \subset \Omega'$, we can choose $b \in \Omega' - \Delta$. If $\mathfrak{P}_1$ is a p-subgroup of $\mathfrak{H}$ for which b is the only fixed point in Ω', Δ is the union of orbits of $\mathfrak{P}_1$ of length divisible by p. Thus $|\Delta| \equiv 0$ (p), a contradiction. Hence $\Delta = \Omega'$ and $\mathfrak{G}_{a_1, \ldots, a_{k-1}}$ is transitive on Ω'. By 6.8, $\mathfrak{G}$ is k-fold transitive. **q.e.d.**

6.10 Lemma. *Suppose that $\mathfrak{G}$ is a permutation group on Ω. Suppose that for some $k > 1$, all stabilisers of $k - 1$ elements of Ω have the same order greater than 1 and that all stabilisers of k elements of Ω have the same order. Then $\mathfrak{G}$ is $(k - 1)$-fold transitive.*

Proof. Suppose that $a_1, \ldots, a_{k-1}$ are distinct elements of Ω. Under $\mathfrak{G}_{a_1, \ldots, a_{k-1}}$, $\Omega - \{a_1, \ldots, a_{k-1}\}$ splits into orbits all of which have the same length

$$|\mathfrak{G}_{a_1, \ldots, a_{k-1}} : \mathfrak{G}_{a_1, \ldots, a_k}| = q.$$

If $q = 1$, $\mathfrak{G}_{a_1, \ldots, a_{k-1}} = 1$, which is not the case by hypothesis. Let p be a prime divisor of q and let $\mathfrak{P}$ be a Sylow p-subgroup of $\mathfrak{G}_{a_1, \ldots, a_{k-1}}$. Thus for any $a_k \in \Omega - \{a_1, \ldots, a_{k-1}\}$, $\mathfrak{P} \not\leq \mathfrak{G}_{a_1, \ldots, a_k}$. Thus $a_1, \ldots, a_{k-1}$ are the only fixed points of $\mathfrak{P}$. By 6.9, $\mathfrak{G}$ is $(k - 1)$-fold transitive. **q.e.d.**

6.11 Theorem (LIVINGSTONE, WAGNER [1]). *Suppose that $\mathfrak{G}$ is a permutation group on Ω, where $|\Omega| = n$. Suppose that $3 \leq k \leq n$. Suppose further that*

(i) *$\mathfrak{G}$ is $(k-2)$-fold transitive but not $(k-1)$-fold transitive on Ω, and*

(ii) *$\mathfrak{G}$ is $(k-1)$-homogeneous and k-homogeneous on Ω.*

Then either $k = n$ and $\mathfrak{G} = \mathfrak{A}_n$ or one of the following occurs.

(1) *$n = 5$, $k = 4$ and $|\mathfrak{G}| = 20$.*

(2) *$n = 6$, $k = 5$, $\mathfrak{G} = PGL(2, 5)$.*

(3) *$n = 9$, $k = 5$, $\mathfrak{G}$ is $PGL(2, 8)$ or $P\Gamma L(2, 8)$.*

Proof. Since $\mathfrak{G}$ is k-homogeneous, the subgroups $\mathfrak{G}_{(\{a_1,\dots,a_k\})}$ for distinct elements $a_1, \dots, a_k$ of Ω, are all conjugate in $\mathfrak{G}$. Thus the $\mathfrak{G}_{(\{a_1,\dots,a_k\})}$ are all similar permutation groups, and

$$|\mathfrak{G}_{(\{a_1,\dots,a_k\})}| = s, \quad |\mathfrak{G}_{a_1,\dots,a_k}| = m$$

for integers s, m independent of $a_1, \dots, a_k$. Similarly, since $\mathfrak{G}$ is $(k-1)$-homogeneous,

$$|\mathfrak{G}_{(\{a_1,\dots,a_{k-1}\})}| = t, \quad |\mathfrak{G}_{a_1,\dots,a_{k-1}}| = l$$

for integers t, l independent of $a_1, \dots, a_{k-1}$. Further

(1) $$|\mathfrak{G}| = s\binom{n}{k} = t\binom{n}{k-1}.$$

Since $\mathfrak{G}$ is not $(k-1)$-fold transitive, it follows from 6.10 that $l = 1$. But since $\mathfrak{G}$ is $(k-2)$-fold transitive,

$$\begin{aligned} |\mathfrak{G}| &= n(n-1)\cdots(n-k+3)|\mathfrak{G}_{c_1,\dots,c_{k-2}}| \\ &= \binom{n}{k-2}(k-2)!\,|\mathfrak{G}_{c_1,\dots,c_{k-2}}|. \end{aligned}$$

Let $h = |\mathfrak{G}_{c_1,\dots,c_{k-2}}|$. If $a \notin \{c_1, \dots, c_{k-2}\}$, $\mathfrak{G}_{c_1,\dots,c_{k-2},a} = 1$, since $l = 1$. Hence the length of any orbit of $\mathfrak{G}_{c_1,\dots,c_{k-2}}$ on $\Omega - \{c_1, \dots, c_{k-2}\}$ is h. Since $\mathfrak{G}$ is not $(k-1)$-fold transitive, h is a proper divisor of $n - k + 2$. Since $|\mathfrak{G}| = \binom{n}{k-2}(k-2)!\,h$, (1) gives

(2) $$h = t\frac{n-k+2}{(k-1)!}$$

and

(3) $$t = s\frac{n-k+1}{k}.$$

Case 1. $k = 3$.

In this case, $\mathfrak{G}$ is 2-homogeneous but not doubly transitive. Thus $|\mathfrak{G}|$ is odd, by 6.2.

Since h divides $n - k + 2 = n - 1$, it follows from (2) that t divides $(k-1)! = 2$. By (1), $|\mathfrak{G}| = t\binom{n}{2}$, so t is odd, $t = 1$ and $|\mathfrak{G}| = \binom{n}{2}$. By (3), $3 = k = s(n - k + 1) = s(n - 2)$. If $s = 1$, then $n = 5$ and $|\mathfrak{G}| = \binom{5}{2} = 10$ is even. Thus $s = 3$ and $n = 3 = k$. Hence $\mathfrak{G} = \mathfrak{A}_3$.

Case 2. $k = 4$.

In this case, (2) and (3) give

$$6h = (n-2)t, \quad 4t = (n-3)s.$$

Since h is a proper divisor of $n - k + 2 = n - 2$, t is a proper divisor of 6. And $4t^2 \geq (n-3)t = 6h - t \geq 6 - t$, so $t > 1$. Thus $t \in \{2, 3\}$. The solutions of the above equations are then as follows.

α) $t = 2,\quad n = 11,\quad s = 1,\quad h = 3,\quad |\mathfrak{G}| = 330.$

β) $t = 2,\quad n = 5,\quad s = 4,\quad h = 1,\quad |\mathfrak{G}| = 20.$

γ) $t = 3,\quad n = 6,\quad s = 4,\quad h = 2,\quad |\mathfrak{G}| = 60.$

δ) $t = 3,\quad n = 4,\quad s = 12,\quad h = 1,\quad |\mathfrak{G}| = 12.$

β) is the conclusion (1) of the theorem, and δ) is the case $n = k$, $\mathfrak{G} = \mathfrak{A}_n$. Thus it is necessary to exclude the cases α), γ).

In case α), $|\mathfrak{G}| = 330$, so $\mathfrak{G}$ has an element g of order 3. If Λ is a non-trivial orbit of g, $|\Lambda| = 3$ and $g \in \mathfrak{G}_{(\Lambda)}$, so $|\mathfrak{G}_{(\Lambda)}|$ is divisible by 3. But $|\mathfrak{G}_{(\Lambda)}| = t = 2$, a contradiction. In case γ), $|\mathfrak{G}_{c_1,c_2}| = h = 2$, so $\mathfrak{G}$ contains an element of the form $(a, b)(c) \cdots$. But then $3 = t = |\mathfrak{G}_{(\{a,b,c\})}|$ is even, a contradiction.

Case 3. $k = 5$.

In this case, $\mathfrak{G}$ is triply but not quadruply transitive. Suppose first that $\mathfrak{G}$ is sharply triply transitive. Thus $h = 1$, and (2), (3) give

$$24 = t(n-3), \quad 5t = s(n-4).$$

Thus $s(n-4)(n-3) = 120$, so $n \leq 14$. The only solutions are the following.

$$\begin{aligned}
&\alpha)\ t = 4, && s = 4, && n = 9, && |\mathfrak{G}| = 504.\\
&\beta)\ t = 6, && s = 10, && n = 7, && |\mathfrak{G}| = 210.\\
&\gamma)\ t = 8, && s = 20, && n = 6, && |\mathfrak{G}| = 120.\\
&\delta)\ t = 12, && s = 60, && n = 5, && |\mathfrak{G}| = 60.
\end{aligned}$$

As $\mathfrak{G}$ is sharply triply transitive, $n - 1$ is a prime-power (XI, 1.4b)). Thus β) cannot occur. If $n = 9$, $\mathfrak{G}$ is $PGL(2, 8)$, by XI, 2.1, and this is part of (3). If $n = 6$, $\mathfrak{G}$ is $PGL(2, 5)$, by XI, 2.6, and this is (2). Finally, if $n = 5$, we have the case $n = k$ and $\mathfrak{G} = \mathfrak{A}_n$.

If $\mathfrak{G}$ is not sharply triply transitive, then since $\mathfrak{G}_{c_1,c_2,c_3,a} = 1$, it follows from 3.6 that $\mathfrak{G}$ is one of the groups $\mathfrak{A}_6$, $\mathfrak{M}_{11}$ or $P\Gamma L(2, 2^p)$ (p prime) in its natural permutation representation. Since $\mathfrak{G}$ is not 4-fold transitive, $\mathfrak{G} = P\Gamma L(2, 2^p)$. Thus $n = 2^p + 1$ and

$$|\mathfrak{G}| = (2^p + 1)2^p(2^p - 1)p.$$

The order h of the stabiliser of any three points is thus p. Thus by (2),

$$24p = t(2^p - 2).$$

Thus

$$12p = t(2^{p-1} - 1).$$

For $p > 7$, $2^{p-1} - 1 > 12p$. As $2^6 - 1 = 63$ does not divide 84, $p \neq 7$. If $p = 5$, then $t = 4$, but this is impossible, since we would have $n = 33$ and $4 = t = s\frac{29}{5}$. Thus $p \leq 3$. The case $p = 2$ is impossible, since $P\Gamma L(2, 4) \cong \mathfrak{S}_5$ is 4-fold transitive. Thus $p = 3$, which is part of (3).

Case 4. $k = 6$.

$\mathfrak{G}$ is 4-fold transitive but not 5-fold transitive, and the stabiliser of any 5 points is 1. Thus by 3.3 and 3.6, $\mathfrak{G}$ is one of the groups $\mathfrak{S}_4$, $\mathfrak{S}_5$, $\mathfrak{A}_6$, $\mathfrak{M}_{11}$, $\mathfrak{A}_7$, $\mathfrak{S}_6$ or $\mathfrak{M}_{12}$. Excluding the cases when $n < k$ or $\mathfrak{G}$ is 5-fold transitive, there remain only $\mathfrak{A}_6$ and $\mathfrak{M}_{11}$. But $\mathfrak{M}_{11}$ is not 5-homogeneous, since $11 \cdot 10 \cdot 9 \cdot 8$ is not divisible by $\dfrac{11 \cdot 10 \cdot 9 \cdot 8 \cdot 7}{5!}$.

Case 5. $k = 7$.

By 3.4 and 3.6, $\mathfrak{G}$ is one of $\mathfrak{S}_5$, $\mathfrak{S}_6$, $\mathfrak{A}_7$, $\mathfrak{M}_{12}$, $\mathfrak{A}_8$, $\mathfrak{S}_7$. The only possibilities are $\mathfrak{A}_7$ and $\mathfrak{M}_{12}$. $\mathfrak{M}_{12}$ is not 6-homogeneous, since $12 \cdot 11 \cdot 10 \cdot 9 \cdot 8$ is not divisible by $\dfrac{12 \cdot 11 \cdot 10 \cdot 9 \cdot 8 \cdot 7}{6!}$.

Case 6. $k \geq 8$.

Theorems 3.5 and 3.6 now show that $\mathfrak{G}$ is one of $\mathfrak{S}_{k-2}$, $\mathfrak{S}_{k-1}$, $\mathfrak{A}_k$, $\mathfrak{A}_{k+1}$ or $\mathfrak{S}_k$. Only $\mathfrak{A}_k$ can occur. **q.e.d.**

6.12 Theorem (LIVINGSTONE, WAGNER [1]). *Suppose that $\mathfrak{G}$ is k-homogeneous on Ω, where $|\Omega| = n$. If $4 \leq 2k \leq n$, then $\mathfrak{G}$ is $(k - 1)$-fold transitive.*

Proof. For $k = 2$ the statement follows from 6.2d).

For $k > 2$, $\mathfrak{G}$ is $(k - 1)$-homogeneous, by 6.7. By the inductive hypothesis, $\mathfrak{G}$ is $(k - 2)$-fold transitive. By 6.11, $\mathfrak{G}$ is $(k - 1)$-fold transitive, since $2k \leq n$. **q.e.d.**

We shall see in 6.16 that a k-homogeneous group is k-fold transitive, provided that $2k \leq n$ and $k \geq 5$.

We can now easily obtain the oldest result on multiply homogeneous permutation groups.

6.13 Theorem (BEAUMONT, PETERSON [1]). *Suppose that $\mathfrak{G}$ is a permutation group of degree n and that $\mathfrak{G}$ is k-homogeneous for all $k = 1, \ldots, n$. Then one of the following holds.*

(1) $\mathfrak{G} = \mathfrak{A}_n$ *or* $\mathfrak{G} = \mathfrak{S}_n$.

(2) $n = 5$, $|\mathfrak{G}| = 20$.

(3) $n = 6$, $\mathfrak{G} = PGL(2, 5)$.

(4) $n = 9$, $\mathfrak{G}$ *is* $PGL(2, 8)$ *or* $P\Gamma L(2, 8)$.

Proof. If $\mathfrak{G}$ is $(n - 2)$-fold transitive, $\mathfrak{G}$ is $\mathfrak{A}_n$ or $\mathfrak{S}_n$. Otherwise, there exists an integer k such that $3 \leq k < n$ and $\mathfrak{G}$ is $(k - 2)$-fold transitive but not $(k - 1)$-fold transitive. The assertion thus follows from 6.11. **q.e.d.**

All the groups in the conclusion of the theorem actually do satisfy the hypothesis. In case (2), $\mathfrak{G}$ is doubly transitive and hence k-homogeneous for $k = 1, 2, 3, 4, 5$. Similarly, in case (3), $\mathfrak{G}$ is triply transitive. Case (4) was handled in 6.4a).

6.14 Lemma. *Suppose that $\mathfrak{G}$ is k-fold transitive on Ω, where $k < |\Omega|$. Suppose that Δ is a minimal subset of Ω such that $|\Delta| > k$ and the group $\mathfrak{H}$ of permutations of Δ induced by $\mathfrak{G}_{(\Delta)}$ is k-fold transitive on Δ. Then for any $k + 1$ distinct elements $a_1, \dots, a_{k+1}$ of Δ, $\mathfrak{H}_{a_1,\dots,a_{k+1}} = 1$.*

Proof. Suppose that there exist distinct elements $a_1, \dots, a_{k+1}$ of Δ such that $\mathfrak{K} = \mathfrak{H}_{a_1,\dots,a_{k+1}} \neq 1$. Let p be a prime divisor of $|\mathfrak{K}|$. Suppose that $\mathfrak{P}^* \in S_p(\mathfrak{K})$ and $\mathfrak{P}^* \leq \mathfrak{P} \in S_p(\mathfrak{H}_{a_1,\dots,a_k})$. Then $1 \neq \mathfrak{P}^* = \mathfrak{P} \cap \mathfrak{K} = \mathfrak{P}_{a_{k+1}}$. By II, 1.13c), $\mathbf{N}_{\mathfrak{H}}(\mathfrak{P})$ is k-fold transitive on the set Γ of all fixed points of $\mathfrak{P}$ (in Δ). Also, $\Gamma \subset \Delta$, since $\mathfrak{P} \neq 1$. Thus by minimality of Δ, $|\Gamma| = k$ and $\Gamma = \{a_1, \dots, a_k\}$. Hence if c is an element of an orbit of $\mathfrak{P}$ on $\Delta - \{a_1, \dots, a_k\}$ of minimal length, $\mathfrak{P}_c < \mathfrak{P}$ and $|\mathfrak{P} : \mathfrak{P}_c| \leq |\mathfrak{P} : \mathfrak{P}_{a_{k+1}}|$. Thus $1 \neq |\mathfrak{P}_{a_{k+1}}| \leq |\mathfrak{P}_c|$ and $\mathfrak{P}_c \neq 1$. Hence if Λ is the set of fixed points of $\mathfrak{P}_c$, $\Lambda \subset \Delta$. But also $|\Lambda| \geq k + 1$, so by minimality of Δ, the group of permutations of Λ induced by $\mathfrak{H}_{(\Lambda)}$ is not k-fold transitive. By 6.9, there exist k distinct points $b_1, \dots, b_k$ of Λ such that any p-subgroup of $\mathfrak{H}_{(\Lambda),b_1,\dots,b_k}$ has a further fixed point in Λ.

Now $\mathfrak{P}_c \leq \mathfrak{H}_{b_1,\dots,b_k}$. Suppose that $\mathfrak{P}_c \leq \mathfrak{S} \in S_p(\mathfrak{H}_{b_1,\dots,b_k})$. Since $\mathfrak{H}$ is k-fold transitive, $\mathfrak{H}_{b_1,\dots,b_k}$ is conjugate to $\mathfrak{H}_{a_1,\dots,a_k}$, so $\mathfrak{S}$ is conjugate to $\mathfrak{P}$ in $\mathfrak{H}$. Since $\mathfrak{P}_c < \mathfrak{P}$, $\mathfrak{P}_c < \mathfrak{S}$. Let $\mathfrak{Q}$ be a subgroup of $\mathfrak{S}$ for which $|\mathfrak{Q} : \mathfrak{P}_c| = p$. Then $\mathfrak{P}_c \trianglelefteq \mathfrak{Q}$, so $\mathfrak{Q} \leq \mathfrak{H}_{(\Lambda)}$ and $\mathfrak{Q} \leq \mathfrak{H}_{(\Lambda),b_1,\dots,b_k}$. Hence from above, there exists $b \in \Lambda - \{b_1, \dots, b_k\}$ such that $\mathfrak{Q} \leq \mathfrak{H}_b$. Then the length of the orbit of $\mathfrak{S}$ containing b is

$$|\mathfrak{S} : \mathfrak{S}_b| \leq |\mathfrak{S} : \mathfrak{Q}| = \frac{|\mathfrak{P}|}{p|\mathfrak{P}_c|} = \frac{1}{p}|\mathfrak{P} : \mathfrak{P}_c|.$$

Since $\mathfrak{S}$ is conjugate to $\mathfrak{P}$ and c lies in an orbit of $\mathfrak{P}$ of minimal length greater than 1, $\mathfrak{S}_b = \mathfrak{S}$. But then, $\mathfrak{P}$ has a fixed point outside Γ, contrary to the above. **q.e.d.**

6.15 Lemma. *Suppose that $\mathfrak{G}$ is k-fold transitive on Ω, where $4 \leq k < |\Omega|$. Then there is a subset Δ of Ω such that $|\Delta| = k + 1$ and the group of permutations of Δ induced by $\mathfrak{G}_{(\Delta)}$ contains the alternating group $\mathfrak{A}_{k+1}$.*

Proof. Let Γ be a minimal subset of Ω such that $|\Gamma| > k$ and the group $\mathfrak{H}$ of permutations of Γ induced by $\mathfrak{G}_{(\Gamma)}$ is k-fold transitive. By 6.14, $\mathfrak{H}_{a_1,\dots,a_{k+1}} = 1$ for any distinct elements $a_1, \dots, a_{k+1}$ of Γ.

First suppose that $k = 4$. Since $|\Gamma| > 4$, it follows from 3.3 and 3.6 that $\mathfrak{H}$ is one of the groups $\mathfrak{S}_5$, $\mathfrak{A}_6$, $\mathfrak{M}_{11}$, $\mathfrak{S}_6$, $\mathfrak{A}_7$ or $\mathfrak{M}_{12}$. The assertion of the theorem follows at once if $\mathfrak{H}$ is $\mathfrak{A}_6$ or a 5-fold transitive group, for then $\mathfrak{H}_{(\Delta)}$ induces at least $\mathfrak{A}_5$ on any Δ for which $|\Delta| = 5$. Thus we may suppose that $\mathfrak{H}$ is $\mathfrak{M}_{11}$. Then by 1.3, we may suppose that $\Gamma = GF(3^2) \cup \{v, \infty\}$ and $\mathfrak{H} = \langle M(3^2), s_2 \rangle$, where

$$\infty s_2 = v, \quad v s_2 = \infty, \quad (x + jy)s_2 = x - jy \quad (x, y \in GF(3)).$$

Let $\Delta = GF(3) \cup \{v, \infty\}$, and let $\mathfrak{K}$ be the permutation group on Δ induced by $\mathfrak{H}_{(\Delta)}$. Since $GF(3)^\times \subseteq GF(3^2)^{\times 2}$, $M(3^2)$ contains all permutations of the form

$$x \to \frac{cx + d}{ex + f}, \quad v \to v$$

for c, d, e, f in $GF(3)$ and $cf - de \neq 0$, and these permutations leave Δ fixed. Thus $\mathfrak{K}_v$ contains all such permutations on Δ. Also $\mathfrak{K}$ contains the restriction of s_2 to Δ, which is simply the transposition (v, ∞). Thus $\mathfrak{K}$ is transitive on Δ and $\mathfrak{K}_v$ contains $PGL(2, 3) \cong \mathfrak{S}_4$. Hence $\mathfrak{K}$ is $\mathfrak{S}_5$, contrary to the minimality of Γ.

For $k = 5$, it follows from 3.4 and 3.6 that $\mathfrak{H}$ is one of the groups $\mathfrak{S}_6$, $\mathfrak{A}_7$, $\mathfrak{M}_{12}$, $\mathfrak{S}_7$ or $\mathfrak{A}_8$. The theorem is clear if $\mathfrak{H}$ is $\mathfrak{A}_7$ or a 6-fold transitive group, so we suppose that $\mathfrak{H}$ is $\mathfrak{M}_{12}$. By 1.3, we may suppose that $\Gamma = GF(3^2) \cup \{v, w, \infty\}$ and $\mathfrak{H} = \langle \mathfrak{M}_{11}, s_3 \rangle$, where

$$\infty s_3 = \infty, \quad v s_3 = w, \quad w s_3 = v, \quad z s_3 = z^3 \quad (z \in GF(3^2)).$$

Let $\Delta = GF(3) \cup \{v, w, \infty\}$, and let $\mathfrak{K}$ be the permutation group on Δ induced by $\mathfrak{H}_{(\Delta)}$. From above $\mathfrak{K}_w$ is the whole symmetric group $\mathfrak{S}_5$. Since s_3 leaves Δ fixed and induces the transposition (v, w) on it, $\mathfrak{K}$ is transitive and is therefore $\mathfrak{S}_6$. Again this contradicts the minimality of Γ.

For $k \geq 6$, it follows from 3.5 and 3.6 that $\mathfrak{H}$ is one of $\mathfrak{S}_{k+1}$, $\mathfrak{A}_{k+2}$, $\mathfrak{S}_{k+2}$ or $\mathfrak{A}_{k+3}$. The assertion is clear in all cases. **q.e.d.**

This brings us to the following beautiful theorem.

6.16 Theorem (LIVINGSTONE, WAGNER [1]). *Suppose that $\mathfrak{G}$ is a k-homogeneous permutation group on Ω, where $10 \leq 2k \leq |\Omega| = n$. Then $\mathfrak{G}$ is k-fold transitive.*

Proof. By 6.12, $\mathfrak{G}$ is $(k - 1)$-fold transitive. By 6.15, there is a subset Γ of Ω such that $|\Gamma| = k$ and the group of permutations of Γ induced by $\mathfrak{G}_{(\Gamma)}$ contains $\mathfrak{A}_k$. Since $\mathfrak{G}$ is k-homogeneous, for any subset Δ of Ω for which $|\Delta| = k$, the group of permutations of Δ induced by $\mathfrak{G}_{(\Delta)}$ contains

$\mathfrak{A}_k$. We show that it is in fact $\mathfrak{S}_k$. To do this it suffices to display an element of $\mathfrak{G}$ which induces an odd permutation on some k-point subset Δ of Ω.

Since $k - 1 \geq 4$, $\mathfrak{G}$ is 4-fold transitive. Since $|\Omega| \geq 2k \geq 10$, it follows from 3.8 that if the order of the stabiliser of three elements of Ω is odd, then $\mathfrak{G}$ is $\mathfrak{M}_{11}$. But the stabiliser of three elements in $\mathfrak{M}_{11}$ is of order 8. Hence the order of the stabiliser in $\mathfrak{G}$ of three elements is even, and $\mathfrak{G}$ contains an involution j such that the number r of fixed points of j is at least 3. Write $n = r + 2s$; thus $r \geq 3$ and $s \geq 1$. Let

$$j = (a_1) \cdots (a_r)(b_1, c_1) \cdots (b_s, c_s)$$

be the cycle decomposition of j. Let $s' = \min(s, [\frac{1}{2}k])$ and let

$$\Delta = \{a_1, \ldots, a_{r'}, b_1, c_1, \ldots, b_{s'}, c_{s'}\},$$

where $r' + 2s' = k$. If s' is odd, j induces an odd permutation on Δ. If s' is even and $r \geq r' + 2$, then j induces an odd permutation on

$$\{a_1, \ldots, a_{r'+2}, b_1, c_1, \ldots, b_{s'-1}, c_{s'-1}\}.$$

If $r < r' + 2$, then $r - 1 \leq r'$ and

$$\begin{aligned} 2s - 2s' &\geq 2s - 4s' - (r - 3) = (r + 1) + 2(1 - r) - 4s' + 2s \\ &\geq r + 1 - 2r' - 4s' + 2s = n - 2k + 1 \geq 1, \end{aligned}$$

so $s \geq s' + 1$. Then j induces an odd permutation on

$$\{a_1, \ldots, a_{r'-2}, b_1, c_1, \ldots, b_{s'}, c_{s'}, b_{s'+1}, c_{s'+1}\}.$$

Now let $(a_1, \ldots, a_k)$, $(b_1, \ldots, b_k)$ be ordered k-tuples of distinct elements of Ω. Since $\mathfrak{G}$ is k-homogeneous, there exists $g \in \mathfrak{G}$ such that $\{a_1, \ldots, a_k\}g = \{b_1, \ldots, b_k\}$. Since $\mathfrak{G}$ induces the symmetric group $\mathfrak{S}_k$ on any k-element subset of Ω, there exists $h \in \mathfrak{G}$ such that $a_i gh = b_i$ $(i = 1, \ldots, k)$. Hence $\mathfrak{G}$ is k-fold transitive. **q.e.d.**

6.17 Remarks. a) 2-homogeneous groups which are not doubly transitive are discussed in 6.5 and 6.6.

b) (KANTOR [3]). If $\mathfrak{G}$ is 3-homogeneous but not 3-fold transitive, then either (i) $n = q + 1$, $q = p^f \equiv 3$ (4) and $PSL(2, q) \leq \mathfrak{G} < P\Gamma L(2, q)$, (ii) $\mathfrak{G}$ is the group of permutations of $GF(8)$ of the form $x \to ax + b$

($a \neq 0$), or (iii) $\mathfrak{G}$ is the group of all semilinear mappings $x \to a(x\alpha) + b$ ($a \neq 0$) on $GF(8)$ or $GF(32)$.

c) (KANTOR [3]). If $\mathfrak{G}$ is 4-homogeneous but not 4-fold transitive, $\mathfrak{G}$ is one of the groups $PGL(2, 8)$, $P\Gamma L(2, 8)$ or $P\Gamma L(2, 32)$ in its natural representation.

§ 7. Doubly Transitive Soluble Permutation Groups

The group $\Gamma(p^n)$ of all semilinear mappings

$$x \to a(x\alpha) + b \quad (a \neq 0, \alpha \in \mathbf{Aut}\, GF(p^n))$$

on $GF(p^n)$ is doubly transitive and soluble. In this section, we shall generalize earlier results (II, 3.13) and show that with a finite number of exceptions, any doubly transitive soluble permutation group of degree p^n is a subgroup of $\Gamma(p^n)$. To do this, we shall need the theorem of Zsigmondy (IX, 8.3). We shall also need the following lemmas.

7.1 Lemma. *Suppose that $\mathfrak{A}$ is a maximal normal Abelian subgroup of the soluble group $\mathfrak{G}$ and that $\mathbf{C}_{\mathfrak{G}}(\mathfrak{A})$ is non-Abelian. Let $\mathfrak{N}$ be a minimal element of the set of normal non-Abelian subgroups of $\mathfrak{G}$ contained in $\mathbf{C}_{\mathfrak{G}}(\mathfrak{A})$.*

a) *$\mathbf{Z}(\mathfrak{N}) = \mathfrak{A} \cap \mathfrak{N}$ and $\mathfrak{N}/\mathbf{Z}(\mathfrak{N})$ is a chief factor of $\mathfrak{G}$. Thus $\mathfrak{N}/\mathbf{Z}(\mathfrak{N})$ is an elementary Abelian p-group for some prime p. $\mathfrak{N}$ is a p-group of class 2 and p divides $|\mathfrak{A}|$.*

b) *If, further, $\mathfrak{A}$ is cyclic, $|\mathfrak{N}'| = p$ and $|\mathfrak{N} : \mathbf{Z}(\mathfrak{N})|$ is a square. Also, if $p = 2$, the exponent of $\mathfrak{N}$ is 4 and $|\mathbf{Z}(\mathfrak{N})|$ is 2 or 4.*

Proof. a) Let $\mathfrak{M}$ be a normal subgroup of $\mathfrak{G}$ for which $\mathfrak{M} < \mathfrak{N}$. By minimality of $\mathfrak{N}$, $\mathfrak{M}$ is Abelian. Since $[\mathfrak{M}, \mathfrak{A}] = 1$, $\mathfrak{M}\mathfrak{A}$ is Abelian. By maximality of $\mathfrak{A}$, $\mathfrak{M} \leq \mathfrak{A}$. Thus $\mathfrak{M} \leq \mathfrak{A} \cap \mathfrak{N}$.

Since $\mathfrak{N}$ is non-Abelian, $\mathbf{Z}(\mathfrak{N}) < \mathfrak{N}$ and we may take $\mathfrak{M} = \mathbf{Z}(\mathfrak{N})$. It follows that $\mathbf{Z}(\mathfrak{N}) \leq \mathfrak{A} \cap \mathfrak{N}$. Since $[\mathfrak{A}, \mathfrak{N}] = 1$, it is clear that $\mathfrak{A} \cap \mathfrak{N} \leq \mathbf{Z}(\mathfrak{N})$. Thus $\mathbf{Z}(\mathfrak{N}) = \mathfrak{A} \cap \mathfrak{N}$. Hence if $\mathfrak{M} \trianglelefteq \mathfrak{G}$ and $\mathfrak{M} < \mathfrak{N}$, then $\mathfrak{M} \leq \mathbf{Z}(\mathfrak{N})$. Thus $\mathfrak{N}/\mathbf{Z}(\mathfrak{N})$ is a chief factor of $\mathfrak{G}$. Hence $\mathfrak{N}/\mathbf{Z}(\mathfrak{N})$ is an elementary Abelian p-group for some prime p and $\mathfrak{N}$ is nilpotent of class 2. If $\mathfrak{S} \in S_p(\mathfrak{N})$, $\mathfrak{S} \trianglelefteq \mathfrak{G}$. But $\mathfrak{S} \nleq \mathbf{Z}(\mathfrak{N})$, so $\mathfrak{S} = \mathfrak{N}$. Thus $\mathfrak{N}$ is a p-group. Since $\mathbf{Z}(\mathfrak{N}) = \mathfrak{N} \cap \mathfrak{A} \neq 1$, p divides $|\mathfrak{N} \cap \mathfrak{A}|$ and $|\mathfrak{A}|$.

b) Since $\mathbf{Z}(\mathfrak{N}) \leq \mathfrak{A}$, $\mathbf{Z}(\mathfrak{N})$ is cyclic. Hence by III, 13.7, $|\mathfrak{N}'| = p$ and $\mathfrak{N}/\mathbf{Z}(\mathfrak{N})$ is a square.

Suppose that $p = 2$. Since $\mathfrak{N}$ has class 2,

$$(xy)^4 = x^4y^4[y, x]^6$$

for all x, y in $\mathfrak{N}$, by III, 1.3. Since $|\mathfrak{N}'| = 2$, $(xy)^4 = x^4y^4$. Thus the set of solutions of $t^4 = 1$ in $\mathfrak{N}$ is a characteristic subgroup $\Omega_2(\mathfrak{N})$. If $\Omega_2(\mathfrak{N}) < \mathfrak{N}$, then $\Omega_2(\mathfrak{N}) \leq \mathbf{Z}(\mathfrak{N})$ and $\Omega_2(\mathfrak{N})$ is cyclic. But then $\Omega_2(\mathfrak{N})$ is the only subgroup of $\mathfrak{N}$ of order 4, and by III, 8.3, $\mathfrak{N}$ is cyclic. Since $\mathfrak{N}$ is non-Abelian, this is not the case, so $\Omega_2(\mathfrak{N}) = \mathfrak{N}$. Thus $\mathfrak{N}$ is of exponent 4 and the order of the cyclic group $\mathbf{Z}(\mathfrak{N})$ is 2 or 4. **q.e.d.**

In 7.1b), every characteristic Abelian subgroup of $\mathfrak{N}$ is cyclic, so III, 13.10 is applicable.

7.2 Lemma. *Let $\mathfrak{G}$ be a nilpotent group of class 2 and let ρ be a faithful complex irreducible representation of $\mathfrak{G}$ of degree n. Then $|\mathfrak{G} : \mathbf{Z}(\mathfrak{G})| = n^2$.*

Proof. By Schur's lemma, there is a representation λ of $\mathbf{Z}(\mathfrak{G})$ of degree 1 such that $\rho(x) = \lambda(x)I_n$ for every $x \in \mathbf{Z}(\mathfrak{G})$. Thus if x, y are elements of $\mathfrak{G}$,

$$\rho(x^y) = \rho(x[x, y]) = \lambda([x, y])\rho(x).$$

Hence if χ is the character of ρ,

$$\chi(x) = \chi(x^y) = \lambda([x, y])\chi(x).$$

Now if $x \notin \mathbf{Z}(\mathfrak{G})$, there exists $y \in \mathfrak{G}$ such that $[x, y] \neq 1$. Since ρ is faithful, we then have $\lambda([x, y]) \neq 1$. Hence $\chi(x) = 0$ for all $x \in \mathfrak{G} - \mathbf{Z}(\mathfrak{G})$. The orthogonality relations now give

$$|\mathfrak{G}| = \sum_{x \in \mathfrak{G}} |\chi(x)|^2 = \sum_{x \in \mathbf{Z}(\mathfrak{G})} |\chi(x)|^2 = \sum_{x \in \mathbf{Z}(\mathfrak{G})} n^2|\lambda(x)|^2 = n^2|\mathbf{Z}(\mathfrak{G})|.$$

Therefore $n^2 = |\mathfrak{G}/\mathbf{Z}(\mathfrak{G})|$. **q.e.d.**

We now prove the main theorem of this section.

7.3 Theorem (HUPPERT [1]). *Any doubly transitive soluble permutation group of degree p^n is similar to a subgroup of $\Gamma(p^n)$, except possibly when p^n is 3^2, 5^2, 7^2, 11^2, 23^2 or 3^4.*

Proof. Suppose that $\mathfrak{G}$ is a doubly transitive soluble group of degree p^n but that $\mathfrak{G}$ is not a group of semilinear mappings over $GF(p^n)$. By II, 3.5, we may suppose that $\mathfrak{G}$ is a permutation group on the n-dimensional $GF(p)$-vector space $V = V(n, p)$ and that the only minimal normal subgroup $\mathfrak{T}$ of $\mathfrak{G}$ is the group of translations on V. Then $\mathfrak{H} = \mathfrak{G}_0$ is a subgroup of $GL(n, p)$ which permutes transitively the set of non-zero elements of V. In particular, V is an irreducible $\mathfrak{H}$-module and $p^n - 1$ is a divisor of $|\mathfrak{H}|$.

a) If $1 \neq \mathfrak{K} \trianglelefteq \mathfrak{H}$, then V is the direct sum of isomorphic irreducible $\mathfrak{K}$-modules.

By V, 17.3, there is a decomposition $V = W_1 \oplus \cdots \oplus W_k$, where each W_i is the direct sum of isomorphic irreducible $\mathfrak{K}$-modules and the W_i are transitively permuted by $\mathfrak{H}$. If $k > 1$, suppose that $0 \neq w_i \in W_i$ for $i = 1$, 2. Then for each $h \in \mathfrak{H}$, $w_1 h$ lies in one of the W_j, so $w_1 h \neq w_1 + w_2$. This contradicts the transitivity of $\mathfrak{H}$ on $V - \{0\}$, so $k = 1$.

b) Suppose that $\mathfrak{A}$ is a maximal normal Abelian subgroup of $\mathfrak{H}$. Then $\mathfrak{A}$ is cyclic, and V is the direct sum of s isomorphic irreducible $\mathfrak{A}$-modules of dimension $k = n/s$ over $GF(p)$. Here $s > 1$ and $|\mathfrak{A}|$ divides $p^k - 1$. The group $\mathfrak{H}$ is similar to a subgroup of the group of all semilinear mappings on the vector space $V' = V(s, p^k)$ of dimension s over $GF(p^k)$, and $\mathbf{C}_{\mathfrak{H}}(\mathfrak{A})$ corresponds to the $GF(p^k)$-linear mappings on V' which lie in $\mathfrak{H}$. Further, $|\mathfrak{H}/\mathbf{C}_{\mathfrak{H}}(\mathfrak{A})|$ divides k.

By a), V is the direct sum of s isomorphic irreducible $\mathfrak{A}$-modules for some $s \geq 1$. By II, 3.11, $\mathfrak{H}$ is similar to a group of semilinear mappings on V', and $\mathbf{C}_{\mathfrak{H}}(\mathfrak{A})$ corresponds to the group of $GF(p^k)$-linear mappings on V' which lie in $\mathfrak{H}$. Since $\mathfrak{H}$ is not a group of semilinear mappings over $GF(p^n)$, $s > 1$. $\mathfrak{A}$ has a faithful irreducible representation of degree k over $GF(p)$, so by II, 3.10, $\mathfrak{A}$ is cyclic and $|\mathfrak{A}|$ divides $p^k - 1$. Finally $\mathfrak{H}/\mathbf{C}_{\mathfrak{H}}(\mathfrak{A})$ is isomorphic to a subgroup of the group of automorphisms of $GF(p^k)$, so $|\mathfrak{H}/\mathbf{C}_{\mathfrak{H}}(\mathfrak{A})|$ divides k.

c) If $n \neq 2$ and $p^n \neq 2^6$, there exists an element g in $\mathbf{C}_{\mathfrak{H}}(\mathfrak{A})$ of prime order q, where q does not divide n, q divides $p^n - 1$ but q does not divide $p^i - 1$ for all $i < n$. Also V is an irreducible $\langle g \rangle$-module.

For $n \neq 2$ and $p^n \neq 2^6$, it follows from IX, 8.3 that there exists a prime q with the stated properties. Since $p^n - 1$ is a divisor of $|\mathfrak{H}|$, q divides $|\mathfrak{H}|$. If q divides $|\mathfrak{H} : \mathbf{C}_{\mathfrak{H}}(\mathfrak{A})|$, then by b), q divides k. But this is impossible, since k divides n and q does not divide n. Hence q divides $|\mathbf{C}_{\mathfrak{H}}(\mathfrak{A})|$. Let g be an element of $\mathbf{C}_{\mathfrak{H}}(\mathfrak{A})$ of order q. By the Maschke-Schur theorem, V is the direct sum of irreducible $\langle g \rangle$-modules, since $q \neq p$. Let V_1 be an irreducible $\langle g \rangle$-submodule of V on which $\langle g \rangle$ is represented faithfully. Then by II, 3.10, q divides $p^m - 1$, where m is the dimension of V_1. Since $p^i - 1$ is not divisible by q for $i < n$, $m = n$ and $V_1 = V$. Hence V is an irreducible $\langle g \rangle$-module.

d) $\mathfrak{A} < \mathbf{C}_{\mathfrak{H}}(\mathfrak{A})$ and $\mathbf{C}_{\mathfrak{H}}(\mathfrak{A})$ is non-Abelian.

If $\mathfrak{A} = \mathbf{C}_{\mathfrak{H}}(\mathfrak{A})$, then $|\mathfrak{H}| = |\mathfrak{H}/\mathbf{C}_{\mathfrak{H}}(\mathfrak{A})||\mathfrak{A}|$ divides $(p^k - 1)k$. Thus $(p^k - 1)k$ is divisible by $p^n - 1 = p^{ks} - 1$, and k is divisible by $1 + p^k \cdots + p^{k(s-1)}$. But then $s = 1$, contrary to b). Thus $\mathfrak{A} < \mathbf{C}_{\mathfrak{H}}(\mathfrak{A})$, and by maximality of $\mathfrak{A}$, $\mathbf{C}_{\mathfrak{H}}(\mathfrak{A})$ is non-Abelian.

e) $\mathfrak{H}$ has a normal r-subgroup $\mathfrak{N}$ with the following properties.
(1) $\mathfrak{N} \leq \mathbf{C}_{\mathfrak{H}}(\mathfrak{A})$.
(2) $\mathfrak{N}' \leq \mathbf{Z}(\mathfrak{N}) \leq \mathfrak{A}$.
(3) $|\mathfrak{N}'|$ is a prime r and r divides $p^k - 1$.
(4) $\mathfrak{N}/\mathbf{Z}(\mathfrak{N})$ is elementary Abelian of order r^{2m} for some m.
(5) If $r = 2$, $|\mathbf{Z}(\mathfrak{N})|$ is 2 or 4.

By d), the set of non-Abelian normal subgroups of $\mathfrak{H}$ contained in $\mathbf{C}_{\mathfrak{H}}(\mathfrak{A})$ is non-empty and therefore has a minimal element $\mathfrak{N}$. By 7.1, $\mathfrak{N}$ has all the stated properties, since $\mathfrak{A}$ is cyclic and $|\mathfrak{A}|$ divides $p^k - 1$.

f) r^m divides s.

Let $GF(p^{kf})$ be a splitting field for $\mathfrak{N}$ over $\mathsf{K} = GF(p^k)$, and let $\mathsf{V}'' = \mathsf{V}' \otimes_\mathsf{K} GF(p^{kf})$. Thus V'' is an $\mathfrak{N}$-module, and by the Maschke-Schur theorem, V'' is the direct sum of absolutely irreducible $\mathfrak{N}$-modules V''_j. Suppose that the representation of $\mathfrak{N}$ on V''_j is not faithful. Then the kernel $\mathfrak{M}$ of this representation is a non-identity normal subgroup of $\mathfrak{N}$. Therefore $1 \neq \mathfrak{M} \cap \mathbf{Z}(\mathfrak{N}) \leq \mathfrak{A}$, and it follows that there exist $a \in \mathfrak{A}$, $w \in \mathsf{V}''$ such that $wa = w$, $w \neq 0$ and $a \neq 1$. Hence there exists $v \in \mathsf{V}$ such that $va = v$ and $v \neq 0$. But by b), this implies that there is a faithful irreducible representation of $\mathfrak{A}$ in which a has a fixed point. This is impossible by II, 3.10, since such a representation is equivalent to multiplication in a field by non-zero elements.

Hence V'' is the direct sum of faithful absolutely irreducible $\mathfrak{N}$-modules, and it is therefore sufficient to show that the degree of any faithful, absolutely irreducible $\mathfrak{N}$-module is r^m. But since $r \neq p$, this follows from V, 12.9 and 7.2.

g) If $n \neq 2$ and $p^n \neq 2^6$, the element g of $\mathbf{C}_{\mathfrak{H}}(\mathfrak{A})$ introduced in c) does not centralize $\mathfrak{N}/\mathbf{Z}(\mathfrak{N})$.

Since $g \in \mathbf{C}_{\mathfrak{H}}(\mathfrak{A})$ and $\mathbf{Z}(\mathfrak{N}) \leq \mathfrak{A}$, $g \in \mathbf{C}_{\mathfrak{H}}(\mathbf{Z}(\mathfrak{N}))$. Suppose that $g \in \mathbf{C}_{\mathfrak{H}}(\mathfrak{N}/\mathbf{Z}(\mathfrak{N}))$. Now $r \neq q$, since by f), r^m divides s and s divides n, whereas by c), q does not divide n. Hence by I, 4.4, it follows from

$$g \in \mathbf{C}_{\mathfrak{H}}(\mathbf{Z}(\mathfrak{N})) \cap \mathbf{C}_{\mathfrak{H}}(\mathfrak{N}/\mathbf{Z}(\mathfrak{N}))$$

that $g \in \mathbf{C}_{\mathfrak{H}}(\mathfrak{N})$. Thus $\mathfrak{N} \leq \mathbf{C}_{\mathfrak{H}}(g)$ and $\mathfrak{N}$ is faithfully represented on V by elements of $\mathfrak{D} = \mathrm{Hom}_{\langle g \rangle}(\mathsf{V}, \mathsf{V})$. By c), V is an irreducible $\langle g \rangle$-module, so by Schur's lemma, $\mathfrak{D}$ is a division ring. Since $\mathfrak{D}$ is finite, it follows from Wedderburn's theorem that $\mathfrak{D}$ is commutative. Hence $\mathfrak{N}$ is commutative, a contradiction.

h) If $n \neq 2$ and $p^n \neq 2^6$, then $r = 2 \neq p$, $n = s = 2^m$, $k = 1$ and $q = 2^m + 1$. The representation of $\mathfrak{N}$ on V is then absolutely irreducible.

By g), there exists an irreducible $\langle g \rangle$-submodule $\mathfrak{B}$ of $\mathfrak{N}/\mathbf{Z}(\mathfrak{N})$ on which $\langle g \rangle$ is faithfully represented. If $|\mathfrak{B}| = r^{m'}$, then $r^{m'} - 1$ is divisible by q, by II, 3.10; write $r^{m'} = 1 + qx > q$. By c), n is the order of p modulo q, so n divides $q - 1$. By f), r^m divides n, so we may put $q = 1 + r^m y > r^m$. Hence

$$r^{m'} = 1 + x + xyr^m.$$

Since $r^m < q < r^{m'}$, $m < m'$, and so r^m divides $1 + x$. Thus $x \geq r^m - 1$. Also

$$r^{m'} = 1 + qx \geq 1 + (r^m + 1)(r^m - 1) = r^{2m} = |\mathfrak{N}/\mathbf{Z}(\mathfrak{N})| \geq |\mathfrak{B}| = r^{m'},$$

so $q = r^m + 1$ and $x = r^m - 1$. Since q is a prime, $r = 2$. Thus n divides $q - 1 = 2^m$. By f), 2^m divides s. Since $n = sk$, it follows that $n = s = 2^m$ and $k = 1$. By b), $|\mathfrak{A}|$ divides $p - 1$, so $p \neq 2$. Since the faithful absolutely irreducible representations of $\mathfrak{N}$ are of degree 2^m, V is an absolutely irreducible $\mathfrak{N}$-module.

Let $\mathfrak{L} = \mathbf{C}_{\mathfrak{H}}(\mathbf{Z}(\mathfrak{N})) \cap \mathbf{C}_{\mathfrak{H}}(\mathfrak{N}/\mathbf{Z}(\mathfrak{N}))$.

i) $\mathbf{C}_{\mathfrak{H}}(\mathbf{Z}(\mathfrak{N}))/\mathfrak{L}$ is isomorphic to a subgroup of the symplectic group $Sp(2m, r)$.

By III, 13.7, there exists a non-singular skew-symmetric bilinear form f on $\mathfrak{N}/\mathbf{Z}(\mathfrak{N})$ given by

$$[x, y] = c^{f(x\mathbf{Z}(\mathfrak{N}),\, y\mathbf{Z}(\mathfrak{N}))},$$

where c is a generator of $\mathfrak{N}'$. If $h \in \mathbf{C}_{\mathfrak{H}}(\mathbf{Z}(\mathfrak{N}))$,

$$c^{f(x^h\mathbf{Z}(\mathfrak{N}),\, y^h\mathbf{Z}(\mathfrak{N}))} = [x^h, y^h] = [x, y]^h = [x, y] = c^{f(x\mathbf{Z}(\mathfrak{N}),\, y\mathbf{Z}(\mathfrak{N}))},$$

so f is invariant under $\mathbf{C}_{\mathfrak{H}}(\mathbf{Z}(\mathfrak{N}))$. Thus there exists a homomorphism of $\mathbf{C}_{\mathfrak{H}}(\mathbf{Z}(\mathfrak{N}))$ into $Sp(2m, r)$ with kernel $\mathfrak{L}$.

j) If V is an irreducible $\mathfrak{N}$-module, then $|\mathbf{C}_{\mathfrak{H}}(\mathbf{Z}(\mathfrak{N}))|$ divides $(p^k - 1)\,|Sp(2m, r)|\,|\mathbf{Z}(\mathfrak{N})|^{2m}$.

Since V is an irreducible $\mathfrak{N}$-module, it follows from Schur's lemma and Wedderburn's theorem that $\mathrm{Hom}_{\mathfrak{N}}(\mathrm{V}, \mathrm{V})$ is a field. Thus $\mathbf{C}_{\mathfrak{H}}(\mathfrak{N})$ is Abelian. But $\mathbf{C}_{\mathfrak{H}}(\mathfrak{N}) \geq \mathfrak{A}$ and $\mathfrak{A}$ is a maximal normal Abelian subgroup of $\mathfrak{H}$. Thus $\mathbf{C}_{\mathfrak{H}}(\mathfrak{N}) = \mathfrak{A}$ and $|\mathbf{C}_{\mathfrak{H}}(\mathfrak{N})|$ divides $p^k - 1$. By i), $|\mathbf{C}_{\mathfrak{H}}(\mathbf{Z}(\mathfrak{N}))/\mathfrak{L}|$ divides $|Sp(2m, r)|$. Hence it suffices to prove that $|\mathfrak{L} : \mathbf{C}_{\mathfrak{H}}(\mathfrak{N})|$ divides $|\mathbf{Z}(\mathfrak{N})|^{2m}$. Suppose that $\mathfrak{N} = \langle x_1, \ldots, x_{2m}, \mathbf{Z}(\mathfrak{N}) \rangle$. The mapping

$$h \to ([x_1, h], \ldots, [x_{2m}, h]) \quad (h \in \mathfrak{L})$$

is a homomorphism of $\mathfrak{L}$ into the direct product of $2m$ groups each isomorphic to $\mathbf{Z}(\mathfrak{N})$, and the kernel is $\mathbf{C}_{\mathfrak{H}}(\mathfrak{N})$. Thus $|\mathfrak{L} : \mathbf{C}_{\mathfrak{H}}(\mathfrak{N})|$ divides $|\mathbf{Z}(\mathfrak{N})|^{2m}$.

k) If $n = 2$, then p is 3, 5, 7, 11 or 23.

Since $2 = n = sk$ and $s > 1$, $k = 1$. Thus $|\mathfrak{A}|$ divides $p - 1$ and $p \neq 2$. By f), r^m divides $s = 2$, so $r = 2$, $m = 1$. Since $\mathfrak{N}$ is non-Abelian and $\dim \mathsf{V} = 2$, V is an irreducible $\mathfrak{N}$-module. Since $k = 1$, $\mathfrak{A}$ and hence $\mathbf{Z}(\mathfrak{N})$ consist of scalar multiples of the identity mapping; thus $\mathbf{C}_{\mathfrak{H}}(\mathbf{Z}(\mathfrak{N})) = \mathfrak{H}$. Hence by j), $|\mathfrak{H}|$ divides

$$(p - 1)|Sp(2, 2)||\mathbf{Z}(\mathfrak{N})|^2 = 6(p - 1)|\mathbf{Z}(\mathfrak{N})|^2.$$

Since $p^2 - 1$ divides $|\mathfrak{H}|$, $p + 1$ divides $6|\mathbf{Z}(\mathfrak{N})|^2$. By e), $|\mathbf{Z}(\mathfrak{N})|$ is 2 or 4, so $p + 1$ divides $6 \cdot 2^4$. If $p \equiv 1$ (4), it follows that $p = 5$. If $p \equiv -1$ (4), $|\mathbf{Z}(\mathfrak{N})| = 2$, since $\mathbf{Z}(\mathfrak{N}) \leq \mathfrak{A}$ and $|\mathfrak{A}|$ divides $p - 1$. Thus $p + 1$ divides 24 and p is 3, 7, 11 or 23.

l) If $n > 2$, either $p^n = 3^4$ or $p^n = 2^6$.

Suppose that $n > 2$ and $p^n \neq 2^6$. By h), $k = 1$, $n = s = 2^m$, $r = 2$ and V is an absolutely irreducible $\mathfrak{N}$-module. Thus $|\mathfrak{A}|$ divides $p - 1$ and $p \neq 2$. Since $\mathbf{Z}(\mathfrak{N}) \leq \mathfrak{A}$ and $\mathfrak{A}$ consists of scalar multiples of the identity mapping, $\mathbf{C}_{\mathfrak{H}}(\mathbf{Z}(\mathfrak{N})) = \mathfrak{H}$. It follows from j) that $p^{2^m} - 1$ divides $(p - 1)|Sp(2m, 2)||\mathbf{Z}(\mathfrak{N})|^{2m}$. Hence $p^{2^{m-1}} + 1$ divides $|Sp(2m, 2)||\mathbf{Z}(\mathfrak{N})|^{2m}$. Since $m > 1$, $p^{2^{m-1}} \equiv 1$ (4), so $p^{2^{m-1}} + 1$ is not divisible by 4. By II, 9.13,

$$|Sp(2m, 2)| = 2^{m^2} \prod_{i=1}^{m} (2^{2i} - 1).$$

It follows that

$$(*) \quad p^{2^{m-1}} + 1 \text{ is a divisor of } 2 \prod_{i=1}^{m} (2^{2i} - 1).$$

For $m = 2$, it follows that $p^2 + 1$ divides $2 \cdot 3^2 \cdot 5 = 90$, so $p = 3$ and $p^n = 3^4$.

For $m > 2$, let l be the greatest integer for which $3^l \leq p$. Since

$$2(2^2 - 1)(2^4 - 1)(2^6 - 1) < 3^8,$$

it follows from $(*)$ that

$$3^{l2^{m-1}} < p^{2^{m-1}} + 1 \leq 3^8 \prod_{i=4}^{m} 2^{2i} < 3^8 \prod_{i=4}^{m} 3^{4i/3} = \prod_{i=1}^{m} 3^{4i/3} = 3^{2(m+m^2)/3}.$$

Thus $l2^{m-1} < 2(m + m^2)/3$, which forces $m \leq 5$ and $l = 1$. Since $p < 3^{l+1}$, p is 3, 5 or 7. By h), $2^m + 1 = q$ is a prime, so the only possibility for $m > 2$ is $m = 4$. It then follows from (∗) that $p^8 + 1$ divides $2\prod_{i=1}^{4}(2^{2i} - 1) = 2 \cdot 3^5 \cdot 5^2 \cdot 7 \cdot 17$. But this is not the case for $p = 3, 5$ or 7.

m) The case $p^n = 2^6$ does not occur.

If $p^n = 2^6$, then $ks = n = 6$. By b), $|\mathfrak{A}|$ divides $2^k - 1$, so $k > 1$. Also $s > 1$, so $k = 2$ or $k = 3$. By f), r^m divides s and by e), r divides $2^k - 1$. This gives $k = 2$, $s = r^m = 3$, $|\mathfrak{A}| = 3$. Since $\mathbf{Z}(\mathfrak{N}) \leq \mathfrak{A}$, $|\mathfrak{N}| = r^{2m}|\mathbf{Z}(\mathfrak{N})| = 3^3$.

Since 7 divides $2^6 - 1$ and $2^6 - 1$ divides $|\mathfrak{H}|$, $\mathfrak{H}$ possesses an element h of order 7. Since the order of the automorphism group of $\mathbf{Z}(\mathfrak{N})$ is 2, h centralizes $\mathbf{Z}(\mathfrak{N})$. Also $|GL(2, 3)| = 48$, so h centralizes $\mathfrak{N}/\mathbf{Z}(\mathfrak{N})$. By I, 4.4, h centralizes $\mathfrak{N}$. Since the degree of any faithful absolutely irreducible representation of $\mathfrak{N}$ is $r^m = 3$, $\mathsf{V}' = \mathsf{V}(3, 2^2)$ is an irreducible $\mathfrak{N}$-module. By b), $\mathbf{C}_{\mathfrak{H}}(\mathfrak{A})$ consists of linear mappings over $GF(2^2)$. Hence $\mathbf{C}_{\mathfrak{H}}(\mathfrak{N})$ consists of linear mappings as $\mathbf{C}_{\mathfrak{H}}(\mathfrak{N}) \leq \mathbf{C}_{\mathfrak{H}}(\mathfrak{A})$. Since $\mathfrak{N}$ operates irreducibly on V', $\mathbf{C}_{\mathfrak{H}}(\mathfrak{N})$ is Abelian and $\mathbf{C}_{\mathfrak{H}}(\mathfrak{N}) = \mathfrak{A}$. Thus $h \in \mathfrak{A}$, which is impossible, since $|\mathfrak{A}| = 3$. **q.e.d.**

7.4 Example (BUCHT [1]). There exist doubly transitive, soluble permutation groups of degree 3^4 and orders $2^5 \cdot 3^4 \cdot 5$, $2^6 \cdot 3^4 \cdot 5$, $2^7 \cdot 3^4 \cdot 5$, which do not consist of mappings on $GF(3^4)$ of the form $x \to a(x\alpha) + b$.

We consider the following matrices over $GF(3)$.

$$A_1 = \begin{pmatrix} 1 & 1 \\ 1 & -1 \end{pmatrix}, \quad A_2 = \begin{pmatrix} 0 & -1 \\ 1 & 0 \end{pmatrix}, \quad A_3 = \begin{pmatrix} 0 & 1 \\ -1 & 0 \end{pmatrix}, \quad A_4 = \begin{pmatrix} 1 & 0 \\ 0 & -1 \end{pmatrix}.$$

We write J for the 2×2 unit matrix and put

$$N_1 = A_1 \otimes J, \quad N_2 = A_2 \otimes J, \quad N_3 = J \otimes A_3, \quad N_4 = J \otimes A_4.$$

Then, denoting the 4×4 unit matrix by I,

$$[N_1, N_3] = [N_1, N_4] = [N_2, N_3] = [N_2, N_4] = I, \quad N_1^4 = I,$$
$$N_1^2 = N_2^2 = N_3^2, \quad N_1^{N_2} = N_1^{-1}, \quad N_3^4 = N_4^2 = I, \quad N_3^{N_4} = N_3^{-1}.$$

Hence $\mathfrak{N} = \langle N_1, N_2, N_3, N_4 \rangle$ is the central product of a quaternion group $\langle N_1, N_2 \rangle$ and a dihedral group $\langle N_3, N_4 \rangle$. Put

$$F = \begin{pmatrix} 1 & 1 & -1 & -1 \\ 0 & 0 & -1 & 1 \\ 0 & 0 & -1 & -1 \\ -1 & 1 & 1 & -1 \end{pmatrix}, \quad G = \begin{pmatrix} -1 & 0 & 1 & 0 \\ 0 & 0 & 0 & -1 \\ -1 & 0 & -1 & 0 \\ 0 & -1 & 0 & 0 \end{pmatrix}.$$

Then the characteristic polynomial of F is $t^4 + t^3 + t^2 + t + 1$, and the following relations hold.

$$\begin{aligned} &F^5 = I, \quad G^4 = N_1^2 N_4, \quad G^{-1}FG = F^2 N_1^{-1} N_2 N_4, \\ &F^{-1}N_1F = N_2N_3N_4, \quad G^{-1}N_1G = N_1N_2N_4, \\ &F^{-1}N_2F = N_1^3, \qquad G^{-1}N_2G = N_2N_4, \\ &F^{-1}N_3F = N_1N_4, \qquad G^{-1}N_3G = N_1N_3N_4, \\ &F^{-1}N_4F = N_1^2N_3N_4, \quad G^{-1}N_4G = N_4. \end{aligned}$$

Put $\mathfrak{H} = \langle \mathfrak{N}, F, G \rangle$. Then $\mathfrak{N} \trianglelefteq \mathfrak{H}$ and $\mathfrak{H}/\mathfrak{N}$ is the split extension of a cyclic group of order 5 by its automorphism group.

We show that the subgroup $\mathfrak{L} = \mathfrak{N}\langle F \rangle$ of $\mathfrak{H}$ permutes transitively the non-zero elements of $\mathsf{V} = \mathsf{V}(4, 3)$. Put $v = (1, 0, 0, 0)$. If $\mathfrak{L}_v$ contains an element of order 5, then 1 is an eigen-value of F. This is not the case, so $\mathfrak{L}_v \leq \mathfrak{N}$. Hence any element of $\mathfrak{L}_v$ is of the form $A \otimes B$, where $A = (a_{ij}) \in \langle A_1, A_2 \rangle$ and $B = (b_{ij}) \in \langle A_3, A_4 \rangle$. Then

$$v = v(A \otimes B) = (a_{11}b_{11}, a_{11}b_{12}, a_{12}b_{11}, a_{12}b_{12}).$$

This forces $a_{12} = b_{12} = 0$. Hence A is $\pm J$ and B is either $\pm J$ or $\pm A_4$. Thus $A \otimes B$ is either $\pm I$ or $\pm N_4$, and it is clear that in fact $A \otimes B$ must be I or N_4. Hence $\mathfrak{L}_v = \langle N_4 \rangle$ and

$$|\mathfrak{L} : \mathfrak{L}_v| = 2^4 \cdot 5 = 3^4 - 1.$$

Therefore $\mathfrak{L}$ is transitive on $\mathsf{V} - \{0\}$.

If we extend the group of translations of V by $\mathfrak{L} = \langle \mathfrak{N}, F \rangle, \langle \mathfrak{N}, F, G^2 \rangle$ or $\langle \mathfrak{N}, F, G \rangle$, we obviously obtain soluble doubly transitive permutation groups of degree 3^4 and orders $2^5 \cdot 3^4 \cdot 5$, $2^6 \cdot 3^4 \cdot 5$, $2^7 \cdot 3^4 \cdot 5$ respectively. It remains to show that these groups are not similar to groups of permutations of $GF(3^4)$ of the form $x \to a(x\alpha) + b$. But the stabiliser of 0 in such a group has a normal Sylow 5-subgroup, whereas $\mathfrak{L}$ does not.

7.5 Remark. a) The following much deeper result (HERING [2]) depends on the classification of finite simple groups.

Let $\mathfrak{H}$ be a subgroup of $GL(n, p)$ which acts transitively on the non-zero vectors of $\mathsf{V} = \mathsf{V}(n, p)$. Let L be a subfield of $\mathrm{Hom}(\mathsf{V}, \mathsf{V})$, containing the identity mapping and maximal with respect to $h^{-1}\mathsf{L}h = \mathsf{L}$ for all $h \in \mathfrak{H}$. If $|\mathsf{L}| = p^m$, then $n = km$ and we can consider V as a vector space $\mathsf{V}(k, p^m)$ of dimension k over L. Then one of the following assertions holds.

(1) $SL(k, p^m) \leq \mathfrak{H} \leq \Gamma L(k, p^m)$.

(2) There exists a non-singular symplectic form on $\mathsf{V}(k, p^m)$, and the symplectic group $Sp(k, p^m)$ is normal in $\mathfrak{H}$.

(3) $k = 6$, $p = 2$ and $\mathfrak{H}$ contains a normal subgroup isomorphic to the Chevalley group $G_2(2^m)$.

(4) $\mathfrak{H}$ contains a normal extraspecial subgroup $\mathfrak{N}$ such that $|\mathfrak{N}| = 2^{n+1}$, $\mathbf{C}_{\mathfrak{H}}(\mathfrak{N}) = \mathbf{Z}(\mathfrak{N})$ and $\mathfrak{H}/\mathfrak{N}\mathbf{Z}(\mathfrak{N})$ is faithfully represented on $\mathfrak{N}/\mathbf{Z}(\mathfrak{N})$.

If $n = 2$, then $k = 2$ and $|\mathsf{L}| = 3, 5, 7, 11, 23$.

If $n > 2$, then $k = n = 4$ and $|\mathsf{L}| = 3$.

(5) If $\mathfrak{H}^{(\infty)}$ is the final term of the derived series of $\mathfrak{H}$, then $\mathfrak{H}^{(\infty)} \cong SL(2, 5)$, $k = 2$, $|\mathsf{L}| = 9, 11, 19, 29, 59$ (see 9.8).

(6) $\mathfrak{H} \cong \mathfrak{A}_6$, $n = 4$, $|\mathsf{L}| = 2$.

(7) $\mathfrak{H} \cong \mathfrak{A}_7$, $n = 4$, $|\mathsf{L}| = 2$ (see II, 2.6).

(8) $\mathfrak{H} \cong SL(2, 13)$, $n = 6$, $|\mathsf{L}| = 3$.

(9) $\mathfrak{H} \cong PSU(3, 3^2)$, $n = 6$, $|\mathsf{L}| = 2$.

b) Let $\mathfrak{G}$ be a doubly transitive permutation group and $\mathfrak{N}$ a minimal normal subgroup of $\mathfrak{G}$. If $\mathfrak{N}$ is regular, we obtain the structure of $\mathfrak{G}$ from a). If $\mathfrak{N}$ is not regular, then by XI, 7.12 $\mathfrak{N}$ is simple non-Abelian and $\mathbf{C}_{\mathfrak{G}}(\mathfrak{N}) = 1$. So we have $\mathfrak{N} \leq \mathfrak{G} \leq \mathbf{Aut}\,\mathfrak{N}$. The problem is now reduced to a question about multiply transitive permutation representations of certain subgroups of the automorphism group of a finite simple group. This question has been solved for Chevalley groups in CURTIS, KANTOR, SEITZ [1]. Their result is the following.

(1) $\mathfrak{N} = PSL(n, q)$ $(n \geq 2)$ in degree $\dfrac{q^n - 1}{q - 1}$.

(2) $\mathfrak{N} = \mathfrak{G} = PSp(2n, 2)$ of degree $2^{n-1}(2^n \pm 1)$, here the stabilisers are orthogonal groups $O^{\pm}(2n, 2)$.

(3) $\mathfrak{N} = PSU(3, q^2)$ of degree $q^3 + 1$ (see II, 10.12).

(4) $\mathfrak{N} = Sz(q)$ of degree $q^2 + 1$ (see XI, 3.3).

(5) $\mathfrak{N} = R(q)$ of degree $q^3 + 1$ (see XI, 13.2).

(6) $\mathfrak{N} = PSL(4, 2) \cong \mathfrak{A}_8$.

(7) Various cases where $\mathfrak{N} \cong PSL(2, q)$ for some special q; these can be obtained easily from II, 8.27.

(8) $\mathfrak{G} \cong P\Gamma L(2, 8)$ of degree 28 (see II, § 8, Exerc. 17).

(9) $\mathfrak{G} \cong PSU(3, 3^2)\ \mathbf{Aut}(GF(3^2))$ and $\mathbf{Aut}\,\mathfrak{G} \cong P\Gamma U(3, 3^2)$.

§ 8. A Characterization of $SL(2, 5)$

If $\mathfrak{G}$ is a group of fixed point free automorphisms of a finite group, the Sylow p-subgroups of $\mathfrak{G}$ have a very special structure, but $\mathfrak{G}$ is not necessarily soluble. For in V, 8.8b), we saw that $SL(2, 5)$ always admits a fixed point free representation on the vector space of dimension 2 over $GF(p)$ when $p^2 \equiv 1$ (5). Our main aim in this section is to prove that $SL(2, 5)$ is the only perfect group with this property (Theorem 8.24). (The group $\mathfrak{G}$ is called *perfect* if $\mathfrak{G}' = \mathfrak{G} \neq 1$.)

8.1 Theorem. *The following conditions on $\mathfrak{G}$ are equivalent.*

a) *$\mathfrak{G}$ is isomorphic to a group of fixed point free automorphisms of some finite group $\mathfrak{H}$.*

b) *$\mathfrak{G}$ has a faithful complex irreducible representation ρ such that if $1 \neq g \in \mathfrak{G}$, 1 is not an eigen-value of $\rho(g)$.*

Proof. a) $\Rightarrow$ b): We may clearly suppose that no proper non identity subgroup of $\mathfrak{H}$ is $\mathfrak{G}$-invariant. Then by V, 8.7, $|\mathfrak{H}|$ is a prime-power p^n and $\mathfrak{H}$ is elementary Abelian. By V, 8.3, p does not divide $|\mathfrak{G}|$. We may think of $\mathfrak{H}$ as a $\mathsf{K}\mathfrak{G}$-module V, where $\mathsf{K} = GF(p)$. Any non-identity element g of $\mathfrak{G}$ operates fixed point freely on V and is therefore represented on V by a linear transformation which does not have the eigen-value 1.

Let L be a $\mathfrak{p}$-adic splitting field for $\mathfrak{G}$ with residue class field $\bar{\mathsf{L}}$ of characteristic p (V, 11.7). Let I be the ring of integers in L. By V, 12.9 there is an $\mathsf{I}\mathfrak{G}$-module W such that $\mathsf{V} \otimes_{\mathsf{K}} \bar{\mathsf{L}} \cong \mathsf{W}/\mathfrak{p}\mathsf{W}$ and W is a free I-module. Suppose that $0 \neq w \in \mathsf{W}$, $1 \neq g \in \mathfrak{G}$ and $wg = w$. If π is a prime element of L, then $w = \pi^r w_0$ for some $r \geq 0$ and $w_0 \in \mathsf{W} - \mathfrak{p}\mathsf{W}$. But then

$$0 = wg - w = \pi^r(w_0 g - w_0).$$

Since W is a free I-module, $w_0 g - w_0 = 0$ and g has a non-zero fixed point on $\mathsf{W}/\mathfrak{p}\mathsf{W} \cong \mathsf{V} \otimes_{\mathsf{K}} \bar{\mathsf{L}}$. This is not the case, so every non-identity element of $\mathfrak{G}$ operates fixed point freely on W and on the $\mathsf{L}\mathfrak{G}$-module $\mathsf{W} \otimes_{\mathsf{I}} \mathsf{L}$. Let ρ be the representation of $\mathfrak{G}$ on an irreducible $\mathsf{L}\mathfrak{G}$-submodule of $\mathsf{W} \otimes_{\mathsf{I}} \mathsf{L}$. Then if $1 \neq g \in \mathfrak{G}$, 1 is not an eigen-value of $\rho(g)$. Since L is a splitting field, ρ remains irreducible in the algebraic closure of L and can be realized in the algebraic closure of the prime subfield of L. Hence ρ can be realized in $\mathbb{C}$.

b) $\Rightarrow$ a): Let ρ be a complex irreducible representation of $\mathfrak{G}$ such that if $1 \neq g \in \mathfrak{G}$, 1 is not an eigen-value of $\rho(g)$. Then ρ can be realized in a finite algebraic extension K of $\mathbb{Q}$. Let p be a prime which does not divide $|\mathfrak{G}|$ and let L be a completion of K with finite residue field $\bar{\mathsf{L}}$ of characteristic p. Then ρ is a representation of $\mathfrak{G}$ in L. By the procedure described in V, 12.7 (Konstantenreduktion), ρ gives rise to a representation σ of $\mathfrak{G}$ on an $\bar{\mathsf{L}}\mathfrak{G}$-module V. The eigen-values of $\sigma(g)$ are the elements $\varepsilon + \mathfrak{p}$, where ε runs through the eigen-values of $\rho(g)$. For $g \neq 1$, $\varepsilon \neq 1$. Suppose that

$$\varepsilon + \mathfrak{p} = 1 + \mathfrak{p}.$$

Let π be a prime element of L and suppose that $\varepsilon - 1 = \pi^r a$, where $r > 0$ and a is a $\mathfrak{p}$-adic integer not in $\mathfrak{p}$. Then

$$1 = \varepsilon^{|\mathfrak{G}|} \equiv 1 + |\mathfrak{G}|\pi^r a \quad (\pi^{2r}).$$

As p does not divide $|\mathfrak{G}|$, this is a contradiction. Hence for $g \neq 1$, 1 is not an eigen-value of $\sigma(g)$, and g operates fixed point freely on the finite group V. Thus a) holds. **q.e.d.**

We shall also need the following lemmas.

8.2 Lemma *(Thompson's transfer lemma). Suppose that $\mathfrak{G}$ is a group with no subgroup of index 2 and that $\mathfrak{K}$ is a subgroup such that $|\mathfrak{G} : \mathfrak{K}|$ is twice an odd number. Then any involution in $\mathfrak{G}$ is conjugate to an involution in $\mathfrak{K}$.*

Proof. Suppose that t is an involution in $\mathfrak{G}$ which is conjugate to no element of $\mathfrak{K}$. Let ρ be the permutation representation of $\mathfrak{G}$ on the set of right cosets of $\mathfrak{K}$ in $\mathfrak{G}$ given by

$$(\mathfrak{K}x)\rho(g) = \mathfrak{K}xg \quad (x \in \mathfrak{G}, g \in \mathfrak{G}).$$

If $\mathfrak{K}x$ is a fixed point of $\rho(t)$, $\mathfrak{K}xt = \mathfrak{K}x$ and $t \in \mathfrak{K}^x$. But this is never the case, so $\rho(t)$ has no fixed points. Hence $\rho(t)$ is the product of $\frac{1}{2}|\mathfrak{G} : \mathfrak{K}|$ transpositions and is an odd permutation. Thus $\{g \mid g \in \mathfrak{G}, \rho(g) \text{ is even}\}$ is a subgroup of $\mathfrak{G}$ of index 2. This is impossible, by hypothesis. **q.e.d.**

8.3 Lemma. *If $\mathfrak{G}$ is a perfect group of order 120, $\mathfrak{G} \cong SL(2, 5)$.*

Proof. If the Sylow 5-subgroup of $\mathfrak{G}$ is normal, then $\mathfrak{G}$ is soluble, contrary to $\mathfrak{G} = \mathfrak{G}'$. Hence $\mathfrak{G}$ has 6 Sylow 5-subgroups and there exists an epimorphism of $\mathfrak{G}$ onto a transitive subgroup $\tilde{\mathfrak{G}}$ of $\mathfrak{S}_6$. Since $\mathfrak{G} = \mathfrak{G}'$, $\tilde{\mathfrak{G}} \leq \mathfrak{A}_6$, and since $|\mathfrak{G}| = 120$, $\tilde{\mathfrak{G}} < \mathfrak{A}_6$. Since $\mathfrak{A}_6$ is simple, $\mathfrak{A}_6$ has no proper subgroup of index less than 6. Hence $|\tilde{\mathfrak{G}}| \leq 60$. Since all groups of order less than 60 are soluble, $|\tilde{\mathfrak{G}}| = 60$ and $\tilde{\mathfrak{G}} \cong \mathfrak{A}_5$ (I, 8.14). Thus $\mathfrak{G}/\mathbf{Z}(\mathfrak{G}) \cong \mathfrak{A}_5$. Since $\mathbf{Z}(\mathfrak{G}) \leq \mathfrak{G}'$, $\mathfrak{G}$ is the representation group of $\mathfrak{A}_5$, which is $SL(2, 5)$ (V, 25.7). **q.e.d.**

We shall need a certain group of order 48 defined as follows.

8.4 Definition. $SL(2, 3)$ is the split extension of a quaternion group of order 8 by the cyclic group of order 3; thus $SL(2, 3) = \langle x, y, a \rangle$ and

$$x^2 = y^2 = (xy)^2, \quad x^4 = a^3 = 1, \quad x^a = y, \quad y^a = xy$$

are defining relations. Hence $SL(2, 3)$ has an automorphism β given by

$$x\beta = x, \quad y\beta = xy, \quad a\beta = a^{-1}xy^{-1},$$

as is easily verified. Further, β^2 is the inner automorphism induced by x^{-1}. Thus there is an extension of $SL(2, 3)$ of the form $\langle SL(2, 3), b \rangle$, where

$$b^2 = x^{-1}, \quad x^b = x, \quad y^b = xy, \quad a^b = a^{-1}xy^{-1}.$$

(see I, 14.8). We denote this extension by $\mathfrak{G}_{48}$. Thus $\mathfrak{G}_{48} = \langle a, b \rangle$ and the following are defining relations:

$$x^2 = y^2 = (xy)^2, a^3 = 1, y^a = xy, b^y = b^{-1}, a^b = a^{-1}xy^{-1},$$

$$b^2 = x^{-1}, x^a = y.$$

Since $\mathfrak{G}'_{48} \geq \langle SL(2, 3)', axy^{-1} \rangle = \langle a, x, y \rangle$, we have

$$\mathfrak{G}'_{48} = \langle a, x, y \rangle \cong SL(2, 3).$$

There is an epimorphism of $\mathfrak{G}_{48}$ onto $\mathfrak{S}_4$ in which

$$a \to (1, 2, 3), \quad b \to (1, 3, 2, 4),$$

so $\mathfrak{G}_{48}/\mathbf{Z}(\mathfrak{G}_{48}) \cong \mathfrak{S}_4$ and $\mathbf{Z}(\mathfrak{G}_{48}) = \langle x^2 \rangle$. Also the Sylow 2-subgroup

$\langle b, y\rangle$ of $\mathfrak{G}_{48}$ is a generalized quaternion group of order 16, so $\mathfrak{G}_{48}$ has only one element of order 2. (In particular, $\mathfrak{G}_{48}$ is not isomorphic to $GL(2, 3)$.)

8.5 Lemma. *Suppose that $\mathfrak{G}$ has a normal subgroup $\mathfrak{N}$ of order 2 and that $\mathfrak{G}$ has only one element of order 2.*

a) *If $\mathfrak{G}/\mathfrak{N} \cong \mathfrak{A}_4$, $\mathfrak{G} \cong SL(2, 3)$.*

b) *If $\mathfrak{G}/\mathfrak{N} \cong \mathfrak{S}_4$, $\mathfrak{G} \cong \mathfrak{G}_{48}$.*

Proof. Suppose that $\mathfrak{N} = \langle j\rangle$. In either case, let $\bar{a}$, $\bar{x}$, $\bar{y}$ be elements of $\mathfrak{G}/\mathfrak{N}$ corresponding to the even permutations $(1, 2, 3)$, $(1, 2)(3, 4)$, $(1, 4)(2, 3)$ respectively. Then $\bar{a} = a\mathfrak{N}$ for some element a of order 3 and $\bar{x} = x\mathfrak{N}$, where $x^2 = j$. Also $(\bar{a}\bar{x})^3 = 1$, so either $(ax)^3 = 1$ or $(ax)^3 = j$. In the latter case, we replace x by xj, then $(ax)^3 = 1$. Putting $y = x^a$, we have $\bar{y} = y\mathfrak{N}$ and, since $\bar{x}\bar{y}$ corresponds to $(1, 3)(2, 4)$,

$$x^2 = y^2 = (xy)^2 = j.$$

Further,

$$y^a = x^{a^2} = axa^2 = x^{-1}a^{-1}x^{-1}a = xa^{-1}xa = xy.$$

Thus if $\mathfrak{G}/\mathfrak{N} \cong \mathfrak{A}_4$, $\mathfrak{G} = \langle a, x\rangle \cong SL(2, 3)$.

If $\mathfrak{G}/\mathfrak{N} \cong \mathfrak{S}_4$, $\mathfrak{G}$ has an element b such that $\bar{b} = b\mathfrak{N}$ corresponds to $(1, 3, 2, 4)$. Then $\bar{a}^{\bar{b}} = \bar{a}^{-1}\bar{x}\bar{y}$ and

$$(a^{-1}xy)^3 = (xy)^a(xy)^{a^2}(xy) = (yxy)(xyx)(xy) = j,$$

so $a^b = a^{-1}xyj$. Hence

$$(ba)^2 = b^2a^ba = b^2a^{-1}xyja = b^2xj = b^2x^{-1}.$$

But $\bar{b}^2 = \bar{x}$, so b^2 is either x or x^{-1}. Since $(ba)^2 \neq 1$, $b^2 = x^{-1}$. Since $(\bar{b}\bar{y})^2 = 1$, $(by)^2 = j$ and $b^y = b^{-1}$. Thus $\mathfrak{G} \cong \mathfrak{G}_{48}$. **q.e.d.**

The finite subgroups of $SL(2, \mathbb{C})$ may be determined as follows.

8.6 Theorem. *A finite subgroup of $SL(2, \mathbb{C})$ is either*

(i) *cyclic,*

(ii) *the non-split extension of a cyclic group $\langle y\rangle$ of even order by the group generated by the automorphism $y \to y^{-1}$, or*

(iii) *isomorphic to $SL(2, 3)$, $\mathfrak{G}_{48}$ or $SL(2, 5)$.*

Proof. Let $\mathfrak{X}$ be a finite subgroup of $SL(2, \mathbb{C})$. Then $\mathfrak{X}$ can be realized in an algebraic number field K, and indeed, the coefficients may be supposed to be in the ring $\mathfrak{o}$ of algebraic integers in K (V, 12.5). Let p be an odd prime which does not divide $|\mathfrak{X}|$, and let $\mathfrak{p}$ be a prime ideal divisor of p. If each coefficient x in any element of $\mathfrak{X}$ is replaced by the corresponding element $x + \mathfrak{p}$ of $\mathfrak{o}/\mathfrak{p}$, we obtain a representation ρ of $\mathfrak{X}$ in $\mathfrak{o}/\mathfrak{p} = GF(p^f)$ for some f.

We show that ρ is faithful. Otherwise the kernel of ρ contains an element g of prime order q, and $q \neq p$. Then there exists an integer $m \geq 1$ such that all the coefficients of $\rho(g) - I$ lie in $\mathfrak{p}^m$ but they are not all in $\mathfrak{p}^{m+1}$. Thus

$$I = \rho(g^q) = \rho(g)^q = (I + \rho(g) - I)^q \equiv I + q(\rho(g) - I) \ (\mathfrak{p}^{m+1}).$$

But this is impossible, since $\mathfrak{p}$ does not divide q and $\mathfrak{p}^{m+1}$ does not divide $\rho(g) - I$.

Thus $\mathfrak{X}$ is isomorphic to a subgroup $\mathfrak{Y}$ of $SL(2, p^f)$, where $|\mathfrak{Y}|$ is not divisible by p. Let $\mathfrak{Z} = \mathbf{Z}(SL(2, p^f)) = \langle -I \rangle$. Then by II, 8.27, $\mathfrak{Y}\mathfrak{Z}/\mathfrak{Z}$ is cyclic, dihedral or isomorphic to one of the groups $\mathfrak{A}_4$, $\mathfrak{S}_4$ or $\mathfrak{A}_5$. Since $-I$ is the only element of $SL(2, p^f)$ of order 2, it follows in all except the cyclic case that $-I \in \mathfrak{Y}$.

If $\mathfrak{Y}\mathfrak{Z}/\mathfrak{Z}$ is cyclic, so is $\mathfrak{Y}$. If $\mathfrak{Y}\mathfrak{Z}/\mathfrak{Z}$ is dihedral, $\mathfrak{Y} = \mathfrak{Y}\mathfrak{Z}$ has a cyclic subgroup of even order and index 2. Since the Sylow 2-subgroup of $\mathfrak{Y}$ is cyclic or generalized quaternion, $\mathfrak{Y}$ is as described in (ii). If $\mathfrak{Y}/\mathfrak{Z}$ is isomorphic to $\mathfrak{A}_4$ or $\mathfrak{S}_4$, the assertion follows from 8.5; if $\mathfrak{Y}/\mathfrak{Z} \cong \mathfrak{A}_5$, it follows from 8.3. **q.e.d.**

The degrees of the faithful irreducible complex representations of $SL(2, 5)$ are 2, 2, 4 and 6, so $SL(2, 5)$ is isomorphic to a subgroup of $SL(2, \mathbb{C})$.

8.7 Lemma. *The group of automorphisms of the alternating group $\mathfrak{A}_5$ is isomorphic to $\mathfrak{S}_5$.*

Proof. It is easy to see that any subgroup of $\mathfrak{A}_5$ of index 5 is the normaliser of a Sylow 2-subgroup of $\mathfrak{A}_5$; thus the set $\mathscr{S}$ of subgroups of index 5 is precisely the set of stabilisers of the symbols. Thus the permutation representation of **Aut** $\mathfrak{A}_5$ on $\mathscr{S}$ is of degree 5. Suppose that α is in the kernel. Any Sylow 3-subgroup of $\mathfrak{A}_5$ is the intersection of two stabilisers, so α leaves every Sylow 3-subgroup fixed. The intersection of the normalisers of $\langle(1, 2, 3)\rangle$ and $\langle(3, 4, 5)\rangle$ is $\langle(1, 2)(4, 5)\rangle$, so α leaves each element of order 2 fixed. Then $\alpha = 1$ and **Aut** $\mathfrak{A}_5$ has a faithful permutation representation of degree 5. Hence **Aut** $\mathfrak{A}_5 \cong \mathfrak{S}_5$. **q.e.d.**

8.8 Lemma. *Suppose that $\mathfrak{G}$ is a group and that $|\mathbf{Z}(\mathfrak{G})| \leq 2$. Let π, σ be sets of primes such that every element of $\mathfrak{G}/\mathbf{Z}(\mathfrak{G})$ is either a π'-element or a σ'-element. If χ is a complex irreducible character of $\mathfrak{G}$, then $\chi(g)$ is rational either for all π-elements g or for all σ-elements g of $\mathfrak{G}$.*

Proof. Suppose that $\mathfrak{G}$ contains a π-element g_0 and a σ-element h_0 for which $\chi(g_0)$, $\chi(h_0)$ are both irrational.

Write $|\mathfrak{G}| = mn$, where all the prime divisors of m are in π and all the prime divisors of n are in π'. Then the order of g_0 is a divisor of m, so $\chi(g_0) \in \mathbb{Q}(\omega^n)$, where ω is a primitive mn-th root of unity. Since $\chi(g_0) \notin \mathbb{Q}$, there is a field automorphism α of $\mathbb{Q}(\omega^n)$ such that $\chi(g_0)\alpha \neq \chi(g_0)$. Suppose that $(\omega^n)\alpha = \omega^{ln}$, where $(l, m) = 1$. Then $\chi(g_0)\alpha = \chi(g_0^l)$, by V, 13.1c). Now there exists an integer k for which

$$k \equiv l \ (m), \quad k \equiv 1 \ (n).$$

Thus $(k, mn) = 1$, so there is an automorphism β of $\mathbb{Q}(\omega)$ such that $\omega\beta = \omega^k$. Then

$$\chi(g_0) \neq \chi(g_0)\alpha = \chi(g_0^l) = \chi(g_0^k) = \chi(g_0)\beta,$$

so $\chi \neq \chi\beta$. We show, however, that $\chi(g)\beta = \chi(g)$ when $\mathbf{Z}(\mathfrak{G})g$ is a π'-element of $\mathfrak{G}/\mathbf{Z}(\mathfrak{G})$. Indeed, we have $g = zg'$, where $z \in \mathbf{Z}(\mathfrak{G})$ and g' is a π'-element of $\mathfrak{G}$; (if $2 \in \pi'$, we can choose $z = 1$). Since χ is irreducible, z is represented by $\pm I$; thus

$$\chi(g)\beta = (\pm\chi(g'))\beta = \pm\chi(g'^k) = \pm\chi(g') = \chi(g).$$

Similarly, there exists an automorphism γ of $\mathbb{Q}(\omega)$ such that $\chi \neq \chi\gamma$ but $\chi(h)\gamma = \chi(h)$ whenever $\mathbf{Z}(\mathfrak{G})h$ is a σ'-element of $\mathfrak{G}/\mathbf{Z}(\mathfrak{G})$. Let

$$\psi = \chi - \chi\beta - \chi\gamma + \chi\beta\gamma.$$

If $\mathbf{Z}(\mathfrak{G})g$ is a π'-element, then

$$\psi(g) = \chi(g) - \chi(g)\beta - (\chi(g) - \chi(g)\beta)\gamma = 0.$$

And if $\mathbf{Z}(\mathfrak{G})h$ is a σ'-element, since $\beta\gamma = \gamma\beta$, we have

$$\psi(h) = \chi(h) - \chi(h)\gamma - (\chi(h) - \chi(h)\gamma)\beta = 0.$$

By hypothesis, every element of $\mathfrak{G}/\mathbf{Z}(\mathfrak{G})$ is either a π'-element or a σ'-element, so $\psi = 0$ and

$$\chi + \chi\beta\gamma = \chi\beta + \chi\gamma.$$

Since decomposition into irreducible characters is unique, it follows that $\chi = \chi\beta$ or $\chi = \chi\gamma$. This is a contradiction. **q.e.d.**

Next we give the elementary properties of perfect groups of fixed point free automorphisms.

8.9 Hypothesis. a) $\mathfrak{G} = \mathfrak{G}'$.

b) For p odd, the Sylow p-subgroups of $\mathfrak{G}$ are cyclic.

c) The Sylow 2-subgroups of $\mathfrak{G}$ are generalized quaternion.

d) $\mathfrak{G}$ has just one element j of order 2.

e) If p, q are distinct primes, any subgroup of $\mathfrak{G}$ of order pq is cyclic.

8.10 Lemma. *If $\mathfrak{G}$ is a perfect group of fixed point free automorphisms of some finite group, $\mathfrak{G}$ satisfies Hypothesis 8.9.*

Proof. a) is clear, and b) follows from V, 8.7b). Since $\mathfrak{G}$ is not soluble, it follows from IV, 2.11 that $\mathfrak{G}$ has at least one non-cyclic Sylow subgroup. Thus c) also follows from V, 8.7b). d) follows from V, 8.18a) and e) from V, 8.12a). **q.e.d.**

8.11 Lemma. *Suppose that $\mathfrak{G}$ satisfies Hypothesis 8.9.*

a) $|\mathbf{Z}(\mathfrak{G})| = 2$ *and* $\mathbf{Z}(\mathfrak{G})$ *is the only non-identity cyclic normal subgroup of* $\mathfrak{G}$.

b) *All the elements of $\mathfrak{G}$ of order* 4 *are conjugate in* $\mathfrak{G}$.

c) *If u is an element of $\mathfrak{G}$ of order* 4, *all the Sylow subgroups of* $\mathbf{C}_{\mathfrak{G}}(u)$ *are cyclic and* $\mathbf{C}_{\mathfrak{G}}(u)$ *is 2-nilpotent.*

d) $|\mathfrak{G}|$ *is divisible by* 3.

Proof. a) By 8.9d), $|\mathbf{Z}(\mathfrak{G})|$ is divisible by 2. Let $\mathfrak{Z}$ be a cyclic normal subgroup of $\mathfrak{G}$ of prime-power order. Since $\mathfrak{G}/\mathbf{C}_{\mathfrak{G}}(\mathfrak{Z})$ is Abelian and $\mathfrak{G} = \mathfrak{G}'$, $\mathbf{C}_{\mathfrak{G}}(\mathfrak{Z}) = \mathfrak{G}$ and $\mathfrak{Z} \leq \mathbf{Z}(\mathfrak{G})$. If $\mathfrak{S}$ is a Sylow subgroup of $\mathfrak{G}$ containing $\mathfrak{Z}$,

$$\mathfrak{Z} \leq \mathfrak{S} \cap \mathbf{Z}(\mathfrak{G}) \cap \mathfrak{G}' \leq \mathfrak{S}' \cap \mathbf{Z}(\mathfrak{S}),$$

by IV, 2.2. By 8.9b), $\mathfrak{S}' = 1$ if $|\mathfrak{S}|$ is odd, and by 8.9c), $|\mathbf{Z}(\mathfrak{S})| = 2$ if $\mathfrak{S}$ is a 2-group. Thus $|\mathfrak{Z}| \leq 2$. In particular, $|\mathbf{Z}(\mathfrak{G})| = 2$.

b) By 8.9d), the element x of $\mathfrak{G}$ is of order 4 if and only if $x\mathbf{Z}(\mathfrak{G})$ is an involution. Let $\mathfrak{H}$ be a Sylow 2-subgroup of $\mathfrak{G}$. By 8.9c), $\mathfrak{H}$ has a cyclic subgroup $\mathfrak{K}$ of index 2, and if a is an element of $\mathfrak{K}$ of order 4, a is conjugate to a^{-1} in $\mathfrak{H}$. If x is of order 4, then by 8.2, $x\mathbf{Z}(\mathfrak{G})$ is conjugate to the unique involution $a\mathbf{Z}(\mathfrak{G})$ in $\mathfrak{K}/\mathbf{Z}(\mathfrak{G})$, so x is conjugate to a or a^{-1}.

c) The centre of a generalized quaternion group has no element of order 4, so the Sylow 2-subgroups of $\mathbf{C}_{\mathfrak{G}}(u)$ are not generalized quaternion. Thus by 8.9, all the Sylow subgroups of $\mathbf{C}_{\mathfrak{G}}(u)$ are cyclic. By IV, 2.8, $\mathbf{C}_{\mathfrak{G}}(u)$ is 2-nilpotent.

d) $\mathfrak{G}$ is not 2-nilpotent and the Sylow 2-subgroups of $\mathfrak{G}$ are metacyclic. Thus by IV, 5.11, $|\mathfrak{G}|$ is divisible by 3. **q.e.d.**

8.12 Lemma. *Suppose that $\mathfrak{G}$ satisfies Hypothesis 8.9, $\mathfrak{U} \leq \mathfrak{G}$, all the Sylow subgroups of $\mathfrak{U}$ are cyclic, $\mathfrak{Q}$ is a nilpotent subgroup of $\mathbf{N}_{\mathfrak{G}}(\mathfrak{U})$ and* $(|\mathfrak{U}|, |\mathfrak{Q}|) = 1$.

a) *$\mathfrak{U}'$ is cyclic, and there exists a cyclic subgroup $\mathfrak{V}$ of $\mathfrak{U}$ for which $\mathfrak{Q} \leq \mathbf{N}_{\mathfrak{G}}(\mathfrak{V})$, $\mathfrak{U} = \mathfrak{U}'\mathfrak{V}$ and $(|\mathfrak{U}'|, |\mathfrak{V}|) = 1$.*

b) *If v is an element of $\mathfrak{V}$ of prime order, $v \in \mathbf{Z}(\mathfrak{U})$.*

c) *If u is an element of $\mathfrak{U}$ of prime order, $\langle u\rangle$ is a characteristic subgroup of $\mathfrak{U}$.*

d) *If $\mathfrak{U}$ is not cyclic, $\mathfrak{V} \cap \mathbf{Z}(\mathfrak{V}\mathfrak{Q}) \neq 1$.*

Proof. a) By IV, 2.11, $\mathfrak{U}'$ and $\mathfrak{U}/\mathfrak{U}'$ are cyclic of coprime orders. Let $\mathfrak{H} = \mathfrak{U}\mathfrak{Q}$. Since $\mathfrak{U}$ and $\mathfrak{Q}$ are soluble, so is $\mathfrak{H}$. Also $\mathfrak{U}'$ is a normal Hall subgroup of $\mathfrak{H}$, so by VI, 1.8, $\mathfrak{Q}$ is contained in a complement $\mathfrak{W}$ of $\mathfrak{U}'$ in $\mathfrak{H}$. Then $\mathfrak{U} = \mathfrak{U}'\mathfrak{V}$, where $\mathfrak{V} = \mathfrak{W} \cap \mathfrak{U}$, and $\mathfrak{U}' \cap \mathfrak{V} = 1$. Thus $\mathfrak{V} \cong \mathfrak{U}/\mathfrak{U}'$ is cyclic.

b) Let q be the order of v. Let $\langle a\rangle$ be a Sylow subgroup of $\mathfrak{U}'$, and let $p^k = |\langle a\rangle|$. By a), $p \neq q$. Then $v \in \mathbf{N}_{\mathfrak{G}}(\langle a^{p^{k-1}}\rangle)$ and $|\langle a^{p^{k-1}}, v\rangle| = pq$. By 8.9e), $\langle a^{p^{k-1}}, v\rangle$ is cyclic. Thus v induces the identity automorphism on $\Omega_1(\langle a\rangle)$. Since $p \neq q$, it follows that v also induces the identity automorphism on $\langle a\rangle$ (IV, 5.12). Hence $\mathbf{C}_{\mathfrak{G}}(v) \geq \langle \mathfrak{U}', \mathfrak{V}\rangle = \mathfrak{U}$ and $v \in \mathbf{Z}(\mathfrak{U})$.

c) If $u \in \mathfrak{U}'$, $\langle u\rangle$ is a characteristic subgroup of $\mathfrak{U}$, since $\mathfrak{U}'$ is cyclic. Otherwise, the order of u divides $|\mathfrak{V}|$ and $u^x \in \mathfrak{V}$ for some $x \in \mathfrak{U}$. Hence by b), $u = u^x \in \mathbf{Z}(\mathfrak{U})$. But $\mathbf{Z}(\mathfrak{U})$ is cyclic, so again $\langle u\rangle$ is a characteristic subgroup of $\mathfrak{U}$.

d) Since $(|\mathfrak{V}|, |\mathfrak{Q}|) = 1$ and $\mathfrak{Q} \leq \mathbf{N}_{\mathfrak{G}}(\mathfrak{V})$, it follows from a theorem of Zassenhaus (III, 13.4) that

$$\mathfrak{V} = (\mathfrak{V} \cap (\mathfrak{V}\mathfrak{Q})') \times (\mathfrak{V} \cap \mathbf{Z}(\mathfrak{V}\mathfrak{Q})).$$

Thus if $\mathfrak{V} \cap \mathbf{Z}(\mathfrak{V}\mathfrak{Q}) = 1$, $\mathfrak{V} \leq (\mathfrak{V}\mathfrak{Q})'$. But now $\mathfrak{H} = \mathfrak{U}\mathfrak{Q}$ is super-soluble, since $\mathfrak{U}'$, $\mathfrak{U}/\mathfrak{U}'$ are cyclic and $\mathfrak{H}/\mathfrak{U}$ is nilpotent. Thus by VI, 9.1b), $\mathfrak{H}'$ is nilpotent. Since $\mathfrak{B} \leq \mathfrak{H}'$ and $(|\mathfrak{B}|, |\mathfrak{U}'|) = 1$, it follows that $[\mathfrak{B}, \mathfrak{U}'] = 1$. Hence $\mathfrak{U} = \mathfrak{U}'\mathfrak{B}$ is cyclic, contrary to hypothesis. **q.e.d.**

8.13 Lemma. *Suppose that $\mathfrak{G}$ satisfies Hypothesis 8.9. If $\mathfrak{M}$ is a maximal subgroup of $\mathfrak{G}$, then $\mathbf{Z}(\mathfrak{M}) = \mathbf{Z}(\mathfrak{G})$. In particular, $|\mathfrak{M}|$ is even.*

Proof. If $\mathbf{Z}(\mathfrak{G}) \not\leq \mathfrak{M}$, then $\mathfrak{G} = \mathfrak{M}\mathbf{Z}(\mathfrak{G})$, so $\mathfrak{M} \geq \mathfrak{G}' = \mathfrak{G}$, a contradiction. Thus $\mathbf{Z}(\mathfrak{G}) \leq \mathbf{Z}(\mathfrak{M})$. Hence if $\mathbf{Z}(\mathfrak{M}) \neq \mathbf{Z}(\mathfrak{G})$, $\mathbf{Z}(\mathfrak{M})$ has an element x of order either an odd prime p or 4. By 8.11, $\langle x \rangle$ is not normal in $\mathfrak{G}$, so $\mathbf{N}_{\mathfrak{G}}(\langle x \rangle) = \mathfrak{M}$ and $\mathbf{C}_{\mathfrak{G}}(x) = \mathfrak{M}$.

Suppose first that the order of x is an odd prime p. If $\mathfrak{P} \in S_p(\mathfrak{G})$ and $x \in \mathfrak{P}$, then $\mathbf{N}_{\mathfrak{G}}(\mathfrak{P}) \leq \mathbf{N}_{\mathfrak{G}}(\langle x \rangle) = \mathfrak{M}$. Hence $\mathbf{N}_{\mathfrak{G}}(\mathfrak{P})$ centralizes $\Omega_1(\mathfrak{P}) = \langle x \rangle$, so by IV, 5.12, $\mathbf{N}_{\mathfrak{G}}(\mathfrak{P}) = \mathbf{C}_{\mathfrak{G}}(\mathfrak{P})$. By IV, 2.6, $\mathfrak{G}$ is p-nilpotent, contrary to $\mathfrak{G} = \mathfrak{G}'$.

Now suppose that the order of x is 4. Then $\mathbf{Z}(\mathfrak{M})$ contains an element of order 4 and the Sylow 2-subgroups of $\mathfrak{M}$ cannot be generalized quaternion. They are therefore cyclic. Now $|\mathfrak{G}|$ is divisible by 8, so $\langle x \rangle$ is normal in a group of order 8. Thus $|\mathfrak{M}|$ is divisible by 8. Let $\langle a \rangle$ be a Sylow 2-subgroup of $\mathfrak{M}$ containing x, and suppose that $\langle a \rangle \leq \mathfrak{Q} \in S_2(\mathfrak{G})$. Since the order of a is at least 8, $|\mathfrak{Q}| \geq 16$. Thus $\langle a \rangle$, being of order at least 8, lies in the unique cyclic maximal subgroup of $\mathfrak{Q}$, and $\langle a \rangle$ is a characteristic subgroup of $\mathfrak{Q}$. Since $\langle x \rangle$ is a characteristic subgroup of $\langle a \rangle$, $\langle x \rangle \trianglelefteq \mathfrak{Q}$ and $\mathfrak{Q} \leq \mathbf{N}_{\mathfrak{G}}(\langle x \rangle) = \mathfrak{M}$. This is a contradiction. Hence $\mathbf{Z}(\mathfrak{M}) = \mathbf{Z}(\mathfrak{G})$. **q.e.d.**

The largest part of the proof of the main theorem of this section consists of a detailed investigation of the maximal subgroups of $\mathfrak{G}$. We begin with the 2-nilpotent ones.

8.14 Theorem. *Suppose that $\mathfrak{G}$ satisfies Hypothesis 8.9 and that $\mathfrak{M}$ is a 2-nilpotent maximal subgroup of $\mathfrak{G}$. Let $\mathfrak{N} = \mathbf{O}_{2'}(\mathfrak{M})$; thus $\mathfrak{M} = \mathfrak{N}\mathfrak{Q}$ for a Sylow 2-subgroup $\mathfrak{Q}$ of $\mathfrak{M}$.*

a) *$\mathfrak{N}$ is cyclic.*

b) *$\mathfrak{N}$ is a Hall subgroup of $\mathfrak{G}$ and $\mathbf{N}_{\mathfrak{G}}(\mathfrak{X}) = \mathfrak{M}$ for any non-identity subgroup $\mathfrak{X}$ of $\mathfrak{N}$.*

c) *If $g \notin \mathfrak{M}$, then $\mathfrak{N} \cap \mathfrak{N}^g = 1$.*

d) *Suppose that $\mathfrak{Q}$ is not cyclic and $\mathfrak{N} \neq 1$. Then $\mathfrak{M}$ is generated by elements b, c satisfying the relations*

$$c^{4k} = 1, \quad c^b = c^{-1}, \quad b^2 = c^{2k},$$

where $|\mathfrak{M}| = 8k$. In particular, $\mathfrak{M} = \mathbf{N}_{\mathfrak{G}}(\langle c^k \rangle)$.

Proof. a) Suppose that $\mathfrak{N}$ is not cyclic. Since $|\mathfrak{N}|$ is odd, all the Sylow subgroups of $\mathfrak{N}$ are cyclic. Hence by 8.12, there exists a subgroup $\mathfrak{L}$ of $\mathfrak{N}$ for which

$$\mathfrak{N} = \mathfrak{N}'\mathfrak{L}, \quad \mathfrak{N}' \cap \mathfrak{L} = 1, \quad \mathfrak{Q} \leq \mathbf{N}_{\mathfrak{G}}(\mathfrak{L}), \quad \mathfrak{L} \cap \mathbf{Z}(\mathfrak{L}\mathfrak{Q}) \neq 1.$$

Let y be an element of $\mathfrak{L} \cap \mathbf{Z}(\mathfrak{L}\mathfrak{Q})$ of prime order. By 8.12b), $y \in \mathbf{Z}(\mathfrak{N})$, so $y \in \mathbf{Z}(\mathfrak{M})$, since $\mathfrak{M} = \mathfrak{N}\mathfrak{Q}$. This is impossible, since the order of y is odd and, by 8.13, $|\mathbf{Z}(\mathfrak{M})| = 2$.

b) If $\mathfrak{X}$ is a non-identity subgroup of $\mathfrak{N}$, $\mathfrak{X} \trianglelefteq \mathfrak{M}$. By 8.11, $\mathfrak{M} = \mathbf{N}_{\mathfrak{G}}(\mathfrak{X})$. Hence any non-identity Sylow subgroup of $\mathfrak{N}$ is also a Sylow subgroup of $\mathfrak{G}$, so $\mathfrak{N}$ is a Hall subgroup of $\mathfrak{G}$.

c) If $\mathfrak{N} \cap \mathfrak{N}^g \neq 1$, then by b), $\mathfrak{M} = \mathbf{N}_{\mathfrak{G}}(\mathfrak{N} \cap \mathfrak{N}^g) = \mathfrak{M}^g$. Hence $g \in \mathfrak{M}$.

d) This is proved in a number of steps.

(i) If $\mathfrak{P}$ is a non-identity Sylow subgroup of $\mathfrak{N}$, $|\mathfrak{Q} : \mathbf{C}_{\mathfrak{Q}}(\mathfrak{P})| = 2$.

Since $|\mathfrak{P}|$ is odd, the group of automorphisms of $\mathfrak{P}$ is cyclic, so $\mathfrak{Q}/\mathbf{C}_{\mathfrak{Q}}(\mathfrak{P})$ is cyclic. Hence $|\mathfrak{Q} : \mathbf{C}_{\mathfrak{Q}}(\mathfrak{P})| \leq 2$. If $\mathfrak{Q} = \mathbf{C}_{\mathfrak{Q}}(\mathfrak{P})$, $\mathfrak{P} \leq \mathbf{Z}(\mathfrak{M})$, contrary to 8.13. Thus $|\mathfrak{Q} : \mathbf{C}_{\mathfrak{Q}}(\mathfrak{P})| = 2$.

(ii) If y is an element of $\mathfrak{Q}$ of order 4, $|\mathbf{C}_{\mathfrak{G}}(y)|$ is divisible by $|\mathfrak{N}|$.

It is sufficient to show that $|\mathbf{C}_{\mathfrak{G}}(y)|$ is divisible by $|\mathfrak{P}|$ for any non-identity Sylow subgroup $\mathfrak{P}$ of $\mathfrak{N}$. By (i), $\mathbf{C}_{\mathfrak{G}}(\mathfrak{P})$ contains an element x of order 4. By 8.11b), $y = x^g$ for some $g \in \mathfrak{G}$, so $\mathfrak{P}^g \leq \mathbf{C}_{\mathfrak{G}}(y)$. Thus $|\mathbf{C}_{\mathfrak{G}}(y)|$ is divisible by $|\mathfrak{P}|$.

(iii) If $\mathfrak{P}$ is a non-identity Sylow subgroup of $\mathfrak{N}$ and g is an element of $\mathbf{C}_{\mathfrak{G}}(\mathfrak{P})$ of order 4, then $g \in \mathbf{C}_{\mathfrak{G}}(\mathfrak{N})$.

By 8.11c), all the Sylow subgroups of $\mathbf{C}_{\mathfrak{G}}(g)$ are cyclic, so by 8.12c), $\Omega_1(\mathfrak{P}) \trianglelefteq \mathbf{C}_{\mathfrak{G}}(g)$. By b), $\mathbf{C}_{\mathfrak{G}}(g) \leq \mathbf{N}_{\mathfrak{G}}(\Omega_1(\mathfrak{P})) = \mathfrak{M}$. Using (ii), it follows that $\mathfrak{N} \leq \mathbf{C}_{\mathfrak{G}}(g)$.

(iv) If $\mathfrak{P}$ is a non-identity Sylow subgroup of $\mathfrak{N}$, $\mathbf{C}_{\mathfrak{Q}}(\mathfrak{P})$ is a cyclic subgroup of $\mathfrak{Q}$ of index 2.

Suppose not. Then by (i), $|\mathfrak{Q}| > 8$ and $\mathbf{C}_{\mathfrak{Q}}(\mathfrak{P}) = \langle \mathfrak{Q}', a \rangle$ for some element a of order 4. By (iii), $\mathfrak{N} \leq \mathbf{C}_{\mathfrak{G}}(a)$. Let $\mathfrak{K}$ be a normal subgroup of $\mathfrak{N}$ of prime order. By 8.11c), all the Sylow subgroups of $\mathbf{C}_{\mathfrak{G}}(a)$ are cyclic, so by 8.12c), $\mathfrak{K}$ is a characteristic subgroup of $\mathbf{C}_{\mathfrak{G}}(a)$. Hence $\mathfrak{K}$ is the only subgroup of $\mathbf{C}_{\mathfrak{G}}(a)$ of order $|\mathfrak{K}|$. But now, $\mathfrak{Q}'$ contains an element h of order 4, and $a = h^g$ for some $g \in \mathfrak{G}$, by 8.11b). Since $\mathfrak{N}$ is cyclic, $\mathfrak{Q}' \leq \mathbf{C}_{\mathfrak{G}}(\mathfrak{N})$, so $\mathfrak{N} \leq \mathbf{C}_{\mathfrak{G}}(h)$. Then $\mathfrak{K}^g \leq \mathfrak{N}^g \leq \mathbf{C}_{\mathfrak{G}}(a)$, and it follows that $\mathfrak{K}^g = \mathfrak{K}$. By b), $g \in \mathfrak{M} = \mathfrak{Q}\mathfrak{N}$. Since $a \in \mathbf{C}_{\mathfrak{G}}(\mathfrak{N})$, it follows that a, h are conjugate in $\mathfrak{Q}$. Thus $a \in \mathfrak{Q}'$, a contradiction. Hence (iv) is proved.

If $|\mathfrak{Q}| = 8$, it follows from (i) and (iii) that $\mathbf{C}_{\mathfrak{Q}}(\mathfrak{N})$ is a cyclic subgroup $\mathfrak{Z}$ of order 4. If $|\mathfrak{Q}| > 8$, $\mathfrak{Q}$ has a unique cyclic subgroup $\mathfrak{Z}$ of index 2.

By (iv), $\mathfrak{Z}$ centralizes every Sylow subgroup of $\mathfrak{N}$, so again $\mathbf{C}_{\mathfrak{Q}}(\mathfrak{N}) = \mathfrak{Z}$. In either case, $\mathbf{C}_{\mathfrak{G}}(\mathfrak{N}) = \mathfrak{N} \times \mathfrak{Z}$ is a cyclic subgroup $\langle c \rangle$ of index 2 in $\mathfrak{M}$. We have $\mathfrak{Q} = \mathfrak{Z}\langle b \rangle$ for some b and $\mathfrak{M} = \mathfrak{N}\mathfrak{Q} = \mathbf{C}_{\mathfrak{G}}(\mathfrak{N})\langle b \rangle$. Then $\mathbf{C}_{\mathfrak{N}}(b) \leq \mathfrak{N} \cap \mathbf{Z}(\mathfrak{M}) = 1$, so b induces on $\mathfrak{N}$ a fixed point free automorphism of order 2. Thus $c^b = c^{-1}$. Since $c^k \notin \mathbf{Z}(\mathfrak{G})$, $\mathbf{N}_{\mathfrak{G}}(\langle c^k \rangle) = \mathfrak{M}$.

q.e.d.

We turn now to maximal subgroups which are soluble but not 2-nilpotent.

8.15 Lemma. *Suppose that $\mathfrak{G}$ satisfies Hypothesis 8.9 and that $\mathfrak{M}$ is a maximal subgroup which is soluble but not 2-nilpotent.*

a) *If $|\mathfrak{G}|$ is divisible by 2^4, $\mathfrak{M}/\mathbf{O}_{2'}(\mathfrak{M})$ is isomorphic to $\mathfrak{G}_{48}$.*

b) *If $|\mathfrak{G}|$ is not divisible by 2^4, $\mathfrak{M} \cong SL(2, 3)$.*

Proof. Let $\mathfrak{N} = \mathbf{O}_{2'}(\mathfrak{M})$, $\mathfrak{L} = \mathbf{O}_{2',2}(\mathfrak{M})$ and $\mathfrak{K}/\mathfrak{N} = \mathbf{Z}(\mathfrak{L}/\mathfrak{N})$. By a lemma of Hall and Higman (IX, 1.4), $\mathbf{C}_{\mathfrak{M}}(\mathfrak{L}/\mathfrak{N}) = \mathfrak{K}$, so $\mathfrak{M}/\mathfrak{K}$ is isomorphic to a group of automorphisms of $\mathfrak{L}/\mathfrak{N}$. Now $\mathfrak{L}/\mathfrak{N}$ is either cyclic or a generalized quaternion group. Thus **Aut** $\mathfrak{L}/\mathfrak{N}$ is a 2-group except in the case when $\mathfrak{L}/\mathfrak{N}$ is a quaternion group of order 8. But $\mathfrak{M}/\mathfrak{K}$ is not a 2-group, since $\mathfrak{M}$ is not 2-nilpotent. Thus $\mathfrak{L}/\mathfrak{N}$ is a quaternion group of order 8 and **Aut** $\mathfrak{L}/\mathfrak{N} \cong \mathfrak{S}_4$. Since $|\mathfrak{L} : \mathfrak{K}| = 4$, $|\mathfrak{M}/\mathfrak{K}|$ is divisible by 12. Hence $\mathfrak{M}/\mathfrak{K}$ is isomorphic to $\mathfrak{S}_4$ or $\mathfrak{A}_4$.

If $\mathfrak{M}/\mathfrak{K} \cong \mathfrak{S}_4$, then by 8.5, $\mathfrak{M}/\mathfrak{N} \cong \mathfrak{G}_{48}$. This is case a).

Now suppose that $\mathfrak{M}/\mathfrak{K} \cong \mathfrak{A}_4$. By 8.5, $\mathfrak{M}/\mathfrak{N} \cong SL(2, 3)$. It is to be shown that $\mathfrak{N} = 1$ and that $|\mathfrak{G}|$ is not divisible by 2^4. Let $\mathfrak{Q}$ be a Sylow 2-subgroup of $\mathfrak{M}$. Thus $\mathfrak{Q}\mathfrak{N} = \mathfrak{L}$. Since $\mathfrak{K}/\mathfrak{N} = (\mathfrak{L}/\mathfrak{N})'$ and $(\mathfrak{M}/\mathfrak{K})' = \mathfrak{L}/\mathfrak{K}$, $\mathfrak{L} = \mathfrak{M}'\mathfrak{N}$. Hence $\mathfrak{Q} \leq \mathfrak{M}'\mathfrak{N}$. Since $|\mathfrak{N}|$ is odd, $\mathfrak{Q} \leq \mathfrak{M}'$. By 8.12, $\mathfrak{N}/\mathfrak{N}'$ and $\mathfrak{N}'$ are cyclic, so it follows that they are centralized by $\mathfrak{Q}$. Since $(|\mathfrak{Q}|, |\mathfrak{N}|) = 1$, $\mathfrak{Q}$ centralizes $\mathfrak{N}$.

Suppose that $\mathfrak{P} \in S_3(\mathfrak{M})$ and put $\mathfrak{R} = \mathfrak{N}\mathfrak{P}$. Then $|\mathfrak{R} : \mathfrak{N}| = 3$ and all the Sylow subgroups of $\mathfrak{R}$ are cyclic. By 8.12, $\mathbf{Z}(\mathfrak{R}) \neq 1$. But $\mathfrak{M} = \mathfrak{L}\mathfrak{R} = \mathfrak{Q}\mathfrak{N}\mathfrak{R} = \mathfrak{Q}\mathfrak{R}$, so $[\mathbf{Z}(\mathfrak{R}) \cap \mathfrak{N}, \mathfrak{M}] = [\mathbf{Z}(\mathfrak{R}) \cap \mathfrak{N}, \mathfrak{Q}\mathfrak{R}] = 1$. Thus $\mathbf{Z}(\mathfrak{R}) \cap \mathfrak{N} \leq \mathbf{Z}(\mathfrak{M})$. But by 8.13, $|\mathbf{Z}(\mathfrak{M})| = 2$, so $\mathbf{Z}(\mathfrak{R}) \cap \mathfrak{N} = 1$. Thus $\mathfrak{N} < \mathbf{Z}(\mathfrak{R})\mathfrak{N} \leq \mathfrak{R}$ and $|\mathfrak{R} : \mathfrak{N}| = 3$, so $\mathfrak{R} = \mathbf{Z}(\mathfrak{R})\mathfrak{N} = \mathbf{Z}(\mathfrak{R}) \times \mathfrak{N}$. Hence $\mathbf{Z}(\mathfrak{N}) = 1$. By 8.12, $\mathfrak{N} = 1$. Thus $\mathfrak{M} \cong SL(2, 3)$. Note that $\mathfrak{Q} \trianglelefteq \mathfrak{M}$.

If $|\mathfrak{G}|$ is divisible by 2^4, $\mathfrak{Q} \lhd \mathfrak{S}$ for some 2-subgroup $\mathfrak{S}$. Then $\mathfrak{S} \nleq \mathfrak{M}$, so $\mathbf{N}_{\mathfrak{G}}(\mathfrak{Q}) > \mathfrak{M}$. Since $\mathfrak{M}$ is maximal, it follows that $\mathfrak{Q} \trianglelefteq \mathfrak{G}$. Then the order of the Sylow 2-subgroups of $\mathfrak{G}/\mathfrak{Q}$ is 2, since the order of the normaliser of a quaternion subgroup of order 8 in a generalized quaternion group of greater order is 16. This contradicts $\mathfrak{G} = \mathfrak{G}' \neq 1$. Thus case b) holds.

q.e.d.

In the proof of the main theorem, we shall have an inductive hypothesis which is needed to deal with insoluble maximal subgroups. We therefore have an additional hypothesis in this case.

8.16 Lemma. *Suppose that $\mathfrak{G}$ satisfies Hypothesis 8.9 and that any perfect proper subgroup of $\mathfrak{G}$ is isomorphic to $SL(2, 5)$. If $\mathfrak{M}$ is an insoluble maximal subgroup of $\mathfrak{G}$, $|\mathfrak{M}/\mathfrak{M}'| \leq 2$ and $\mathfrak{M}' \cong SL(2, 5)$. Further* $\mathbf{C}_{\mathfrak{M}}(\mathfrak{M}') = \mathbf{Z}(\mathfrak{M}) = \mathbf{Z}(\mathfrak{G})$.

Proof. Put $\mathfrak{N} = \mathfrak{M}^{(i)} = \mathfrak{M}^{(i+1)} \neq 1$. Then $\mathfrak{N} = \mathfrak{N}'$. Hence by hypothesis, $\mathfrak{N} \cong SL(2, 5)$.

Let $\mathfrak{C} = \mathbf{C}_{\mathfrak{M}}(\mathfrak{N}/\mathbf{Z}(\mathfrak{N}))$. Thus $\mathfrak{C} \trianglelefteq \mathfrak{M}$ and $[\mathfrak{C}, \mathfrak{N}, \mathfrak{N}] = 1$. It follows from III, 1.10b) that $[\mathfrak{N}', \mathfrak{C}] = 1$. Thus $[\mathfrak{N}, \mathfrak{C}] = 1$ and $\mathfrak{C} = \mathbf{C}_{\mathfrak{M}}(\mathfrak{N})$.

Now $\mathfrak{M}/\mathfrak{C}$ is isomorphic to a subgroup of the automorphism group of $\mathfrak{N}/\mathbf{Z}(\mathfrak{N})$ containing all the inner automorphisms of $\mathfrak{N}/\mathbf{Z}(\mathfrak{N})$. Since $\mathfrak{N}/\mathbf{Z}(\mathfrak{N}) \cong \mathfrak{A}_5$, it follows from 8.7 that $|\mathfrak{M} : \mathfrak{M}_0| \leq 2$, where $\mathfrak{M}_0 = \mathfrak{C}\mathfrak{N}$.

Note that $\mathfrak{C} \cap \mathfrak{N} = \mathbf{Z}(\mathfrak{N})$ is of order 2. Also,

$$\mathfrak{M}_0/\mathbf{Z}(\mathfrak{N}) = (\mathfrak{C}/\mathbf{Z}(\mathfrak{N})) \times (\mathfrak{N}/\mathbf{Z}(\mathfrak{N})).$$

Since $\mathfrak{N}/\mathbf{Z}(\mathfrak{N}) \cong \mathfrak{A}_5$ has an elementary Abelian subgroup of order 4 and non-identity subgroups of orders 3 and 5, it follows from the structure of the Sylow subgroups of $\mathfrak{G}$ that $|\mathfrak{C}/\mathbf{Z}(\mathfrak{N})|$ is not divisible by 2, 3 or 5. Thus $(|\mathfrak{C}/\mathbf{Z}(\mathfrak{N})|, |\mathfrak{N}|) = 1$.

In particular, $|\mathfrak{C}/\mathbf{Z}(\mathfrak{N})|$ is odd, so by the Schur-Zassenhaus theorem (I, 18.1), there exists a subgroup $\mathfrak{K}$ of $\mathfrak{C}$ such that $\mathfrak{C} = \mathfrak{K} \times \mathbf{Z}(\mathfrak{N})$. Thus $|\mathfrak{K}| = |\mathfrak{C} : \mathbf{Z}(\mathfrak{N})|$ is coprime to $|\mathfrak{N}|$. Also, $\mathfrak{M}_0 = \mathfrak{K} \times \mathfrak{N}$, since $\mathfrak{K} \leq \mathfrak{C} = \mathbf{C}_{\mathfrak{M}}(\mathfrak{N})$. Thus $\mathfrak{K}$ is a characteristic subgroup of $\mathfrak{M}_0$ and a normal subgroup of $\mathfrak{M}$.

Suppose that $\mathfrak{K} \neq 1$. Then by 8.12, $\mathbf{Z}(\mathfrak{K}) \neq 1$. By 8.11a), $|\mathbf{Z}(\mathfrak{G})| = 2$, so $\mathbf{Z}(\mathfrak{K}) \nleq \mathbf{Z}(\mathfrak{G})$ and $\mathbf{Z}(\mathfrak{K})$ is not normal in $\mathfrak{G}$. Hence neither is $\mathfrak{K}$. Since $\mathfrak{K} \trianglelefteq \mathfrak{M}$ and $\mathfrak{M}$ is a maximal subgroup, $\mathbf{N}_{\mathfrak{G}}(\mathfrak{K}) = \mathfrak{M}$. Again, $\mathbf{Z}(\mathfrak{M}_0) = \mathbf{Z}(\mathfrak{K}) \times \mathbf{Z}(\mathfrak{N})$, so $|\mathbf{Z}(\mathfrak{M}_0)| > 2$. But by 8.13, $|\mathbf{Z}(\mathfrak{M})| = 2$, so $\mathfrak{M}_0 < \mathfrak{M}$. Thus $|\mathfrak{M} : \mathfrak{M}_0| = 2$. Hence $\mathfrak{M}/\mathfrak{C} = \mathbf{Aut}\, \mathfrak{A}_5 \cong \mathfrak{S}_5$. Thus $\mathfrak{M}/\mathfrak{C}$ contains an involution $x\mathfrak{C}$ and an element $y\mathfrak{C}$ of order 3 such that $x\mathfrak{C}$, $y\mathfrak{C}$ commute; further, we may suppose that x is a 2-element and that the order of y is 3. Since $|\mathfrak{M} : \mathfrak{N}| = 2|\mathfrak{M}_0 : \mathfrak{N}| = 2|\mathfrak{K}|$ is prime to 3, $y \in \mathfrak{N}$. We have $y^x = yz$ for $z \in \mathfrak{C} \cap \mathfrak{N} = \mathbf{Z}(\mathfrak{N})$. Thus y, z commute, and since the order of y is 3, it follows that $z = 1$. Thus $y^x = y$.

Now $x^2 \in \mathfrak{C} = \mathbf{Z}(\mathfrak{N}) \times \mathfrak{K}$, so $x^2 \in \mathbf{Z}(\mathfrak{N})$. Since the only involution in $\mathfrak{G}$ is central, $x^2 \neq 1$. Thus the order of x is 4. Also, $\mathfrak{N}$ has an element x' of order 4. By 8.11b), there exists $g \in \mathfrak{G}$ such that $x' = x^g$. If $y' = y^g$,

$y' \in \mathbf{C}_{\mathfrak{G}}(x')$. By 8.11c), all the Sylow subgroups of $\mathbf{C}_{\mathfrak{G}}(x')$ are cyclic, so by 8.12c), $\langle y' \rangle \trianglelefteq \mathbf{C}_{\mathfrak{G}}(x')$. But $\mathfrak{K} \leq \mathbf{C}_{\mathfrak{G}}(\mathfrak{N}) \leq \mathbf{C}_{\mathfrak{G}}(x')$, so $\mathfrak{K} \leq \mathbf{N}_{\mathfrak{G}}(\langle y' \rangle)$. Since $|\mathfrak{K}|$ is odd and the order of y' is 3, it follows that $\mathfrak{K} \leq \mathbf{C}_{\mathfrak{G}}(y')$. Hence $y' \in \mathbf{C}_{\mathfrak{G}}(\mathfrak{K}) \leq \mathbf{N}_{\mathfrak{G}}(\mathfrak{K}) = \mathfrak{M}$. Since $|\mathfrak{M} : \mathfrak{N}|$ is prime to 3, $y' \in \mathfrak{N}$. Hence $x'y'$ is an element of $\mathfrak{N}$ of order 12. But this is impossible, since $\mathfrak{A}_5$ has no element of order 6.

Thus $\mathfrak{K} = 1$, $\mathfrak{C} \leq \mathfrak{N}$ and $|\mathfrak{M} : \mathfrak{N}| \leq 2$. Thus $\mathfrak{N} = \mathfrak{M}' \cong SL(2, 5)$. And $\mathbf{C}_{\mathfrak{M}}(\mathfrak{M}') = \mathfrak{C} = \mathbf{Z}(\mathfrak{M}') = \mathbf{Z}(\mathfrak{M}) = \mathbf{Z}(\mathfrak{G})$. **q.e.d.**

We shall now show that if, in 8.15, $\mathfrak{M}/\mathbf{O}_{2'}(\mathfrak{M}) \cong \mathfrak{G}_{48}$, then $\mathbf{O}_{2'}(\mathfrak{M}) = 1$. The proof of this, however, is very long.

8.17 Lemma. *Suppose that $\mathfrak{G}$ satisfies Hypothesis 8.9 and that any perfect proper subgroup of $\mathfrak{G}$ is isomorphic to $SL(2, 5)$. If $\mathfrak{M}$ is a maximal subgroup of $\mathfrak{G}$ and $\mathfrak{M}/\mathbf{O}_{2'}(\mathfrak{M}) \cong \mathfrak{G}_{48}$, then $\mathbf{O}_{2'}(\mathfrak{M}) = 1$.*

Proof. Put $\mathfrak{N} = \mathbf{O}_{2'}(\mathfrak{M})$ and suppose that $\mathfrak{N} \neq 1$.

a) $\mathfrak{N}$ is cyclic, $\mathbf{C}_{\mathfrak{G}}(\mathfrak{N}) = \mathfrak{M}'$ and $|\mathfrak{M} : \mathfrak{M}'| = 2$. Also, $\mathbf{N}_{\mathfrak{G}}(\mathfrak{X}) = \mathfrak{M}$ for any non-identity subgroup $\mathfrak{X}$ of $\mathfrak{N}$.

Suppose that $\mathfrak{N}$ is not cyclic. Let $\mathfrak{Q} \in S_2(\mathfrak{M})$, $\mathfrak{P} \in S_3(\mathfrak{M})$. All the Sylow subgroups of $\mathfrak{N}$ are cyclic, so by 8.12, there exists a subgroup $\mathfrak{L}$ of $\mathfrak{N}$ for which $\mathfrak{N} = \mathfrak{N}'\mathfrak{L}$, $\mathfrak{N}' \cap \mathfrak{L} = 1$ and $\mathfrak{Q} \leq \mathbf{N}_{\mathfrak{G}}(\mathfrak{L})$; also $\mathfrak{L} \cap \mathbf{Z}(\mathfrak{L}\mathfrak{Q})$ has an element x of odd prime order, $x \in \mathbf{Z}(\mathfrak{N})$ and $\langle x \rangle$ is a characteristic subgroup of $\mathfrak{N}$. Thus $\langle x \rangle \trianglelefteq \mathfrak{M}$ and $\mathfrak{M}' \leq \mathbf{C}_{\mathfrak{G}}(x)$. By the definition of $\mathfrak{G}_{48}$ (8.4), $\mathfrak{P}\mathfrak{N}/\mathfrak{N} \leq (\mathfrak{M}/\mathfrak{N})'$, so $\mathfrak{P} \leq \mathfrak{M}'\mathfrak{N} \leq \mathbf{C}_{\mathfrak{G}}(x)$. Then $\mathbf{C}_{\mathfrak{G}}(x) \geq \langle \mathfrak{N}, \mathfrak{P}, \mathfrak{Q} \rangle = \mathfrak{M}$. This is impossible, since $|\mathbf{Z}(\mathfrak{M})| = 2$. Hence $\mathfrak{N}$ is cyclic and $\mathfrak{M}' \leq \mathbf{C}_{\mathfrak{G}}(\mathfrak{N})$. Since $|\mathfrak{M} : \mathfrak{M}'\mathfrak{N}| = 2$, it follows that $|\mathfrak{M} : \mathbf{C}_{\mathfrak{M}}(\mathfrak{N})| = 2$. If b is an element of $\mathfrak{M} - \mathbf{C}_{\mathfrak{M}}(\mathfrak{N})$, $\mathbf{C}_{\mathfrak{N}}(b) = \mathfrak{N} \cap \mathbf{Z}(\mathfrak{M}) = 1$, so b induces a fixed point free automorphism of $\mathfrak{N}$ of order 2. Hence $x^b = x^{-1}$ for all $x \in \mathfrak{N}$ and $x^{-2} = [x, b] \in \mathfrak{M}'$. Thus $\mathfrak{M}' \geq \mathfrak{N}$, $|\mathfrak{M} : \mathfrak{M}'| = 2$ and $\mathfrak{M}' = \mathbf{C}_{\mathfrak{M}}(\mathfrak{N})$. Finally, if $\mathfrak{X}$ is a non-identity subgroup of $\mathfrak{N}$, $\mathfrak{X} \trianglelefteq \mathfrak{M}$ and, by 8.11, $\mathfrak{X} \ntrianglelefteq \mathfrak{G}$, so $\mathbf{N}_{\mathfrak{G}}(\mathfrak{X}) = \mathfrak{M}$ and hence $\mathbf{C}_{\mathfrak{G}}(\mathfrak{N}) = \mathbf{C}_{\mathfrak{M}}(\mathfrak{N})$.

b) Suppose that $|\mathfrak{N}|$ is not divisible by 3 and that $\mathfrak{P} \in S_3(\mathfrak{M})$. Then $\mathbf{C}_{\mathfrak{G}}(\mathfrak{P}) \leq \mathfrak{M}$ and indeed, $\mathfrak{M}$ is the only maximal subgroup of $\mathfrak{G}$ which contains $\mathbf{C}_{\mathfrak{G}}(\mathfrak{P})$.

To prove this, observe that by a), $\mathfrak{N} \leq \mathbf{C}_{\mathfrak{G}}(\mathfrak{P}) < \mathfrak{G}$. Thus there is a maximal subgroup $\mathfrak{M}^*$ of $\mathfrak{G}$ such that $\mathbf{C}_{\mathfrak{G}}(\mathfrak{P}) \leq \mathfrak{M}^*$. Suppose that $\mathfrak{M}^*$ is insoluble. Then by 8.16, $|\mathfrak{M}^* : \mathfrak{M}^{*\prime}| \leq 2$ and $\mathfrak{M}^{*\prime} \cong SL(2, 5)$. Thus $|\mathfrak{N}| = 5$ and $\mathfrak{M}^*$ has no element of order 15. But this contradicts $\mathfrak{N} \leq \mathbf{C}_{\mathfrak{G}}(\mathfrak{P}) \leq \mathfrak{M}^*$, so $\mathfrak{M}^*$ is soluble. Since $|\mathfrak{G}|$ is divisible by 2^4, it follows from 8.15 that either $\mathfrak{M}^*/\mathbf{O}_{2'}(\mathfrak{M}^*) \cong \mathfrak{G}_{48}$ or $\mathfrak{M}^*$ is 2-nilpotent. In either case, $\mathbf{O}_{2'}(\mathfrak{M}^*)$ is cyclic; this follows from a) (applied to $\mathfrak{M}^*$) in the former case and from 8.14 in the latter. But $\mathfrak{N} \leq \mathbf{O}_{2'}(\mathfrak{M}^*)$, so $\mathfrak{N} \trianglelefteq \mathfrak{M}^*$ and $\mathfrak{M}^* \leq \mathbf{N}_{\mathfrak{G}}(\mathfrak{N}) = \mathfrak{M}$.

c) $\mathfrak{M}$ is a Hall subgroup of $\mathfrak{G}$; in particular, $|\mathfrak{G}|$ is not divisible by 2^5.

Suppose that p divides $|\mathfrak{N}|$. If $\mathfrak{S} \in S_p(\mathfrak{N})$, then $\mathbf{N}_{\mathfrak{G}}(\mathfrak{S}) = \mathfrak{M}$, by a). Hence $\mathfrak{M}$ contains a Sylow p-subgroup of $\mathfrak{G}$ and $|\mathfrak{G} : \mathfrak{M}|$ is not divisible by p.

We show next that $|\mathfrak{G} : \mathfrak{M}|$ is odd. Since $|\mathfrak{M} : \mathbf{C}_{\mathfrak{M}}(\mathfrak{N})| = 2$, $\mathbf{C}_{\mathfrak{M}}(\mathfrak{N})$ contains an element g of order 4. Thus

$$\mathfrak{N} \leq \mathbf{C}_{\mathfrak{G}}(g) \leq \mathbf{N}_{\mathfrak{G}}(\langle g \rangle).$$

Now $|\mathbf{N}_{\mathfrak{G}}(\langle g \rangle) : \mathbf{C}_{\mathfrak{G}}(g)| \leq 2$ and, by 8.11c), $\mathbf{C}_{\mathfrak{G}}(g)$ is 2-nilpotent. Hence $\mathbf{N}_{\mathfrak{G}}(\langle g \rangle)$ is 2-nilpotent, and since $|\mathfrak{N}|$ is odd, $\mathfrak{N} \leq \mathfrak{K} = \mathbf{O}_{2'}(\mathbf{N}_{\mathfrak{G}}(\langle g \rangle))$. Since all elements of $\mathfrak{G}$ of order 4 are conjugate and the Sylow 2-subgroups of $\mathfrak{G}$ have a normal cyclic subgroup of order 4, $\mathbf{N}_{\mathfrak{G}}(\langle g \rangle)$ contains a Sylow 2-subgroup $\mathfrak{Q}$ of $\mathfrak{G}$. Thus $\mathbf{N}_{\mathfrak{G}}(\langle g \rangle) = \mathfrak{K}\mathfrak{Q}$. Now if $\mathfrak{T}$ is a normal subgroup of $\mathfrak{N}$ of prime order, $\mathfrak{T} \leq \mathfrak{N} \leq \mathfrak{K}$, and by 8.12c), $\mathfrak{T}$ is a characteristic subgroup of $\mathfrak{K}$. Since $\mathfrak{Q} \leq \mathbf{N}_{\mathfrak{G}}(\mathfrak{K})$, it follows that $\mathfrak{Q} \leq \mathbf{N}_{\mathfrak{G}}(\mathfrak{T}) = \mathfrak{M}$. Thus $|\mathfrak{G} : \mathfrak{M}|$ is odd.

It remains to show that $|\mathfrak{G} : \mathfrak{M}|$ is not divisible by 3, and from above, we may suppose that 3 does not divide $|\mathfrak{N}|$. Let $\mathfrak{P} \in S_3(\mathfrak{M})$. By b), $\mathbf{C}_{\mathfrak{G}}(\mathfrak{P}) \leq \mathfrak{M}$. Since $\mathbf{C}_{\mathfrak{G}}(\mathfrak{P})$ contains a Sylow 3-subgroup of $\mathfrak{G}$, $|\mathfrak{G} : \mathfrak{M}|$ is not divisible by 3. Thus $\mathfrak{M}$ is a Hall subgroup of $\mathfrak{G}$.

d) If $g \notin \mathfrak{M}$, the order of any element of $\mathfrak{M} \cap \mathfrak{M}^g$ is a divisor of 4.

Suppose that x is an element of $\mathfrak{M} \cap \mathfrak{M}^g$ of prime order p. If $p > 3$ or if $p = 3$ and 3 divides $|\mathfrak{N}|$, then $x \in \mathfrak{N}$ and by a), $\mathfrak{M} = \mathbf{N}_{\mathfrak{G}}(\langle x \rangle) = \mathfrak{M}^g$. Thus $g \in \mathfrak{M}$. If $p = 3$ and 3 does not divide $|\mathfrak{N}|$, $\mathfrak{P} = \langle x \rangle$ and $\mathfrak{P}^{g^{-1}}$ are Sylow 3-subgroups of $\mathfrak{M}$. By b), $\mathbf{C}_{\mathfrak{G}}(\mathfrak{P}^{g^{-1}}) \leq \mathfrak{M}$, so $\mathbf{C}_{\mathfrak{G}}(\mathfrak{P}) \leq \mathfrak{M}^g$. By b) again, $\mathfrak{M}^g = \mathfrak{M}$ and $g \in \mathfrak{M}$.

Now suppose that y is an element of $\mathfrak{M} \cap \mathfrak{M}^g$ of order 8. By a), $y^2 \in \mathbf{C}_{\mathfrak{G}}(\mathfrak{N}) \cap \mathbf{C}_{\mathfrak{G}}(\mathfrak{N}^g)$. Suppose that $\mathfrak{K}$ is a subgroup of $\mathfrak{N}$ of prime order p; thus $\mathfrak{K} \leq \mathbf{C}_{\mathfrak{G}}(y^2)$ and $\mathfrak{K}^g \leq \mathbf{C}_{\mathfrak{G}}(y^2)$. By 8.11, all the Sylow subgroups of $\mathbf{C}_{\mathfrak{G}}(y^2)$ are cyclic, so by 8.12c), $\mathfrak{K}$ and $\mathfrak{K}^g$ are normal in $\mathbf{C}_{\mathfrak{G}}(y^2)$. Since the Sylow p-subgroups of $\mathbf{C}_{\mathfrak{G}}(y^2)$ are cyclic, this forces $\mathfrak{K} = \mathfrak{K}^g$. By a), $\mathbf{N}_{\mathfrak{G}}(\mathfrak{K}) = \mathfrak{M}$, so $g \in \mathfrak{M}$.

e) The order of the centralizer of any element of $\mathfrak{G}$ of order 4 is $8|\mathfrak{N}|$. Further, there exists $g \in \mathbf{C}_{\mathfrak{M}}(\mathfrak{N})$ such that g is of order 4 and $\mathbf{C}_{\mathfrak{G}}(g) \leq \mathfrak{M}$.

Let g be a generator of the derived group of a Sylow 2-subgroup of $\mathfrak{M}$. Thus $g \in \mathbf{C}_{\mathfrak{M}}(\mathfrak{N})$ and g is of order 4. Also, $|\mathbf{C}_{\mathfrak{M}}(g)|$ is divisible by 8. In $\mathfrak{M}/\mathfrak{N}$, no element of $\mathfrak{M}'/\mathfrak{N}$ of order 4 is centralized by an element of order 3, so $|\mathbf{C}_{\mathfrak{M}}(g)| = 8|\mathfrak{N}|$. By 8.11c), $\mathbf{C}_{\mathfrak{G}}(g)$ is 2-nilpotent and the Sylow 2-subgroups of it are cyclic. If $\mathfrak{T}$ is a subgroup of $\mathfrak{N}$ of prime order,

$\mathfrak{T} \leq \mathbf{C}_{\mathfrak{G}}(g)$. By 8.12, $\mathfrak{T} \trianglelefteq \mathbf{C}_{\mathfrak{G}}(g)$, so $\mathbf{C}_{\mathfrak{G}}(g) \leq \mathbf{N}_{\mathfrak{G}}(\mathfrak{T})$. By a), $\mathbf{N}_{\mathfrak{G}}(\mathfrak{T}) = \mathfrak{M}$, so $\mathbf{C}_{\mathfrak{G}}(g) \leq \mathfrak{M}$.

Since all elements of $\mathfrak{G}$ of order 4 are conjugate, the order of the centralizer of any element of order 4 is $8|\mathfrak{N}|$.

f) $\mathfrak{M}$ has exactly $3|\mathfrak{N}|$ Sylow 2-subgroups and $12|\mathfrak{N}|$ elements of order 8. Also all elements of $\mathfrak{M} - \mathfrak{M}'$ are 2-elements.

By definition of $\mathfrak{G}_{48}$, $\mathfrak{M}/\mathfrak{N}$ has a normal quaternion subgroup $\mathfrak{L}/\mathfrak{N}$ of order 8, and $\mathfrak{M}/\mathfrak{L} \cong \mathfrak{S}_3$. Thus $\mathfrak{L} \leq \mathfrak{M}' = \mathbf{C}_{\mathfrak{G}}(\mathfrak{N})$. If $\mathfrak{R} \in S_2(\mathfrak{L})$, then $\mathfrak{L} = \mathfrak{R} \times \mathfrak{N}$, so $\mathfrak{R} \trianglelefteq \mathfrak{M}$. Since $\mathfrak{L}/\mathfrak{R} \leq \mathbf{Z}(\mathfrak{M}'/\mathfrak{R})$ and $\mathfrak{M}'/\mathfrak{L}$ is cyclic, $\mathfrak{M}'/\mathfrak{R}$ is Abelian. Now a 2-element b of $\mathfrak{M} - \mathfrak{M}'$ induces fixed point free automorphisms on both $\mathfrak{M}'/\mathfrak{L}$ and $\mathfrak{L}/\mathfrak{R}(\cong \mathfrak{N})$, so b induces a fixed point free automorphism of order 2 on $\mathfrak{M}'/\mathfrak{R}$. Hence $\mathfrak{M}/\mathfrak{R}$ is a dihedral group and all the elements of $\mathfrak{M} - \mathfrak{M}'$ are 2-elements. Also the number of Sylow 2-subgroups of $\mathfrak{M}$ is the same as that of $\mathfrak{M}/\mathfrak{R}$, which is $|(\mathfrak{M}/\mathfrak{R}) - (\mathfrak{M}'/\mathfrak{R})| = 6|\mathfrak{N}| - 3|\mathfrak{N}| = 3|\mathfrak{N}|$. Each Sylow 2-subgroup of $\mathfrak{M}$ has 4 elements of order 8 and is generated by $\mathfrak{R}$ and any one of them. Thus $\mathfrak{M}$ has just $12|\mathfrak{N}|$ elements of order 8.

g) $|\bigcup_{g \in \mathfrak{G}} \mathfrak{M}^g| = \frac{3}{4}|\mathfrak{G}| + 2 - 2|\mathfrak{G} : \mathfrak{M}|$.

Let $\mathsf{M} = \bigcup_{g \in \mathfrak{G}} \mathfrak{M}^g$. Let M_1, M_2 be the sets of those elements of M which are of order 8, 4 respectively, and let M_3 be the set of elements of M which are not of order a power of 2. Then M is the disjoint union of M_1, M_2, M_3 and $\mathbf{Z}(\mathfrak{G})$.

By d), $\mathfrak{M} \cap \mathfrak{M}^g$ has no element of order 8 if $g \notin \mathfrak{M}$, so by f),

$$|\mathsf{M}_1| = |\mathfrak{G} : \mathfrak{M}| \cdot 12|\mathfrak{N}| = \frac{|\mathfrak{G}|}{2^4 \cdot 3 \cdot |\mathfrak{N}|} \cdot 12|\mathfrak{N}| = \frac{1}{4}|\mathfrak{G}|.$$

Since all elements of order 4 are conjugate, M_2 is the set of elements of order 4 in $\bigcup_{g \in \mathfrak{G}} \mathfrak{M}'^g$. Thus $\mathsf{M}_2 \cup \mathsf{M}_3$ is the set of all elements of order greater than 2 in $\bigcup_{g \in \mathfrak{G}} \mathfrak{M}'^g$. If $y \in (\mathfrak{M}' \cap \mathfrak{M}'^g) - \mathbf{Z}(\mathfrak{G})$, then y is of order 4 and $\langle \mathfrak{N}, \mathfrak{N}^g \rangle \leq \mathbf{C}_{\mathfrak{G}}(y)$. If $\mathfrak{K}$ is a subgroup of $\mathfrak{N}$ of prime order, $\mathfrak{K} \leq \mathbf{C}_{\mathfrak{G}}(y)$ and $\mathfrak{K}^g \leq \mathbf{C}_{\mathfrak{G}}(y)$. By 8.11 and 8.12, $\mathfrak{K}$ and $\mathfrak{K}^g$ are characteristic subgroups of $\mathbf{C}_{\mathfrak{G}}(y)$, so $\mathfrak{K}^g = \mathfrak{K}$. By a), $\mathbf{N}_{\mathfrak{G}}(\mathfrak{K}) = \mathfrak{M}$, so $g \in \mathfrak{M}$ and $\mathfrak{M}'^g = \mathfrak{M}'$. It follows that

$$|\mathsf{M}_2 \cup \mathsf{M}_3| = |\mathfrak{G} : \mathfrak{M}||\mathfrak{M}' - \mathbf{Z}(\mathfrak{G})| = \tfrac{1}{2}|\mathfrak{G}| - 2|\mathfrak{G} : \mathfrak{M}|.$$

Thus

$$|\mathsf{M}| = \tfrac{3}{4}|\mathfrak{G}| - 2|\mathfrak{G} : \mathfrak{M}| + 2.$$

Hence g) is proved.

By c), $\mathfrak{M}$ is a Hall subgroup of $\mathfrak{G}$, so there exist primes which divide $|\mathfrak{G}|$ but not $|\mathfrak{M}|$.

h) Let p be a prime which divides $|\mathfrak{G}|$ but does not divide $|\mathfrak{M}|$. Suppose that $\mathfrak{P} \in S_p(\mathfrak{G})$. Then $\mathbf{N}_{\mathfrak{G}}(\mathfrak{P})$ is 2-nilpotent with a cyclic 2-complement $\mathfrak{Z}$ and $|\mathbf{N}_{\mathfrak{G}}(\mathfrak{P})|$ is not divisible by 2^4. Either $\mathbf{N}_{\mathfrak{G}}(\mathfrak{P})$ is a 2-nilpotent maximal subgroup of $\mathfrak{G}$ or $|\mathfrak{P}| = |\mathfrak{Z}| = 5$ and $|\mathbf{N}_{\mathfrak{G}}(\mathfrak{P})|$ is 20 or 40. In either case, $\mathbf{N}_{\mathfrak{G}}(\mathfrak{P}) = \mathbf{N}_{\mathfrak{G}}(\mathfrak{Z})$.

Let $\mathfrak{M}^*$ be a maximal subgroup of $\mathfrak{G}$ which contains $\mathbf{N}_{\mathfrak{G}}(\mathfrak{P})$. First suppose that $\mathfrak{M}^*$ is insoluble. By 8.16, $|\mathfrak{M}^*/\mathfrak{M}^{*\prime}| \leq 2$ and $\mathfrak{M}^{*\prime} \cong SL(2, 5)$. Since $p > 3$, $p = 5$ and $\mathfrak{P} \in S_5(\mathfrak{M}^{*\prime})$. But the normaliser of a Sylow 5-subgroup of $SL(2, 5)$ is 2-nilpotent of order 20, so $\mathbf{N}_{\mathfrak{G}}(\mathfrak{P}) = \mathbf{N}_{\mathfrak{M}^*}(\mathfrak{P})$ is 2-nilpotent of order 20 or 40. If $\mathfrak{Z} = \mathbf{O}_{2'}(\mathbf{N}_{\mathfrak{G}}(\mathfrak{P}))$, $\mathfrak{Z} = \mathfrak{P}$ and $\mathbf{N}_{\mathfrak{G}}(\mathfrak{Z}) = \mathbf{N}_{\mathfrak{G}}(\mathfrak{P})$. Thus h) is proved in this case.

Suppose then that $\mathfrak{M}^*$ is soluble. If $\mathfrak{M}^*$ is not 2-nilpotent, then by 8.15, $\mathfrak{M}^*/\mathbf{O}_{2'}(\mathfrak{M}^*) \cong \mathfrak{G}_{48}$, for $|\mathfrak{G}|$ is divisible by 2^4. Since $p > 3$, $\mathfrak{P} \leq \mathbf{O}_{2'}(\mathfrak{M}^*)$, so $\mathbf{O}_{2'}(\mathfrak{M}^*) \neq 1$ and we may apply e) to $\mathfrak{M}^*$; hence there is an element g^* of order 4 in $\mathbf{C}_{\mathfrak{M}^*}(\mathbf{O}_{2'}(\mathfrak{M}^*))$. Since $\mathfrak{P} \leq \mathbf{O}_{2'}(\mathfrak{M}^*)$, $\mathfrak{P} \leq \mathbf{C}_{\mathfrak{G}}(g^*)$. By e), $|\mathfrak{P}|$ divides $8|\mathfrak{N}|$, a contradiction. Thus $\mathfrak{M}^*$ is 2-nilpotent. By 8.14, the 2-complement $\mathfrak{Z}$ of $\mathfrak{M}^*$ is cyclic. Since $\mathfrak{Z} \geq \mathfrak{P}$, $\mathbf{N}_{\mathfrak{G}}(\mathfrak{P}) = \mathfrak{M}^* = \mathbf{N}_{\mathfrak{G}}(\mathfrak{Z})$ and $\mathbf{N}_{\mathfrak{G}}(\mathfrak{P})$ is a 2-nilpotent maximal subgroup. If $|\mathbf{N}_{\mathfrak{G}}(\mathfrak{P})|$ is divisible by 2^4, the Sylow 2-subgroup $\mathfrak{Q}$ of $\mathbf{N}_{\mathfrak{G}}(\mathfrak{P})$ is also a Sylow 2-subgroup of $\mathfrak{G}$ (by c)). So $\mathfrak{Q}'$ contains an element h of order 4 and $h \in \mathbf{N}_{\mathfrak{G}}(\mathfrak{P})' \leq \mathbf{C}_{\mathfrak{G}}(\mathfrak{P})$. By e), $|\mathfrak{P}|$ divides $8|\mathfrak{N}|$, a contradiction. Thus h) is proved in this case also.

i) Let p be a prime which divides $|\mathfrak{G}|$ but does not divide $|\mathfrak{M}|$, let $\mathfrak{P} \in S_p(\mathfrak{G})$ and $\mathfrak{Z} = \mathbf{O}_{2'}(\mathbf{N}_{\mathfrak{G}}(\mathfrak{P}))$. Then $(|\mathfrak{Z}|, |\mathfrak{M}|) = 1$.

If not, let q be a prime divisor of $(|\mathfrak{Z}|, |\mathfrak{M}|)$ and let x be an element of $\mathfrak{M}$ of order q. Then $x^h \in \mathfrak{Z}$ for some h. By h), $\mathfrak{Z}$ is cyclic, so $\mathfrak{P} \leq \mathbf{N}_{\mathfrak{G}}(\langle x^h \rangle)$. If $q > 3$ or if $q = 3$ and 3 divides $|\mathfrak{N}|$, then $x \in \mathfrak{N}$, so $\mathbf{N}_{\mathfrak{G}}(\langle x \rangle) = \mathfrak{M}$ and $\mathfrak{M} \geq \mathfrak{P}^{h^{-1}}$, a contradiction. Thus $q = 3$ and 3 does not divide $|\mathfrak{N}|$. Hence $|\mathfrak{Z}| \neq 5$ and by h), $\mathbf{N}_{\mathfrak{G}}(\mathfrak{P})$ is a 2-nilpotent maximal subgroup of $\mathfrak{G}$. Since $\mathbf{N}_{\mathfrak{G}}(\mathfrak{P}) = \mathbf{N}_{\mathfrak{G}}(\mathfrak{Z}) \leq \mathbf{N}_{\mathfrak{G}}(\langle x^h \rangle) < \mathfrak{G}$, $\mathbf{N}_{\mathfrak{G}}(\langle x^h \rangle) = \mathbf{N}_{\mathfrak{G}}(\mathfrak{P})$. But $x \in \mathfrak{M}' = \mathbf{C}_{\mathfrak{G}}(\mathfrak{N})$, so $\mathfrak{N}^h \leq \mathbf{C}_{\mathfrak{G}}(x^h) \leq \mathbf{N}_{\mathfrak{G}}(\mathfrak{P})$. Since $|\mathfrak{N}|$ is odd, $\mathfrak{N}^h \leq \mathbf{O}_{2'}(\mathbf{N}_{\mathfrak{G}}(\mathfrak{P})) = \mathfrak{Z}$. Since $\mathfrak{Z}$ is cyclic, $\mathfrak{N}^h \trianglelefteq \mathbf{N}_{\mathfrak{G}}(\mathfrak{Z})$ and $\mathbf{N}_{\mathfrak{G}}(\mathfrak{P}) \leq \mathbf{N}_{\mathfrak{G}}(\mathfrak{N}^h) = \mathfrak{M}^h$. Since $|\mathfrak{P}|$ does not divide $|\mathfrak{M}|$, this is impossible.

j) Let p be a prime which divides $|\mathfrak{G}|$ but does not divide $|\mathfrak{M}|$ and let $\mathfrak{P} \in S_p(\mathfrak{G})$. Then $|\mathbf{N}_{\mathfrak{G}}(\mathfrak{P})|$ is divisible by 8 but not by 16. Also, if $\mathfrak{Z} = \mathbf{O}_{2'}(\mathbf{N}_{\mathfrak{G}}(\mathfrak{P}))$,

$$\left|\bigcup_{g \in \mathfrak{G}} (\mathfrak{Z} \times \mathbf{Z}(\mathfrak{G}) - \mathbf{Z}(\mathfrak{G}))^g\right| = \frac{1}{4}|\mathfrak{G}| - \frac{1}{4}|\mathfrak{G} : \mathfrak{Z}|.$$

By h), $|\mathbf{N}_{\mathfrak{G}}(\mathfrak{P})|$ is not divisible by 16. Also $\mathbf{N}_{\mathfrak{G}}(\mathfrak{P})$ is 2-nilpotent and $\mathfrak{Z}$ is cyclic. Thus $\mathbf{N}_{\mathfrak{G}}(\mathfrak{P})/\mathbf{C}_{\mathfrak{G}}(\mathfrak{P})$ is a 2-group. Since $\mathfrak{G} = \mathfrak{G}'$, it follows from

Burnside's transfer theorem (IV, 2.6) that $\mathbf{N}_{\mathfrak{G}}(\mathfrak{P}) > \mathbf{C}_{\mathfrak{G}}(\mathfrak{P})$. Thus $|\mathbf{N}_{\mathfrak{G}}(\mathfrak{P}) : \mathbf{C}_{\mathfrak{G}}(\mathfrak{P})|$ is divisible by 2. Let 2^j be the highest power of 2 which divides $|\mathbf{N}_{\mathfrak{G}}(\mathfrak{P})|$. Since $\mathbf{C}_{\mathfrak{G}}(\mathfrak{P}) \geq \mathbf{Z}(\mathfrak{G})$ and $|\mathbf{Z}(\mathfrak{G})| = 2$, $j \geq 2$. By h), $|\mathbf{N}_{\mathfrak{G}}(\mathfrak{Z})| = |\mathbf{N}_{\mathfrak{G}}(\mathfrak{P})| = 2^j|\mathfrak{Z}|$. Now $\mathfrak{Z} \cap \mathfrak{Z}^g = 1$ if $g \notin \mathbf{N}_{\mathfrak{G}}(\mathfrak{Z})$; this is trivial if $|\mathfrak{Z}| = 5$ and follows from 8.14c) otherwise. Hence if $g \notin \mathbf{N}_{\mathfrak{G}}(\mathfrak{Z})$,

$$(\mathfrak{Z}^g \times \mathbf{Z}(\mathfrak{G})) \cap (\mathfrak{Z} \times \mathbf{Z}(\mathfrak{G})) = \mathbf{Z}(\mathfrak{G}).$$

Hence

$$\left|\bigcup_{g \in \mathfrak{G}} (\mathfrak{Z} \times \mathbf{Z}(\mathfrak{G}) - \mathbf{Z}(\mathfrak{G}))^g\right| = |\mathfrak{G} : \mathbf{N}_{\mathfrak{G}}(\mathfrak{Z})|\,|\mathfrak{Z} \times \mathbf{Z}(\mathfrak{G}) - \mathbf{Z}(\mathfrak{G})|$$

$$= \frac{|\mathfrak{G}|}{2^j|\mathfrak{Z}|}(2|\mathfrak{Z}| - 2) = \frac{1}{2^{j-1}}(|\mathfrak{G}| - |\mathfrak{G} : \mathfrak{Z}|).$$

It remains to prove that $j = 3$. Otherwise $j = 2$. By i),

$$\mathfrak{M}^g \cap (\mathfrak{Z} \times \mathbf{Z}(\mathfrak{G}))^h = \mathbf{Z}(\mathfrak{G})$$

for all g, h in $\mathfrak{G}$. Therefore

$$|\mathfrak{G}| \geq \left|\bigcup_{g \in \mathfrak{G}} \mathfrak{M}^g\right| + \left|\bigcup_{g \in \mathfrak{G}} (\mathfrak{Z} \times \mathbf{Z}(\mathfrak{G}) - \mathbf{Z}(\mathfrak{G}))^g\right|.$$

Hence by g),

$$|\mathfrak{G}| \geq \tfrac{3}{4}|\mathfrak{G}| + 2 - 2|\mathfrak{G} : \mathfrak{M}| + \tfrac{1}{2}(|\mathfrak{G}| - |\mathfrak{G} : \mathfrak{Z}|).$$

This gives

$$\frac{|\mathfrak{G}|}{4} < 2\,|\mathfrak{G} : \mathfrak{M}| + \tfrac{1}{2}\,|\mathfrak{G} : \mathfrak{Z}|,$$

so

$$\frac{1}{4} < \frac{2}{|\mathfrak{M}|} + \frac{1}{2|\mathfrak{Z}|} = \frac{1}{24|\mathfrak{N}|} + \frac{1}{2|\mathfrak{Z}|} \leq \frac{1}{24 \cdot 3} + \frac{1}{2 \cdot 5} < \frac{1}{4},$$

since $\mathfrak{N} \neq 1$ gives $|\mathfrak{N}| \geq 3$ and $(|\mathfrak{M}|, |\mathfrak{Z}|) = 1$ gives $|\mathfrak{Z}| \geq 5$.

k) If $g \in \mathfrak{G}$ and the order of g is divisible by 4, then some conjugate of g lies in $\mathfrak{M}$.

If the order of g is $4k$, then by e), there exists $x \in \mathfrak{G}$ such that $\mathbf{C}_{\mathfrak{G}}((g^k)^x) \leq \mathfrak{M}$. Thus $g^x \in \mathfrak{M}$.

l) Let p be a prime which divides $|\mathfrak{G}|$ but does not divide $|\mathfrak{M}|$, let $\mathfrak{P} \in S_p(\mathfrak{G})$ and $\mathfrak{Z} = \mathbf{O}_{2'}(\mathbf{N}_{\mathfrak{G}}(\mathfrak{P}))$. If $g \in \mathfrak{G}$ and the order of g is divisible by a prime divisor q of $|\mathfrak{Z}|$, some conjugate of g lies in $\mathfrak{Z} \times \mathbf{Z}(\mathfrak{G})$.

Suppose that the order of g is qk. Then $g^k \in \mathfrak{Z}^x - \{1\}$ for some x. If $\mathfrak{P} = \mathfrak{Z}$ and $|\mathbf{N}_{\mathfrak{G}}(\mathfrak{P})| = 40$, then $q = 5$ and $\langle g^k \rangle = \mathfrak{Z}^x$, whence $g \in \mathbf{C}_{\mathfrak{G}}(\mathfrak{Z})^x \leq \mathbf{N}_{\mathfrak{G}}(\mathfrak{Z})^x = \mathbf{N}_{\mathfrak{G}}(\mathfrak{P})^x$. Otherwise, by h), $\mathbf{N}_{\mathfrak{G}}(\mathfrak{P})$ is maximal and in this case also $g \in \mathbf{C}_{\mathfrak{G}}(g^k) \leq \mathbf{N}_{\mathfrak{G}}(\langle g^k \rangle) = \mathbf{N}_{\mathfrak{G}}(\mathfrak{P})^x$. Since the order of g is divisible by q, g is not conjugate to any element of $\mathfrak{M}$, so by k), the order of g is not divisible by 4. It follows that $g \in (\mathfrak{Z} \times \mathbf{Z}(\mathfrak{G}))^x$.

m) If $g \in \mathfrak{G}$ and the order of g is divisible by an odd prime divisor q of $|\mathfrak{M}|$, some conjugate of g lies in $\mathfrak{M}$.

Suppose that the order of g is qk. If $q > 3$ or $q = 3$ and 3 divides $|\mathfrak{N}|$, $g^k \in \mathfrak{N}^x - \{1\}$ for some x. Hence

$$g \in \mathbf{C}_{\mathfrak{G}}(g^k) \leq \mathbf{N}_{\mathfrak{G}}(\langle g^k \rangle) = \mathfrak{M}^x.$$

Suppose then that $q = 3$ and 3 does not divide $|\mathfrak{N}|$. Then g^k is of order 3 and is conjugate to an element h of $\mathfrak{M}$ for which $\mathfrak{N} \leq \mathbf{C}_{\mathfrak{G}}(h)$. Since $g \in \mathbf{C}_{\mathfrak{G}}(g^k)$, it is sufficient to show that $\mathbf{C}_{\mathfrak{G}}(h) \leq \mathfrak{M}$. Let $\mathfrak{M}^*$ be a maximal subgroup of $\mathfrak{G}$ containing $\mathbf{C}_{\mathfrak{G}}(h)$. If $\mathfrak{M}^*$ is insoluble, then $|\mathfrak{M}^*/\mathfrak{M}^{*\prime}| \leq 2$ and $\mathfrak{M}^{*\prime} \cong SL(2, 5)$, by 8.16. Since the centralizer in the alternating group $\mathfrak{A}_5$ of an element of order 3 is the group generated by it, $|\mathbf{C}_{\mathfrak{M}^*}(h)|$ divides 12. Hence $|\mathbf{C}_{\mathfrak{G}}(h)|$ divides 12. However, this is impossible, since $\mathfrak{N} \leq \mathbf{C}_{\mathfrak{G}}(h)$, $\mathfrak{N} \neq 1$ and $(|\mathfrak{N}|, 6) = 1$. Hence $\mathfrak{M}^*$ is soluble. By 8.15, $|\mathfrak{M}^* : \mathbf{O}_{2'}(\mathfrak{M}^*)|$ is an integer of the form $2^m 3^n$. But $\mathfrak{N} \leq \mathfrak{M}^*$ and $(|\mathfrak{N}|, 6) = 1$, so $\mathfrak{N} \leq \mathbf{O}_{2'}(\mathfrak{M}^*)$. By 8.14, 8.15 and a) (applied to $\mathfrak{M}^*$), $\mathbf{O}_{2'}(\mathfrak{M}^*)$ is cyclic, so $\mathfrak{M}^* \leq \mathbf{N}_{\mathfrak{G}}(\mathfrak{N}) = \mathfrak{M}$. Thus $\mathfrak{M}^* = \mathfrak{M}$ and $\mathbf{C}_{\mathfrak{G}}(h) \leq \mathfrak{M}$.

To complete the proof of 8.17, let p be a fixed prime divisor of $|\mathfrak{G}|$ which does not divide $|\mathfrak{M}|$. Let $\mathfrak{P} \in S_p(\mathfrak{G})$, $\mathfrak{Z} = \mathbf{O}_{2'}(\mathbf{N}_{\mathfrak{G}}(\mathfrak{P}))$. By g), i) and j), the number of elements in

$$\left(\bigcup_{g \in \mathfrak{G}} \mathfrak{M}^g\right) \cup \left(\bigcup_{g \in \mathfrak{G}} (\mathfrak{Z} \times \mathbf{Z}(\mathfrak{G}) - \mathbf{Z}(\mathfrak{G}))^g\right)$$

is

$$\begin{aligned} &\tfrac{3}{4}|\mathfrak{G}| + 2 - 2|\mathfrak{G} : \mathfrak{M}| + \tfrac{1}{4}|\mathfrak{G}| - \tfrac{1}{4}|\mathfrak{G} : \mathfrak{Z}| \\ &= |\mathfrak{G}| + 2 - 2|\mathfrak{G} : \mathfrak{M}| - \tfrac{1}{4}|\mathfrak{G} : \mathfrak{Z}| < |\mathfrak{G}|. \end{aligned}$$

Thus there exists an element y not in this set. By k), l) and m), the order of y cannot be divisible by either 4 or an odd prime divisor of $|\mathfrak{M}||\mathfrak{Z}|$. Nor can it be 2, so it is divisible by some odd prime r which does not divide $|\mathfrak{M}||\mathfrak{Z}|$. Let $\mathfrak{R} \in S_r(\mathfrak{G})$, $\mathfrak{Y} = \mathbf{O}_{2'}(\mathbf{N}_{\mathfrak{G}}(\mathfrak{R}))$. By h), $\mathbf{N}_{\mathfrak{G}}(\mathfrak{R})$ is 2-nilpotent and $\mathfrak{Y}$ is cyclic. By i), $(|\mathfrak{Y}|, |\mathfrak{M}|) = 1$. Further, $(|\mathfrak{Y}|, |\mathfrak{Z}|) = 1$, since otherwise it would follow from l) that some conjugate of $\mathfrak{Y}$ is contained in $\mathfrak{Z} \times \mathbf{Z}(\mathfrak{G})$, which is impossible since r does not divide $|\mathfrak{Z}|$.

Thus by g), i) and j),

$$|\mathfrak{G}| \geq \left|\bigcup_{g \in \mathfrak{G}} \mathfrak{M}^g\right| + \left|\bigcup_{g \in \mathfrak{G}} (\mathfrak{Z} \times \mathbf{Z}(\mathfrak{G}) - \mathbf{Z}(\mathfrak{G}))^g\right| + \left|\bigcup_{g \in \mathfrak{G}} (\mathfrak{Y} \times \mathbf{Z}(\mathfrak{G}) - \mathbf{Z}(\mathfrak{G}))^g\right|$$

$$\geq \tfrac{3}{4}|\mathfrak{G}| + 2 - 2|\mathfrak{G} : \mathfrak{M}| + \tfrac{1}{2}|\mathfrak{G}| - \tfrac{1}{4}|\mathfrak{G} : \mathfrak{Z}| - \tfrac{1}{4}|\mathfrak{G} : \mathfrak{Y}|.$$

This gives

$$\frac{1}{4} < \frac{2}{|\mathfrak{M}|} + \frac{1}{4|\mathfrak{Z}|} + \frac{1}{4|\mathfrak{Y}|}.$$

Since $|\mathfrak{M}| = 48|\mathfrak{N}| \geq 48 \cdot 3$ and $|\mathfrak{Z}|, |\mathfrak{Y}|$ both have prime divisors which do not divide $|\mathfrak{M}|$, it follows that

$$\frac{1}{4} < \frac{1}{24 \cdot 3} + \frac{1}{20} + \frac{1}{20} < \frac{3}{20},$$

a contradiction. **q.e.d.**

We summarize these results as follows.

8.18 Theorem. *Suppose that $\mathfrak{G}$ satisfies Hypothesis 8.9 and that any perfect proper subgroup of $\mathfrak{G}$ is isomorphic to $SL(2, 5)$. Every maximal subgroup $\mathfrak{M}$ of $\mathfrak{G}$ satisfies one of the following.*

a) *$\mathfrak{M}$ is 2-nilpotent.*

b) *$\mathfrak{M} \cong SL(2, 3)$.*

c) *$\mathfrak{M} \cong \mathfrak{G}_{48}$.*

d) *$|\mathfrak{M}/\mathfrak{M}'| \leq 2$ and $\mathfrak{M}' \cong SL(2, 5)$.*

Proof. This follows at once from 8.15, 8.16 and 8.17. **q.e.d.**

The next step in the characterization of $SL(2, 5)$ is to investigate the maximal cyclic subgroups of $\mathfrak{G}$. We begin with the following strengthening of Lemma 8.11.

8.19 Lemma. *Suppose that $\mathfrak{G}$ satisfies Hypothesis 8.9 and that any perfect proper subgroup of $\mathfrak{G}$ is isomorphic to $SL(2, 5)$. Let g be an element of $\mathfrak{G}$ of order 4. Then $\mathbf{C}_{\mathfrak{G}}(g)$ is cyclic and $|\mathbf{N}_{\mathfrak{G}}(\langle g\rangle):\mathbf{C}_{\mathfrak{G}}(g)| = 2$. Also either $\mathbf{C}_{\mathfrak{G}}(g)$ is a 2-group or $\mathbf{N}_{\mathfrak{G}}(\langle g\rangle)$ is a 2-nilpotent maximal subgroup.*

Proof. Since all elements of order 4 are conjugate, there exists a Sylow 2-subgroup $\mathfrak{Q}$ of $\mathfrak{G}$ such that $g \in \mathbf{Z}_2(\mathfrak{Q})$. Since $g \notin \mathbf{Z}(\mathfrak{Q})$, $|\mathbf{N}_{\mathfrak{G}}(\langle g\rangle):\mathbf{C}_{\mathfrak{G}}(g)| = 2$.

If $\mathbf{C}_{\mathfrak{G}}(g)$ is a 2-group, $\mathbf{C}_{\mathfrak{G}}(g)$ is cyclic, since any subgroup of a generalized quaternion group having a central element of order 4 is cyclic. Suppose then that $\mathbf{C}_{\mathfrak{G}}(g)$ is not a 2-group, and let $\mathfrak{M}$ be a maximal subgroup of $\mathfrak{G}$ for which $\mathbf{N}_{\mathfrak{G}}(\langle g\rangle) \leq \mathfrak{M}$. Thus $\mathfrak{Q} \leq \mathfrak{M}$.

If $\mathfrak{M} \cong SL(2, 3)$ or $\mathfrak{M} \cong \mathfrak{G}_{48}$, then $\mathfrak{M}/\mathbf{Z}(\mathfrak{G})$ is isomorphic to $\mathfrak{A}_4$ or $\mathfrak{S}_4$. But in both $\mathfrak{A}_4$ and $\mathfrak{S}_4$, the centralizer of an involution is a 2-group, so $\mathbf{C}_{\mathfrak{G}}(g)$ is a 2-group.

If $\mathfrak{M}$ is insoluble, either $\mathfrak{M} \cong SL(2, 5)$ or $|\mathfrak{M}:\mathfrak{M}'| = 2$ and $\mathfrak{M}' \cong SL(2, 5)$. In the first case $\mathfrak{M}/\mathbf{Z}(\mathfrak{G}) \cong \mathfrak{A}_5$ and the centralizer of an involution in $\mathfrak{A}_5$ is a 2-group. In the second case, observe that $|\mathfrak{G}|$ is divisible by 16, so $g \in \mathfrak{Q}' \leq \mathfrak{M}'$. Thus $\mathbf{C}_{\mathfrak{M}'}(g)$ and $\mathbf{C}_{\mathfrak{M}}(g) = \mathbf{C}_{\mathfrak{G}}(g)$ are 2-groups.

Therefore $\mathfrak{M}$ is 2-nilpotent. Since $\mathbf{C}_{\mathfrak{G}}(g)$ is not a 2-group, $\mathbf{O}_{2'}(\mathfrak{M}) \neq 1$. And since $\mathfrak{M} \geq \mathfrak{Q}$, $\mathfrak{M}/\mathbf{O}_{2'}(\mathfrak{M})$ is not cyclic. Hence by 8.14d), $\mathfrak{M}$ has a cyclic subgroup $\mathfrak{Y}$ of index 2 which is inverted by all elements of $\mathfrak{M} - \mathfrak{Y}$. Since $\mathbf{C}_{\mathfrak{G}}(g)$ is not a 2-group, $g \in \mathfrak{Y}$. Hence $\mathbf{C}_{\mathfrak{G}}(g) = \mathfrak{Y}$ is cyclic and $\mathfrak{M} = \mathbf{N}_{\mathfrak{G}}(\langle g\rangle)$. **q.e.d.**

Note that since $|\mathbf{Z}(\mathfrak{G})| = 2$, any maximal cyclic subgroup of $\mathfrak{G}$ is of even order.

8.20 Lemma. *Suppose that $\mathfrak{G}$ satisfies Hypothesis 8.9 and that any perfect proper subgroup of $\mathfrak{G}$ is isomorphic to $SL(2, 5)$. Let $\mathfrak{Z}$ be a maximal cyclic subgroup of $\mathfrak{G}$.*

a) *If $|\mathfrak{Z}|$ is divisible by 4, $\mathfrak{Z} = \mathbf{C}_{\mathfrak{G}}(h)$ for some element h of $\mathfrak{G}$ of order 4. Further, $|\mathbf{N}_{\mathfrak{G}}(\mathfrak{Z}):\mathfrak{Z}| = 2$, and there exists a Sylow 2-subgroup $\mathfrak{Q}$ of $\mathfrak{G}$ for which $\mathfrak{Q} \leq \mathbf{N}_{\mathfrak{G}}(\mathfrak{Z})$ and $h \in \mathbf{Z}_2(\mathfrak{Q})$. Also, $\mathfrak{Z} \cap \mathfrak{Z}^g = \mathbf{Z}(\mathfrak{G})$ whenever $g \notin \mathbf{N}_{\mathfrak{G}}(\mathfrak{Z})$.*

b) *If $|\mathfrak{Z}|$ is twice an odd number, then $\mathbf{O}_{2'}(\mathfrak{Z})$ is a Hall subgroup of $\mathfrak{G}$, $\mathfrak{Z} \cap \mathfrak{Z}^g = \mathbf{Z}(\mathfrak{G})$ whenever $g \notin \mathbf{N}_{\mathfrak{G}}(\mathfrak{Z})$. Also $\mathbf{N}_{\mathfrak{G}}(\mathfrak{Z})/\mathfrak{Z}$ is a non-identity 2-group.*

Proof. a) Let h be an element of $\mathfrak{Z}$ of order 4. Then $\mathfrak{Z} \leq \mathbf{C}_{\mathfrak{G}}(h)$ and, by 8.19, $\mathbf{C}_{\mathfrak{G}}(h)$ is cyclic. Thus $\mathfrak{Z} = \mathbf{C}_{\mathfrak{G}}(h)$. Hence $\mathbf{N}_{\mathfrak{G}}(\langle h\rangle)$ normalises $\mathfrak{Z}$. But trivially, $\mathbf{N}_{\mathfrak{G}}(\mathfrak{Z})$ normalises $\langle h\rangle$, so $\mathbf{N}_{\mathfrak{G}}(\mathfrak{Z}) = \mathbf{N}_{\mathfrak{G}}(\langle h\rangle)$. Hence by 8.19, $|\mathbf{N}_{\mathfrak{G}}(\mathfrak{Z})/\mathfrak{Z}| = |\mathbf{N}_{\mathfrak{G}}(\langle h\rangle)/\mathbf{C}_{\mathfrak{G}}(h)| = 2$. Since all elements of order 4 are conjugate, $h \in \mathbf{Z}_2(\mathfrak{Q})$ for some $\mathfrak{Q} \in S_2(\mathfrak{G})$. Thus $\mathfrak{Q} \leq \mathbf{N}_{\mathfrak{G}}(\mathfrak{Z})$.

If $\mathfrak{Z} \cap \mathfrak{Z}^g$ contains an element a of odd order, then $\mathbf{C}_{\mathfrak{G}}(h)$ is not a 2-group, so by 8.19, $\mathfrak{M} = \mathbf{N}_{\mathfrak{G}}(\langle h\rangle)$ is a 2-nilpotent maximal subgroup. Then $\langle a\rangle \trianglelefteq \mathfrak{M}$, so $\mathfrak{M} = \mathbf{N}_{\mathfrak{G}}(\langle a\rangle)$. Similarly, $\mathfrak{M}^g = \mathbf{N}_{\mathfrak{G}}(\langle a\rangle)$, so $g \in \mathfrak{M} = \mathbf{N}_{\mathfrak{G}}(\mathfrak{Z})$.

Thus, if $g \notin \mathbf{N}_{\mathfrak{G}}(\mathfrak{Z})$, $\mathfrak{Z} \cap \mathfrak{Z}^g$ is a 2-group. If $|\mathfrak{Z} \cap \mathfrak{Z}^g| > 2$, the subgroup of $\mathfrak{Z} \cap \mathfrak{Z}^g$ of order 4 is both $\langle h\rangle$ and $\langle h^g\rangle$, so again $g \in \mathbf{N}_{\mathfrak{G}}(\mathfrak{Z})$. Hence $\mathfrak{Z} \cap \mathfrak{Z}^g = \mathbf{Z}(\mathfrak{G})$.

b) In this case, $\mathfrak{Z} = \mathfrak{A} \times \mathbf{Z}(\mathfrak{G})$, where $\mathfrak{A} = \mathbf{O}_{2'}(\mathfrak{Z})$. It follows that $\mathbf{N}_{\mathfrak{G}}(\mathfrak{Z}) = \mathbf{N}_{\mathfrak{G}}(\mathfrak{A})$ and that $\mathfrak{A}$ is a maximal cyclic subgroup of odd order. Let $\mathfrak{M}$ be a maximal subgroup of $\mathfrak{G}$ for which $\mathbf{N}_{\mathfrak{G}}(\mathfrak{Z}) \leq \mathfrak{M}$.

Note that if $|\mathfrak{A}|$ is prime, then $\mathfrak{A}$ is a Sylow subgroup of $\mathfrak{G}$, since the Sylow subgroups of odd order are cyclic. And clearly, $\mathfrak{Z} \cap \mathfrak{Z}^g = (\mathfrak{A} \cap \mathfrak{A}^g) \times \mathbf{Z}(\mathfrak{G}) = \mathbf{Z}(\mathfrak{G})$ in this case.

If $\mathfrak{M}$ is insoluble, then $|\mathfrak{A}|$ is 3 or 5, since the alternating group $\mathfrak{A}_5$ has no element of order 15. Also, $\mathbf{N}_{\mathfrak{G}}(\mathfrak{Z})/\mathfrak{Z} = \mathbf{N}_{\mathfrak{M}}(\mathfrak{Z})/\mathfrak{Z}$ is of order 2 or 4 in either case. Similarly, if $\mathfrak{M} \cong \mathfrak{G}_{48}$, then $|\mathfrak{A}| = 3$ and $|\mathbf{N}_{\mathfrak{G}}(\mathfrak{Z})/\mathfrak{Z}| = 2$. Thus all the assertions are proved in these cases. If $\mathfrak{M} \cong SL(2, 3)$, then $|\mathfrak{A}| = 3$ and $\mathbf{N}_{\mathfrak{G}}(\mathfrak{Z}) = \mathfrak{Z}$. But this is impossible, since it would follow from Burnside's transfer theorem that $\mathfrak{G}/\mathbf{Z}(\mathfrak{G})$ has a normal 3-complement.

If $\mathfrak{M}$ is 2-nilpotent, $\mathbf{O}_{2'}(\mathfrak{M})$ is cyclic, by 8.14. Then $\mathfrak{A} \leq \mathbf{O}_{2'}(\mathfrak{M})$ and, by maximality, $\mathfrak{A} = \mathbf{O}_{2'}(\mathfrak{M})$. Thus $\mathfrak{M} = \mathbf{N}_{\mathfrak{G}}(\mathfrak{A})$ and $\mathbf{N}_{\mathfrak{G}}(\mathfrak{A})/\mathfrak{A}$ is a 2-group. By 8.14, $\mathfrak{A}$ is a Hall subgroup of $\mathfrak{G}$ and $\mathfrak{A} \cap \mathfrak{A}^g = 1$ if $g \notin \mathfrak{M}$. If $\mathbf{N}_{\mathfrak{G}}(\mathfrak{Z}) = \mathfrak{Z}$, then $\mathfrak{M} = \mathbf{N}_{\mathfrak{G}}(\mathfrak{A}) = \mathfrak{Z}$ and Burnside's transfer theorem, applied to a Sylow subgroup of $\mathfrak{A}$, gives a contradiction. Thus $\mathbf{N}_{\mathfrak{G}}(\mathfrak{Z})/\mathfrak{Z}$ is a non-identity 2-group. **q.e.d.**

8.21 Theorem. *Suppose that $\mathfrak{G}$ satisfies Hypothesis 8.9 and that any perfect proper subgroup of $\mathfrak{G}$ is isomorphic to $SL(2, 5)$. Let $\mathfrak{Z}_0, \mathfrak{Z}_1, \dots, \mathfrak{Z}_r$ be representatives of all the conjugacy classes of maximal cyclic subgroups of $\mathfrak{G}$, so numbered that $\mathfrak{Z}_0$ contains an element of order 4. Let $|\mathfrak{Z}_i| = 2m_i$.*

a) *For $i > 0$, m_i is odd, and $2m_0 = 2^n m'$, where 2^{n+1} is the order of a Sylow 2-subgroup of $\mathfrak{G}$ and m' is odd.*

b) *For $i \neq j$, $(m_i, m_j) = 1$.*

c) $|\mathfrak{G}| = 4m_0 m_1 \cdots m_r$.

d) $|\mathbf{N}_{\mathfrak{G}}(\mathfrak{Z}_i)/\mathfrak{Z}_i| = 2^{n_i}$, *where $n_0 = 1$ and $n_i \geq 1$ for $i = 1, \dots, r$. If $\mathfrak{P}$ is a subgroup of $\mathfrak{Z}_i$ of odd prime order, $\mathbf{N}_{\mathfrak{G}}(\mathfrak{Z}_i)/\mathfrak{Z}_i$ is faithfully represented on $\mathfrak{P}$. In particular, $n_i = 1$ if 3 divides $|\mathfrak{Z}_i|$.*

e) $|\mathfrak{G}| = 2 + |\mathfrak{G}|\sum_{i=0}^{r}\left(1 - \dfrac{1}{m_i}\right)2^{-n_i}$.

f) $r \geq 2$.

Proof. a) If m_i is even, then by 8.20, $\mathfrak{Z}_i$ is the centralizer of an element of order 4. Since all elements of order 4 are conjugate, $\mathfrak{Z}_i$ is conjugate to $\mathfrak{Z}_0$ and $i = 0$. The fact that $2m_0 = 2^n m'$ follows from the assertions $\mathfrak{Q} \leq \mathbf{N}_{\mathfrak{G}}(\mathfrak{Z})$ and $|\mathbf{N}_{\mathfrak{G}}(\mathfrak{Z})/\mathfrak{Z}| = 2$ of 8.20.

b) Suppose that p is a prime divisor of (m_i, m_j), where $i < j$. Then p is odd, and for some $y \in \mathfrak{G}$, $\mathfrak{Z}_i \cap \mathfrak{Z}_j^y$ contains an element x of order p. Then $\langle \mathfrak{Z}_i, \mathfrak{Z}_j^y \rangle \leq \mathbf{C}_{\mathfrak{G}}(x) \leq \mathbf{N}_{\mathfrak{G}}(\langle x \rangle) \leq \mathfrak{M}$ for some maximal subgroup $\mathfrak{M}$ of $\mathfrak{G}$. Now any subgroup of $\mathfrak{A}_4$, $\mathfrak{S}_4$ or $\mathfrak{A}_5$ of odd prime order is self-centralizing, so if $\mathfrak{M}$ is not 2-nilpotent, $|\mathbf{C}_{\mathfrak{G}}(x)|$ divides $p \cdot 2^2$ and $\mathbf{C}_{\mathfrak{G}}(x)$ is cyclic. Thus $\mathfrak{Z}_i = \mathfrak{Z}_j^y$ and $i = j$. Hence $\mathfrak{M}$ is 2-nilpotent and, by 8.14, $\mathbf{O}_{2'}(\mathfrak{M})$ is cyclic. Also $\mathbf{O}_{2'}(\mathfrak{Z}_i) \leq \mathbf{O}_{2'}(\mathfrak{M})$ and $\mathbf{O}_{2'}(\mathfrak{Z}_j^y) \leq \mathbf{O}_{2'}(\mathfrak{M})$. Since $j > 0$, $\mathbf{O}_{2'}(\mathfrak{Z}_j)$ is a maximal cyclic subgroup of odd order, so $\mathbf{O}_{2'}(\mathfrak{Z}_j^y) = \mathbf{O}_{2'}(\mathfrak{M}) \geq \mathbf{O}_{2'}(\mathfrak{Z}_i)$. Hence $\mathfrak{Z}_j^y \geq \mathbf{Z}(\mathfrak{G}) \times \mathbf{O}_{2'}(\mathfrak{Z}_i)$. Since $i < j$, it follows that $i = 0$. By 8.20, $\mathbf{N}_{\mathfrak{G}}(\mathfrak{Z}_0)$ contains a Sylow 2-subgroup $\mathfrak{Q}$ of $\mathfrak{G}$. Then $\mathfrak{Q} \leq \mathbf{N}_{\mathfrak{G}}(\mathfrak{Z}_0) \leq \mathbf{N}_{\mathfrak{G}}(\langle x \rangle) \leq \mathfrak{M}$. Thus by 8.14d), $\mathfrak{M}$ has a cyclic subgroup $\mathfrak{Y}$ of index 2 which is inverted by all elements of $\mathfrak{M} - \mathfrak{Y}$. Thus $\mathbf{C}_{\mathfrak{M}}(x) \leq \mathfrak{Y}$, so

$$\mathfrak{Z}_0 \leq \mathbf{C}_{\mathfrak{G}}(x) = \mathbf{C}_{\mathfrak{M}}(x) \leq \mathfrak{Y}$$

and $\mathfrak{Z}_0 = \mathfrak{Y}$. Hence $\mathfrak{Z}_j^y = \mathbf{O}_{2'}(\mathfrak{Z}_j^y) \times \mathbf{Z}(\mathfrak{G}) = \mathbf{O}_{2'}(\mathfrak{M}) \times \mathbf{Z}(\mathfrak{G}) \leq \mathfrak{Y} = \mathfrak{Z}_0$. This is impossible.

c) Suppose that p is an odd prime. If p^a divides $|\mathfrak{G}|$, $\mathfrak{G}$ has a cyclic subgroup of order p^a. Thus p^a divides $|\mathfrak{Z}_i|$ for some i and p^a divides $4m_0m_1 \cdots m_r$. If p^a divides $4m_0m_1 \cdots m_r$, then p^a divides m_i for some i, by b). Hence p^a divides $|\mathfrak{G}|$. If 2^{n+1} divides $|\mathfrak{G}|$, then 2^{n+1} divides $4m_0$, by a). By 8.20a), $|\mathbf{N}_{\mathfrak{G}}(\mathfrak{Z}_0)| = 4m_0$, so $4m_0$ divides $|\mathfrak{G}|$. Thus $|\mathfrak{G}| = 4m_0m_1 \cdots m_r$.

d) By 8.20, $|\mathbf{N}_{\mathfrak{G}}(\mathfrak{Z}_0)/\mathfrak{Z}_0| = 2$ and $\mathbf{N}_{\mathfrak{G}}(\mathfrak{Z}_i)/\mathfrak{Z}_i$ is a non-identity 2-group.

To show that $\mathbf{N}_{\mathfrak{G}}(\mathfrak{Z}_i)/\mathfrak{Z}_i$ operates faithfully on $\mathfrak{P}$, we may suppose that $\mathbf{C}_{\mathfrak{G}}(\mathfrak{P})$ is not cyclic, for otherwise, $\mathbf{C}_{\mathfrak{G}}(\mathfrak{P}) = \mathfrak{Z}_i$. Let $\mathfrak{M}$ be a maximal subgroup of $\mathfrak{G}$ for which $\mathbf{N}_{\mathfrak{G}}(\mathfrak{P}) \leq \mathfrak{M}$. If $\mathfrak{M}$ is not 2-nilpotent, $\mathbf{C}_{\mathfrak{G}}(\mathfrak{P})$ is cyclic because any subgroup of $\mathfrak{A}_4$, $\mathfrak{S}_4$ or $\mathfrak{A}_5$ of odd prime order is self-centralizing. If $\mathfrak{M}$ is 2-nilpotent and the Sylow 2-subgroups of $\mathfrak{M}$ are not cyclic, then by 8.14d), $\mathfrak{M}$ has a normal cyclic subgroup $\mathfrak{Y}$ of index 2, so $\mathfrak{P} \leq \mathfrak{Y}$ and $\mathbf{C}_{\mathfrak{G}}(\mathfrak{P}) = \mathbf{C}_{\mathfrak{M}}(\mathfrak{P}) \leq \mathfrak{Y}$. In the remaining case, $i > 0$, since $\mathbf{N}_{\mathfrak{G}}(\mathfrak{Z}_0) \leq \mathfrak{M}$ and the Sylow 2-subgroups of $\mathbf{N}_{\mathfrak{G}}(\mathfrak{Z}_0)$ are not cyclic, by 8.20. Since $(m_0, m_i) = 1$, $\mathfrak{P} \not\leq \mathfrak{Z}_0^x$ for any $x \in \mathfrak{G}$. Thus by 8.20, $\mathfrak{P}$

centralizes no element of order 4, and $|\mathbf{C}_{\mathfrak{G}}(\mathfrak{P})| \equiv 2$ (4). Thus $|\mathbf{C}_{\mathfrak{G}}(\mathfrak{P})/\mathfrak{Z}_i|$ is odd. Since $\mathbf{N}_{\mathfrak{G}}(\mathfrak{Z}_i)/\mathfrak{Z}_i$ is a 2-group, it follows that $\mathbf{C}_{\mathfrak{G}}(\mathfrak{P}) \cap \mathbf{N}_{\mathfrak{G}}(\mathfrak{Z}_i) \leq \mathfrak{Z}_i$.

e) Each element of $\mathfrak{G}$ lies in at least one conjugate of a $\mathfrak{Z}_i$. By b), $\mathfrak{Z}_i^g \cap \mathfrak{Z}_j^h = \mathbf{Z}(\mathfrak{G})$ for $i \neq j$. By 8.20, $\mathfrak{Z}_i \cap \mathfrak{Z}_i^g = \mathbf{Z}(\mathfrak{G})$ for $g \notin \mathbf{N}_{\mathfrak{G}}(\mathfrak{Z}_i)$. Thus

$$\begin{aligned}|\mathfrak{G}| &= |\mathbf{Z}(\mathfrak{G})| + \left|\bigcup_{i=0}^{r} \bigcup_{g \in \mathfrak{G}} (\mathfrak{Z}_i - \mathbf{Z}(\mathfrak{G}))^g\right| \\ &= 2 + \sum_{i=0}^{r} |\mathfrak{G} : \mathbf{N}_{\mathfrak{G}}(\mathfrak{Z}_i)|(|\mathfrak{Z}_i| - 2) \\ &= 2 + |\mathfrak{G}| \sum_{i=0}^{r} \frac{2m_i - 2}{2^{n_i+1}m_i} \\ &= 2 + |\mathfrak{G}| \sum_{i=0}^{r} \left(1 - \frac{1}{m_i}\right) 2^{-n_i}.\end{aligned}$$

f) Certainly $r > 0$, for a finite group is never covered by the conjugates of a proper subgroup. If $r = 1$, e) gives

$$|\mathfrak{G}| = 2 + \frac{|\mathfrak{G}|}{2} - \frac{|\mathfrak{G}|}{2m_0} + \frac{|\mathfrak{G}|}{2^{n_1}} - \frac{|\mathfrak{G}|}{2^{n_1}m_1}.$$

Since $n_1 \geq 1$, it follows that

$$2 - \frac{|\mathfrak{G}|}{2m_0} - \frac{|\mathfrak{G}|}{2^{n_1}m_1} = \frac{|\mathfrak{G}|}{2} - \frac{|\mathfrak{G}|}{2^{n_1}} \geq 0.$$

Since $|\mathfrak{G}| > 2m_0$ and $|\mathfrak{G}| > 2^{n_1}m_1$, this is impossible. **q.e.d.**

8.22 Corollary. *If, in 8.21, $n_1 = n_2 = \cdots = n_r = 1$, then $\mathfrak{G} \cong SL(2, 5)$.*

Proof. By 8.21e),

$$1 = \frac{2}{|\mathfrak{G}|} + \frac{1}{2} \sum_{i=0}^{r} \left(1 - \frac{1}{m_i}\right).$$

By 8.21f), $r \geq 2$. By 8.21b), $(m_i, m_j) = 1$ for $i \neq j$, so we may suppose that $m_0 \geq 2$, $m_1 \geq 3$, $m_2 \geq 5$ and, if $r > 2$, $m_3 \geq 7$. Thus if $r > 2$,

$$1 > \frac{1}{2}\left\{\left(1 - \frac{1}{2}\right) + \left(1 - \frac{1}{3}\right) + \left(1 - \frac{1}{5}\right) + \left(1 - \frac{1}{7}\right)\right\},$$

which is not the case. Thus $r = 2$. If m_0, m_1, m_2 are all at least 3, we get

$$1 \geq \frac{2}{|\mathfrak{G}|} + \frac{3}{2}\cdot\frac{2}{3},$$

a contradiction. Hence $m_0 = 2$. By 8.21a), m_1 and m_2 are both odd. If they are both greater than 3, we again get a contradiction, since $1 - \dfrac{1}{m_i} \geq \dfrac{4}{5}$. Thus $m_1 = 3$ and

$$\frac{5}{12} = \frac{2}{|\mathfrak{G}|} + \frac{1}{2}\left(1 - \frac{1}{m_2}\right) > \frac{1}{2} - \frac{1}{2m_2}.$$

It follows that $m_2 < 6$, so $m_2 = 5$. By 8.21c),

$$|\mathfrak{G}| = 4m_0m_1m_2 = 120.$$

By 8.3, $\mathfrak{G} \cong SL(2, 5)$. **q.e.d.**

We shall now use character theory to obtain the decisive result for the characterization.

8.23 Theorem. *Suppose that $\mathfrak{G}$ is perfect and that $\mathfrak{G}$ is isomorphic to a group of fixed point free automorphisms of a finite group. If every perfect proper subgroup of $\mathfrak{G}$ is isomorphic to $SL(2, 5)$, then $\mathfrak{G}$ is isomorphic to a subgroup of $SL(2, \mathbb{C})$.*

Proof. By 8.1, we may suppose that $\mathfrak{G}$ is an irreducible subgroup of $GL(k, \mathbb{C})$ for some k and that, if $1 \neq g \in \mathfrak{G}$, 1 is not an eigen-value of g. Let χ be the character of $\mathfrak{G}$; $\chi(g) = \operatorname{tr} g$ for all $g \in \mathfrak{G}$.

By 8.10, $\mathfrak{G}$ satisfies Hypothesis 8.9. Let $\mathfrak{Z}_0, \mathfrak{Z}_1, \ldots, \mathfrak{Z}_r$ be representatives of all the conjugacy classes of maximal cyclic subgroups of $\mathfrak{G}$, so numbered that $\mathfrak{Z}_0$ contains an element of order 4. Let $|\mathfrak{Z}_i| = 2m_i$, $|\mathbf{N}_{\mathfrak{G}}(\mathfrak{Z}_i)/\mathfrak{Z}_i| = 2^{n_i}$.

a) k is even.

Since $\mathfrak{G}$ is irreducible, the non-identity element of $\mathbf{Z}(\mathfrak{G})$ is $-I$. But since $\mathfrak{G} = \mathfrak{G}'$, $\det g = 1$ for all $g \in \mathfrak{G}$. Hence $\det(-I) = (-1)^k = 1$.

b) There is at most one $\mathfrak{Z}_i$ such that χ is not rational-valued on $\mathfrak{Z}_i$.

For suppose that χ is not rational-valued on both $\mathfrak{Z}_i$ and $\mathfrak{Z}_j$, where $i \neq j$. Let π, σ be the sets of prime divisors of m_i, m_j respectively. Since the order of any element of $\mathfrak{G}/\mathbf{Z}(\mathfrak{G})$ divides m_l for some l, every element of

$\mathfrak{G}/\mathbf{Z}(\mathfrak{G})$ is either a π'-element or a σ'-element. By 8.8, χ is rational-valued on either $\mathfrak{Z}_i$ or $\mathfrak{Z}_j$. This is a contradiction.

c) If χ is rational-valued on $\mathfrak{Z}_i$, then k is divisible by $\phi(2m_i)$.

Let ψ be the restriction to $\mathfrak{Z}_i$ of χ. If $1 \neq h \in \mathfrak{Z}_i$, 1 is not an eigen-value of h, so ψ is the sum of faithful irreducible characters of $\mathfrak{Z}_i$. The irreducible characters of the cyclic group $\mathfrak{Z}_i$ form a cyclic group, generated, say, by λ. Thus the faithful irreducible characters of $\mathfrak{Z}_i$ are the $\lambda\alpha$, where α runs through the Galois group $\mathfrak{J}$ of $\mathbb{Q}(\omega)$ and ω is a primitive $2m_i$-th root of unity. Since ψ is rational-valued,

$$(\psi, \lambda\alpha)_{\mathfrak{Z}_i} = (\psi\alpha, \lambda\alpha)_{\mathfrak{Z}_i} = ((\psi, \lambda)_{\mathfrak{Z}_i})\alpha = (\psi, \lambda)_{\mathfrak{Z}_i}.$$

Hence

$$\psi = b \sum_{\alpha \in \mathfrak{J}} \lambda\alpha$$

for some b, and $k = \psi(1) = b|\mathfrak{J}| = b\phi(2m_i)$. Thus c) is proved.

By 8.11, $|\mathfrak{G}|$ is divisible by 3, so there exists s such that 3 divides m_s. Suppose that χ has a algebraic conjugates over $\mathbb{Q}$.

d) $ak^2 < 6\phi(2m_s)$ and $k < 6$.

Let $\chi_1, \ldots, \chi_a$ be the algebraic conjugates of χ over $\mathbb{Q}$ and put $\eta = \chi_1 + \cdots + \chi_a$. Thus η is rational-valued.

Write $\mathbf{Z}(\mathfrak{G}) = \langle j \rangle$. Then since $(\eta, \eta)_{\mathfrak{G}} = a$,

$$\begin{aligned} a|\mathfrak{G}| = \sum_{g \in \mathfrak{G}} |\eta(g)|^2 &\geq \eta(1)^2 + \eta(j)^2 + |\mathfrak{G} : \mathbf{N}_{\mathfrak{G}}(\mathfrak{Z}_s)| \sum_{h \in \mathfrak{Z}_s - \mathbf{Z}(\mathfrak{G})} |\eta(h)|^2 \\ &> |\mathfrak{G} : \mathbf{N}_{\mathfrak{G}}(\mathfrak{Z}_s)|(|\mathfrak{Z}_s|(\psi, \psi)_{\mathfrak{Z}_s} - \eta(1)^2 - \eta(j)^2), \end{aligned}$$

where ψ is the restriction to $\mathfrak{Z}_s$ of η. By 8.21d), $|\mathbf{N}_{\mathfrak{G}}(\mathfrak{Z}_s) : \mathfrak{Z}_s| = 2$, so

$$a|\mathfrak{G}| > \frac{|\mathfrak{G}|}{4m_s}(2m_s(\psi, \psi)_{\mathfrak{Z}_s} - 2(ak)^2). \tag{1}$$

Since h does not have the eigen-value 1 for $1 \neq h \in \mathfrak{Z}_s$, ψ is the sum of faithful irreducible characters of $\mathfrak{Z}_s$. As in c), these are $\lambda\alpha$, where α runs through $\mathfrak{J}$, so

$$(\psi, \lambda\alpha)_{\mathfrak{Z}_s} = (\psi, \lambda)_{\mathfrak{Z}_s}$$

and

$$\psi = b \sum_{\alpha \in \mathfrak{J}} \lambda\alpha$$

for some b. Thus

$$(2) \qquad ak = \eta(1) = \psi(1) = b|\mathfrak{I}| = b\phi(2m_s).$$

Also

$$(\psi, \psi)_{\mathfrak{Z}_s} = b^2|\mathfrak{I}| = b^2\phi(2m_s).$$

It follows from (1) that

$$a|\mathfrak{G}| > \frac{|\mathfrak{G}|}{4m_s}(2m_s b^2\phi(2m_s) - 2a^2k^2) = \frac{|\mathfrak{G}|}{4m_s}\left(\frac{2m_s a^2 k^2}{\phi(2m_s)} - 2a^2k^2\right)$$
$$= \frac{|\mathfrak{G}|}{2m_s} a^2k^2\left(\frac{m_s}{\phi(2m_s)} - 1\right).$$

Since $2m_s \equiv 0$ (6),

$$\phi(2m_s) = 2m_s(1 - \tfrac{1}{2})(1 - \tfrac{1}{3}) \cdots \leq \tfrac{2}{3}m_s.$$

Hence $m_s - \phi(2m_s) \geq \frac{1}{3}m_s$ and

$$a > \frac{a^2k^2}{2m_s}\frac{m_s}{3\phi(2m_s)} = \frac{a^2k^2}{6\phi(2m_s)}.$$

Thus $6\phi(2m_s) > ak^2$. By (2), $6\phi(2m_s) > kb\phi(2m_s)$ and $k \leq kb < 6$.

e) $k = 2$.

Suppose that $k \neq 2$. By a), k is even, and by d), $k < 6$. Hence $k = 4$.

Suppose first that χ is rational-valued. By c), 4 is divisible by $\phi(2m_i)$ for all $i = 0, 1, \ldots, r$. This shows that $m_i \in \{2, 3, 4, 5, 6\}$. Since $r \geq 2$ (8.21f)) and $(m_i, m_j) = 1$ for $i \neq j$ (8.21b)), we can put $m_1 = 3$, $m_2 = 5$, and m_0 is 2 or 4. By 8.21c), $|\mathfrak{G}| = 4m_0m_1m_2 = 60m_0$. If $m_0 = 2$, then by 8.3, $\mathfrak{G} \cong SL(2, 5)$. If s_j ($j = 1, 2, \ldots$) are the degrees of the faithful irreducible characters of $SL(2, 5)$, then $s_j = 2t_j$ (as $SL(2, 5)$ is perfect) and

$$120 = SL(2, 5) = 60 + 4\sum_j t_j^2.$$

This forces $s_1 = 2$, so $SL(2, 5)$ has a faithful irreducible representation over $\mathbb{C}$ of degree 2. If $m_0 = 4$, $|\mathfrak{G}| = 240$. By 8.21e),

$$1 = \frac{2}{|\mathfrak{G}|} + \sum_{i=0}^{2}\left(1 - \frac{1}{m_i}\right)2^{-n_i}$$
$$= \frac{1}{120} + \left(1 - \frac{1}{4}\right)\frac{1}{2} + \left(1 - \frac{1}{3}\right)\frac{1}{2} + \left(1 - \frac{1}{5}\right)2^{-n_2}.$$

This gives $2^{n_2} = \frac{48}{17}$, a contradiction.

Now suppose that χ is not rational-valued. Then in d), $a \neq 1$, so

$$\phi(2m_s) > \frac{k^2}{3} = \frac{16}{3} > 4.$$

Thus by c), χ is not rational-valued on $\mathfrak{Z}_s$. If $i \neq s$, then by b), c), 4 is divisible by $\phi(2m_i)$. Also m_i is not divisible by 3, so $m_i \in \{2, 4, 5\}$. Since $r \geq 2$ and m_i is odd for $i > 0$, $s \neq 0$. Thus we may suppose that m_1 is divisible by 3, $m_2 = 5$ and m_0 is either 2 or 4. Since m_1 is divisible by 3, $n_0 = n_1 = 1$ by 8.21d). Since $m_2 = 5$, 8.21d) gives $n_2 \leq 2$. Thus

$$\begin{aligned} 1 &= \frac{2}{|\mathfrak{G}|} + \sum_{i=0}^{2} \left(1 - \frac{1}{m_i}\right) 2^{-n_i} \\ &= \frac{2}{20 m_0 m_1} + \left(1 - \frac{1}{m_0}\right)\frac{1}{2} + \left(1 - \frac{1}{m_1}\right)\frac{1}{2} + \left(1 - \frac{1}{5}\right) 2^{-n_2}. \end{aligned}$$

If $n_2 = 2$, this gives $m_1 = -9$ for $m_0 = 2$ and $m_1 = \frac{19}{3}$ for $m_0 = 4$. Both are impossible, so $n_2 = 1$. Thus by 8.22, $\mathfrak{G} \cong SL(2, 5)$ and $\mathfrak{G}$ has a faithful representation of degree 2.

By e), $\mathfrak{G}$ is a subgroup of $GL(2, \mathbb{C})$. Since $\mathfrak{G} = \mathfrak{G}'$, $\mathfrak{G} \leq SL(2, \mathbb{C})$.
q.e.d.

The main theorem of this section is now very easy to prove.

8.24 Theorem (ZASSENHAUS [3]). *Suppose that $\mathfrak{G}$ is a group of fixed point free automorphisms of some finite group and that $\mathfrak{G} = \mathfrak{G}' \neq 1$. Then $\mathfrak{G} \cong SL(2, 5)$.*

Proof. Suppose that $\mathfrak{G}$ is a counterexample of minimal order. Then any perfect proper subgroup of $\mathfrak{G}$ is isomorphic to $SL(2, 5)$. By 8.23, $\mathfrak{G}$ is isomorphic to a subgroup of $SL(2, \mathbb{C})$. Since $\mathfrak{G} = \mathfrak{G}'$, it follows from 8.6 that $\mathfrak{G} \cong SL(2, 5)$. **q.e.d.**

§ 9. Sharply Doubly Transitive Permutation Groups

In XI, § 2 and XII, § 3, we determined all sharply k-fold transitive permutation groups for $k \geq 3$. We now turn to the sharply doubly transitive permutation groups.

9.1 Theorem. *Suppose that* $\mathfrak{G}$ *is a sharply doubly transitive group of degree n. Then n is a prime-power* p^f *and* $\mathfrak{G}$ *is a Frobenius group with elementary Abelian Frobenius kernel. Also,* $\mathfrak{G}$ *is similar to a permutation group* $\mathfrak{F}\mathfrak{H}$ *on the vector space* $\mathsf{V} = \mathsf{V}(f, p)$, *where* $\mathfrak{F}$ *is the group of all translations* $v \to v + a$ *of* V *and* $\mathfrak{H}$ *is the stabiliser of 0; also* $\mathfrak{H}$ *is regular on* $\mathsf{V} - \{0\}$ *and* $\mathfrak{H} \leq GL(V)$. *In particular,* $|\mathfrak{H}| = p^f - 1$.

Proof. For any distinct elements a, b of the set of symbols permuted, $\mathfrak{G}_{a,b} = 1$, so $\mathfrak{G}$ is a Frobenius group. Since $\mathfrak{G}$ is primitive, it follows from V, 8.19 that the Frobenius kernel is elementary Abelian. Hence $n = p^f$ for some prime p. By II, 3.5, it follows that $\mathfrak{G}$ is similar to the group $\mathfrak{F}\mathfrak{H}$. Since $\mathfrak{H}$ is regular on $\mathsf{V} - \{0\}$, $|\mathfrak{H}| = p^f - 1$. **q.e.d.**

Henceforth we shall retain the notation of 9.1 and assume that $\mathfrak{G} = \mathfrak{F}\mathfrak{H}$. Of course $\mathfrak{F}$ is the Frobenius kernel of $\mathfrak{G}$.

9.2 Definition. We distinguish between three types of sharply doubly transitive permutation groups.

(1) $\mathfrak{G}$ is said to be of *type I* if $\mathfrak{H}$ has a cyclic normal subgroup which is irreducible on V. (In this case, by II, 3.12, $\mathfrak{G}$ is similar to a subgroup of the group $\Gamma(p^n)$ of semilinear mappings $x \to a(x\alpha) + b$ on $GF(p^f)$.)

(2) $\mathfrak{G}$ is said to be of *type II* if $\mathfrak{G}$ is soluble but $\mathfrak{G}$ is not of type I.

(3) $\mathfrak{G}$ is said to be of *type III* if $\mathfrak{G}$ is insoluble.

9.3 Lemma. *Suppose that* $\mathfrak{G}$ *is of type II or III and let* $\mathfrak{A}$ *be a maximal normal Abelian subgroup of* $\mathfrak{H}$.

a) $\mathfrak{A}$ *is cyclic and reducible on* V. *The* $\mathfrak{A}$*-module* V *is the direct sum of isomorphic* $\mathfrak{A}$*-submodules of dimension k, say. Also* $|\mathfrak{A}|$ *divides* $p^k - 1$.

b) $\mathfrak{A} < \mathbf{C}_{\mathfrak{H}}(\mathfrak{A})$. *Thus* $\mathbf{C}_{\mathfrak{H}}(\mathfrak{A})$ *is non-Abelian.*

c) *Let* $\mathfrak{N}$ *be a minimal non-Abelian normal subgroup of* $\mathfrak{H}$ *contained in* $\mathbf{C}_{\mathfrak{H}}(\mathfrak{A})$. *Then either* $\mathfrak{N}$ *is a quaternion group of order 8 or* $\mathfrak{N} \cong SL(2, 5)$.

d) V *is an irreducible* $\mathfrak{N}$*-module.*

e) $\mathbf{C}_{\mathfrak{H}}(\mathfrak{N}) = \mathfrak{A}$.

Proof. a) If $\mathfrak{A}$ were irreducible, $\mathfrak{G}$ would be of type I. Thus $\mathfrak{A}$ is reducible. By V, 17.3, there is a decomposition $\mathsf{V} = \mathsf{W}_1 \oplus \cdots \oplus \mathsf{W}_l$, where each W_i is the direct sum of isomorphic irreducible $\mathfrak{A}$-modules and the W_i are permuted transitively by $\mathfrak{H}$. If $l > 1$, suppose that $0 \neq w_i \in \mathsf{W}_i$ $(i = 1, 2)$. Then for each $h \in \mathfrak{H}$, $w_1 h$ lies in one of the W_j, so $w_1 h \neq w_1 + w_2$. This contradicts the transitivity of $\mathfrak{H}$ on $\mathsf{V} - \{0\}$, so $l = 1$. Hence V is the direct sum of isomorphic irreducible $\mathfrak{A}$-submodules, say of dimension k. Thus $\mathfrak{A}$ has a faithful irreducible representation in $GF(p)$ of degree k. By II, 3.10, $\mathfrak{A}$ is cyclic and $|\mathfrak{A}|$ divides $p^k - 1$.

b) Since $\mathfrak{A}$ is cyclic, $|\mathbf{Aut}\,\mathfrak{A}| = \phi(|\mathfrak{A}|) \leq |\mathfrak{A}|$, so $|\mathfrak{H} : \mathbf{C}_{\mathfrak{H}}(\mathfrak{A})| \leq |\mathfrak{A}|$ and $|\mathbf{C}_{\mathfrak{H}}(\mathfrak{A})| \geq |\mathfrak{H}|/|\mathfrak{A}|$. Since f is a multiple of k,

$$\frac{|\mathfrak{H}|}{|\mathfrak{A}|} \geq \frac{p^f - 1}{p^k - 1} = 1 + p^k + \cdots > p^k - 1 \geq |\mathfrak{A}|,$$

so $\mathbf{C}_{\mathfrak{H}}(\mathfrak{A}) > \mathfrak{A}$. By maximality of $\mathfrak{A}$, $\mathbf{C}_{\mathfrak{H}}(\mathfrak{A})$ is non-Abelian.

c) If $\mathfrak{N}' = \mathfrak{N}$, then $\mathfrak{N} \cong SL(2, 5)$, by 8.24. Suppose that $\mathfrak{N}' < \mathfrak{N}$ (cf. 7.1). Then $\mathfrak{N}'$ is Abelian, and since $\mathfrak{N}' \leq \mathbf{C}_{\mathfrak{G}}(\mathfrak{A})$, $\mathfrak{N}'\mathfrak{A}$ is Abelian. Hence $\mathfrak{N}' \leq \mathfrak{A}$ on account of the maximality of $\mathfrak{A}$. Thus

$$\mathfrak{N}' \leq \mathfrak{N} \cap \mathfrak{A} \leq \mathbf{Z}(\mathfrak{N}).$$

Hence $\mathfrak{N}$ is nilpotent of class 2. The minimality of $\mathfrak{N}$ forces $\mathfrak{N}$ to be a q-group for some prime q. By V, 8.7b), the Sylow q-subgroups of $\mathfrak{H}$ are cyclic for q odd and generalized quaternion or cyclic for $q = 2$. Thus $q = 2$ and $\mathfrak{N}$, being of class 2, is the quaternion group of order 8.

d) Suppose that V is a reducible $\mathfrak{N}$-module and let U be an irreducible $\mathfrak{N}$-submodule of V. Then for each $h \in \mathfrak{H}$, $\mathsf{U}h$ is also an irreducible $\mathfrak{N}$-submodule of V. Let

$$\mathscr{U} = \{\mathsf{U}h | h \in \mathfrak{H}\}.$$

Then $\mathscr{U}$ is a partition of V. For if $0 \neq v \in \mathsf{V}$, then since $\mathfrak{H}$ is transitive on $\mathsf{V} - \{0\}$, there exists $h \in \mathfrak{H}$ such that $vh^{-1} \in \mathsf{U}$ and $v \in \mathsf{U}h$. And on account of the irreducibility of the $\mathsf{U}h$, $\mathsf{U}h_1 \cap \mathsf{U}h_2 \neq 0$ implies that $\mathsf{U}h_1 = \mathsf{U}h_2$.

Consider

$$\mathfrak{C} = \{\gamma | \gamma \in \mathrm{Hom}_{\mathsf{K}}(\mathsf{V}, \mathsf{V}), (\mathsf{U}h)\gamma \subseteq \mathsf{U}h \quad \text{for all } h \in \mathfrak{H}\}.$$

We show that $\mathfrak{C}$ is a division ring. Clearly, $\mathfrak{C}$ is a subring of $\mathrm{Hom}_{\mathsf{K}}(\mathsf{V}, \mathsf{V})$. If $\gamma \in \mathfrak{C}$ and γ is non-singular, then $\gamma^{-1} \in \mathfrak{C}$. Let γ be a singular element in $\mathfrak{C}$. Suppose that $0 \neq v \in \ker\gamma$. Then $v \in \mathsf{U}h_1$ for some $h_1 \in \mathfrak{H}$. Suppose that $w \notin \mathsf{U}h_1$. Then $w \in \mathsf{U}h_2$ and $v + w \in \mathsf{U}h_3$ for some h_2, h_3 in $\mathfrak{H}$. Thus

$$(v + w)\gamma = w\gamma \in \mathsf{U}h_2 \cap \mathsf{U}h_3.$$

Hence either $w\gamma = 0$ or $\mathsf{U}h_2 = \mathsf{U}h_3$. In the latter case, $v + w \in \mathsf{U}h_2$ and $w \in \mathsf{U}h_2$, so $v \in \mathsf{U}h_2 \cap \mathsf{U}h_1$, $\mathsf{U}h_1 = \mathsf{U}h_2$ and $w \in \mathsf{U}h_1$, a contradiction. Hence $w\gamma = 0$ for all $w \notin \mathsf{U}h_1$ and $\mathsf{V} = (\mathsf{U}h_1) \cup (\ker\gamma)$. Since $\mathsf{U}h_1 \neq \mathsf{V}$, $\ker\gamma = \mathsf{V}$ and $\gamma = 0$.

Hence $\mathfrak{C}$ is a division ring. But $\mathfrak{C}$ is finite, so by Wedderburn's theorem, $\mathfrak{C}$ is commutative. But $\mathfrak{N} \subseteq \mathfrak{C}$ and $\mathfrak{N}$ is non-Abelian. This is a contradiction, so V is an irreducible $\mathfrak{N}$-module.

e) Since $\mathfrak{N}$ is irreducible, it follows from Schur's lemma that

$$\mathfrak{D} = \mathrm{Hom}_{K\mathfrak{N}}(\mathsf{V}, \mathsf{V})$$

is a division ring. By Wedderburn's theorem, $\mathfrak{D}$ is commutative. But $\mathbf{C}_{\mathfrak{H}}(\mathfrak{N}) \subseteq \mathfrak{D}$, so $\mathbf{C}_{\mathfrak{H}}(\mathfrak{N})$ is Abelian. Since $\mathfrak{A} \leq \mathbf{C}_{\mathfrak{H}}(\mathfrak{N})$ and $\mathfrak{A}$ is maximal, $\mathfrak{A} = \mathbf{C}_{\mathfrak{H}}(\mathfrak{N})$. **q.e.d.**

9.4 Theorem. *If $\mathfrak{G}$ is of type II, p^f is $5^2, 7^2, 11^2$ or 23^2.*

If $\mathfrak{G}$ is of type III, p^f is $11^2, 29^2$ or 59^2 and $\mathfrak{H} \cong SL(2, 5) \times \mathfrak{Z}$, where $\mathfrak{Z}$ is a cyclic group of order 1, 7 or 29 respectively.

Proof. We use the notation of 9.3.

a) First suppose that $\mathfrak{N}$ is a quaternion group of order 8. By 9.3d), V is an irreducible $\mathfrak{N}$-module. Thus $p \neq 2$. Hence the group-ring $\mathfrak{R}$ of $\mathfrak{N}$ over $GF(p)$ is semisimple. Since $\mathfrak{N}$ has four distinct representations of degree 1 in $GF(p)$, $\mathfrak{R}$ has four ideals of dimension 1, so to within equivalence $\mathfrak{N}$ has only one faithful irreducible representation. This is of degree 2, for if i, j are elements of $GF(p)$ satisfying $i^2 + j^2 = -1$ (see II, 10.6),

$$\begin{pmatrix} i & j \\ j & -i \end{pmatrix}, \begin{pmatrix} 0 & 1 \\ -1 & 0 \end{pmatrix}$$

generate a quaternion group of order 8. Hence $f = 2$ and $|\mathfrak{H}| = p^2 - 1$.

By 9.3a), $k = 1$ and $|\mathfrak{A}|$ divides $p - 1$. By 9.3e), $\mathfrak{A} = \mathbf{C}_{\mathfrak{H}}(\mathfrak{N})$, so $\mathfrak{H}/\mathfrak{A}$ is isomorphic to a subgroup of **Aut** $\mathfrak{N}$. Since **Aut** $\mathfrak{N} \cong \mathfrak{S}_4$, $\mathfrak{H}$ is soluble and $\mathfrak{G}$ is of type II. Also $|\mathfrak{H}/\mathfrak{A}|$ divides $|\mathfrak{S}_4| = 24$ and $|\mathfrak{H}|$ divides $24(p - 1)$. Hence $p + 1$ divides 24. If $p = 3$, $|\mathfrak{H}| = 8 = |\mathfrak{N}|$, so $\mathfrak{H}$ has a cyclic normal subgroup of order 4 which is irreducible on V. This shows that $\mathfrak{G}$ is of type I, a contradiction. Hence p is 5, 7, 11 or 23.

b) Now suppose that $\mathfrak{N} \cong SL(2, 5)$. Thus $\mathfrak{G}$ is of type III. By 9.3e), $\mathfrak{A} = \mathbf{C}_{\mathfrak{H}}(\mathfrak{N})$. Let $\mathfrak{Z}$ be the subgroup of **Aut** $\mathfrak{N}$ consisting of automorphisms which induce the identity mapping on $\mathfrak{N}/\mathbf{Z}(\mathfrak{N})$. Then

$$[\mathfrak{N}, \mathfrak{Z}] = [\mathfrak{N}, \mathfrak{N}, \mathfrak{Z}] \leq [\mathfrak{N}, \mathfrak{Z}, \mathfrak{N}] \leq [\mathbf{Z}(\mathfrak{N}), \mathfrak{N}] = 1,$$

so $\mathfrak{Z} = 1$. Hence **Aut** $\mathfrak{N}$ operates faithfully on $\mathfrak{N}/\mathbf{Z}(\mathfrak{N}) \cong \mathfrak{A}_5$. By 8.7, **Aut** $\mathfrak{A}_5 \cong \mathfrak{S}_5$, so $|\mathbf{Aut}\,\mathfrak{N}|$ divides 120. Thus $|\mathfrak{H}/\mathfrak{A}|$ divides 120. Since

$|\mathfrak{H}| = p^f - 1$ and $|\mathfrak{A}|$ divides $p^k - 1$, $p^f - 1$ divides $120(p^k - 1)$. Now $f = kl$ with $l > 1$, so

$$\frac{p^f - 1}{p^k - 1} = 1 + p^k + \cdots + p^{k(l-1)}$$

divides 120. And since 120 divides $p^f - 1$, $p \geq 7$. If $l > 2$, $1 + p^k + p^{2k} \leq 120$, so $p^k = 7$; but then $1 + p^k + p^{2k} + \cdots$ does not divide 120. Hence $l = 2$. Also $p^k < 120$. Thus if $k > 1$, $p = 7$ and $k = 2$; but then $1 + p^k = 50$ does not divide 120. Hence $k = 1$ and $f = kl = 2$. Thus $1 + p$ divides 120 and $\dfrac{120}{1 + p}$ divides $p - 1$. The only possibilities are 11, 19, 29 or 59.

Since $\mathfrak{A} \cap \mathfrak{N} = \mathbf{Z}(\mathfrak{N})$, $\mathfrak{A}\mathfrak{N}/\mathbf{Z}(\mathfrak{N})$ is the direct product of $\mathfrak{A}/\mathbf{Z}(\mathfrak{N})$ and $\mathfrak{N}/\mathbf{Z}(\mathfrak{N})$. But the Sylow subgroups of $\mathfrak{H}$ of odd order are cyclic and the Sylow 2-subgroups of $\mathfrak{H}/\mathbf{Z}(\mathfrak{N})$ have no elementary Abelian subgroup of order 8, so $|\mathfrak{A}/\mathbf{Z}(\mathfrak{N})|$ is not divisible by 2, 3 or 5.

It follows that $p \neq 19$. For if $p = 19$, $|\mathfrak{A}|$ divides $p - 1 = 18$ and, since $(3, |\mathfrak{A}|) = 1$, $|\mathfrak{A}| = 2$. Thus $|\mathfrak{H}/\mathfrak{A}| = \frac{1}{2}(19^2 - 1) = 180$ does not divide 120. Hence p^f is 11^2, 29^2 or 59^2.

Now $|\mathfrak{H}/\mathfrak{A}|$ divides 120 and $|\mathfrak{N}\mathfrak{A}/\mathfrak{A}| = |\mathfrak{N}/\mathbf{Z}(\mathfrak{N})| = 60$, so $|\mathfrak{H} : \mathfrak{N}\mathfrak{A}| \leq 2$. If $|\mathfrak{H} : \mathfrak{N}\mathfrak{A}| = 2$, $\mathfrak{H}/\mathfrak{A} \cong \mathfrak{S}_5$ and there exist elements x, y in $\mathfrak{H}$ such that the order of $y\mathfrak{A}$ is 5 and

$$y^x \equiv y^2 \mod \mathfrak{A};$$

further we may suppose that $y^5 = 1$. Since $k = 1$, $\mathfrak{A}$ consists of scalar multiplications, so $y^x = \lambda y^2$ for some $\lambda \in GF(p)$. In a suitable extension of the ground field, y is represented by a non-scalar diagonal matrix, so x is represented by a matrix of the form

$$\begin{pmatrix} 0 & \mu \\ \nu & 0 \end{pmatrix}.$$

Thus $x^2 \in \mathbf{Z}(\mathfrak{H})$, contrary to $y^{x^2} \equiv y^4 \mod \mathfrak{A}$. Hence $\mathfrak{H} = \mathfrak{N}\mathfrak{A}$. Since $|\mathfrak{A}|$ is not divisible by 4, $\mathfrak{A} = \mathbf{Z}(\mathfrak{N}) \times \mathfrak{Z}$ and $\mathfrak{H} = \mathfrak{N} \times \mathfrak{Z}$ for some cyclic group $\mathfrak{Z}$. Finally,

$$|\mathfrak{Z}| = \frac{|\mathfrak{H}|}{|\mathfrak{N}|} = \frac{p^f - 1}{120}. \qquad \textbf{q.e.d.}$$

9.5 Remark. It can be shown that there exist sharply doubly transitive permutation groups of type II or III of each of the degrees $5^2, 7^2, 11^2, 23^2, 29^2$ and 59^2. Indeed there is just one such group of each degree except 11^2, and there are two such groups of degree 11^2. We state what these groups are without proof; we describe the groups by giving matrices which generate $\mathfrak{H}$.

a) $p^f = 5^2$, $\mathfrak{H} \cong SL(2, 3)$ and $\mathfrak{H}$ can be generated by the matrices

$$\begin{pmatrix} 2 & 0 \\ 0 & -2 \end{pmatrix}, \begin{pmatrix} 0 & -1 \\ 1 & 0 \end{pmatrix}, \begin{pmatrix} 1 & 2 \\ 1 & -2 \end{pmatrix}.$$

b) $p^f = 7^2$, $\mathfrak{H} \cong GL(2, 3)$ and $\mathfrak{H}$ can be generated by

$$\begin{pmatrix} 2 & 3 \\ 3 & -2 \end{pmatrix}, \begin{pmatrix} 0 & -1 \\ 1 & 0 \end{pmatrix}, \begin{pmatrix} 0 & -2 \\ -3 & -1 \end{pmatrix}, \begin{pmatrix} 3 & -1 \\ 3 & -3 \end{pmatrix}.$$

c) $p^f = 11^2$, $\mathfrak{H} \cong SL(2, 3) \times \mathfrak{Z}_5$ and $\mathfrak{H}$ can be generated by

$$\begin{pmatrix} 1 & 3 \\ 3 & -1 \end{pmatrix}, \begin{pmatrix} 0 & -1 \\ 1 & 0 \end{pmatrix}, \begin{pmatrix} -5 & 4 \\ 3 & 4 \end{pmatrix}, \begin{pmatrix} 4 & 0 \\ 0 & 4 \end{pmatrix}.$$

d) $p^f = 23^2$, $\mathfrak{H} \cong GL(2, 3) \times \mathfrak{Z}_{11}$ and $\mathfrak{H}$ can be generated by

$$\begin{pmatrix} 3 & 6 \\ 6 & -3 \end{pmatrix}, \begin{pmatrix} 0 & -1 \\ 1 & 0 \end{pmatrix}, \begin{pmatrix} 1 & -4 \\ -5 & -2 \end{pmatrix}, \begin{pmatrix} 4 & -6 \\ -1 & -4 \end{pmatrix}, \begin{pmatrix} 2 & 0 \\ 0 & 2 \end{pmatrix}.$$

e) $p^f = 11^2$, $\mathfrak{H} \cong SL(2, 5)$ and $\mathfrak{H}$ can be generated by

$$\begin{pmatrix} 0 & -1 \\ 1 & 0 \end{pmatrix}, \begin{pmatrix} 2 & 4 \\ 1 & -3 \end{pmatrix}.$$

f) $p^f = 29^2$, $\mathfrak{H} \cong SL(2, 5) \times \mathfrak{Z}_7$ and $\mathfrak{H}$ can be generated by

$$\begin{pmatrix} 0 & -1 \\ 1 & 0 \end{pmatrix}, \begin{pmatrix} 1 & -7 \\ -12 & -2 \end{pmatrix}, \begin{pmatrix} -13 & 0 \\ 0 & -13 \end{pmatrix}.$$

g) $p^f = 59^2$, $\mathfrak{H} \cong SL(2, 5) \times \mathfrak{Z}_{29}$ and $\mathfrak{H}$ can be generated by

$$\begin{pmatrix} 0 & -1 \\ 1 & 0 \end{pmatrix}, \begin{pmatrix} 9 & 15 \\ -10 & -10 \end{pmatrix}, \begin{pmatrix} 4 & 0 \\ 0 & 4 \end{pmatrix}.$$

We prepare the way for the description of the sharply doubly transitive permutation groups of type I by a number-theoretical lemma.

9.6 Lemma. *Let q be a prime-power and let n be an integer greater than* 1. *Then the following are equivalent.*

a) *Every prime divisor of n divides $q - 1$, and if 4 divides n, 4 divides $q - 1$.*

b) $q^n \equiv 1\ (n(q-1))$, *but* $q^i \not\equiv 1\ (n(q-1))$ *for* $0 < i < n$.

Proof. a) $\Rightarrow$ b): Let p be a prime divisor of $n(q-1)$. By a), p divides $q - 1$; write $q = 1 + rp^a$, where $(p, r) = 1$ and $a \geq 1$. Also write $n = n'p^b$ with $(p, n') = 1$. Then

$$q^n - 1 = (1 + rp^a)^{n'p^b} - 1 = rn'p^{a+b} + r^2p^{2a}\binom{n'p^b}{2} + \cdots + (rp^a)^{n'p^b}.$$

By I, 13.18,

$$\binom{n'p^b}{i}p^i \equiv 0 \quad (p^{b+1})$$

for $i \geq 1$, so

$$\binom{n'p^b}{i}p^{ia} \equiv 0 \quad (p^{b+1+ia-i}).$$

Since $b + 1 + ia - i \geq b + 1 + a - 1 = a + b$, $q^n - 1$ is divisible by p^{a+b}. Since this holds for all prime divisors p of $n(q-1)$,

$$q^n - 1 \equiv 0 \quad (n(q-1)).$$

If $0 < i < n$, there is a prime p such that if p^k is the highest power of p which divides i, p^{k+1} divides n. By a), p divides $q - 1$. Write $q = 1 + rp^a$, where $(p, r) = 1$, and $i = p^k i'$. Again,

$$q^i - 1 = (1 + rp^a)^{p^k i'} - 1 = \sum_{j=1}^{p^k i'}\binom{p^k i'}{j}r^j p^{aj}.$$

First suppose that p is odd. Then by I, 13.18a),

$$\binom{p^k i'}{j}p^{aj} \equiv 0 \quad (p^{k+2+aj-j})$$

for $j \geq 2$. Since $k + 2 + aj - j \geq k + 2a \geq k + a + 1$,

$$q^i - 1 \equiv p^{k+a}i'r \; (p^{k+a+1}).$$

Since p^{k+a+1} divides $n(q - 1)$, $n(q - 1)$ does not divide $q^i - 1$.

For $p = 2$ and $j \geq 1$, I, 13.18b) gives

$$\binom{2^k i'}{j} 2^{aj} \equiv 0 \quad (2^{k+1+aj-j}).$$

Thus if $a > 1$, we have

$$k + 1 + aj - j \geq k + 1 + 2(a - 1) = k + a + (a - 1) \geq k + a + 1$$

for $j \geq 2$, and the result follows as before. Finally, suppose that $a = 1$. By a), n is not divisible by 4, so $k = 0$ and i is odd. Then

$$q^i - 1 = (1 + 2r)^i - 1 \equiv 2ir \equiv 2 \quad (4),$$

so $q^i - 1$ is not divisible by $n(q - 1)$.

b) $\Rightarrow$ a). We show first that each prime divisor of n divides $q - 1$. If this is false, let r be the greatest prime divisor of n which does not divide $q - 1$. Let $n = rs$. We show that $q^s \equiv 1 \; (n(q - 1))$, which contradicts b).

Let p^a be the highest power of the prime p which divides $n(q - 1)$, with $a \geq 1$. If p divides $q - 1$, then $p \neq r$, so

$$\frac{q^n - 1}{q^s - 1} = 1 + q^s + \cdots + q^{s(r-1)} \equiv r \not\equiv 0 \; (p).$$

Thus, since $n(q - 1)$ divides $q^n - 1$, p^a divides $q^s - 1$. If p does not divide $q - 1$, then $p \leq r$ by choice of r. Also p^a divides n; hence p^a divides $n(q - 1)$ and $q^n - 1$. Since $a \geq 1$, p does not divide q. Hence

$$q^{p^{a-1}(p-1)} = q^{\phi(p^a)} \equiv 1 \quad (p^a).$$

Since $p \leq r$, $p - 1$ is not divisible by r, so $(n, p^{a-1}(p - 1))$ divides $\frac{n}{r} = s$. Hence $q^s - 1 \equiv 0 \; (p^a)$. Altogether, then,

$$q^s \equiv 1 \quad (n(q - 1)),$$

a contradiction.

Finally, we show that if $q \equiv 3\ (4)$, then $n \not\equiv 0\ (4)$. Suppose that $n = 2^a n'$, where $a \geq 2$ and n' is odd. We have

$$q^n - 1 = (q^{\frac{1}{2}n} - 1)(q^{\frac{1}{2}n} + 1) \equiv 0 \quad (n(q-1)).$$

Let p be a prime divisor of $n(q-1)$. We have already shown that this implies that p divides $q - 1$. Thus for p odd, $q^{\frac{1}{2}n} + 1 \not\equiv 0\ (p)$. Hence if p is odd and $n(q-1) = p^b m$ with $(p, m) = 1$,

$$q^{\frac{1}{2}n} \equiv 1 \quad (p^b).$$

For $p = 2$, 2^{a+1} is the highest power of 2 which divides $n(q-1)$, since $q \equiv 3\ (4)$. Putting $q = -1 + 4c$,

$$\begin{aligned} q^{\frac{1}{2}n} - 1 &= (-1 + 4c)^{2^{a-1}n'} - 1 \\ &= (-1)^{2^{a-1}n'} - 1 + \sum_{i=1}^{2^{a-1}n'} \pm \binom{2^{a-1}n'}{i}(4c)^i. \end{aligned}$$

But for $i \geq 1$, I, 13.18b) gives

$$\binom{2^{a-1}n'}{i} 4^i \equiv 0 \quad (2^{a+i}).$$

Since $a \geq 2$, this gives $q^{\frac{1}{2}n} - 1 \equiv 0\ (2^{a+1})$, and altogether,

$$q^{\frac{1}{2}n} - 1 \equiv 0 \quad (n(q-1)),$$

contrary to b). **q.e.d.**

9.7 Theorem (ZASSENHAUS [3]). a) *Suppose that q is a prime-power and n is a natural number with the following properties.*

(1) *Each prime divisor of n divides $q - 1$.*

(2) *If 4 divides n, then 4 divides $q - 1$.*

Let $GF(q^n)^\times = \langle \omega \rangle$ and suppose that $(n, t) = 1$. Let a, b be the following elements of the group of semilinear mappings on $GF(q^n)$.

$$xa = \omega^n x, \quad xb = \omega^t x^q.$$

Then $\mathfrak{H} = \langle a, b \rangle$ is a regular permutation group on $GF(q^n)^\times$ and $\langle a \rangle$ is an irreducible normal subgroup of $\mathfrak{H}$. If $\mathfrak{T}$ is the group of all translations on $GF(q^n)$, $\mathfrak{H}$ normalises $\mathfrak{T}$ and $\mathfrak{G}(n, q, t) = \mathfrak{T}\mathfrak{H}$ is a sharply doubly transitive permutation group of type I and degree q^n.

b) *If* $\mathfrak{G}$ *is a sharply doubly transitive permutation group of type I,* $\mathfrak{G}$ *is similar to* $\mathfrak{G}(n, q, t)$ *for some* n, q, t.

Proof. a) By 9.6, $q^n - 1 = n(q - 1)k$ for some integer k. The order of a is $\frac{q^n - 1}{n} = (q - 1)k$, and $b^{-1}ab = a^q$. Since

$$xb^i = \omega^{t(1+q+\cdots+q^{i-1})}x^{q^i},$$

$b^i \notin \langle a \rangle$ for $0 < i < n$. But $xb^n = \omega^{tnk}x$ and $b^n = a^{tk}$. Hence $|\mathfrak{H}| = n\frac{q^n - 1}{n} = q^n - 1$.

We show that the stabiliser of 1 is the unit subgroup of $\mathfrak{H}$. For suppose that $1a^ib^j = 1$, where $0 \leq j < n$. Then

$$1 = \omega^{ni}b^j = \omega^{t(1+q+\cdots+q^{j-1})+niq^j}$$

and

$$t(1 + q + \cdots + q^{j-1}) + niq^j \equiv 0 \quad (n(q - 1)k).$$

Thus n divides $t(1 + q + \cdots + q^{j-1})$. Since $(n, t) = 1$, it follows that $n(q - 1)$ divides $q^j - 1$. By 9.6, $j = 0$, so $1a^i = 1$, $\omega^{ni} = 1$ and $a^ib^j = 1$.

Since $|\mathfrak{H}| = q^n - 1$, it follows that $\mathfrak{H}$ is transitive and regular on $GF(q^n)^\times$.

Suppose that p is the prime divisor of q and $q = p^f$. Since the order of a is prime to p, $GF(q^n)$ is a completely reducible module for $\langle a \rangle$ over $GF(p)$, and $\langle a \rangle$ is faithfully represented on each irreducible submodule. Thus by II, 3.10, the number of elements in each irreducible submodule is p^i, where i is the smallest positive integer for which $p^i - 1$ is divisible by the order $(q - 1)k$ of a. Hence i divides the degree fn. Since $p^i - 1$ is divisible by $q - 1$, $i = fj$ for some integer j, and j divides n. Thus if $j < n$, $j \leq \frac{1}{2}n$ and

$$q^n - 1 = (q^{\frac{1}{2}n} - 1)(q^{\frac{1}{2}n} + 1) \geq (q^{\frac{1}{2}n} - 1)(2^{\frac{1}{2}n} + 1) > (q^j - 1)n,$$

so $(q - 1)k = \frac{q^n - 1}{n}$ cannot divide $p^i - 1 = q^j - 1$. Hence $j = n$, $i = nf$ and $\langle a \rangle$ is irreducible. The remaining assertions are clear.

b) By 9.1, we may suppose that $\mathfrak{G} = \mathfrak{F}\mathfrak{H}$, where $\mathfrak{F}$ is the group of all translations on $\mathsf{V} = \mathsf{V}(f, p)$ and $\mathfrak{H} = \mathfrak{G}_0$; further $\mathfrak{H}$ is regular on

$\mathsf{V} - \{0\}$ and $\mathfrak{H} \leq GL(\mathsf{V})$. By 9.2, $\mathfrak{H}$ has a cyclic normal subgroup which is irreducible on V. By II, 3.11, we may suppose that $\mathsf{V} = GF(p^f)$, and $\mathfrak{H}$ is a group of semilinear mappings $x \rightarrow c(x\alpha)$. The field automorphisms which occur form a group $\mathfrak{A}$. By the fundamental theorem of Galois theory, there is a subfield L of $GF(p^f)$ such that $\mathfrak{A}$ is the group of all automorphisms of $GF(p^f)$ over L. Suppose that $|\mathsf{L}| = q$ and $p^f = q^n$. Then $\mathfrak{A}$ is generated by the automorphism $x \rightarrow x^q$ and $\mathfrak{H}$ contains an element b given by $xb = dx^q$ for some $d \in GF(q^n)$. The kernel of the obvious epimorphism of $\mathfrak{H}$ onto $\mathfrak{A}$ is a cyclic normal subgroup $\langle a \rangle$ and $\mathfrak{H} = \langle a, b \rangle$. Also $|\mathfrak{H} : \langle a \rangle| = |\mathfrak{A}| = n$, so $|\langle a \rangle| = \dfrac{q^n - 1}{n}$. Hence with suitable choice of ω,

$$xa = \omega^n x.$$

Put $d = \omega^t$. Then the $\mathfrak{H}$-orbit of 1 lies in $\langle \omega^n, \omega^t \rangle$. Thus $\langle \omega^n, \omega^t \rangle = GF(q^n)^\times$ and $(n, t) = 1$. Further,

$$xb^i = \omega^{t(1+q+\cdots+q^{i-1})} x^{q^i}.$$

Since $b^n \in \langle a \rangle$, it follows that $n(q - 1)$ divides $q^n - 1$. But if $0 < i < n$, $b^i \notin \langle a \rangle$ and $1b^i \notin \{1a^j | j \in \mathbb{Z}\}$. Thus

$$\omega^{t(1+q+\cdots+q^{i-1})} \neq \omega^{jn}$$

for all j, and $1 + q + \cdots + q^{i-1}$ is not divisible by n. Hence $q^i - 1$ is not divisible by $n(q - 1)$. By 9.6, q and n have the required properties. Hence $\mathfrak{G}$ is similar to $\mathfrak{G}(n, q, t)$. **q.e.d.**

We put these results together as follows.

9.8 Theorem (ZASSENHAUS [3]). *Let $\mathfrak{G}$ be a sharply doubly transitive permutation group. Then $\mathfrak{G}$ is a Frobenius group with elementary Abelian Frobenius kernel $\mathfrak{F}$. To within similarity, $\mathfrak{G}$ permutes $GF(p^f)$, and $\mathfrak{F}$ is the group of translations on $GF(p^f)$. If p^f is distinct from 5^2, 7^2, 11^2, 23^2, 29^2 and 59^2, $\mathfrak{G}$ is similar to one of the groups $\mathfrak{G}(n, q, t)$ defined in 9.7.*

An alternative formulation is as follows.

9.9 Definition. A set F satisfying the following conditions is called a (complete) *near-field*.

(1) F is an additively written Abelian group.

(2) Multiplication is defined in F and $\mathsf{F} - \{0\}$ is a group with respect to multiplication.

(3) $a(b + c) = ab + ac$ for all a, b, c in F.

(It is the other distributive law which is not required.)

Finite near-fields and sharply doubly transitive permutation groups are closely related.

9.10 Theorem (ZASSENHAUS [3]). a) *Suppose that* F *is a finite near-field. Let* $\mathfrak{G}$ *be the set of all permutations of the form*

$$x \to ax + b \quad (a, b, x \in \mathsf{F}, a \neq 0).$$

Then $\mathfrak{G}$ *is a sharply doubly transitive permutation group.*

b) *Let* $\mathfrak{G}$ *be a sharply doubly transitive permutation group on* Ω. *Then addition and multiplication can be defined in* Ω *in such a way that* Ω *becomes a near-field and* $\mathfrak{G}$ *is the group of all permutations of* Ω *of the form*

$$x \to ax + b.$$

Proof. a) Suppose that $x\alpha = ax + b$, $x\beta = cx + d$. Then

$$x(\alpha\beta) = c(x\alpha) + d = c(ax) + cb + d = (ca)x + cb + d$$

and $x\alpha^{-1} = a^{-1}x - a^{-1}b$. Thus $\mathfrak{G}$ is a permutation group on F. Clearly $\mathfrak{G}$ is transitive and $\mathfrak{G}_0$ is transitive on $\mathsf{F} - \{0\}$, so $\mathfrak{G}$ is doubly transitive. Since $|\mathfrak{G}| = |\mathsf{F}|(|\mathsf{F}| - 1)$, $\mathfrak{G}$ is sharply doubly transitive.

b) By 9.1, Ω can be given the structure of an additive Abelian group in such a way that $\mathfrak{G} = \mathfrak{F}\mathfrak{H}$, where $\mathfrak{F}$ is the group of translations on Ω and $\mathfrak{H}$ is regular on $\Omega - \{0\}$. Choose an arbitrary $1 \in \Omega - \{0\}$. Then given $a \in \Omega - \{0\}$, there is a unique $h_a \in \mathfrak{H}$ such that $ah_a = 1$. We put

$$ab = bh_a^{-1} \quad (b \in \Omega)$$

and $0 \cdot b = 0$. Then $(ab)h_a = b$, so if also $b \neq 0$,

$$(ab)h_a h_b = bh_b = 1 = (ab)h_{ab}.$$

Hence $h_a h_b = h_{ab}$. It follows that $\Omega - \{0\}$ is a group isomorphic to $\mathfrak{H}$. Also $\mathfrak{H}$ is linear with respect to addition, so

$$(b + c)h_a^{-1} = bh_a^{-1} + ch_a^{-1}.$$

Thus $a(b + c) = ab + ac$ for all a, b, c in Ω. Since $\mathfrak{G} = \mathfrak{H}\mathfrak{F}$, every element of $\mathfrak{G}$ is of the form $h_a^{-1}g$, with $g \in \mathfrak{F}$. If $xg = x + b$,

$$xh_a^{-1}g = ax + b,$$

so $\mathfrak{G}$ is the group of all permutations of this form. **q.e.d.**

§ 10. Permutation Groups of Prime Degree

In V, § 21, we considered transitive permutation groups of prime degree. We shall now prove these earlier results by simpler means and to some extent sharpen them.

10.1 Definition. Let $\mathfrak{G}$ be a permutation group on a finite set Ω and let K be a field.

a) Let $\mathsf{F}(\Omega)$ be the set of all mappings of Ω into K. Thus $\mathsf{F}(\Omega)$ is a vector space over K, where

$$i(f_1 + f_2) = if_1 + if_2, \quad i(\lambda f) = \lambda(if) \quad (i \in \Omega).$$

For $i \in \Omega$, define $h_i \in \mathsf{F}(\Omega)$ by $jh_i = \delta_{ij}$; then the h_i form a K-basis of $\mathsf{F}(\Omega)$.

b) $\mathsf{F}(\Omega)$ becomes a $\mathsf{K}\mathfrak{G}$-module if we put

$$i(fg) = (ig^{-1})f \quad (i \in \Omega, f \in \mathsf{F}(\Omega), g \in \mathfrak{G}).$$

We have $h_i g = h_{ig}$.

c) We define a non-singular symmetric bilinear form $(\ ,\ )$ on $\mathsf{F}(\Omega)$ by putting

$$(f_1, f_2) = \sum_{i \in \Omega} (if_1)(if_2).$$

Clearly

$$(f_1 g, f_2 g) = (f_1, f_2)$$

for all $g \in \mathfrak{G}$. If U is a $\mathsf{K}\mathfrak{G}$-submodule of $\mathsf{F}(\Omega)$, we put

$$\mathsf{U}^\perp = \{f \mid (u, f) = 0 \quad \text{for all} \quad u \in \mathsf{U}\}.$$

Then $\mathsf{U}^\perp$ is a $\mathsf{K}\mathfrak{G}$-submodule of $\mathsf{F}(\Omega)$.

10.2 Lemma. *If* V, W *are* $K\mathfrak{G}$*-modules and* U *is a* $K\mathfrak{G}$*-submodule of* V, *then*

$$\dim_K \mathrm{Hom}_{K\mathfrak{G}}(V, W) \leq \dim_K \mathrm{Hom}_{K\mathfrak{G}}(U, W) + \dim_K \mathrm{Hom}_{K\mathfrak{G}}(V/U, W).$$

Proof. From the exact sequence

$$0 \to U \to V \xrightarrow{\pi} V/U \to 0,$$

we obtain the sequence

$$\mathrm{Hom}_{K\mathfrak{G}}(V/U, W) \xrightarrow{\pi'} \mathrm{Hom}_{K\mathfrak{G}}(V, W) \xrightarrow{\iota'} \mathrm{Hom}_{K\mathfrak{G}}(U, W).$$

Here ι' is restriction from V to U and π' is inflation, defined by

$$v(\phi\pi') = (v\pi)\phi = (v + U)\phi \quad (\phi \in \mathrm{Hom}_{K\mathfrak{G}}(V/U, W)).$$

It is easily verified that π' is a monomorphism and $\operatorname{im}\pi' = \ker\iota'$, for if $\phi \in \mathrm{Hom}_{K\mathfrak{G}}(V, W)$, ϕ induces a homomorphism on V/U if and only if $U \subseteq \ker\phi$. (This is the right exactness of the functor Hom.) Hence

$$\dim_K \mathrm{Hom}_{K\mathfrak{G}}(V, W) \leq \dim_K \mathrm{Hom}_{K\mathfrak{G}}(U, W) + \dim_K \mathrm{Hom}_{K\mathfrak{G}}(V/U, W).$$

q.e.d.

10.3 Lemma. a) *Suppose that* $\mathfrak{G}$ *has permutation representations on* Ω_1, Ω_2. *Let* K *be a field, and form the* $K\mathfrak{G}$*-modules* $F(\Omega_1)$, $F(\Omega_2)$. *Then*

$$\dim_K \mathrm{Hom}_{K\mathfrak{G}}(F(\Omega_1), F(\Omega_2))$$

is the number of orbits of $\mathfrak{G}$ *on the Cartesian product* $\Omega_1 \times \Omega_2$.

b) *Let* $\mathfrak{G}$ *be a transitive permutation group on* Ω. *Then* $\dim_K \mathrm{Hom}_{K\mathfrak{G}}(F(\Omega), F(\Omega))$ *is the number of orbits of the stabiliser of an element of* Ω. *(This is called the rank of* $\mathfrak{G}$.*)*

Proof. a) Let $\{h_i | i \in \Omega_1\}$ be the K-basis of $F(\Omega_1)$ described in 10.1a); thus $h_i g = h_{ig}$ for $g \in \mathfrak{G}$. Given $\phi \in F = \mathrm{Hom}_K(F(\Omega_1), F(\Omega_2))$ and $g \in \mathfrak{G}$, we define $\phi^g \in F$ by putting

$$f\phi^g = fg^{-1}\phi g \quad (f \in F(\Omega_1)).$$

This gives F the structure of a $K\mathfrak{G}$-module. We define a K-linear mapping ρ of F into $F(\Omega_1 \times \Omega_2)$ by putting

$$(i, j)(\phi\rho) = j(h_i\phi) \quad (i \in \Omega_1, j \in \Omega_2).$$

Obviously ρ is injective, and by comparison of dimensions, ρ is bijective. In fact ρ is a $\mathsf{K}\mathfrak{G}$-isomorphism, for if $\phi \in \mathsf{F}$ and $g \in \mathfrak{G}$,

$$\begin{aligned}(i, j)(\phi^g\rho) &= j(h_i\phi^g)\\ &= j(h_ig^{-1}\phi g)\\ &= (jg^{-1})(h_ig^{-1}\phi)\\ &= (jg^{-1})(h_{ig^{-1}}\phi)\\ &= (ig^{-1}, jg^{-1})(\phi\rho)\\ &= (i, j)((\phi\rho)g)\end{aligned}$$

for all $i \in \Omega_1, j \in \Omega_2$. Now $\mathrm{Hom}_{\mathsf{K}\mathfrak{G}}(\mathsf{F}(\Omega_1), \mathsf{F}(\Omega_2))$ is the set of elements of F fixed under all elements of $\mathfrak{G}$, so

$$\dim_{\mathsf{K}}\mathrm{Hom}_{\mathsf{K}\mathfrak{G}}(\mathsf{F}(\Omega_1), \mathsf{F}(\Omega_2))$$

is the dimension d of the space F_0 of $\mathfrak{G}$-invariant elements of $\mathsf{F}(\Omega_1 \times \Omega_2)$. But if $f \in \mathsf{F}(\Omega_1 \times \Omega_2)$, then $f \in \mathsf{F}_0$ if and only if f is constant on all the $\mathfrak{G}$-orbits of $\Omega_1 \times \Omega_2$. Hence d is the number of such orbits.

b) By a), $\dim_{\mathsf{K}}\mathrm{Hom}_{\mathsf{K}\mathfrak{G}}(\mathsf{F}(\Omega), \mathsf{F}(\Omega))$ is the number r of orbits of $\mathfrak{G}$ on $\Omega \times \Omega$. Choose $1 \in \Omega$. Since $\mathfrak{G}$ is transitive, each orbit B of $\mathfrak{G}$ on $\Omega \times \Omega$ contains elements of the form $(1, j)$; indeed

$$B \cap (\{1\} \times \Omega) = \{1\} \times \Gamma$$

for some orbit Γ of $\mathfrak{G}_1$. The correspondence between B and Γ is bijective, so r is also the number of orbits of $\mathfrak{G}_1$ on Ω. **q.e.d.**

10.4 Lemma. *Suppose that $\mathfrak{G}$ is a transitive permutation group on a set Ω, where $|\Omega|$ is a prime-power p^f. If* char $\mathsf{K} = p$, $\mathsf{K}\sum_{i\in\Omega} h_i$ *is the unique irreducible* $\mathsf{K}\mathfrak{G}$*-submodule of* $\mathsf{F}(\Omega)$.

Proof. Let V be an irreducible $\mathsf{K}\mathfrak{G}$-submodule of $\mathsf{F}(\Omega)$. Let $\mathfrak{P}$ be a Sylow p-subgroup of $\mathfrak{G}$ and let U be an irreducible $\mathsf{K}\mathfrak{P}$-submodule of V. By V, 5.16, $\mathfrak{P}$ operates trivially on U. But if $1 \in \Omega$, $|\mathfrak{G} : \mathfrak{G}_1| = p^f$, so $\mathfrak{G} = \mathfrak{G}_1\mathfrak{P}$. It follows that $\mathfrak{P}$ is transitive on Ω. Suppose that $x_i \in \mathfrak{P}$ and $1x_i = i$. If $\sum_{j\in\Omega}\lambda_j h_j \in \mathsf{U}$, then

$$\begin{aligned}\sum \lambda_j h_j &= (\sum \lambda_j h_j) x_i \\ &= \lambda_1 h_{1x_i} + \cdots \\ &= \lambda_1 h_i + \cdots.\end{aligned}$$

Hence $\lambda_i = \lambda_1$ for all $i \in \Omega$, and $\mathsf{U} = \mathsf{K}\sum_{i\in\Omega} h_i$. Since $\mathsf{K}\sum_{i\in\Omega} h_i$ is a $\mathsf{K}\mathfrak{G}$-submodule of $\mathsf{F}(\Omega)$, it follows that $\mathsf{V} = \mathsf{K}\sum_{i\in\Omega} h_i$, since V is irreducible.

q.e.d.

10.5 Lemma (KLEMM [1]). *Suppose that $\mathfrak{G}$ is transitive on Ω and that* char $\mathsf{K} = p$ *is a divisor of $|\Omega|$. Let*

$$\begin{aligned}\mathsf{U}_1 &= \mathsf{K}\sum_{i\in\Omega} h_i \\ &= \{f \mid f \in \mathsf{F}(\Omega),\ f \text{ is constant on } \Omega\}\end{aligned}$$

and

$$\mathsf{U}_2 = \{f \mid f \in \mathsf{F}(\Omega),\ \sum_{i\in\Omega} if = 0\}.$$

Then U_1, U_2 are $\mathsf{K}\mathfrak{G}$-submodules of $\mathsf{F}(\Omega)$ and U_1, $\mathsf{F}(\Omega)/\mathsf{U}_2$ are trivial $\mathsf{K}\mathfrak{G}$-modules. Also $\mathsf{U}_1 \subseteq \mathsf{U}_2$; $\mathsf{H} = \mathsf{U}_2/\mathsf{U}_1$ is called the heart of $\mathsf{F}(\Omega)$.

a) *Suppose that $|\Omega| = p^f$, where $f \geq 1$. If H has no trivial $\mathsf{K}\mathfrak{G}$-composition factor, $\mathfrak{G}$ is doubly transitive.*

b) *Suppose that $\mathfrak{G}$ is doubly transitive and that $|\mathfrak{G}_i|$ is even ($i \in \Omega$). Let*

$$\Omega^{(2)} = \{\{a, b\} \mid a \in \Omega,\ b \in \Omega,\ a \neq b\}.$$

Then $\mathfrak{G}$ operates on $\Omega^{(2)}$ in a natural way. If $\mathrm{Hom}_{\mathsf{K}\mathfrak{G}}(\mathsf{H}, \mathsf{F}(\Omega^{(2)})) = 0$, *$\mathfrak{G}$ is triply transitive.*

Proof. It is clear that U_1, $\mathsf{F}(\Omega)/\mathsf{U}_2$ are trivial $\mathsf{K}\mathfrak{G}$-modules. Also $\mathsf{U}_1 \subseteq \mathsf{U}_2$, since

$$\sum_{i\in\Omega} i\left(\sum_{j\in\Omega} h_j\right) = |\Omega| 1_{\mathsf{K}} = 0.$$

a) Let

$$0 = \mathsf{F}_0 \subset \mathsf{F}_1 \subset \cdots \subset \mathsf{F}_{k-1} \subset \mathsf{F}_k = \mathsf{F}(\Omega)$$

be a composition series of $\mathsf{F}(\Omega)$, with $\mathsf{F}_1 = \mathsf{U}_1$, $\mathsf{F}_{k-1} = \mathsf{U}_2$. By 10.2,

$$\dim_{\mathsf{K}} \mathrm{Hom}_{\mathsf{K}\mathfrak{G}}(\mathsf{F}(\Omega), \mathsf{F}(\Omega)) \leq \sum_{i=1}^{k} \dim_{\mathsf{K}} \mathrm{Hom}_{\mathsf{K}\mathfrak{G}}(\mathsf{F}_i/\mathsf{F}_{i-1}, \mathsf{F}(\Omega)).$$

By 10.4, the only irreducible $\mathsf{K}\mathfrak{G}$-submodule of $\mathsf{F}(\Omega)$ is a trivial $\mathsf{K}\mathfrak{G}$-module, so

$$\dim_{\mathsf{K}} \mathrm{Hom}_{\mathsf{K}\mathfrak{G}}(\mathsf{F}_i/\mathsf{F}_{i-1}, \mathsf{F}(\Omega)) = \begin{cases} 1 & \text{if } \mathsf{F}_i/\mathsf{F}_{i-1} \cong \mathsf{K}, \\ 0 & \text{otherwise.} \end{cases}$$

On account of the hypothesis, it follows that

$$\dim_{\mathsf{K}} \mathrm{Hom}_{\mathsf{K}\mathfrak{G}}(\mathsf{F}(\Omega), \mathsf{F}(\Omega)) \leq 2.$$

Thus by 10.3b), the number of orbits of the stabiliser of an element of Ω is at most 2, so $\mathfrak{G}$ is doubly transitive.

b) Since $\mathfrak{G}$ is doubly transitive, $\mathfrak{G}$ is transitive on $\Omega^{(2)}$. Hence $\mathsf{F}(\Omega^{(2)})$ has only one trivial $\mathsf{K}\mathfrak{G}$-submodule and

$$\dim_{\mathsf{K}} \mathrm{Hom}_{\mathsf{K}\mathfrak{G}}(\mathsf{K}, \mathsf{F}(\Omega^{(2)})) = 1.$$

By 10.2, it follows from the hypothesis that

$$\begin{aligned}\dim_{\mathsf{K}} \mathrm{Hom}_{\mathsf{K}\mathfrak{G}}(\mathsf{F}(\Omega), \mathsf{F}(\Omega^{(2)})) &\leq 2 + \dim_{\mathsf{K}} \mathrm{Hom}_{\mathsf{K}\mathfrak{G}}(\mathsf{H}, \mathsf{F}(\Omega^{(2)})) \\ &= 2.\end{aligned}$$

By 10.3a), the number of orbits of $\mathfrak{G}$ on $\Omega \times \Omega^{(2)}$ is at most 2. But

$$\Delta_1 = \{(a, \{b, c\}) | a \notin \{b, c\}\}$$

and

$$\Delta_2 = \{(a, \{b, c\}) | a \in \{b, c\}\}$$

are $\mathfrak{G}$-invariant subsets of $\Omega \times \Omega^{(2)}$. Hence $\mathfrak{G}$ is transitive on Δ_1.

Now since $|\mathfrak{G}_i|$ is even, $\mathfrak{G}$ contains an involution

$$h = (i)(j, k) \cdots.$$

Given distinct elements a, b, c of Ω, there exists $g \in \mathfrak{G}$ such that $ig = a$

and $\{j, k\}g = \{b, c\}$. Thus either

$$(i, j, k)g = (a, b, c) \quad \text{or} \quad (i, j, k)hg = (a, b, c).$$

Hence $\mathfrak{G}$ is triply transitive. **q.e.d.**

10.6 Lemma. *Let $\mathfrak{G}$ be a transitive permutation group on $\Omega = GF(p)$. We choose the notation so that $(0, 1, \ldots, p-2, p-1)$ generates a Sylow p-subgroup $\mathfrak{P}$ of $\mathfrak{G}$. Let K be a field of characteristic p.*

a) *If $f \in \mathsf{F}(\Omega)$, there is a unique polynomial $\sum_{r=0}^{p-1} a_r t^r$ in $\mathsf{K}[t]$ such that*

$$jf = \sum_{r=0}^{p-1} a_r j^r \quad (j \in \Omega).$$

We put $f = \sum_{r=0}^{p-1} a_r t^r$ and define the degree of f to be that of $\sum_{r=0}^{p-1} a_r t^r$. (The degree of 0 is -1.)

b) *For $r = 0, \ldots, p$, let*

$$\mathsf{F}_r = \{f \mid f \in \mathsf{F}(\Omega), \text{degree of } f < r\}.$$

Then $\mathsf{F}_0, \mathsf{F}_1, \ldots, \mathsf{F}_p$ are all the $\mathsf{K}\mathfrak{P}$-submodules of $\mathsf{F}(\Omega)_{\mathfrak{P}}$. We have

$$0 = \mathsf{F}_0 \subset \mathsf{F}_1 \subset \cdots \subset \mathsf{F}_p = \mathsf{F}(\Omega)$$

and $\dim_{\mathsf{K}} \mathsf{F}_r = r$.

c) *F_1 is the set of $f \in \mathsf{F}(\Omega)$ which are constant on Ω and $\mathsf{F}_{p-1} = \{f \mid \sum_{j \in \Omega} jf = 0\}$.*

Proof. a) No non-zero polynomial of degree less than p is identically zero on $GF(p)$, so the dimension of the subspace of functions defined by polynomials is p. Since $\dim_{\mathsf{K}} \mathsf{F}(\Omega) = p$, the assertions in a) follow at once.

b) Let $x = (0, 1, \ldots, p-2, p-1)$. If $f \in \mathsf{F}(\Omega)$,

$$j(fx) = (jx^{-1})f = (j-1)f \quad (j \in \Omega).$$

Thus if $f = \sum_{r=0}^{p-1} a_r t^r$, we have

$$fx = \sum_{r=0}^{p-1} a_r (t-1)^r.$$

Hence the F_r are $\mathsf{K}\mathfrak{P}$-submodules and, since the degree of $fx - f$ is less than that of f, $\mathsf{F}_r/\mathsf{F}_{r-1}$ is a trivial $\mathsf{K}\mathfrak{P}$-module ($i = 1, \ldots, p$). Hence the $\mathsf{K}\mathfrak{P}$-composition factors of $\mathsf{F}(\Omega)$ are all trivial modules.

If $\mathsf{F}(\Omega)$ has other $\mathsf{K}\mathfrak{P}$-submodules, let V be one of smallest dimension. Then $\mathsf{V} \neq 0$; let U be a maximal $\mathsf{K}\mathfrak{P}$-submodule of V. Then V/U is a composition factor of $\mathsf{F}(\Omega)$ and is thus a trivial $\mathsf{K}\mathfrak{P}$-module of dimension 1. By minimality of $\dim_{\mathsf{K}}\mathsf{V}$, $\mathsf{U} = \mathsf{F}_s$ for some $s < p - 1$. Then $\mathsf{V} \nsubseteq \mathsf{F}_{s+1}$, so V contains an element $f = \sum_{r=0}^{k} a_r t^r$ with $a_k \neq 0$ and $s < k < p$. Hence V contains f_1, where

$$\begin{aligned} f_1 &= fx - f \\ &= \sum_{r=0}^{k} a_r((t-1)^r - t^r) \\ &= -ka_k t^{k-1} + \cdots. \end{aligned}$$

Since $ka_k \neq 0$ and $k - 1 \geq s$, $f + \mathsf{F}_s$ and $f_1 + \mathsf{F}_s$ are linearly independent elements of V/U. This is impossible, since $\dim_{\mathsf{K}}\mathsf{V}/\mathsf{U} = 1$.

c) This follows at once from b). **q.e.d.**

10.7 Theorem (KLEMM [1], WIELANDT [7]). *Let $\mathfrak{G}$ be an insoluble transitive permutation group on Ω. Suppose that $|\Omega| = p$ and that K is a field of characteristic p. Then the heart of $\mathsf{F}(\Omega)$ is an irreducible $\mathsf{K}\mathfrak{G}$-module of dimension $p - 2$.*

Proof. As in 10.6, we write the elements of $\mathsf{F}(\Omega)$ as polynomials. Thus it is to be shown that $\mathsf{F}_{p-1}/\mathsf{F}_1$ is an irreducible $\mathsf{K}\mathfrak{G}$-module.

Let U/F_1 be a proper non-zero $\mathsf{K}\mathfrak{G}$-submodule of $\mathsf{F}_{p-1}/\mathsf{F}_1$. Since U is *a fortiori* a $\mathsf{K}\mathfrak{P}$-module, it follows from 10.6b) that $\mathsf{U} = \mathsf{F}_k$, where $1 < k < p - 1$. Also $\mathsf{U}^\perp$ is a $\mathsf{K}\mathfrak{G}$-submodule and

$$\dim_{\mathsf{K}} \mathsf{U}^\perp = p - \dim_{\mathsf{K}} \mathsf{U} = p - k.$$

Hence $\mathsf{U}^\perp = \mathsf{F}_{p-k}$. Thus we may suppose that $k \leq \frac{1}{2}p$ by replacing U by $\mathsf{U}^\perp$ if necessary.

Suppose that $g \in \mathfrak{G}$. Since $\Omega = GF(p)$ and $GF(p) \subseteq \mathsf{K}$, g may be regarded as an element of $\mathsf{F}(\Omega)$. Suppose that

$$g = \sum_{r=0}^{j} b_r t^r,$$

where $b_j \neq 0$ ($1 \leq j < p$). The identity mapping 1_Ω on Ω is represented by the polynomial t and thus lies in F_2. Since $2 \leq k$ and $\mathsf{F}_k = \mathsf{U}$ is a

$\mathsf{K}\mathfrak{G}$-module, $1_\Omega g^{-1} \in \mathsf{F}_k$. Since

$$s(1_\Omega g^{-1}) = sg = \sum_{r=0}^{j} b_r s^r,$$

we have $j < k$. Suppose that i is the least integer such that $ij \geq k$. Then $(i - 1)j < k$ and

$$ij < j + k < 2k \leq p.$$

Let f be the element of $\mathsf{F}(\Omega)$ having polynomial t^i. Then if $s \in \Omega$,

$$s(fg^{-1}) = (sg)f = \left(\sum_{r=0}^{j} b_r s^r\right) f = \left(\sum_{r=0}^{j} b_r s^r\right)^i.$$

Since $ij < p$, the polynomial of fg^{-1} is

$$\left(\sum_{r=0}^{j} b_r t^r\right)^i,$$

so $fg^{-1} \notin \mathsf{F}_{ij}$. Since $ij \geq k$, $fg^{-1} \notin \mathsf{F}_k$. Since F_k is $\mathfrak{G}$-invariant, $f \notin \mathsf{F}_k$. Hence $i \geq k$. Thus $(k - 1)j < k$ and $j = 1$. Therefore

$$xg = b_0 + b_1 x \quad (x \in GF(p)),$$

and $\mathfrak{G}$ is soluble, contrary to hypothesis. **q.e.d.**

This brings us to a very elementary proof of the theorem of Burnside (V, 21.3).

10.8 Theorem (BURNSIDE). *Suppose that $\mathfrak{G}$ is transitive of prime degree p. If $\mathfrak{G}$ is insoluble, $\mathfrak{G}$ is doubly transitive.*

Proof. (WIELANDT [7]). Suppose that $\mathfrak{G}$ is not doubly transitive. By 10.5a), the heart H has a trivial $\mathsf{K}\mathfrak{G}$-composition factor if Char $\mathsf{K} = p$. Since $p > 3$, H is reducible. By 10.7, $\mathfrak{G}$ is soluble. **q.e.d.**

Next we strengthen part of the assertion in V, 21.6.

10.9 Theorem (P. M. NEUMANN [1]). *Suppose that $\mathfrak{G}$ is an insoluble transitive permutation group of degree p. Let $\mathfrak{P}$ be a Sylow p-subgroup of $\mathfrak{G}$. If $|\mathbf{N}_{\mathfrak{G}}(\mathfrak{P})|$ is even, $\mathfrak{G}$ is triply transitive.*

Proof (GREEN [2]). We may suppose that $\Omega = GF(p)$ and that $\mathfrak{P} = \langle x \rangle$, where $x = (0, 1, \ldots, p-1)$. Then $\mathbf{N}_{\mathfrak{G}}(\mathfrak{P})$ consists of mappings of the form $t \to at + b$. Since $|\mathbf{N}_{\mathfrak{G}}(\mathfrak{P})|$ is even, $\mathbf{N}_{\mathfrak{G}}(\mathfrak{P})$ contains an involution, which must be of the form $t \to -t + b$. Hence $\mathbf{N}_{\mathfrak{G}}(\mathfrak{P})$ contains the involution j given by $tj = -t$ $(t \in \Omega)$.

We use the notation U_1, U_2, H, as defined in 10.5. Since $j \in \mathfrak{G}_0$, $|\mathfrak{G}_0|$ is even. By 10.8 and 10.5b), it is sufficient to prove that

$$\mathrm{Hom}_{\mathsf{K}\mathfrak{G}}(\mathsf{H}, \mathsf{F}(\Omega^{(2)})) = 0.$$

To do this consider in $\mathsf{F}(\Omega)/\mathsf{U}_1$ the element

$$v = \sum_{i \in \Omega} ih_i + \mathsf{U}_1.$$

(Here ih_i denotes i times the function h_i, not h_i evaluated at i.)

Since $\sum_{i \in \Omega} i = \frac{1}{2}p(p-1)1_{\mathsf{K}} = 0$, $v \in \mathsf{U}_2/\mathsf{U}_1 = \mathsf{H}$. Also $v \neq 0$. By 10.7, H is irreducible, so H is the $\mathsf{K}\mathfrak{G}$-module generated by v. Hence it is sufficient to prove that if $\alpha \in \mathrm{Hom}_{\mathsf{K}\mathfrak{G}}(\mathsf{H}, \mathsf{F}(\Omega^{(2)}))$, then $v\alpha = 0$. Note that

$$vx = \sum_{i \in \Omega} ih_{i+1} + \mathsf{U}_1 = \sum_{i \in \Omega} (i+1)h_{i+1} - \sum_{i \in \Omega} h_{i+1} + \mathsf{U}_1 = v$$

and

$$vj = \sum_{i \in \Omega} ih_{-i} + \mathsf{U}_1 = -\sum_{i \in \Omega} ih_i + \mathsf{U}_1 = -v.$$

Since $vx = v$, $(v\alpha)x = v\alpha$, that is, $v\alpha$ is an element of $\mathsf{F}(\Omega^{(2)})$ invariant under $\langle x \rangle$. The orbits of $\langle x \rangle$ in $\Omega^{(2)}$ are the sets

$$\Delta_a = \{\{b, c\} \mid c - b = a\} \quad (0 < a \leq \tfrac{1}{2}(p-1)),$$

so

$$v\alpha = \sum_{a=1}^{\frac{1}{2}(p-1)} \lambda_a f_a,$$

where f_a is the element of $\mathsf{F}(\Omega^{(2)})$ which takes the value 1 on Δ_a and 0 elsewhere. But by definition of Δ_a,

$$\{b, c\}(f_a j) = (\{b, c\}j^{-1})f_a = \{-b, -c\}f_a = \{-c, -b\}f_a = \{b, c\}f_a,$$

so $f_a j = f_a$. Hence $(v\alpha)j = v\alpha$. Since $vj = -v$, this gives $v\alpha = 0$. **q.e.d.**

10.10 Theorem (WIELANDT [5]). *Suppose that $\mathfrak{G}$ is a transitive permutation group of prime degree p. Let $\mathfrak{P}$ be a Sylow p-subgroup of $\mathfrak{G}$. If $|\mathbf{N}_{\mathfrak{G}}(\mathfrak{P})|$ is even, $\mathfrak{G}$ has only one conjugacy class of subgroups of index p, namely, the set of stabilisers of the elements of Ω.*

Proof. (P. M. NEUMANN [1]). Suppose that $|\mathfrak{G}:\mathfrak{H}| = p$ and that $\mathfrak{H}$ is distinct from all the stabilisers of elements of Ω. Since all the p-complements of a soluble group are conjugate, $\mathfrak{G}$ is not soluble. By 10.8, $\mathfrak{G}$ is doubly transitive.

Let Δ be an orbit of $\mathfrak{H}$. Since $|\mathfrak{H}|$ is not divisible by p, $\Delta \neq \Omega$; hence $2 \leq |\Delta| \leq p - 2$. The stabiliser of Δ is $\mathfrak{H}$, since $\mathfrak{H}$ is a maximal subgroup of $\mathfrak{G}$, so Δ has p images under $\mathfrak{G}$, which we denote by $\Delta_1, \dots, \Delta_p$. Clearly $\mathfrak{G}$ is represented transitively on $\Omega' = \{\Delta_1, \dots, \Delta_p\}$. But this representation is also faithful, since the kernel is a normal p'-subgroup of $\mathfrak{G}$ and is thus contained in the stabiliser of any element of Ω.

Both on Ω and on Ω', $\mathbf{N}_{\mathfrak{G}}(\mathfrak{P})$ is represented as a Frobenius group. By hypothesis $|\mathbf{N}_{\mathfrak{G}}(\mathfrak{P})|$ is even, so $\mathbf{N}_{\mathfrak{G}}(\mathfrak{P})$ possesses an involution j. Thus j leaves fixed precisely one element of Ω', which we may suppose to be Δ_1. Since $|\Delta_1| > 1$, Δ_1 contains two elements a, b interchanged by j. Hence j permutes those elements of Ω' which contain both a and b. It follows that if k is the number of elements of Ω' containing both a and b, k is odd, for Δ_1 is the only such element left fixed by j.

But again, since $|\Delta_1| \leq p - 2$, there exist two elements a', b' outside Δ_1 which are interchanged by j, and j permutes the elements of Ω' which contain both a' and b'. It follows that the number k' of elements of Ω' containing both a' and b' is even. But since $\mathfrak{G}$ is doubly transitive, $k = k'$. This is a contradiction. **q.e.d.**

10.11 Remark. For each prime $p > 3$, there are among the transitive groups of degree p the symmetric and alternating ones and the soluble ones; the number of types of the latter is the number of divisors of $p - 1$ (see II, 3.6). In addition, only the following are known at present.

(1) $PSL(2, 11)$ in its representation of degree 11 (see II, 8.28).

(2) The Mathieu groups $\mathfrak{M}_{11}$ and $\mathfrak{M}_{23}$.

(3) Consider the projective space $\mathsf{P}(n, q)$ of dimension $n - 1$ over $GF(q)$. If the number $\dfrac{q^n - 1}{q - 1}$ of points of $\mathsf{P}(n, q)$ is a prime p, then each group $\mathfrak{G}$ for which

$$PSL(n, q) \leq \mathfrak{G} \leq P\Gamma L(n, q)$$

is a transitive group of degree p. For example, this yields the following groups.

$PSL(3, 2)$ of degree 7.

$PSL(3, 3)$ of degree 13.

$PSL(2, 16)$ of degree 17.

$PSL(5, 2)$ and $PSL(3, 5)$ of degree 31.

It is not known whether there are infinitely many primes of the form $\frac{q^n - 1}{q - 1}$; it is easy to see that for this number to be prime, n must also be prime.

All these groups have just one conjugacy class of subgroups of index p, except for the groups in (3) with $n \geq 3$; in these cases the stabilisers of the points and of the hyperplanes form two classes of subgroups of index p, and these are the only ones.

We determine all the transitive permutation groups of degree 7 and 11.

10.12 Theorem. *If $\mathfrak{G}$ is an insoluble transitive group of degree 7 and $\mathfrak{A}_7 \not\leq \mathfrak{G}$, then $\mathfrak{G} = PSL(3, 2)$.*

Proof. Let $\mathfrak{P}$ be a Sylow 7-subgroup of $\mathfrak{G}$ and write $|\mathbf{N}_{\mathfrak{G}}(\mathfrak{P})| = 7d$. Thus d is a divisor of 6, and by V, 21.1, $d > 2$. Thus $d = 3$ or $d = 6$.

a) Suppose that $d = 3$. Then

$$|\mathfrak{G}| = 7 \cdot 3(1 + 7k)$$

for some integer $k > 0$. Since $\mathfrak{A}_7 \not\leq \mathfrak{G}$, it follows from II, 4.6 that

$$|\mathfrak{S}_7 : \mathfrak{G}| \geq \left[\frac{7 + 1}{2}\right]! = 24.$$

Thus $|\mathfrak{G}| \leq 7 \cdot 6 \cdot 5$ and $1 + 7k \leq 10$. Thus $k = 1$ and $|\mathfrak{G}| = 168$. If $1 < \mathfrak{N} \trianglelefteq \mathfrak{G}$, then $|\mathfrak{N}|$ is divisible by 7, $\mathfrak{G} = \mathfrak{N}\mathbf{N}_{\mathfrak{G}}(\mathfrak{P})$, $\mathfrak{N}$ is insoluble and $|\mathfrak{N}| = 168$, so $\mathfrak{N} = \mathfrak{G}$. Thus $\mathfrak{G}$ is simple. By II, 6.15, $\mathfrak{G} \cong PSL(2, 7) \cong PSL(3, 2)$. This group has two classes of subgroups of order 24 and thus has two inequivalent transitive representations of degree 7.

b) If $d = 6$, $|\mathfrak{S}_7 : \mathfrak{G}| \geq 24$ gives $|\mathfrak{G}| = 7 \cdot 6(1 + 7k)$ with $1 + 7k \leq 5$. This is impossible, since $\mathfrak{G}$ is insoluble. **q.e.d.**

10.13 Theorem. *If $\mathfrak{G}$ is an insoluble transitive group of degree 11 and $\mathfrak{A}_{11} \not\leq \mathfrak{G}$, then either $\mathfrak{G} = \mathfrak{M}_{11}$ or $\mathfrak{G} = PSL(2, 11)$.*

Proof. Let $\mathfrak{P}$ be a Sylow 11-subgroup of $\mathfrak{G}$ and write $|\mathbf{N}_{\mathfrak{G}}(\mathfrak{P})| = 11d$. Thus d is a divisor of 10. By V, 21.1, $d > 2$, so $d = 5$ or $d = 10$. Also

$$|\mathfrak{G}| = 11d(1 + 11k)$$

for some k.

a) Since $\mathfrak{G}$ is primitive and $\mathfrak{A}_{11} \not\leq \mathfrak{G}$, II, 4.6 gives

$$|\mathfrak{S}_{11} : \mathfrak{G}| \geq \left[\frac{11 + 1}{2}\right]! = 6!.$$

Thus

$$(1) \qquad |\mathfrak{G}| \leq 11 \cdot 10 \cdot 9 \cdot 8 \cdot 7.$$

If $|\mathfrak{G}|$ is divisible by 7, then $\mathfrak{G}$ contains a cycle of length 7 and, by II, 4.5, $\mathfrak{G}$ is 5-fold transitive. By (1), $\mathfrak{G}$ is sharply 5-fold transitive. But by 3.4, there is no sharply 5-fold transitive group of degree 11. Thus

$$(2) \qquad |\mathfrak{G}| \text{ is not divisible by } 7.$$

If $|\mathfrak{G}|$ is divisible by 5^2, $\mathfrak{G}$ contains a cycle of length 5 and, by II, 4.5, $\mathfrak{G}$ is 7-fold transitive. By (2), this is not possible. Thus

$$(3) \qquad |\mathfrak{G}| \text{ is not divisible by } 5^2.$$

Altogether, then, $1 + 11k$ is not divisible by 5, 7, 11 or any larger prime. Hence

$$|\mathfrak{G}| = 11 \cdot 5 \cdot 2^a \cdot 3^b.$$

If $\mathfrak{G}$ contains a cycle of length 2 or 3, $\mathfrak{G} \geq \mathfrak{A}_{11}$. Hence $a \leq 7$ and $b \leq 3$.

b) Suppose that $d = 5$. Then

$$|\mathfrak{G}| = 11 \cdot 5(1 + 11k) \leq 11 \cdot 10 \cdot 9 \cdot 8 \cdot 7,$$

so

$$1 + 11k = 2^a \cdot 3^b \leq 9 \cdot 8 \cdot 7 \cdot 2.$$

The only possibilities are that $1 + 11k$ is 12 or 144. Thus either $|\mathfrak{G}| = 12 \cdot 11 \cdot 5$ or $|\mathfrak{G}| = 11 \cdot 10 \cdot 9 \cdot 8$. In the first case, it is not difficult to see that $\mathfrak{G}$ is faithfully represented on its Sylow 11-subgroups by a Zassenhaus group of degree 12. Thus $\mathfrak{G} \cong PSL(2, 11)$, by XI, 9.1.

c) Suppose next that $d = 5$ and $|\mathfrak{G}| = 11 \cdot 10 \cdot 9 \cdot 8$. Choose elements 1, 2 in Ω and let $\mathfrak{H}$ denote the permutation group induced on $\Omega' = \Omega - \{1, 2\}$ by $\mathfrak{G}_{1,2}$. Thus $\mathfrak{H}$ is of degree 9 and order 72, and $\mathfrak{H}$ contains no cycle of length 2 or 3.

Suppose that $\mathfrak{H}$ is intransitive. Then $\mathfrak{H}$ has no element of order 9, so $\mathfrak{H}$ has an elementary Abelian subgroup $\langle a, b\rangle$ of order 9. Let

$$a = (i, j, k)(i', j', k') \cdots .$$

Suppose first that i, i' lie in distinct $\mathfrak{H}$-orbits Δ, Δ'. If $|\Delta| = |\Delta'| = 3$, a Sylow 2-subgroup of $\mathfrak{H}$ is also a Sylow 2-subgroup of the group of permutations of Ω' which leave Δ, Δ' fixed; but this contains a transposition. Thus $|\Delta| + |\Delta'| > 6$. If $|\Delta| \leq 5$ and $|\Delta'| \leq 5$, b induces cycles on Δ, Δ' which commute with those induced by a; but this implies that $\mathfrak{H}$ contains a cycle of length 3. Hence $\mathfrak{H}$ has an orbit Δ_1 of length at least 6, and this is also the case if i, i' lie in the same orbit. If $\Delta_2 = \Omega' - \Delta_1$, the order of the group of permutations induced on Δ_2 by $\mathfrak{H}$ is divisible by 3, since the Sylow 3-subgroup of the group of all permutations of Δ_1 is of order 3^2 and contains cycles of length 3. Thus $|\Delta_1| = 6$ and Δ_2 is an orbit of length 3. Hence $\mathfrak{H}$ is faithfully represented on Δ_1. Since $\mathfrak{H}$ is soluble, the representation of $\mathfrak{H}$ on Δ_1 is imprimitive. But $|\mathfrak{H}|$ does not divide $(2!)^3 \cdot 3!$ and $|\mathfrak{H}| = (3!)^2 \cdot 2!$. Hence the group of permutations induced on Δ_1 by $\mathfrak{H}$ is $\mathfrak{S}_3 \wr \mathfrak{S}_2$. Since $\mathfrak{S}_3 \wr \mathfrak{S}_2$ contains transpositions, $\mathfrak{H}$ induces the whole of $\mathfrak{S}_3$ on Δ_2. This implies that there exists an epimorphism of $\mathfrak{S}_3 \wr \mathfrak{S}_2$ onto $\mathfrak{S}_3$. This is not the case, since the Sylow 3-subgroup is the only minimal normal subgroup of $\mathfrak{S}_3 \wr \mathfrak{S}_2$. Hence $\mathfrak{H}$ is transitive.

It follows that the stabiliser of an element in $\mathfrak{H}$ is a Sylow 2-subgroup of $\mathfrak{H}$, so $\mathbf{O}_2(\mathfrak{H}) = 1$. Thus by IX, 1.4, $\mathbf{O}_3(\mathfrak{H})$ is self-centralizing in $\mathfrak{H}$; therefore $\mathbf{O}_3(\mathfrak{H})$ is a Sylow 3-subgroup of $\mathfrak{H}$. The Sylow 3-subgroup of $\mathfrak{H}$ is thus a regular normal subgroup. Also, $\mathfrak{G}_1$ is doubly transitive since $\mathfrak{H}$ is transitive, so $\mathfrak{G}_1$ cannot have a regular normal subgroup on $\Omega - \{1\}$, since this would be of order 10. Hence $\mathfrak{G}_1$ is a group of Suzuki type in the sense of 2.1. It follows from 2.6 that $\mathfrak{G}_{1,2,3}$ has only one involution j. Let Γ be an orbit of $\mathfrak{G}_{1,2,3}$ on $\Omega - \{1, 2, 3\}$ on which j is represented non-trivially. Then $\mathfrak{G}_{1,2,3}$ is represented faithfully on Γ. Hence $|\Gamma| \geq 4$, and if $|\Gamma| = 4$, $\mathfrak{G}_{1,2,3}$ is a dihedral group of order 8. Since $\mathfrak{G}_{1,2,3}$ has only one involution, it follows that $|\Gamma| = 8$ and $\mathfrak{G}_{1,2,3}$ is transitive on $\Omega - \{1, 2, 3\}$. Hence $\mathfrak{G}$ is sharply 4-fold transitive. By 3.3, $\mathfrak{G} \cong \mathfrak{M}_{11}$.

d) Now suppose that $d = 10$. By a),

$$|\mathfrak{G}| = 11 \cdot 5 \cdot 2^a \cdot 3^b \leq 11 \cdot 10 \cdot 9 \cdot 8 \cdot 7,$$

where $2^{a-1}3^b \equiv 1$ (11) and $b \leq 3$. Again, the only possibilities are that $1 + 11k$ is 12 or 144. But by 10.9, $\mathfrak{G}$ is triply transitive, so $|\mathfrak{G}|$ is divisible by 9. Thus

$$|\mathfrak{G}| = 11 \cdot 10 \cdot 144 = 2|\mathfrak{M}_{11}|.$$

$\mathbf{N}_{\mathfrak{G}}(\mathfrak{P})$ contains a cycle of length 10, so the even permutations in $\mathfrak{G}$ form a normal subgroup $\mathfrak{H}$ of index 2. Since $|\mathbf{N}_{\mathfrak{H}}(\mathfrak{P})| = 11 \cdot 5$ and $|\mathfrak{H}| = |\mathfrak{M}_{11}|$, it follows from the above that $\mathfrak{H} \cong \mathfrak{M}_{11}$. Let 1, 2, 3 be elements of Ω. Then $\mathfrak{H}_{1,2}$ is a Frobenius group of order 72. Thus $\mathfrak{H}_{1,2}$ has a regular normal Sylow 3-subgroup. Hence so does $\mathfrak{G}_{1,2}$. Since $\mathfrak{G}_1$ is doubly transitive, $\mathfrak{G}_1$ is a group of Suzuki type. By 2.6, $\mathfrak{G}_{1,2,3}$ has just one involution. On the other hand, the Sylow 3-subgroup of $\mathfrak{G}_{1,2}$ is self-centralizing in $\mathfrak{S}_9$, so $\mathfrak{G}_{1,2,3}$ is faithfully represented on it. Hence $\mathfrak{G}_{1,2,3}$, which is of order 16, is isomorphic to a Sylow 2-subgroup of $GL(2, 3)$. But this has more than one involution, so we have a contradiction. Hence the case $d = 10$ does not occur. **q.e.d.**

Notes on Chapter XII

§ 1: The constructions of the Mathieu groups given in 1.3 and 1.4 originated in WITT [2]. Earlier constructions were difficult to check. The reader is referred to WITT [2] and LÜNEBURG [2] for the geometrical treatment of the Mathieu groups by Steiner systems.
§ 2: This section follows SUZUKI [9].
§ 3: The main results of this section, 3.3 and 3.4, are due to C. Jordan. The proofs are considerably simplified by using the determination of all sharply triply transitive permutation groups in XI, § 2. We emphasize that only elementary results from Chapter XI are used.
§ 4: The methods of this section go back to WIELANDT [3], where it was shown that if the Schreier hypothesis always holds, every 8-fold transitive group is alternating or symmetric. The extension of this result to 7-fold transitive groups comes from SUZUKI [9].
§ 5: This section follows W. WONG [1]. The procedures followed are typical of many of the characterizations of simple groups in terms of the centralizers of involutions.
§ 6: This section rests predominantly on LIVINGSTONE and WAGNER [1]. An essential simplification is brought about by Lemma 6.3, from WIELANDT [6]. There are some extensions in papers of Kantor.
§ 7: The main result, 7.3, of this section is from HUPPERT [1]. The use of the theorem of Zsigmondy (see 7.3, part c) of the proof) brings about considerable simplifications and is due to HERING [1].
§ 8: This section is based on ZASSENHAUS [3], with many small alterations.
§ 9: This section also basically follows ZASSENHAUS [3].
§ 10: The elementary proof of the theorem of Burnside in 10.8 comes essentially from Wielandt. Theorem 10.9 was first proved by P. M. Neumann with the help of modular representation theory.

Bibliography

Books and lecture notes are distinguished by *

ADYAN, S. I.:* [1] The Burnside problem and identities in groups. Springer-Verlag 1978.

ALPERIN, J. L.: [1] Automorphisms of solvable groups. Proc. Amer. Math. Soc. *13*, 175–180 (1962).

— [2] Sylow intersections and fusion. J. Algebra *6*, 222–241 (1967).

ALPERIN, J. L., and D. GORENSTEIN: [1] The multiplicators of certain simple groups. Proc. Amer. Math. Soc. *17*, 515–519 (1966).

— [2] Transfer and fusion in finite groups. J. Algebra *6*, 242–255 (1967).

AMAYO, R. K., and I. STEWART:* [1] Infinite-dimensional Lie algebras. Leyden: Noordhoff International Publishing 1974.

ARAD, Z., and G. GLAUBERMAN: [1] A characteristic subgroup of a group of odd order. Pacific J. Math. *56*, 305–319 (1975).

ARTIN, E.: [1] The order of the linear groups. Comm. Pure Appl. Math. *8*, 355–366 (1955).

— [2] The order of the classical simple groups. Comm. Pure Appl. Math. *8*, 455–472 (1955).

ASCHBACHER, M.: [1] Thin finite simple groups. Bull. Amer. Math. Soc. *82*, 484 (1976).

BAŠEV, V. A.: [1] Representations of $Z_2 \times Z_2$ in the field of characteristic 2. Dokl. Akad. Nauk *141*, 1015–1018 (1961).

BEAUMONT, R. A. and R. P. PETERSON: [1] Set transitive permutation groups. Canad. J. Math. *7*, 35–42 (1955).

BENDER, H.: [1] Über den grössten p'-Normalteiler in p-auflösbaren Gruppen. Arch. Math. (Basel) *18*, 15–16 (1967).

— [2] Endliche zweifach transitive Permutationsgruppen, deren Involutionen keine Fixpunkte haben. Math. Z. *104*, 175–204 (1968).

— [3] On the uniqueness theorem. Illinois J. Math. *14*, 376–384 (1970).

— [4] On groups with Abelian Sylow 2-subgroups. Math. Z. *117*, 164–176 (1970).

— [5] Transitive Gruppen gerader Ordnung, in denen jede Involution genau einen Punkt festlässt. J. Algebra *17*, 527–554 (1971).

— [6] A group-theoretic proof of Burnside's $p^a q^b$-theorem. Math. Z. *126*, 327–338 (1972).

— [7] Finite groups with large subgroups. Illinois J. Math. *18*, 223–228 (1974).

— [8] The Brauer-Suzuki-Wall theorem. Illinois J. Math. *18*, 229–235 (1974).

— [9] On the normal p-structure of a finite group and related topics I. Hokkaido Math. J. *7*, 271–288 (1978).

BERGER, T. R.: [1] Hall-Higman type theorems, I. Canad. J. Math. *29*, 513–531 (1974); II. Trans. Amer. Math. Soc. *205*, 47–69 (1975); III. Trans. Amer. Math. Soc. *228*, 47–83 (1977); IV. Proc. Amer. Math. Soc. *37*, 317–325 (1973); V. Pacific J. Math. *73*, 1–62 (1977); VI. J. Algebra *51*, 416–424 (1978); VII. Proc. London Math. Soc. (3) *31*, 21–54 (1975).

— [2] Nilpotent fixed point free automorphism groups of solvable groups. Math. Z. *131*, 305–312 (1973).

BERGER, T. R. and F. GROSS: [1] 2-length and the derived length of a Sylow 2-subgroup. Proc. London Math. Soc. (3) *34*, 520–534 (1977).

BERMAN, S. D.: [1] The number of irreducible representations of a finite group over an arbitrary field. Dokl. Akad. Nauk *106*, 767–769 (1956).

BLACKBURN, N.: [1] Über Involutionen in 2-Gruppen. Arch. Math. (Basel) 35, 75–78 (1980).

BLACKBURN, N. and L. EVENS: [1] Schur multipliers of p-groups. J. Reine Angew. Math. *309*, 100–113 (1979).

BOMBIERI, E.: [1] Thompson's problem ($\sigma^2 = 3$). Invent. Math. *58*, 77–100 (1980).

BOURBAKI, N.:* [1] Algèbre, Chap. 4, 5. Paris: Hermann 1959.

— *[2] Algèbre Commutative, Chap. 2. Paris: Hermann 1961.

— *[3] Groupes et algèbres de Lie, Chap. 2: Algèbres de Lie libres. Paris: 1972.

BRAUER, R.: [1] Über Systeme hyperkomplexer Zahlen. Math. Z. *30*, 79–107 (1929).

— [2] Über die Darstellungen von Gruppen in Galoisschen Feldern. Act. Sci. No. 135, Paris, Hermann 1935.

— [3] On groups whose order contains a prime number to the first power I, II. Amer. J. Math. *64*, 401–420 and 421–440 (1942).

— [4] On the representation of a group of order g in the field of g-th roots of unity. Amer. J. Math. *67*, 461–471 (1945).

— [5] On the structure of groups of finite order. Proc. International Congress, Amsterdam 1954, vol. 1, 209–217.

— [6] Zur Darstellungstheorie der Gruppen endlicher Ordnung. Math. Z. *63*, 406–444 (1956).

— [7] Some applications of the theory of blocks of characters of finite groups. J. Algebra *1*, 152–167 (1964).

— [8] Some applications of the theory of blocks of characters of finite groups IV. J. Algebra *17*, 489–521 (1971).

BRAUER, R. and P. FONG: [1] A characterization of the Mathieu group M_{12}. Trans. Amer. Math. Soc. *122*, 18–47 (1966).

BRAUER, R. and K. A. FOWLER: [1] Groups of even order. Ann. of Math. *62*, 565–583 (1955).

BRAUER, R. and C. NESBITT: [1] On the modular characters of groups. Ann. of Math. *42*, 556–590 (1941).

BRAUER, R., M. SUZUKI and G. E. WALL: [1] A characterization of the one-dimensional unimodular projective groups over finite fields. Illinois J. Math. *2*, 718–745 (1958).

BRENNER, S.: [1] Modular representations of p-groups. J. Algebra *15*, 89–102 (1970).

BRYANT, R. M. and L. C. KOVÁCS: [1] Lie representations and groups of prime-power order. J. London Math. Soc. (2) *17*, 415–421 (1978).

BRYCE, N.: [1] On the Mathieu group M_{23}. J. Austral. Math. Soc. *12*, 385–392 (1971).

BUCHT. G.: [1] Die umfassendsten primitiven metazyklischen Kongruenzgruppen in drei oder vier Variablen. Ark. Mat. Astronom. Fys. *11*, (92 pages, 1917).

BURGOYNE, N. and P. FONG: [1] Multipliers of the Mathieu groups. Nagoya Math. J. *27*, 733–745 (1966). Corrections, Nagoya Math. J. *31*, 297–304 (1968).

BURKHARDT, R.: [1] Die Zerlegungsmatrizen der Gruppen PSL(2, p^f). J. Algebra *40*, 75–96 (1976).

BURNSIDE, W.:* [1] Theory of groups of finite order, 2nd edn. Cambridge, 1911; Dover Publications, 1955.

CARTIER, P.: [1] Remarques sur le théorème de Birkhoff-Witt. Ann. Scuola Norm. Sup. Pisa (3) *12*, 1–4 (1958).

COLLINS. M. J.: [1] The characterization of the Suzuki groups by their Sylow 2-subgroups. Math. Z. *123*, 432–48 (1971).

CONLON, S.: [1] The modular representation algebra of groups with Sylow 2-subgroups $Z_2 \times Z_2$. J. Austral. Math. Soc. *6*, 76–88 (1966).

COSSEY, J. and W. GASCHÜTZ: [1] A note on blocks. Proc. Second Intern. Conf. Theory of Groups, 238–240. Canberra 1973.

COXETER, H. and W. MOSER:* [1] Generators and relations for discrete groups. 2nd edition, Ergebnisse der Mathematik 14, Springer 1964.

CURTIS, C., W. KANTOR and G. SEITZ: [1] The 2-transitive permutation representations of the finite Chevalley groups. Trans. Amer. Math. Soc. *218*, 1–59 (1976).

CURTIS, C. and I. REINER:* [1] Representation theory of finite groups and associative algebras. New York: Interscience Publishers Inc. 1962.
DADE, E. C.: [1] Some p-solvable groups. J. Algebra *2*, 395–401 (1965).
— [2] Blocks with cyclic defect groups. Ann. of Math. (2) *84*, 20–48 (1966).
— [3] Degress of modular irreducible representations of p-solvable groups. Math. Z. *104*, 141–143 (1968).
— [4] Carter subgroups and Fitting heights of finite solvable groups. Illinois J. Math. *13*, 347–369 (1972).
— [5] Une extension de la théorie de Hall et Higman. J. Algebra *20*, 570–609 (1972).
DEURING, M.: [1] Galoissche Theorie und Darstellungstheorie. Math. Ann. *107*, 140–144 (1932).
— *[2] Algebren. Ergebn. der Math. 41, 2. Aufl., Springer 1968.
DIEUDONNÉ, J.:*[1] La géométrie des groupes classiques, 2nd edition. Springer-Verlag 1963.
DOLAN, S. W.: [1] Some problems in the theory of finite groups, D. Phil. dissertation, Oxford, 1975.
DORNHOFF, L.: [1] The rank of primitive solvable permutation groups. Math. Z. *109*, 205–210 (1969).
FEIT, W.: [1] On a class of doubly transitive permutation groups. Illinois J. Math. *4*, 170–186 (1960).
— [2] Group characters, exceptional characters. Summer Institute of Finite Groups, Pasadena (1960).
— *[3] Representation theory of finite groups. Lecture Notes, Yale University 1969.
FEIT, W. and J. G. THOMPSON: [1] Solvability of groups of odd order. Pacific J. Math. *13*, 773–1029 (1963).
FELSCH, W., J. NEUBÜSER and W. PLESKEN: [1] Space groups and groups of prime-power order IV. J. London Math. Soc. (2) *24*, 113–122 (1981).
FISCHER, I. and R. R. STRUIK: [1] Nil algebras and periodic groups. Amer. Math. Monthly *75*, 611–623 (1968).
FONG, P.: [1] On the characters of p-solvable groups. Trans. Amer. Math. Soc. *98*, 263–284 (1961).
— [2] On decomposition numbers of J_1 and $R(q)$. Sympos. Math. Vol. XIII, 414–422. London, New York. Acad. Press 1974.
FONG, P. and W. GASCHÜTZ: [1] A note on the modular representations of solvable groups. J. Reine Angew. Math. *208*, 73–78 (1961).
FOULSER, D.: [1] Solvable primitive permutation groups of low rank. Trans. Amer. Math. Soc. *143*, 1–54 (1969).
FOX, R. H.: [1] Free differential calculus I. Ann. of Math. *57*, 547–560 (1953).
FRASCH, H.: [1] Die Erzeugenden der Hauptkongruenzgruppen für Primzahlstufen. Math. Ann. *108*, 229–252 (1933).
FROBENIUS, G.: [1] Über Gruppencharaktere. Sitz. preuss. Akad. 1896, 985–1021.
— [2] Über Gruppen des Grades p oder $p + 1$. Sitz. preuss. Akad. 1902, 351–369.
— [3] Über die Charaktere der mehrfach transitiven Gruppen. Sitz. preuss. Akad. Berlin 1904, 558–571.
FROBENIUS, G. and I. SCHUR: [1] Über die reellen Darstellungen endlicher Gruppen. Sitz. preuss. Akad. 1906, 186–208.
GAGEN, T. M.:* [1] Topics in Finite Groups. London Mathematical Society Lecture Note Series 16, CUP 1976.
GASCHÜTZ, W.: [1] Über den Fundamentalsatz von Maschke zur Darstellungstheorie endlicher Gruppen. Math. Z. *56*, 376–387 (1952).
GASCHÜTZ, W. and T. YEN: [1] Groups with an automorphism group which is transitive on the elements of prime order. Math. Z. *86*, 123–127 (1964).
GLAUBERMAN, G.: [1] On the automorphism group of a finite group having no non-identity normal subgroups of odd order. Math. Z. *93*, 154–160 (1966).
— [2] Prime-power factor groups of finite groups. Math. Z. *107*, 159–172 (1968).
— [3] A characteristic subgroup of a p-stable group. Canad. J. Math. *20*, 1101–1135

(1968).
— [4] On a class of doubly transitive permutation groups. Illinois J. Math. *13*, 394–399 (1969).
— [5] Prime-power factor groups of finite groups II. Math. Z. *117*, 46–56 (1970).
— [6] A sufficient condition for p-stability. Proc. London Math. Soc. (3) *25*, 253–287 (1972).
— [7] Failure of factorization in p-solvable groups. Quart. J. Math. Oxford (2) *24*, 71–77 (1973); II, Quart. J. Math. Oxford (2) *26*, 257–261 (1975).
— [8] On Burnside's other $p^a q^b$ theorem. Pacific J. Math. *56*, 469–476 (1975).
— [9] On solvable signalizer functors in finite groups. Proc. London Math. Soc. (3) *33*, 1–27 (1976).
— [10] Factorizations in local subgroups of finite groups. Providence, R. I. (1977).
GLEASON, A. M.: [1] Finite Fano planes. Amer. J. Math. *78*, 797–807 (1956).
GLOVER, D. J.: [1] A study of certain modular representations. J. Algebra *51*, 425–475 (1978).
GOLDSCHMIDT, D. M.: [1] A conjugation family for finite groups. J. Algebra *16*, 138–142 (1970).
— [2] A group theoretic proof of the $p^a q^b$ theorem for odd primes. Math. Z. *113*, 373–375 (1970).
— [3] Solvable signalizer functors on finite groups. J. Algebra *21*, 137–148 (1972).
— [4] 2-fusion in finite groups. Ann. of Math. (2) *99*, 70–117 (1974).
— [5] Elements of order two in finite groups. Delta (Waukesha) *4*, 45–58 (1974/5).
GORENSTEIN, D.:* [1] Finite groups. New York, Harper and Row 1968.
GORENSTEIN, D. and I. N. HERSTEIN: [1] Finite groups admitting a fixed-point-free automorphism of order 4. Amer. J. Math. *83*, 71–78 (1961).
GORENSTEIN, D. and D. R. HUGHES: [1] Triply transitive groups in which only the identity fixes four letters. Illinois J. Math. *5*, 486–491 (1961).
GORENSTEIN, D. and J. WALTER: [1] The π-layer of a finite group. Illinois J. Math. *15*, 555–564 (1971).
— [2] Balance and generation in finite groups, J. Algebra *33*, 224–287 (1975).
GOW, R.: [1] Extensions of modular representations for relatively prime operator groups. J. Algebra *36*, 492–494 (1975).
GREEN, J. A.: [1] On the indecomposable representations of a finite group. Math. Z. *70*, 430–445 (1959).
— [2] On a theorem of P. M. Neumann. Workshop on permutation groups and indecomposable modules. Giessen, 1975.
GREEN, J. A. and R. HILL: [1] On a theorem of Fong and Gaschütz. J. London Math. Soc. (2) *1*, 573–576 (1969).
GREEN, J. A. and S. E. STONEHEWER: [1] The radicals of some group algebras. J. Algebra *13*, 137–142 (1969).
GROSS, F.: [1] The 2-length of a finite solvable group. Pacific J. Math. *15*, 1221–1237 (1965).
— [2] The 2-length of groups whose Sylow 2-groups are of exponent 4. J. Algebra *2*, 312–314 (1965).
— [3] Solvable groups admitting a fixed-point-free automorphism of prime power order. Proc. Amer. Math. Soc. *17*, 1440–1446 (1966).
— [4] A note on fixed-point-free solvable operator groups. Proc. Amer. Math. Soc. *19*, 1363–1365 (1968).
— [5] 2-automorphic 2-groups. J. Algebra *40*, 348–353 (1976).
HALL, M.: [1] Solution of the Burnside problem for exponent 6. Proc. Nat. Acad. Sci. U.S.A. *43*, 751–753 (1957); Illinois J. Math. *2*, 764–786 (1958).
— *[2] The theory of groups. New York: Macmillan Company 1959.
HALL, P.: [1] A contribution to the theory of groups of prime-power order. Proc. London Math. Soc. (2) *36*, 29–95 (1933).
HALL, P. and G. HIGMAN: [1] The p-length of a p-soluble group, and reduction theorems for Burnside's problem. Proc. London Math. Soc. (3) *6*, 1–42 (1956).

HARTLEY, B.: [1] Sylow p-subgroups and local p-solubility. J. Algebra *23*, 347–369 (1972).
HARTLEY, B. and D. J. S. ROBINSON: [1] On finite complete groups. Arch. Math. (Basel) *35*, 67–74 (1980).
HAWKES, T. O.: [1] On the automorphism group of a 2-group. Proc. London Math. Soc. (3) *26*, 207–225 (1973).
HEINEKEN, H. and H. LIEBECK: [1] The occurrence of finite groups in the automorphism group of nilpotent groups of class 2. Arch. Math. (Basel) *25*, 8–16 (1974).
HELD, D.: [1] Eine Kennzeichnung der Mathieu Gruppe M_{22} und der alternierenden Gruppe A_{10}. J. Algebra *8*, 436–449 (1968).
— [2] A characterization of some multiply transitive permutation groups I. Illinois J. Math. *13*, 224–240 (1969).
— [3] The simple groups related to M_{24}. J. Algebra *13*, 253–296 (1969).
HELD, D. and U. SCHOENWAELDER: [1] A characterization of the simple group $\mathfrak{M}_{24}$. Math. Z. *117*, 289–308 (1970).
HELLER, A. and I. REINER: [1] Indecomposable representations. Illinois J. Math. *5*, 314–323 (1961).
HERING, C.: [1] Zweifach transitive Permutationsgruppen, in denen 2 die maximale Anzahl von Fixpunkten von Involutionen ist. Math. Z. *104*, 150–174 (1968).
— [2] Transitive linear groups and linear groups which contain irreducible subgroups of prime order. Geom. Dedic. *2*, 425–460 (1974).
HERING, C., W. M. KANTOR and G. M. SEITZ: [1] Finite groups with a split BN-pair of rank 1, I. J. Algebra *20*, 435–475 (1972).
HIGMAN, D. G.: [1] Focal series in finite groups. Canad. J. Math. *5*, 477–497 (1953).
— [2] Modules with a group of operators. Duke Math. J. *21*, 369–376 (1954).
— [3] Indecomposable representations at characteristic p. Duke Math. J. *21*, 377–381 (1954).
— [4] Induced and produced modules, Canad. J. Math. *7*, 490–508 (1955).
HIGMAN, D. G. and C. SIMS: [1] A simple group of order 44,352,000. Math. Z. *105*, 110–113 (1968).
HIGMAN, G.: [1] Groups and rings which have automorphisms without nontrivial fixed elements. J. London Math. Soc. *32*, 321–334 (1957).
— [2] Suzuki 2-groups. Illinois J. Math. *7*, 79–96 (1963).
— [3] On the simple group of D. G. Higman and C. C. Sims. Illinois J. Math. *13*, 74–80 (1969).
HILL, E. T.: [1] The annihilator of radical powers in the modular group ring of a p-group. Proc. Amer. Math. Soc. *25*, 811–815 (1970).
HOARE, A. H. M.: [1] A note on 2-soluble groups. J. London Math. Soc. *35*, 193–199 (1960).
HOLT, D. F.: [1] On the local control of Schur multipliers. Quart. J. Math. Oxford (2) *28*, 495–508 (1977).
HOPKINS, C.: [1] Metabelian groups of order p^m, $p > 2$. Trans. Amer. Math. Soc. *37*, 161–195 (1935).
HUGHES, D. R.: [1] Extensions of designs and groups: Projective, symplectic and certain affine groups. Math. Z. *89*, 199–205 (1965).
HUPPERT, B.: [1] Zweifach transitive, auflösbare Permutationsgruppen. Math. Z. *68*, 126–150 (1957).
— [2] Scharf dreifach transitive Permutationsgruppen. Arch. Math. (Basel) *13*, 61–72 (1962).
— [3] Singer-Zyklen in klassischen Gruppen. Math. Z. *117*, 141–150 (1970).
— [4] Bemerkungen zur modularen Darstellungstheorie 1. Absolut unzerlegbare Moduln. Arch. Math. (Basel) *26*, 242–249 (1975).
— [5] Zur Konstruktion der reellen Spiegelungsgruppe $\mathfrak{H}_4$. Acta Math. Szeged *26*, 331–336 (1975).
HUPPERT, B. and H. WIELANDT: [1] Normalteiler mehrfach transitiver Permutationsgruppen. Arch. Math. (Basel) *9*, 18–26 (1958).
HUPPERT, B. and W. WILLEMS: [1] Bemerkungen zur modularen Darstellungstheorie 2. Darstellungen von Normalteilern. Arch. Math. (Basel) *26*, 486–496 (1975).

HURLEY, T. C.: [1] On a problem of Fox. Invent. Math. *21*, 139–141 (1973).

ISAACS, I. M.: [1] Extensions of group representations over arbitrary fields. J. Algebra *68*, 54–74 (1981).

ITÔ, N.: [1] On a theorem of H. F. Blichfeldt. Nagoya Math. J. *5*, 75–77 (1954).

— [2] Normalteiler mehrfach transitiver Permutationsgruppen. Math. Z. *70*, 165–173 (1958).

— [3] On a class of doubly transitive permutation groups. Illinois J. Math. *6*, 341–352 (1962).

— [4] Normal subgroups of quadruply transitive permutation groups. Hokkaido Math. J. *1*, 1–6 (1972).

JACOBSON, N.:* [1] Lie algebras. New York: Interscience Publishers Inc. 1962.

JAMES, G. D.: [1] The modular characters of the Mathieu groups. J. Algebra *27*, 57–111 (1973).

JANKO, Z.: [1] A new finite simple group with Abelian Sylow 2-subgroups and its characterization. J. Algebra *3*, 147–186 (1966).

— [2] A characterization of the smallest group of Ree associated with the simple Lie algebra of type (G_2). J. Algebra *4*, 293–299 (1966).

— [3] A characterization of the Mathieu simple groups I. J. Algebra *9*, 1–19 (1968): II, J. Algebra *9*, 20–41 (1968).

JANKO, Z. and J. G. THOMPSON: [1] On a class of finite simple groups of Ree. J. Algebra *4*, 274–292 (1966).

JENNINGS, S.: [1] The structure of the group ring of a p-group over a modular field. Trans. Amer. Math. Soc. *50*, 175–185 (1941).

KANTOR, W. M.: [1] 4-homogeneous groups. Math. Z. *103*, 67–68 (1968). Corrections, Math. Z. *109*, 86 (1969).

— [2] Automorphism groups of designs. Math. Z. *109*, 246–252 (1969).

— [3] k-homogeneous groups. Math. Z. *124*, 261–265 (1972).

KAPLANSKY, I.:* [1] Fields and Rings. Chicago: University of Chicago Press 1969.

KASCH, F., KNESER, M. and H. KUPISCH: [1] Unzerlegbare modulare Darstellungen endlicher Gruppen mit zyklischer p-Sylowgruppe. Arch. Math. (Basel) *8*, 320–321 (1957).

KLEMM, M.: [1] Primitive Permutationsgruppen von Primzahlpotenzgrad. Comm. Algebra *5*, 193–205 (1977).

KOSTRIKIN, A. I.: [1] The Burnside problem. Izv. Akad. Nauk SSSR Ser. Mat. *23*, 3–34 (1959), translated in Amer. Math. Soc. Trans. (2nd ser.) *36*, 63–100 (1964).

KREKNIN, V. A.: [1] Solvability of Lie algebras with a regular automorphism of finite period. Dokl. Akad. Nauk SSSR *150*, 467–469 (1963), Soviet Mat. Dokl. *4*, 683–685 (1963).

KREKNIN, V. A. and A. I. KOSTRIKIN: [1] Lie algebras with a regular automorphism. Dokl. Akad. Nauk SSSR *149*, 249–251 (1963), Soviet Math. Dokl. *4*, 355–358 (1963).

KURZWEIL, H.: [1] p-Automorphismen von auflösbaren p'-Gruppen. Math. Z. *120*, 326–354 (1971).

LANDROCK, P.: [1] Finite groups with a quasisimple component of type $PSU(3, 2^n)$ on elementary Abelian form. Illinois J. Math. *19*, 198–230 (1975).

LANDROCK, P. and G. MICHLER: [1] Block structure of the smallest Janko group. Math. Ann. *232*, 205–238 (1978).

LAZARD, M.: [1] Sur les groupes nilpotents et les anneaux de Lie. Ann. Sci. École Norm. Sup. *71*, 101–190 (1954).

LEEDHAM-GREEN, C. R., P. M. NEUMANN and J. WIEGOLD: [1] The breadth and the class of a finite p-group. J. London Math. Soc. (2) *1*, 409–420 (1969).

LIVINGSTONE, D.: [1] On a permutation representation of the Janko group. J. Algebra *6*, 43–55 (1967).

LIVINGSTONE, D. and A. WAGNER: [1] Transitivity on unordered sets. Math. Z. *90*, 393–403 (1965).

LORENZ, F.: [1] A remark on real characters of compact groups. Proc. Amer. Math. Soc. *21*, 391–393 (1969).

LÜNEBURG, H.:* [1] Die Suzuki-Gruppen und ihre Geometrien. Springer Lecture Notes 10

(1965).
— [2] Über die Gruppen von Mathieu. J. Algebra *10*, 194–210 (1968).
— *[3] Transitive Erweiterungen endlicher Permutationsgruppen. Springer Lecture Notes 84 (1969).
MAC LANE, S.:* [1] Homology. Springer 1963.
MCDERMOTT, J. P. J.: [1] On ($t + \frac{1}{2}$)-transitive permutation groups. Math. Z. *148*, 61–62 (1976).
MACDONALD, I. D.: [1] Groups of breadth four have class five. Glasgow Math. J. *19*, 141–148 (1978).
MCLAUGHLIN, J. F.: [1] Some groups generated by transvections. Arch. Math. (Basel) *18*, 364–368 (1967).
MAGNUS, W.: [1] Beziehungen zwischen Gruppen und Idealen in einem speziellen Ring. Math. Ann. *111*, 259–280 (1935).
— [2] Über Beziehungen zwischen höheren Kommutatoren. J. Reine Angew. Math. *177*, 105–115 (1937).
MARTINEAU, R. P.: [1] Solubility of groups admitting a fixed-point-free automorphism group of type (p, p). Math. Z. *124*, 67–72 (1972).
— [2] Elementary Abelian fixed point free automorphism groups. Quart. J. Math. Oxford (2) *23*, 205–212 (1972).
— [3] Solubility of groups admitting certain fixed-point-free automorphism groups. Math. Z. *130*, 143–147 (1973).
MATSUYAMA, H.: [1] Solvability of groups of order $2^a p^b$. Osaka J. Math. *10*, 375–8 (1973).
MAZET, P.: [1] Sur le multiplicateur de Schur du groupe de Mathieu M_{22}. C. R. Acad. Sci. Paris *289*, Série A, 659–661 (1979).
— [2] Some subgroups of $SL_n(F_2)$. Illinois J. Math. *13*, 108–115 (1969).
MEIXNER, T.: [1] Über endliche Gruppen mit Automorphismen, deren Fixpunktgruppen beschränkt sind. Doctoral thesis, University of Erlangen-Nürnberg.
MICHLER, G.: [1] The kernel of a block of a group algebra. Proc. Amer. Math. Soc. *37*, 47–49 (1973).
— [2] The blocks of p-nilpotent groups over arbitrary fields. J. Algebra *27*, 303–315 (1973).
NAGAO, H.: [1] On multiply transitive permutation groups I. Nagoya Math. J. *27*, 15–19 (1966).
NAKAYAMA, T.: [1] Some studies on regular representations, induced representations and modular representations. Ann. of Math. *39*, 361–369 (1938).
— [2] On Frobeniusean algebras I. Ann of Math. *40*, 611–633 (1939).
— [3] Finite groups with faithful irreducible and directly indecomposable modular representations. Proc. Japan Acad. *23*, 22–25 (1947).
NESBITT, C.: [1] On the regular representation of algebras. Ann of Math. *39*, 634–658 (1938).
NEUMANN, P. M.: [1] Transitive permutation groups of prime degree. J. London Math. Soc. (2) *5*, 202–207 (1972).
NEUMANN, P. M. and M. R. VAUGHAN-LEE: [1] An essay on BFC groups. Proc. London Math. Soc. (3) *35*, 213–237 (1977).
ONO, T.: [1] An identification of Suzuki groups with groups of generalized Lie type. Ann. of Math. *75*, 251–259 (1962); Corrigendum, Ann. of Math. *77*, 413 (1963).
OSIMA, M.: [1] Note on blocks of group characters. Math. J. Okayama Univ. *4*, 175–188 (1955).
PAHLINGS, H.: [1] Über die Kerne von Blöcken einer Gruppenalgebra. Arch. Math. (Basel) *25*, 121–124 (1974).
— [2] Groups with faithful blocks. Proc. Amer. Math. Soc. *51*, 37–40 (1975).
— [3] Minimale Kerne von Darstellungen. Arch. Math. (Basel) *32*, 431–435 (1979).
PARROTT, D.: [1] On the Mathieu groups M_{22} and M_{11}. J. Austral. Math. Soc. *11*, 69–81 (1970).
PASSMAN, D.: [1] Solvable 3/2-transitive permutation groups. J. Algebra 7, 192–207 (1967).
— [2] p-solvable doubly transitive permutation groups. Pacific J. Math. *26*, 555–577

(1968).
— *[3] Permutation Groups. New York, Harper and Row (1968).
— [4] Exceptional 3/2-transitive permutation groups. Pacific J. Math. *29*, 669–713 (1969).
— [5] Central idempotents in group rings. Proc. Amer. Math. Soc. *22*, 555–556 (1969).
PETTET, M. R.: [1] A sufficient condition for solvability in groups admitting elementary Abelian operator groups. Canad. J. Math. *29*, 848–855 (1977).
POLLATSEK, H.: [1] First cohomology groups of some linear groups over fields of characteristic 2. Illinois J. Math. *15*, 393–417 (1971).
POWELL, M. B. and G. HIGMAN:* [1] Finite Simple Groups. London and New York: Academic Press 1971.
RAE, A.: [1] Sylow p-subgroups of finite p-soluble groups. J. London Math. Soc. (2) *7*, 117–123 (1973); Corrigendum, J. London Math. Soc. (2) *11*, 11 (1975).
RALSTON, E. W.: [1] Solvability of finite groups admitting fixed point free automorphisms of order rs. J. Algebra *23*, 164–180 (1972).
RAZMYSLOV, YU. P.: [1] On Engel Lie algebras. Algebra i Logika *10*, 33–44 (1971) translated in Algebra and Logic *10* (1971).
REE, R.: [1] A family of simple groups associated with the simple Lie algebra of type (G_2). Amer. J. Math. *83*, 432–462 (1961).
— [2] Sur une famille des groups de permutations doublement transitifs. Canad. J. Math. *16*, 797–820 (1964).
REINER, I.: [1] On the number of irreducible modular representations of a finite group. Proc. Amer. Math. Soc. *15*, 810–812 (1964).
RICKMAN, B.: [1] Groups which admit a fixed-point-free automorphism of order p^2. J. Algebra *59*, 77–171 (1959).
RINGEL, C. M.: [1] The indecomposable representations of dihedral 2-groups. Math. Ann. *214*, 19–34 (1975).
RIPS, I. A.: [1] On the fourth integer dimension subgroup. Israel J. Math. *12*, 342–346 (1972).
ROITER, A. V.: [1] Unboundedness of the dimensions of the indecomposable representations of an algebra which has infinitely many indecomposable representations. Izv. Akad. Nauk. SSSR Ser. Mat. *32*, 1275–1282 (1968).
ROTH, R. L.: [1] A dual view of Clifford theory of characters of finite groups. Canad. J. Math. *23*, 857–865 (1971).
ROWLEY, P. J.: [1] Solubility of finite groups admitting a fixed-point-free Abelian automorphism group of square-free exponent rs. Proc. London Math. Soc. (3) *37*, 385–421 (1978).
SANDLING, R.: [1] The dimension subgroup problem. J. Algebra *21*, 216–231 (1972).
SANOV, I. N.: [1] Solution of Burnside's problem for exponent 4. Leningrad State University Ann. *10*, 166–170 (1940).
SCHWARZ, W.: [1] Die Struktur modularer Gruppenringe endlicher Gruppen der p-Länge 1. J. Algebra *60*, 51–75 (1979).
SERRE, J. P.:* [1] Corps locaux. Paris: Hermann 1962.
SHAW, D.: [1] The Sylow 2-subgroups of finite soluble groups with a single class of involutions. J. Algebra *16*, 14–26 (1970).
SHULT, E. E.: [1] On groups admitting fixed point free Abelian operator groups. Illinois J. Math. *9*, 701–720 (1965).
— [2] On finite automorphic algebras. Illinois J. Math. *13*, 625–653 (1969).
— [3] On the triviality of finite automorphic algebras. Illinois J. Math. *13*, 654–659 (1969).
SIBLEY, D. A.: [1] Coherence in finite groups containing a Frobenius section. Illinois J. Math. *20*, 434–442 (1978).
SJOGREN, J. A.: [1] Dimension and lower central subgroups. J. Pure App. Algebra *14*, 175–194 (1979).
SMITH, S. D. and A. P. TYRER: [1] On finite groups with a certain Sylow normalizer. I. J. Algebra *26*, 343–365 (1973); II. J. Algebra *26*, 366–367 (1973); III, J. Algebra *29*, 489–503 (1974).

SRINIVASAN, B.: [1] On the indecomposable representations of a certain class of groups. Proc. London Math. Soc. (3) *10*, 497–513 (1960).
STANTON, R. G.: [1] The Mathieu groups. Canad. J. Math. *3*, 164–174 (1951).
SUZUKI, M.: [1] Finite groups with nilpotent centralizers. Trans. Amer. Math. Soc. *99*, 425–470 (1961).
— [2] On a finite group with a partition. Arch. Math. (Basel) *12*, 241–254 (1961).
— [3] On a class of doubly transitive groups. Ann. of Math. *75*, 104–145 (1962).
— [4] On the characterization of linear groups III. Nagoya Math. J. *21*, 159–183 (1962).
— [5] On a class of doubly transitive groups II. Ann. of Math. *79*, 514–589 (1964).
— [6] Finite groups of even order in which Sylow 2-subgroups are independent. Ann. of Math. *80*, 58–77 (1964).
— [7] A characterization of the 3-dimensional projective unitary group over a finite field of odd characteristic. J. Algebra *2*, 1–14 (1965).
— [8] Finite groups in which the centralizer of any element of order 2 is 2-closed. Ann. of Math. *82*, 191–212 (1965).
— [9] Transitive extensions of a class of doubly transitive groups. Nagoya Math. J. *27*, 159–169 (1966).
SWAN, R. G.: [1] The Grothendieck ring of a finite group. Topology *2*, 85–110 (1963).
THOMPSON, J. G.: [1] Automorphisms of solvable groups. J. Algebra *1*, 259–267 (1964).
— [2] Fixed points of p-groups acting on p-groups. Math. Z. *86*, 12–13 (1964).
— [3] Factorizations of p-soluble groups. Pacific J. Math. *16*, 371–372 (1966).
— [4] Vertices and sources. J. Algebra *6*, 1–6 (1967).
— [5] Towards a characterization of $E_2^*(q)$. J. Algebra *7*, 406–414 (1967); II. J. Algebra *20*, 610–621 (1972).
— [6] Nonsolvable finite groups all whose local subgroups are solvable, I. Bull. Amer. Math. Soc. *74*, 383–437 (1968); II. Pacific J. Math. *33*, 451–536 (1970); III. Pacific J. Math. *39*, 483–534 (1971); IV. Pacific J. Math. *48*, 511–592 (1973); V. Pacific J. Math. *50*, 215–297 (1974); VI. Pacific J. Math. *51*, 573–630 (1974).
— [7] A replacement theorem for p-groups and a conjecture. J. Algebra *13*, 149–151 (1969).
TITS, J.: [1] Généralisation des groupes projectifs, Acad. Roy. Belgique Bull. Cl. Sci. *35*, 197–208, 224–233, 568–589, 756–773 (1949).
— [2] Les groupes simples de Suzuki et de Ree. Sem. Bourbaki, décembre, 1960.
— [3] Ovoïdes et groupes de Suzuki. Arch. Math. (Basel) *13*, 187–198 (1962)
— [4] Une propriété caractéristique des ovoides associés aux groupes de Suzuki. Arch. Math. (Basel) *17*, 136–153 (1966).
TSUSHIMA, Y.: [1] On the annihilator ideals of the radical of a group algebra. Osaka J. Math. *8*, 91–97 (1971).
VAUGHAN-LEE, M. R.: [1] Metabelian BFC-groups. J. London Math. Soc. (2) *5*, 673–680 (1972).
— [2] Breadth and commutator subgroups of p-groups. J. Algebra *32*, 278–285 (1976).
VILLAMAYOR, O. E.: [1] On the semisimplicity of group algebras II. Proc. Amer. Math. Soc. *10*, 27–31 (1959).
van der WAERDEN, B. L.:* [1] Gruppen von linearen Transformationen. Ergebn. der Math. 4, Springer 1935.
— *[2] Algebra 1, 7. Aufl., Springer 1966.
— *[3] Algebra 2, 5. Aufl., Springer 1967.
WAGNER, A.: [1] Normal subgroups of triply transitive permutation groups of odd degree. Math. Z. *94*, 219–222 (1966).
WALL, G. E.: [1] On the Lie ring of a group of prime exponent. Proceedings of the second International Conference on the Theory of Groups, Canberra 1973, 667–690.
— [2] Secretive prime-power groups of large rank, Bull. Austral. Math. Soc. *12*, 363–369 (1975).
WALTER, J.: [1] Finite groups with Abelian Sylow 2-subgroups of order 8. Invent. Math. *2*, 332–376 (1967).
— [2] The characterization of finite groups with Abelian Sylow 2-subgroups. Ann. of

Math. *89*, 405–514 (1969).
WARD, H. N.: [1] On Ree's series of simple groups. Trans. Amer. Math. Soc. *121*, 62–89 (1966).
— [2] The analysis of representations induced from normal subgroups. Michigan Math. J. *15*, 417–428 (1968).
WARD, J. N.: [1] Automorphisms of finite groups and their fixed-point groups. J. Austral. Math. Soc. *9*, 467–477 (1969).
WIEGOLD, J.: [1] Groups with boundedly finite classes of conjugate elements. Proc. Roy. Soc. Ser. A. *238*, 389–401 (1957).
WIELANDT, H.: [1] Primitive Permutationsgruppen vom Grad $2p$. Math. Z. *63*, 478–485 (1956).
— [2] Beziehungen zwischen den Fixpunktzahlen von Automorphismengruppen einer endlichen Gruppe. Math. Z. *73*, 146–158 (1960).
— [3] Über den Transitivitätsgrad von Permutationsgruppen. Math. Z. *74*, 297–298 (1960).
— *[4] Finite permutation groups. New York and London, Academic Press, 1964.
— [5] On automorphisms of doubly transitive permutation groups. Proc. Intern. Conf. Theory of Groups, Canberra 1965, 389–393.
— [6] Endliche k-homogene Permutationsgruppen. Math. Z. *101*, 142 (1967).
— [7] Permutation groups through invariant relations and invariant functions. Ohio State University, Columbus 1969.
WILLEMS, W.: [1] Bemerkungen zur modularen Darstellungstheorie 3. Induzierte und eingeschränkte Moduln. Arch. Math. (Basel) *26*, 497–503 (1976).
— [2] Metrische G-Moduln über Körpern der Charakteristik 2. Math. Z. *157*, 131–139 (1977).
— [3] On the projectives of a group algebra. Math. Z. *171*, 163–174 (1980).
WITT, E.: [1] Treue Darstellung Liescher Ringe. J. Reine Angew. Math. *177*, 152–160 (1973).
— [2] Die 5-fach transitiven Gruppen von Mathieu. Abh. Math. Sem. Univ. Hamburg *12*, 256–264 (1938).
— [3] Über Steinersche Systeme. Abh. Math. Sem. Univ. Hamburg *12*, 265–275 (1938).
WONG, W.: [1] On finite groups whose 2-Sylow subgroups have cyclic subgroups of index 2. J. Austral. Math. Soc. *4*, 90–112 (1964).
— [2] A characterization of the Mathieu group $\mathfrak{M}_{12}$. Math. Z. *84*, 378–388 (1964).
ZASSENHAUS, H.: [1] Über transitive Erweiterungen gewisser Gruppen aus Automorphismen endlicher mehrdimensionaler Geometrien. Math. Ann. *111*, 748–756 (1935).
— [2] Kennzeichnung endlicher linearer Gruppen als Permutationsgruppen. Abh. Math. Sem. Univ. Hamburg *11*, 17–44 (1936).
— [3] Über endliche Fastkörper. Abh. Math. Sem. Univ. Hamburg *11*, 187–220 (1936).
— [4] Ein Verfahren, jeder endlichen p-Gruppe einen Lie-Ring mit der Charakteristik p zuzuordnen. Abh. Math. Sem. Univ. Hamburg *13*, 200–207 (1939).
ZSIGMONDY, K.: [1] Zur Theorie der Potenzreste. Monatsh. Math. Phys. *3*, 265–284 (1892).

Index of Names

Index

B. Huppert

Endliche Gruppen I

Nachdruck. 1979. 15 Abbildungen.
XII, 796 Seiten (Grundlehren der mathematischen Wissenschaften, Band 134)
ISBN 3-540-03825-6

Inhaltsverzeichnis: Grundlagen. - Permutationsgruppen und lineare Gruppen. - Nilpotente Gruppen und *p*-Gruppen. - Verlagerung und *p*-nilpotente Gruppen. - Darstellungstheorie. - Auflösbare Gruppen. - Literaturverzeichnis. - Namenverzeichnis. - Sachverzeichnis. - Errata.

From the reviews:

"The aim of this impressive book is to present in an organized manner a large part of the theory of finite groups. ...

The author began his task in 1958, which is about the same time the present surge in activity in the field began. Thus his goal of a degree of comprehensiveness has become difficult to achieve; the author has met this problem by carefully selecting the material he presents. ...

In a sense, the reader is presented with the equivalent of what could be two or three books on, say, nilpotent groups, representation theory, and solvable groups. This makes this volume long and makes it appear formidable. However, there is no need for anyone to cover the entire book in one effort. The author has carefully noted in the later sections the back references to material presented earlier. Hence without too much difficulty, one can take any topic presented and work through that without burdening oneself with extraneous material. ... it will take its place as a classic in the field. ...

More than any other work, this book reflects the status of research in a large part. ..."

Mathematical Reviews

Springer-Verlag
Berlin
Heidelberg
New York

M. Suzuki

Group Theory I

1982. XIV, 434 pages. (Grundlehren der mathematischen Wissenschaften, Band 247)
ISBN 3-540-10915-3

Group Theory I is the author's translation of the first volume from his highly successful Japanese original. It provides the knowledge basic to an understanding of finite group theory, from which the reader acquainted only with rudimentary linear algebra can move on to his specialized field of interest.

The author presents all of the topics necessary for a thorough understanding of the subject, including cohomology theory, the theory of Schur multipliers, and the generalized Sylow theorems, as well as a consideration of the classification of finte reflection groups (by the Witt method), an introduction to the theory of Tits systems, and a chapter on recent advances in the theory of finite simple groups. In presenting this material, the text emphasizes groups of matrices (mainly 2 x 2 nonsingular matrices with entries in some field F). Groups consisting of permutations are introduced at an early stage, followed later by computations with cycles.

While the scope is much broader than other presentations of the subject, there is a clear distinction between essentials and secondary concepts. In some cases, inessential results and conjectures are briefly mentioned and linked to bibliographical references. However, most readers will find Suzuki's overview to be amply detailed and functionally self-contained.

Springer-Verlag
Berlin
Heidelberg
New York